Lecture Notes in Computer Science 4735

Commenced Publication in 1973
Founding and Former Series Editors:
Gerhard Goos, Juris Hartmanis, and Jan van Leeuwen

Gregor Engels   Bill Opdyke
Douglas C. Schmidt   Frank Weil (Eds.)

Model Driven Engineering Languages and Systems

10th International Conference, MODELS 2007
Nashville, USA, September 30 - October 5, 2007
Proceedings

 Springer

Volume Editors

Gregor Engels
University of Paderborn
Department of Computer Science
Warburger Str. 100, 33098 Paderborn, Germany
E-mail: engels@upb.de

Bill Opdyke
Motorola, Inc., Technology Office
Home & Networks Mobility
1303 E. Algonquin Rd; Schaumburg, IL 60196, USA
E-mail: Bill.Opdyke@motorola.com

Douglas C. Schmidt
Vanderbilt University
Department of Electrical Engineering and Computer Science
2015 Terrace Place, Nashville, TN 37203, USA
E-mail: schmidt@dre.vanderbilt.edu

Frank Weil
Motorola, Inc., Software Design
Automation Center
1303 E. Algonquin Rd, Schaumburg, IL 60196, USA
E-mail: Frank.Weil@motorola.com

Library of Congress Control Number: 2007936008

CR Subject Classification (1998): D.2, D.3, K.6, I.6

LNCS Sublibrary: SL 2 – Programming and Software Engineering

ISSN   0302-9743
ISBN-10   3-540-75208-0 Springer Berlin Heidelberg New York
ISBN-13   978-3-540-75208-0 Springer Berlin Heidelberg New York

Springer is a part of Springer Science+Business Media

springer.com

© Springer-Verlag Berlin Heidelberg 2007

Typesetting: Camera-ready by author, data conversion by Scientific Publishing Services, Chennai, India
Printed on acid-free paper   SPIN: 12164127   06/3180   5 4 3 2 1 0

Preface

MODELS 2007 was the tenth incarnation of the series of conferences on Model Driven Engineering Languages and Systems. The conference was held in Nashville (TN), USA, during the week of September 30 to October 5, 2007. The local arrangements were provided by the Department of Electrical Engineering and Computer Science of Vanderbilt University.

The program of the week comprised three keynote presentations, technical paper presentations, two panels, as well as several workshops and tutorials on key topics in the field. The invited keynote speakers were Mary Shaw (Carnegie Mellon University), Kevin Sullivan (University of Virginia), and Patrick Lardieri (Lockheed Martin Advanced Technology Lab).

This volume contains the final versions of the papers accepted for presentation at the conference. These papers cover topics from the field including metamodeling, model transformations, model analysis, aspect-oriented modeling, and modeling process support.

We received 158 full paper submissions (including 18 experience reports) for review from 36 different countries. Of these, 31 papers were submitted with authors from more than one country. The top countries submitting papers were France (72), USA (64), and Germany (59). A total of 45 papers were accepted for inclusion in the proceedings, including five experience papers. The acceptance rate was therefore 28 %, which is comparable to those of previous MODELS conferences.

At least three Program Committee members reviewed each paper. Reviewing was thorough, and most authors received detailed comments on their submissions. Conflicts of interest were taken very seriously. No one participated in any way in the decision process of any paper where a conflict of interest was identified. In particular, PC members who submitted papers did not have access to information concerning the reviews of their papers.

We would like to thank everyone who submitted a paper or a proposal for a workshop or a tutorial. We would also like to thank the large number of volunteers who contributed to the success of the conference. Richard van de Stadt deserves special thanks for his prompt and gracious service in supporting special requests for CyberChairPRO, the conference management system used to manage papers submissions and the virtual PC meeting. Finally, we would like to thank our sponsors, ACM and IEEE Computer Society, for their support of the MODELS 2007 conference.

October 2007

Gregor Engels
Bill Opdyke
Douglas C. Schmidt
Frank Weil

Organization

General Chair
Douglas C. Schmidt (Vanderbilt University, USA)

Program Chair
Gregor Engels (University of Paderborn, Germany)

Experience Track Chairs
Bill Opdyke (Motorola, Inc., USA)
Frank Weil (Motorola, Inc., USA)

Workshop Chair
Holger Giese (University of Paderborn, Germany)

Tutorial Chair
Jeff Gray (University of Alabama at Birmingham, USA)

Panel Chair
Jean-Michel Bruel (Université de Pau et des Pays de l'Adour, France)

Doctoral Symposium Chair
Claudia Pons (Universidad Nacional de La Plata, Argentina)

Educators' Symposium Chair
Miroslaw Staron (IT University of Göteborg, Sweden)

Vendor Exhibition Chair
Gunther Lenz (Siemens Corporate Research, USA)

Local Arrangements Chair
Aniruddha Gokhale (Vanderbilt University, USA)

Academic Poster and Demo Chair
Daniel Varro (Budapest University of Technology and Economics, Hungary)

Publicity Chairs

Laurence Tratt (King's College London, UK)
Jules White (Vanderbilt University, USA)

Web Chair

Eric Hall (Vanderbilt University, USA)

Program Committee

Aditya Agrawal (USA)
Gabriela Arévalo (Argentina)
Colin Atkinson (Germany)
Thomas Baar (Switzerland)
Brian Berenbach (USA)
Jean Bezivin (France)
Ruth Breu (Austria)
Lionel Briand (Norway)
Siobhán Clarke (Ireland)
Shigeru Chiba (Japan)
Krzysztof Czarnecki (Canada)
Serge Demeyer (Belgium)
Christophe Dony (France)
Stéphane Ducasse (France)
Gregor Engels (Germany)
Jean-Marie Favre (France)
Harald Gall (Switzerland)
Geri Georg (USA)
Sudipto Ghosh (USA)
Martin Gogolla (Germany)
Susanne Graf (France)
Øystein Haugen (Norway)
Robert Hirschfeld (Germany)
Seongsoo Hong (Korea)
Gabor Karsai (USA)
Jörg Kienzle (Canada)

Thomas Kühne (Germany)
Michele Lanza (Switzerland)
Timothy C. Lethbridge (Canada)
Radu Marinescu (Romania)
Tom Mens (Belgium)
Dragan Milicev (Serbia)
Ana Moreira (Portugal)
Pierre-Alain Muller (France)
Bill Opdyke (USA)
Rob Pettit (USA)
Alexander Pretschner (Switzerland)
Gianna Reggio (Italy)
Bernhard Rumpe (Germany)
Laurent Safa (Japan)
Jean-Guy Schneider (Australia)
Andy Schürr (Germany)
Bran Selic (Canada)
Perdita Stevens (UK)
Juha-Pekka Tolvanen (Finland)
Aswin A. van den Berg (USA)
Markus Völter (Germany)
Alain Wegmann (Switzerland)
Thomas Weigert (USA)
Frank Weil (USA)
Jon Whittle (USA)
Clay E. Williams (USA)

Steering Committee

Thomas Baar (Switzerland)
Jean Bézivin (France)
Lionel Briand (Norway)
Steve Cook (UK)
Gregor Engels (Germany)

Andy Evans (UK)
Robert France (USA)
Geri Georg (USA)
Martin Gogolla (Germany)
Heinrich Hussmann (Germany)

Jean-Marc Jézéquel (France)
Stuart Kent (UK)
Cris Kobryn (USA)
Ana Moreira (Portugal)
Pierre-Alain Muller (France)
Oscar Nierstrasz (Switzerland)
Gianna Reggio (Italy)

David Rosenblum (UK)
Bernhard Rumpe (Germany)
Douglas C. Schmidt (USA)
Bran Selic (Canada)
Perdita Stevens (UK)
Jon Whittle (USA)

Sponsors

ACM Special Interest Group on Software Engineering
(www.sigsoft.org)

IEEE Computer Society
(www.computer.org)

Additional Referees

Khalid Allem
Carsten Amelunxen
Michal Antkiewicz
Joao Araujo
Thiago Bartolomei
Christian Basarke
Benoit Baudry
Hanna Bauerdick
Bernhard Beckert
Stanislav Belogolov
Alexandre Bergel
Kirsten Berkenkötter
Bart Du Bois
François Bronsard
Fabian Büttner
Damien Cassou
Cecilia Challiol
Natalia Correa
Pascal Costanza
Thomas Cottenier
Jose Diego de la Cruz
Duc-Hanh Dang
Helmut Degen
Roland Ducournau

Cedric Dumoulin
Peter Ebraert
Maged Elasaar
Michael Felderer
Bernd Fischer
Franck Fleurey
Joris Van Geet
Pieter Van Gorp
Hans Groenniger
Michael Hafner
Ulrich Hannemann
Michael Haupt
Christoph Herrmann
Berthold Hoffmann
Karsten Hölscher
Marianne Huchard
Andrew Jackson
Eshref Januzaj
Frederic Jouault
Anne Keller
Felix Klar
Jacques Klein
Anneke Kleppe
Sascha Konrad

Holger Krahn
Mirco Kuhlmann
Alexander Königs
Lam-Son Le
Chris Lenz
Tihamér Levendovszky
Lingling Liao
Jintack Lim
Alejandra Lliteras
Jose Magno Lopes
Cristina Marinescu
Sasa Misailovic
Darko Marinov
Aleksandar Milicevic
Sasa Misailovic
Olaf Muliawan
Sadaf Mustafiz
Clémentine Nebut
Jon Oldevik
Joanna Chimika-Opoka
Ellen Van Paesschen
Jiyong Park
Eduardo Piveta
Damien Pollet

Claudia Pons
Chetan Raj
Gil Regev
Irina Rychkova
Dirk Reiss
Filippo Ricca
Matthias Rieger
Ragnhild Kobro Runde
Irina Rychkova
Filip Van Rysselberghe
Salah Sadou

Tim Schattkowski
Martin Schindler
Hans Schippers
Lionel Seinturier
Karsten Sohr
Taehyo Song
Mike Sowka
Mark Stein
Ashley Sterritt
Mathieu Suen
Gabriele Taentzer

Christelle Urtado
Sylvain Vauttier
Steven Voelkel
Andrzej Wasowski
Seungmin We
Ingo Weisemoeller
Richard Wettel
Clinton Woodward
Xiaolin Xi
Jonghun Yoo
Wooseok Yoo

Table of Contents

Model Transformation

Modeling Constraints

Meta-Modeling

Consistent Models

Modeling Support

User Interface Design

Language Definition

Modeling Methods

Service and Process Modeling

Model Analysis

Modeling Process

Aspects

New Language Features

Debugging Support

Statecharts

Workshops, Tutorials and Panels

Bidirectional Model Transformations in QVT: Semantic Issues and Open Questions

Perdita Stevens*

School of Informatics, University of Edinburgh
Fax: +44 131 667 7209
perdita@inf.ed.ac.uk

Abstract. We consider the OMG's Queries, Views and Transformations (QVT) standard as applied to the specification of bidirectional transformations between models. We discuss what is meant by bidirectional transformations, and the model-driven development scenarios in which they are needed. We analyse the fundamental requirements on tools which support such transformations, and discuss some semantic issues which arise. We argue that a considerable amount of basic research is needed before suitable tools will be fully realisable, and suggest directions for this future research.

Keywords: bidirectional model transformation, QVT.

1 Introduction

The central idea of the OMG's Model Driven Architecture is that human intelligence should be used to develop models, not programs. Routine work should be, as far as possible, delegated to tools: the human developer's intelligence should be used to do what tools cannot. To this end, it is envisaged that a single platform independent model (PIM) might be created and transformed, automatically, into various platform specific models (PSMs) by the systematic application of understanding concerning how applications are best implemented on each specific platform. The OMG's Queries, Views and Transformations (QVT) standard [12] defines languages in which such transformations can be written.

In this paper we will discuss *bidirectional* transformations, focusing on basic requirements which such transformations should satisfy.

The structure of the paper is as follows. In the remainder of this section, we motivate bidirectional transformation, and especially, the need for non-bijective bidirectional transformations; we then discuss related work. Section 2 briefly summarises the most relevant aspects of the QVT standard. Section 3 discusses key semantic issues that arise. Section 4 proposes and motivates a framework and a definition of "coherent transformation". Finally Section 5 concludes.

In order to justify the considerable cost of developing a model transformation, it should ideally be reused; perhaps a vendor might sell the same transformation

---

* Corresponding author.

G. Engels et al. (Eds.): MoDELS 2007, LNCS 4735, pp. 1–15, 2007.

to many customers. However, in practice a transformation will usually have to be adapted to the needs of a particular application. Similarly, whilst we might like to argue that only the PIM would ever need to be modified during development, with model transformation being treated like compilation, the transformed model never being directly edited, nevertheless in practice it will be necessary for developers to modify both the source and the target models of a transformation and propagate changes in both directions. The interesting albeit unfinished document [14] makes these and other points, emphasising especially that bidirectional transformations are a key user requirement on QVT, and that ease of use of the transformation language is another key requirement.

Even in circumstances where it is in principle possible to make every change in a single source model, and roll changes down to target models by reapplying unidirectional transformations, in practice this is not desirable for a number of reasons. A human reason is that different developers are familiar with different models, and even different modelling languages. Developers are less likely to make mistakes if they change models they are comfortable with. A technical reason is that some changes are most simply expressed in the vocabulary, or with respect to the structure, of one model. For example, a single change to one model might correspond, semantically, to a family of related changes in the other.

Given the need for transformations to be applied in both directions, there are two possible approaches: write two separate transformations in any convenient language, one in each direction and ensure "by hand" that they are consistent, or use a language in which one expression can be read as a transformation in either direction. The second is very much preferable, because the checking required to ensure consistency of two separate transformations is hard, error-prone, and likely to cause a maintenance problem in which one direction is updated and the other not, leaving them inconsistent. QVT Relational takes this second approach: a transformation written in QVT Relational is intended to be able to be read as a specification of a relation which should hold between two models, or as a transformation function in either direction.

1.1 Bidirectional Versus Bijective Transformations

A point which is vital for the reader to understand is that bidirectional transformations need not be bijective. A transformation between metamodels M and N given by a relation R is *bijective* if for every model m conforming to M there exists exactly one model n conforming to N such that m and n are related by R, and vice versa (for every model n conforming to N there exists exactly one ...). This is an important special case because there is then no choice about what the transformation must do: given a source model, it must return the unique target model which is correctly related to the source. Ideally, the developer writing a bijective transformation does not have to concern herself with how models should be transformed: it should suffice to specify the relation, which will in fact be a bijective function. (In practice, depending on exactly how the relation is expressed, it might be far from trivial for a tool to extract the functions, however.)

Modulo information encoded in the transformation itself, both source and target models contain exactly the same information; they just present it differently. The classic example in the QVT standard of transformation between a UML class diagram and a database schema is a case where both models contain almost (but not quite) the same information, so it happens not to be a clear illustration of the inadequacy of bijective transformations. More realistically we might express the requirement as the synchronisation of a full UML model, including dynamic diagrams, with a database schema, which makes it obvious that there will be many UML models which might be related to a given schema. More generally, bijective transformations are the exception rather than the rule: the fact that one model contains information not represented in the other is part of the reason for having separate models. The QVT standard [12] is somewhat ambivalent about whether it intends all bidirectional QVT transformations to be bijective. On the one hand, the requirements of MDD clearly imply that it should be possible to write non-bijective transformations (see also [14]): for example, in general, the development of a PSM will involve the addition of information concerning design decisions on a particular platform. On the other hand, it is technically a consequence of statements made in the QVT Relations chapter that all "valid" transformations expressed in that language must be bijective, as we will show below. We take the latter to be a bug in the document, or at least, a restriction which needs to be relaxed for the promise of MDD to be fully realised.

1.2   Related Work

In the context of model transformations, almost all formal work on bidirectional transformations is based on graph grammars, especially triple graph grammars (TGGs) as introduced by Schürr (see, for example, [7]). Indeed, the definition of the QVT core language was clearly influenced by TGGs. A master's thesis by Greenyer [4] studies the relationship between QVT (chiefly QVT core) and TGGs, defining a translation from (a simplified subset of) QVT core into a version of TGGs that can be input into a TGG tool. More broadly, the field of model transformations using graph transformation is very active, with several groups working and tools implemented. We mention in particular [8,13]. Most recently, the paper [2] by Ehrig et al. addresses questions about the circumstances in which a set of TGG rules can indeed be used for forward and backward transformations which are information preserving in a certain technical sense. It is future work to relate our approach to TGGs.

   In this context, it may seem foolhardy to write a paper which approaches semantic issues in bidirectional model transformations from first principles. However, there is currently a wide gap between what is desired for the success of MDD and what is known to be soundly supportable by graph transformations; the use of QVT-style bidirectional transformations has not spread fast, despite the early availability of a few tools, partly (we think) because of uncertainty among users over fundamental semantic issues; and moreover, there is a large body of quite different work from which we may hope to gain important insights. Here we give a few pointers.

Benjamin Pierce and colleagues in the Harmony group have explored bidirectional transformations extensively in the context of trees [3], and more recently in the context of relational databases [1]. In their framework, a *lens*, or bidirectional transformation, is a pair of functions (a "get" function and a "putback" function) which are required to satisfy certain properties to ensure their coherence. They define a number of primitive lenses, and combinators with which to build more complex lenses. Thus, they define a programming language operating on trees in which a program can be read either forwards or backwards. Coherence of the forward and backward readings of the program follows from properties of the primitives and combinators. Their framework is asymmetric, however: their forward transformation is always a function on the source model only, which, in conjunction with their coherence conditions, implies that the target model is always an abstraction of the source model: it contains less information. This has advantages and disadvantages. It is insufficiently flexible to serve as a framework for MDA-style model transformations in general, but the restriction permits certain constructs, especially composition, to work in a way which does not seem to be possible in the more general setting. We shall return to this work later in the paper.

Bidirectional programming languages have been developed in various areas, and a survey can be found in [3]. Notably Lambert Meertens' paper [9] addresses the question of developing "constraint maintainers" for use in user interface design, but his approach is far more general. His maintainers are essentially model transformations which, in terms we shall introduce below, are required to be *correct* and *hippocratic*, but not *undoable*. In [6], Kawanaka and Hosoya develop a bidirectional programming language for XML. In Tokyo, Masato Takeichi and colleagues Shin-Cheng Mu and Zhenjiang Hu have also worked extensively on an algebraic approach to bidirectional programming: see [10,11,5].

2   QVT

The OMG's Queries, Views and Transformations (QVT) standard [12] addresses a family of related problems which arise in tool-supported model driven development. Not all information which is modelled is relevant at any one time, so there is a need to be able to abstract out the useful information; and models need to be held in meaningful relationships with one another. Provided that we permit non-bijective transformations (required to support model views), transformations subsume views.

The QVT standard describes three languages for transformations: the Operational, Relational and Core languages. The Relational language is the most relevant here. In the Operational language, someone wishing to specify a bidirectional transformation would have to write a pair of transformations and make them consistent by hand, which we have already said is undesirable. QVT Core is a low level language into which the others can be translated; an example translation from QVT Relational to QVT Core is given in the standard. Since we are concerned with transformations as expressed by users, we will work with QVT Relational.

The issue of whether transformations expressed in QVT Relational are supposed to be bijective is not explicitly discussed, but it seems to be a – possibly unintended – consequence of statements made in [12] that they must be. Specifically, QVT transformations are given a "check then enforce" semantics which means that a transformation must not modify a target model if it is already correctly related to the source model. At the same time, [12] page 18 states:

> In relations, the effect of propagating a change from a source model to a target model is semantically equivalent to executing the entire transformation afresh in the direction of the target model.

This seems to imply that if a transformation is to propagate changes made in a source model m to a target model n, the new target model that results must be independent of the old one: the result of the transformation must depend only on m, since it is equivalent to "executing the entire transformation" on m. In other words given m, there is a unique target model n which must result from executing the transformation. Now suppose that there is also a different model n' which is correctly related to m. Of course, this is quite compatible with the functional interpretation of transformation given in the above quotation: it could happen that even though n' would be a correct target model, n is the one which happens to be produced when the transformation is run on m. In this case, if the transformation is run in a situation where the source model is m and the target model is n', the target model must be transformed into n, even though n' was already correctly related to m. This, however, is exactly what is forbidden by the "check then enforce" semantics: given that n' is already correctly related to m, it must not be modified by running the transformation. If the situation of running a transformation on models which are already correctly related seems too artificial, the reader may prefer to consider a situation in which a target model may be put in correct relation with a source model in two different ways: either by making a tiny change to turn it into one correctly related model, or by making a large change to turn it into a different correctly related model. It will be natural to want the transformation to make the minimal possible changes to ensure relatedness (and, indeed, the text in [12] immediately following the above quotation suggests that this is intended). Interpreting the quoted text literally, though, forbids the transformation to give different results depending on how close the existing target model is to each correctly related target model.

It might be possible to interpret "semantically equivalent" in the above quotation so as to resolve this problem, but this seems forced (since it would require being able to regard models which contained very different information as being "semantically equivalent"). A better solution seems to be to assume that the above quotation is unintentionally restrictive, and disregard it.

3   Semantic Issues

In this section we raise a variety of issues which we consider to need further study: they are settled neither by the QVT standard, nor as far as we know by existing related work.

3.1   What Exactly It Makes Sense to Write

The QVT Relational language is designed to be easy for someone familiar with related OMG standards such as OCL to learn and use; this has clearly been a higher design goal than ensuring that only safe transformations can be written. There are several places (when and where clauses, among others) where arbitrary OCL expressions are permitted, even though only certain expressions make sense as part of a bidirectional transformation. For example, a transformation may in one direction give an attribute a value using an non-invertible expression.[1] Specifying exactly what language subset is permitted, however, is likely to run quickly into a familiar problem: that any reasonably easy-to-define language subset which is provably safe will also exclude many safe expressions of the full language. It may well be preferable to be permissive, and rely on users not to choose to write things that don't make sense. They will, however, require a clear understanding of what it means for a transformation to "make sense". In Section 4 we propose first steps in this direction and give simple postulates which, we argue, any bidirectional model transformation should obey.

3.2   Determining Validity of a Transformation

Let us suppose that the reader and the developer accept that model transformations will be written in an expressive, unsafe language, but that the transformations written should obey our proposed postulates (even though this has to be verified on a case-by-case basis, lacking a language in which any transformation is guaranteed to satisfy the postulates). How can developers become confident that their transformations do indeed obey these postulates? Ideally, the language and the postulates should be so clearly understandable that the developers can be confident in their intuition: tool support is no substitute for this kind of clarity. However, it is also desirable that a tool should be able to check, given the text of a transformation and the metamodels to which it applies, that it obeys the postulates. That is, this transformation should be able to be verified statically at the time of writing it, as opposed to failing when it is applied to arguments which expose a problem. Whether or to what extent this can be done is an open question.

A major danger with bidirectional transformations is that one direction of the transformation may be a seldom used but very important "safety net". It will be unfortunate if the user only finds out that their transformation cannot be executed in the less usual direction long after the transformation has been written, in circumstances where the reverse transformation is really needed.

---

[1] Note that permitting non-bijective transformations does not make this unproblematic: since transformations are to be deterministic, where there are several relationally possible choices of value the language needs to provide a way to specify which should be chosen.

3.3   Composition of Relations in QVT: *When* and *Where* Clauses

Most of this paper takes a high level view of transformations, in which a whole
transformation text specifies a relation and a pair of transformational functions.
We have not yet considered the details of how simpler relations are combined
and built up into transformations in QVT. This is interesting, however, and not
least because it gives another justification for considering non-bijective transformations. A QVT relational transformation has an overall structure something
like this:

```
transformation ... {
top relation R {
  domain a...
  domain b...
  when {...}
  where {...}
}
top relation S ...
relation ...
relation ...
...
}
```

In order to understand when and where clauses, note that [12] uses two different
notions of a relation holding. At the top level, a relation holds of a pair of models – checking will return TRUE – if they are consistent. E.g. if a UML model
m is consistent with an RDBMS model s according to relation ClassToTable,
we will write ClassToTable$^+$(m,s). The + is intended to distinguish this notion from the following: the consistency between m and s is *demonstrated* by
matching individual classes in m to individual tables in s: then the class c and
table t (or more formally, the corresponding valid bindings of domain variables
in the text of ClassToTable) may also be said to be related. We will write
ClassToTable(c,t). Note that the relation on models ClassToTable$^+$ is a lifted
version of the relation on bindings ClassToTable: a UML model is related to an
RDBMS model by ClassToTable$^+$ iff for every class there is a table related to
it by ClassToTable and vice versa.

The when and where clauses can contain arbitrary OCL, but are typically
expected to contain (if anything) statements about relations satisfied by variables of the domain patterns. Thus in fact, the relation R holds if for every
valid match of the first domain, there exists a valid match of the second domain *such that the where clause holds*. The when clause "specifies the conditions
under which the relationship needs to hold". The example used in the standard is the relation ClassToTable with domains binding c:Class (and hence
p:Package etc.) and t:Table (and hence s:Schema etc.), the when clause being
PackageToSchema(p,s) and the where clause being AttributeToColumn(c,t).

Now, what does this mean in relational terms, and specifically, what is the difference between the when clause and the where clause, both of which

appear at first sight to impose extra conditions on valid matches of bindings, thus forming an intersection of relations? Unfortunately, this is not straightforward to express relationally. Operationally, the idea is that the variables in the when clause "are already bound" "at the time this relation is invoked". Roughly speaking, when a relation `ClassToTable` has domain patterns with variables including `p : Package` and `s : Schema`, and a when clause which states `PackageToSchema(p,s)`, the QVT engine is supposed to have already processed the `PackageToSchema` relation (if not, it will postpone consideration of the `ClassToTable` relation). The matchings calculated for `PackageToSchema` provide bindings for variables `p` and `s` in `ClassToTable`. Evaluation of `ClassToTable` now proceeds, looking for compatible valid bindings of all the other variables.

We have sketched the operational view of what happens in one example, but an open problem is to give a clean compositional account of even the relation (let alone the transformation) defined by a whole QVT transformation. Making this precise would involve a full definition of R^+ taking account of when and where clauses, and an account of the relationship between properties of R and properties of R^+. As an example of the complications introduced by dependencies between relations, suppose that there are two ways of matching pairs of valid bindings (skolem functions) for one relation, one of which leads to a compatible matching for a later-considered relation and one of which does not. If a QVT engine picks "the wrong" matching for the first relation considered, is it permitted to return the result that the models are inconsistent, even though a different choice by the tool would have given a different result? Surely not: but then there is a danger that the tool will need to do global constraint solving or arbitrarily deep backtracking to ensure that it is not missing a solution. Not only is this inefficient, but it will be very hard for the human user to understand. Now, looking at the examples in [12], it seems clear that this kind of problem is not supposed to arise, because when clauses are used in very restricted circumstances. However, it is an open question what can be permitted, and we can expect to encounter the usual problem of balancing expressivity against safety.

For a simple example of "spatial" composition of relations where we *can* lift good properties of simple relations to good properties of a more complex relation, see the next section.

3.4 Sequential Composition of Transformations

We have discussed the ways in which relations are composed in QVT to make up transformations. A different issue is the sequential composition of whole transformations. We envisage a bidirectional QVT tool which does not retain information between uses: it simply expects to be given a pair of models, a transformation, and a command telling it in which direction to apply the transformation and whether to check or enforce.[2]

---

[2] If the tool is allowed to retain trace information – the correspondence graph in TGG terms – between executions, the problem becomes more tractable. But this is a severe pragmatic limitation.

We naturally expect to be able to define a transformation to be the sequential composition of two other transformations, and then treat the composition as a first-class transformation in its own right. In this case, the pair of models given to the tool will be the source of the first transformation and the target of the second: the tool will not receive a version of the intermediate model, the one which acts as target of the first transformation and source of the second. In order to define a general way to compose transformations, we need to suppose that we are given transformations R from M to N and S from N to P and show how to construct a composed transformation $T = R; S$, giving the relational and functional parts of the composed relation in terms of the parts of the constituent relations.

We will return to this issue in the next section, after introducing appropriate notation.

4   Requirements for Bidirectional Model Transformations

In this section we discuss bidirectional model transformations which are not necessarily bijective, and discuss under what circumstances these will make sense. We will give postulates which are clearly satisfied by bijective transformations, but also by certain non-bijective transformations.

First let us set some basic notation. We will use capital letters such as R, S, T for the relations which transformations are supposed to ensure. That is, if M and N are metamodels (or sets of models) to be related by a model transformation, the relation $R \subseteq M \times N$ holds of a pair of models – and we write $R(m,n)$ – if and only if the pair of models is deemed to be consistent. Associated with each relation will be the two directional transformations:

$$\overrightarrow{R} : M \times N \longrightarrow N$$

$$\overleftarrow{R} : M \times N \longrightarrow M$$

The idea is that $\overrightarrow{R}$ looks at a pair of models (m,n) and works out how to modify n so as to enforce the relation R: it returns the modified version. Similarly, $\overleftarrow{R}$ propagates changes in the opposite direction.

In practical terms, what we expect is that the programmer writes a single text (or set of diagrams) defining the transformation in the QVT relational language (or indeed, in another appropriate language). This same text can be read in three ways: as a definition of a relation which must hold between pairs of models; as a "forward" transformation which explains how to modify the right-hand model so as to make it relate to the left-hand model; as a "backward" transformation which explains how to modify the left-hand model so as to make it relate to the right-hand model. By slight abuse of notation, we will use capital letters R, S etc. to refer to the whole transformation, including both transformational functions as well as the relation itself, when no confusion can result.

Our notation already incorporates some assumptions, or rather assertions, which need justification.

First, and most importantly, that the behaviour of a transformation should be deterministic, so that modelling it by a mathematical function is appropriate. The same transformation, given the same pair of models, should always return the same proposed modification. This is a strong condition: it proscribes, for example, transformation texts being interpreted differently by different tools. An alternative approach, which we reject, would have been to permit a tool to modify the target model by turning it into *any* model which is related to the source model by the relation encoded in the transformation. There are several good reasons to reject that approach. Most crucially, the model transformation does not take place in isolation but in the presence of the rest of the development process. Even though certain aspects of one model may be irrelevant to users of the other – so that the transformation will deliberately abstract away those aspects – this does not imply that the abstracted away aspects are not important to other users! Usually, it will be unacceptable for a tool to "invent information" in any way, e.g. by making the choice of which related model to choose. The developer needs full control of what the transformation does. Even in rare cases where certain aspects of the transformation's behaviour (say, the choice of name for a newly created model element) might be thought of as unimportant, we claim that determinism is necessary in order to ensure, first, that developers will find tool behaviour predictable, and second, that organisations will not be unacceptably "locked in" to the tool they first use. Experience shows that even when a choice is arbitrary, people find it important that the way the arbitrary choice is made be consistent. One example of this is the finding that, even though the spatial layout of UML diagrams does not (generally) carry semantic information, it is important for UML tools to preserve the information.

Our second assertion is that the behaviour of a transformation may reasonably depend on the current value of the target model which will be replaced, as well as on the source model. This follows from our argument in Section 1 that restricting bidirectional transformations to be bijective is too restrictive. Of course, the fact that we choose a formalism which permits the behaviour of a transformation to depend on both arguments does not force it to do so. In the special case of a bijective transformation, the result of $\vec{R}$ may be independent of its second argument, and the result of $\overleftarrow{R}$ independent of its first argument. Another important special case is when the transformation in one direction uses only one of its arguments, while the reverse transformation uses both. Pierce et al.'s lenses fall into this category, and we will discuss how they fit into this framework below.

A technical point is that we require transformations to be total, in the sense that $\vec{R}$ and $\overleftarrow{R}$ are total functions, defined on the whole of $M \times N$. We may want to define, for a metamodel M, a distinguished "content-free" model ϵ_M to be used as a dummy argument e.g. in the case that a target model is created afresh from a source model. Note that since the model containing no model elements might not conform to M, ϵ_M might not literally be empty.

Correctness. Our notation is chosen to suggest that the job of $\overrightarrow{R}$ and $\overleftarrow{R}$ is to enforce the relation R, and our first postulates state that they actually do this. We will say that a transformation T is *correct* if

$$\forall m \in M \ \forall n \in N \qquad T(m, \overrightarrow{T}(m, n))$$

$$\forall m \in M \ \forall n \in N \qquad T(\overleftarrow{T}(m, n), n)$$

These postulates clearly have to be satisfied by any QVT-like transformation.

Hippocraticness, or "check-then-enforce". The QVT standard very clearly states that a QVT transformation must not modify either of a pair of models if they are already in the specified relation. That is, even if models n_1 and n_2 are both properly related to m by R, it is not acceptable for $\overrightarrow{R}$, given pair (m, n_1), to return n_2. Formally, we say that a transformation is *hippocratic*[3] if for all $m \in M$ and $n \in N$, we have

$$T(m, n) \quad \Longrightarrow \quad \overrightarrow{T}(m, n) = n$$

$$T(m, n) \quad \Longrightarrow \quad \overleftarrow{T}(m, n) = m$$

These postulates imply that if the relation T is *not* bijective, then (at least one of) the transformations must look at both arguments. As a consequence, applying a transformation to a source model in the presence of an existing target model will not in general be equivalent to applying it in the presence of an empty target model.

Undoability. Our final pair of postulates is motivated by thinking about the following scenario. The developer, beginning with a consistent pair of models m (the source) and n (the target, perhaps produced by a model transformation), makes a modification to the source model, producing m', and propagates it using the model transformation tool (so that target model n is replaced by $\overrightarrow{T}(m', n)$). Immediately, without making any other changes to either model, our developer realises that the modification was a mistake. She reverts the modified model to the original version m, and propagates the change. The developer reasonably expects that the effect of the modification has been completely undone: just as the modified model has been returned to its original state m, so has the target model been returned to its original state n.

Formally, we will say that transformation T is *undoable* if for all $m, m' \in M$ and $n, n' \in N$, we have

$$T(m, n) \quad \Longrightarrow \quad \overrightarrow{T}(m, \overrightarrow{T}(m', n)) = n$$

$$T(m, n) \quad \Longrightarrow \quad \overleftarrow{T}(\overleftarrow{T}(m, n'), n) = m$$

---

[3] First, do no harm. Hippocrates, 450-355BC

It turns out that this requirement is hard to meet in general, and arguably too strong. However, we find the scenario compelling: a model transformation which did not allow one's changes to be undone in this way would be quite confusing. Therefore, in the present paper, we take it to be essential (we shall shortly show that we can still write non-bijective transformations).

Definition 1. *Let R be a transformation between metamodels M and N, consisting of a relation $R \subseteq M \times N$ and transformation functions $\overrightarrow{R} : M \times N \longrightarrow N$ and $\overleftarrow{R} : M \times N \longrightarrow M$. Then R is a* coherent transformation *if it is correct, hippocratic and undoable.*

4.1 Examples and Consequences

Having presented a framework for bidirectional transformations and argued for a set of postulates that they should obey, let us explore the consequences of our choices. (Proofs, all easy, are omitted for space reasons; as are various other small results.) First we state two reassuring trivialities:

Lemma 1. *Let M be any metamodel. Then the trivial transformation, given by:*

- *$R(m, n)$ if and only if $m = n$*
- *$\overrightarrow{R}(m, n) = m$*
- *$\overleftarrow{R}(m, n) = n$*

is a coherent transformation.

Lemma 2. *Let M and N be any metamodels. Then the universal transformation, given by:*

- *$R(m, n)$ always*
- *$\overrightarrow{R}(m, n) = n$*
- *$\overleftarrow{R}(m, n) = m$*

is a coherent transformation.

Note that the latter lemma already proves that our postulates permit bidirectional transformations which are not bijective. We would of course expect that any bijective transformation is coherent, and so it is:

Lemma 3. *Let M and N be any metamodels. Then any bijective transformation, given by:*

- *$R(m, n)$ if and only if $n = r(m)$*
- *$\overrightarrow{R}(m, n) = r(m)$*
- *$\overleftarrow{R}(m, n) = r^{-1}(n)$*

where $r : M \longrightarrow N$ is a bijective function, is a coherent transformation.

The relationship between our framework and that of [3] is close. Note that it is our undoability postulates which prevent more general lenses being coherent.

Lemma 4. *Any very well behaved lens l can be regarded as a coherent transformation.*

The reader familiar with [3] may be surprised that our postulates do not include analogues of the GETPUT and PUTGET laws from that work. They are in fact immediate consequences of correctness and hippocraticness:

Lemma 5. *Let M and N be any metamodels, and let R be a correct and hippocratic (but not necessarily undoable) transformation. Then for any $m \in M$, $n \in N$:*

- $\overleftarrow{R}(m, \overrightarrow{R}(m,n)) = m$
- $\overrightarrow{R}(\overleftarrow{R}(m,n), n) = n$

4.2   Composition of Metamodels

Let us say that a metamodel M is the *direct product* of metamodels M_1 and M_2 if any model m conforming to M can be written in exactly one way as a pair of a model m_1 conforming to M_1 and a model m_2 conforming to M_2, and conversely, any such pair conforms to M. For example, perhaps M_1 and M_2 comprise disjoint sets of metaclasses, with no relationships or constraints between the two sets. (This is admittedly an artificially constraining scenario: we will discuss relaxations in a moment.)

Now suppose that we have coherent transformations R on $M_1 \times N_1$ and S on $M_2 \times N_2$. We can construct a transformation which we will call $R \oplus S$ on $M \times N$ pointwise as follows:

- $(R \oplus S)(m_1 \oplus m_2, n_1 \oplus n_2)$ if and only if $R(m_1, n_1)$ and $S(m_2, n_2)$
- $\overrightarrow{(R \oplus S)}(m_1 \oplus m_2, n_1 \oplus n_2) = (\overrightarrow{R}(m_1, n_1)) \oplus (\overrightarrow{S}(m_2, n_2))$
- $\overleftarrow{(R \oplus S)}(m_1 \oplus m_2, n_1 \oplus n_2) = (\overleftarrow{R}(m_1, n_1)) \oplus (\overleftarrow{S}(m_2, n_2))$

Then

Lemma 6. *If R and S are coherent transformations, $R \oplus S$ is also a coherent transformation.*

The proof is a straight application of the definitions. This captures the intuition that transformations on parts of models which are completely independent ought to be able to be combined without difficulty. One would expect to be able to extend this result to cover carefully-defined simple dependencies between the metamodel parts, perhaps sufficient to justify, for example, applying a transformation defined only for class diagrams to a complete UML model, rolling the resulting changes to the class diagram through to the rest of the model. Even here, though, the issues are not entirely trivial.

4.3   Sequential Composition Revisited

The relation part of the sequential composition of transformations must be given by the usual mathematical composition of relations: $(R; S)(m, p)$ if and only if *there exists some* n such that $R(m, n)$ and $S(n, p)$. Mathematically this is a fine definition, but we already see the core of the problem: a tool has no obvious way to find a relevant n. What about the associated transformations? For example, $\overrightarrow{T}$ may be given models m and p such that there does not exist any n such that $R(m, n)$ and $S(n, p)$. It is required to calculate an update of p; that is, to find a new model p' such that such an intermediate model does exist, and in general the choice of intermediate model will depend on both m and p. However, $\overrightarrow{R}$ "does not understand" p, etc., so there does not appear to be any way to do this in general.

We may consider two special cases in which it is possible to define composition of transformations.

1. If R and S are bijective transformations, then the intermediate model is unique, and is found by applying $\overrightarrow{R}$ just to the first argument m. Composition of transformations in this case is just the usual composition of invertible functions.
2. More interestingly, the Harmony group considers transformations in which $\overrightarrow{R}$ is a function of the source model only, even though $\overleftarrow{R}$ still uses both source and target model. Here $\overrightarrow{R; S}$ must clearly be defined to be $\overrightarrow{R}; \overrightarrow{S}$, and we can define $\overleftarrow{R; S}$ using a trick: use the function $\overrightarrow{R}$ to bring the source model forward into the middle in order to use it to push the changes back. Formally (and translating into our notation)

$$\overleftarrow{R; S}(m, n) = \overleftarrow{R}(m, \overleftarrow{S}(\overrightarrow{R}(m), n))$$

5   Conclusion

We have explored some fundamental issues which arise when we consider relationally defined transformations between models which are bidirectional and not necessarily bijective. We have motivated our work from the current QVT standard, and some of the issues we raise are specific to it, but most are more general. We have suggested a framework and a set of postulates which ensure that bidirectional transformations will behave reasonably for some definition of "reasonable", and explored some consequences of our choice. Future work includes relating our framework to triple graph grammars, and further exploration of the relation with bidirectional programming.

Acknowledgements. The author would like to thank Reiko Heckel, Conrad Hughes, Gabriele Taentzer and especially Benjamin Pierce for helpful discussions.

References

1. Bohannon, A., Vaughan, J.A., Pierce, B.C.: Relational lenses: A language for updateable views. In: Principles of Database Systems (PODS), Extended version available as University of Pennsylvania technical report MS-CIS-05-27 (2006)

2. Ehrig, H., Ehrig, K., Ermel, C., Hermann, F., Taentzer, G.: Information preserving bidirectional model transformations. In: FASE 2007. LNCS, vol. 4422, pp. 72–86. Springer, Heidelberg (2007)
3. Foster, J.N., Greenwald, M.B., Moore, J.T., Pierce, B.C., Schmitt, A.: Combinators for bi-directional tree transformations: A linguistic approach to the view update problem. ACM Transactions on Programming Languages and Systems (to appear, 2007) (preprint), available from http://www.cis.upenn.edu/~bcpierce/papers/index.shtml
4. Greenyer, J.: A study of technologies for model transformation: Reconciling TGGs with QVT. Master's thesis, University of Paderborn, Department of Computer Science, Paderborn, Germany (July 2006)
5. Hu, Z., Mu, S.-C., Takeichi, M.: A programmable editor for developing structured documents based on bidirectional transformations. In: PEPM'04. Proceedings of the 2004 ACM SIGPLAN Workshop on Partial Evaluation and Semantics-based Program Manipulation, pp. 178–189. ACM Press, New York (2004)
6. Kawanaka, S., Hosoya, H.: biXid: a bidirectional transformation language for XML. In: ICFP'06. Proceedings of the International Conference on Functional Programming, pp. 201–214 (2006)
7. Königs, A., Schürr, A.: Tool Integration with Triple Graph Grammars - A Survey. In: Heckel, R. (ed.) Proceedings of the SegraVis School on Foundations of Visual Modelling Techniques. ENTCS, vol. 148, pp. 113–150. Elsevier Science Publ., Amsterdam (2006)
8. Königs, A.: Model transformation with triple graph grammars. In: In Proceedings, Workshop on Model Transformations in Practice (September 2005)
9. Meertens, L.: Designing constraint maintainers for user interaction. (unpublished manuscript) (June 1998), available from http://www.kestrel.edu/home/people/meertens/
10. Mu, S.-C., Hu, Z., Takeichi, M.: An algebraic approach to bi-directional updating. In: Chin, W.-N. (ed.) APLAS 2004. LNCS, vol. 3302, pp. 2–20. Springer, Heidelberg (2004)
11. Mu, S.-C., Hu, Z., Takeichi, M.: An injective language for reversible computation. In: Kozen, D. (ed.) MPC 2004. LNCS, vol. 3125, pp. 289–313. Springer, Heidelberg (2004)
12. OMG. MOF2.0 query/view/transformation (QVT) adopted specification. OMG document ptc/05-11-01 (2005), available from www.omg.org
13. Taentzer, G., Ehrig, K., Guerra, E., de Lara, J., Lengyel, L., Levendovsky, T., Prange, U., Varro, D., Varro-Gyapay, S.: Model transformation by graph transformation: A comparative study. In: Workshop on Model Transformations in Practice (September 2005)
14. Witkop, S.: MDA users' requirements for QVT transformations. OMG document 05-02-04 (2005), available from www.omg.org

Reconciling TGGs with QVT

Joel Greenyer and Ekkart Kindler

University of Paderborn, Department of Computer Science,
33100 Paderborn, Germany
{jgreen,kindler}@upb.de

Abstract. The Model Driven Architecture (MDA) is an approach to develop software based on different models. There are separate models for the business logic and for platform specific details. Moreover, code can be generated automatically from these models. This makes transformations a core technology for MDA. QVT (Query/View/Transformation) is the transformation technology recently proposed for this purpose by the OMG.

TGGs (Triple Graph Grammars) are another transformation technology proposed in the mid-nineties, used for example in the FUJABA CASE tool. In contrast to many other transformation technologies, both QVT and TGGs declaratively define the relation between two models. With this relation definition, a transformation engine can execute a transformation in both directions and, based on the same definition, can also propagate changes from one model to the other.

In this paper, we compare the concepts of QVT and TGGs. It turns out that TGGs and QVT have many concepts in common. In fact, fundamental parts of QVT-Core can be implemented by a TGG transformation engine. Moreover, we discuss how both technologies could profit from each other.

Keywords: MDA, model based software engineering, model transformation, model synchronization, Query/View/Transformation (QVT), Triple Graph Grammar (TGG).

1 Introduction

In the recent years, several approaches to *model based software engineering* have been proposed. One of the most prominent approaches is the *Model Driven Architecture (MDA)* of the OMG [1]. The main idea of all these approaches is that software should no longer be programmed, but developed by a stepwise refinement and extension of models. In the MDA, the focus is on separating the models for the business logic and for platform and implementation specific details. In the end, the code can be generated from these models. This makes technologies for transforming, integrating, and synchronizing models a core technology within model based software engineering.

Today, there are many different technologies for transforming one model into another. Most of these technologies are defined in a more or less operational way; i.e. they basically define instructions how the elements from the source model must be

G. Engels et al. (Eds.): MoDELS 2007, LNCS 4735, pp. 16–30, 2007.

transformed into elements of the target model. This implies that forward and backward transformations between two models are defined more or less independently of each other. Moreover, the operational definition of a transformation makes it very hard to verify its correctness. By contrast, QVT[1] (Query/View/ Transformation) [2] and TGGs (Triple Graph Grammars) [3] allow us to declaratively define the relation between two or more models. Such a declarative definition of a relation can be used by a transformation engine in different ways which we call *application scenarios*: Firstly, there are the *forward* and *backward transformations* of one model into another. Secondly, once we have transformed one model into another, the engine can keep track of changes in either model and propagate those changes to the other model and change it accordingly. This is called *model synchronization* and is one of the most crucial application scenarios in round-trip engineering. Since QVT and TGGs have only a single definition of the relation between two classes of models, inconsistencies among the different transformation scenarios are avoided.

QVT is the transformation technology recently proposed by the OMG for the MDA. Actually, QVT has different parts: There are declarative and operational languages. Here, we focus on QVT-Core, which forms the basic infrastructure of the declarative part of QVT. TGGs were introduced in the mid-nineties, and are now used in the FUJABA Tool Suite, which is a CASE tool supporting round-trip engineering. TGGs are at the core of FUJABA, for transforming back and forth between UML diagrams and Java code [4]. Another implementation of TGGs exists in the MOFLON tool set [5]. In addition to being declarative, QVT-Core and TGGs have many concepts in common and – upon closer investigation – have striking similarities. In this paper, we investigate these similarities for several reasons. Firstly, the common concepts of QVT-Core and TGGs identify the essential concepts of a declarative approach toward specifying the relationship between two classes of models. Secondly, the analysis shows that QVT-Core can be mapped to the concepts of TGGs so that QVT-Core can be implemented by an engine for executing TGG transformations. This mapping was worked out in a master thesis [6] and is briefly discussed in this paper. Thirdly, the differences between QVT and TGGs are analyzed and we discuss how both technologies can benefit from the concepts provided by the other technology. This will help to improve both transformation technologies – in particular this could provide valuable input for QVT as a standard.

This paper is structured as follows: Section 2 introduces the main concepts of QVT and TGGs with the help of an example. Section 3 identifies the similar concepts and shows how QVT can be mapped to TGGs, which provides an implementation of QVT based on a TGG engine. Section 4 gives a more detailed comparison of the philosophical, conceptual, and technical differences between QVT and TGGs.

2   QVT and TGG Transformation Rule Examples

In the following, we introduce QVT and TGG transformation rules along a small example. The example is picked from the ComponentTools project where component-

---

[1] Actually, QVT has different ways for defining transformations. Here, we refer to QVT-Core only, which is the basic infrastructure of the declarative part of QVT.

based material-flow systems can be designed and analyzed by the help of formal methods [7]. For instance, transportation or manufacturing systems can be designed by placing and connecting components, such as tracks, switches and stoppers, inside a *project*. Then, the project and its interconnected components are transformed into a formal model, for example a Petri net. Figure 1 shows how two connected Tracks in a Project should be transformed into a corresponding Petri net: A Track is represented by a Place, an Arc, and a Transition. The Connection simply corresponds to an Arc.

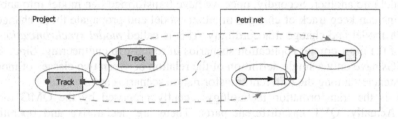

Fig. 1. An example of how two connected tracks are mapped to the corresponding Petri net

Figure 2 shows the way a rule can express how model structures correspond to each other. Here, it is specified that a Track relates to the particular Petri net construct in a certain context. The required context here is an existing relation between the parent model elements, the Project and the Petri net.

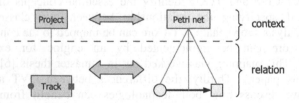

Fig. 2. A rule expressing the relation of model structures

Such rules, which express the relation between model structures, can be classified as *relational* rules and both QVT and TGG rules follow this archetype. Note that such relations can also be expressed between more than two models [8]. However, for simplicity, we focus on relations between only two models in this paper. The advantage of such rules over other transformation approaches is that they can be applied in different ways. They can be used to bidirectionally transform models, to check if two given models are equivalent, or, after an initial transformation, to incrementally propagate changes made in one model to the other. But, the rules do not describe operationally how to transform models. They are *declarative* and it is up to a transformation engine to make them operational.

For example, transforming a ComponentTools Project into its corresponding Petri net would involve the following steps. Firstly, an *axiom* or *start rule* is needed to map the Project root model object to a Petri net root model object. This provides the context necessary to apply the rule shown in Figure 2. The application of the above rule is visualized in Figure 3: For every Track in the Project, the corresponding Place-Arc-Transition-construct is created in the Petri net.

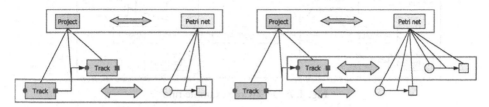

Fig. 3. Two rule applications in the example transformation

The model structures which are newly created during such rule applications provide the context for applying further rules. For example, a further rule would now be involved to transform the Connection between the components.

After introducing the concepts of relational rules and how they are used for model transformation, the following subsections will inspect the details of QVT and TGGs.

2.1 QVT-Core Mappings

The QVT specification defines two declarative transformation languages which form two layers of abstraction. Firstly, there is the more abstract and more user friendly QVT-Relations. QVT-Relations is then mapped to a more concrete language, QVT-Core, for which the semantics for performing transformations is defined in more detail. Although the concepts of QVT-Relations and its mapping to QVT-Core are very interesting, this paper focuses on QVT-Core, because it is the foundation of the declarative QVT and structurally more similar to TGGs.

In both QVT-Relations and QVT-Core, model patterns are described by OCL expressions. Although there is no graphical syntax specified for QVT-Core, Figure 4 illustrates the QVT-Core representation of the example rule from Figure 2 in a graphical way. Instead of using the concrete syntax of components and Petri nets, the model patterns are now shown in a notation similar to object diagrams. Each object node shown here represents an OCL variable. In the following, the terms *variable* and *node* are used interchangeably.

A single transformation rule in QVT-Core as shown here is called a *mapping*. The patterns in a mapping are structured in three columns[2], called *areas*, each consisting of a *guard pattern* and a *bottom pattern*. The bottom patterns represent the model elements which are actually brought into relationship by this mapping. The guard patterns specify the context which is required for this relation to hold. The outer columns, which contain the patterns belonging to the different involved models, are called the *domain areas*. In this example, the ComponentTools domain is abbreviated as `ctools` and the Petri net domain is abbreviated as `pnet`.

---

[2] There can be more columns in QVT-Core to specify relations between more than two models.

Mapping TrackToPlaceArcTransition

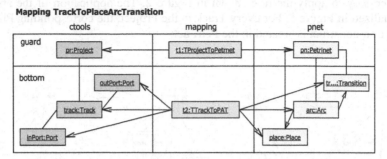

Fig. 4. A QVT-Core example rule

The center column, called the *mapping area*, contains additional elements, which embody the mapping of the involved domain patterns. In particular, there is one[3] mapping node in the middle-bottom pattern of a rule, which references all nodes in the domain-bottom patterns. In the course of a transformation, these "mapping nodes" are instantiated and keep track of the corresponding model structures. These objects are therefore called *trace objects*. In the rule, we will refer to those "mapping nodes" also as *trace nodes* or *trace variables*. In addition to the middle-bottom pattern, the middle-guard pattern can contain an arbitrary number of trace nodes, depending on the complexity of the context specified by this mapping.

Next, we have a look at the textual representation of this mapping, which is shown in Listing 1. First of all, we see the two domain areas represented by "**check ctools**" and "**check enfo□ce**" pnet". The keywords *check* and *enforce* determine whether the model patterns should just be matched in an existing model or whether model elements should also be created when they are missing. So, the mapping as shown below can just be applied in certain application scenarios, also called application *modes*: The rule can be applied to transform from a ctools model to a pnet model, but not backwards. Furthermore, if the ctools and pnet models both exist, the rule can check for a valid correspondence between the models. In particular, the enforce-keyword denotes that parts of the pnet model, which do not correspond to the ctools model, can be altered to establish a valid correspondence.

The domain areas are then structured in the way that the guard pattern is specified inside the parentheses following the domain identifier and the bottom pattern is specified in braces. The domain-guard patterns are fairly simple in this example, since they each contain just a single variable. But, in the bottom pattern, we see how the variable declaration is followed by a number of OCL expressions. These expressions describe the model structure, i.e. how the model objects should be referencing each other. Additionally, we see that there are two constraints formulated regarding a "type"-string of the Track's ports. In this example, these constraints are necessary to determine the in-Port and out-Port of the Track. For simplicity, these constraints were not reflected in the rule's graphical representation in Figure 4.

---

[3] QVT-Core is not restricted to just one mapping node in the bottom pattern of a rule. However, the QVT specification does not seem to intend the use of more than one.

Listing 1. The QVT-Core example rule in its textual notation

```
map"TrackToPlaceArcTransition{
""check"ctools(project:Project){
""""track:Track,"portIn:Port,"portOut:Port|
""""track.project"="project;"track.port"="portIn;"track.port"="portOut;
""""portIn.type"=""in";"portOut.type"=""out";
""}

""check"enfo□ce"pnet(petrinet:Petrinet){
""""□ealize"place:Place,"□ealize"arc:Arc,"□ealize"trans:Transition|
""""arc.arcToPetrinet":="petrinet;"arc.arcToPlace":="place;
""""arc.arcToTransition":="trans;"place.placeToPetrinet":="petrinet;
""""trans.transitionToPetrinet":="petrinet;
""}

""whe□e(t1:TProjectToPetrinet|
""""t1.project=project,t1.petrinet=petrinet){
""""□ealize"t2:TTrackToPlaceArcTransition|
""""t2.track":="track;"t2.inPort":="inPort;"t2.outPort":="outPort;
""""t2.place":="place;"t2.arc":="arc;"t2.transition":="trans;
""}
}
```

Looking at the bottom pattern of the enforceable `pnet` domain, we see some differences to the previous domain. Firstly, we see that the variables are marked as *realizable*. This means that, when missing, they can be created when the rule is applied. Secondly, we see that the OCL expressions contain an assignment symbol `:="` instead of a normal equals symbol. Since OCL is just a language to formulate constraints and queries, QVT introduces additional operations, called *assignments*, to assign reference or attribute values to model objects.

The mapping area is specified in the **whe□e**-section. Here, we see the trace variables of the guard and bottom pattern. The expressions in these patterns specify how they reference the variables in the domain areas.

2.2 TGG Rules

The actual idea of Triple Graph Grammars (TGGs) is to specify how two types of graphs relate to each other. Because software models can be considered graphs, this theory can be applied to models as well. We assume that we have two types of graphs given and their structure is specified by (single) graph grammars. Then, TGG rules allow us to specify how these single graph grammar rules structurally correspond to each other. This correspondence is expressed by inserting a third graph grammar, where the nodes provide a mapping by referencing the nodes in the other two graph grammars. This is illustrated in Figure 5 for the example rule from Figure 2. The generation of graphs through the simultaneous application of these corresponding graph grammars always results in structurally corresponding graphs.

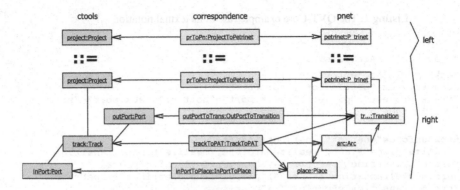

Fig. 5. The three graph grammar rules in a Triple Graph Grammar rule

Now, this formalism can be used to transform one graph into another. This is achieved by *parsing* an existing graph with the graph grammar on one side of the TGG. Then, during this process, the other graph grammar and the correspondence grammar of the TGG are applied to *create* the target graph and the correspondence graph. Another application of TGGs is to check two given graphs for their correspondence by parsing the two graphs simultaneously with the particular graph grammars in the TGG rules. In this process, the correspondence graph is built up and, in the end, represents the detailed correspondence of the nodes in the two graphs.

TGG rules are structurally very similar to QVT-Core rules. Figure 6 shows a collapsed representation of the above TGG rule. This collapsing is possible, because TGGs use only non-deleting graph grammars. This means that every element on the left-hand (top) side of the rule also appears on the right-hand (bottom) side. Note that we also introduced two attribute value constraints which were not present in Figure 7.

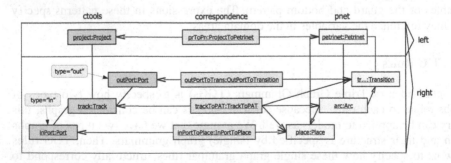

Fig. 6. A TGG example rule

Similar to QVT, there are three columns. The two columns containing the `ctools` and `pnet` patterns are called the *domains*. In fact, TGGs can also relate more than two models and may thus also be called Multi Graph Grammars (MGGs) [8]. In TGGs, the mapping nodes are called *correspondence nodes* and the middle column is therefore called the *correspondence domain*.

One difference between the above QVT-Core mappings and the TGG rules is that TGG rules have a graphical syntax. In Figure 4, we have just visualized the pattern structure made up by the OCL expressions in the QVT-Core rule. A TGG rule, in contrast, actually consists of *nodes*, representing the model objects, and *edges*, representing the references between these objects. Additionally, TGG rules may specify the attribute values of objects with *attribute value constraints*, as expressed by the rounded boxes containing labels attached to the nodes. Similar to QVT, OCL expressions can also be used for this purpose to specify literal values or values calculated from attribute values somewhere in the involved models.

After illustrating the structural similarities of QVT-Core and TGGs, the following section shows how the constructs of QVT-Core can be mapped to TGGs. The semantics of these rules are also very similar due to the fact that both QVT and TGGs are relational rules as illustrated in Figure 2. Remaining issues concerning slight differences in the semantics, or rather philosophies, are discussed in Section 4.

3 Mapping QVT to TGGs

As we have seen in the previous section, there are some apparent similarities between QVT-Core and TGGs. Therefore, a mapping can be specified from QVT-Core to TGGs. The full details have been worked out and implemented in a master thesis [6]. In the following, we informally describe the major steps of this mapping.

In the previous section, we observed that a variable in QVT-Core is essentially the same as a node in TGGs. Both a variable and a node are typed by a class in one of the involved models. Actually, in the redesigned TGG model proposed in [6], TGG nodes are also variables in terms of OCL, because the nodes should be reused as variables in OCL expressions as shown in Figure 6. Secondly, expressions which specify the reference values of objects in QVT are mapped to edges in TGGs. This holds for both OCL equality expressions, as `track.port=portIn,` and for assignments in an enforceable domain, as `arc.arcToPlace:=place.` Note that in TGGs, there is no distinction between an expression which is *enforceable* and one which is not. It is up to the transformation engine to decide the enforcement in a particular transformation scenario – but, we will come back to that in Section 4.

Apart from specifying reference values of objects, there are expressions in QVT which specify attribute values. For example equality expressions as `portIn.type"` `="in"` or assignments like `place.name:=track.name+"_"+portIn.name.` These expressions are not mapped to edges in TGGs, but to *attribute value constraints* as shown in Figure 6. Figure 7 summarizes the mapped constructs.

The overall structure of a single rule in QVT-Core is also quite similar to the structure of a TGG rule. The QVT-Core domain areas are mapped to TGG domain sides and, accordingly, the mapping area is mapped the correspondence column of a TGG. Then, the guard and bottom patterns in QVT-Core mappings are mapped to TGGs in such a way that the variables in the guard pattern of a QVT-Core mapping are mapped to such nodes in the TGG rule, which belong to the right-hand *and* the left-hand side. Variables which belong to the bottom pattern in the QVT-Core rule are mapped to nodes which belong to the right-hand side of the TGG rule *only*.

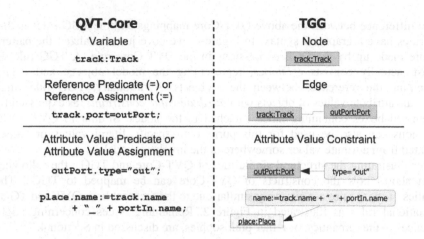

Fig. 7. Mapping the constructs of QVT and TGG transformation rules

Before designing a set of rules in QVT or TGGs, we need to specify a general transformation *setting*. This setting consists of references to the packages of the domain models involved in the transformation as well as the package of the correspondence or trace model. In the redesign of the TGG model proposed in [6], this description of a transformation setting was adopted from what is called the QVT-Base package in QVT. Therefore, technically, there is a one-to-one mapping at this level between QVT and TGGs.

We have actually specified the above mapping from QVT-Core to TGGs by a set of TGG rules. So, we use TGGs to transform QVT-Core mappings into TGG rules and, thus, we can perform model transformations specified in QVT-Core. One example of such a QVT-Core-to-TGG rule is shown in Figure 8. Here, a variable in the guard-pattern of a QVT-Core mapping is related to a node which belongs to the right-hand side and left-hand side pattern of a TGG rule. It is made sure that the QVT variable and TGG node belong to the correct domain column by referring to the corresponding CoreDomain and DomainGraphPattern in the context of the rule.

However, we do not explain the details of this rule, but rather illustrate the steps required to actually implement this transformation. Firstly, to create and transform QVT and TGG rules, both a QVT and TGG metamodel is needed. Because there are yet no implemented metamodels of the declarative QVT languages available, we implemented them anew, according to the QVT specification. A TGG metamodel and transformation engine was available from the ComponentTools project [9]. However, due to insights gained during the comparison of TGGs with QVT, we decided to conduct a redesign and reimplementation of the TGG technology. For the implementation, we chose the modeling framework of the Eclipse platform, EMF [10], as a basis. For one reason, The ComponentTools project and its TGG technology was already based on EMF. Another argument for using EMF is that it provides many useful features and that there are many interesting projects now based on EMF. One particularly interesting project is the Graphical Modeling Framework, GMF [11], where graphical editors can be generated from EMF models. Figure 8 actually shows a screenshot of a TGG rule in the generated graphical GMF-editor.

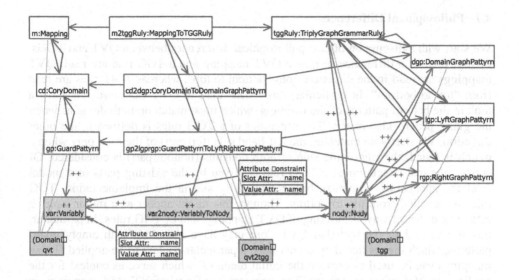

Fig. 8. Mapping the constructs of QVT and TGG transformation rules

Our TGG engine supports the interpretation of the TGG rules and is thus also called *TGG interpreter*. The integration of OCL is not yet completed, so that for now, only simple attribute value constraints are supported.

The transformation from TGGs back to QVT-Core is only possible under certain conditions. The backward mapping works if, firstly, we can assume that QVT-Core supports multiple trace variables in the bottom pattern. Secondly, because TGG rules do not specify if any domain is enforceable or not, the resulting QVT-Core would need to consider every domain area to be enforceable. A backward mapping from QVT to TGGs could nevertheless be interesting to evaluate advantages or disadvantages between the TGG interpreter and upcoming implementations of QVT-Relations or QVT-Core.

4 Comparing Concepts of QVT and TGGs

In the previous sections, we have discussed the basic concepts of QVT and TGGs. In this section, we relate the concepts of QVT and TGGs and discuss how the concepts of both technologies could benefit from each other.

The starting point of our work was that there are striking similarities and analogies between QVT and TGGs. We have seen in Section 3 that the basic structure of QVT and TGGs coincide and the rough meaning of these concepts are the same. But, there are some more or less significant differences, on the technical, the conceptual, and the philosophical level. Firstly, we discuss some philosophical differences. Then we compare the definitions of the semantics and some conceptual differences. In the end, we discuss some advanced features.

4.1 Philosophical Differences

We start with a discussion of two philosophical differences between QVT and TGGs. Firstly, there is a difference in how a QVT mapping and a TGG rule are read: QVT mappings are read in the direction from "bottom to top", whereas TGG rules are read from "top to bottom". In particular, this means that, in QVT-Core, we first have a look at the bottom patterns of the mapping, which must match on both domains when the guard pattern also match. The semantics of a TGG rules is defined in the other direction. We start with matching the left-hand side (top part) of the rule; only when a match is found, the right-hand side of both domains (bottom part) is considered. Of course, the real execution of a TGG rule is driven by the existing parts of a model (and the application scenario); but, conceptually and in the implementation, TGG rules are applied from top to bottom. Though, this might appear as a minor issue, it reflects a different way of thinking of QVT mappings and of TGG rules. In particular, this is reflected in the fact that QVT-Core may have mappings with empty guard patterns, which means that they do not need a particular context to be applied. These mappings can be used to provide the initial bindings, which serve as context for the application of further mappings. The counterpart of such QVT start mappings in TGGs is the axiom, which provides the start context for applying further rules in TGGs. The difference is that QVT can start with several start mappings in different parts of the model, whereas the matching of a TGG always starts with a single axiom.

Secondly, there is a fundamental difference in the way QVT and TGGs are applied in a concrete transformation scenario. In QVT, the application scenario, i.e. whether we perform a consistency check only or a transformation in one direction or the other, is partly encoded in the QVT mappings. The directives *check* and *enforce* within a QVT mapping state in which domain patterns will be created, deleted, or modified. Therefore, the QVT mappings are written with an application scenario in mind. By contrast, TGG rules do not refer to the application scenario. The same set of TGG rules can be used for checking consistency, for transformations in either direction, as well as for synchronizing both models after they have been changed independently of each other. The benefit is that we have a single set of TGG rules with the same semantics for all application scenarios—the only thing changing is the application scenario itself. This guarantees that the models are and remain consistent with respect to the single TGG specification, even when switching the application scenario.

This second philosophical difference has an important implication. In QVT, there is no explicit scenario which synchronizes two models that have been changed independently of each other. This can be achieved only by two subsequent incremental transformations in both directions[4]. By contrast, TGGs can be interpreted to synchronize two models in both directions in a single run at the same time.

4.2 Semantical Comparison

In Section 4.1, we have discussed the philosophical differences between QVT and TGGs already. Next, we ask a more technical question concerning the semantics of

---

[4] Actually, most of today's synchronization technologies use this approach; they incrementally transform the changes of one model into the other and vice versa subsequently.

QVT and TGGs. Let us assume that we have two models which are in the relation specified by the QVT mappings, resp. TGG rules. Now, we have a set of applications of these rules which prove that the models are in relation. Then, each variable or node is mapped or *bound* to the actual model elements. Now, let us consider these bindings in the opposite direction.

In TGGs, every object in the two models has exactly one binding to a *creating* node of an *applied* TGG rule (i.e. an application of a TGG rule where the node occurs on the right-hand side of this rule only). This results from the definition of the semantics of TGG rules which generates legal pairs of corresponding models. Conceptually, these pairs of models are defined by generating pairs of corresponding models starting from the axiom by applying some TGG rules: This way, both models are generated simultaneously. In the end, each object of the two generated models corresponds to exactly one creating node (right-hand side only) in an application of a TGG rule and vice versa. For more details, see [12]. In practical cases, however, we may want to describe only the relation between parts of the models. Therefore, the TGG rules will cover only these relevant parts and the rest of the models will be ignored. In these cases, there is exactly one binding of the relevant model objects to a creating node of an application of a TGG rule and there is no such binding for a model object which is not considered by the TGG.

The QVT standard, in contrast, does not make it clear whether there is at most one binding of each model object to a variable in a bottom pattern of exactly one application of a QVT mapping. Some explanations in the standard as well as the examples suggest that this is true and, thus, the interpretation seems to be very similar to the TGG interpretation. But, the mathematical formalization does not guarantee that. However, under the assumption that, at least for enforced variables, QVT also implies such a one-to-one correspondence between the relevant objects of the model and variables in the bottom patterns, the semantics of QVT-Core mappings can be mapped to TGG rules as discussed in Section 3. We consider this an open issue and a final justification of this assumption will have to be discussed.

This assumption is however supported by other considerations which concern verification. As pointed out in [13] and proved in a case study [14], TGGs can be used for verifying the semantical correctness of models that are transformed by TGGs. This kind of verification is possible, because of the one-to-one semantics of TGGs, which nicely reflects the definition of a semantics in SOS-style (structural operational semantics).

4.3 Conceptual Differences

In addition to the philosophical differences, there are some conceptual differences between QVT and TGGs. But, we will see that these can be easily aligned.

Though there is a graphical notation for QVT, QVT is conceptually more closely related to defining a set of *variables* and the definition of relations among these variables. These relations are defined in terms of OCL *expressions* and *assignments*. This makes QVT very flexible and expressive—due to the expressive power of OCL. By contrast, TGGs are graphical in nature and originate from the realm of graph grammars and graph transformations. A model is considered as a graph with *typed nodes* and edges between them. On the one hand, this results in a simple, precise and intuitive semantics (see Section 4.2); on the other hand, this imposes some restrictions. Some restrictions have been overcome by introducing different kinds of expressions to different variants of TGGs. Most of these extensions are straightforward, but not quite in the spirit of TGGs. However, inspired by QVT, we have shown how to introduce OCL constraints to TGGs without spoiling the "spirit of TGGs" and their graphical nature [12]. By equipping TGGs with OCL constraints, the essence of a relation between two models can be specified in a graphical way; still the expressive power of OCL is at hand when necessary. In particular, we do not need to distinguish between (querying) expressions and assignments in TGGs.

As mentioned above, QVT has the concepts of *check* and *enforce*. In fact, these keywords occur in two quite different ways in the QVT specification. So they actually constitute two different concepts. The first concepts defines the *mode* of application, i.e. it defines whether the mapping should be interpreted as a consistency check, or as a transformation in different directions, or as a synchronization. In TGGs, we call this the *application scenario*. So, *mode* and *application scenario* are just two different names for essentially the same concept. In QVT, however, *check* and *enforce* also occur within a mapping—with a similar but still different meaning. Some areas can be marked with **check** and **enfo□ce**. The idea of check and enforce in this context is that, during a transformation, an existing object of a model can be reused in such a mapping. Only if this object does not exist, it will be created. This increases the efficiency in transformations where parts of models already exist. Though there are many different extensions of TGGs, the basic form of TGGs does not have such a concept of *reusable nodes*. A node is either required (when it occurs on the left-hand side of the rule) or it is created (when it occurs on the right-hand side only). Based on the practical experience with TGGs and inspired by QVT, we have introduced a concept of *reusable nodes* to TGGs. There meaning is that these nodes can be reused if a node with the required properties already exists. Note that reusable nodes in TGGs do allow to reuse nodes, but they do not require their reuse. If we want to force the reuse of a node, this needs to be specified by additional global constraints, which will be explained shortly. Altogether, the concept of check and enforce of QVT can be expressed in TGGs by the concepts of reusable nodes and global constraints. Since reusable elements are identified on the level of individual nodes, TGGs can express this concepts on a finer level of granularity.

As mentioned above, we also introduced the concept of *global constraints* to TGGs. A global constraint allows us to enforce that a node with specific properties exists at most once in a model. This way, the reuse of a node can be enforced. Note that a global constraint is very similar to the concept of *keys* of QVT-Relations, which is used to uniquely identify an object by some of its attributes. This guarantees that a transformation does not generate duplicates of objects that have the same key.

A last difference between QVT and TGG is of technical nature: QVT was proposed in the context of the MDA. Therefore, QVT is defined based on MOF and uses the underlying concepts. TGGs were introduced long before the existence of MOF, and there are different implementations for different technologies— independently from MOF. But, there are implementations based on the Eclipse Modeling Framework (EMF), which in turn is an implementation of MOF.

4.4 Advanced Features

In addition to the differences discussed above, there are some advanced features of QVT and TGGs that are briefly discussed here.

First of all, QVT-Core allows to nest mappings within mappings. According to the QVT-Specification [2], this nesting of mappings helps to avoid inefficient and iterated deletion and creation of objects. Thus, nested mappings do not increase the expressive power of QVT mappings, but do increase the efficiency of their application. By contrast, it is not suitable to adopt the concept of nested rules to TGG rules. As pointed out before, this would not increase the expressive power, but only the efficiency. Efficiency, however, is an implementation matter. In TGGs, efficiency might be achieve in a simpler way, due to the top down interpretation from a single start context. Still, TGGs can be designed in such way that an implementation works efficiently. One of the main concerns for efficiency is to reduce the size of each individual TGG rule (because applying a rule basically means applying a graph-matching algorithm between this rule and the models).

A useful feature in QVT-Core is the refinement of rules by others, similar to inheritance in object orientation. This is a feature which is not yet present in TGGs. We feel that this is useful for better maintaining sets of TGGs and for making them more understandable. But, this can be built on top of the concepts of TGGs and it is not necessary to make this a core feature of TGGs.

As pointed out earlier, TGGs allow many different correspondence nodes in a single TGG rule. In QVT is seems to be possible to use more than one trace node in the bottom pattern, but it does not seem to be strongly encouraged. In TGGs, multiple correspondence nodes allow us to keep track of relations between individual model elements in a very detailed way. This helps to design TGG rules in a local way, which in turn results in simpler and smaller TGG rules. We identified examples where this clearly reduces the number and size of rules. In this way, multiple correspondence nodes compensate the efficiency of nested mappings in QVT. But, they do more: they also help us to design clearer and better understandable TGG rules.

5 Conclusion

In this paper, we have discussed the similarities of QVT and TGGs. Due to the similar structure and concepts, QVT mappings can be transformed into TGG rules. This way, a TGG engine can execute transformations specified in QVT-Core.

In addition, we have discussed the differences in the philosophy and concepts between QVT and TGGs. This improves our understanding of how model transformations and model synchronizations work with relational rules. The insights gained here have inspired the extension of TGGs and might provide valuable input for QVT as a standard. The goal is to have a clear and simple semantics to support a straightforward employment of model transformation technologies and to facilitate the use of validation and verification techniques.

Furthermore, it could be interesting to inspect the relation between QVT-Relations and TGGs. QVT-Relations provides more structure in order to simplify and better organize a set of transformation rules We believe that this could result in some concepts on top of TGGs, which could improve the comprehensibility of TGGs. Also we believe that there are additional concepts needed for the efficient synchronization of models, which go beyond transforming back and forth. This will need a closer investigation in the future.

References

1. Object Management Group (OMG): Model Driven Architecture - A Technical Perspective (July 2001) (last accessed 2nd of April 2007), http://www.omg.org/docs/ormsc/01-07-01.pdf
2. Object Management Group (OMG): MOF QVT Final Adopted Specification (November 2005) (last accessed 2nd of April 2007), http://www.omg.org/docs/ptc/05-11-01.pdf
3. Schürr, A.: Specification of Graph Translators with Triple Graph Grammars. In: Mayr, E.W., Schmidt, G., Tinhofer, G. (eds.) WG 1994. LNCS, vol. 903, Springer, Heidelberg (1995)
4. Wagner, R.: Developing Model Transformations with Fujaba. In: Proceedings of the 4th International Fujaba Days 2006, Bayreuth, Germany, pp. 79–82 (September 2006)
5. The MOFLON Tool Set (last accessed 30th of June 2007), http://www.moflon.org
6. Greenyer, J.: A Study of Model Transformation Technologies - Reconciling TGGs with QVT. University of Paderborn, Department of Computer Science, Master/Diploma thesis (July 2006)
7. Gepting, A., Greenyer, J., Kindler, E., Maas, A., Munkelt, S., Pales, C., Pivl, T., Rohe, O., Rubin, V., Sander, M., Scholand, A., Wagner, C., Wagner, R.: Component Tools: A vision for a tool. In: Kindler, E. (ed.) Algorithmen und Werkzeuge für Petrinetze (AWPN) - Algorithms and Tools for Petri nets. Proceedings of the Workshop AWPN, September 30th, October 1st 2004, pp. 37–42 (2004)
8. Königs, A., Schürr, A.: MDI - a Rule-Based Multi-Document and Tool Integration Approach Special Section on Model-based Tool Integration in Journal of Software&System Modeling. Academic Press, San Diego (2006)
9. Rohe, O.: Model Transformation by Interpreting Triple Graph Grammars: Evaluation and Case Study. Bachelor thesis, University of Paderborn (January 2006)
10. The Eclipse Project: The Eclipse Modeling Framework (last accessed 2nd of April 2007), http://www.eclipse.org/emf/
11. The Eclipse Project: The Graphical Modeling Framework (last accessed 2nd of April 2007), http://www.eclipse.org/gmf/
12. Kindler, E., Wagner, R.: Triple Graph Grammars: Concepts, Extensions, Implementations, and Application Scenarios. Technical Report, University of Paderborn, Department of Computer Science (June 2007)
13. Giese, H., Glesner, S., Leitner, J., Schäfer, W., Wagner, R.: Towards Verified Model Transformations. In: MoDeV^{2}a. Proceedings of the 3rd International Workshop on Model Development, Validation and Verification, Genova, Italy, pp. 78–93. Le Commissariat à l'Energie Atomique - CEA (October 2006)
14. Leitner, J.: Verifikation von Modelltransformationen basierend auf Triple Graph Grammatiken. Diploma thesis, University of Karlsruhe/TU Berlin (March 2006)

UniTI: A Unified Transformation Infrastructure*

Bert Vanhooff, Dhouha Ayed, Stefan Van Baelen, Wouter Joosen, and Yolande Berbers

Department of Computer Science, K.U.Leuven, Celestijnenlaan 200A, 3001 Leuven, Belgium
{bert.vanhooff,dhouha.ayed,stefan.vanbaelen,wouter.joosen,
yolande.berbers}@cs.kuleuven.be

Abstract. A model transformation can be decomposed into a sequence of sub-transformations, i.e. a transformation chain, each addressing a limited set of concerns. However, with current transformation technologies it is hard to (re)use and compose subtransformations without being very familiar with their implementation details. Furthermore, the difficulty of combining different transformation technologies often thwarts choosing the most appropriate technology for each subtransformation. In this paper we propose a model-based approach to reuse and compose subtransformations in a technology-independent fashion. This is accomplished by developing a unified representation of transformations and facilitating detailed transformation specifications. We have implemented our approach in a tool called UniTI, which also provides a transformation chain editor. We have evaluated our approach by comparing it to alternative approaches.

1 Introduction

Model transformations are a key ingredient of Model Driven Development (MDD). They can for example be used to add details to a model, incorporate non-functional concerns, convert between different types of models, and refactor certain constructs within a model. Model transformations quickly become complex when they need to address many concerns at once. Monolithic transformations, as most non-modularized software entities [1], have some inherent problems: little reuse opportunities, bad scalability, bad separation-of-concerns, sensitivity to requirement changes, etc. A number of these problems can be solved by decomposing a transformation into a sequence of smaller subtransformations, i.e. a transformation chain.

Currently, most transformation technologies do not very well support reuse and composition of subtransformations as high-level building blocks. One of the causes is the fuzzy distinction between specification, implementation and execution of transformations [2]. We define these terms as follows:

Implementation is the transformation source code, as seen by the developer. For many common transformation languages this comes down to a set of mapping rules.
Specification is the documentation that describes how a transformation behaves, independently of its concrete implementation. We focus on the functional interface

---

* The described work is part of the EUREKA-ITEA MARTES project, and partly funded by the Flemish government institution IWT (Institute for the Promotion of Innovation by Science and Technology in Flanders).

G. Engels et al. (Eds.): MoDELS 2007, LNCS 4735, pp. 31–45, 2007.

(similar to programming languages) of a transformation in terms of input and output model types.

Execution is the runtime instance of a transformation that has concrete input models and produces one or more output models.

Furthermore, current transformation technologies offer little to no support for combining transformation technologies even though in practice different technologies may be suitable for different parts of a transformation chain. Lastly, the focus of transformation technologies on local implementations prevents them from adressing issues that go beyond the boundary of a single transformation such as end-to-end traceability.

In this paper we propose a technology-neutral view on model transformations: the Unified Transformation Representation (UTR). We have implemented a tool called UniTI (Unified Transformation Infrastructure) based on UTR that facilitates transformation composition and execution without having to know underlying implementation details. UniTI provides the groundwork to incorporate cross-transformation services such as traceability, mentioned in the previous paragraph.

The remainder of this paper is structured as follows. In Section 2 we give a concise overview of current transformation techniques and identify a number of concrete problems concerning (re)usability and composability. We then list a set of characteristics that a solution to these problems must possess (Section 3). In Section 4 we present the Unified Transformation Representation; the implementation of the latter – UniTI – is discussed in Section 5. We summarize related work in Section 6 and evaluate our approach by comparing it to alternative solutions in Section 7. Finally, we round up by drawing conclusions and identifying future work in Section 8.

2 Using and Composing Transformations

Current transformations technologies all have their own specific vision on model transformations. In Section 2.1 we give a concise overview of a selection of available transformation technologies. We then zoom in on a number of shortcomings of transformation technologies related to the creation of transformation chains in Section 2.2.

2.1 Characteristics of Current Transformation Technologies

A lot of effort has already been spent on the development of suitable languages to express model transformations. Key examples are ATL [3], MTF [4], VIATRA [5], UM-LAUT [6] and OMG's QVT [7]. Note that generic programming languages such as JAVA can also be used to implement transformations. The main objective of specialized languages is to provide a number of powerful yet easy to use constructs to express relations between model elements. Many of the mentioned transformation languages take a different approach to implementing transformations. They differ in notational style (graphical:VIATRA; textual: ATL, MTF or a combination:QVT), specification style (imperative:JAVA; declarative:MTF or hybrid:ATL, QVT), directionality (unidirectional:ATL or omnidirectional:MTF), supported model types (UML [8]:UMLAUT; MOF [9]; Ecore [10]), etc. The examples given here cover only a small subset of available transformation languages; a more complete discussion of these and other differences between transformation languages can be found in [11].

In this subsection we describe to what extent one must be familiar with a transformation's implementation in order to use it in a meaningful way. We take ATL, MTF and JAVA as reference languages; we believe these cover a substantial part of the existing transformation approaches.

MTF enables the implementation of transformations by specifying an arbitrary number of declarative mapping rules. Each one of these rules denotes a relationship between one or more model elements and can call out to other rules. An MTF transformation is always omnidirectional, which means that any of the (meta)models involved can be designated both as input or as output. Moreover, a transformation can be initiated from any of the mapping rules, possibly producing a different result. In order to execute an MTF transformation we need to select an appropriate *initial rule* and select the *direction* of the transformation. One should be very careful when choosing a direction. It is not because a transformation turns model A into B that applying the same transformation to B yields A. We will not go into the intricate details of omnidirectional transformations but we emphasize that it is far from trivial to choose a specific direction and initial rule without being very familiar with its implementation details.

In ATL we use transformation rules that have both declarative and imperative characteristics to relate a number of models. The transformations implemented as such are unidirectional. ATL transformations are, to some extent, metamodel independent – references to metamodels are not included in the implementation. Only when we execute a transformation, we need to select concrete metamodels. The range of valid metamodels directly depends on the set of transformation rules. For example, if rules only refer to 'Class' and 'StructuralProperty' we can execute the transformation for any version of the UML metamodel. When we add an additional rule that refers to 'Port', we can only choose UML2 since that element is not present in lower versions. Hence, the set of valid metamodels can only be derived by studying the implementation.

A transformation written in JAVA, or any other general purpose language, accesses models directly via the model repository. Hence, the developer is not bound to any rules to implement a transformation. There is no standard technique to specify the input and output models; this can be done through method parameters, command-line arguments, files, etc. The same goes for specifying the entry point of the implementation (which method should be invoked?). So, in case of JAVA, we need complete knowledge of the implementation in order to use it.

2.2 Limitations

In this subsection we identify a number of deficiencies in the area of current transformation technologies with respect to using transformations as (reusable) building blocks in transformation chains.

Incomplete Specification. It is clear from the previous subsection that an intimate knowledge of a transformation's implementation is required in order to use it. The following elements are not clear from the implementation alone:

- MTF : direction; initial rule
- ATL : concrete metamodels
- JAVA : concrete metamodels; direction; entry point, ...

In order to be able to (re)use, compose and execute transformations we must clearly fix the possible values of these enumerated properties.

Imprecise Specification. When we have filled the specification gaps of each technology, we can say that a transformation is mainly characterized by its input and output model types. Metamodels are used to type models: a model *conforms to* a metamodel. This implies that if we know the metamodel we automatically know what kind of models to expect. This is indeed true, but the sheer amount of possible models that conform to a (complex) metamodel often make this an inadequate way to type models (from a transformation point of view). We give examples in three categories:

– *Metamodel variation* In order to allow a greater flexibility, a metamodel can deliberately be left incomplete. For example, the UML has a number of so-called semantic variation points, which need to be resolved before using the metamodel.
– *Metamodel delimitation* Complex metamodels such as the UML are not often used completely; a transformation usually only considers a subset of UML. For example, a UML to RDB (Relational DataBase) transformation, only takes structural elements such as classes and associations into account and ignores other elements such as actions and states.
– *Model structure* Next to global metamodel concerns, transformations may also make assumptions at the model level. For example, requiring specific model elements such as a 'Car' class, a directed association named 'fuel', etc.

Mind that the above issues are not due to shortcomings of particular metamodeling languages. Metamodels such as the UML are conceived to accommodate a wide range of modeling possibilities, making them reusable in many domains. A transformation, however, often only makes sense on a small subset of all the possible instances of a metamodel.

Current transformation languages cannot directly express such subsets. The best they can do is check whether a model belongs to the expected subset at execution time, which requires additional code that pollutes the implementation. This also means that the requirements on the involved models are hidden inside the implementation. In order to check whether we can successfully execute a transformation or whether transformations can be connected in a chain, these subsets must explicitly be defined as part of its specification.

Technology Lock-in. Different technologies can be suited to implement different types of transformations. Figure 1 suggests a technology for each subtransformation in an imaginary transformation chain.

To go from *Modela* to *Modelb* we use a generic transformation language that supports any metamodel (e.g. ATL). To make minor changes to *Modelb* we prefer an in-place (source=target) technology (e.g. JAVA). In order to combine *Modelb* and *Modelc* we use a model weaving engine (e.g. AMW [12]). Finally, we use a domain specific transformation language (XSTL) to transform XML models.

If we choose a different technology for each subtransformation, we must ensure the ability to combine the technologies. Since current transformation technologies offer no (or only very basic by QVT) possibilities to call out to other independently defined

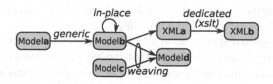

Fig. 1. Different types of transformations in a transformation chain

transformations, it is very hard to mix different technologies. In order to realize a cross-technology transformation composition we need to write additional, external *glue code*.

Megamodel Concerns. Megamodel management is the term introduced by Bézivin et al [12] to indicate the need to establish and use global relations between macroscopic MDD entities such as (meta)models and transformations while ignoring the internal details of these entities. Reusing and composing transformations in transformation chains is one of the use cases within this area.

Current transformation languages focus on implementation details of individual transformations and cannot express much at the level of complete models or transformations. A higher level transformation infrastructure is needed to offer additional functionality on top of a transformation chain, such as end-to-end traceability [13].

3   Characteristics of Reusable and Composable Transformations

Since we are using transformations as building blocks, our point of view is related to the field of component based software engineering (CBSE) [14]. We will use the main principles of CBSE as guidelines to solve the issues raised in the previous section. These principles, reformulated to fit the transformation domain, are summarized below. We also describe deficiencies of current transformation technologies with respect to each principle.

Black-Box principle. *The black-box principle is the strict separation of a transformation's public behavior and its internal implementation. Implementation hiding makes any technique an eligible candidate to implement the transformation.*

As discussed in the previous section, transformation specifications in the considered languages are both incomplete and imprecise. We need to look into their implementation in order to use them, which violates the black-box principle.

Composition. *Constructing a complete transformation with reusable transformation building blocks should be considerably less work than writing it from scratch. A transformation component should be a self-contained unit and composition should be as easy as possible without the need for much glue code.*

Neither ATL or MTF have an explicit notion of (external) transformation composition; they only support low-level internal composition of individual transformation rules (see Section 2.2). Some other transformation languages have limited high-level composition support (QVT, VMT [15]) but cross-technology composition is generally hard to accomplish.

External specification. *Each transformation should clearly specify what it requires from the environment and what it provides.*

An ideal external specification should only provide detailed information about the expected input and provided output models of a transformation (its environment). All other information such as metamodel choice, initial rule, invocation method, etc. is not relevant at this level and should be hidden in the specification. The external specification is only weakly defined in current transformation technologies (at many points implementation knowledge is required) and differs profoundly among the different technologies (see Section 2.1).

4 Unified Transformation Representation

In this section we describe a common, model-based, representation for transformations (Unified Transformation Representation – UTR) to establish a common ground among different transformation technologies. The UTR separates implementation, specification and execution of transformations and contains concepts to type and compose transformations (see Section 4.1). We also show how the UTR metamodel gives rise to a number of transformation roles in Section 4.2 and discuss a practical usage scenario that involves these roles in Section 4.3.

4.1 UTR Metamodel

Figure 2 shows an extract of the UTR metamodel. We clarify this metamodel in the following paragraphs.

Basic Transformation Concepts. We make a distinction between the *specification* of a transformation, which encapsulates a concrete *implementation*, and the *execution* or instance of a transformation, which is subject to composition.

Each *TFSpecification* is characterized by an implementation and an external specification. An *AtomicTFSpecification* defines the transformation directly in terms of a specific transformation technology. Alternatively, a *CompositeTFSpecification* is itself expressed as a chain of subtransformations (inherits from *TFChain*). Implementation details, whether atomic or composite, are hidden by subclassing (*ATLImpl, MTFImpl* and *JavaImpl*). Notice that it is not our intention to allow multiple implementations for one *TFSpecification* but rather to provide a clear specification that hides intricate technology and implementation details.

A *TFSpecification* is represented in terms of input and output models, denoted by *TFFormalParameters* and their respective model types, denoted by *ModelingPlatforms*. We explain the concept of *ModelingPlatform* in greater detail later in this section, for now it suffices to see it as a constrained metamodel. The combination of *TFFormalParameters* and *ModelingPlatforms* specify exactly what a *TFSpecification* requires from its inputs and provides on its outputs and hence defines the context in which a transformation can be meaningfully executed. *ModelingPlatforms* and *TFSpecifications* can be grouped and organized in *TFLibraries*.

The lower part of Figure 2 represents the execution level. *TFExecution* and *TFActualParameter* are the runtime counterparts (or instances) of *TFSpecification* and *TFFormalParameter*. *TFActualParameter* is a container for a concrete *Model*, so only at this

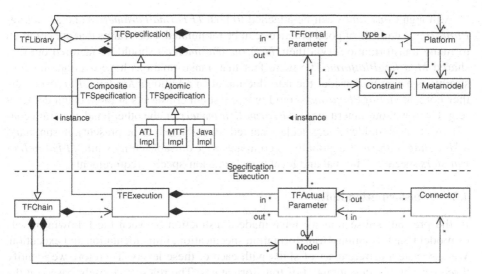

Fig. 2. Extract of the UTR metamodel

level we can execute a transformation. We introduce the *Connector* element to intercon-nect two *TFExecutions* through their *TFActualParameters*. The *Connector* is modeled as a class so that it can provide additional behavior such as verifying models, pause execution, etc. A connection is valid only if the *TFActualParameters* have the same type (equal *ModelingPlatforms*). Finally, the *TFChain* class represents a transformation chains that is composed of *TFExecutions*, *Connectors* and *Models*.

Model Types. As discussed in Section 2.2, a metamodel is not always sufficient to accurately specify inputs and outputs of a *TFSpecification*. We need to be able to tune existing metamodels such as UML to our needs (metamodel variation and delimitation) and to enforce local transformation requirements (model structure).

In classical programming languages, similar problems can be encountered when specifying the signature of an operation. Some programming languages offer as a so-lution preconditions and invariants (Design by Contract [16]) that limit the range of allowed parameter values. We propose a similar approach to solve the limitations of typing by metamodel. We allow additional constraints on each transformation input so that we can exactly describe what is expected from the input model, beyond the struc-ture imposed by the metamodel.

The specification of input and output models is accomplished in two stages (see Figure 2): (1) through the transformation-specific *TFFormalParameters* and (2) through cross-transformation reusable *ModelingPlatforms*. Both can impose *Constraints* on the models. We must stress that we do not refer to execution platforms such as J2EE or .NET in any way. A *ModelingPlatform* is composed of a *Metamodel* along with a num-ber of *Constraints* that describe our expectations on the structure of the model beyond those captured in the metamodel. Such constraints are typically expressed using OCL [17]. In this way *ModelingPlatforms* allow for a more controlled reuse of existing meta-models. A *TFFormalParameter* is typed by a *ModelingPlatform* and can impose addi-tional, transformation-local, constraints on the global *ModelingPlatform*.

Although *Constraints* can be attached to both *TFFormalParameters* and *Modeling-Platforms*, a number of good practices can be formulated. Constraints that are common to many transformations (via their *TFFormalParameters*) should be factored out to a shared *ModelingPlatform*. The mere fact that transformations have overlapping constraints is a sign to consider the introduction of an additional *ModelingPlatform*. Furthermore, *ModelingPlatforms* should only constrain constructs at the metamodel level (e.g. UML without inheritance). *TFFormalParameters* on the other hand constrain constructs at the model level (e.g. a class named 'Proxy' needs to be present). In summary, a *ModelingPlatform* is a globally or company-wide reusable entity while *TFFormalParameters* specify additional and local transformation-specific requirements.

4.2 Transformation Roles

In the previous subsection we have made a distinction between the following levels of model transformations: transformation specification, implementation and execution. We associate a different set of skills with each of these levels. Therefore we identify three roles in the development of transformations. The roles are loosely based on the roles of Knowledge Builder, Facilitator and User as proposed in [18].

Transformation Developer. A person in this role is responsible for implementing transformations using the most appropriate technique (ATL, JAVA, etc.). In many cases this role will be anonymous, for example when reusing third-party transformations.

Transformation Specifier. The Transformation Specifier is responsible for the external specification of a transformation. Depending on the context, this role performs a different task, hence we split it up in two subroles:

The **designer** subrole is used in the classical specification-before-implementation sequence. The designer gathers the requirements and defines the external specification of the required transformations. This information is then passed on to a Developer, who implements the given *TFSpecification*.

Because many transformation implementations are readily available we also introduce the more pragmatic role of the **harvester**. In this subrole the Specifier searches for available and appropriate transformation implementations that could be of use in a project. The harvester thoroughly studies and tests the implementations (a reverse engineering activity) so that appropriate *TFSpecifications* can be created. Basically this involves creating an explicit manifestation of the implicit assumptions that are present in the implementation.

Transformation Assembler. This role selects appropriate *TFSpecifications* and composes *TFExecutions* into a transformation chain that realizes an overall transformation goal. The Assembler is shielded from low level implementation details because only *TFSpecifications* (provided by the Specifier) must be looked at.

4.3 Usage Scenario

To get a better view on how UTR concepts can be used and where the different roles come in, we present a typical usage scenario (see Figure 3). We describe the creation

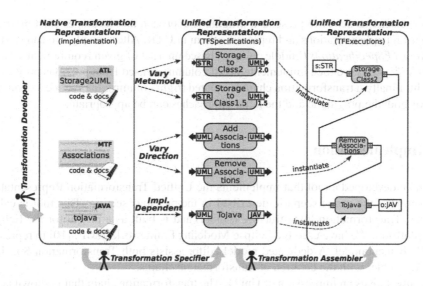

Fig. 3. A typical usage scenario. Part of the notation is derived from UML Activity diagrams.

of a transformation chain (TFChain) that takes a *Storage* model as input and produces a corresponding JAVA model. The input model specifies a storage application such as a library or warehouse in a domain specific modeling language.

The process of building the transformation chain consists of three steps. **First**, the Transformation Specifier/Harvester searches for candidate transformation implementations that can contribute to the chain. We have selected three relevant transformation implementations (left column of Figure 3): one converts Storage to UML models (ATL), one changes Associations within the UML model (MTF) and one transforms UML to equivalent JAVA models (JAVA). The only information that is available at this point is the source code of the transformation and textual documentation. As explained in Section 2, we need to fix many variables before we can execute any of the transformations.

In a **second step** we translate the transformation implementations into UTR Transformation Specifications. The Transformation Specifier is again responsible for this step. Due to the variability points of the implementations (e.g. direction, metamodel), there are often several alternative but valid Specifications possible. For example, in the middle column of Figure 3, we have defined a separate Specification for each direction of the omnidirectional MTF transformation *Associations*: *AddAssociations* and *RemoveAssociations*. In the case of ATL (*Storage2UML*) we have varied the metamodels to target UML2.0 or UML1.5. The latter two are materialized as *ModelingPlatforms*.

The **final step** is to create the actual transformation chain from the building blocks provided by the previous steps. We are completely shielded from transformation implementation details at this stage. The Transformation Assembler instantiates the appropriate *TFSpecifications*, which yield *TFExecutions*, and composes them using *Connectors* (see right of Figure 3). Mind that Connectors can only connect *TFActualParameters* of the same type: their constraints must be compatible and their *ModelingPlatforms* must be the same. Finally we provide the input model *s* and a container for the result *o*.

The discussed scenario presents a pragmatic approach that leverages existing transformation implementations and represents them in UTR. Alternatively, we can start by defining *TFSpecifications* (middle column) based on a set of given requirements. These *TFSpecifications* are then implemented (left column), reused from previous projects or bought. Finally a transformation chain is created (right column). In practice a combination of both the pragmatic and the latter approaches can be appropriate.

5   Implementation

We have developed a tool that implements the Unified Transformation Representation and supports the usage scenario described in the previous section. This tool is called Unified Transformation Infrastructure (UniTI) and is built as a plugin for the Eclipse [19] platform. We have used the Eclipse Modeling Framework (EMF) [10] to represent the UTR metamodel. A dedicated model editor assists both Transformation Specifier and Assembler with the creation of transformation chains.

Figure 4 gives an impression of UniTI. The transformation chain that is shown is the same as in the usage scenario of Section 4.3. The left part of the figure shows a library of *TFSpecifications*. A number of so-called *wizards* assist the Transformation Specifier by asking for the necessary implementation details of each technology. *TFSpecifications* are then automatically generated. Note that the Transformation Specifier can now alter the generated *TFSpecifications* to his likings without destroying the coupling with the

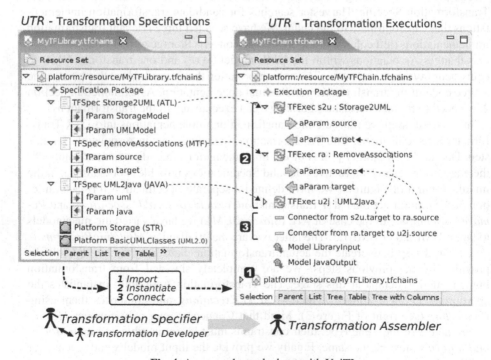

Fig. 4. An example worked out with UniTI

underlying implementation. Typically, *ModelingPlatforms* are refined at this point – e.g. the *BasicUMLClasses ModelingPlatform* in the figure.

If we shift focus to the right of Figure 4, we see the actual transformation chain model. Before modeling can start we need to import the necessary *TFLibraries* – only 'MyTfLibrary' in this case. We can now instantiate the necessary *TFExecutions* and connect them appropriately. The transformation chain can then be executed. Execution of each transformation is taken care of behind the scenes and hides technology-specific details. All intermediate models are automatically saved and are available for inspection during and after execution. UniTI supports parallel execution of transformations by allowing multiple *Connectors* on the same *TFActualParameter*. Conditional execution is not yet supported.

We allow extensions of UniTI with new transformation technologies through two mechanisms. A *lightweight* extension mechanism is provided by the standard support for JAVA transformations. A simple JAVA interface that represents JAVA transformations can be reused to encapsulate any transformation technology that provides a JAVA API. A more *heavyweight* approach is by extending the metamodel by subclassing the Atomic Transformation Specification class. The main advantage of the heavyweight approach is better integration in UniTI; for example, we can provide a technology-specific wizard that simplifies the creation of Transformation Specifications.

6   Related Work

In our work we consider models and transformations as course-grained building blocks of an MDD approach. In the following paragraphs we discuss other approaches that apply a similar point of view.

In [2], transformation composition is seen as the composition of different tools that support a number of dedicated transformations or provide a generic transformation specification facility. An Eclipse plugin, the Model Control Center (MCC) allows the creation of transformation *networks* by using a simple scripting language that allows sequential, parallel and conditional execution. Our approach is similar to [2] by making a clear distinction between the specification and execution of transformations and allowing different transformation technologies to work together. Our approach differs in the way we specify transformations. They reuse JAVA interfaces to encapsulate model types while we have the dedicated concept of Transformation Parameters and Platforms. We make a more detailed comparison of UniTI and MCC in the next section.

Another approach similar to ours is described in [20]. They use a classical component system as a Transformation Composition Framework (TCF). Transformations and models are encapsulated in components and component interfaces act as model types, execution entry points and context information nodes. They provide a basic transformation language for defining simple transformations; composite transformations are facilitated by the existing component framework. As in [2], transformations/models are typed by regular interfaces which limits their preciseness. They are also limited to their own transformation language, although this limitation seems easy to circumvent. Most of the limitations of this approach are a direct cause of reusing an existing component framework. UniTI avoids many of these limitations by offering more dedicated constructs. TCF will be compared to our approach in more detail in the next section.

In [21] a Domain Specific Language (DSL) is described to compose JAVA metamodel-specific transformation classes and automatically choose the right transformation instance depending on the preceding transformations. Our approach is similar to this one in that we also use a kind of DSL, which is defined by the UTR metamodel.

OMG's QVT [7] supports chaining of internal (QVT native) and external (implemented in another language) transformations at the transformation language level. It is, however, up to the tool implementer to provide the mechanisms to call out to external transformations so the composition of subtransformations is limited in practice. QVT supports precise model types by metamodel and compliance kind (strict, effective). This approach is comparable to, but more restricted than, our notion of Platform.

A number of approaches make use of classical build tools such as ANT [22] to manage transformation compositions. Examples are AM3/ATL [12] and OMELET [23]; the latter tries to extend build tools with the notion of metamodel as data type in order to fit them better to transformations.

Finally the MDDI ModelBus [24] provides a middleware that enables the execution of all kinds of model services provided by different tools in a technology-neutral fashion. Transformation services offered through the ModelBus could be integrated in UniTI through the provided extension mechanisms.

In summary, we proposed a model-based solution for transformation reuse and composition while most other approaches reuse existing technology to facilitate transformation chaining. Hence, the main advantage of our approach is that the abstractions that we use map very well to the transformation domain, while this is not always the case for the other approaches. A more detailed comparison of UniTI, MCC, TCF and ANT is made in the next section.

7 Evaluation

In this section we evaluate UniTI by comparison with other approaches that offer support for composition and reuse of transformations. We also look at how regular transformation languages support modularization and composition of transformations.

Table 1 shows a comparison of selected characteristics of UniTI, ANT, TCF and MCC. Although ANT cannot be considered as a dedicated transformation composition tool we have included it because it is used for that purpose in practice. For each of the characteristics we have indicated the corresponding element/value in each of the approaches. For example, the transformation module of UniTI is the TFSpecification while for TCF it is a regular component and for ANT there is no such concept.

In order to get a measure of the effort required for defining a transformation chain in each of these tools we have expressed our running example (see Subsection 4.3) in each of them. The amount of necessary glue code was then recorded as lines of code. Because UniTI does not offer a textual syntax we have recorded the amount of model elements that needed to be created instead. We looked at the transformation chain from the perspective of a Transformation Assembler. Therefore we never counted the specification of a transformation in the glue code if it was possible to leave it out.

Table 1. Qualitative comparison of UniTI, ANT, TFC and MCC

	UniTI	ANT	TCF	MCC
Tf Module	TF Spec. (wizard)	N/A	Component	Eclipse plugin
Tf exchange/reuse	import library	N/A	copy component	install plugin
Model Typing	Platform	N/A	JAVA Interface	JAVA Interface
Tf Composition	model-based	script	component composition	script
Cross-technology	built-in	manual	manual?	manual
Glue code	7 model elements	~60 lines	~20 lines	~6 lines

From this comparison we can conclude that ANT offers the least support for the Transformation Assembler. UniTI, TCF and MCC offer substantial advantages in the area of model typing, chaining and exchange. Notice that UniTI is the only approach that takes a model-based approach. It offers transformation concepts as first-class entities (e.g. Transformation Specification) while others reuse existing notions such as components and Eclipse plugins to represent transformations. Therefore, UniTI has very expressive model typing, easy exchange of transformations and concise glue code. Furthermore, UniTI has built-in support for different transformation languages and is thus the only approach that also supports the task of the Transformation Specifier by automatically generating Transformation Specifications.

Table 2. Modularization alternatives for transformations

Transformation chain	*Rule-based*	*Modularized rule-based*
transf GET **transf** SET **connect** GET.out,SET.in	**transf** getset **rule** getset **from** Property **to** Operation 'get' **to** Operation 'set'	**transf** getset **rule** get **from** Property **to** Operation 'get' **rule** set **from** Property **to** Operation 'set'

In this paper we have focussed on composition of complete transformations as building blocks. Since the modularization mechanisms offered by transformation languages are usually situated at a lower level, e.g. transformation rules, it is hard to make a meaningful comparison. Instead, we believe that both techniques have to be used in conjunction. In some cases it might be more opportune to modularize a transformation at the implementation level, while in others, composition of larger building blocks might be a good solution. In Table 2, we show alternative possibilities to modularize a transformation that adds accessor and mutator operations (i.e. getters and setters) to a Class, expressed in pseudo-code. The leftmost definition makes use of transformation chaining while the other definitions are expressed in a rule-based transformation language. The middle definition shows a naive implementation and the rightmost definition decomposes the transformation into two separate rules. For the transformation chaining approach, one line of code corresponds with one model element in UniTI.

At this point it is not clear which (de)composition strategy is best for different situations. We believe that both approaches can live in harmony, but more experience/research is required in order to define a set of guidelines that guide the developer in choosing the right strategy for each situation. A more thorough study is out of the scope of this paper. For a more detailed evaluation of the composition possibilities offered by rule-based transformation languages we refer to [25].

8   Conclusions and Future Work

A possible approach to implement complex transformations is by composing many simple subtransformations. We showed that this is not so easy to accomplish with current transformation technologies, certainly not when mixing different technologies. The main problem is the weak separation of specification and implementation of transformations, which requires a deep implementation knowledge both when reusing and composing transformations. Therefore we have proposed a model-based approach that offers a unified view on transformations with a clear distinction between implementation and specification. Transformation chains can thus be created without bothering with the technological or implementation details of each subtransformation.

The core concepts of our approach are defined in the Unified Transformation Representation (UTR) metamodel, which fulfills the following characteristics for transformations: the *black-box principle*, subject to *composition* and clear *external specification*. In order to realize these we made a clear distinction between implementation, specification and execution of transformations. Our approach gave rise to different usage roles: Transformation Developer, Specifier and Assembler. Each of these roles has a restricted view on transformations and is assigned with a clear set of responsibilities. The Developer implements a transformation in the most appropriate technology, the Assembler composes transformations without having to know anything about the underlying implementation and the Specifier mediates between the former two by providing Transformation Specifications. UniTI implements the UTR metamodel and assists both Transformation Specifier and Assembler to create transformation chains in a technology-transparent fashion.

In future work we will investigate how to offer cross-transformation services such as end-to-end traceability. We will explore how this traceability information can be used throughout the transformation chain to improve subsequent transformations.

References

1. Parnas, D.L.: On the criteria to be used in decomposing systems into modules. Commun. ACM 15, 1053–1058 (1972)
2. Kleppe, A.: Mcc: A model transformation environment. In: ECMDA-FA, pp. 173–187 (2006)
3. Jouault, F., Kurtev, I.: Transforming models with atl. In: Briand, L.C., Williams, C. (eds.) MoDELS 2005. LNCS, vol. 3713, Springer, Heidelberg (2005)
4. IBM Alphaworks: Model transformation framework. Misc (2004),
   http://www.alphaworks.ibm.com/tech/mtf
5. Csertán, G., Huszerl, G., Majzik, I., Pap, Z., Pataricza, A., Varró, D.: VIATRA Visual Automated Transformations for Formal Verification and Validation of UML Models. In: Proceedings of the 17th IEEE international conference on Automated software engineering, IEEE Computer Society Press, Los Alamitos (2002)

6. Ho, W.M., Jezequel, J.M., Pennanc'h, F., Plouzeau, N.: A toolkit for weaving aspect-oriented uml designs. In: Proceedings of the 1st Conference on Aspect-Oriented Software Development, pp. 99–105. ACM Press, New York (2002)
7. Object Management Group: Qvt-merge group submission for mof 2.0 query/view/transformation. Misc (2005)
8. Object Management Group: Uml 2.0 superstructure ftf convenience document. Misc (2004)
9. Object Management Group: Meta object facility (mof) 2.0 core specification. Misc (2004)
10. Budinsky, F., Brodsky, S.A., Merks, E.: Eclipse Modeling Framework. Pearson Education (2003)
11. Czarnecki, K., Helsen, S.: Classification of model transformation approaches. In: OOPSLA 2003 Workshop on Generative Techniques in the context of MDA (2003)
12. Bézivin, J., Jouault, F., Rosenthal, P., Valduriez, P.: The AMMA platform support for modeling in the large and modelling in the small. Technical Report 04.09, LINA (2004)
13. Barbero, M., Fabro, M.D.D., Bézivin, J.: Traceability and provenance issues in global model management. In: 3rd ECMDA-Traceability Workshop (2007)
14. Szyperski, C.: Component Software: Beyond Object-Oriented Programming. Addison-Wesley Professional, Reading (1997)
15. Sendall, S., Perrouin, G., Guelfi, N., Biberstein, O.: Supporting model-to-model transformations: The vmt approach. Technical report (2003)
16. Meyer, B.: Applying "design by contract". Computer 25, 40–51 (1992)
17. Object Management Group: Uml 2.0 ocl final adopted specification. Misc (2003)
18. Gavras, A., Belaunde, M., Almeida, L.F.: Towards an mda-based development methodology. In: Oquendo, F., Warboys, B.C., Morrison, R. (eds.) EWSA 2004. LNCS, vol. 3047, pp. 230–240. Springer, Heidelberg (2004)
19. Beaton, W., d. Rivieres, J.: Eclipse platform technical overview. Technical report, The Eclipse Foundation (2006)
20. Marvie, R.: A transformation composition framework for model driven engineering. Technical Report LIFL-2004-10, LIFL (2004)
21. Wagelaar, D.: Blackbox composition of model transformations using domain-specific modelling languages. In: Rensink, A., Warmer, J. (eds.) ECMDA-FA 2006. LNCS, vol. 4066, Springer, Heidelberg (2006)
22. Moodie, M.: Pro Apache Ant (Pro). Apress, Berkely, CA, USA (2005)
23. Willink, E.D.: Omelet:exploiting meta-models as type systems. In: 2nd European Workshop on MDA with an emphasis on Methodologies and Transformations (2004)
24. Blanc, X., Gervais, M.P., Sriplakich, P.: Model bus: Towards the interoperability of modelling tools. In: MDAFA, pp. 17–32 (2004)
25. Kurtev, I., van den Berg, K., Jouault, F.: Evaluation of rule-based modularization in model transformation languages illustrated with atl. In: Proceedings of the 2006 ACM symposium on Applied computing, pp. 1202–1209. ACM Press, New York (2006)

Guided Development with Multiple Domain-Specific Languages

Anders Hessellund[1], Krzysztof Czarnecki[2], and Andrzej Wąsowski[1]

[1] IT University of Copenhagen, Denmark
{hessellund,wasowski}@itu.dk
[2] University of Waterloo, Canada
kczarnec@swen.uwaterloo.ca

Abstract. We study the Apache Open for Business (OFBiz), an industrial-strength platform for enterprise applications. OFBiz is an example of a substantial project using model-driven development with multiple domain-specific languages (DSLs). We identify consistency management as one of its key challenges. To address this challenge, we present SmartEMF, which is an extension of the Eclipse Modeling Framework that provides support for representing, checking, and maintaining constraints in the context of multiple loosely-coupled DSLs. SmartEMF provides a simple form of *user guidance* by computing the valid set of editing operations that are available in a given context. We evaluate the prototype by applying it to the OFBiz project.

1 Introduction

Successful development and customization of ever more complex enterprise systems depends on effective collaboration between several stakeholders as well as on a flexible and coherent conceptualization of the problem domain. Among the different approaches towards tackling this challenge, *domain-specific modeling* seems especially promising. Domain-specific modeling can be defined as the systematic application of *domain-specific languages* (DSLs) in the design and programming phases of a development project. In complex projects, multiple DSLs are usually necessary in order to cope with different concerns. This requirement raises the need to manage the consistency among several models in multiple DSLs, which is the focus of this paper.

We give an example of an industrial-strength enterprise application framework that uses multiple DSLs, namely Apache Open for Business (OFBiz) [1]. We analyze the use of multiple DSLs in OFBiz applications by studying the OFBiz documentation, issue tracking system, developer forums, and the OFBiz implementation artifacts. We identify consistency management, and in particular ensuring referential integrity across models, as one of the key challenges of multi-DSL development. We want to address these challenges in a non-invasive way that can be incorporated in an existing development process and system architecture.

G. Engels et al. (Eds.): MoDELS 2007, LNCS 4735, pp. 46–60, 2007.

To address the problems identified in the OFBiz study, we introduce SmartEMF, which is an extension of the Eclipse Modeling Framework (EMF) [2]. SmartEMF provides support for representing, checking, and maintaining constraints using Prolog. SmartEMF can represent and check four kinds of constraints that we identified in OFBiz DSLs. Furthermore, it provides a simple form of *user guidance* by computing the valid set of editing operations that can be applied in a given context based on the current state of all models.

We believe that our study of OFBiz offers a valuable example of how multiple DSLs are used in industry today and the challenges that arise from such use. We are not aware of other quantitative studies of using multiple DSLs to describe a single system. Furthermore, although Prolog has been previously used to represent models and provide constraint checking and editing guidance, three aspects of SmartEMF are novel: (i) the use of Prolog to compute *multiple* valid operations; (ii) the exposition of how Prolog's *higher-order queries* can elegantly support constraint checking and the computation of valid operations; (iii) the support for loosely coupled DSLs by defining the valid target domain for *name-based references* through an annotation mechanism. The last capability makes it possible for SmartEMF to automatically support existing DSLs represented as XML Schemas that use name-based references to cross-link elements of individual models without the need to create a single, integrated metamodel.

The paper is structured as follows. Section 2 introduces OFBiz—our case study. Section 3 elaborates on issues in the typical process of developing applications in OFBiz. Section 4 describe the different kinds of consistency constraints that we have identified in OFBiz. Section 5 presents our tool SmartEMF, which addresses the issues by guided development with multiple DSLs. Section 6 discusses the solution and examines possible alternative approaches. Section 7 describes the related work and finally, section 8 concludes the paper and suggests possible future work.

2   Motivating Example: Apache Open for Business

The Apache Open for Business (OFBiz) framework [1] is an open source platform for building enterprise automation software, such as Enterprise Resource Planning (ERP), Content Management System (CMS), Customer Relationship Management (CRM), and Electronic Commerce systems. OFBiz is a top-level project at the Apache Foundation. Its users include both large companies, such as British Telecom and United Airlines [3], and a range of small and medium-sized ones. The framework is an excellent example of state-of-the-art, industrial-strength application development with multiple DSLs.

From a technical viewpoint, OFBiz is a J2EE framework that delivers a service-oriented architecture with persistent business objects, its own web application framework, and support for business rules, workflow, role-based security, and localization. OFBiz based applications are expressed using multiple DSLs. The core of OFBiz is an engine that can load and interpret more than fifteen DSLs (Figure 1). Each DSL covers a different aspect of application

Table 1. Overview of the OFBiz DSLs

Tier	DSL	Description	No. of Elements
Data	Entity Model	Define business objects, attributes, and relations	23
	Fieldtype Model	Define attribute types	3
	Entity Config	Configure data sources, files, and transactions	19
	Entity Group	Configure active models and entities	2
	Entity ECA	Define events, conditions, and actions for entities	5
Service	Service Def.	Define service interfaces and permissions	18
	Service Group	Configure active models and services	3
	Service Config	Configure security, threading, and service engine	13
	Service ECA	Define events, conditions, and actions for services	6
	Minilang	Implement services	154
	XPDL	Define workflows	89
UI	Screen	Implement screens and layout	65
	Form	Implement user forms and data binding	57
	Menu	Implement menus	27
	Tree	Implement visual tree structures and data binding	38
WWW	Site Config	Define web controller behaviour	15
	Regions Def.	Define screen regions	3

development such as defining business objects, services, graphical user interfaces, and workflows. Each DSL is defined using an XML Schema, and individual models are represented simply as XML documents. Table 1 provides a list of the DSLs. The table also specifies the number of elements in the schema of each DSL as an estimate of its size.

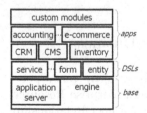

Fig. 1. The architecture

OFBiz applications are implemented as *modules* on top of the engine and the DSL layer (Figure 1). Each module typically consists of 20 to 60 models expressed in different DSLs and sometimes also custom Java code. The framework includes predefined modules such as Inventory, Customer Service, Product Catalogs, Order Entry, Accounting, and other ERP functions. Table 2 lists the artifacts constituting two of the predefined modules, including artifact sizes and numbers of cross-references among the artifacts. The framework is highly extensible allowing custom modules to be build on existing ones.

3   Application Development with Multiple DSLs in OFBiz

To understand how multiple DSLs are used in OFBiz, we analyzed the freely available documentation, including tutorials, Wiki sites, user forums, the project's issue tracking system [4], and the actual source code (stable build, September 2, 2006).

Table 2. DSL usage statistics for selected OFBiz modules

DSLs in 'Accounting' module	No. of Models	No. of Elements	No. of Cross-refs	DSLs in 'Content' module	No. of Models	No. of Elements	No. of Cross-refs
Entity Model	2	2105	723	Entity Model	1	1005	271
Entity Group	1	140	138	Entity Group	1	71	70
Service Def.	18	1726	433	Service Def.	6	1334	389
Service Group	1	15	10	Service ECA	1	10	9
Service ECA	2	59	57	Minilang	9	2718	506
Minilang	16	2127	277	Form	13	3487	1699
Form	11	2400	1141	Site Config	1	1443	284
Site Config	1	1087	228	Screen	12	2303	796
Screen	11	1889	648	Tree	2	122	11
Tree	1	25	4	Menu	9	366	107
Menu	2	268	45				

Table 3. Summary of a sample OFBiz customization

Reference	OFBIZ-93: Support BillingAcct & PaymentMethod for Payment
Link	http://issues.apache.org/jira/browse/OFBIZ-93
Module	Accounting
Problem	The requirement is that a customer be able to use a billing account plus another form of payment, such as a credit card, for a payment on an order. The billing account is to be used first.
Solution	• New service declaration captureBillingAccountPayment • Java implementation of this service • An extra parameter in the calcBillingAccountBalance service definition • Minor changes to 3 existing service implementations and a few utility methods • Minor changes to a single screen definition

The recommended OFBiz application development process involves a bottom-up development of new models or customization of existing ones according to the tiered architecture of OFBiz [5]. The first step is to define business objects and data models using the data-tier DSLs (Table 1). Then services are defined using the service-tier DSLs and, in complex cases, also Java and scripting languages. The third step is to implement the user interface using the user interface DSLs and possibly HTML/CSS code. A particular customization employs one or more of these steps depending on its purpose and requirements.

Multiple DSLs are involved not only in the development of complete OFBiz applications, but even in small customizations of the existing ones. Descriptions of customizations are available in the OFBiz issue tracking system. In our study we have selected a sample set of eleven completed customizations of predefined applications, all categorized as *new features* or *improvement requests*. Table 3 shows an example of such a customization. The number of affected artifacts for each of the eleven customizations are listed in Table 4. The average number of affected artifacts in the selected set was five, which approximates well the number of DSLs used in an average customization.

We have examined the discussions in the OFBiz issue forum [4] related to the issues in Table 4 and have found that the customizations typically

Table 4. Number of affected artifacts per customization

Issue	Affected modules	No. of artifacts	Issue	Affected modules	No. of artifacts
OFBIZ-16	ECommerce	3	OFBIZ-361	Webtools	5
OFBIZ-93	Accounting	13	OFBIZ-435	Marketing	4
OFBIZ-113	Order	1	OFBIZ-540	WorkEffort, Catalog, Product	14
OFBIZ-188	ECommerce	2	OFBIZ-557	Product	4
OFBIZ-338	Manufacturing	7	OFBIZ-580	WorkEffort	6
OFBIZ-339	WorkEffort	6			

required several iterations of changes to the involved artifacts before they were correctly implemented. A very common problem is inconsistency among the new or modified artifacts and the existing artifacts, mainly caused by dangling references. Currently developers use ordinary XML and Java editors to implement their customizations. These tools offer little help to keep the artifacts consistent. To check for inconsistencies, the developers start up the application and run test scenarios, which is time-consuming and error-prone. According to the OFBiz forum [6], one of the main future tool requirements is better consistency checks and editing guidance that could visualize how different artifacts are related.

4   Consistency Constraints in OFBiz

Our survey has revealed that inconsistency was one of the main development problems. We will now illustrate this problem with some concrete examples taken from the OFBiz framework. We cover both the problem of consistency within a single artifact and consistency among multiple artifacts. On the surface, these cases do not seem to differ: in either the goals are to avoid dangling references, to enforce typing, and to satisfy other constraints. In practice, the mechanisms for expressing references and enforcing constraints within and across artifacts are likely to differ. Different artifacts need to support independent editing and storage and also may belong to different technical spaces, e.g., XML and Java. In the following we identify four kinds of constraints that need to be maintained in application development. Unfortunately, the current OFBiz tools cannot represent, check, and maintain these constraints.

(1) Well-formedness of individual artifacts. Currently, all OFBiz DSLs are XML-based, which means that well-formedness can be established by checking whether a model conforms to its schema. Unfortunately, XML Schemas have serious limitations. In particular, element and attribute declarations are context insensitive and therefore cannot express whether their presence depends on the presence of other elements or attributes in their context [7, Sec 4.3-4.4]. However, OFBiz requires expressing such constraints. For example, according to the OFBiz documentation, if the `alias` element in the `Entity` DSL contains the `group-by` attribute, it should also contain the `function` attribute.

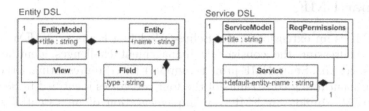

Fig. 2. Simplified excerpt from the metamodels of two DSLs

(2) Simple referential integrity across artifacts. A serious and frequent problem is
that of referential integrity across models. Multiple DSLs often refer to each other
because they represent different views of the same system. We have identified
more than 50 such references across the OFBiz DSLs. For example, each `Service`
in the `Service Definition` language needs to refer to an `Entity` in the `Entity`
`Model` language since services operate on entities. In OFBiz, as illustrated in
Figure 2, all such references across DSLs are *name-based*: the value of the
`default-entity-name` attribute in `Service` should match the `name` attribute
of the corresponding `Entity`. Sadly, there is no mechanism in XML Schema to
enforce this. Observe that *typed references*, absent from XML Schema, would
offer only a partial solution to the problem since name-based references between
XML and Java still need to be enforced. Cross-model name references are used
also in other approaches, e.g., in Microsoft DSL Software Factories [8].

(3) References with additional constraints. One can also find more complex
constraints imposed on references among OFBiz models, for example, between
models expressed in the `Form` and the `Entity` languages. A typical `Form` on a
webpage possesses fields linked to attributes of some `Entity`. A `Registration`
form may, for instance, have fields `Firstname`, `Lastname` and `Password` generated
using a reference to a `Person`. However, it is sometimes necessary to override these
generated fields. We may want to create a password text widget instead of the default
textfield for the `Password` field. In this case, the reference from the `Registration`
form to the `Person` entity has the additional constraint that the overridden field
must correspond to an attribute on the entity. More generally, all overriden form
fields should correspond to attributes on the entity that the form refers to. If this
constraint is violated, the engine will not be able to render the overriden form fields
correctly since it can not determine their origin in the entity layer.

(4) Style constraints. A fourth class of constraints suitable for OFBiz instal-
lations are style constraints. Enforcement of such constraints is not necessary
to execute OFBiz applications but facilitates maintenance in the long run. For
instance, the OFBiz designers have consciously adopted typical J2EE design
patterns. An example of a style constraint is to require that entity models
conform to the *ObjectRole* pattern as discussed in the OFBiz forums [9]: all
entities with a name that ends with *Role* should connect entities that do not
end with *Role*. This constraint ensures that relationships in the entity layer are
only specified between entities and not between relationships.

5   SmartEMF

SmartEMF is an extension of the Eclipse Modeling Framework (EMF) [2] that aims at addressing the consistency management challenges identified in the previous sections. SmartEMF provides support for (i) representing, (ii) checking, and (iii) maintaining constraints of the four categories identified in Section 4.

SmartEMF builds on EMF—an implementation of an essential subset of the Meta Object Facility [10]. EMF is a platform for defining DSLs that has several advantages over XML. In contrast to XML, EMF supports typed references, proper many-to-many relationships, and a standard cross-model reference mechanism. EMF has an editing and rendering API with a command framework and an adapter layer for integration with model editors. A generator of tree-based editors is included, while graphical editors are supported via the Graphical Modeling Framework. In contrast to the string-based Document Object Model (DOM) of XML, the EMF editing API is strongly typed.

SmartEMF achieves constraint checking and editing guidance using a logical representation of EMF models. The logical representation is maintained in parallel to the model. Constraints are expressed as Prolog rules and a Prolog inference engine is used to evaluate them. For a given model a set of valid operations is inferred and presented, guiding the user to select valid targets for references. The following sections explain each of these aspects.

5.1   Ecore-to-Prolog Mapping

SmartEMF assumes that a metamodel of each DSL is given in EMF's *Ecore* notation, which closely resembles MOF [10]. EMF offers bi-directional bridges between Ecore and other technical spaces, such as XML and Java. In particular, XML Schema Definition (XSD) files of OFBiz DSLs and the corresponding XML documents can be automatically imported into EMF, which makes them accessible as Ecore models and instances.

Figure 3 shows an excerpt of the mapping from Ecore to Prolog for a fragment of the Entity model from Figure 2. The mapping is directly inspired by the GEMS project [11]. Similar to GEMS, all elements of an Ecore model representing the DSL metamodel and all elements of an Ecore model instance representing a concrete model in the DSL are declared as facts in the fact base. For example, an Ecore class is represented as a fact using the `eclass` predicate with a unique identifier of the class as an argument. N-ary predicates are used to assert relations such as between an attribute and its containing class or between an integer attribute and its upper bound.

Upon startup, our prototype initializes the fact base by traversing and asserting model elements from Ecore, EMF's embedded XML metamodel, and all relevant DSLs and their instances. The resulting fact base then serves as the underlying representation of a reflective Ecore editor, which manipulates and queries both the EMF object model and the fact base. SmartEMF extends the standard EMF editing commands, such as *add*, *set*, and *delete*, to propagate changes of the model to the Prolog fact base.

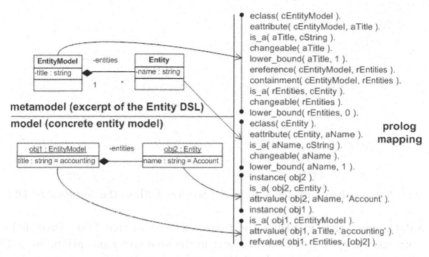

Fig. 3. Mapping from Ecore to Prolog

5.2 Representing Constraints

Consistency constraints from all of the categories discussed in Section 4 can be represented as Prolog rules. Since all the DSLs and models are represented in the Prolog fact base, constraints spanning one or more DSLs are expressed naturally.

A simple example of a well-formedness constraint (1) is *required_value_present*: every mandatory feature should have a value. Every constraint consist of two parts: a name (`required_value_present`) and a rule. The rule expresses a negation of the constraint, so that the rule is satisfied whenever the constraint is violated. In our example this happens if a mandatory feature, i.e., a feature with a lower bound of 1, has value id_UNSET. As shown below such constraints are relatively simple to read and write.

```
% name
constraint( required_value_present ) .
% rule representing the negation of the constraint
required_value_present( Object, Feature ) :-
lower_bound( Feature, 1 ) ,
( attrvalue( Object, Feature, id_UNSET ) ;
  refvalue( Object, Feature, id_UNSET ) ) .
```

Another class of constraints (2) considers consistency relations across distinct DSLs. A DSL can refer to another one in two ways. Either by using types from the other language or by name-based references.

Typed references are natively supported by Ecore and our mapping to Prolog. However, name-based references require additional information in the Ecore model, which SmartEMF supports with the *modelref* annotation. Figure 4 shows how the sample reference from *Service* to *Entity* from Figure 2 is represented using the annotation. The annotation consists of two key/value pairs: a *model* key which denotes the target DSL by its namespace and an *xpath* key which is

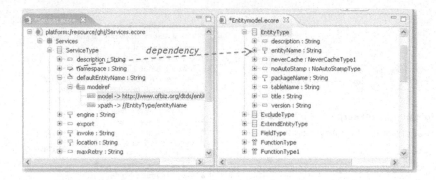

Fig. 4. Name-based reference from the *Services* DSL to the *Entitymodel* DSL

an XPath query that identifies the target element in that DSL. Provided with the corresponding values of the two keys in the *modelref* annotation, SmartEMF can determine the set of legal values (a *valid domain*) of an annotated model element. In the example, the annotation on the *defaultEntityName* attribute of the service shows that the valid values for this attribute are names of entities in the *Entitymodel* DSL. The rule expressing the negation of this constraint follows:

```
% name
constraint( no_dangling_modelrefs ).
% rule
no_dangling_modelrefs( Object, AnnotatedFeature ) :-
  modelref( AnnotatedFeature, DomainFeature ) ,
  attrvalue( Object, AnnotatedFeature, Value ) ,
  not( attrvalue( _ , DomainFeature, Value ) ) .
```

5.3   Constraint Checking Using Higher-Order Queries

Constraint checking utilizes Prolog's support for higher-order queries. The meta-logical `call` predicate facilitates such queries. The `call` predicate invokes a goal with an optional set of extra arguments. Since all our constraints are declared using the custom `constraint` predicate, we can easily compute the set of all constraints in the fact base. By using the `call` predicate, we can then determine which constraints are violated, i.e., evaluate to true for a given binding of their variables. The `check` rule states this query:

```
check( Object, Violations ) :-
  findall( [ Goal, Object, Feature ] ,
    ( constraint( Goal ) ,
      call( Goal, Object, Feature ) ) ,
    ViolationsUnsorted ) ,
  sort( ViolationsUnsorted, Violations ) .
```

If the `check` predicate is evaluated with the `Object` variable bound to an object then the result is a binding of the `Violations` variable to an empty list in case of no constraint violations or a list of tuples. In the latter case each tuple consists of the violated constraint (a goal), the concrete object, and the feature of that object. If the check predicate is evaluated with two variables, the query produces *all* constraint violations in the *entire* fact base in one shot.

5.4   Preconditions of Editing Operations

Simple editing guidance can be offered by computing the set of editing operations (and possibly some or all of their arguments) that are available in a given context based on the current state of the fact base. This facility is achieved by representing the preconditions of editing operations such as *add*, *set*, and *delete* as Prolog rules and querying them using a higher-order query.

```
% name
operation( add ) .
% rule
add( Object, Feature, AddableTypes ) :-
    instance( Object ) ,
    is_a( Object, ObjectType ) ,
    containment( ObjectType, Feature ) ,
    upper_bound( Feature, Upper ),
    refvalue( Object, Feature, CurrentValues ) ,
    not( length( CurrentValues, Upper ) ) ,
    is_a( Feature, AddableTypes ) .
```

The above listing shows a declaration of the *add* operation, which adds a child element to a containment reference. Similarly to constraints declarations, the precondition consists of a fact declaring the rule name and the rule. The rule states that only the instances of the feature's type can be added and only as long as the number of instances in the containment list does not exceed the upper bound. Similar preconditions are declared for other operations.

We determine valid editing operations in a context by using the higher-order query operations shown below. Depending on whether the Object and/or the Feature variables are bound, we can either determine all valid operations, all valid operations for a given object, or all valid values for a given feature.

```
operations( Object, Feature, Operations ) :-
    findall( [ Goal, Object, Feature, Value ] ,
      ( operation( Goal ) ,
        call( Goal, Object, Feature, Value ) ) ,
      OperationsUnsorted ) ,
    sort( OperationsUnsorted, Operations ) .
```

5.5   Reflective Editor

SmartEMF provides a reflective editor that exploits the underlying representation and previously described queries. It is a form-based editor implemented as an Eclipse EMF plugin (Figure 5). It enables users to access and modify

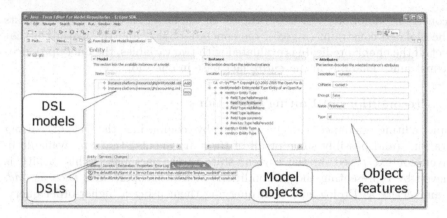

Fig. 5. SmartEMF's reflective editor

instances of different DSLs in a uniform way. Each DSL is represented in a tab containing three columns. The first column lists different models in the selected DSL. The second column contains a hierarchical view of the model elements in the currently selected model. The third column displays the features of a selected model element. All objects and feature values can be loaded, edited, and serialized respecting the individual file formats of their DSLs. Specifically, in the OFBiz case, every modified model is saved in XML conforming to the original DSL-defining XSDs.

The editor uses reflective capabilities of the regular EMF object model in order to structure the user interface. When the user selects an object or a feature, the framework queries the underlying representation for valid editing operations using the `operations` predicate from Section 5.4. The resulting tuples are presented in the form of various visual or textual hints as shown in Figure 6. Since the framework simultaneously queries the representation using the *check* predicate, a list of inconsistencies is available, which is both shown in the bottom as well as using other visual hints.

The reflective nature of the SmartEMF editor is one of its main advantages. Most EMF editors are generated and hence customized for a particular set of models. In contrast, the SmartEMF editor allows the user to quickly include new DSLs, new instances, and new constraints just by changing the loading configuration of the editor. Upon loading, the editor automatically adapts to the current selection of DSLs while still providing guidance and consistency management. Of course, the query facilities of SmartEMF could also be used from specialized and generated editors.

6   Discussion and Suggestions for Future Work

Experience in applying SmartEMF to OFBiz. OFBiz applications are traditionally developed using built-in XML and Java editors of IDEs like Eclipse. As

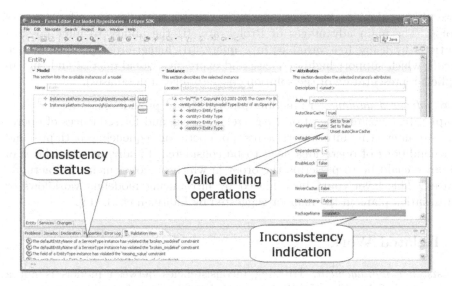

Fig. 6. Guidance and consistency management in the SmartEMF editor

described in the previous section, these tools do not offer any cross-language support or editing guidance. In our experiments, we annotated the `Entity`, `Service definition`, and `Form` languages and loaded all models in these languages from the *accounting* module. The setup comprised 31 models with a total of 6231 model elements and 2297 potentially broken cross-references. We implemented a set of simple customizations, such as extending an entity, revising a service, and displaying the results in the user interface, experiencing no performance problems with checking and guidance.

We have not yet performed any user studies apart from the customizations that we ourselves have done. We do, however, expect a significant drop in undetected inconsistencies when SmartEMF is systematically applied to OFBiz projects. The guidance facilities should also speed up development since they provide very concrete hints on how to complete the models in a given installation. Empirical studies need to be performed in order to validate these claims.

Applicability to other DSL-based infrastructures. A growing number of XML-based infrastructures (e.g., Struts and Spring) use similar mechanisms as OFBiz and we expect that these infrastructures could benefit equally well from SmartEMF's guidance and consistency management facilities. Also, we wish to examine how Java-customizations can be related to DSL artifacts. One possible approach would be to extract Ecore-based models from Java code as it has been done in framework-specific modeling languages [12].

Prolog as a constraint language. An object-oriented language such as OCL, which supports a concise expression of constraints and navigation, would seem to be a better choice than Prolog. Alternatively an approach based on a relational database and SQL queries could have been adopted. While both solutions are more elegant than a pure Java approach, they do not offer the advantages of Prolog. The main advantages of Prolog are: (i) the ability to infer possible solutions by using free variables in a query, (ii) higher-order predicates such as *call*, which support highly concise and expressive global queries and (iii) an implicit representation of the solution space (unlike in databases).

More advanced guidance. Extending SmartEMF to provide guidance for more complex (composite) operations, e.g., refactorings, and for sequences of operations are interesting future work items. For operation sequencing, both pre- and post-conditions of operations need to be considered. Finally, an extra layer of guidance could be to prescribe the order in which model elements are actually created. This layer could be achieved by introducing modeling workflows or instantiation plans in the manner suggested by Lahtinen et al. [13].

7   Related Work

Consistency management. Tracking inconsistencies between models is not a new research field. The *ViewPoint* method [14] is one of the best treatments of the subject. Recently several authors have addressed the problem, also considering repairs. Mens et al. [15] propose a graph-based approach where different models are represented as a single graph. The graph can be searched for erroneous patterns. Each pattern has a corresponding repair that can be applied to the graph. Similarly, other approaches support repairs by annotating OCL-constraints with repair actions [16] or even generating repair actions automatically [17]. More advanced repairs and diagnostics may be offered by implementing model checks as transformations as suggested by Bézivin and Jouault [18]. The checking aspect of all these approaches is impressive, but the guidance is limited to the predefined repairs. SmartEMF's ability to search for possible editing operations is not present in any of these.

Mapping models to a logical fact base. We have adopted the idea of mapping an Ecore-based model to Prolog from the GEMS project [11]. The primary emphasis in the GEMS project is on creating a graphical, guided editor for one or more DSLs based on a single, thightly-integrated meta-metamodel and allowing automatic configuration of models. SmartEMF draws on this approach, but uses pure Ecore as its meta-metamodel rather than a custom Ecore-based meta-metamodel. This choice allows SmartEMF to leverage existing languages by loading, editing, and saving different models in a schema-conformant way. Furthermore, in the case of multiple DSLs, GEMS would require a composite metamodel with multiple *aspects*, or views. SmartEMF handles this case by using a generic editor which can display both regular cross-model references as well as name-based references. Another Prolog-based approach is the Design Critiquing facility of the ArgoUML tool described by Robbins et al. [19]. This approach is focused on critiquing rather than suggesting possible edits as SmartEMF.

Multiple DSL development. Development scenarios with multiple DSL is sparsely described in the existing literature. One of the best empirical studies of DSL usage to date is Tolvanen and Kelly's work [20], which is primarily concerned with single DSLs. There are, as far as we know, no comprehensive empirical studies of industrial instances of multiple DSL applications like OFBiz. An early theoretical reference is the DRACO system [21], which contains a systematic approach for dealing with multiple DSLs. Recently, Warmer and Kleppe [8] have described an approach based on multiple DSLs, but that work does not provide empirical data such as the size of the involved DSLs and the number of cross-references among them.

Partial models and name-based references. Warmer and Kleppe [8] advocate the use of loosely coupled, *partial models*. Name-based references are used to integrate multiple DSLs. Their approach uses a consistency checking mechanism tailored to a specific set of languages.

8 Conclusion

We have examined the problem of working with multiple DSLs in domain-specific modeling. Our main contributions are: (i) a qualitative study of the architecture and development problems of an industrial-strength, multiple DSL application, OFBiz, and (ii) SmartEMF—an EMF-based framework offering consistency checking and editing guidance. We have tested our prototype within OFBiz development scenarios, applying it to the problems that were identified in the study as common issues in OFBiz development. Preliminary experiments suggest that a guidance tool can significantly help in maintaining consistency during development with multiple DSLs.

References

1. The Apache Software Foundation: The Apache Open for Business Project (March 8, 2007) (2007), http://ofbiz.apache.org/
2. Budinsky, F., Steinberg, D., Merks, E., Ellersick, R., Grose, T.J.: Eclipse Modeling Framework: a Developer's Guide. Addison-Wesley, Reading (2004)
3. Chen, S.: Opening Up Enterprise Software: Why Enterprises are Adopting Open Source Applications (2006), http://www.opensourcestrategies.com/slides/
4. The Apache Software Foundation: The Open for Business Project. Issue Tracking System (March 22, 2007), https://issues.apache.org/jira/browse/OFBIZ
5. Undersun Consulting LLC: OFBiz Framework Quick Reference Book, ver. 1.5.1 (March 26, 2007) (2004),
   http://bigfiles.ofbiz.org/FrameworkIntro/01MainDiagram.pdf
6. Jones, D.E.: Requirements for an OFBiz IDE (March 27, 2007) (2006),http://www.nabble.com/Re%3A-requirements-for-an-OFBiz-IDE-p8066093.html
7. Møller, A., Schwartzbach, M.I.: An Introduction to XML and Web Technologies. Addison-Wesley, Reading (2006)
8. Warmer, J., Kleppe, A.: Building a Flexible Software Factory Using Partial Domain Specific Models. In: Proc. of The 6th OOPSLA Workshop on Domain-Specific Modeling (2006), http://www.dsmforum.org/events/DSM06/

9. Howe, C.: Party Relationship Best Practices (March 27, 2007) (2006),
   http://www.nabble.com/Party-Relationship-Best-Practices-p5453154.html
10. Object Management Group: Meta-Object Facility (March 12, 2007) (2007),
    http://www.omg.org/mof/
11. White, J., Schmidt, D., Nechypurenko, A., Wuchner, E.: Domain-Specific
    Intelligence Frameworks for Assisting Modelers in Combinatorically Challenging
    Domains. In: GPCE4QoS (2006)
12. Antkiewicz, M., Czarnecki, K.: Framework-Specific Modeling Languages with
    Round-Trip Engineering. In: Nierstrasz, O., Whittle, J., Harel, D., Reggio, G. (eds.)
    MoDELS 2006. LNCS, vol. 4199, pp. 200–214. Springer, Heidelberg (2006)
13. Lahtinen, S., Peltonen, J., Hammouda, I., Koskimies, K.: Guided Model Creation:
    A Task-Driven Approach. In: VLHCC '06: Proc. of the Visual Languages and
    Human-Centric Computing, pp. 89–94 (2006)
14. Nuseibeh, B., Kramer, J., Finkelstein, A.: Expressing the relationships between
    multiple views in requirements specification. In: ICSE '93: Proc. of the 15th Int'l.
    Conf. on Software Engineering, pp. 187–196 (1993)
15. Mens, T., Van Der Straeten, R., D'Hondt, M.: Detecting and Resolving Model
    Inconsistencies Using Transformation Dependency Analysis. In: Nierstrasz, O.,
    Whittle, J., Harel, D., Reggio, G. (eds.) MoDELS 2006. LNCS, vol. 4199, pp.
    200–214. Springer, Heidelberg (2006)
16. Kolovos, D.S., Paige, R.F., Polack, F.A.: On the Evolution of OCL for Capturing
    Structural Constraints in Modelling Languages. In: Proc. Dagstuhl Workshop on
    Rigorous Methods for Software Construction and Analysis (2007)
17. Nentwich, C., Emmerich, W., Finkelstein, A.: Consistency Management with
    Repair Actions. In: Proc. of the 25th Int'l. Conf. on Software Engineering, May
    3-10, 2003, Portland, Oregon, USA, pp. 455–464 (2003)
18. Bézivin, J., Jouault, F.: Using ATL for Checking Models. In: GraMoT workshop,
    4th Int'l. Conf. on Generative Programming and Component Engineering (2005)
19. Robbins, J.E., Hilbert, D.M., Redmiles, D.F.: Software Architecture Critics in
    Argo. In: IUI '98: Proc. of the 3rd Int'l. Conf. on Intelligent User Interfaces, pp.
    141–144. ACM Press, New York (1998)
20. Tolvanen, J.P., Kelly, S.: Defining Domain-Specific Modeling Languages to
    Automate Product Derivation: Collected Experiences. In: Obbink, H., Pohl, K.
    (eds.) SPLC 2005. LNCS, vol. 3714, pp. 198–209. Springer, Heidelberg (2005)
21. Neighbors, J.M.: Software Construction using Components. PhD thesis, UC Irvine,
    Tech. Report UCI-ICS-TR-160 (1980)

Model-Driven, Network-Context Sensitive Intrusion Detection

Frederic Massicotte[1], Mathieu Couture[1], Lionel Briand[2], and Yvan Labiche[2]

[1] Communication Research Centre, 3701 Carling Avenue, P.O. Box 11490, Stn. H,
Ottawa, ON K2H 8S2, Canada
http://www.crc.ca

[2] Department of Systems and Computer Eng., Carleton University, 1125 Colonel By Drive
Ottawa, ON K1S 5B6, Canada
{briand, labiche}@sce.carleton.ca

Abstract. Intrusion Detection Systems (IDSs) have the reputation of generating many false positives. Recent approaches, known as stateful IDSs, take the state of communication sessions into account to address this issue. A substantial reduction of false positives, however, requires some correlation between the state of the session, known vulnerabilities, and the gathering of more network context information by the IDS than what is currently done (e.g., configuration of a node, its operating system, running applications). In this paper we present an IDS approach that attempts to decrease the number of false positives by collecting more network context and combining this information with known vulnerabilities. The approach is model-driven as it relies on the modeling of packet and network information as UML class diagrams, and the definition of intrusion detection rules as OCL expressions constraining these diagrams. The approach is evaluated using real attacks on real systems, and appears to be promising.

Keywords: Intrusion Detection, UML modeling, OCL constraints.

1 Introduction

An Intrusion Detection System (IDS) monitors network traffic in real-time, identifies undesirable utilization of computer resources, and reports on such utilization to network administrators. Most available (signatures-based) IDSs use packets and protocol headers, i.e., they rely on specific characteristics of packets and headers (i.e., their signatures) to define intrusion detection rules. This is currently the most efficient approach. However, these IDSs have the reputation of generating many false positives, i.e., false alarms. More recent IDSs, called stateful IDSs, additionally rely on the state of the communication sessions in order to reduce the number of false positives, i.e., an attack can only be successful during specific states of a session.

We performed an experiment [13] during which we evaluated the response to well-known attacks of two of the most advanced and widely used stateful IDSs to date, namely Snort 2.3.2 [7] and Bro 0.9a9 [16]. We used 92 vulnerability programs, implementing 57 vulnerabilities. Results show that Snort, in particular, has a high rate

G. Engels et al. (Eds.): MoDELS 2007, LNCS 4735, pp. 61–75, 2007.
© Springer-Verlag Berlin Heidelberg 2007

of false positives (83%). This experiment confirmed our intuition that despite the adoption of stateful approaches, much remains to be done in the area of intrusion detection. Specifically, we identified that: (1) the false positives problem partly lies in the inability of current IDSs to fully exploit network context information in order to correctly interpret intrusion alarms. For instance, if a packet is recognized as being a threat to a Windows machine but is sent to a Unix machine (network context), a context-based IDS can be designed to be silent (if this is acceptable to the network administrator); and (2) IDSs mostly rely on one or two packets within one session to identify attacks, whereas detecting modern attacks requires monitoring series of packets in multiple sessions to distinguish between successful and failed attacks.

The Passive Network Monitoring approach we describe in this paper attempts to address these issues. It can monitor complex communication patterns that involve multiple packets in multiple sessions, and analyze these patterns within the particular network context in which they occur. As a result, it is possible to correlate attacks and vulnerabilities of a particular system. By adopting the Model Driven Architecture (MDA) approach [8], whereby the level of abstraction is raised to the level of models and model transformations are automated, our approach models the monitored (packet and network) information in UML and defines detection rules in OCL.

The rest of the article is structured as follows. Section 2 describes the related work on current IDS technologies. Section 3 describes our approach. Section 4 presents the results we obtained, and a comparative analysis with a well-known IDS, with real attacks and systems. Conclusions are drawn in Section 5.

2 Related Work

We performed a systematic review of approaches currently used in IDSs. The most common approaches we discovered, among the ten IDSs we studied, include Temporal Logic [17], Petri Nets [9], State Machines [24, 25], Expert Systems [1, 10], Event Calculus [14], Regular Expressions [19] and Ad hoc paradigms [3, 7, 16]. In this article, due to space constraints, we only report on the most salient conclusions of this study, referring to only two well-known, mostly used IDSs, namely Snort [7] and Bro [16], and refer the reader interested in more details to [2].

The two main technological drawbacks we identified are that:

1. Network monitoring engines often restrict their analysis of network traffic to multiple packets from the same session. However, (modern) attacks involve several packets over several sessions, which call for an IDS with a multi-packet, multi-session monitoring engine.

   Snort and Bro have originally been built to monitor multiple packets in one session. The plugin-based extension mechanisms of Snort (plugins are programmed in C) can be used to address this issue to some extent though. However, each time a new plugin is created (or modified), the whole Snort IDS must be rebuilt. Similarly, it is theoretically possible to detect a multi-session attack with Bro, using the Bro customization language (similar to C). However, in both cases, this functionality is not built-in and has to be programmed at a low-level using a non-declarative language such as C. This is one of the issues further discussed below (issue 4).

2. IDSs do not fully benefit from the information that can be gathered about the state (or context) of the network such as the configuration of a host, its operating system, its role on the network, its active services and known vulnerabilities. Context information includes static and dynamic data [12]. The former do not depend on a specific network infrastructure (e.g., a given operating system is vulnerable to a specific attack) and typically come from vulnerability databases. The latter are collected from monitored packets, e.g., a computer on the network has an open port.

Though Snort has not been originally built to be context-aware, a recent approach has introduced limited network context awareness to Snort rules [26] (limited to server reactions), and Snort plugins (programmed in C) can be added and re-built to become more context-aware. For instance, sfPortScan is a preprocessor for Snort to detect scans on ports. Bro offers certain ability to capture and to represent network context (e.g. server reaction, product version). However, the extent of the network context that can be reached is not clear (see issue 3 below), and the context verification is limited to the same session as the attack.

From a methodological standpoint, we additionally found that it was often difficult to precisely compare existing IDSs, and to identify the root causes of technological drawbacks for two main reasons:

3. It is often impossible to precisely understand what packet and network information is monitored, and how this information is structured. What is often missing is a description of a formal packet model (specifying what information related to packets is used) and a formal network model (specifying what information on the network context is used).

For instance, the packet data used by Snort and Bro is specified in English. Although most of this relates to the communication protocols being monitored (e.g., TCP), which is standardized information, natural language tends to be imprecise. Similarly, since both Snort and Bro have not been originally built to be network context-aware, the network context data they use is scattered in all the rules/scripts they provide.

4. IDSs typically infer information as specified in the (often implicit) network model from data collected according to the (often implicit) packet model. Ideally, IDS rules should be defined in a declarative language that is independent of the monitoring engine. As a result, those two aspects of the IDS are able to evolve (and likely improve) independently.

As already mentioned, Snort and Bro rely on an imperative programming language for the definition of advanced intrusion detection rules (e.g., C in the case of Snort). Although a flexible option (one can potentially do anything), this results in a strong coupling between the monitoring engine and the rule evaluation engine and this requires more effort to write rules.

We acknowledge that these last two issues are not mandatory technical requirements for designing and reporting on an IDS. We however believe that it is important that researchers precisely specify the packet and network models they rely on, using an unambiguous specification language, to easily compare the IDS approaches and capabilities.

3 The PNMT Approach

To help leverage IDS technology, we developed a Passive Network Monitoring approach and Tool (PNMT) that specifically addresses the four issues listed previously. The approach is illustrated in Figure 1. In the Monitoring & Reporting swimlane, the packet capturing engine (Sniffer), collects packet data that is used by the Dispatcher to populate a packet data model. The Rule Evaluation Engine evaluates compiled OCL rules against packet information to derive network information. Both the packet and network information are formally defined by a packet and network data model, respectively (issue 3). The Rule Evaluation Engine is able to monitor a series of packets and relate monitored packets to each other, thus capturing the network context and detecting multi-packet attack scenarios in possibly multiple sessions (issue 1). The Reporting Engine is then used by the network administrator to identify attack attempts. This requires that OCL rules be defined and processed, which is the purpose of the Rule Specification swimlane: OCL rules created by an engineer well versed into OCL and the two data models are automatically processed by the OCL Rule Compiler to generate pieces of code that can directly be used by the Monitoring Engine. To increase the accuracy of the context-based analysis of the OCL rules, additional network information data is derived from vulnerability databases (issue 2). We have shown [12] how this can be done by combining Snort rules [7] and Nessus scripts [22] (for the identification of vulnerabilities) with Bugtraq data [18] (a well-known database of identified systems' vulnerabilities). Typically, by using the Bugtraq number reported in a Snort rule or in a Nessus script, we are able to know the systems that are vulnerable to an attack.

To address issue 3, we specify a packet data model and a network data model under the form of UML class diagrams (Sections 3.1 and 3.2, respectively). This allows us to precisely specify the packet and network information used in our approach and PNMT tool. This also allows us to define logical rules relating to either intrusion detection or network context discovery with the OCL (Section 3.3). OCL provides PNMT with a declarative rule specification language that is independent of the monitoring engine (issue 4). Another advantage of using UML is that the class diagrams provide a solid starting point for the design of the tool supporting our approach (PNMT): This is shortly described in Section 3.4.

Our choice of UML was obvious as it is now the de-facto standard for modeling domain concepts and for designing object-oriented software systems. Because OCL is a declarative constraint language which is already part of UML, our choice of OCL for the definition of intrusion detection rules was also immediate. Other technical aspects, further discussed in Section 3.4, also strengthened our decision to use OCL. As an alternative, extending Snort (or Bro) was not considered, since we wanted to use a higher level of abstraction than an imperative programming language.

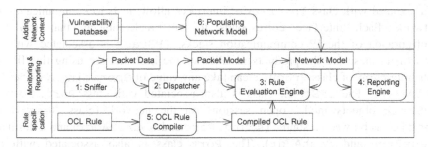

Fig. 1. PNMT Activity Diagram

3.1 Network Data Model

Few methods have been proposed in the literature to model a computer network from a perspective useful to IDSs. The most important contributions are reported in [6, 15, 23], where the model suggested in [15], referred to as M2D2, encompasses concepts from the other two. M2D2 models network topology, host configuration (including products and vulnerabilities) and packets. M2D2 seems the most appropriate for modeling networks from an intrusion detection perspective.

To model the network, we extended the M2D2 model and made some adjustments. For example, we went from a functional specification to an object-oriented specification using the UML and OCL. Second, we made the model suitable for modeling missing network context information. For example, we changed the minimum multiplicity of some relations to 0, as opposed to 1 for M2D2. In particular, this modification was applied to the Interface - IPStack relationship because it is possible to have an Interface object without any associated IPStack object at some moment in time. Finally, we added/removed some data (i.e., class relationships, attributes) to/from the M2D2 model. For example, we added a state attribute to class Port and we added a relationship between Product and Port.

Whenever, possible our models abstracted away from implementation details. For instance, we could have added an IPAddress class with four attributes of type Integer (representing the four number composing an IP address), associated with the class IPStack. However, we did not consider this level of details necessary for our model, and instead represented IP addresses as strings.

Figure 2 shows our network data model, where the classes can be split into three categories. The *network configuration classes* represent the network components: Host, Interface, IPStack, Session, and Port. The *host configuration classes* represent how each network component is configured: Product and Vendor. The *information security classes* represent security information that needs to be modeled, in relation with the host network configuration and the host configuration classes: Alarm, Exploit, Vulnerability, and Reference.

The Host class models each component communicating on the network. This class represents infrastructure equipment (routers, switches, workstations, servers). It is associated with class `Interface`: interfaces allow hosts to communicate on the network. Each interface corresponds to `IPStack` objects that model the configuration of the IP communication stacks. `IPStack` is associated with class `Product`: these are the products installed on the host that are using this IP stack. Attribute `type` of class `Product` can take any value of `OperatingSystem`, `FTP`, `Telnet`, ... The `IPStack` class is also associated with class `Port`: `Port` and `IPStack` objects model open sessions on the network (class `Session`). A `Session` is between a source (`sourcePort` and `sourceAddr`) and a destination (`destPort` and `destAddr`). The `Port` class is also associated with class `Product` to capture information such as the fact that a port is associated with a specific product (e.g., a Microsoft IIS FTP server) running on a particular port of a particular host. Each `Product` object is also associated to a `Vendor` object that represents this product's vendor, and may be associated to other `Product(s)`. The recursive association on class `Product` models the fact that a given service (a `Product` instance) is only available on a specific operating system (another `Product` instance). The `Alarm` class represents the alarms generated by PNMT. The `Alarm` objects are associated to `Exploit` objects that represent known computer system exploits. Each `Exploit` object is associated to a particular `Vulnerability` object, which are known software vulnerabilities of `Product` running on computer systems. The `Reference` class links vulnerabilities to existing databases recording exploits and systems' vulnerabilities.

3.2 Packet Data Model

An intrusion detection system analyses packet traffic. Hence, it is important to properly represent network packets. In [6, 15], packets are represented as tuples, where each element of the tuple corresponds to a protocol header. Although this is a very natural way to proceed, because of the encapsulated nature of headers in the OSI model [21], we have opted for a model based on compositions. This is similar to Snort [7], though based on a UML class diagram instead of plain language descriptions. Our

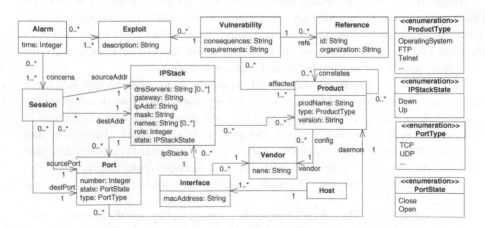

Fig. 2. Network data model

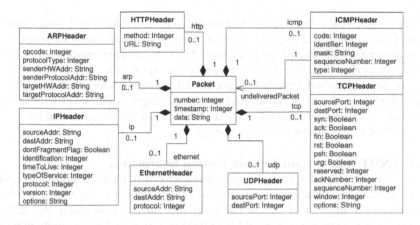

Fig. 3. Packet data model

packet data model is depicted in Figure 3. (Note that some attributes have been omitted and a complete description of the attributes can be found in [11].) A `Packet` instance describes a complete packet collected by the sniffer (Figure 1). Attribute `number` is a unique identifier of the packet instance and attribute `timestamp` indicates when the packet was collected. A unique `number` is necessary since, depending on the precision of the clock, two different packets may have the same `timestamp`. On the other hand, a `timestamp` attribute is required to measure delay among packets. Each packet is composed of any of the following headers, at most once: `ARPHeader` for the Address Resolution Protocol (ARP); `ICMPHeader` for the Internet Control Message Protocol (ICMP); `TCPHeader` for the Transmission Control Protocol (TCP); `EthernetHeader` for the Ethernet protocol; `IPHeader` for the Internet Protocol (IP); `UDPHeader` for the User Datagram Protocol (UDP); `HTTPHeader` for the Hypertext Transfer Protocol (HTTP). Note that other application protocols than HTTP can be implemented, but so far it is the only application protocol implemented in PNMT.

3.3 Rule Specification

In the case of context-based intrusion detection, we drew on the same idea as model transformation in the context of MDA [8] and used OCL to express how to translate packet header information into network context information such as alarms (i.e., intrusion detection), network configuration and host configuration information. Instead of resorting to OCL extensions that have not yet been standardized (e.g., [4, 27]), we decided to add the notion of time to the packet model, thus attribute `timestamp` in class `Packet` (Figure 3). This way, as the `Sniffer` (Figure 1) collects packet data and forwards them to the `Dispatcher`, the `Dispatcher` creates a sequence of `Packet` instances, ordered according to their arrival time (i.e., `timestamp`). We can then define OCL rules that specify characteristics of `Packet` instances in a sequence of packets.

We describe below three representative examples of PNMT (OCL) rules. Note that the first rule cannot be expressed using the Snort rule specification language. Only a Snort plugin (a.k.a., *flow* plugin) can provide the same functionality. Also, it is not possible to write the other two rules using the Snort rule specification language because Snort does not have a network model capturing network context. In these cases, there is no plugin available. Additionally, a plugin would be difficult to build as it would have to gather network context which is not the default behavior of Snort. (This will be further illustrated in Section 4.)

3.3.1 PNMT Rule to Gain Network Context Information

The OCL rule in Figure 4 is used to gain more information on the monitored network, i.e., to increase contextual information. This rule infers that a given TCP port is open on a given computer. In order to come up with such a conclusion, the IDS must observe, within a specific period of time (2000 ms), two packets traveling in opposite directions, one being a SYN packet and the other being a SYN ACK packet. The rule specifies that if such a condition is satisfied (left-hand part of the `implies` operator) for any two observed packets `p1` and `p2` (`Packet.allInstances->forAll(...)`), p2 being the SYN ACK response, then the source of p2 has an open port (right-hand side of the `implies` operator). The OCL expression is simplified thanks to four operations defined using `def` expressions in the context of `Packet` (Figure 5), but not shown in Figure 4. These expressions define variables and operations that can be used in other OCL expressions in the context of `Packet`. This illustrates that a PNMT user can define other utility operations that can be used in the PNMT rules. A tool supporting the PNMT approach would therefore not need to be re-built (e.g., recompiled) if new rules (or supporting functions) are needed.

Similar context gathering rules to identify open sessions, operating systems and products running on hosts' ports can be found in [11].

```
Context Packet
inv: Packet.allInstances->forAll(p1, p2 | ( p1.syn and p2.synAck
    and p1.oppositeTCPFlow(p2) and p2.occuredWithin(2000,p1))
  implies IPStack.allInstances->exists(i | i.ipAddr = p2.ip.sourceAddr
    and i.ports->exists(po:Port | po.state = PortState::Open
      and po.type = PortType::TCP and po.number = p2.tcp.sourcePort)))
```

Fig. 4. OCL expression modeling an Open Port

```
Context Packet
def: syn: Boolean = self.tcp.syn = true and  self.tcp.ack = false
def: synAck: Boolean  = self.tcp.syn = true and  self.tcp.ack = true
def: oppositeIPFlow(p:Packet):Boolean = self.ip.sourceAddr =
       p.ip.destAddr  and  self.ip.destAddr = p.ip.sourceAddr
def: oppositeTCPFlow(p:Packet):Boolean = self.oppositeIPFlow(p)
       and  self.tcp.sourcePort = p.tcp.destPort
       and  self.tcp.destPort = p.tcp.sourcePort
def: occuredWithin(t:Integer,p:Packet):Boolean =
       self.timeStamp > p.timeStamp
       and ((self.timeStamp - p.timeStamp) < t)
```

Fig. 5. Utility Operations of class `Packet`

```
Context Packet
inv: Packet.allInstances->forAll(p:Packet |
  ( p.http.url.contains(".printer") and p.tcp.destPort = 80
    and sessionOpen(p)
    and IPStack.hasDaemonOnPort(p, PortType::TCP, "IIS", 5.0))
  implies
    Alarm.allInstances->exists(p, "WEB-IIS ISAPI .printer access"))
```

Fig. 6. PNMT rule for vulnerability 2674 (Bugtraq reference)

3.3.2 Microsoft IIS Server 5.0 Printer ISAPI Extension Buffer Overflow

The Microsoft IIS Web Server 5.0 is vulnerable to the ".printer ISAPI Exten-sion Buffer Overflow". An unchecked buffer in a dll file could allow attackers to execute arbitrary code. This vulnerability corresponds to Bugtraq number 2674 [18], and only affects version 5.0 of the server. A successful attack consists in sending an HTTP ".printer" request containing approximately 420 bytes in the "Host:" field.

Figure 6 shows the PNMT OCL intrusion detection rule for this vulnerability. It is simple thanks to other utility operations, specified in Figure 7: sessionOpen() returns true if the packet passed as a parameter is part of an open session (i.e., the parameter is a TCP packet and the source/destination of the packet corresponds to a session in the model); hasDaemonOnPort() returns true if a specific port, on a specific computer (to which the Packet passed as a parameter is sent) is running a specific daemon; exists() is used to describe the kind of alarm that is sent by the PNMT rule. The rule in Figure 6 specifies that if PNMT finds a packet p that (i) contains an HTTP header with string ".printer" in its url field, (ii) has a TCP destination port of 80, (iii) is part of an open session, and (iv) there is a computer in

```
Context Packet
def: sessionOpen(p:Packet):Boolean =
Session.allInstances->exists(s | s.sourcePort.number = p.tcp.sourcePort
  and s.sourcePort.type = PortType::TCP
  and s.destPort.type = PortType::TCP
  and s.destPort.number = p.tcp.destPort
  and s.sourceAddr.ipAddr = p.ip.sourceAddr
  and s.destAddr.ipAddr = p.ip.destAddr)
Context IPStack
def: hasDaemonOnPort(p:Packet, t:PortType, n:String, v:String):Boolean =
IPStack.allInstances->exists(i | i.ipAddr = p.ip.destAddr
    and i.port->exists(po:Port| po.number = p.tcp.destPort
              and po.type = t and po.daemon.prodName = n
              and po.daemon.version = v)  )
Context Alarm
def: exists(p:Packet, descr:String):Boolean =
Alarm.allInstances->exists(a | a.exploit.description = descr
  and a.concerns->exists(s | s.sourceAddr.ipAddr = p.ip.sourceAddr
    and s.time = p.timeStamp  and s.destAddr.ipAddr = p.ip.destAddr
    and s.destPort.type = PortType::TCP
    and s.sourcePort.type = PortType::TCP
    and s.sourcePort.number = p.tcp.sourcePort
    and s.destPort.number = p.tcp.destPort  )  )
```

Fig. 7. Utility Operations of class Packet (cont.)

the network to which p is sent (IP address and port identical to p's data) and that runs a daemon for a product called "IIS" with version equal to "5.0", then there must be an Alarm instance in the network model that is characterized by information from packet p with its message value equal to "WEB-IIS ISAPI .printer access".

Our PNMT rules, like the one in Figure 6, detect when attacks are made against vulnerable systems. It could be debated that a network administrator could also be very interested in knowing that an intruder made an attempt against a non-vulnerable system. Distinguishing such attempts is possible with PNMT (see the discussion at the beginning of Section 3.3) by creating an OCL rule that is a slight variation of the one in Figure 6: the left part of the implies would check for versions different from version 5.0, and the right part would raise different kind of alarms (e.g., different severity of alarms) depending on the identified versions.

3.3.3 Wu-ftpd File Globbing Heap Corruption
A well-known vulnerability against the WU-FTP server is the Wu-ftpd File Glob-bing Heap Corruption vulnerability, which allows attackers to execute arbitrary code on a server remotely. This vulnerability corresponds to Bugtraq number 3581 [18].

Figure 8 shows the PNMT rule for this vulnerability. It specifies that if PNMT finds a packet p with a data field that matches regular expression "~. [", that is part of an open session, and if p's target (destination IP address and port) is running a daemon with a product name equal to "WU-FTP" and a version number equal to 2.5.0, 2.6.0 or 2.6.1, then the network model must contain a corresponding Alarm instance. Note that the rule uses a String operation called match() to match a string with a regular expression.

```
Context Packet
inv:Packet.allInstances->forAll(p | p.data.match("~.*[")
    and p.tcp.destinationPort = 21 and  sessionOpen(p)
    and  (IPStack.hasDaemonOnPort(p, PortType::TCP, "WU-FTP", "2.5.0")
      or  IPStack.hasDaemonOnPort(p, PortType::TCP, "WU-FTP", "2.6.0")
      or  IPStack.hasDaemonOnPort(p, PortType::TCP, "WU-FTP", "2.6.1"))
   implies Alarm.allInstances->exists(p, "FTP wu-ftp bad file completion
attempt ["))
```

Fig. 8. PNMT rule for vulnerability 3581 (Bugtraq reference)

3.4 Tool Design

In order to quickly develop a prototype version of PNMT for the experiment we present in Section 4, we decided to implement the OCL rules by hand instead of building a complete OCL rule compiler (Figure 1). However, we developed a framework to help specify the OCL rules. This framework is not described here due to space constraints and the interested reader is referred to [11]. We are considering, as an alternative to building an OCL rule compiler, the use of recent technology developed by IBM in the framework of the Eclipse platform. An EMF model of our packet and network models could be queried using the EMF Query Framework [5] which supports OCL queries.

We have created a prototype version of PNMT in Java (52 classes and 9 KLOC) for the Windows platform [11]. The sniffer uses the Windows version of the *libpcap* library included in *tcpdump*, thus capturing all the packets it sees on the network.

4 Case Study

In this section, we report on an experiment we performed to assess our approach and tool (PNMT). This experiment has two objectives: (1) Evaluate whether the (packet and network) models are sufficient to handle a variety of situations, i.e., to evaluate the completeness and correctness of our approach; (2) Quantitatively and qualitatively evaluate the reduction of false positives by comparing PNMT with a well-known and widely used IDS, namely Snort (version 2.3.2) [7]. Section 4.1 describes the setting of the experiment and Section 4.2 discusses the results.

4.1 Case Study Setting

To test Snort and PNMT we developed a data set of attack scenarios. To make this data set similar to attacks perpetrated on the Internet, trying to be as general as possible, we carefully picked different Vulnerability Exploitation Programs (VEP) that exploit vulnerabilities in the most popular key services (HTTP, FTP, SMTP) that are running on the most common operating systems (e.g., Linux, Windows, FreeBSD). We also used VEPs that affect services directly related to the operating systems. We therefore selected the following vulnerabilities, all referred to by their Bugtraq number (BID) [18]: BID 2674 is an HTTP vulnerability against the IIS web server running on Windows; BID 3581 (resp. BID 4482) is an FTP vulnerability against *wu-ftpd* (resp. IIS FTP server) which runs on Linux (resp. Windows); BID 6991 is an SMTP vulnerability against *sendmail* that can run on Linux or FreeBSD; BID 8205 (resp. BID 4006) is a vulnerability of RPC (resp. MSDTC) on Windows operating systems. The complete descriptions of these vulnerabilities, as well as the corresponding VEPs, can be found on the SecurityFocus web site [18].

To generate this data set, we used the following testing strategy: we launched every VEP against vulnerable and non-vulnerable services and operating systems and we recorded each attack scenario and documented them. The resulting data set is therefore composed of traffic traces that each contain an attack scenario. Each traffic trace is then analyzed by each IDS as a test case to verify first whether it is able to detect the attack scenarios and second whether it is able to distinguish between an attack to a vulnerable target or not (to reduce false positives).

Table 1 summarizes the case study setting. The first column identifies the vulnerabilities we used, referred to by their Bugtraq number. The following two columns report the number of VEPs used for each vulnerability and the number of (non-)vulnerable systems against which these VEPs were executed. (The rest of the table is discussed in the next section.) For instance, we used 6 different VEPs for vulnerability BID 2674, and those 6 different attack programs were used to attack 9 vulnerable systems (9 Windows systems) and 26 non-vulnerable systems (6 Windows systems and 20 Linux systems), thus resulting in 210 traffic traces (54 traces for vulnerable targets and 156 traces for non-vulnerable targets) to be analyzed by both

Snort and PNMT. We only report on a representative subset of our results in this paper as we possess a larger number of traffic traces: in [13], we used 92 VEPs implementing 57 vulnerabilities. To build such a large set, we developed a controlled network infrastructure [13] that allows network traffic recording, prevention of information leakage and noise (to obtain clean traffic traces), control of attack propagation (i.e., attack confinement), usage of real and heterogeneous system configurations and fast attack recovery to initial conditions.

The number of VEPs in Table 1 varies from one BID to another for two main reasons. First, there are not always many VEPs for a given vulnerability on the SecurityFocus web site. Second, we only selected VEPs that actually attack the target systems. In other words, we discarded VEPs that are smart enough to recognize that the target system would not be vulnerable to their attack and therefore do not perform any attack. The number of (non-)vulnerable systems also varies from one BID to another. One main reason is that some vulnerabilities require a specific service and a specific operating system version (one possible vulnerable combination) whereas others require a specific service but can work on several operating system versions (several possible vulnerable combinations).

Table 1. Case study setting and results

BID	Number of		Alarms raised Vuln. / not Vuln.		Successful / non Successful attacks	
	VEPs	Vulnerable / not Vulnerable	Snort	PNMT	Snort	PNMT
2674	6	9 / 26	54 / 156	54 / 0	16 / 194	16 / 38
3581	1	6 / 69	6 / 69	6 / 0	6 / 69	6 / 0
4482	2	10 / 68	20 / 136	20 / 0	8 / 148	8 / 12
4006	1	3 / 1	3 / 1	3 / 0	0 / 4	0 / 3
6991	2	1 / 1	2 / 2	2 / 0	2 / 2	2 / 0
8205	5	14 / 3	378 / 52 [1]	378 / 0	87 / 343	87 / 291

The main observation we can draw from this initial (rule modeling) step is that we were able to use the packet and network models to define intrusion detection rules for this initial set of representative (services, systems, operating systems) vulnerabilities. (The rules for the two first vulnerabilities in Table 1 have been discussed in Sections 3.3.2 and 3.3.3, respectively.) This suggests that our packet and network models are complete and correct based on the case study.

4.2 Results

The fourth and fifth columns of Table 1 show the number of alarms raised by Snort and PNMT on the attack scenarios for vulnerable and non-vulnerable targets. For example, in the case of BID 2674, Snort raised an alarm for all the 54 (6*9) attacks on vulnerable targets, as well as for all the 156 (6*26) attacks on non-vulnerable targets (156 false positives). On the other hand, for the same vulnerability, PNMT only raised

---

[1] As opposed to the other VEPs that have only one configuration, the five VEPs for BID 8205 have various configurations, specifically 5, 11, 7, 2, and 2 configurations for the five VEPs, respectively. This results in (5+11+7+2+2)*14=378 attack traces for vulnerable targets.

an alarm for the 54 attacks on vulnerable targets (0 false positive). This observation applies for the six selected vulnerabilities (except for BID 8205 that we discuss below), and is representative of other vulnerabilities we have tried: Snort raised an alarm, regardless of the vulnerability of the target, but PNMT is able to distinguish attacks that will not succeed because sent on non-vulnerable targets. This confirms that using (more) network context allows an IDS to reduce the number of false positives.

There is one exception to the trend we just discussed: vulnerability 8205. In this case, the reaction of some non-vulnerable targets differs from the one of vulnerable targets. At the same time, the Snort rule for BID 8205 only triggers an alarm if the attack is detected and the target system has a specific reaction. However, the reaction part of the rule is not accurate enough to detect all possible reactions that correspond to a failed attack. In other words, Snort uses some network context, but this is not enough to reduce the number of false positives to the level of PNMT.

Snort was able to take some of the systems reaction into context for BID 8205 and it generated 52 false positives (out of 81 attack scenarios against non-vulnerable targets) against the 3 non-vulnerable systems. However, PNMT performed better than Snort in this situation.

The last two columns of Table 1 report on additional qualitative analysis we performed on the attacks. Specifically, we looked at whether the attacks we used were successful or not on the vulnerable targets. For instance (Table 1), for BID 2674, only 16 of the 54 attacks on vulnerable targets succeeded. Therefore, Snort really produces 194 false positives (156 non-vulnerable targets, and 54-16=38 vulnerable targets), and PNMT really produces 38 false positives. An example of unsuccessful attack is the following: some VEPs execute shell code by exploiting a vulnerability, but the shell code is sometimes operating system dependent, thus a shell code for FreeBSD used against a vulnerable Linux operating systems will not succeed. This experimentally confirms that future work will have to improve IDSs by incorporating even more network context (e.g., by looking at the target reaction to the attack scenario) [20, 26].

5 Conclusion

Intrusion Detection Systems (IDSs) are known to generate large numbers of false positives, i.e., to raise alarms from monitored packets that are not actual attacks or attacks that have no chance of being successful. Though different techniques have been investigated to reduce the number of false positives, one promising solution is to have a so-called stateful approach: to account for some communication session context in the intrusion detection rules. In existing solutions, this context information is limited and does not entail for instance the characteristics of the system under attack (e.g., the actual operating system and its known vulnerabilities).

In this paper, we present a proof of concept IDS called Passive Network Monitoring Tool (PNMT) that is able to passively acquire network context information, allowing the inclusion of such context in network intrusion detection rules. Our approach is model-driven and based on the UML/OCL standard in order to achieve two additional objectives. First, the packet and network information on which the detection rules rely is precisely and formally modeled, thus facilitating any future comparisons with other IDSs and providing a good starting point for the design of other, similar IDSs. The UML class diagrams are a starting point to derive an object-oriented IDS design. Second, using UML and OCL, the specification of the intrusion detection rules are independent from any specific monitoring engine, so that those two aspects of an IDS can evolve independently. The Object Constraint Language (OCL) is a natural choice to express constraints on class diagrams, and thus on our data models for packets and network context. We developed a set of network context gathering rules and intrusion detection rules in OCL that are shown, on a case study, to reduce the number of false positives when compared to Snort, a well-known and widely used IDS.

Considerable improvements can be made in the future by investigating two main avenues of research. First, we should improve the automation for creating checkable OCL intrusion detection rules (and study the performance) and, as discussed, some open source technologies are promising to that end. Second, during our experimentation with PNMT and Snort, while using attack programs, we discovered that even more context is needed, for instance, to infer that an attack has succeeded or not. For instance we have discovered that some attacks conducted on vulnerable systems do not necessarily succeed although the traffic trace contains all the elements for Snort and PNMT to raise an alarm. This calls for an improved approach where even more network context data is collected and used (e.g., the fact that a machine is no longer responding because of a denial of service attack indicates a successful attack). Last, we recognize that a comprehensive list of Bugtraq vulnerabilities used to test PNMT and Snort may not be sufficient to validate that PNMT produces a small rate of false positive. Based on the results presented in this article, we are considering to evaluate PNMT on a larger, more complete set of vulnerability exploit programs, and target systems. Finally, an alternative to OCL for specifying detection rules could be to embed the security expert's knowledge into a domain specific language.

Acknowledgements

This work results from a collaboration between Carleton University and the Communications Research Center of Canada (CRC), an Agency of Industry Canada. Lionel Briand and Yvan Labiche are further supported by NSERC Operational Grants.

References

[1] Anderson, D., Frivold, T., Valdes, A.: Next-generation Intrusion Detection Expert System (NIDES): A Summary. SRI International, Technical Report SRI-CSL-95-07 (May 1995), http://www.sdl.sri.com/nides/reports/4sri.pdf
[2] Couture, M., Massicotte, F.: Systèmes et Languages de Détection d'Intrusion. CRC, Technical Report CRC-RP-2005-001 (July 2005)
[3] Deraison, R., Gula, R., Hayton, T.: Passive Vulnerability Scanning - An Introduction to NeVO. Tenable Network Security, White Paper (2003), www.tenablesecurity.com/
[4] Distefano, D., Katoen, J.-P., Rensink, A.: On A Temporal Logic For Object-Based Systems. In: Proc. IFIP Formal Methods for Open Object-Based Distributed Systems, pp. 305–326 (2000)
[5] Eclipse Foundation, Tutorial: Querying EMF Models with OCL, http://help.eclipse.org/

[6] Goldman, R.P., Heimerdinger, W., Geib, C.W., Thomas, V., Carter, R.L.: Information modeling for intrusion report aggregation. In: Proc. DARPA Information Survivability Conference and Exposition, pp. 329–342 (2001)

[7] Green, C., Roesch, M.: The Snort Project: version 2.3.2., User Manual (2003), www.snort.org

[8] Kleppe, A., Warmer, J., Bast, W.: MDA Explained. Addison-Wesley, Reading (2003)

[9] Kumar, S., Spafford, E.: A Software Architecture to Support Misuse Intrusion Detection. In: Proc. National Information Security Conference, pp. 194–204 (1995)

[10] Lindqvist, U., Porras, P.A.: Detecting Computer and Network Misuse through the Prediction-Based Expert System Toolset (P-BEST). In: Proc. IEEE Symposium on Security and Privacy, pp. 146–161 (1999)

[11] Massicotte, F.: Using Object-Oriented Modeling for Specifying and Designing a Network-Context Sensitive Intrusion Detection System, Masters Thesis, Carleton University, Systems and Computer Engineering (2005)

[12] Massicotte, F., Couture, M., Briand, L.C., Labiche, Y.: Context-Based Intrusion Detection Using Snort, Nessus and Bugtraq Databases. In: Proc. Annual Conference on Privacy, Security and Trust (2005)

[13] Massicotte, F., Gagnon, F., Labiche, Y., Briand, L., Couture, M.: Automatic Evaluation of Intrusion Detection Systems. In: Proc. Annual Computer Security Applications Conference (2006)

[14] Morin, B., Debar, H.: Correlation of Intrusion Symptoms: an Application of Chronicles. In: Vigna, G., Krügel, C., Jonsson, E. (eds.) RAID 2003. LNCS, vol. 2820, pp. 94–112. Springer, Heidelberg (2003)

[15] Morin, B., Mé, L., Debar, H., Ducassé, M.: M2d2: A formal data model for ids alert correlation. In: Wespi, A., Vigna, G., Deri, L. (eds.) RAID 2002. LNCS, vol. 2516, pp. 177–198. Springer, Heidelberg (2002)

[16] Paxson, V.: BRO: A System for Detecting Network Intrusion in Real-Time. Computer Networks 31(23-24), 2435–2463 (1999)

[17] Roger, M., Goubault-Larrecq, J.: Log Auditing though Model Checking. In: Proc. IEEE Computer Security Foundations Workshop, pp. 220–236 (2001)

[18] Security Focus, Bugtraq Homepage, http://www.securityfocus.com/

[19] Sekar, R., Guang, Y., Verma, S., Shanbhag, T.: A High-Performance Network Intrusion Detection System. In: Proc. ACM Symposium on Computer and Communication Security, pp. 8–17 (1999)

[20] Sommer, R., Paxson, V.: Enhancing byte-level network intrusion detection signatures with context. In: Proc. ACM Conference on Computer and Communications Security (2003)

[21] Stallings, W.: Data and Computer Communications. Addison-Wesley, Reading (1996)

[22] Tenable Network Security, Nessus Scripts, www.nessus.org/plugins/

[23] Vigna, G.: A topological characterization of tcp/ip security. Politecnico di Milano, Technical Report TR-96.156 (1996)

[24] Vigna, G., Kemmerer, R.A.: Netstat: A network-based intrusion detection approach. In: Proc. IEEE Annual Computer Security Applications Conference, pp. 25–34 (1998)

[25] Vigna, G., Valeur, F., Kemmerer, R.: Designing and implementing a family of intrusion detection systems. In: Proc. ACM SIGSOFT European Software Engineering Conference, pp. 88–97 (2003)

[26] Zhou, J., Carlson, A.J., Bishop, M.: Verify Results of Network Intrusion Alerts Using Lightweight Protocol Analysis. In: Srikanthan, T., Xue, J., Chang, C.-H. (eds.) ACSAC 2005. LNCS, vol. 3740, Springer, Heidelberg (2005)

[27] Ziemann, P., Gogolla, M.: An Extension of OCL with Temporal Logic. In: Proc. Workshop on Critical Systems Development with UML, in conjunction with the UML conference (2002)

An Empirical Study of the Impact of OCL Smells and Refactorings on the Understandability of OCL Specifications

Alexandre Correa[1], Cláudia Werner[2], and Márcio Barros[3]

[1,2] COPPE/UFRJ – Computer Science Department
Federal University of Rio de Janeiro (UFRJ)
C.P. 68511, Rio de Janeiro, RJ, Brazil – 21945-970
{alexcorr, werner}@cos.ufrj.br
[3] UNIRIOTEC – Computer Science Department
Federal University of the State of Rio de Janeiro (UNIRIO)
Av. Pasteur 458, Rio de Janeiro, RJ, Brazil – 22290-240
marcio.barros@uniriotec.br

Abstract. The Object Constraint Language (OCL) is a OMG standard that plays an important role in the elaboration of precise models. However, it is not hard to find models and metamodels containing overly complex OCL expressions. Refactoring is a technique that can be used in this context since its goal is to reduce complexity by incrementally improving the internal software quality. Indeed several refactorings have already been proposed to improve the quality of OCL expressions. This paper presents the results of an empirical study that investigates the impact of poor OCL constructs, also known as *OCL Smells,* and OCL refactorings on the understandability of OCL expressions. Current results show that most refactorings significantly improve the understandability of OCL specifications.

1 Introduction

The Model Driven Architecture (MDA) is gradually becoming an important element of software development. MDA is a framework for model-based software development sponsored by the Object Management Group (OMG) [1]. In MDA, platform independent models, platform dependent models and automatic model transformations have a key importance in the software development process. The Object Constraint Language (OCL) [2] plays an important role in this context, because it allows the elaboration of precise, consistent and computer-processable models. Besides, OCL can be used in the context of model transformation languages, as is the case of OMG QVT (Query, Views and Transformations) standard [3].

Although OCL was designed to be both formal and simple when compared to formal specification languages such as Z [4] and VDM-SL [5], specifications written in OCL may be difficult to understand and evolve, particularly when they contain overly complex or duplicate OCL expressions. *OCL smell* was defined in [6] as a hint that some parts of a specification written in OCL, or even of its underlying model,

G. Engels et al. (Eds.): MoDELS 2007, LNCS 4735, pp. 76–90, 2007.
© Springer-Verlag Berlin Heidelberg 2007

contain constructs that might generate a negative impact on its understandability and extensibility. In [7], we presented a catalogue of OCL smells often found in OCL specifications, including the UML 2.0 specifications [8, 9] and several conference papers.

Refactoring is considered an essential technique for handling software evolution [10]. It is defined as changes made to the internal structure of a software, without modifying its observable semantics, aiming at improving quality factors such as understandability, modularity and extensibility of a software model or implementation [11]. In [7], we presented a catalogue of refactorings which can be applied to remove OCL smells from a specification. Automated support for OCL refactorings have recently appeared in research tools such as RocIET [12] and Odyssey-PSW [7]. Experienced programmers easily acknowledge and realize the effectiveness of code refactorings. Although there is anecdotal evidence on their usefulness, few quantitative evaluations of software refactorings have been done so far [13].

This paper reports on a controlled experimental study that was performed with the purpose of evaluating whether the understandability of constraints written in OCL can be affected by the structure of their expressions. In particular, our goal was to verify whether the presence of OCL smells impacts the understandability of OCL expressions when compared to refactored versions of the same expressions. We chose to perform a controlled study in order to control for extraneous factors which could have affected the results.

The rest of this paper is structured as follows: section 2 briefly presents the OCL smells involved in the study. Section 3 describes some OCL refactorings that can be used to remove those OCL smells. Section 4 describes the objectives, design and the instrumentation of the study. Section 5 reports on the results obtained from the study and the threats to their validity. In section 6, we draw some conclusions in terms of practical significance and future work.

2 OCL Smells

This section briefly presents the OCL smells involved in the empirical study reported in this paper. The main session of the study was designed to take no longer than 90 minutes, not only due to the limited availability of subjects, but also because longer sessions might introduce undesirable effects caused by factors such as fatigue, for example. Therefore, although there are twelve OCL smells catalogued in [7], we limited our study to the following five OCL smells, which correspond to the ones that we have found most often in OCL specifications.

a) **Implies Chain**
    Implies chain corresponds to OCL expressions of the form *b1 implies (b2 implies bn)*, where *b1, b2* and *bn* are boolean expressions.

b) **Verbose Expression**
    This OCL smell corresponds to expressions that are larger than necessary.

Two usual forms of this smell are:

- Expressions containing more operation calls than needed. Ex: X->select(x | P(x))->size() > 0   vs. X->exists(x | P(x)).
- Invariants defined in the wrong context class: since an invariant can be described in many ways depending on its context class, attaching an invariant to the wrong context usually makes it harder to specify and maintain [2].

c) **Forall Chain**
ForAll chain is a special case of the *Verbose Expression* smell corresponding to expressions containing a chain of *forAll* operation calls. This chain can be replaced by a single *forAll* applied to a navigation expression:
Ex: A ->forAll(a1 | a1.B->forAll(b1 | b1.C->forAll(c1 | P(c1))))

d) **Downcasting**
Downcasting is a well-known smell in the object oriented programming community. In OCL, it corresponds to the use of expressions of the form *x.oclAsType(Y).z*, usually preceded by an expression of the form *x.oclIsKindOf(Y)*.

e) **Type Related Conditionals**
This smell occurs in expressions of the form *if x.oclIsKindOf(A) then <exp1> else if x.oclIsKindOf(B) then <exp2> else ... endif*, i.e., the result of the expression depends on the type of a given object *x* which is obtained through calls to *oclIsKindOf* or *oclIsTypeOf* operations.

3 OCL Refactorings

A number of refactorings can be applied to OCL expressions and their underlying model in order to remove OCL smells such as the ones described in the previous section. These refactorings can be classified into three categories:

a) *OCL-exclusive refactorings* refer to changes that only affect OCL expressions, i.e., the underlying model remains the same and no new OCL helper attributes or operations are defined. Examples of this kind of refactoring are:

- *Replace Implies Chain by a Single Implication*: transform an *Implies Chain* - A implies (B implies C)) - into an expression of the form (A and B) implies C.

- *Replace ForAll Chain by Navigations*: transform a *ForAll Chain* A->forAll (a1 | a1.B->forAll (b1 | b1.C->forAll (c1 | P(c1)))) into a navigation expression of the form A.B.C->forAll(c1 | P(c1)).

- *Simplify Operation Calls*: Rewrites an expression using less operation calls. Ex: X->select(x | P(x))->size() > 0 and its refactored version: X-> exists(x | P(x)).

- *Change Context*: This refactoring is usually motivated by the presence of the *Verbose Expression* smell, and corresponds to writing an invariant in a shorter form by using a different context class.

b) **OCL definition constraint refactorings** correspond to changes related to OCL definition constraints, i.e., addition, renaming, removal or use of OCL helper attributes or operations. Examples of this kind of refactoring are:

- *Add Operation Definition* and *Replace Expression by Operation Call:* The main motivation for these two refactorings is to promote encapsulation and reuse across a specification. By using operation definitions, one can hide complex expressions from other parts of a specification and avoid the duplication of expressions that are used in many parts of a specification.

- *Add Property Definition and Replace Expression by Property Call Expression* are analogous to the refactorings based on operation definitions previously described. Instead of creating and using OCL helper operations, they create and use OCL helper properties.

c) **Underlying model refactorings** are the ones that demand changes in the underlying model. Addition, removal or renaming of classes, attributes, associations or operations are examples of model refactorings. Another specific example of this type of refactoring is:

- *Introduce Polymorphism*: this refactoring replaces complex *if-then-else-endif* expressions that makes considerable use of operations such as *oclIsKindOf, oclIsTypeOf, oclAsType*, by a combination of generic and specific operations defined in a hierarchy of classes.

4 Study Design

This section describes the design of the experimental study. Section 4.1 defines the study and describes its context and material. Section 4.2 formulates the hypothesis. The study design and instrumentation are detailed in section 4.3.

4.1 Definition

The purpose of the study was to evaluate whether the understandability of constraints written in OCL can be affected by the structure of their expressions. In particular, our goal was to verify whether the presence of OCL smells impacts the understandability of OCL expressions when compared to their refactored versions.

The study consists of reading and interpretating OCL constraints associated to elements defined in the same UML model. Those activities were not performed in the context of a real industry project. All instruments were specially prepared in laboratory.

Each subject answered ten questions. Each question presents an OCL invariant and a small object diagram corresponding to a snapshot of objects of a given UML model. Subjects should answer whether the given snapshot violates the invariant, and justify their answers. They were divided in two groups (GI and GII) and each group answered a different set of questions (Set I and Set II, respectively). Both sets contain five questions of two types: *S-Type* questions are interpretation of constraints containing OCL smells; *R-Type* questions are interpretation of constraints corresponding to refactored versions of the constraints present in the *S-type* questions answered by the other group.

The subjects were 23 graduated software developers who have attended to a 40-hour course in UML/OCL offered by Federal University of Rio de Janeiro. An invitation was sent by e-mail to everyone who has attended to this course in the last 3 years. We invited about 100 attendees, from which 23 volunteered to participate in the study.

Most subjects have experience on using UML in real projects, but their previous experience with OCL was limited to simple examples and small case studies. Subjects were told they would not be rewarded on performance, but that they were expected to perform theirs tasks in a professional manner. They were asked to answer the questions as quickly as possible, without sacrificing the quality of their answers. The subjects were aware that we were attempting to evaluate some issues related to OCL, but they were not aware of the exact hypotheses we were testing or what results we were hoping to obtain. Their knowledge of OCL was restricted to basic OCL syntax and semantics, i.e., they were not aware of concepts such as OCL smells and OCL refactorings.

4.2 Hypothesis

The type of each question answered by the subjects is the independent variable of this study. There are two possible values for this variable: S (*S-Type* questions) and R (*R-type* questions).

We investigated the impact of the independent variable on the following dependent variables:

- Question Score (QS): the score of a subject in a specific question. Each question has a two-part answer, and QS is computed as follows: one point for the yes/no part of the answer regarding a possible violation of the constraint in a given snapshot of objects, and one point for the explanation part. Two additional scores are computed for each subject: SS (Smell Score) = sum of QS in questions of type S; and RS (Refactored Score) = sum of QS in questions of type R.
  Therefore, $0 <= QS <= 2$; $0 <= SS <= 10$; $0 <= RS <= 10$.
- Question Time to Answer (QT): time in seconds spent by a subject to give the two-part answer to a question. Two other variables are computed for each subject: ST (smell time) = sum of QT in questions of type S; and RT (refactored time) = sum of QT in questions of type R.

The Null Hypothesis (H_0) is formulated as follows: "The understandability of OCL constraints is not influenced by the structure of their expressions, i.e., there is no difference in accuracy and time to answer interpretation questions on constraints containing OCL smells, when compared to their refactored versions". Therefore, H_0: $\mu SS = \mu RS$ and $\mu ST = \mu RT$.

The alternative hypothesis (H_1) is that OCL smells affect accuracy or time to answer those interpretation questions. To be more precise, H_1 should be one-tailed: we expect OCL smells to have one or both of the following effects: decreased accuracy or increased time to answer. Therefore, **H_1: μSS < μRS or μST > μRT**.

4.3 Study Design and Instrumentation

This study involves only one factor of interest: whether or not the expressions used in OCL invariants contain OCL smells. In order to increase the number of observations given the number of participants, all subjects were submitted to both treatments, i.e., they analyzed OCL constraints containing OCL smells (*S-type* questions) and refactored constraints (*R-type* questions). The study was organized in the following activities:

a) Enrollment
A pre-session survey questionnaire was distributed to obtain information about the background of the participants, e.g., their academic background, their UML and OCL experience and their industrial experience in software development and modeling.

b) Self-Study
After answering the survey, the participants received a self-study tutorial on OCL, authored by us, which describes the main concepts of OCL and all operations defined in the OCL standard library. The description of each concept and operation was accompanied with examples and self assessment exercises. Subjects were given two weeks to study this tutorial. The goal of this tutorial was to give the participants enough knowledge to answer the questions presented in the main session of the study.

c) Initial Assessment
After studying the OCL tutorial, the knowledge in OCL of each subject was assessed by a test containing a UML model, a set of objects of the classes defined in that model, and a set of OCL expressions whose values should be manually evaluated by the subjects. Although there might be more robust forms to assess the knowledge of each participant in OCL, we judged this assessment as the most adequate given the constraints on the time available of each participant.

Given the results of this assessment, subjects were grouped in two blocks: High (grades above the median) and Low (grades below the median). Each group (GI and GII) was then randomly assigned subjects from both blocks (High and Low) in nearly identical proportions. GI was assigned with 6 subjects from High and 5 subjects from Low; GII was composed of 5 subjects from High and 6 subjects from Low. One subject was allocated to test the instrumentation, and therefore did not take part in the main session of the study. The goal of this test was to avoid a possible bias on the assignment of subjects in the two groups regarding their knowledge of OCL.

d) Instructions and Warm-Up
Each participant received detailed written instructions about the main session of the study. Before the start of the main session, they answered a warm-up question

in order to make them familiar with the format of the questions to be answered in the main session. This warm-up question also helped the participants on understanding and practicing the procedure that they were expected to follow during the session.

e) Main session

In this session, each participant answered the set of ten questions assigned to his group (SI or SII). In order to avoid biasing the results due to the presentation order of the questions, each subject had to fully answer one question to proceed to the next one. If the current question were of type S, the next must be of type R and vice versa. The type of the first question was selected randomly for each subject. This strategy allowed us to collect the time spent by the subjects on each question. They were not allowed to change the answer of any previous question.

We used an adaptation from the Royal & Loyal case study [2] as the underlying model for the sets of questions answered by the participants during the main session of the study. The adapted model contains 12 classes, 13 associations and 2 generalizations. Each subject was given 10 minutes to study the model before starting to answer the set of ten questions. The participants were expected to take 60 minutes to answer all ten questions, but there was no time limit to answer each of the questions.

R-type questions of one set (Set I, for example) use the same object diagrams present in S-type questions of the other set (Set II). The constraints of R-type questions of one set are equivalent to the constraints of S-type questions in the other set. The main difference is that the constraints of R-type questions are refactored versions of the expressions present in S-type questions of the other set. Figure 1 shows an example of the invariant used in a S-type question containing the Verbose Expression smell and its corresponding R-type question of the other set which contains a refactored version of the same expression.

The two sets of questions tried to expose all subjects to expressions containing the same number of OCL smells at the same difficulty level. Therefore, if the set SI has a question S_m with an expression that has an OCL smell X, the set SII must have a question S_n containing that same OCL smell. Questions S_m and S_n have different OCL expressions but they are at a similar level of difficulty.

Question S7
context Client
inv: self.cards.transactions
       ->select(x | x.oclIsKindOf(EarnTransaction))
      ->forAll(t | t.value > 10.0)

Question R7
context EarnTransaction
def: minValue : Real = 10.0
inv: self.value > minValue

Fig. 1. Example of OCL invariants present in R-type and S-type questions

Table 1 presents a summary of the structure of each set of questions and the OCL smells present in each *S-type* question. The relation between questions containing the same OCL smells is given by the following pairs: (S1-S10); (S2-S5); (S3-S4); (S6-S9) and (S7-S8).

Table 1. Composition of each set of questions

Set SI		Set SII	
Question	*OCL Smells*	*Question*	*OCL Smells*
S1	Implies Chain	R1	Refactored version of S1
R2	Refactored version of S2	S2	Forall Chain
S3	Downcasting	R3	Refactored version of S3
R4	Refactored version of S4	S4	Downcasting
S5	Forall Chain	R5	Refactored version of S5
R6	Refactored version of S6	S6	Type Rel. Contitionals
S7	Verbose Expression (Wrong Context)	R7	Refactored version of S7
R8	Refactored version of S8	S8	Verbose Expression (Wrong Context)
S9	Type Rel. Conditionals	R9	Refactored version of S9
R10	Refactored version of S10	S10	Implies Chain

f) Subjective Evaluation

After answering the set of questions, the participants were asked to classify each question according to two aspects: the difficulty level and the perceived quality of the OCL expressions.

For the evaluation of the difficulty level of each question, we used a Likert scale from 1 to 5 (*1-very easy, 2-easy, 3-medium, 4-difficult, 5-very difficult*). The quality of the OCL expressions present in each question was evaluated according to a 1-3 scale: *1- constraint is badly written; 2- not sure whether the constraint is well or badly written; 3- constraint is well written*. The participants were also asked to describe possible improvements that could be applied to the OCL expressions present in each question. Those subjective evaluations were collected in order to verify a possible influence of the instruments on the result, and also to supplement the analysis of the objective results obtained from the sets of questions SI and SII.

5 Results

This section reports the results of the study. Section 5.1 describes the analysis of the instruments. The analysis of the scores and time to answer the questions are discussed in sections 5.2 and 5.3 respectively. Section 5.4 presents the results of the subjective evaluation of the questions and section 5.5 discusses threats to the validity of the study.

5.1 Instruments

The first analysis that we conducted was related to the instruments used in the main session. Our goal was to evaluate whether the two sets of questions had a significant influence on the results, since they were designed to give a similar experience to all subjects. In this evaluation, we applied the ANOVA test, with a significant threshold (α-level) of 0.05, comparing the following data: mean score of each set of questions; mean score of questions of the same type in each set of questions; mean time spent on each set of questions; mean time spent on questions of the same type in each set of questions; frequency of each difficulty level (subjective evaluation) in each set of questions; frequency of each difficulty level (subjective evaluation) in questions of the same type.

Table 2 presents the results regarding the mean score in each set of questions. Since F(ANOVA) < F_{CRIT}, we conclude that there is no significant difference between the mean score obtained by the subjects in both sets of questions. The results obtained in the other data sets also indicate that there was no significant difference in the performance and difficulty perceived by the participants in both sets of questions. Therefore, we concluded that there was no need to analyze the results of each set separately.

Table 2. ANOVA table: mean score in each set of questions

Set	Size (N)	Sum of Squares	Mean Square	Individual Mean
SI	110	368	321	17,09
SII	110	373	332	17,36
F (ANOVA)	0,10	F_{CRIT}	3,88	

5.2 Scores

Table 3 shows the mean score of the subjects considering the two sets of questions. Compared to *S-Type* questions, the mean score was higher and the standard deviation was lower in *R-type* questions. 59% of the subjects correctly answered all *R-type* questions, while only 22% answered all *S-type* questions correctly.

Table 3. Descriptive statistics of the scores

	Type S	Type R	Total
Mean	7,95	9,27	17,22
Standard Deviation	1,65	0,98	2,04
% with highest score (20 points)	22%	59%	18%

Figure 2 graphically shows the total score obtained by all subjects in each question. The highest score for each question is 22 (11 subjects x 2 points). Nine questions were correctly answered by all subjects (six *R-type* questions and three *S-type* questions). The bottom four scores correspond to *S-type* questions (S5, S10, S6 and S9). This graph also shows that the total scores in all R-type questions were greater than or equal to their respective S-type questions.

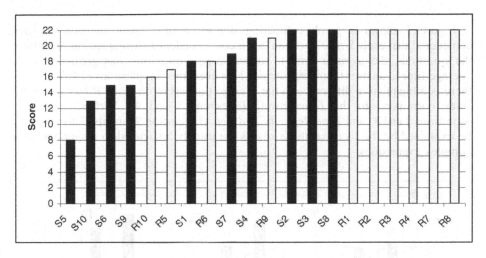

Fig. 2. Total score by question

The ANOVA test (α-level = 0.05) was applied to the score of each type of question. The results shown in Table 4 rejected the null hypothesis H_0, in favor of the alternative hypothesis H_1: $\mu SS < \mu RS$, i.e., the mean score in *S-type* questions is lower than the mean score in *R-type* questions.

Table 4. ANOVA table: score in each type of question

Question Type	Size (N)	Sum of Squares	Mean Square	Individual Mean
S	110	339	278	7,95
R	110	402	378	9,27
F (ANOVA)	9,89	F_{CRIT}	6,75	

Therefore, the results show that, at least in the sample analyzed in the study, the presence of OCL smells in expressions negatively impacts the understanding of OCL constraints.

5.3 Time to Answer

Results show that most subjects spent more time answering *S-type* questions than *R-type* questions. Table 5 shows the mean time spent by the subjects in each type of question, and the mean time they took to answer all questions.

Table 5. Descriptive statistics of the time to answer

	S-Type	R-Type	Total
Mean	26:12	20:00	46:12
Standard Deviation	8:56	5:55	14:11

Figure 3 shows a graph comparing the mean time spent in each question, considered that each *S-type* question has a correspondent *R-type* question. In 90% of the questions, the average time to answer *S-type* questions was greater than the average time to answer their respective *R-type* questions. In questions 3, 4, 7, 8 and 9, there is a big relative difference in the average time spent on *S* and *R-type* questions. They correspond to the *Downcasting* (questions 3 and 4), *Verbose Expression* (questions 7 and 8) and *Type Related Conditionals* (question 9) smells. This result can be explained by the fact that those smells are usually associated to longer and more complex expressions.

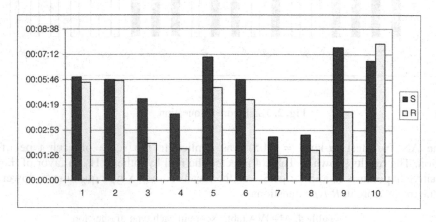

Fig. 3. Time to answer each question (left bars = S-Type)

On the other hand, there was only a small difference in the average time to answer questions 1 and 10, which correspond to the *Implies Chain* smell. A possible explanation for this result is that the expressions present in those questions are of almost the same size. However, as shown in Figure 2, the total score in questions R1 and R10 (38) was significantly higher than the score in questions S1 and S10 (31). This latter result suggests that it might be easier to misunderstand an expression containing an *implies chain*, but more specific studies need to be done in order to draw such conclusions.

The ANOVA test (α-level = 0.05) was applied to the time to answer each type of question. The result shown in Table 6 rejected the null hypothesis H_0, in favor of the alternative hypothesis H_1: $\mu ST > \mu RT$, i.e., the mean time to answer *S-type* questions is greater than the mean time to answer *R-type* questions.

Table 6. ANOVA table: time to answer each type of question

Question Type	Size (N)	Sum of Squares	Mean Square	Individual mean time spent in each question
S	105	13.236.000	10.047.146,67	05:14
R	103	8.525.400	5.942.403,88	04:00
F (ANOVA)	8,87	F_{CRIT}	3,89	

Therefore, the results show that, at least in the sample analyzed in the study, the presence of OCL smells in expressions can negatively impact the time needed to understand OCL constraints.

5.4 Subjective Evaluation

a) Perceived Level of Difficulty
Data collected from the subjective evaluation made by the subjects were analyzed in order to investigate whether they perceived some difference in the difficulty level of *S-type* questions compared to *R-type* questions. Figure 4 shows that more than 60% of *R-type* questions were perceived as easy or very easy, and only 10% of *R-type* questions were perceived as difficult or very difficult. On the other hand, less than 30% of *S-type* questions were perceived as easy or very easy, while 30% of them were perceived as difficult or very difficult.

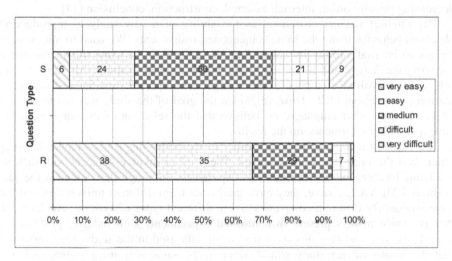

Fig. 4. Subjective evaluation of the difficulty level by the question type

After mapping the difficulty level from the ordinal scale to an interval scale using a monotonical function that preserves the order of the elements, we found a negative correlation (-0.71) between the difficulty perceived by the subjects and their score, i.e, the higher the difficulty level of a question, the lower the score was. We also found a positive correlation (0.73) between the difficulty level of a question and the time spent by the subjects to answer it.

b) Perceived Quality of the Expressions
Table 7 shows the overall evaluation of *S-type* and *R-type* questions regarding the quality perceived in the OCL expressions present in those questions. The results show that there is a significant difference in the perceived quality of expressions present in S-type questions and R-type questions.

While only 4% of the evaluations of R-type questions classified their expressions as of poor quality, that number raised to 36% in S-type questions.

80% of the evaluations indicate that R-type questions were well written. However, a significant number of evaluations (44%) perceived expressions containing OCL smells (S-type) as of good quality.

Table 7. Descriptive statistics of the perceived quality of the expressions

Evaluation	S-type	R-type
Well Written	44%	80%
Neutral (not sure whether good or poor)	20%	16%
Badly Written	36%	4%

5.5   Threats to Validity

This section discusses threats to the validity of the results found in this study, in decreasing priority order: internal, external, construction, conclusion [14].

The *internal validity* is defined as the ability of a new study to reproduce the observed behavior using the same subjects and instruments. We tried to minimize the threats to internal validity by submitting all subjects to the same treatments and by applying the *R-type* and *S-type* questions in an alternate fashion. All subjects were selected from volunteered people and most of them reported some interest in using or learning more about OCL. However, since the goal of the study was not to compare OCL with any other language, we believe that the selection of the subjects did not have a significant influence on the results.

The *external validity* reflects the ability to reproduce the same behavior in groups other than the ones that were analyzed. We tried to involve subjects with different academic background and professional experience. Although they attended the same 40-hour UML/OCL course, they have graduated from different universities and they have worked for different companies and in different types of systems. Another issue that is almost always present in controlled experiments with industry professionals regards the size and complexity of the constraints used in the study: we cannot state that the results of this study would occur in the same way using bigger and more complex models and OCL constraints.

The *construction validity* refers to the relationship between the instruments / subjects and the theory under study. We designed the study so that all subjects could have comparable and similar experiences. The results presented in section 5.1 indicate that this goal was achieved. It is not possible to state that we used the best strategy to evaluate the understandability of OCL specifications. However, we followed an approach similar to those used in other empirical studies that evaluated some aspect related to program or specification understanding [15, 16 and 17].

The *conclusion validity* relates the treatments and the results, defining the ability of the study in generating some conclusion. Threats to conclusion validity were mitigated by carefully designing all instruments and by using objective measures and statistical parametric tests. We also used subjective evaluations in order to support the quantitative results. Although the number of subjects could be considered low, we tried to increase the number of data points by submitting all subjects to both treatments.

6 Conclusions

In this paper, we investigate an important issue regarding the specification of constraints in OCL: whether the structure of the expressions used in OCL constraints has a negative impact on their understandability. The results indicate that the presence of OCL smells in OCL expressions may have a negative impact in both the correctness and the time necessary to understand a constraint written in OCL.

Subjects scored better and took less time to answer *R-type* questions than *S-type* questions. *S-type* questions were perceived as being more difficult than *R-type* questions. Moreover, we found a correlation of this level of difficulty perceived by the subjects and their performance: lower score and increased time to answer *S-type* questions. Although some insights on the impact of each OCL smell were presented in section 5, the focus of this study was on the global performance of the subjects considering the whole set of questions. We plan to conduct more specific assessments of the impact of each OCL smell in a future work.

The subjective evaluation of the perceived quality of the expressions reflects somehow the lack of experience of the subjects with OCL. Although a significant part was able to see that some expressions are more complex than necessary, few subjects were able to correctly explain how they could be made simpler. Besides the lack of experience with OCL, we believe that an additional reason for such results is that their knowledge is restricted to basic OCL syntax and semantics. Therefore, the results suggest that a catalogue of OCL smells and their respective refactorings is an important asset that the OCL community should consider to continually use and evolve.

References

[1] OMG: Model Driven Architecture (MDA) Guide, document number omg/2003-06-01 – (June 2003)
[2] Warmer, J., Kleppe, A.: The Object Constraint Language – Getting Your Models Ready for MDA. Addison-Wesley, Reading (2003)
[3] OMG: MOF QVT Final Adopted Specification. Document ptc/05-11-01 – (November 2005)
[4] Woodcock, J., Davis, J.: Using Z. Specification, Refinement and Proof. Prentice-Hall, Englewood Cliffs (1996)
[5] Jones, C.B.: Systematic Software Development Using VDM. Prentice-Hall, Englewood Cliffs (1989)
[6] Correa, A., Werner, C.: Applying Refactoring Techniques to UML/OCL Models. In: Baar, T., Moreira, A., Strohmeier, A., Mellor, S. (eds.) UML 2004. LNCS, vol. 3273, pp. 173–187. Springer, Heidelberg (2004)
[7] Correa, A., Werner, C.: Refactoring Object Constraint Language Specifications. In: Journal of Software and Systems Modeling (SoSyM), Springer, Heidelberg (2006)
[8] OMG: Unified Modeling Language (UML) Infrastructure Specification, version 2.0, Object Management Group, ptc/03-09-15 (2003)
[9] OMG: Unified Modeling Language (UML) Superstructure Specification, version 2.0, Object Management Group, ptc/03-08-02 (2003)

[10] Roberts, D.B.: Practical Analysis for Refactoring. PhD thesis, University of Illnois at Urbana-Champaign (1999)

[11] Mens, T., Tourwe, T.: A Survey of Software Refactoring. IEEE Transactions on Software Engineering 30(2), 126–139 (2004)

[12] Jeanneret, C., Eyer, L., Markovic, S., Baar, T.: RocIET – Refactoring OCL Expressions by Transformations. In: 19th International Conference on Software & Systems Engineering and their Applications (2006)

[13] Kataoka, Y., Imai, T., Andou, H., Fukaya, T.: A Quantitative Evaluation of Maintainability Enhancement by Refactoring. In: Proceedings of the International Conference on Software Maintainance (ICSM), Montreal, Canada (2002)

[14] Wohlin, C., Runeson, P., Host, M., Ohlsson, M.C., Regnell, B., Wesslen, A.: Experimentation in Software Engineering – An Introduction. Kluwer, Dordrecht (2000)

[15] Finney, K., Fenton, N., Fedorec, A.: Effects of structure on the comprehensibility of formal specifications. IEE Proceedings – Software 146(4), 193–202 (1999)

[16] Briand, L., Labiche, Y., Penta, M., et al.: An Experimental Investigation of Formality in UML-Based Development. IEEE Transactions on Software Engineering 31(10), 833–849 (2005)

[17] Snook, C., Harrison, R.: Experimental Comparison of the Comprehensibility of a Z Specification and its Implementation. In: Proceedings of the Conference on Empirical Assessment in Software Engineering – EASE 01, England (April 2001)

On Metamodeling in Megamodels

Dragan Gašević[1], Nima Kaviani[2], and Marek Hatala[2]

[1] Athabasca University, Canada
[2] Simon Fraser University Surrey, Canada
dragang@athabascau.ca, {nkaviani, mhatala}@sfu.ca

Abstract. Model-Driven Engineering (MDE) introduced the notion of metamodeling as the main means for defining modeling languages. As a well organized engineering discipline, MDE should also have its theory clearly defined in terms of the relationships between key MDE concepts. Following the spirit of MDE, where models are first class citizens, even the MDE theory can be defined by models, or so called megamodels. In this paper, we use Favre's megamodel that was already used for defining linguistic metamodeling. Starting from the premise that this megamodel can also be used for defining other MDE concepts, we use it to specify the notion of ontological metamodeling. Here, we show that in order for this megamodel to be able to fully capture all the concepts of ontological metamodeling, some refinements should be applied to its definition. We also show how these new changes are in the same direction with the work of Kühne in defining linguistic and ontological metamodels.

1 Introduction

The idea of Model Driven Engineering (MDE) stems from software engineering, and more specifically, from the recent research in software development. MDE evolved as the paradigm shifted from the object-oriented approach where the main principle is that *everything is an object* into the model engineering paradigm based on the principle that *everything is a model* [6]. The object-oriented technology is about classes and objects, and main relations are *instantiation* (an object is an instance of a class) and *inheritance* (a class inherits from another class). MDE is about models, but it is also about relations between a model and the system under study (which can be a software artifact or a real world domain), metamodels, and model transformations. Similar to the object-oriented technology, MDE can be characterized by two main relations, namely, *representation* (a model represents a software artifact or real world domain) and *conformance* (a model conforms to a metamodel). Generally speaking, MDE is a field of system engineering in which the process heavily relies on the use of models and model engineering. Here, model engineering is considered as a disciplined and rationalized production of models [11].

For MDE to be popularized and accepted by the software engineering community, the theory behind it should be precise and easy to grasp. Furthermore, this theory must be comprehensive enough to address all the phenomena related to the languages

G. Engels et al. (Eds.): MoDELS 2007, LNCS 4735, pp. 91–105, 2007.
© Springer-Verlag Berlin Heidelberg 2007

and metamodels used in MDE (e.g. OWL, XML, and UML). It makes the study of MDE's theory critical. Modeling, as the spirit of MDE, is arguably the best method to study the theory of MDE, especially because of its natural ability in simplifying the specification and description of the intended goal in mind [7].

Atkinson & Kühne [1, 2, 15, 16] and Favre [11, 12] have taken important steps in clarifying the theory of MDE through using models. Atkinson & Kühne have distinguished between two types of metamodeling, namely *ontological* and *linguistic* (cf. Section 3), and Favre has used a megamodel to represent the concepts of modeling and metamodeling in MDE. Our aim in this paper is to compare these two research works and bring clarification inputs, in the form of metamodeling concepts, into both of them. Relying on the generality of Favre's megamodel to cover the concepts of both linguistic and ontological metamodeling, we apply his megamodel, currently only applied to linguistic metamodels, to ontological metamodels and discuss the raised problems. We try to solve these problems by expanding the megamodel, so that it can support the modeling of ontological metamodels as well. Since MDE theory should be able to cover any modeling language, in this paper we are exercising Favre's megamodel on the example of the Web Ontology Language (OWL) [23]. As both Favre's megamodel and OWL are based on *set theory*, applying the megamodel to OWL can help with refining and increasing the generality of the megamodel and thus the MDE theory. The composition of the worlds represented by Semantic Web ontology languages and MDE is followed as a goal by the Object Management Group (OMG) through defining the Ontology Definition Metamodel [4, 8, 13, 20]. Our work can also be regarded as another step towards reconciliation of these two worlds.

The rest of the paper has been organized as follows. In Section 2, we review the notions of megamodel and also describe ontologies and models and how they are inter-related. Section 3 further describes the definition of Favre's megamodel which is used to provide linguistic metamodeling and ontological metamodeling for OWL as a modeling language in MDE. Section 4 discusses the problems of the current megamodel with modeling the concepts of OWL and gives suggestions on how it can be improved. Finally, Section 5 will be a conclusion to this work followed by planning the future research in this area.

2   Megamodels and Models in Model Driven Engineering

Favre has introduced the notion of *megamodel* as a way to formally define the theory of MDE [10]. The term megamodel is selected to avoid the confusion with the basic meanings of the terms *model* and *metamodel*. Defining the formal theory in the form of a model representing the basic MDE concepts and their relations helps us infer new facts about MDE that are not explicitly represented in the megamodel. From this perspective the megamodel can also be regarded as an ontology of MDE.

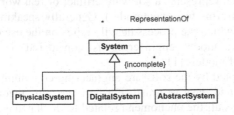

Fig. 1. The megamodel: Classification of systems

Fig. 1 shows an excerpt of Favre's megamodel [11] as a UML class diagram. The diagram defines the most abstract concept of discourse in MDE – *System*. The (incomplete) classification of systems shown in Fig. 1 distinguishes between physical systems, digital systems, and abstract systems. PhysicalSystem represents things from the reality such as 'a travel agency'. AbstractSystem is an abstraction in human minds that can be processed by human brains, for example, concepts and their relations from the biological domain. Finally, DigitalSystem is a digital representation that can be processed by computers, such as an XML document with an OWL representation of biological classes and their properties.

In constructing the megamodel, the following definition of models is used: *A model is a set of statements about some system under study* [22]. This definition introduces the notions of a model, the system under study, and their relations. The fact that being a model or a system under study (aka subject [15, 16]) is a relative notion, not an intrinsic property of an artifact, is represented by the *non-transitive* relation *RepresentationOf* (or simply μ) between different systems. This means that, one system can be a model of another system, and this can be represented by *RepresentationOf*. For example, an ontology of the biological domain (abstract system) is a representation of the conceptualization used by a biologist. A file in an OWL XML format (digital system) is a representation of the ontology of the biological domain. The same ontology can also be represented in an XML format of UML (i.e., UML XML Metadata Interchange, XMI).

2.1 On Meaning of Models and Ontologies

Since in this paper we are using the OWL language to experiment with Favre's megamodel, in this subsection, we discuss the relations between ontologies and models in order to motivate why the MDE theory should be applicable to the ontology languages. We have already defined a model as a set of statements about some system under study [22]. Looking strictly from the software development perspective, models are used to *specify (or prescribe)* a software system being developed. Having models formally defined, we can use them to check the validity of developed systems with respect to these models. Furthermore, we can check the consistency of well-defined models. In addition to specifying systems, models can also be used to *describe* systems under study. This is the role of conceptual models that model concepts and their relations within the system under study. In a megamodel, the role of models is defined relative to the roles that reflect relations between systems they represent. A model can be an abstract system that is a *RepresentationOf* another system (that can be physical, abstract, or digital, see Fig. 1), expressed in some modeling language. As one system is a *RepresentationOf* another system, the meaning of a model is always relative. For example, a UML model is an abstract system that is used to describe a specific domain (either physical or abstract system) or to specify a software system (a digital system).

On the other hand, we have ontologies based on the knowledge representation languages. Initially, ontology was defined as a formal specification of a conceptualization. More recent definition states that an ontology is a set of knowledge terms, including the vocabulary, the semantic interconnections, and some simple rules of inference and logic for some particular topic [14]. This definition implies the *descriptive* nature of ontologies, which is equivalent to the descriptive nature of models. Thus, the *RepresentationOf* relation from the megamodel can be applied to ontologies as well. However, nothing can prevent us from building specification models (e.g., of a software system) using ontologies, as nothing can prevent us from representing ontologies using UML constructs and later transform that representation into ontology languages [2]. One of the main advantages of ontologies over UML is that ontologies have formally defined semantics which makes them better suited for checking the validity of the systems they specify. This also proves the usefulness of the *RepresentationOf* relation from the megamodel for ontologies when they are used as specification models.

Based on the above discussion, we can conclude that although UML and OWL originate from different domains they have a lot in common. Furthermore, we should not neglect the fact that the design of UML and object-oriented programming languages is generally based on frame-based systems. In addition, object-oriented technology has had an important impact on ontology development methodologies and the construction of ontology languages. Some types of description logics adopted the basic object-oriented concepts, such as instantiation and inheritance[1]. This resulted in a proposal for unification of software modeling and knowledge representation technologies [2]. Therefore, in the rest of the paper, we use OWL to exemplify the metamodeling phenomena that are defined by using the megamodeling approach.

3   Metamodeling

We start our discussion on metamodeling with the definition of metamodels given in the OMG's MOF specification [19]: *A metamodel is a model that defines the language for expressing a model.* This definition emphasizes the linguistic nature of metamodeling where metamodeling is used for defining modeling languages. Seidewitz [22] gives the most commonly used definition of metamodels in MDE. *A metamodel is a specification model for a class of systems under study where each system under study in the class is itself a valid model expressed in a certain modeling language.* In fact, a metamodel is a specification model, which leads us to the conclusion that metamodels should be used to validate models represented in a specific language. That is, a metamodel makes statements about what can be expressed in the valid models of a certain modeling language. Generally speaking, a metamodel is any language specification written in English, such as the W3C's OWL language specification or the OMG's UML specification. However, in order to automatically check the validity of models w.r.t. their metamodels, we need to formalize the definition of the metamodel. In the scope of the MDA's pile of standards, MOF is defined as a metamodeling language for defining other languages (see [17] for further details about the MDA). In common understanding of the MDA, metamodels and models are connected by the *instanceOf* relation meaning that a metamodel element (e.g. the Class metaclass from the UML metamodel) is instantiated at the model level (e.g. a UML class Collie).

---

[1] The discussion about different design practices, such as the use of inheritance in software designs and knowledge representation is out of scope of this paper – see [3] for details.

Earlier research in MDE recognized two different types of instantiation relations in various metamodeling architectures: i) instantiation of elements residing at different modeling layers; and ii) instantiation of elements residing at the same modeling layer [5]. This is further explored in [1] where the example shown in Fig. 2 is used. We can say that *linguistic metamodeling* is equivalent to the definitions cited above [22] and [19]. In fact, linguistic metamodeling is a way for defining modeling languages and their primitives (e.g., Object, Class, MetaClass) on the layer of a metamodel (L2 in Fig. 2 or M2 in terms of the MDA). These language concepts are *linguistically* instantiated on the layer of models (L1 in Fig. 2 or M1 in terms of the MDA).

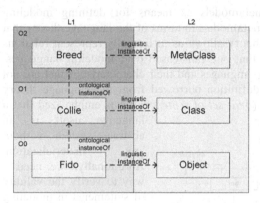

Fig. 2. Two types of metamodeling [1]: ontological and linguistics

Ontological metamodeling is concerned with domain definition and uses ontological instantiation. For example, a UML object (Fido) is an ontological instance of a UML class (Collie). It is important to say that a UML class and a UML object are on different ontological layers (O0 and O1, respectively), but still they are on the same linguistic metamodeling layer (L1 or M1 in terms of the MDA). We can go further and provide more abstract domain types or domain metaclasses (Breed) on higher ontological layers (O2) without the need to leave the model layer L1.

This distinction of metamodeling types had two important implications on the MDA:

1. The existence of the M0 layer in the classical four-layer MDA architecture has been discarded, as there is no need to separate the M0 and M1 layers. In fact, M0 layer comprising UML objects which are instances of UML classes is now a part of the M1 layer, but UML objects and UML classes are still on different ontological layers (O0 and O1).
2. The UML2 specification enables designing domain concepts whose instances can be both classes and objects.

Looking at the OWL language, we can identify both types of metamodeling. The OWL specification can be regarded as a model of the OWL modeling language, thus it is a linguistic metamodel. To allow for the automatic checking of ontology validity, this linguistic metamodel of OWL has to be specified by using a metamodeling language. Indeed, this is the role of the RDF schema of the OWL language that is a part of the official OWL specification. This schema can be regarded as a linguistic metamodel of the OWL language. To bring OWL ontologies to the MDA, a linguistic metamodel of OWL needs to be represented by an MDA language (i.e. MOF). This, at a more general level, is the goal of the ODM initiative [20], namely, to use MOF as a metamodeling language to develop a linguistic metamodel for representing ontological languages.

In terms of ontological metamodeling, OWL Full allows an instance of a class to be another class, while Boris Motik recently described one OWL Full semantic that is decidable [18].

Taking into account the above examples, we can identify an important issue that should be answered: "How can we formally specify the difference between the ontological and linguistic metamodeling?" In the remainder of this section, we will try to answer this question by using the megamodel [12].

3.1 The Megamodel: Metamodeling Relations

Having in mind the purpose of metamodels as means for defining modeling languages, Favre suggests analyzing metamodels in terms of the language theory that is based on sets [12]. The language theory allows us to formally represent relations between models and metamodels, so that one can reason over modeling languages, and thus deduce facts about modeling languages and their characteristics that are not explicitly asserted. In a very broad definition borrowed from the language theory where *a language is a set of systems (or a set of sentences)*, it is emphasized that a language itself is an abstract system without any materialization. Additionally, we need a way to verify the validity of statements in modeling languages. Models seem as one way to do so due to their specification nature.

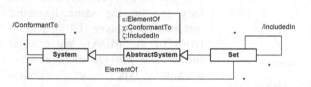

Fig. 3. Elements of the megamodel to define metamodeling

Based on these facts, the megamodel is extended with new elements in order to explain the meaning of metamodeling (Fig. 3). Besides adding the concept *Set* into the megamodel, the relation *ElementOf* (ε) is added as a means to assert that a system belongs to a set. In terms of OWL, this relation allows us to state that an individual (i.e. an Abstract System) is an *ElementOf* a class (i.e., a Set). This relation can also relate two sets stating that a set is *ElementOf* (or subset of) another set. However, *ElementOf* is not a transitive relation. This implies that the elements of a set's subset are not elements of its superset. It is obvious that this relation can not be used to model inheritance. To enable inheritance, the megamodel defines the *IncludedIn* (ζ) relation which is transitive. The transitivity of this relation means that the elements of a set's subset are also elements of its superset, which is the common meaning of this relation in UML (generalization) and OWL (subClassOf). However, verifying the validity of statements in modeling languages can not be done by using only sets and their relations. For this purpose, the megamodel defines the relation *ConformantTo* (χ): a non-transitive relation between systems in the megamodel. In the rest of this section, we use these relations to define two types of metamodeling.

3.2 Linguistics Metamodeling

Let us consider now how we can express relations between some constructs of the OWL language in terms of the extended megamodel (see Fig. 4). As an illustration, we use the concepts from Fig. 2. In OWL, we can use an individual to represent a

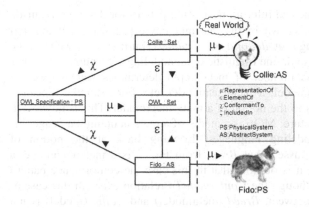

Fig. 4. Relations between models, metamodels, and modeling languages

physical system. In Fig. 4, an OWL individual is an abstract system (Fido:AS) that is a representation of (μ) a physical system (our pet collie Fido:PS). Additionally, we can use an OWL class, e.g. Collie, to generalize the characteristics of all collies. According to the definitions of the megamodel the class Collie should be defined as a set in order to be a representation of the set of all individuals. Thus, the Collie:Set class is a representation of (or a model of) the abstract system Collie:AS. Note that we leave the analysis of the relation between the class Collie:Set and the individual Fido:AS for the next subsection. According to the definition borrowed from the language theory that a *language is a set of systems*, we infer that both the individual Fido:AS and the class Collie:Set are elements of (ε) the OWL language (i.e. OWL:Set). Furthermore, we can say that a *modeling language* (in this case OWL:Set) *is a set of models*. Since a set is an abstract system, we need to represent (μ) a modeling language by a physical system in order to develop tools for checking the validity of expressions in that language. In fact, such a representation of (μ) a modeling language (OWL:Set) in the form of a physical system (OWL Specification:PS) is actually *a model of a modeling language*, i.e. a *metamodel*. This is why in this case we can say that the official W3C's OWL specification can play the role of a metamodel. In the same way, the ODM specification is a metamodel of OWL but specified in MOF. Finally, we can conclude that a metamodel is *a model of a set of models* [12].

There is one important relation between elements of (e.g. Collie:Set and Fido:AS) a modeling language (e.g. OWL:Set) and the metamodel of the modeling language (e.g. OWL Specification:PS). This relation is *ConformantTo* (χ), which means that all elements (classes and individuals) of a modeling language have to conform to (χ) the metamodel of the modeling language. It is very important to note that the relation between models (Collie:Set and Fido:AS) and their metamodel *is not ElementOf*, hence the metamodel does not contain them. Another observation is that the relation *ConformantTo* is equivalent to the *linguistic instanceOf* relation from Fig. 2 and this type of metamodeling is *linguistic metamodeling*. In terms of the OWL language this relation is represented by the *rdf:type* property relating the OWL language construct *owl:class* with a concrete ontology class, such as the class Collie:Set. We refer readers to [12] where the notion of linguistic metamodeling is defined for the UML language in a similar way.

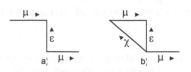

Fig. 5. The $\mu\varepsilon\mu$ (a) and conformant (b) patterns

From the above analysis, one can infer an interesting pattern for defining linguistic metamodeling. This pattern is shown in Fig. 5a, and Favre [12] named it the μεμ pattern or a pattern representing *a model of a set of models*. Furthermore, we used the conformance pattern (Fig. 5b) to define linguistic metamodeling. The number of times we apply this pattern (i.e. the number of meta-steps) determines whether we are dealing with a model, a metamodel, a metametamodel, etc. For example, repeating this pattern two times leads us to the notion of a meta-metamodel. This is how MOF is defined in the MDA architecture: MOF is a model of a metamodeling language, i.e. a language for defining modeling languages. Looking back at the notion of metamodeling used in [18] to assert that *Breed* from Fig. 2 is a metamodel (i.e. a metaclass) of the *Collie* class, it is obvious that in this case the conformance pattern (Fig. 5b) is not applicable. Although the *ConformantTo* relation exists in this case as well, the *ElementOf* relation between *Breed* (metamodel) and *Collie* (model) is not present, although it should be defined according to the definition of the OWL class from the OWL specification. The existence of the *ElementOf* relation would imply that the metamodel (*Breed*) contains specific elements of a modeling language (in this case the class *Collie*), which is not the case in linguistic metamodeling. Instead, it is the case in ontological metamodeling as explained in the next subsection[2].

3.3 Ontological Metamodeling

To explain ontological metamodeling, let us first clarify the relations between the elements of the modeling language. Fig. 6a shows an extended UML static structure diagram from Fig. 4 with the relations between the set Collie:Set and the abstract system Fido:AS shown. The notion of a set for Collie has been chosen correspondent to the definitions of Favre's megamodel in which ontological concepts have been considered as sets [12]. The set Collie:Set in modeling languages (such as OWL and UML) can be best modeled by using the term *class*. Such a set is typically composed of elements that are just abstract systems (or facts in terms of the OWL specification). This explains why we have the *ElementOf* relation between Collie:Set and Fido:AS with the meaning that Fido:AS is an *ElementOf* Collie:Set. However, the set of abstract systems groups them also based on their common characteristics. Consequently, according to Favre's megamodel, abstract systems have to be *ConformantTo* their sets. This implies having a *ConformantTo* relationship from the Fido:AS to Collie:Set, as it has been shown in Fig. 6a.

In Fig. 6b, we analyze the relation between the meta-class Breed:Set and the class Collie:Set from Fig. 2 using the megamodel. The meaning of the set Collie:Set is the same as above (Fig. 6a): *RepresentationOf* (μ) the abstract system Collie:AS; *ConformantTo* (χ) the OWL specification; *ElementOf* (ε) the OWL:Set (i.e. OWL language); and Fido is *ElementOf* (ε) and *ConformantTo* (χ) the set Collie. Note that Breed also has been chosen as a set by following the same rationality that was expressed for Collie. This way we can keep up with the notions of megamodel that has been suggested by Favre.

---

[2] This also reflects the strict separation (a basis for strict metamodeling architectures) of a set of elements or symbols belonging to models from a set of elements belonging to the metamodel, i.e., having strict separation between meta-layers in metamodeling architectures [22]. This idea has also inspired [19] to develop their metamodeling architecture for the Semantic Web.

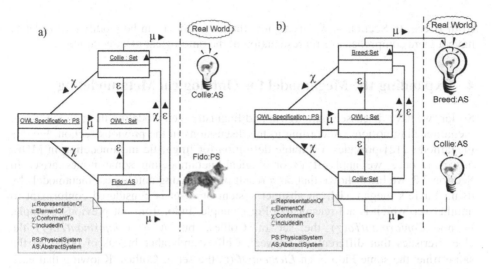

Fig. 6. Relations between abstract systems and sets of abstract systems represented in a modeling language (a); Ontological metamodel in the megamodel: Relations between a set and sets of models (b)

Let us look at Breed. Breed:AS is the real meaning of Breed in human's mind, while the set Breed:Set (according to Favre's megamodel) is considered as a *RepresentationOf* (μ) the abstract system Breed:AS, i.e Breed:Set is a model of the abstract system Breed:AS. Furthermore, the Breed:Set contains the Collie:Set meaning that the Collie:Set is an *ElementOf* (ε) the Breed:Set. In terms of OWL, the Breed:Set is a class that contains another class, i.e. the Collie:Set. Since the set Breed:Set defines characteristics of its elements (instantiation relation from Fig. 2), the Collie:Set should be also *ConformantTo* (χ) the Breed:Set. Due to the non-transitive definition of the *ElementOf* relation, the elements of the Collie:Set (i.e. Fido:AS) *are not* elements of the Breed:Set. In this case, we can say that Breed:Set is a superset of the Collie:Set whose element is the abstract system Fido:AS. In the OWL language, it can be said that the class Collie is of type Breed, which means that the Breed class is a meta-class for Collie. The individual Fido is of type Collie, but Fido is not of type Breed.

Here we have a case of metamodeling, which is not linguistic metamodeling as defined in the previous section; it is ontological metamodeling. In this case, we can identify another pattern that is based on the χε relation between two sets, and both sets are models of abstract systems. However, the *ConformantTo* relationship between Fido as an abstract system and Collie as a set, and also Collie as a set and Breed as another set, opens some points for discussion. Although it is reasonable to consider an abstract system conformant to another abstract system, it is not reasonable to consider an abstract system or a set conformant to another set. Looking at Collie as a set of all collies in the world, there is nothing in common between our Fido and a community of Collies, but the commonality comes from the properties that all these dogs share as Collies. Similarly the representation and definition of the concept Breed makes Collie a Breed, but the Breed set is just to identify the individuals that have been instantiated from Breed. In Section 4, we argue how the megamodel can be modified in order to have a more comprehensive representation of the ontological metamodeling.

4 Expanding the Megamodel for Ontological Metamodeling

So far, we have discussed ontological and linguistic metamodeling through the use of megamodeling notions. According to the discussions in the previous section, Favre's megamodel [12] provides a suitable definition for linguistic metamodeling in MDE, however, once we apply it to ontological metamodeling some issues arise. In Section 3.3, we have shown that, as a result of expressing ontological metamodels by using Favre's megamodel, an abstract system or a set is considered conformant to another set, which is not reasonable. For example, Fido, from our previous example, is not *ConformantTo*(χ) the set of Collies, but it is *ConformantTo*(χ) the characteristics that differentiate between Collies and other breeds of dogs. At the same time, the same Fido is an *ElementOf* (ε) the set of Collies. Knowing that each ontological concept can be modeled with a class, the question is how we can precisely define the sets and the characteristics of a class, and its correspondent concept, in the megamodel.

Let us start with the definition of a class in the OWL specification. According to W3C's specification for OWL [23], "Classes provide an abstraction mechanism for grouping resources with similar characteristics. Every OWL class is associated with a set of individuals, called the *class extension*. The individuals in the class extension are called the *instances* of the class. A class has an *intensional* meaning (the underlying concept) which is related but not equal to its class extension. Thus, two classes may have the same class extension, but still be different classes."

The main problem with the definition of Favre's megamodel is that there is no distinction between the *intension* and the *extension* of a class. While our Fido is *ConformantTo*(χ) the characteristics of the class Collie, i.e. the intension of the class Collie, it is an *ElementOf* (ε) the extension of this class. In fact, in ontological metamodeling the instantiation relationship (*instanceOf*) happens between two concepts when one concept is *ConformantTo*(χ) the intension and is an *ElementOf* (ε) the extension of the other concept. Consequently, the relations between Breed, Collie, and Fido of Fig. 2 should be modeled similar to Fig. 7, which is different from what we obtained in Fig. 6. Note that the <<instanceOf>> relationship of Fig. 7 is not part of the megamodel and we are using the link between the super concept and its instance only for the sake of clarification.

We claim that a set is equivalent to the extensional part of a concept, and considering the concept equal to a set means dropping the intensional part of that concept. Especially, it has been also pointed out in the OWL definition of a class that the extensional part might be shared among a couple of classes with each class representing different intensions. Referring to a class as a set makes the desired intensional meaning ambiguous. Also, according to the defintion of a class in OWL, the extension of a class is a set of its instances or individuals. It means that this extension is in the same ontological level as one single individual of the class (see Fig. 7). Additionally, for each of the intensional and extensional properties of a concept, there is a description in the form of an abstract system in the real world which can be connected to the intension or extension of the system with a *RepresentationOf* (μ) relation as shown in Fig. 7. In OWL, the intensional part of a concept is defined in the form of class properties, also each object instantiated from a class is an element of the extensional part of the class. Consequently, a class can be considered an abstract system composed of intensional and extensianl parts which are also abstract systems.

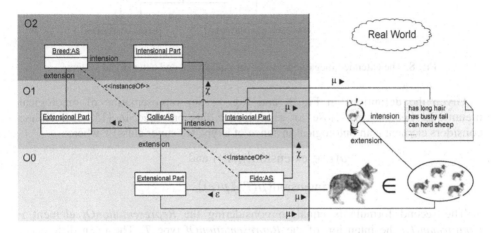

Fig. 7. The relations between the instantiated elements and the intensional and extensional properties of their super classes

Based on the discussions that we have made to this point, the megamodel introduced by Favre should be further refined to cover the intensional and extensional meanings of a concept. We argued that intensional and extensional properties of a class are also abstract systems. We believe, the term *Set*, coined in Favre's megamodel as an abstract system, is nothing but the extensional meaning of the ontological concept, and thus this extensional meaning of the concept is an abstract system as well. Moreover, the intensional meaning of a concept is fully in line with the definition of the abstract system in Favre's megamodel (cf. Section 2) and can be considered an abstract system. This leads us to improving Favre's megamodel by extending the abstract system according to Fig. 8. It is worth mentioning that Favre in his megamodel represented the identified types of a system as *incomplete*, which has left the room for further expansion and improvement. Our expansion of Favre's system is yet an incomplete representation of the system as well.

As Fig. 8 shows, an abstract system composes a set (as the extensional part) and an instensional part which both are also abstract systems. However, a set is an abstract system for which there is no intensional part, while the intensional system is an abstract system that can have both the extensional and intensional parts. This definition of an intensional system makes it possible to have a set of intensions for an abstract system, which is generally the case when defining the ontological concepts.

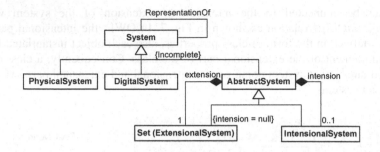

Fig. 8. The extended megamodel to cover intensional and extensional systems

Given the definitions in Fig. 7 and Fig. 8, our representation of ontological metamodel becomes so close to what Kühne has provided in [15, 16]. Kühne considers element *e* an ontological instance of a type (or super class) *T* where:

$$\mu(e) \in \text{extension}(\mu(T)) \text{ and}$$

$$\text{intension}(\mu(T))(\mu(e))$$

The second formula is equal to considering the *RepresentationOf* element *e* *ConformantTo* the intension of the *RepresentationOf* type *T*. The main difference, however, is that in Kühne's representation of ontological metamodel there is no intensional or extensional system in the ontological layers (e.g. O_0 or O_1) and these two systems are only considered as the meanings in the real world (referred to in the form of *RepresentationOf*). Nonetheless, we believe that extensional and intensional meanings also exist as systems within the modeling space [9]. The witness to this claim is the representation of the properties for a class which can be considered as the intension of the class. Furthermore, referring to the definition of a class in OWL, provided in the beginning of this section, one can recognize that the set of individuals for an OWL class are not judged based on their presence in the real world, but these individuals are the objects instantiated from the OWL class in the knowledge base. We can even consider the representation of a class in OWL as a *DigitalSystem* which is a model of the *AbstractSystem* intended in the program. However, as the notion of model is relative; this chain of finding a model for another model will never stop unless we decide to stop it at a certain point. As a result, we consider the definition of a class in OWL as an *AbstractSytem*.

Now that we have made a clear understanding of the ontological metamodeling by refining the definitions of the megamodel, we can provide a definition for the ontological metamodel, based on our notions of megamodel. We define *an ontological metamodel as a set (or extension) of models of abstract systems classified based on some instensions, and as an element of a modeling language*. At the same time, *the ontological metamodel is a model of an abstract system*. Given the definition above, the next section compares the implications of ontological and linguistic metamodeling.

4.1 Ontological and Linguistic Metamodeling: Implications

There are several important consequences of the definitions above:

1. An ontological metamodel is a model of an abstract system, which means that ontological metamodel can only model abstract systems (not physical and digital systems). This further means that an ontological metamodel can only be a representation of systems that are sets of either other sets or abstract systems. Consequently, an ontological metamodel in the OWL Full language can not contain individuals, only classes.
2. Being an ontological metamodel means being only a model of an abstract system, but not a model of models of systems like it is the case with linguistic metamodel. Consequently, an ontological metamodel *is not* a model of a modeling language, and thus it is not supposed to define a set of valid constructs of a modeling language.
3. The definition of an ontological metamodel is surprising at first, since an ontological metamodel is not just a model, but it is also a set. Having in mind that the main purpose of ontological metamodel is to define libraries of high level domain types [1] it seems quite natural that an ontological metamodel is indeed a set of (predefined domain specific) models of abstract systems. That is to say, a set of classes, meta-classes, meta-meta-classes and so forth in terms of OWL or UML.
4. An ontological metamodel should be conformant to the linguistic metamodel of the modeling language like any other element of the modeling language. This is again compliant with the notion of ontological metamodeling from Fig. 2 where the libraries of high level domain types are linguistic instances of the linguistic metamodel.

   There are also some inferred facts about relations between ontological and linguistic metamodeling.
5. The first concern with regards to the linguistic metamodel is whether the change to the megamodel affects the definition of linguistic metamodeling or not. Fig. 9a shows the exact pattern that we introduced in Fig. 5b. Fig. 9b is the rotation of that same pattern, and finally, Fig. 9c is the pattern that we obtained for ontological metamodeling based on intensional and extensional expansion of abstract system, which is really close to Fig. 9b. It can be said that in ontological metamodeling, because we are mostly dealing with the *AbstractSystems*, the notions of intensional and extensional systems become more important, so that they need to be clearly defined. However, in linguistic metamodeling, because the metamodel is usually a *PhysicalSystem* or *DigitalSystem*, it can play as both intensional and extensional systems and thus these two systems can be omitted (in Fig. 8 we consider intensional and extensional systems only for *AbstractSystems*). The possibility of having an *AbstractSystem* as a linguistic metamodel is open for future research.

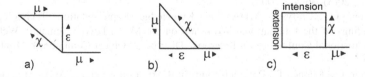

Fig. 9. The effect of the change to the megamodel on the pattern of Fig. 5

6. Linguistic metamodeling is used to specify a set of valid statements of a modeling language, and thus linguistic metamodeling (a linguistic metamodel of the OWL or UML languages) specifies whether ontological metamodeling is supported by a modeling language. Defining a modeling language by a linguistic metamodel does not necessarily mean that the modeling language supports ontological metamodeling.
7. Metamodeling languages, such as MOF, may have also support for ontological metamodeling according to the meta-step definition for linguistic metamodeling. This means that we can define meta-classes that are ontological instances of other meta-classes on the same linguistic layer. However, the pragmatics for the presence of such features in metamodeling languages is out of scope of this paper (see discussion on minimal reflective metamodel [22]).

5 Conclusion

Favre by representing megamodels and Kühne by identifying linguistic and ontological metamodeling have had important contributions to clarifying the theory of MDE. In this paper, we investigated the generality and comprehensiveness of Favre's megamodel, once it is applied to the ontological metamodels. We have shown that for this megamodel to be able to fully capture the intended meaning of the ontological metamodeling, the concept of *AbstractSystems* should be further expanded to entail *intensional* and *extensional* systems which are also *AbstractSystems*. We also argued that the extensional system is what we know as a *Set*, while the intensional system is the characteristic or the set of characteristics (as an intensional system can entail a set itself as a result of being an abstract system) that classify the objects in the set. As a consequence of applying the change to the megamodel, we have shown how the ontological classification becomes more meaningful by making each concept *ConformantTo* the intensional system of its super class and an *ElementOf* the extensional system of the super class. We have also shown that this change does not really interfere with Favre's representation of linguistic metamodeling using megamodels and the whole systems remains consistent. As a future plan for this research, we decide to consider the relations between different intensional systems that have the share the same extensional system (e.g. a group of planets clustered in one single set based on both their biological and botanical characteristics) and see how the intensional and extensional relations affect the definition of the megamodel.

References

1. Atkinson, C., Kühne, T.: Model-Driven Development: A Metamodeling Foundation. IEEE Software 20(5), 36–41 (2003)
2. Atkinson, C.: Unifying MDA and Knowledge Representation Technologies. In: Proceedings of the International Workshop on the Model-Driven Semantic Web (At the 8th International Conference on Enterprise Distributed Object Computing), Monterey, CA (2004)
3. Atkinson, C., Kühne, T.: Rearchitecting the UML infrastructure. ACM Transactions on Modeling and Computer Simulation 12(4), 290–321 (2002)

4. Baclawski, K., Kokar, M., Kogut, P., Hart, L., Smith, J.E., Letkowski, J., Emery, P.: Extending the Unified Modeling Language for ontology development. Software and Systems Modeling 1(2), 142–156 (2002)
5. Bézivin, J., Lemesle, R.: Ontology-Based Layered Semantics for Precise OA&D Modeling. In: Proc. of the WSh on Object-Oriented Tech., Jyväskylä, Finland, pp. 151–154 (1998)
6. Bézivin, J.: On the unification power of models. Software and System Modeling 4(2), 171–188 (2005)
7. Bézivin, J., Grebe, O.: Towards a Precise Definition of the OMG/MDA Framework. In: Proceedings of ASE'01 (November 2001)
8. Bodoff, D., Ben-Menachem, M., Hung, P.C.K.: Web Metadata Standards: Observations and Prescriptions. IEEE Software 22(1), 78–85 (2005)
9. Djurić, D., Gašević, D., Devedžić, V.: The Tao of Modeling Spaces. Journal of Object Technology 5(8), 125–147 (2006)
10. Favre, J.M.: Towards a Basic Theory to Model Driven Engineering. In: WISME 2004. Proceedings of the UML2004 International Workshop on Software Model Engineering, Lisbon, Portugal (2004)
11. Favre, J.M.: Foundations of Model (Driven) (Reverse) Engineering: Models - Episode I, Stories of the Fidus Papyrus and of the Solarus, Dagstuhl Seminar 04101 on Language Engineering for Model-Driven Software Development, Dagsthul, Germany (2004)
12. Favre, J.M.: Foundations of the Meta-pyramids: Languages and Metamodels - Episode II, Story of Thotus the Baboon, Dagstuhl Seminar 04101 on Language Engineering for Model-Driven Software Development, Dagsthul, Germany (2004)
13. Gašević, D., Djurić, D., Devedžić, V.: Model Driven Architecture and Ontology Development. Springer, Berlin (2006)
14. Hendler, J.: Agents and the Semantic Web. IEEE Intelligent Systems 16(2), 30–37 (2001)
15. Kühne, T.: Clarifying matters of (meta-) modeling: an author's reply. Software and Systems Modeling 5(4), 395–401 (2006)
16. Kühne, T.: Matters of (Meta)- Modeling. Software and Systems Modeling 5(4), 369–385 (2005)
17. Miller, J., Mukerji, J.: MDA Guide Version 1.0., OMG Document: omg/2003-05-01 (2003), Available: http://www.omg.org/mda/mda_files/ MDA_Guide_Version1-0.pdf
18. Motik, B.: On the Properties of Metamodeling in OWL. In: Proceedings of the 4th International Semantic Web Conference, Galway, Ireland, pp. 548–562 (2005)
19. OMG MOF OMG Meta Object Facility Specification v1.4, OMG Document formal/02-04-03 (2002), Available: http://www.omg.org/cgi-bin/apps/doc?formal/02-04-03.pdf
20. OMG ODM: Ontology Definition Metamodel, OMG Document ad/05-08-01 (2005), Available: http://www.omg.org/cgi-bin/apps/doc?ad/05-08-01.pdf
21. Pan, J.Z., Horrocks, I.: Metamodeling architecture of Web ontology languages. In: Proceedings of the 1st Semantic Web Working Symposium, Stanford Univ., USA, pp. 131–149 (2001)
22. Seidewitz, E.: What Models Mean. IEEE Software 20(5), 26–32 (2003)
23. W3C Specification for OWL: OWL Web Ontology Language Reference: (February 2004), Available: http://www.w3.org/TR/owl-ref/

Magritte – A Meta-driven Approach to Empower Developers and End Users

Lukas Renggli[1], Stéphane Ducasse[2], and Adrian Kuhn[1]

[1] Software Composition Group, University of Bern, Switzerland
{renggli,akuhn}@iam.unibe.ch
[2] LISTIC, University of Savoie & INRIA Futurs Lille, France
stephane.ducasse@free.fr

Abstract. Model-driven engineering is a powerful approach to build large-scale applications. However, an application's metamodel often remains static after the initial development phase and cannot be changed unless a new development effort occurs. Yet, end users often need to rapidly adapt their applications to new needs. In many cases, end users would know how to make the required adaptations, if only the application would let them do so. In this paper we present how we built a runtime-dynamic meta-environment into Smalltalk's reflective language model. Our solution offers the best of both worlds: developers can develop their applications using the same tools they are used to and gain the power of meta-programming. We show in particular that our approach is suitable to support end user customization without writing new code: the adaptive model of Magritte not only describes existing classes, but also lets end users build their own metamodels on the fly.

Keywords: Meta-Modeling, Meta-Data, Adaptive Object Model, Business Application Development, Smalltalk.

1 Introduction

As a result of our experience with developing dynamic web applications at an industrial scale[1], we recognized the need to introduce a meta-layer to provide us with more flexibility. Describing domain entities is not a new idea [1,2,3,4,5]. However, often meta-descriptions remain static after the initial development phase and cannot be changed unless a new development effort occurs. Yet, end users often need to rapidly adapt their applications to new business needs [6] and in many cases, they would know how to make the required adaptations, if only the application would let them do so [7].

Application requirements usually do not remain static after the initial development phase. Changing business plans typically boils down to minor modifications to domain objects and behavior, for example new input fields have to be added, configured differently, rearranged or removed. Unfortunately most of today's applications don't provide this ability to their end users. The situation is even more striking in the context of web applications that are typically built

G. Engels et al. (Eds.): MoDELS 2007, LNCS 4735, pp. 106–120, 2007.
© Springer-Verlag Berlin Heidelberg 2007

for a lot of different people with varying needs. Furthermore it is often the case that software systems have a static object model: one that has been defined by the software architect at implementation time and that cannot be changed later without changing and recompiling the source-code.

Generative techniques should be avoided, as they prevent the metamodel from being dynamically changed at runtime. Also, the introduction of meta-descriptions should not disrupt the normal way of programming and the tools used to program. The development tools (refactorings, version control, unit testing, debugger, etc.) should continue to work as if there were no meta-descriptions [9]. The approach should be integrated as closely as possible into the object-oriented paradigm, the tools and the programming environment. In our case we use Squeak[2], an open-source Smalltalk [10,11], and Seaside[3], an open-source web application framework [12].

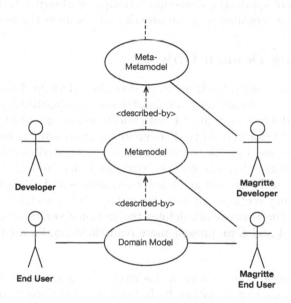

Fig. 1. Magritte is self-described and features metamodel changes at runtime. This allows end users not only to interact with the application data, but also change the metamodel without having to write code.

This publication reports on our experience with using the Magritte meta-descriptions framework. Magritte has been originally developed for web applications, but its applicability goes beyond that context. The Magritte meta-descriptions are integrated into the reflective metamodel of Smalltalk to support

---

[1] The first author of this paper is an independent consultant and software architect. In the context of his Master thesis [8] he invented and developed the Magritte framework, which is used in several large-scale industry and open-source projects.

[2] http://www.squeak.org

[3] http://www.seaside.st

the development of flexible applications. As the Magritte metamodel is self-described, it is possible to apply the same editors for both domain data and its corresponding metamodel. As illustrated on Figure 1 this enables a Magritte user to work on two meta-levels at the same time. This applies to both the end user and the developer. With Magritte we can reap the benefit of the two worlds: On the one hand we keep our efficient and dynamic object-oriented programming with an excellent tooling context, and at the same time we gain the flexibility and compactness of meta-descriptions to factor repetitive tasks of our application development.

This paper is structured as follows: Section 2 introduces the Magritte framework and presents an example how Magritte descriptions are specified. In Section 3 we present different interpreters that have been written for Magritte. Section 4 explains how Magritte is self-described and how this enables end users to customize their applications. Section 5 compares Magritte to related frameworks and Section 6 evaluates our approach and discusses the lessons learnt.

2   Describing Domain Objects

Magritte is a meta-description framework, describing domain classes and their respective attributes, relationships and constraints [4]. Magritte augments the reflective metamodel of Smalltalk [13] with additional means to reason about the structure and behavior of objects. The Smalltalk programming language is used to define Magritte meta-entities and their behavior. An attribute description contains the type information, the way the attribute is accessed, and some optional information such as a comment and label, relationships and validation conditions.

In the following sections we use the example of a meta-described person domain-model. The Person class defines the instance variables name and birthday. In Sections 4 and 6 we present more realistic examples used in productive applications.

Example. To describe the entities in this model we need corresponding description instances, that can be either built from the source-code at development time, dynamically at run-time, or a combination of the two approaches. Either way, the code to build the descriptions looks the same. To describe the name, we create an instance of StringDescription, define an access strategy (in this case the getter method #name is used), provide a label and add the constraint that this is a required value[4].

```
(StringDescription new)
    selectorAccessor: #name;
    label: 'Name';
    beRequired
```

---

[4] In Smalltalk messages follow the pattern receiver methodName: argument, which is equivalent to the Java syntax receiver.methodName(argument). Hence String-Description new sends the message new to the class StringDescription that returns a new instance of the receiving class. Subsequently the messages selectorAccessor:, label: and beRequired are sent to this instance.

Note that descriptions provide much more information than just type information. A date description, for example, knows how the attribute should be displayed (June 11, 1980, 11 June 1980, 06/11/1980), edited (text-input fields, drop-down boxed, date-picker), and validated. Moreover descriptions do not necessarily describe instance variable attributes, but might also describe derived attributes that are dynamically calculated on demand.

2.1 Structural Descriptions

The Essential Meta-Object Facility (EMOF) is a standard for model driven engineering defined by the Object Management Group (OMG). Similar to EMOF Magritte is not designed as a layered architecture. Magritte descriptions live in a flat world and there is no distinction drawn between objects in the meta-metamodel (M3), the metamodel (M2), the model (M1) and the instances (M0).

Contrary to EMOF Magritte has no notion of instantiation, inheritance and classes. We describe objects that have already been instantiated. Magritte is tightly embedded into the Smalltalk object model. Smalltalk is used to instantiate, configure and compose the descriptions, as well as to model the behavior of the meta-descriptions. In Magritte objects are not tightly connected with a single description. Descriptions can be shared, exchanged and applied to different instances and classes.

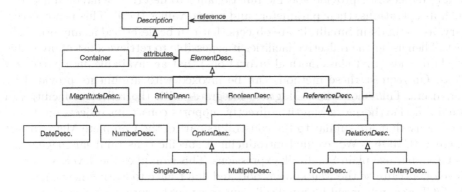

Fig. 2. The Description Hierarchy of Magritte

As seen in Figure 2 the description classes define a type hierarchy. This is similar to the subclasses of Type in EMOF, where a distinction between classes and primitive types is made. An instantiated Magritte description is similar to an EMOF property.

Magritte defines multiplicities using the Composite design pattern. The class ReferenceDescription knows another description, that is used to describe the referenced object. Whether the elements are *ordered* and/or *unique* is determined as a property in ReferenceDescription. Upper and lower bounds of are specified using constraints. In EMOF multiplicities are part of the type information. Our approach has shown to be more straightforward when automatically building editors and reports.

Option Descriptions. The SingleOptionDescription models an 1 : 1 relationship. The class MultipleOptionDescription models a 1 : n relationship. In both cases the referenced objects must be chosen from a list of existing objects satisfying the reference description.

Relationship Descriptions. The ToOneRelationshipDescription models an 1 : 1 relationship. The ToManyRelationshipDescription models an 1 : * relationship. In both cases any object can be referenced that satisfies the reference description.

The architecture of Magritte, *i.e.,* describing Smalltalk class with descriptions, is not new and can be seen as a validation of the nowadays well-known distinction between two conceptually different kinds of instance-of relationships: (1) a traditional and implementation driven one where an instance is an instance of its class, and (2) a representation one where an instance is described by another entity [14]. Atkinson and Kühne named these two forms: form vs. contents or linguistic and logical [15,16].

2.2   Executability and Constraints

Magritte does not provide specific functionality to describe behavioral aspects, such as operations, their parameters and return values [17,9]. This is not necessary, as methods in Smalltalk are objects that can be described as any other object. Then using the reflective facilities it is possible to retrieve a list of invokable method sends (first class method invocations) that are available on a particular class. On request these methods can be invoked with arguments provided by end users. This shows how Magritte integrates with the reflective facilities of Smalltalk. Furthermore Magritte directly supports constraint objects on its descriptions, that are similar to the constraints part of the Complete Meta-Object Facility (CMOF). We avoided introducing a specific constraint language, such as OCL, but use plain Smalltalk expressions. This simplifies the development, as developers can use the well known tools and don't have to learn a new language. As OCL was influenced by Smalltalk, our constraint expressions resemble those of OCL.

Example. To add a size constraint to a string description we use a block closure (anonymous function) to ensure a maximal size of 5 characters. In case the condition is not satisfied the error message "too long" is displayed:

```
aDescription addCondition: [ :value | value size <= 5 ] label: 'too long'
```

3   Interpreting Descriptions

Magritte descriptions can be interpreted in many different ways. Simple interpreters just iterate over the descriptions and perform different tasks on the associated model. In more generic cases we exploit the Visitor design pattern to walk trough the description graph. The most immediate use case is the one to automatically build views, editors and reports.

3.1 Building a View

The simplest interpreter that can be written is one that iterates over all descriptions of a domain model and prints the label and the current values onto a text stream. The following code shows everything that is needed to accomplish this task on any described domain-model as aModel in the following:

```
aModel description do: [ :desc |
  aStream
      nextPutAll: (desc label);
      nextPutAll: ': ';
      nextPutAll: (desc toString: (desc accessor readFrom: aModel));
      cr ].
```

First we ask the model for its description, then we iterate over its individual attributes. Within the loop, we first print the label, then we ask the accessor of the description to return the associated attributes from aModel and transform this value to a string, so that it can be appended to the output. The resulting output might look like:

```
Name: John Lennon
Birthday: 9 October 1940
```

Since every description knows how to convert its values to strings, we get a readable list of all the described attributes of our domain-model. By defining a different string-conversion strategy in descriptions, we are able to change the way values are printed. When adding, removing or changing descriptions in the domain-model, the above code will still print the correct output without having to change a single line of the interpretation code.

3.2 Building an Editor

Most business applications today consist of a large number of input-dialogs that need to be built and validated manually. One of the goals of Magritte was that developers could specify how their domain objects can be modified, so that it becomes possible to automatically build editors for different user-interfaces frameworks, as seen in Figure 3.

Sending the message asComponent to a domain model returns a ready-to-use Seaside component that can be plugged into the web application. As in Section 3.1, Magritte will iterate over the descriptions and compose an editor. The default interpreter creates XHTML markup that is annotated with a variety of CSS classes, so that the layout and look can adapted to most needs by only using a different style-sheet. For specific cases it is always possible to subclass the interpreter or to define a different XHTML generation strategy on a per meta-description bases.

Fig. 3. Interpreting descriptions for different GUI frameworks: the web (left) and Morphic Squeak (right)

During an edit operation, Magritte works on copies of the values being edited, so that the original data remains untouched. Before actually committing the changes, Magritte checks if the model satisfies all its validation conditions. Moreover the framework ensures that there are no edit conflicts caused by other people editing the same objects at the same time, and, if necessary, shows a warning.

All this is very convenient for software developers, as they don't have to do the caching, the validation and the conflict detection for every editor manually. Not only does this increase the development speed, but it also makes the software more robust, since all editing concerns are handled at a single place and are not spread across all editors in the system.

3.3   Other Interpreters

Over the past few years many Magritte interpreters have been written:

Validate, Verify and Setup Objects. Whenever user input is requested incoming data has to be validated. Existing graphs of objects need to be verified from time to time to ensure validity. Complex graphs of objects need to be built and initialized with default values. All these tasks can be accomplished by walking trough a description graph and validate or build these objects on the fly.

Persistency, Indexing and Querying. Making objects persist is one of the most daunting tasks. Magritte description are able to tell a interpreter how a graph of domain models should be stored and loaded. For example Magritte can generate SQL statements to retrieve and update objects in a relational database. In the context of object databases it is crucial to build indexes to be able to efficiently query that data. With Magritte these tasks can be automated.

Introspection, Reflection. The metamodel of Magritte provides additional information that can be used to improve the development processes, for example in the debugger and in the inspector a high level view can be provided instead of a straight memory dump of the object layout.

4   End Users Customizability

Often dialogs in applications remain static after the initial development phase and cannot be changed unless a new development effort occurs. Yet end users

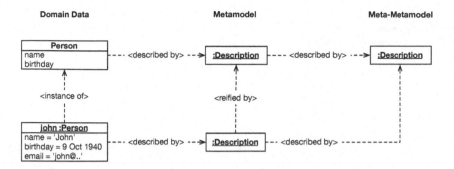

Fig. 4. The meta-levels of the Magritte metamodel

often need to rapidly adapt their applications to new business needs. In many cases they would know how to make the required adaptations, if only the application would let them do so.

As shown in Figure 4 there are different spots to specify, reify and interpret the Magritte metamodels. The domain class Person is written by the application developer. The class is described by a set of Magritte descriptions that are common to all instances of Person. These descriptions are either hardcoded into the source-code of the application or have been specified at runtime by an end user.

john is an instance of the class Person. The instance itself references a set of instance specific descriptions used to reify the class-based descriptions. These descriptions are either dynamically built from the application logic or have been manually specified by an end user using an editor as seen in Figure 5. Instances that do not use instance-specific descriptions simply reference the set of class descriptions. Furthermore to avoid the need to introduce an instance variable to hold the instance specific descriptions on all objects, we propose the use of an adaptive model as presented in the next section.

Figure 5 shows a description editor that is part of a commercial workflow definition and runtime engine. The editor opened on a specific workflow task allows

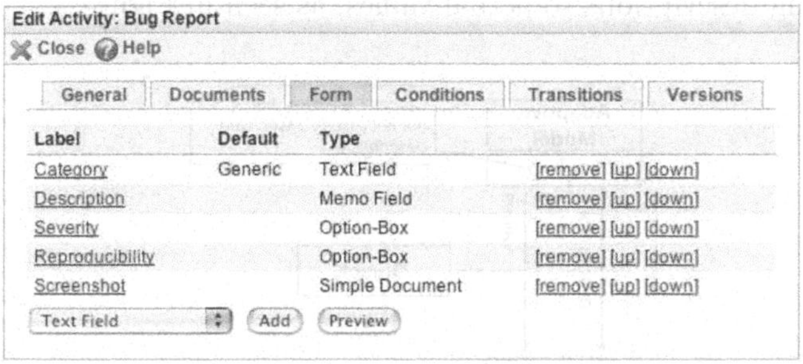

Fig. 5. A Magritte description editor that allows end users to change the metamodel without writing code

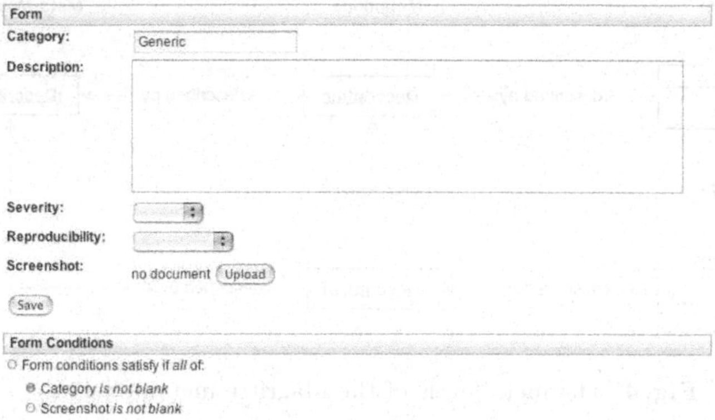

Form

Category: Generic

Description:

Severity: [▼]
Reproducibility: [▼]
Screenshot: no document (Upload)

(Save)

Form Conditions
○ Form conditions satisfy if *all* of:
  ⊕ Category *is not blank*
  ○ Screenshot *is not blank*

Fig. 6. An automatically built editor from runtime customized meta-descriptions displaying if the form conditions are satisfied

end users to customize the existing metamodel to suit their particular needs. Moreover, the end user is able to specify validation and transition conditions in different sections of the user interface. When interpreted by the runtime engine, see Figure 6, the specified metamodel is used to collect the data from the users and to operate the workflow execution. Therefore, an end user can adapt forms on the fly and see its effects directly. Furthermore the customized metamodel is exploited to operate reporting and querying facilities on running workflows. It is Magritte too, that is responsible to make all the meta-data and data persistent.

As we have illustrated in Figure 1, the possibility to work on two meta-level applies to both end users and developers. This, and the fact that Magritte describes itself, are the key concepts to enable end users to modify the metamodel on their own.

4.1   Adaptive Model: Enabling End User Editable Meta-descriptions

To enable instance specific metamodels, Magritte introduces a generic object model mapping descriptions to actual values, as seen in Figure 7. The Adaptive-

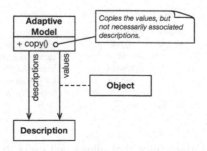

Fig. 7. An adaptive model, mapping a set of descriptions to actual model values

Model has two instance variables, the first being used to refer to the descriptions of the instance and the other one to keep a list of the actual values of the model. Transforming the class of the adaptive model into a Trait [18] allows us turn any existing class into an adaptive model and to combine the descriptions defined in the class with the ones provided by the instance.

End users are able to edit the adaptive model at two different levels, at the model and at the metamodel level:

Domain Data Editing. Since the adaptive model is described, an editor can be built automatically (see Figure 6). The only difference is that the described values are not stored in instance variables of the model, but are kept within a hash table inside the adaptive model, mapping descriptions to their actual values. This gives much better flexibility when descriptions are added and removed.

Metamodel Editing. The descriptions of an adaptive model can change on the fly, since they are stored as part of the model-data. The descriptions can be either changed programmatically by the developer, or through end user interactions from a description editor. Since descriptions are described as well (see Figure 4), it is possible to let Magritte build a meta-editor (see Figure 5).

Descriptions can be shared among different adaptive model instances or can be unique to every instance. Therefore when copying an adaptive model one has to specify if the descriptions should be copied as well. If descriptions are shared, editing the metamodel affects all its associated instances.

5   Related Work

Yoder *et al* propose the type-square design pattern [19], based on the type object that separates the entity from its entity type [20]. Magritte uses these patterns as well, but it makes some generalizations, as seen in Figure 8: the distinction between components and properties is not made. A component and a property are just any kind of object. It is the same for component-types and property-types. They are all descriptions with the same superclass.

JavaBeans [5] includes a property framework similar to the description hierarchy of Magritte. However JavaBeans properties are solely based on the static type signature of the instance variable. Other settings, such as if the value is read-only, is determined implicitly through the absence of a write-accessor. JavaBeans properties do not describe themselves.

One reason that most frameworks do not describe themselves is that they all tend to be very domain-specific: some concentrate on the modeling of a specific business model, others concentrate on a specific output format, such as for a web framework. Unfortunately this leads to a model that is not able to describe itself. Therefore a lot of additional work is required if end users should be able to modify the adaptive-models. Magritte tries to consolidate everything by enabling meta-editing using itself.

Muller et al [21] present an approach to platform-independent web application modeling and development in the context of model-driven engineering.

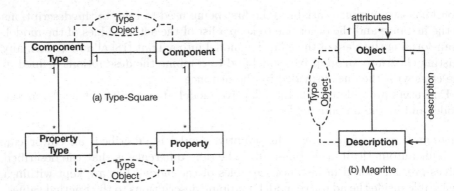

Fig. 8. (a) The type-square, and (b) the meta-recursive model of Magritte are both making extensive use of the type-object design pattern

A specific metamodel (and associated notation) is introduced and motivated for the modeling of dynamic web specific concerns. Web applications are represented via three independent but related models (business, hypertext and presentation). A kind of action language (based on OCL and Java) is used on these models to write methods and actions, specify constraints and express conditions.

WebML [22] enables the high-level description of a web site according to distinct orthogonal dimensions: its data content (structural model), the pages that compose it (composition model), the topology of links between pages (navigation model), the layout and graphic requirements for page rendering (presentation model), and the customization features for one-to-one content delivery (personalization model). WebML goes in the same direction as Netsilon: An application is modeled using different perspectives and generated. Our approach is different. Our object-oriented applications are implemented in Smalltalk but meta-described, and this connected meta-description is used to support the generation of web user interface, queries and persistency. There is no automatic code generation involved in our approach, therefore if the metamodel changes, all the users of the metamodel behave the new way automatically.

6   Evaluation and Lessons Learned

Figure 9 compares two web-based content management systems, SmallWiki[5] and Pier[6]. Both open-source systems have been written by the first author and make it possible to collaboratively build web sites. SmallWiki does not have a metamodel, all its features, such as the different views, the search engine, and the persistency, are hardcoded. Conversely Pier, the successor version of SmallWiki, is built from ground up using the Magritte metamodel. While the code base of

---

[5] http://smallwiki.unibe.ch/smallwiki
[6] http://www.lukas-renggli.ch/smalltalk/pier

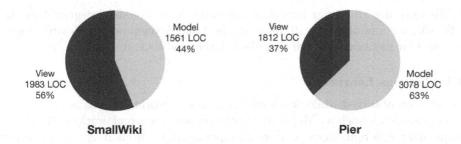

Fig. 9. Comparison of two web applications: SmallWiki and Pier (meta-described)

Pier is noticeably bigger, it also provides functionality that was not possible in SmallWiki:

- As opposed to SmallWiki, Pier has a low coupling between model and view. Different views are interchangeable and their implementations are relatively small, as they only consist of Magritte glue code and some view specific functionality.
- Pier is easily extensible as the entities of the model are specified declaratively. The search engine, the persistency layer and user interface builders all take advantage of the Magritte descriptions.
- Most aspects of Pier can be customized by end users at runtime without having to write code. Additional data fields can be added to any page, to make it simple to collect and display structured data on the web.

As stated by Ralph Johnson [6] a metamodel introduces additional complexity to an application and therefore inexperienced developers might have conceptual problems. Another problem might be a reduced execution speed, as there are additional indirections introduced through the interpretation of the metamodel. Comparing the execution speed of two systems like SmallWiki and Pier is difficult, as their features and implementation details don't exactly match. Most of the time other factors such as the network connection and persistency back-ends are more critical than the use of an underlying metamodel.

To evaluate the speed penalty when using a metamodel we benchmarked the text search of the two frameworks. Both frameworks are using the Visitor pattern to walk over the object graph: in SmallWiki this is hard coded, while in Pier this makes extensive use of meta-descriptions. For the benchmark we created a test setup of 100 pages and run 100 queries on both implementation. As expected SmallWiki performed better with a cumulated search time of 2456 ms. In Pier the search took 8190 ms, so the meta-driven search is about 30% slower than the hard coded one. Given the number of involved objects (a single page consists of hundreds of described objects) text search is a hard task for the meta-driven approach, as many descriptions have to be traversed and matched on the fly. We expect a much better performance for other use-cases and we plan to perform compile-time caching if necessary.

We have described our experience of using a metamodel integrated in the reflective metamodel of Smalltalk to support the development of flexible application. Our metamodel is self-described which enables end user customization.

6.1   Lessons Learned

As we have observed while developing several real world applications, having a meta-framework such as Magritte greatly reduces recurrent work, such as implementing different views, editors and persistency. Often it is much simpler to write a generic interpreter of the metamodel than to manually build specific implementations of the functionality in different places of the application. Developers only change the description at one single place in the source-code without having to refactor all places that deal with the object. More important end users are enabled to reify the choices of the developer through a convenient interface without having to know anything about the implementation and the underlying programming language. Hence, the use of Magritte not only supports the developers, but it makes the application more adaptable to changing needs of end users and reduces the need for development iterations.

Extending the existing Smalltalk metaclass does not allow to keep the metamodel independent of the actual implementation of the class. It should be possible to exchange the metamodel on the fly, and even use multiple metamodels at the same time for the same underlying domain object. Moreover we would like to let end users customize these model, without that they have to know the underlying programming language.

The fact that descriptions are used to describe Magritte itself, makes the system even more versatile: it gives end users the possibility to customize existing models or to build new ones, without having to write a single line of code. The interpreting software system can easily control how far this meta-customization should go. We observed that exposing a small subset of Magritte to end users greatly reduces complexity and increases productivity. Having adaptive models is the key for customizable applications, to allow end users build their own data-models.

As future work, we would like to investigate how the control flow of applications could be meta-described with Magritte. Especially in the context of web application it would be interesting to model the flow between pages as a meta-described graph, and again end user should be empowered to customize it on the fly.

Acknowledgments

We thank Oscar Nierstrasz for his careful review of a draft of this paper. We gratefully acknowledge the financial support of the Swiss National Science Foundation for the project "Analyzing, Capturing and Taming Software Change" (SNF Project No. 200020-113342, Oct. 2006 - Sept. 2008) and of the french ANR (National Research Agency) for the project "COOK: Réarchitecturisation des applications industrielles objets" (JC05 42872).

References

1. Group, O.M.: Common warehouse metamodel. Technical report, Object Management Group (2003)
2. Group, O.M.: Meta object facility (MOF) 2.0 core final adopted specification. Technical report, Object Management Group (2004)
3. Budinsky, F., Steinberg, D., Merks, E., Ellersick, R., Grose, T.: Eclipse Modeling Framework. Addison Wesley Professional, Reading (2003)
4. ITU: Abstract syntax notation one (ASN.1). Technical report, International Telecommunication Union (2002)
5. Hamilton, G.: Javabeans. Technical report, Sun Microsystems (1997)
6. Yoder, J.W., Johnson, R.: The adaptive object model architectural style. In: WICSA3 '02. Proceeding of The Working IEEE/IFIP Conference on Software Architecture 2002 (2002)
7. Atkinson, B.: Hypercard. Hypercard (1987)
8. Renggli, L.: Magritte – meta-described web application development. Master's thesis, University of Bern (2006)
9. Ducasse, S., Gîrba, T.: Using Smalltalk as a reflective executable meta-language. In: Nierstrasz, O., Whittle, J., Harel, D., Reggio, G. (eds.) MoDELS 2006. LNCS, vol. 4199, pp. 604–618. Springer, Heidelberg (2006)
10. Goldberg, A., Robson, D.: Smalltalk 80: the Language and its Implementation. Addison Wesley, Reading (1983)
11. Ingalls, D., Kaehler, T., Maloney, J., Wallace, S., Kay, A.: Back to the future: The story of Squeak, A practical Smalltalk written in itself. In: Proceedings OOPSLA '97, ACM SIGPLAN Notices, pp. 318–326. ACM Press, New York (1997)
12. Ducasse, S., Lienhard, A., Renggli, L.: Seaside — a multiple control flow web application framework. In: Zhang, K., Zheng, Y. (eds.) ISC 2004. LNCS, vol. 3225, pp. 231–257. Springer, Heidelberg (2004)
13. Rivard, F.: Smalltalk: a reflective language. In: Proceedings of REFLECTION '96, pp. 21–38 (1996)
14. Bézivin, J., Gerbé, O.: Towards a precise definition of the OMG/MDA framework. In: Proceedings Automated Software Engineering (ASE 2001), pp. 273–282. IEEE Computer Society, Los Alamitos CA (2001)
15. Atkinson, C., Kuehne, T.: Concepts for comparing modeling tool architecture. In: Proceedings of the UML Conference. LNCS, vol. 3713, pp. 19–33. Springer, Heidelberg (2005)
16. Atkinson, C., Kuehne, T.: The essence of multilevel metamodeling. In: Gogolla, M., Kobryn, C. (eds.) UML 2001. LNCS, vol. 2185, pp. 19–33. Springer, Heidelberg (2001)
17. Muller, P.A., Fleurey, F., Jézéquel, J.M.: Weaving executability into object-oriented meta-languages. In: Briand, L.C., Williams, C. (eds.) MoDELS 2005. LNCS, vol. 3713, pp. 264–278. Springer, Heidelberg (2005)
18. Schärli, N., Ducasse, S., Nierstrasz, O., Black, A.P.: Traits: Composable units of behavior. Technical Report IAM-02-005, Institut für Informatik, Universität Bern, Switzerland (2002) (Also available as Technical Report CSE-02-014, OGI School of Science & Engineering, Beaverton, Oregon, USA)
19. Yoder, J., Balaguer, F., Johnson, R.: Architecture and design of adaptive object models. In: OOPSLA '01. Conference on Object-Oriented Programming Systems, Languages, and Applications, pp. 50–60 (2001)

20. Johnson, R., Wolf, B.: Type object. In: Martin, R.C., Riehle, D., Buschmann, F. (eds.) Pattern Languages of Program Design 3, Addison Wesley, Reading (1998)
21. Muller, P.A., Studer, P., Fondement, F., Bézivin, J.: Independent web application modeling and development with netsilon. Software and System Modeling 4, 424–442 (2005)
22. Ceri, S., Fraternali, P., Bongio, A.: Web modeling language (WebML): a modeling language for designing web sites. In: Ninth International World Wide Web Conference (2000)

Matching Model-Snippets

Rodrigo Ramos[1,2], Olivier Barais[2], and Jean-Marc Jézéquel[2]

[1] Centre of Informatics, Federal University of Pernambuco
P.O. Box 7851, CEP 50732970, Recife, Brazil
[2] IRISA / INRIA, Campus de Beaulieu, 35042 Rennes Cedex, France

Abstract. An important demand in Model-Driven Development is the simple and efficient expression of model patterns. Current approaches tend to distinguish the language they use to express patterns from the one for modelling. Consequently, productivity is reduced by dealing with a distinct new language, and new intermediate steps are introduced in order to support pattern-matching. In this paper we propose a framework for expressing patterns as *model-snippets*. We present how model-snippets are specified upon concepts in a given domain (meta-model), and how we perform pattern-matching with model-snippets, whatever the meta-model. We also provide an implementation which is well integrated with existing technologies, such as Eclipse Modelling Framework.

1 Introduction

Recently, hopes that modelling might play a more important role in the software engineering process have been lived up by Model Driven Development initiatives. These initiatives have been mainly advanced by the Object Management Group and IBM through the Eclipse foundation. The main proposition of this approach consists of considering models as first-class artefacts within the software development process. The recent focus on these assets has raised several new issues [1]. An important demand is the introduction of more automatic ways for searching into model repositories. This demand can be rephrased into the ability to perform efficient pattern matching at the model level.

Pattern-matching at the design level is also used, for example, in order to perform declarative transformations [2], to recover design-patterns in an object-oriented software design [3], or to identify join points in a model from a pointcut expression to weave aspects into models [4,5]. In all these cases, we can imagine patterns as *model snippets* with some conditions, which are possibly expressed by some predicates over the models that they wish to match. In fact, several languages have been proposed to express these patterns [4,5] in this way. Unfortunately, most of these languages were designed to a specific domain, defined by a specific meta-model. Moreover, approaches that implement pattern matching over these languages tend to also be specific to these meta-models. This limitation has been obstructing the application of pattern matching over new meta-models, since each of them requires new support for pattern matching. For instance, despite of the good results of current methods for aspect-oriented

G. Engels et al. (Eds.): MoDELS 2007, LNCS 4735, pp. 121–135, 2007.

modelling, it is still difficult to play with its concepts over non object-oriented paradigms or, even more, meta-models that do not extend UML.

In this paper, we propose a generic pattern framework for expressing pattern at the model level and for performing pattern-matching. The language for expressing patterns is built on demand, conforming to an input meta-model of the model that we wish to match. The main goal of simplifying the expression of patterns at the model level, avoiding any textual regular expressions, is to assist the design of these patterns with existing model editors for the user, whatever the meta-model. Moreover, we consistently define model patterns upon the concepts of *model-typing* [6], which permits us to better manage such patterns during model evolution. The results of this work are also integrated with existing technologies, such as EMF[1] (Eclipse Modelling Framework), Kermeta[1] (a meta-model engineering environment within Eclipse) and Flora-2[1] (An object-oriented knowledge base language that extends Prolog).

The next section complements this introduction with a motivating example, which will also guild us throughout the paper. Then, Section 2 presents our approach for *meta pattern-matching* in details, and Section 3 describes how this is implemented and integrated with Kermeta and Flora-2. Finally, Section 4 relates this work with existing approaches, and Section 5 concludes this paper.

1.1   Motivating Example

In order to motivate this work, a small state-machine meta-model is presented in this section, in which we intend to perform a pattern-matching. As any typical state-machine, it is mainly formed by a set of states and transitions among the states. Additionally, the meta-model might include some constraints (expressed in OCL, for example), which among other things may specify that the state-machine must have a unique initial state.

Now, assume that we have a state-machine obeying this meta-model (left-hand side of Fig. 1), which is composed by several complex states and transitions and detailed with actions to be executed during the state-machine life-cycle.

Moreover, suppose that we want to find some simple patterns over this model, such as cyclic-transitions or, in the opposite way, any transition that links two distinct states. By the simple nature of these patterns, we wish to avoid dealing with complex details of the state-machine meta-model, specifying just what is needed for performing pattern-matching. Moreover, it makes sense to express patterns in a similar language in order to make their specification easier and avoid learning new languages. Another concern is to use the same mechanism for each meta-model, instead of recreate them to each new meta-model. We can summarise these concerns, as follows:

- *How to have a simple way to express a valid pattern?*
- *How can we be guided by our meta-model in the specification of a pattern?*
- *Can we express patterns using existing editors for our meta-model?*
- *Are our models already supported by any existing pattern-matching tool?*

---

[1] see www.eclipse.org/emf, www.kermeta.org and flora.sourceforge.net

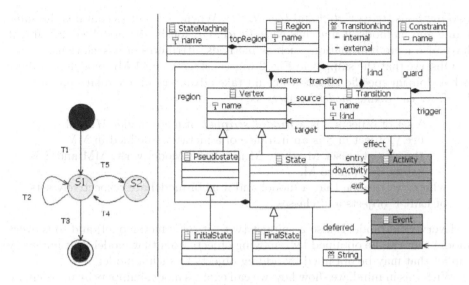

Fig. 1. A model and a meta-model of state-machine

All things considered, we have identified, as other authors [4], that probably the simplest way to express patterns is using the same concepts that we find in our meta-model, i.e. expressing patterns as *snippets* of a valid model in our meta-model. In order to realise this idea, and also to support its application in new meta-models, we discuss throughout this paper a pattern framework for expressing pattern as model snippets and for performing pattern-matching. The questions raised in this section are also addressed during the paper, using the small example of the state-machine to illustrate our discussion.

2   Patterns as Model-Snippets

As suggested in the previous section, patterns might be expressed as *model snippets*, such as the UML templates [4] in Fig. 2. Having patterns expressed in this way, it is possible to allow a user to draw patterns using editors that he is used to, when drawing the models that he intents to match.

Each model-snippet defines a set of information existing in the model that we wish to match. For example, in Fig. 2, a class named *Trace* is declared with

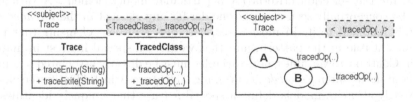

Fig. 2. UML Templates of a Class and State Diagram

two methods (*traceEntry* and *traceExit*). Whenever, a class contains the same name and methods, it matches with *Trace*. Obviously, the matched class might have also more information, such as methods, attributes or associations.

Observe that the snippet in Fig. 2 was constructed for UML models [4]. Many other domain specific languages could take advantage of a similar approach. In brief, we can define *model snippets* as:

> A set of objects S is a **model snippet** of meta-model MM iff:
> - every object in S is an instance of a metaclass defined in MM;
> - there exists a set M where M is a Valid Model w.r.t. MM and S is subset or equal to M.
>
> where, we assume that a model and a meta-model are respectively sets of EMOF objects and classes.

Every valid model is also a snippet (w.r.t its own metamodel) and so is every model that can be obtained by removing objects from that model. But not every model that may be obtained by adding objects to such a model.

With this in mind, we show how we can express model-snippets in any domain. We present in Sect. 2.1 a pattern-framework with the minimal elements that forms a pattern. In Sect. 2.2, we show how to create *model snippets* and to customise this framework according to a target meta-model. Finally, on Sect. 2.3 we formally define what we mean by pattern-matching using model-snippets.

2.1   Pattern-Framework Meta-model

Taking a closer look at the model-snippet in Fig. 2, we can see the snippet as an instance of the UML meta-model. All elements in the snippets are instances of a UML classifier, and furthermore inherit from a superclass *NamedElement*. For instance, *Trace* is an instance of a UML *Class* identified with a feature *name* equal to *'Trace'*, and the method *traceEntry* is an instance of *Operator* with *name* equal to *'traceEntry'*. The same happens in the model that we want to match. An important finding from this observation is that a snippet specifies a subset of instances, and of associations among them, in the model that we want to match. This set of instances is the information that we use to match models. However, a pattern seems to be a little more complex than just a set of instances of a meta-model.

It is clear from Fig. 2 that a pattern is mainly formed by a *snippet* part (in each package of the figure) and a sequence of free-variables over this *snippet* part (in rectangle on the top-right side of each package). The purpose of variables is to define the selection criteria for a particular model element. A variant can also be conceptually seen as a placeholder for any element in the *intended model* that is matched to it. In most cases, variables represent elements that play a significant role in the pattern, and that we have a special interest in matching with. Contrary to variables, non-variables must be directly associated to a unique element in the *intended model*, containing all features which identify the element.

Nearly all meta-models define a special feature that uniquely identifies each element of their models. As we wish to be able to match variables with more than one element in the model, we do not take into account this identifier during pattern-matching of variables. For instance, *TracedClass* is a variable in Fig. 2. So it matches to any class, with any name, that has the same methods than *TracedClass* and an association to a class named *'Trace'*. *Trace* is a non-variable, and, furthermore, we take into account its *name* during pattern-matching. As in most of the cases, UML uses a feature *name* as an identifier. However, the feature can change in other meta-models.

The more information is expressed in the structural part of the pattern, the more precise is the pattern-matching. However, an excessive and detailed *snippet* might also uncover all positive matches. For this reason, as any model, a pattern can have additional constraints, which help to better describe the pattern and to relate variables with other elements in the model.

Note that constraints between variables and non-variables in a pattern should still be valid after they have been matched with elements in the *intended model*. From this standpoint, constraints might also help to describe false positives in the set of possible matchings. In doing so, constraints can represent *Negative Application Conditions* (NAC) in a pattern-matching, improving the accuracy of the matching process.

Based on the concepts presented above, a possible generic meta-model for patterns could be expressed as the one in Fig. 3. In the meta-model, *Pattern* represents the whole pattern, and *PatternStructure* represents its structure. The structural part contains a *PModel* with a set of instances of classes in a given meta-model, which we wish to be related to the metamodel that describes the models in which we look for matches (*intended model*). *PatternStructure* has also a set of *Role*, which express pattern variables in the structural part. Additionally, the pattern can also have some constraints, or *invariants*, that, here, might be expressed in OCL or Kermeta.

Fig. 3 presents what we call a *Pattern Framework*. *PModel* is the point (*hot spots*) where the framework can be adapted or specialised by the developer. The specialisation of our framework to the meta-model that describes the *intended models* is described in the next section.

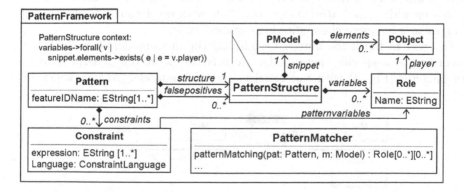

Fig. 3. Meta-model of the pattern framework

2.2   Constructing Model-Snippets

Most of the time, the meta-model (MM) of the *intended model* is too restrictive to represent patterns. The reason for that is very simple; patterns have to be expressed in a higher level of abstraction, such as *model-snippets* (see definition in the beginning of Sect. 2). For example in the state machine meta-model, it is totally understandable that someone does not want to provide the mandatory *event* of a *Transition*, or even want to instantiate a *Vertex*, which is defined as abstract, in order to match over instances of *State* or *PseudoState*.

We wish that a snippet relies as much as possible on the same concepts of MM. In order to do that, we construct by demand a more flexible meta-model (MM') that allows us to represent abstract patterns with all concepts of the meta-model of the *intended model*. If we imagine a flexible meta-model MM' that is equals to MM, except that:

- No invariant or pre-condition is defined in MM';
- All features of all classes in MM' are optional;
- MM' has no abstract element.

Then, we can notice that all concepts in MM are also represented in MM'. MM' describe a wider range of models, including all models described by MM (see Fig. 4). This is obtained by removing all restrictions that exist in MM: invariants, mandatory features, nonexistence of instances of certain class. To allow features to be optional, we just set its lower bound as zero. These restrictions can be taken rewritten as invariants over a group of classes S, and any group of classes with a weaker invariant could be taken as a generalisation of S [6].

Note that any model that conforms to MM also conforms to MM', and, furthermore, any model-snippet that conforms to MM also conforms to MM'. A detailed discussion about this conformance is presented in Sect. 2.4. This is an important result, it shows that we can still use existing graphical editors to draw pattern snippets and use them for pattern-matching. It also says that any meta-model that generalises MM can be used to specify more abstract patterns.

For example, we can generate a new and flexible meta-model (MM') from the state machine meta-model in Fig. 1 (MM). MM' might describe, for instance, a *Region* with zero *InitialStates* or a *State* with no *Activity*. It also describes *Vertex* as a concrete class, allowing instances of it.

In order to finish the automatic building the meta-model of model snippets according to a specific meta-model, we need to merge our general framework for patterns (Fig. 3) with this restrictiveness meta-model (MM'), which takes

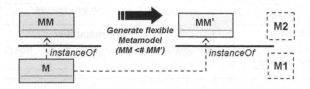

Fig. 4. Process for deriving a meta-model for pattern *snippets*

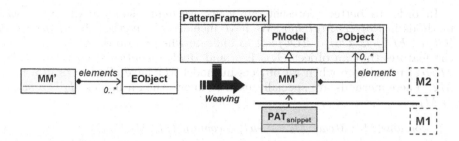

Fig. 5. Process for customising the pattern framework

into account concepts in the intended domain (MM). This composition is called
a weaving because this transformation integrates $PObject$, which comes from
the pattern framework ($PatternFramework$), as a superclass of all meta-classes
in MM'. The transformation can be compared to an interface introduction in
AspectJ[1] that adds a new superclass to a type. Our weaving process is equivalent to this mechanism. The implicit pointcut used in our weaving applies the
introduction of the $PObject$ superclass into all the metaclasses that do not have
any superclass. The whole process to derive MM' from MM and to permit the
specification of valid pattern $snippet(PAT_{snippet})$ is presented in Fig. 5.

For example, we can generate a new meta-model that weaves our *pattern
framework* and the state machine meta-model. It includes classes from both
framework and meta-model, and additionally hook all classes from the latter
with $PObject$. Then, all classes that do not have a super class in MM' inherit
from $PObject$ after the weaving process. Fig. 6 shows the inheritance between
some of these classes and $PObject$. As a result, we obtain a meta-model that
can be used to express model-snippets and also can be taken as an input of the
pattern-matching mechanism.

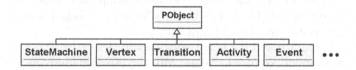

Fig. 6. Weaved state machine meta-model

2.3 Pattern-Matching Behaviour

In this section, we describe what we mean by pattern-matching using *model
snippets*. We make use of a formal notation that looks like OCL, but with explicit
use of quantifier operators of the first order logic. We also assume that any
instance of MM' and MM is also an instance of *Model*, as it is presented in
Fig. 7. Where, *Model* contains a *elements* reference with a set of EMOF object
instances.

---

[1] see www.eclipse.org/aspectj.

In order to better represent our ideas, we define some auxiliary (private) methods in *PatternMatcher*. We have included the method *Booleancontains* ($Obj_1 : EObject, Obj_2 : EObject$) to indicate when an object Obj_1 contains all the features that an object Obj_2 has, and *Booleancontains*($M_1 : Model, M_2 : Model$) to indicate when all objects in model $M1$ have all objects in a model $M2$. These methods are specially used to know when a model M_2 is a snippet of M_1.

$$contains(M_1 : Model, M_2 : Model) = contains(M_1, M_2, \{\}, \{\}) \tag{1}$$

$$contains(Obj_1 : EObject, Obj_2 : EObject) = contains(Obj_1, Obj_2, \{\}, \{\}) \tag{2}$$

$$contains(M_1 : Model, M_2 : Model, fIds : EString[0..*], roles : Role[0..*]) = \tag{3}$$

$$\forall e_2 \in M_2.elements \bullet \exists e_1 | contains(e_1, e_2, fIds, roles) \tag{4}$$

$$contains(Obj_1 : EObject, Obj_2 : EObject, fIds : EString[0..*], \tag{5}$$

$$roles : Role[0..*]) = \tag{6}$$

$$\forall r_2 \in Obj_2.getAllReferences() \mid Obj_2.eGet(r_2).isDefined() \bullet \tag{7}$$

$$\exists r_1 :\in Obj_1.getAllReferences() | r_2.name = r_1.name \wedge \tag{8}$$

$$contains(Obj_1.eGet(r_1), Obj_2.eGet(r_2), fIds, roles) \wedge \tag{9}$$

$$\forall a_2 \in Obj_2.getAllAttributes() \mid Obj_2.eGet(a_2).isDefined() \bullet \tag{10}$$

$$\exists a_1 :\in Obj_1.getAllAttributes() | \tag{11}$$

$$(a_2.name = a_1.name \wedge Obj_1.eGet(a_1) = Obj_2.eGet(a_2)) \vee \tag{12}$$

$$(\exists ro : roles | ro.player = Obj_2 \wedge a_2.name \in fIds) \tag{13}$$

We also included two auxiliary methods that take into account the list of features used as identifiers ($fIds$) in the model type (as we said before, this list changes from a domain to another) and a list of variables ($roles$). The method *Booleancontains*($M_1, M_2, fIds, roles$) is used to indicate that M_2 is a snippet of M_1 considering a set of feature identifiers and pattern variables. With the help of a method with similar signature for objects, it checks if every objects in M_2 has a counterpart in M_1, checking every association (lines 7-9) and comparing every attribute (lines 10-13). While the attributes are checked, feature identifiers of a pattern variable are ignored (line 13).

However, pattern-matching is not solely about detecting that a pattern snippet exists in a model. We need to cope with the possible bindings of the pattern variables with the *intended model*. For our purposes, we define pattern-matching

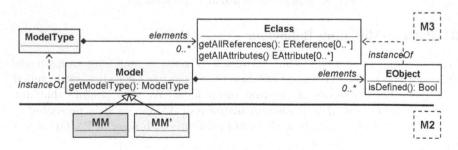

Fig. 7. Abstraction level of MM and MM'

as the set of possible bindings (tuples with variable name and object) from the pattern variables into the objects of the *intended model*.

$$patternMatching(pat : Pattern, m : Model) : Role[0..*][0..*] = \qquad (14)$$

$$\textbf{lets } patsnpt = pat.structure.snippet \wedge vars = pat.structure.variables \wedge \quad (15)$$

$$ids = pat.featureIDName \qquad (16)$$

$$\forall snpt : Model \mid contains(m, snpt) \wedge contains(snpt, patsnpt, ids, vars) \wedge \quad (17)$$

$$\nexists minorsnpt : Model \mid minorsnpt! = snpt \wedge contains(snpt, minorsnpt) \wedge (18)$$

$$contains(minorsnpt, patsnpt, ids, vars) \bullet \qquad (19)$$

$$\exists setroles \in \textbf{result} \mid \qquad (20)$$

$$\forall e_s \in snpt.elements, ro \in vars \mid contains(e_s, ro.player, ids, vars) \wedge \quad (21)$$

$$(\forall(ro.name, e) \in setroles \mid e = e_s) \bullet (ro.name, e_s) \in setroles \qquad (22)$$

It the above expression, the *patternMatching* method is formed by two steps, first, in the lines 17-19, the smallest snippets of the *intended model* that contains the pattern structure are found. Then, in lines 20-22, the bindings with the objects of these snippets are finally defined. This is a very general and inefficient definition. We use it here for explanation purposes. The way that we address its implementation is described in the next section.

2.4 Model-Snippets During Model Evolution

During a software evolution, an important question that might raises is:

– How useful is a model-snippet (or pattern) when definitions in the meta-model are changed ?

The need for identifying relationships between models, including model-snippets, suggests that we might examine *model-typing* [6]. Indeed, we need to identify a type for the model-snippet in order to verify if a pattern can still be applied in a certain domain, after changes in the meta-model.

For this purpose, we can think in *patternMatching()* (see Fig. 3) as a parametrized type operation, that can be performed to any model that conforms to a criterion. A good starting point for this criterion is to use type checking based on the type relationship *matching* [6] ($< \#$), where any operation $Op[X < \#MM_T]$ parametrized with the group of types X can be successfully performed when X matches to a group of types MM_T. Generally speaking, matching[2] in this context is satisfied when all class c_t in MM_T has a respective class c_x in X that matches to it. A class c_x matches to another c_t if: 1) c_x includes all features of c_t, 2) and, for all associations of c_t with other class c_t' there is an association in c_x to a class c_x', which contains c_t'.

Despite we have presented a pattern-framework that is independent of any meta-model, we can use model type as a pre-step criterion for verifying the validity of a pattern in a specific domain, comparing the model types of the

---

[2] Do not mix *pattern-matching* with the type relationship called *matching*. We use only *matching* when we refer to the latter in this paper.

pattern *snippet* and the intended model. This criterion can be expressed as a pre-condition of the pattern-matching operation, as follows.

$$patternMatching(pat : Pattern, m : Model) : Role[0..*][0..*]$$
$$\mathbf{pre} : pat.structure.snippet.getModelType() < \#m.getModelType()$$

From the expression above, we identify that we can use any meta-model MM' to describe a pattern snippet, provided that $MM' < \#MM$. So, in other words, a model-snippets is still useful if the changes applied to a meta-model MM does not make it less abstract than the MM'.

3   Implementation

From Sect. 2, we can see how pattern-matching is performed. However, an efficient implementation might not be so easy to construct. Fortunately, this is an extensive topic of research, which has produced several existing languages and APIs with embedded pattern-matching mechanisms [2,7]. For that reason, we have decided to rely on these existing tools as much as possible in order to integrate our ideas and to contribute with existing tools in this research topic.

Our implementation relies on the Kermeta language [8], an executable and object-oriented DSL (Domain Specific Language) for meta-model engineering. Kermeta is built as a conservative extension of EMOF, giving special attention to the specification of abstract syntax, static semantic (OCL) and operational semantics as well with connexion to the concrete syntax. Consequently, an EMF model is seen as a Kermeta model without operational semantics. Through our implementation, we contribute with pattern-matching mechanisms to the meta-model engineering environment available with Kermeta, which includes model transformations, aspect weaving and loading of EMF models.

For our purpose, we have implemented a pattern-matching front-end in Kermeta. This front-end behaves as an abstract interface between our framework for pattern-matching and existing engines with embedded pattern-matching mechanisms. In order to delegate computation to these engines, we require the implementation of a specialised back-end for each engine.

As a proof of concept, we have constructed a back-end that uses a Prolog engine to perform pattern-matching. Using this approach, facts are derived from

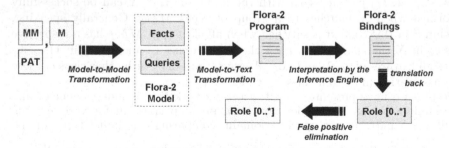

Fig. 8. The workflow of pattern-matching implementation using flora2

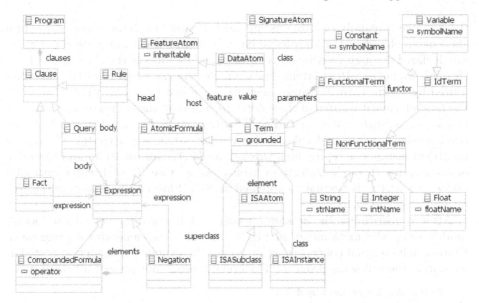

Fig. 9. A simplified meta-model of Flora-2

the model that we intent to match and inserted in a knowledge base of the engine. Then, queries are generated to search for a subset of the facts that matches with the pattern. An abstract workflow of this back-end is presented in Fig. 8.

Initially, the front-end inputs the pattern and the *intended model* into the back-end in Kermeta. The later transforms both into an intermediate model in Flora-2, an object-oriented Prolog dialect that is suitable to represent concepts in EMOF [9]; a simplified meta-model for Flora-2 is presented next. In order to interpret this model in Flora-2, we serialise it in the concrete textual syntax of Flora-2. Then, our back-end implementation in Kermenta sends the Flora-2 program to the Prolog Engine trough a Java proxy layer; Kermeta supports a seamless integration with Java programs. The Engine interprets the program and returns the result for our query. From the analysis of this result, we obtain a set of bindings from pattern variables into elements in the *intended model*.

Being object-oriented, Flora-2 offers all necessary counterparts for concepts in EMOF. However, as it is a conservative extension of Prolog, Flora-2 still is a fact-based language. All information about a class or instance is distributed in several atoms[1] (*AtomicFormula*), as it is presented in Fig. 9. Fig. 9 shows a simplified and incomplete meta-model for Flora-2, which only addresses concepts that Flora-2 takes from F-Logic.

Any concept in Flora-2 starts being represented by the most basic atom, a *Term*, which might be classified as a primitive type, a constant or a variant;

---

[1] Despite *Molecule* constructions permit the direct representation of object and classes and their features, it still is a syntax sugar for a set of related atom in the Knowledge base. Furthermore, it is not represented in our meta-model.

the last one is usually used in queries. Initially any new term might represent a class or an object in the knowledge base. In a class declaration, its super-types are declared through *ISASubClass* atoms, and features of the class are declared through *SignatureAtom* atoms. For example, the class *State* in Fig. 1 is represented by a term with the same name, its superclass by a term *Vertex*, and it is the *host* of a feature *entry* of class *Activity*. These atoms are textually expressed in Flora-2 as *State :: Vertex* and *State[entry *=> Activity]*; they could also be condensed in a unique molecule *State :: Vertex[entry *=> Activity]*. The type of an object instance is represented through an *ISAInstance* atom, and the value of its features through *DataAtom* atoms. So, a state *S1* with entry *entryAct* and *exit exitAct*, is textually expressed as *S1 :: State*, *S1[entry -> entryAct]*, *S1[entry -> entryAct]*.

After transforming the *intended model* and its meta-model into a Flora-2 model, using the meta-model in Fig. 9, the back-end generates a program in Flora-2 in its textual concrete syntax. So, for the example presented in Sect. 1.1, we obtain the following program. For brevity, we do not present all program.

```
// Facts about the meta-model
PseudoState :: Vertex,  State :: Vertex.
FinalState :: State, InitialState :: PseudoState.
...
//Facts about the model
SM1:StateMachine [name -> "sm1", topRegion -> Reg1].
Reg1 [transition => {T1, T2, T3, T4, T5}, vertex => {S0, S1, S2, S3}].
T1:Transition[name -> "T1", source -> S0, target -> S1, trigger -> event1].
T2:Transition[name -> "T2", source -> S1, target -> S1, trigger -> event2].
T3:Transition[name -> "T3", source -> S1, target -> S3, trigger -> event3].
T4:Transition[name -> "T4", source -> S2, target -> S1, trigger -> event4].
T5:Transition[name -> "T5", source -> S1, target -> S2, trigger -> event5].
S0:InitialState[name -> "S0", owner -> Reg1].
S1:State[name -> "S1", entry -> Activity1].
S2:State[name -> "S2", entry -> Activity2].
S3:FinalState[name -> "S3"owner -> Reg1].
```

In the program above, we summarised some facts about subtyping in the meta-model, which we use next for pattern matching. Subtyping is expressed in the form *Class :: SuperClass*. Classes are themselves expressed using atomic clauses (*atoms*) or *molecules*. And, objects are represented in the form *ObjectID* : *Class[feature₁-> value, ..., featureₙ => setOfValues]*, where the operators -> and => map a feature into a single and multiple values, respectively.

In this way, if we wish to detect all transitions from two states, excepting self-transitions, we would use the pattern presented in Fig. 10. The transition, its source and target states are presented in the left-hand side of the figure, while the false-positive pattern with a cyclic transition is shown in the right-hand side.

The Flora-2 textual representation of this pattern, considering the false-positive pattern in the right-hand side of figure, is expressed bellow. The query is composed of clauses, conjuncted by commas, with Flora-2 variables (symbols

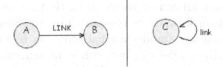

Fig. 10. Patterns of state machines

with a ? suffix) to represent all pattern variables. The false-positive is represented
with a negation clause *not((...)))*.

```
?link : Transition[source -> ?SourceState, target -> ?TargetState],
not ((?link : Transition [source -> ?SameState, target -> ?SameState])),
?SourceState : State, ?TargetState : State, ?SameState : State.
```

The query result is presented as follows. Note that the transition from *S*1
to the final state (*S*3) is also returned, since the type *FinalState* inherits from
State in the state-machine meta-model (see Fig. 1).

?source = S1	?source = S2	?source = S1
?target = S2	?target = S1	?target = S3
?link = T5	?link = T4	?link = T3

After analysing the set of bindings from the inference engine, which maps
variables into elements of the *intended model*, we check constraints over the
pattern considering each new set of binding for roles in the pattern. This final
checking is useful to eliminate false positives in the pattern. Note that part of
these constraints might be anticipated through the representation of patterns
for false positives, using the *falsepositive* feature of the pattern (see Fig. 3).
Elements in *falsepositive* can also be conceptually seen as an negative expression
in the query in Prolog, restricting the possible results in the search. Despite the
possible use of negative patterns, constraints are always checked, as a post-
procedure to filter false positives from the results of the pattern-matching.

After the elimination of false positives, a final set of bindings are presented to
the user, or program which invoked the pattern-matching, who can chose among
the bindings the most convenient one.

4   Related Work

Pattern-matching has been analysed theoretically in various contexts. For metamodel engineering, the most successful results have been achieved in approaches
that use graph transformation, such as GReAT[1] and AGG[1]. They rely on graph
theories to perform searches in an intermediate graph model that represents the
source model. Patterns are also constructed as graphs, which are used to match
with model graphs. Differently, but still using an intermediate language, MOLA
has an efficient approach for pattern matching [7] through SQL queries in a
model repository located in a database.

The unique exception, known by the authors, that proposes a concrete syntax for pattern for graph transformations is reported in [10]. Similar to our approach, this work discuss how a metamodel for patterns can be generated conforming to the meta-model of the model that we wish to match, and how these patterns can visualised using a concrete syntax. Besides its originality, this work does not address negative application conditions and constraints (like in OCL) as we do.

In ATL [2], pattern-matching is used to identify a source element for declarative transformations. Opposite to real pattern-matching, so called pattern-matching in ATL is used in a simplistic way to identify a unique element, rather that a *snippet* of the model. Features related to pattern-matching in all these languages are encompassed in the approach presented in Sect. 2, which solves a common deficiency in non-graph-like transformation languages, as Kermeta.

Concerning the implementation in Sect. 3, pattern-matching implemented via Prolog have been already studied in [3]. However, contrary to our approach, this work has focused in the detection of design patterns in object-oriented programs. While the process steps are very similar, our implementation concerns the application of Prolog in pattern-matching at the model level. The derivated facts from the meta-models and models in this section are also aligned to the formal mapping of class and object into Flora-2 in [9] and other transformations using F-Logic [11]. However, we take as input EMOF classes and objects.

5   Conclusion

Nearly all of graph model transformation languages use pattern-matching as the main functional element for defining how the source model components must be detected in model transformations or weavings. For this reason, this work brings an important contribution to Kermeta, permitting the use of pattern-matching mechanisms in its environment. It also brings an alternative between so many existing pattern-matchings using graph-transformations.

In Sect. 2, we also present a simple meta-model framework for patterns, which is beyond any language or implementation. We have shown that this framework might be used in conjunct with any meta-models in order to represent concepts in existing or new domains. We have also discussed how pattern *snippets* might be validated according to the model type of models in these domains. This have shown that indeed, although its limitations, we can use these meta-models to guide the specification of pattern snippets; this also brings some advantages in the use of existing model editors to express model-snippets. This consideration of model type in pattern-matching, and the relation between patterns and the models that they intent to match, have been neglected so far.

An efficient implementation of these ideas is proposed in Sect. 3, through the integration of Kemeta and Flora-2. This is by itself an original contribution, concerning model transformations from arbitrary EMOF model to Flora-2 and the presentation of a, despite simplified, meta-model for Flora-2.

---

[1] see tfs.cs.tu-berlin.de/agg and repo.isis.vanderbilt.edu/tools

As future work, we plan to match semantic equivalent models. Examples of its need are mainly found in the pattern-matching of behavioural models, since too much semantic information is required. Another future work is to take advantage of transaction logic to directly introduce OCL constraints as Flora-2 facts, in order to improve the overall performance of our implementation. Indeed, such optimisation contributes to a better integration with the Prolog engine, which uses execution strategies with *backtracking* to reduce the exhaustive search.

References

1. Kent, S.: Model driven engineering. In: Butler, M., Petre, L., Sere, K. (eds.) IFM 2002. LNCS, vol. 2335, pp. 286–298. Springer, Heidelberg (2002)
2. Jouault, F., Kurtev, I.: Transforming Models with ATL. In: Model Transformations in Practice Workshop at MoDELS/UML (2005)
3. Prechelt, L., Krämer, C.: Functionality versus practicality: Employing existing tools for recovering structural design patterns. Journal of Universal Computer Science (J.UCS) 4, 866–882 (1998)
4. Clarke, S., Walker, R.J.: Composition patterns: an approach to designing reusable aspects. In: ICSE'01. 23rd International Conference on Software Engineering, Washington, DC, USA, pp. 5–14. IEEE Computer Society, Los Alamitos (2001)
5. Barais, O., Duchien, L., Meur, A.F.L.: A framework to specify incremental software architecture transformations. In: 31st EUROMICRO Conf. on Software Engineering and Advanced Applications, IEEE Computer Society, Los Alamitos (2005)
6. Steel, J., Jézéquel, J.M.: On model typing. Journal of Software and Systems Modeling (SoSyM) (to appear, 2007)
7. Kalnins, A., Celms, E., Sostaks, A.: Simple and efficient implementation of pattern matching in mola tool. In: Baltic DB&IS2006, Vilnius, Lithuania, pp. 159–167 (2006)
8. Muller, P.A., Fleurey, F., Jézéquel, J.M.: Weaving executability into object-oriented meta-languages. In: Briand, L.C., Williams, C. (eds.) MoDELS 2005. LNCS, vol. 3713, pp. 264–278. Springer, Heidelberg (2005)
9. Ramalho, F., Robin, J.: Mapping uml class diagrams to object-oriented logic programs for formal. In: 3rd UML Workshop in Software Model Engineering (WiSME 2004) at MODELS/UML'2004, Lisbon, Portugal, pp. 11–15 (2004)
10. Baar, T., Whittle, J.: On the usage of concrete syntax in model transformation rules. In: Virbitskaite, I., Voronkov, A. (eds.) PSI 2006. LNCS, vol. 4378, Springer, Heidelberg (2007)
11. Gerber, A., Lawley, M., Raymond, K., Steel, J., Wood, A.: Transformation: The missing link of mda. In: Corradini, A., Ehrig, H., Kreowski, H.-J., Rozenberg, G. (eds.) ICGT 2002. LNCS, vol. 2505, pp. 90–105. Springer, Heidelberg (2002)

Improving Inconsistency Resolution with Side-Effect Evaluation and Costs

Jochen M. Küster and Ksenia Ryndina

IBM Zurich Research Laboratory, Säumerstr. 4
8803 Rüschlikon, Switzerland
{jku,ryn}@zurich.ibm.com

Abstract. Consistency management is a major requirement in software engineering. Although this problem has attracted significant attention in the literature, support for inconsistency resolution is still not standard for modeling tools. In this paper, we introduce explicit side-effect expressions for each inconsistency resolution and costs for each inconsistency type. This allows a fine-grained evaluation of each possible inconsistency resolution for a particular inconsistent model. We further show how an inconsistency resolution module for a modeling tool can be designed and implemented based on our approach. We demonstrate the applicability of our approach for resolution of inconsistencies between object life cycles and process models.

1 Introduction

Consistency management is a major requirement in software engineering [6]. It requires one to establish consistency constraints that can be checked to identify inconsistencies in models [5,11]. For resolving a particular inconsistency, it is common practice to specify one or more suitable inconsistency resolutions that can transform the model so that the consistency constraint is no longer violated.

Although many solutions addressing various aspects of inconsistency management have been proposed [7,13,20], most modeling tools currently do not offer adequate support to the user for resolution of inconsistencies, in particular for behavioral models such as activity diagrams, statecharts and sequence diagrams. One reason for this is that inconsistency resolution for these models requires transformations that often have side-effects. Such side-effects include both, introduction of new inconsistencies and expiration of existing inconsistencies [12]. As a consequence, in the presence of numerous inconsistencies in a model, many alternative resolutions can be applicable and it is often not obvious which resolution is most appropriate to apply. One technique that has been proposed to tackle this problem is the detection of potential dependencies between inconsistency resolutions using dependency analysis [12,13,22]. This analysis is performed without taking a particular inconsistent model into account. However, only some of the discovered dependencies are usually relevant for a given inconsistent model and these must be precisely identified for comparing alternative resolutions.

G. Engels et al. (Eds.): MoDELS 2007, LNCS 4735, pp. 136–150, 2007.

In this paper, we propose an approach where inconsistency resolutions are associated with explicit side-effect expressions. Such side-effect expressions are evaluated given a concrete inconsistent model to determine whether or not a resolution will have side-effects. This leads to precise knowledge of both expired and induced inconsistencies before applying a particular resolution. Further, we introduce the concept of costs for inconsistency types that allows one to prioritize resolution of different inconsistencies and calculate the total inconsistency cost for a given model. In combination with side-effect expressions, cost reductions for a resolution can be calculated in advance. Overall, our approach leads to an improved way of inconsistency resolution, because it allows a fine-grained comparison of alternative resolutions. In addition, explicit side-effects are also used to avoid re-checking the whole model for inconsistencies after applying a resolution.

We demonstrate our approach using a case study that deals with inconsistency of object life cycles and process models in the context of IBM Insurance Application Architecture (IAA) [1]. For showing the feasibility, we briefly present a design of an inconsistency resolution module based on the proposed approach. Using this design, we have implemented a prototype extension to IBM Web-Sphere Business Modeler [2] for resolving inconsistencies between object life cycles and process models.

The paper is structured as follows: Section 2 presents our case study by introducing object life cycles, process models and inconsistencies that can occur between these models. In Sect.3, we introduce the concept of explicit resolution side-effects and costs for inconsistency types. Design and implementation of tool support for inconsistency resolution are discussed in Sect.4. We compare our approach with existing work in Sect.5 and conclude the paper in Sect.6.

2   Inconsistency of Object Life Cycles and Process Models

In this section, we first introduce object life cycles and process models, together with an example inspired by IAA. We then discuss inconsistencies that can occur in these models.

An object life cycle [4,9] captures all possible *states* and *state transitions* for a particular *object type* and can be represented as a state machine. Figure 1(a) uses the UML2 State Machine notation [3] for modeling a life cycle for objects of type Claim in an insurance company. The object life cycle shows that after a claim has been registered, it goes through an evaluation process and can be either granted or rejected. Rejected claims are closed directly, while granted claims are first settled and then closed. According to this model, every claim is created in state Registered and passes through state Closed to the only final state.

Our case study deals with *reference object life cycles* that are prescriptive models capturing how objects should be evolved by business processes. Consistency with such reference object life cycles may be an internal business policy or an external legal requirement.

A process model captures the coordination of individual actions in a particular process and the exchange of objects between these actions. Figure 1(b) shows

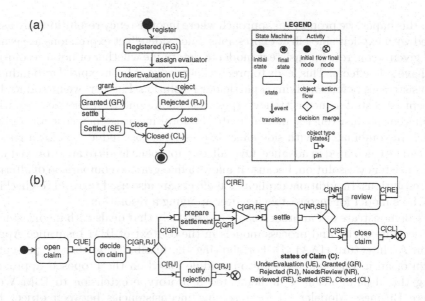

Fig. 1. (a) Reference object life cycle for claims (b) Claim handling process model

a process model for a simplified Claim handling process in the UML2 Activity Diagram notation [3]. In the beginning of this process, a claim is opened and then an evaluation decision is made. If the claim is granted, a settlement is prepared for it. During settlement of a claim, the claimant may appeal for a review of the settlement amount and conditions, in which case these are reviewed and the adjusted settlement is then carried out. Only settled claims are closed in this process.

A process model can also depict how an object changes state as it moves through the process, which is explained in detail in [17]. In our example, the open claim action creates a claim in state UE (UnderEvaluation) and passes it to action decide on claim that changes the claim's state to either GR (Granted) or RJ (Rejected). The following decision node passes claims in state GR to prepare settlement and those in state RJ to notify rejection.

In order to define consistency of a given process model with respect to an object life cycle, we first consider the relation between these models in the UML2 metamodel. Figure 2 shows an extract from the UML2 metamodel that contains relevant classes for State Machine and Activity modeling. The inState association of a Pin to a State makes it possible to specify input and output object states for ActivityNodes in an Activity. We assume that control nodes also have pins that are not explicitly shown in Fig.1(b). For a more convenient means of referring to input and output object states of ActivityNodes, we define additional input-state() and outputstate() operations, shown as Object Constraint Language [10] definitions at the bottom of Fig.2.

We assume that when executing a process model, state transitions of objects are induced. We also identify states in which the process creates objects and states that objects can be in upon termination of the process:

Definition 1 (Induced transition, first state and last state). *Let a process model P (instance of Activity) and an object type o (instance of Class) be given.*

- *An* induced transition *of o in P is a triple* (a, s_{src}, s_{tgt}) *such that there is an Action a in P with* $s_{src} \in a.inputstate(o)$ *and* $s_{tgt} \in a.outputstate(o)$;
- *A state* s_{first} *is a* first state *of o in P such that* $s_{first} \in a.outputstate(o)$ *for some Action a that has no input pins of type o;*
- *A state* s_{last} *is a* last state *of o in P such that* $s_{last} \in n.inputstate(o)$ *for some ActivityNode n that has no output pins of type o.*

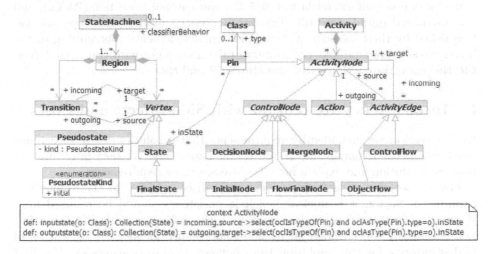

context ActivityNode
def: inputstate(o: Class): Collection(State) = incoming.source->select(oclIsTypeOf(Pin) and oclAsType(Pin).type=o).inState
def: outputstate(o: Class): Collection(State) = outgoing.target->select(oclIsTypeOf(Pin) and oclAsType(Pin).type=o).inState

Fig. 2. UML2 metamodel extract

For consistency of process models and object life cycles, we distinguish between *conformance*[1] and *coverage* [17]: *Conformance* requires that a given process model does not manipulate objects in a way that is not defined in the given life cycle. *Coverage* requires that objects used in the process model cover the entire scope of their life cycle. In the following, we directly define inconsistency types:

Definition 2 (Inconsistency types). *Given an object life cycle OLC (instance of State Machine) for object type o (instance of Class) and a process model P (instance of Activity), we define the following inconsistency types:*

- non-conformant transition: *an induced transition* (a, s_{src}, s_{tgt}) *of o in P, such that OLC contains no transition from state* s_{src} *to state* s_{tgt};
- non-conformant first state: *a first state* s_{first} *of o in P, such that OLC contains no transition from the initial state to state* s_{first};
- non-conformant last state: *a last state* s_{last} *of o in P, such that OLC contains no transition from* s_{last} *to a final state;*

---

[1] Called compliance in [17].

- non-covered transition: *a transition* (s_{src}, s_{tgt}) *in OLC where* s_{src} *is not the initial state, such that there is no induced transition* (a, s_{src}, s_{tgt}) *of o in P for any action a;*
- non-covered initial state: *a state* s_i *in OLC that has an incoming transition from the initial state, such that* s_i *is not a first state of o in P;*
- non-covered final state: *a state* s_f *in OLC that has an outgoing transition to a final state, such that* s_f *is not a last state of o in P.*

We discover the following inconsistencies in the example claim handling process with respect to the reference life cycle for claims (Fig.1): non-conformant transitions (settle, GR, NR), (settle, RE, SE), (review, NR, RE), a non-conformant first state UE, a non-conformant last state RJ, a non-covered transition (RJ, CL), and a non-covered initial state RG. Two inconsistencies of the same type are distinguished by their *contexts* that comprise model elements contributing to the inconsistency. For example, the context of the non-conformant transition (settle, GR, NR) comprises action settle, and states GR and NR.

3   Inconsistency Resolution with Side-Effects and Costs

In this section we first identify requirements that a solution for inconsistency resolution needs to satisfy, motivated by the case study. We then introduce our proposed solution and explain how it addresses these requirements.

Given a set of inconsistencies such as those discovered for the claims handling example in Sect.2, it is unclear which inconsistency should be resolved first. If our goal is to achieve conformance with the reference claim life cycle, but not necessarily full coverage of it, non-conformance inconsistencies will be of higher priority for the resolution than non-coverage inconsistencies. The first requirement is therefore that **priorities of different inconsistency types must be made explicit in the resolution process (Req1).**

In order to support the user in resolving a particular inconsistency, we define a number of alternative resolutions for each inconsistency type, following existing work [21,22]. In this paper, we selected three resolutions for resolving a non-conformant transition (an induced transition (a, s_{src}, s_{tgt}) of o in P, such that OLC contains no transition from state s_{src} to state s_{tgt}). These are informally specified in Fig.3: (a) r_1 removes state s_{src} from $a.inputstate(o)$, (b) r_2 removes s_{tgt} from $a.outputstate(o)$ and (c) r_3 removes the entire action a from the process model. We assume that removing the last input or output state also involves removing the associated pin.

Suppose that from the set of discovered inconsistencies for the claims handling process, we choose to first resolve the non-conformant transitions (settle, GR, NR) and (review, NR, RE). This gives rise to three alternative resolutions for each inconsistency, shown in Fig.4. To assist the user in choosing how to resolve a particular inconsistency, **advantages and disadvantages of each available resolution should be identified and used to rank the resolutions (Req2).**

Some resolutions may have *side-effects*, in other words an application of a resolution may not only resolve its target inconsistency, but also introduce new incon-

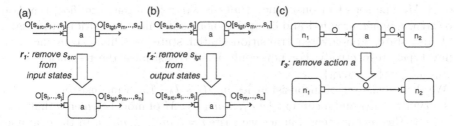

Fig. 3. Resolutions for a non-conformant transition

sistencies (*induced inconsistencies*) or remove other existing inconsistencies (*expired inconsistencies*) [12]. For example, if we apply r_1 to the non-conformant transition (settle, GR, NR) as shown in Fig.4 (b), we will introduce a new non-covered transition (GR, SE). Applying r_1 to the non-conformant transition (review, NR, RE) introduces a new non-covered first state RE, see Fig.4 (e). This shows that applying the same resolution to different inconsistencies may yield different side-effects. Generally, the user will become aware of the side-effects of a resolution only after applying it. To improve this situation, **side-effects of a resolution must be calculated before the resolution is applied (Req3)**.

In the following, we explain our approach to inconsistency resolution that satisfies the identified requirements.

Let $M = \{me_1, ..., me_n\}$ be a model comprising model elements $me_1, ..., me_n$. We denote the set of inconsistency types defined for M as T_M, where each inconsistency type $t \in T_M$ is associated with a set of resolutions R_t. In our example, the claims handling process together with the claim life cycle form the

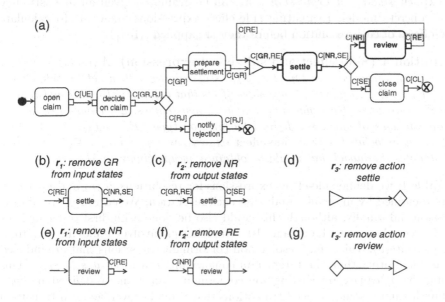

Fig. 4. Resolving non-conformant transitions in the example

model M. The set of inconsistency types is $T_M = \{$ncnf_tran, ncnf_first, ncnf_last, ncov_tran, ncov_init, ncov_fin$\}$ and comprises non-conformant transition, first state, and last state, non-covered transition, initial state and final state inconsistency types, respectively. For non-conformant transitions, the resolution set is $R_{ncnf\_tran} = \{r_1, r_2, r_3\}$.

We denote the set of inconsistencies in M as I_M. Each inconsistency $i \in I_M$ has a type $t \in T_M$ and a context $\{me_j, ...me_k\} \subseteq M$ of model elements that contribute to this inconsistency, hence we write $i = t(me_j, ..., me_k)$. In the example, the set of inconsistencies is $I_M = \{$ ncnf_tran(settle, GR, NR), ncnf_tran(review, NR, RE), ncnf_first(UE), ncnf_last(RJ), ncov_tran(RJ, CL), ncov_init(RG)$\}$.

We introduce *costs* for different inconsistency types to reflect that resolving inconsistencies of some types has a higher priority than of others (Req1).

Definition 3 (Cost of an inconsistency type). *The cost of an inconsistency type is defined by a function cost : $T_M \to \mathbb{N}$ that maps an inconsistency type to a natural number.*

Costs can either be assigned to inconsistency types once and then used for inconsistency resolution in every model, or different costs for each model can be assigned to reflect a specific resolution goal. For our example, we assume that our main goal is to achieve conformance of the claim handling process. We further consider that conformance of transitions and last states is more important than that of first states. Therefore, we assign the following costs to the different inconsistency types: cost(ncnf_tran) = 3, cost(ncnf_first) = 2, cost(ncnf_last) = 3, cost(ncov_tran) = 1, cost(ncov_init) = 1 and cost(ncov_fin) = 1.

We further associate each resolution with one or more *side-effects*, each having an explicit *side-effect expression* that can be evaluated given an inconsistency in a concrete model. Evaluating side-effect expressions allows us to calculate side-effects of each resolution before they are applied (Req3).

Definition 4 (Side-effect and side-effect expression). *A resolution $r \in R_t$ for a given inconsistency type $t \in T_M$ is associated with a set of side-effects $E_r = E_r^- \cup E_r^+$, where E_r^- are side-effects that expire existing inconsistencies and E_r^+ are side-effects that induce new inconsistencies. Each side-effect affects inconsistencies of one type, defined by function type : $E_r \to T_M$. A side-effect $e \in E_r$ is associated with a side-effect expression $exp_e : I_M \to \mathcal{P}(I_{M'})$, where M' denotes the model that would be obtained by applying r to M.*

In Table 1, we define side-effect expressions for resolutions r_1, r_2 and r_3, which were identified by manually analyzing each resolution. We currently specify expressions informally, although this could also be done using first-order logic or the Object Constraint Language [10]. Applying r_1 involves removing s_{src} from $a.inputstate(o)$, which may resolve non-conformant transitions that are induced by action a other than the target non-conformant transition (a, s_{src}, s_{tgt}). This means that r_1 can expire existing non-conformant transitions, as defined in exp_{e_1}.

Furthermore, some induced transitions that provide coverage for a transition in the life cycle may no longer be induced after r_1 is applied. To capture this in a side-effect expression, we introduce the concept of a coverage set: A *coverage set* of a transition (s_k, s_l) in the object life cycle contains all induced transitions of the form (a', s_k, s_l) in the process model. If an induced transition (a', s_k, s_l) is the only element of a coverage set for (s_k, s_l), then removing this induced transition will introduce a new non-covered transition (s_k, s_l). For r_1, induced non-covered transitions are defined in exp_{e_2}.

Table 1. Side-effect expressions, where $i = ncnf\_tran(a, s_{src}, s_{tgt})$

Res	Side-effect	Side-effect expression
r_1	$e_1 \in E_{r_1}^{-}$	$exp_{e_1}(i) = \{ncnf\_tran(a, s_{src}, s_i) \mid s_i \in a.outputstate(o)$ and $ncnf\_tran(a, s_{src}, s_i) \in I_M\}$
	$e_2 \in E_{r_1}^{+}$	$exp_{e_2}(i) = \{ncov\_tran(s_{src}, s_i) \mid (a, s_{src}, s_i)$ is an induced transition of o in P and (a, s_{src}, s_i) is the only element in the coverage set of $(s_{src}, s_i)\}$
	$e_3 \in E_{r_1}^{+}$	$exp_{e_3}(i) = \{ncnf\_first(s_i) \mid s_i \in a.outputstate(o)$ and s_{src} is the only state in $a.inputstate(o)$ and there is no transition from the initial state to s_i in $OLC\}$
r_2	$e_4 \in E_{r_1}^{-}$	$exp_{e_4}(i) = \{ncnf\_tran(a, s_i, s_{tgt}) \mid s_i \in a.inputstate(o)$ and $ncnf\_tran(a, s_i, s_{tgt}) \in I_M\}$
	$e_5 \in E_{r_2}^{+}$	$exp_{e_5}(i) = \{ncov\_tran(s_i, s_{tgt}) \mid (a, s_i, s_{tgt})$ is an induced transition of o in P and (a, s_i, s_{tgt}) is the only element in the coverage set of $(s_i, s_{tgt})\}$
	$e_6 \in E_{r_2}^{+}$	$exp_{e_6}(i) = \{ncnf\_last(s_i) \mid s_i \in a.inputstate(o)$ and s_{tgt} is the only state in $a.outputstate(o)$ and there is no transition from s_i to a final state in $OLC\}$
r_3	$e_7 \in E_{r_3}^{-}$	$exp_{e_7}(i) = \{ncnf\_tran(a, s_i, s_j) \mid s_i \in a.inputstate(o)$ and $s_j \in a.outputstate(o)$ and $ncnf\_tran(a, s_i, s_j) \in I_M\}$
	$e_8 \in E_{r_3}^{+}$	$exp_{e_8}(i) = \{ncov\_tran(s_i, s_j) \mid (a, s_i, s_j)$ is an induced transition of o in P and (a, s_i, s_j) is the only element in the coverage set of $(s_i, s_j)\}$

For resolving non-conformant transitions (settle, GR, NR) and (review, NR, RE), we get the following details about the effect of each resolution if we evaluate the defined side-effect expressions (see Fig.4(b)-(g)):

Resolutions for ncnf_tran(settle, GR, NR)
r_1 resolves ncnf_tran(settle, GR, NR) and introduces ncov_tran(GR,SE);
r_2 resolves ncnf_tran(settle, GR, NR), ncnf_tran(settle, RE, NR);
r_3 resolves ncnf_tran(settle, GR, NR), ncnf_tran(settle, RE, NR) and introduces ncov_tran(GR,SE).

Resolutions for ncnf_tran(review, NR, RE)

r_1 resolves ncnf_tran(review, NR, RE) and introduces ncnf_first(review, RE);
r_2 resolves ncnf_tran(review, NR, RE) and introduces ncnf_last(review, NR);
r_3 resolves ncnf_tran(review, NR, RE).

This detailed information about the effect of each resolution helps the user to decide which resolution to apply in each case. Generally, the most beneficial resolution would be the one that overall removes the greatest number of inconsistencies. In this example, we would choose r_2 to resolve ncnf_tran(settle, GR, NR) and r_3 to resolve ncnf_tran(review, NR, RE).

To satisfy Req2, our approach goes further to provide a more fine-grained comparison of resolutions based on *cost reduction* values calculated for each resolution.

Definition 5 (Cost reduction of a resolution). *Given a resolution* $r \in R_t$ *that can resolve an inconsistency* $i \in I_M$ *of type* $t \in T_M$, *with side-effects* $E_r^- = \{e_{11}, ..., e_{1p}\}$ *and* $E_r^+ = \{e_{21}, ..., e_{2q}\}$, *the* cost reduction *of resolution* r *is denoted by* $costred_r$ *and calculated as follows:*

$$costred_r = cost(t) + \sum_{j=1}^{p} \left(\mid exp_{e_{1j}}(i) \mid \times cost(type(e_{1j})) \right)$$

$$- \sum_{k=1}^{q} \left(\mid exp_{e_{2k}}(i) \mid \times cost(type(e_{2k})) \right)$$

We now calculate cost reduction values for the resolutions in our example:

ncnf_tran(settle, GR, NR)	ncnf_tran(review, NR, RE)
$costred_{r_1} = 3 - (1 \times 1) = 2$	$costred_{r_1} = 3 - (1 \times 2) = 1$
$costred_{r_2} = 3 + (1 \times 3) = 6$	$costred_{r_2} = 3 - (1 \times 3) = 0$
$costred_{r_3} = 3 + (1 \times 3) - (1 \times 1) = 5$	$costred_{r_3} = 3$

It can be seen that based on the calculated cost reduction values, we can perform a more fine-grained comparison of the resolutions that takes into account the priorities or costs of different inconsistency types.

With our approach, we could also introduce more automation into the resolution process by applying resolutions without user intervention whenever there is one resolution that has a highest cost reduction value for a particular inconsistency. However, in our scenario, approving the choice of a resolution needs to be done by an expert who is aware of what impact the change in the process model has on the business. Provided that we are working with a model of an existing business process, removing an action from this model translates to removing a step in the process and may be difficult to implement in practice, even though the cost reduction value indicates that this is the best resolution.

4   Design and Implementation of Tool Support

In this section we present a design for an inconsistency resolution module based on our approach, which leads to an efficient implementation of inconsistency resolution.

Figure 5 shows the fundamental elements of the resolution module design, comprising several abstract classes (labeled in italics) that need to be extended to arrive at an executable implementation.

The central element of the module is the InconsistencyResolver that has references to all InconsistencyTypes that it can handle and a Model that is the subject of inconsistency handling. The checkAndResolve() method of the InconsistencyResolver is the entry-point to the inconsistency checking and resolution process.

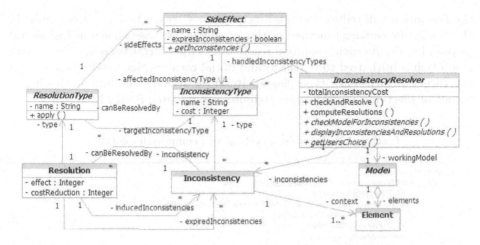

Fig. 5. Design for an inconsistency resolution module

Listing 1.1. InconsistencyResolver : checkAndResolve()

```
1  checkAndResolve()
2      inconsistencies = checkModelForInconsistencies ();
3      while ( inconsistencies . size > 0) do
4          totalInconsistencyCost = 0;
5          for each ( Inconsistency i in  inconsistencies ) do
6              totalInconsistencyCost += i.type.cost;
7          computeResolutions();
8          displayInconsistenciesAndResolutions ();
9          Resolution r = getUsersChoice();
10         if (r == null) then
11             return;
12         else
13             r.type.apply (),
14         inconsistencies .remove(r. inconsistency );
15         inconsistencies .removeAll(r. expiredInconsistencies );
16         inconsistencies .add(r. inducedInconsistencies );
```

As shown in the method in Listing 1.1, the working model is first checked
for inconsistencies and then the total inconsistency cost for the model is computed (lines 4-6). For each discovered inconsistency, possible resolutions are identified and these are communicated to the user (lines 7,8). After a resolution of
user's choice is applied (lines 9-13), the model is not re-checked, but rather the
inconsistency set is directly updated by removing the resolved inconsistency and
those that expired and adding induced inconsistencies (lines 14-16). This added
efficiency is valuable in practice, as in reality many models tend to get very large
and as soon as re-checking them takes a noticeable amount of time, unnecessary
interruptions are introduced into the resolution process.

As shown in Fig.5, each Resolution is associated with sets of inconsistencies that
it will induce or expire if it is applied. An effect of a Resolution is the overall change
in number of inconsistencies that it will inflict and its costReduction reflects how

the resolution will reduce the total inconsistency cost. These are determined in the computeResolutions() method of the InconsistencyResolver shown in Listing 1.2 as pseudocode. As each resolution resolves its target inconsistency, effect of each resolution is initialized to -1 (line 8) and initial costReduction is set to the cost of the target inconsistency type (line 9). Iterating over the resolution side-effects, values of effect and costReduction attributes are updated (lines 13-14,17-18).

Listing 1.2. InconsistencyResolver : computeResolutions()

```
1  computeResolutions()
2      for each ( Inconsistency  i  in  inconsistencies ) do
3          i . canBeResolvedBy.clear ();
4          for each (ResolutionType rt  in  i.type.canBeResolvedBy) do
5              Resolution  r = new Resolution();
6              i . canBeResolvedBy.add(r);
7              r . inconsistency  = i;
8              r . effect  = −1;
9              r .costReduction = i.type. cost ;
10             for each ( SideEffect  se  in  rt . sideEffects ) do
11                 Set  inconsistencies  = se. getInconsistencies ( i ,workingModel);
12                 if (se. expiresInconsistencies ) then
13                     r . effect  −= inconsistencies . size ();
14                     r .costReduction += inconsistencies . size () ∗ se. affectedInconsistencyType . cost ;
15                     r . expiredInconsistencies .add( inconsistencies );
16                 else
17                     r . effect  += inconsistencies . size ();
18                     r .costReduction −= inconsistencies . size () ∗ se. affectedInconsistencyType . cost ;
19                     r . inducedInconsistencies .add( inconsistencies );
```

Our proposed design can be directly used to derive the implementation core of a resolution module, after which all abstract classes need to be extended to complete the implementation. For implementing support for inconsistency resolution between object life cycles and process models for example, we extend InconsistencyResolver with the concrete class ObjectLifeCycleProcessModelResolver that provides an implementation for finding inconsistencies in a model and communication with the user. For each resolution side-effect like the ones shown in Table 1, we create an extension of the SideEffect class and implement the getInconsistencies() method that evaluates the associated side-effect expression. Our approach does not place any restrictions on how transformations associated with resolutions are to be implemented, i.e. this can be done directly in a conventional programming language such as Java or using one of the existing model transformation approaches.

We have implemented a prototype for resolution of inconsistencies between object life cycles and process models, based on the presented approach. Our prototype is an extension to IBM WebSphere Business Modeler [2] that natively supports process modeling and has been extended for modeling object life cycles (Fig.6).

Detected inconsistencies are shown in the inconsistencies view below the model editors, where inconsistencies with higher costs are displayed higher in the list. The total number and cost of inconsistencies are shown at the top of this view, so that the effect on these numbers can be monitored during the resolution process.

After an inconsistency is selected, resolution side-effects and cost reduction values are calculated by the tool and can be taken into account for choosing the most appropriate resolution.

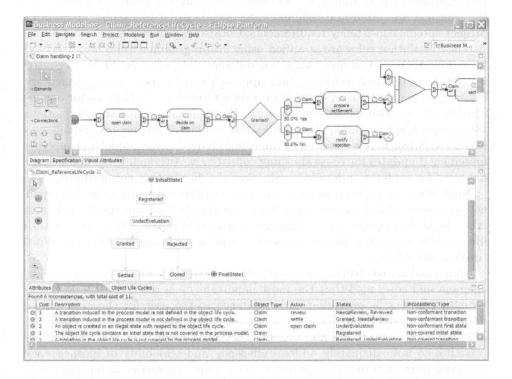

Fig. 6. Inconsistency resolution prototype in IBM WebSphere Business Modeler

We have used the prototype for resolving inconsistencies between several example models inspired by IAA [1]. This initial study has shown that the proposed approach provides powerful assistance for the user in selecting a resolution for each inconsistency.

5   Related Work

Recent work on inconsistency resolution includes work by Mens et al. [12,13] that uses the AGG [23] graph transformation tool to detect potential dependencies between different inconsistency resolutions. Inconsistency detection and resolution rules are expressed as graph transformation rules in AGG and are then analyzed using critical pair analysis. Analysis results point to potentially conflicting resolutions, resolutions that may induce or expire other types of inconsistencies and potential cycles between resolutions. In this paper, we propose explicit specification of side-effect expressions that can be evaluated given a model inconsistency and hence provide the user with a basis for comparison of alternative resolutions. It would be interesting to investigate how automated dependency analysis can assist in the specification of side-effect expressions.

The FUJABA tool suite [15] supports both manual and automatic incremental inconsistency resolution [24]. Consistency checking rules can be configured by the user and organized into different categories in order to support domain- or project-specific consistency requirements. Consistency checking rules and inconsistency resolution rules are specified using graph grammar rules and executed by a FUJABA rule engine. Although different categories could also be used for obtaining different priorities, our approach can be seen as complementary because we focus on the evaluation of several alternative resolutions for one inconsistency based on side-effects and costs. Work on incremental transformations using triple graph grammars [19] studies the problem of keeping two models synchronized [8,18]. This is achieved by analyzing changes in one model and applying incremental updates for re-establishing consistency. Although these updates are analyzed for conflicts, a detailed evaluation of side-effects is not addressed.

Nentwich et al [14] propose to generate inconsistency resolutions (called repair actions) automatically from consistency constraints that are specified in first order logic. As opposed to our approach, generated repair actions do not take into account a concrete model violating consistency constraints and also do not consider side-effects.

Spanoudakis and Zisman [20] conducted a survey about inconsistency management and concluded that the most important open research issue in inconsistency handling is providing more guidance to the user for choosing among multiple alternative resolutions. The authors argue that resolutions should be ordered based on cost, risk and benefit. They further conclude that existing approaches do not adequately address efficiency and scalability of inconsistency detection in models that change during the resolution process. In our approach, we use side-effects and costs for evaluating alternative resolution and avoid rechecking the whole model after a resolution is applied.

Nuseibeh et al. [16] present a framework for managing inconsistency in software development. This framework comprises monitoring, identification and measuring of inconsistencies. Measuring inconsistencies includes attaching priorities to different inconsistencies. In our work, we use costs to reflect priorities of inconsistency types and also to calculate cost reduction for each resolution.

6   Conclusions and Future Work

In this paper, we introduce the concept of side-effect expressions that can be evaluated for a given inconsistent model to determine whether or not a resolution leads to new or expired inconsistencies. This allows the user to compare alternative resolutions for the same inconsistency. We attach costs to each inconsistency type, which enables us to calculate cost reduction values for each resolution and therefore provide a more fine-grained comparison of resolutions. Finally, we show how our concepts can be used to implement an efficient inconsistency resolution module for integration with an existing modeling tool.

Our case study and application of the prototype to various examples have shown that the approach adds significant value for the user during the resolution process. To enhance our solution further, we will next investigate more sophisticated cost models and study how cycles can be detected during the resolution process. Another area of future work is the application of our approach to domain-specific languages. Here, automated resolution dependency analysis and reusable side-effect specifications can be of interest in order to decrease the manual overhead of our approach.

Acknowledgements. We would like to thank Harald Gall, Jana Koehler, Cesare Pautasso, Hagen Völzer and Michael Wahler for their valuable feedback on an earlier version of this paper.

References

1. IBM Insurance Application Architecture, http://www.ibm.com/industries/financialservices/doc/content/solution/278918103.html
2. IBM WebSphere Business Modeler, http://www.ibm.com/software/integration/wbimodeler/
3. UML2.0 Superstructure, formal/05-07-04. OMG Document (2005)
4. Ebert, J., Engels, G.: Specialization of Object Life Cycle Definitions. Fachberichte Informatik 19/95, University of Koblenz-Landau (1997)
5. Engels, G., Küster, J.M., Groenewegen, L., Heckel, R.: A Methodology for Specifying and Analyzing Consistency of Object-Oriented Behavioral Models. In: ESEC'01. Proceedings of the 8th European Software Engineering Conference, pp. 186–195. ACM Press, New York (2001)
6. Finkelstein, A., Gabbay, D., Hunter, A., Kramer, J., Nuseibeh, B.: Inconsistency Handling in Multi-Perspective Specifications. IEEE Transactions on Software Engineering 20(8), 569–578 (1994)
7. Ghezzi, C., Nuseibeh, B.A.: Special Issue on Managing Inconsistency in Software Development (1). IEEE Transactions on Software Engineering 24(11) (November 1998)
8. Giese, H., Wagner, R.: Incremental Model Synchronization with Triple Graph Grammars. In: Nierstrasz, O., Whittle, J., Harel, D., Reggio, G. (eds.) MoDELS 2006. LNCS, vol. 4199, pp. 543–557. Springer, Heidelberg (2006)
9. Kappel, G., Schrefl, M.: Object/Behavior Diagrams. In: Proceedings of the 7th International Conference on Data Engineering, Washington, DC, USA, pp. 530–539. IEEE Computer Society, Los Alamitos (1991)
10. Kleppe, A., Warmer, J.: The Object Constraint Language, 2nd edn. Addison-Wesley, Reading (2003)
11. Küster, J.M.: Consistency Management of Object-Oriented Behavioral Models. PhD thesis, University of Paderborn (March 2004)
12. Mens, T., Van Der Staeten, R., Warny, J.-F.: Graph-Based Tool Support to Improve Model Quality. In: Proceedings of the 1st Workshop on Quality in Modeling co-located with MoDELS 2006, Technical report 0627, Technische Universiteit Eindhoven, pages 47–62 (2006)

13. Mens, T., Van Der Straeten, R., D'Hondt, M.: Detecting and Resolving Model Inconsistencies Using Transformation Dependency Analysis. In: Nierstrasz, O., Whittle, J., Harel, D., Reggio, G. (eds.) MoDELS 2006. LNCS, vol. 4199, pp. 200–214. Springer, Heidelberg (2006)
14. Nentwich, C., Emmerich, W., Finkelstein, A.: Consistency Management with Repair Actions. In: Proceedings of the 25th International Conference on Software Engineering, Portland, Oregon, USA, May 3-10, 2003, pp. 455–464. IEEE Computer Society, Los Alamitos (2003)
15. Nickel, U.A., Niere, J., Zündorf, A.: Tool Demonstration: The FUJABA Environment. In: ICSE. Proceedings of the 22nd International Conference on Software Engineering, Limerick, Ireland, pp. 742–745. ACM Press, New York (2000)
16. Nuseibeh, B., Easterbrook, S., Russo, A.: Making Inconsistency Respectable in Software Development. Journal of Systems and Software 58(2), 171–180 (2001)
17. Ryndina, K., Küster, J.M., Gall, H.: Consistency of Business Process Models and Object Life Cycles. In: Kühne, T. (ed.) Workshops and Symposia at MoDELS 2006. LNCS, vol. 4364, pp. 80–90. Springer, Heidelberg (2007)
18. Lohmann, S., Westfechtel, B., Becker, S., Herold, S.: A Graph-Based Algorithm for Consistency Maintenance in Incremental and Interactive Integration Tools. Journal of Software and Systems Modeling (to appear, 2007)
19. Schürr, A.: Specification of Graph Translators with Triple Graph Grammars. In: Mayr, E.W., Schmidt, G., Tinhofer, G. (eds.) WG 1994. LNCS, vol. 903, pp. 151–163. Springer, Heidelberg (1995)
20. Spanoudakis, G., Zisman, A.: Handbook of Software Engineering and Knowledge Engineering, chapter Inconsistency Management in Software Engineering: Survey and Open Research Issues, pp. 329–380. World Scientific Publishing Co. (2001)
21. Van Der Straeten, R.: Inconsistency Management in Model-Driven Engineering. PhD thesis, Vrije Universiteit Brussel (September 2005)
22. Van Der Straeten, R., D'Hondt, M.: Model Refactorings through Rule-Based Inconsistency Resolution. In: SAC. Proceedings of the 2006 ACM Symposium on Applied Computing, Dijon, France, April 23-27, 2006, pp. 1210–1217. ACM, New York (2006)
23. Taentzer, G.: AGG: A Graph Transformation Environment for Modeling and Validation of Software. In: Pfaltz, J.L., Nagl, M., Böhlen, B. (eds.) AGTIVE 2003. LNCS, vol. 3062, pp. 446–453. Springer, Heidelberg (2004)
24. Wagner, R., Giese, H., Nickel, U.: A Plug-In for Flexible and Incremental Consistency Management. In: Proceedings Workshop on Consistency Problems in UML-based Software Development, San Francisco, USA, Technical Report. Blekinge Institute of Technology, San Francisco (October 2003)

Model Composition in Product Lines and Feature Interaction Detection Using Critical Pair Analysis

Praveen Jayaraman, Jon Whittle, Ahmed M. Elkhodary, and Hassan Gomaa

Department of Information & Software Engineering,
George Mason University, Fairfax VA 22030
praveenjayaraman@yahoo.com, {jwhittle,aelkhoda,hgomaa@gmu.edu}

Abstract. Software product lines (SPL) are an established technology for developing families of systems. In particular, they focus on modeling commonality and variability, that is, they are based on identifying features common to all members of the family and variable features that appear only in some members. Model-based development methods for product lines advocate the construction of SPL requirements, analysis and design models for features. This paper describes an approach for maintaining feature separation during modeling using a UML composition language based on graph transformations. This allows models of features to be reused more easily. The language can be used to compose the SPL models for a given set of features. Furthermore, critical pair analysis is used to detect dependencies and conflicts between features during analysis and design modeling. The approach is supported by a tool that allows automated composition of UML models of features and detection of some kinds of feature interactions.

Keywords: Software product lines, model transformation, feature interaction.

1 Introduction

A software product line (SPL) consists of a family of software systems where family members have some common functionality and some variable functionality [1-3]. Software product line engineering involves developing the requirements, architecture, and component implementations for a family of systems, from which applications (family members) are derived and configured. A better understanding of a product line can be obtained by considering the multiple views, such as requirements models, static models, and dynamic models of the product line, which can be depicted in UML. A key view of a software product line is the feature modeling view [4]. The feature model is crucial for managing variability and product derivation as it describes the product line requirements in terms of common and variable features, as well as defining feature dependencies [5]. SPL model-based development methods, such as PLUS [6] and Orthogonal Variability Modeling [7], typically advocate feature-based approaches for developing SPL requirements, analysis and design models.

Model-based SPL development methods would benefit considerably from automated support for maintaining clear and traceable feature separation throughout the modeling lifecycle. In PLUS, common, optional, and alternative features are

G. Engels et al. (Eds.): MoDELS 2007, LNCS 4735, pp. 151–165, 2007.

determined during feature modeling, which is part of the requirements phase. The common features are called the kernel, and the optional and alternative features are called variants. Impact analysis is used during dynamic modeling to determine each variant feature's impact on kernel interaction and state machine models. This is a manual process. Conventional UML tools do not provide automated feature-based navigation and traceability between multiple view SPL models. In particular, current methods like PLUS do not maintain separation of features at the analysis modeling stage of the modeling lifecycle. This can lead to inconsistencies in software product line models and ultimately to errors in implementations.

To address these issues, this paper describes an automated approach for modeling features in UML such that the UML analysis and design models for each feature are maintained separately and are combined on-demand at any point during the modeling lifecycle. A model composition language, MATA (Modeling Aspects using a Transformation Approach) is used to specify how features relate to each other. MATA is based on the graph transformation formalism. This has two major advantages. Firstly, existing graph rewrite tools can be used to compose the models for a given set of features. In our case, AGG (Attributed Graph Grammar) [8] is used. Secondly, critical pair analysis (CPA) [9] can be used to detect static interactions between models of features. Hence, if the modeler (accidentally) introduces a new feature dependency or conflict during the modeling phase, CPA will inform her. She can then choose to modify the models to avoid the dependency or to update the feature dependency diagram. In either case, the consistency between feature dependency diagram and models of features is maintained. After resolving the conflicts or dependencies, the modeler can execute the actual model composition to generate the combined model consisting of behavior for multiple features.

The remainder of this paper is structured as follows. Section 2 describes existing tool support for SPL engineering. In section 3, we describe the necessary background concepts. In section 4, our approach to modeling SPLs and the detection of feature interactions is explained. Section 5 provides an application on an example case study. Section 6 surveys the related work in the area of model compositions. Finally we present future threads of research and conclude the paper in section 7.

2   Tool Support for Software Product Line Engineering

Automated tool support is highly desirable for managing the complexity and variability inherent in software product lines. Previous research [10-12] provided tool support for representing the multiple graphical views (i.e., static and dynamic views) of a SPL using various CASE tools, including IBM Rational Rose. Using the open architecture provided by these CASE tools, plug-ins were developed to extract the underlying representation of each view and store this information in a SPL repository, which consisted of an integrated set of database relations. A multiple view consistency checking tool was then developed to check for consistency among the multiple views and report any inconsistencies to the user [13]. Automated tool support [12] was also provided for deriving application models for an individual SPL member from the SPL repository. This was achieved by developing a knowledge-based requirements elicitation and application derivation tool, which interacts with the user to ensure selection of a consistent set of application features and then derives the application model [10, 11].

While these tools all rely on the concept of commonality and variability in SPL engineering, they do not maintain separation of features throughout the modeling lifecycle. The novelty of the research described in this paper is providing an approach for modeling individual features, detecting feature relationships at the modeling level and then providing automated composition of models of features.

This paper describes an SPL process and tool which can be used to model features independently of other features, compose models of features and also to detect structural interactions between features automatically. The technique is based on graph transformations, which are described in the next section.

3 Background

This section presents background on graph transformations which form the formal basis of our approach. A graph consists of a set of nodes and a set of edges. A typed graph is a graph in which each node and edge belongs to a type defined in a type graph. An attributed graph is a graph in which each node and edge may contain attributes where each attribute is a (value, type) pair giving the value of the attribute and its type. A graph transformation is a rule used to modify a host graph, G, and is defined by a left hand graph, L, and a right hand graph, R. The process of applying the rule to a graph G involves finding a graph monomorphism, h, from L to G and replacing $h(L)$ in G with $h(R)$. Further details can be found in [14].

We give a simple example of using graph transformations for composing UML class diagrams in a product line. The example is for a family of microwave ovens. The microwave oven product line contains a variant feature, called Language, to select the display language for the microwave display prompts. The kernel class diagram of the product line—i.e., the classes that are common to all family members—contains an abstract class *DisplayPrompts*. The Language feature adds the implementation class *FrenchPrompts* to the kernel to allow user messages to be displayed in another language. The kernel class diagram is shown in Figure 1. The composed class diagram, after the Language feature has been added to the kernel, is shown in Figure 2. The Language feature can be represented as a graph transformation, as shown in Figure 3. Figure 3 defines the left and right graphs of the transformation to the left and right of the arrow, respectively.

One may, of course, define multiple graph transformation rules on a given host graph. For product lines, this corresponds to adding multiple variant features.

In general, there are two ways of introducing non-determinism when applying a set of graph rules—either the same rule may match in different ways, or the rules may be applied in a different order to yield a different result. The latter case of non-determinism arises because of relationships between rules. These relationships are either conflicts or dependencies. A conflict relation arises when one rule modifies an element of the host graph in such a way that another rule can no longer be applied. A dependency relation arises when one rule requires the existence of elements in the host graph that must be produced by another rule. Critical pair analysis [9] is a well-established technique for discovering conflict and dependency relations for a set of

Fig. 1. Kernel class diagram

Fig. 2. Class diagram of kernel with language feature

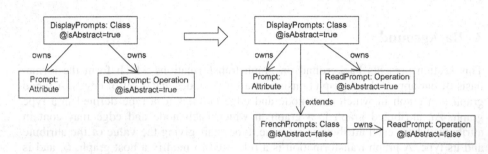

Fig. 3. Graph transformation of language feature

graph rules. Dependency relations can be resolved by imposing an ordering on the application of the rules. Some conflicts can be resolved by choosing an ordering, but, in general, conflicts may require that the models be rewritten.

There already exist a number of tools for applying graph transformations—for example, AGG [8], FUJABA [15], GREAT [16]. AGG, in particular, supports critical pair analysis. We will apply graph transformations and critical pair analysis in the context of modeling software product lines. Hence, AGG is used as the underlying tool for executing graph transformations and detecting conflicts and dependencies.

4   Model Composition and Feature Interaction for Product Lines

Figure 4 illustrates our approach for SPL feature-based modeling, which can be used with existing methods that capture dependencies between features on a feature dependency diagram. Once an initial set of features has been decided on, a feature dependency model is defined showing which features require other features[1] (see Figure 4). Next, the kernel feature set is modeled using UML. Currently, our tool supports class diagrams, sequence diagrams and statecharts. Each variant feature (or feature group if a set of features are always used together) is modeled in the MATA language in terms of how it relates to the kernel UML models. This allows a single feature (or feature group) to be modeled separately from other features. This is in

---

[1] We allow richer relationships between features but only consider <<requires>> in this paper.

contrast to other approaches like PLUS where kernel and variant features are mixed on the same set of diagrams. In particular, PLUS incrementally adds features to a set of kernel models without maintaining any traceability as to where the feature elements come from.

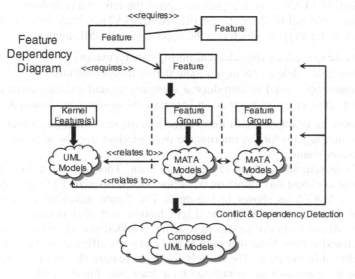

Fig. 4. Feature-based product line development with MATA

Since each variant feature is modeled independently of other variant features, the user can choose a subset of the available features and automatically generate the models (class, sequence and state diagrams) for the combined set of features. Furthermore, new features can be added relatively easily. The user may specify an ordering of features or may leave it unspecified. In any case, critical pair analysis (CPA) is applied to detect any structural conflicts and dependencies between models of features. These results are fed back to the user who either modifies the ordering or the UML models to ensure a consistent composition. The user may also discover an inconsistency with the feature dependency diagram and update it accordingly.

4.1 A Language for Composing Variant Features with the Kernel (MATA)

The MATA specifications shown in Figure 4 are transformations detailing how to modify kernel models. Transformations for UML such as QVT (Queries/Views/Transformations) [17] are typically specified at the meta-level using abstract syntax. (Note, for example, that abstract syntax is used in Figure 3). For product line modeling, this approach is insufficient because it requires the writer of the transformation to have an intimate knowledge of the UML metamodel. Most model developers do not have this knowledge. Therefore, it would be inadvisable to force them to use the abstract syntax of the metamodel. Graph transformations for UML, however, can also be specified directly over UML's concrete syntax. The only drawback of this is that meta-level concepts cannot be referred to.

MATA is essentially a compact representation of graph transformations which allows the user to describe model transformations graphically in UML concrete syntax. MATA is not meant to be a general model transformation language but is instead geared towards feature composition in product lines.

Usually, graph rules have a left- and a right-hand side. For a more compact representation, MATA uses stereotypes to capture the left- and right-hand sides on the same diagram. We follow the notation offered in VIATRA [18]. MATA allows the following three stereotypes, which may be applied to any UML model element:

- <<create>> – create new element during transformation.
- <<delete>> – delete existing element during transformation.
- <<context>> – used to introduce a container around existing elements in the host graph—eg., match a state and create a composite state around it.

We illustrate an example use of these directives for state diagrams. Figure 5 shows the kernel state diagram for the microwave oven product line. Now consider how to specify a variant feature.

The Light feature adds a light object to the oven. This light should be turned off when the door is closed and turned on when the door is opened. The Light feature can be defined in MATA as shown in Figure 6. The figure describes elements of the kernel state diagram affected by the Light feature and also describes how those elements are affected—in this case, a new action is added to each of the Door Opened and Door Closed events. Note that only kernel elements affected by the Light feature need be included in Figure 6. The models for Light feature shown in Figure 6 match any two distinct states that are connected by at least one transition which is triggered by the Door Closed or the Door Opened event.

There are special composition situations when the creation of new elements requires existing elements to be contained in newly created elements. <<context>> is a special directive used when a new element is created such that it encloses an existing element. It follows that <<context>> has to be specified within a <<create>> directive. The use of <<create>> is "optimized" in the sense that if a state is stereotyped as <<create>> then any of its substates or transitions are also created. This optimization reduces the number of stereotypes a user must specify. To stop substates from being created, they are stereotyped as <<context>>. Figure 7 illustrates using a variant feature called Recipe. The Recipe feature requires the creation of a new composite state Recipe around existing states 'Ready to Cook', 'Door Open with Item' and 'Cooking'.

The resultant state diagram obtained from composing the Light and Recipe features with the kernel is shown in Figure 8. Our tool automates the composition of the kernel for a chosen set of variant features. The tool first converts each MATA definition to a standard graph rule. The graph rule is then executed in AGG to produce a composed model and the composed model is converted back to UML. The equivalent meta-level graph rule for the Recipe feature is shown in Figure 9. Note that the Light and Recipe features are orthogonal, i.e. they do not conflict or depend on each other. Hence the ordering between them can be ignored at this point.

MATA allows modular specification of a variant feature with minimal knowledge of the kernel. In other words, only the kernel elements directly affected by the variant need be included in the MATA specification for the variant. Also, each variant feature is represented independently from other variant features and hence the features can be reused more easily.

MATA allows a slice-based approach to describe features. Each feature consists of a "slice" described by class diagrams, sequence diagrams and state diagrams. For example, the slice for the Recipe feature consists of the state diagram slice (Figure 8) and the sequence diagram slice (Figure 12). Each MATA model represents an incremental change to the kernel models. Once the UML and MATA models have been specified, the modeler chooses any subset of the feature set to compose. Our tool translates the MATA models to their equivalent graph transformation rules and uses AGG to execute the transformations on the kernel models. Finally, the composed graph is translated from its graph representation back to UML and is presented to the user. The result is a set of multiple view UML models that describe the composition of the models for the selected features.

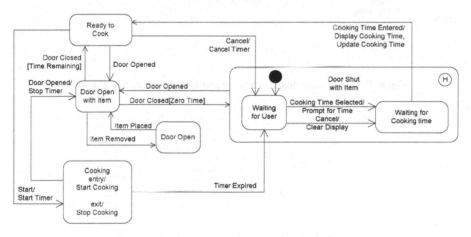

Fig. 5. Kernel state diagram of microwave product line

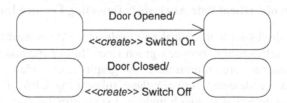

Fig. 6. MATA representation of light feature

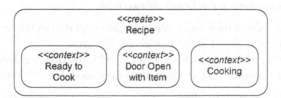

Fig. 7. MATA representation of recipe feature

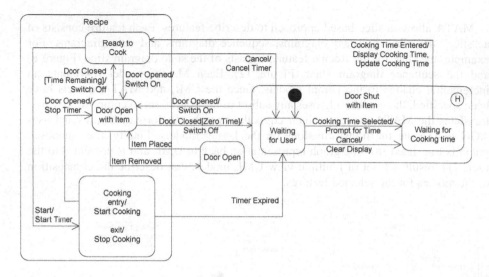

Fig. 8. Resulting state diagram after applying light and recipe feature

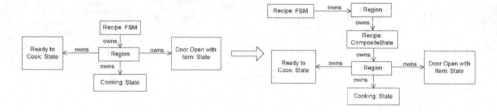

Fig. 9. Equivalent graph rule for recipe feature

4.2 Application of Critical Pair Analysis to Detecting Feature Interactions

In section 3, we discussed how critical pair analysis (CPA) is used to detect conflict and dependency relationships between graph rules. In our methodology, models of product line features are represented as graph rules. Hence, conflicts and dependencies can be detected between these rules using CPA. This is a form of feature interaction detection in which undesired static structural interactions between features are detected automatically. These interactions can be verified for consistency with the relations captured in the feature dependency diagram. The purpose of consistency checking between relations is two fold.

- Detection of redundant feature dependencies captured in the feature dependency diagram but not realized in the models for features.
- Detection of new dependencies not specified in the feature dependency diagram but which are introduced (accidentally) in the models for features.

After detection, the feature interactions are resolved by ordering the features in a way that allows all selected features to be added successfully. In some cases, the detected interactions correspond to design flaws and the models must be changed.

Ordering information is assigned as integers starting from 0 (highest). If no critical pairs are detected, then this means that the feature specifications are orthogonal, i.e., there are no relationships in the feature dependency diagram. In a case like this, no ordering is needed for execution as there is no interaction that needs to be resolved.

We look at an example of dependency interaction between two features for sequence diagrams in the microwave product line. The features are specified using the MATA syntax explained above.

- Multi-Line display (Figure 10) – Add the Multi-Line Display object so that it can be used to display multiple lines of data sent as updates to the microwave controller. (Note that the object sending the data to the controller is deliberately left unspecified).
- Recipe (Figure 12) – Add a Recipes lifeline and add the interactions for the selection of pre-stored recipes and display the recipe.

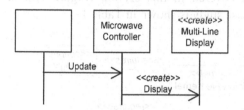

SequenceDiagram Conflict-None
SequenceDiagram Dependencies-
Found critical pairs from Multi-Line
Display to Recipe: Class Multi-Line
Display is produced.

Fig. 10. Multi-line display feature   **Fig. 11.** Critical pair analysis

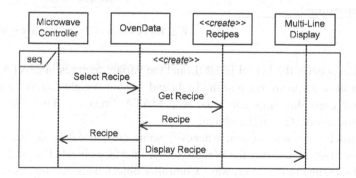

Fig. 12. Recipe feature

The results of critical pair analysis for this example are shown in Figure 11. The results show that no conflicts exist between the between the Multi-Line Display and the Recipe features. However one dependency is found to exist between the two features. The tool detects and describes the dependency as the production of a class called Multi-Line Display by the Multi-Line Display feature. To resolve this dependency, we assign an execution order such that the Multi-Line Display feature is composed before the Recipe feature.

5 MATA Evaluation and Tool Support

To evaluate our software product line composition approach, we used the full microwave oven case study introduced in [6], which was developed using the PLUS product line engineering method. We specified the variant features using MATA in terms of how they impact the kernel class, sequence and state diagrams. Then, we analyzed the relationships among the features using CPA in order to determine the order of composition. We also injected a feature conflict intentionally to the case study since the case study did not already contain a conflict.

In the PLUS SPL method, requirements are captured through use case modeling and feature modeling. Use case variability is addressed by optional and alternative use cases, as well as through variation points, which are mapped to optional and alternative features in the feature model. Due to space limitations, a small subset of the features of the microware oven is covered in this section (Figure 13). A summary of the features in the original case study is shown in Table 1.

Table 1. Microwave case study

Complexity	Count
Kernel Features	1
Variant Features	19
Feature groups	4
Feature Relationships	14

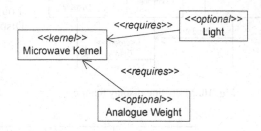

Fig. 13. Features of the microwave oven

We need to specify the kernel in UML and the variant features in MATA. Figures 14a and 14b show a specification of the Light and Analog Weight features using state, sequence, and class diagrams annotated with MATA directives. The state diagrams are for the Microwave Controller object.

In the kernel microwave oven, whenever items are placed inside the oven, the Weight Sensor Device informs the Weight Sensor interface object. The latter sends an Item Placed message to the Microwave Controller object. This message is critical for the Light and Analog Weight features.

Now consider the variant features. The Light feature requires that a weight indicator should be switched on when items are placed in the oven (Figure 14a). The Analog Weight feature, on the other hand, passes the weight of the placed item to the control object. This new logic is captured by replacing the Item Placed message with the Weight Change message (Figure 14b).

When CPA was applied on these models, it detected a conflict between Analog Weight and Light. The Analog Weight feature deletes the Item Placed message that is required by the Light feature. So, if we were to change the order of composition to (1)

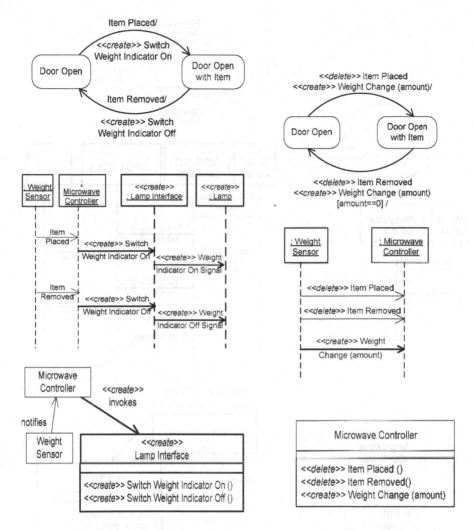

Fig. 14a. Light feature slice   **Fig. 14b.** Weight feature slice

Analog Weight and then (2) Light, the latter would not find the Item Placed message. With manual feature impact analysis, this simple conflict could easily be overlooked CPA automates the detection process and, therefore, it is an important step towards developing consistent models of features in software product line engineering.

To create a new member of the product line that includes both Light and Analog Weight features, the MATA tool composes the models automatically with the kernel. Figure 15a shows the kernel models before composition with the two variant features (in the order (2) then (1)), and Figure 15b shows the result after composition.

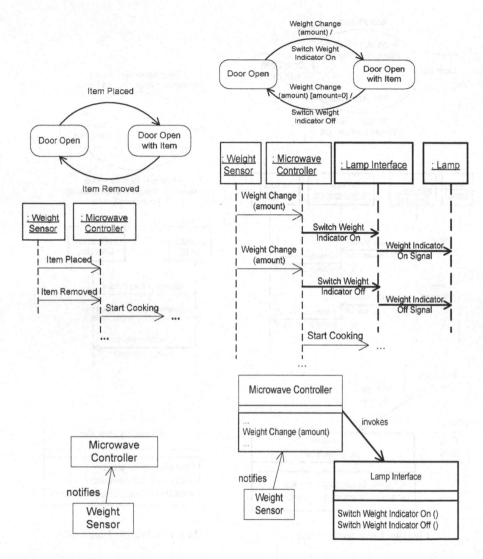

Fig. 15a. Kernel feature before composition

Fig. 15b. Kernel feature after composition

6   Related Work

Hyper/UML [19] is an attempt to bring the composition mechanisms defined for Hyper/J [20] to the modeling level. The composition specification language in Hyper/UML is much more limited than MATA. Modelers are restricted to a predefined set of four composition relations—merge, override, summary and order. Merge, for example, simply takes two models and combines identical elements. As shown in [21], the merge relation is not sufficient to define practical compositions.

Ziadi and Jézéquel [22] describe an approach that addresses composition of static and behavioral models for product lines. Static models of features are restricted in that they must be implemented as subclasses of an abstract kernel class. Feature relationships are enforced on the static models using OCL. Composition for interaction models is by using sequential (seq), alternate (alt) and looping (loop) composition operators. Since only a predefined number of operators is allowed, the approach is not as expressive as in this paper, where the full behavioral model syntax is available for composition. A significant issue not addressed by Ziadi and Jézéquel is the extension of features by other features. Our approach does allow this and also supports richer composition semantics like pattern matches, contextual creation and deletion of base elements. There is no inheritance hierarchy restriction for MATA features and the model composition is automated.

FMP2RSM [23] is an editor for specifying feature relationships and UML models for features. However, in FMP, features are all described on the same UML diagram and each model element is color-coded to show which feature it contributes to. MATA provides a cleaner separation of features and hence better reuse. However, a combination of these two approaches might prove beneficial.

Feature composition is very similar to aspect composition. There is a large body of work on aspect-oriented modeling, although much of this has been restricted to structural models. Work of note that considers behavioral models is the Motorola WEAVR tool for state machines [24], Song et al.'s work on weaving security aspects into sequence diagrams [25], and Jézéquel et al.'s work on semantic composition for interaction diagrams [26]. UML models of features are also very similar to use case slices [27] and themes [28].

AHEAD [29] considers feature interactions formally by using algebraic constructs. The approach follows the Principle of Uniformity in that all representations of a system are converted to a structured model such as object oriented code. AHEAD then applies techniques like aspectual mixins to refine code blocks and compose features. This approach is neither graphical nor does it maintain consistency between feature dependency diagrams and the models of features. It would be interesting, however, to see if our approach can be mapped to AHEAD.

The work described in this paper maintains consistency between a feature dependency diagram and the definitions of the features themselves. A significant amount of work has been done on detecting inconsistencies in UML models [13, 30]. In this research, an inconsistency arises when the same information is represented in different ways in different views. Our work instead looks at dependencies between models in the same view.

7 Future Work and Conclusion

In this paper, we presented an approach for maintaining feature separation throughout the modeling lifecycle, and an approach for detecting undesired structural interactions between models of different features. Kernel features are expressed in UML class diagrams, sequence diagrams and state diagrams. Variant features are specified in MATA, a UML representation of graph transformations, in terms of how they modify the kernel models. Features are composed automatically using the graph rewrite engine AGG and critical pair analysis is used to detect structural feature interactions. These interactions manifest themselves when the feature dependency diagram, developed during requirements engineering, is inconsistent with the UML models of the features. The entire approach is supported by a tool, which is a plug-in to IBM Rational Software Modeler.

As future work, we plan to implement code generators to map models of features to code. This could be done either by composing the features and generating code, or by mapping each feature independently to code in an aspect-oriented language. We are investigating the latter approach for a number of aspect-oriented languages.

We also plan to extend the MATA language itself by adding support for the specification of expressive patterns that would allow, for instance, a modeler to match against an arbitrary sequence of messages, or an arbitrary chain of states. Some other extensions such as element matches using name wildcards and ignoring the rule application for specific matches by a negative constraint are also being investigated. Currently critical pair analysis only checks interactions between models of the same type that belong to different features. A higher level abstraction that relates heterogeneous models needs to be investigated.

References

1. Weiss, D.M., Lai, C.T.R.: Software Product-Line Engineering: A Family-Based Software Development Process. Addison-Wesley, Reading (1999)
2. Parnas, D.: Designing Software for Ease of Extension and Contraction. IEEE Transactions on Software Engineering SE- 5(2), 128–138 (1979)
3. Clements, P., Northrop, L.: Software Product Lines: Practices and Patterns. Addison-Wesley, Reading (2001)
4. Kang, K., Cohen, S., Hess, J., Nowak, W., Peterson, S.: Feature-oriented domain analysis (FODA) feasibility study. Software Engineering Institute Technical Report CMU/SEI-90-TR-21. Carnegie Mellon University, Pittsburgh, PA (1990)
5. Gomaa, H.: A Software Modeling Odyssey: Designing Evolutionary Architecture-Centric Real-Time Systems and Product Lines. In: Nierstrasz, O., Whittle, J., Harel, D., Reggio, G. (eds.) MoDELS 2006. LNCS, vol. 4199, pp. 1–15. Springer, Heidelberg (2006)
6. Gomaa, H.: Designing Software Product Lines With UML: From Use Cases to Pattern-Based Software Architectures. Addison-Wesley, Reading (2005)
7. Pohl, K., Böckle, G., van der Linden, F.J.: Software Product Line Engineering: Foundations, Principles and Techniques. Springer, Heidelberg (2005)
8. Taentzer, G.: AGG: A Graph Transformation Environment for Modeling and Validation of Software. In: Pfaltz, J.L., Nagl, M., Böhlen, B. (eds.) AGTIVE 2003. LNCS, vol. 3062, pp. 446–453. Springer, Heidelberg (2004)
9. Heckel, R., Küster, J.M., Taentzer, G.: Confluence of Typed Attributed Graph Transformation Systems. In: Corradini, A., Ehrig, H., Kreowski, H.-J., Rozenberg, G. (eds.) ICGT 2002. LNCS, vol. 2505, pp. 161–176. Springer, Heidelberg (2002)
10. Gomaa, H., Saleh, M.: Feature Driven Dynamic Customization of Software Product Lines. In: Morisio, M. (ed.) ICSR 2006. LNCS, vol. 4039, pp. 58–72. Springer, Heidelberg (2006)
11. Gomaa, H., Kerschberg, L., Sugumaran, V., Bosch, C., Tavakoli, I., O'Hara, L.: A Knowledge-Based Software Engineering Environment for Reusable Software Requirements and Architectures. Automated Software Engineering 3(3-4), 285–307 (1996)

12. Gomaa, H., Shin, M.E.: Automated Software Product Line Engineering and Product Derivation. In: HICSS. 40th Annual Hawaii International Conference on System Sciences, Waikoloa, HI, p. 285 (2007)
13. Gomaa, H., Shin, M.E.: A Multiple-View Meta-Modeling Approach for Variability Management in Software Product Lines. In: Bosch, J., Krueger, C. (eds.) ICOIN 2004 and ICSR 2004. LNCS, vol. 3107, pp. 274–285. Springer, Heidelberg (2004)
14. Rozenberg, G.: A Handbook of Graph Grammars and Computing by Graph Transformation: Application Languages and Tools. World Scientific Publishing Company, Singapore (1997)
15. Nickel, U., Niere, J., Zuendorf, A.: The FUJABA Environment. In: International Conference on Software Engineering, Limerick, Ireland, pp. 742–745 (2000)
16. Agrawal, A.: Graph Rewriting And Transformation (GReAT): A Solution For The Model Integrated Computing (MIC) Bottleneck. In: ASE'03. 18th IEEE International Conference on Automated Software Engineering, Montréal, Quebec, Canada, pp. 364–368. IEEE Computer Society Press, Los Alamitos (2003)
17. OMG: MOF^{TM} Query / Views / Transformations Specification (2005)
18. Csertan, G., Huszerl, G., Majzik, I., Pap, Z., Pataricza, A., Varro, D.: VIATRA - visual automated transformations for formal verification and validation of UML models. In: Automated Software Engineering, 2002 (ASE), Edinburgh, UK, p. 267 (2002)
19. Philippow, I., Riebisch, M., Boellert, K.: The Hyper/UML approach for feature based software design. In: The 4th AOSD Modeling with UML Workshop, San Francisco, CA, USA (2003)
20. Ossher, H., Tarr, P.L.: Hyper/J: Multi-Dimensional Separation of Concerns for Java. In: International Conference on Software Engineering (ICSE), Limerick, Ireland, pp. 737–737 (2000)
21. Whittle, J., Moreira, A., Araújo, J., Jayaraman, P., Elkhodary, A., Rabbi, R.: An Expressive Aspect Composition Language for UML State Diagrams. In: MODELS. International Conference on Model Driven Engineering Languages and Systems, Nashville, TN (2007)
22. Ziadi, T., Jézéquel, J.-M.: Software Product Line Engineering with the UML: Deriving Products. In: Käköla, T., Duenas, J.C. (eds.) Software Product Lines, pp. 557–588. Springer, Heidelberg (2006)
23. Czarnecki, K., Antkiewicz, M.: Mapping Features to Models: A Template Approach Based on Superimposed Variants. In: Glück, R., Lowry, M. (eds.) GPCE 2005. LNCS, vol. 3676, pp. 422–437. Springer, Heidelberg (2005)
24. Cottenier, T., van den Berg, A., Elrad, T.: Motorola WEAVR: Model Weaving in a Large Industrial Context. In: Aspect-Oriented Software Development, Vancouver, Canada (2007)
25. Song, E., Reddy, R., France, R.B., Ray, I., Georg, G., Alexander, R.: Verifiable Composition of Access Control and Application Features. In: SACMAT. ACM Symposium on Access Control Models and Technologies, Stockholm, Sweden, pp. 120–129 (2005)
26. Klein, J., Helouet, L., Jézéquel, J.: Semantic-Based Weaving of Scenarios. In: Aspect-Oriented Software Development (AOSD), Vancouver, Canada, pp. 27–38 (2006)
27. Jacobson, I., Ng, P.-W.: Aspect Oriented Software Development with Use Cases. Addison Wesley, Reading (2004)
28. Clarke, S., Baniassad, E.: Aspect-Oriented Analysis and Design: The Theme Approach. Addison Wesley, Reading (2005)
29. Batory, D., Sarvela, J.N., Rauschmayer, A.: Scaling step-wise refinement. In: ICSE. 25th International Conference on Software Engineering, Portland, OR, pp. 187–197 (2003)
30. Egyed, A.: Instant consistency checking for the UML. In: International Conference on Software Engineering, pp. 381–390 (2006)

Automated Semantic Analysis of Design Models

Frank Weil, Brian Mastenbrook, David Nelson, Paul Dietz,
and Aswin van den Berg

Motorola, Schaumburg, IL 60196, USA
{frank.weil, brian.mastenbrook, david.c.nelson,
paul.f.dietz, aswin.vandenberg}@motorola.com

Abstract. Based on several years of experience in generating code from large SDL and UML models in the telecommunications domain, it has become apparent that model analysis must be used to augment more traditional validation and testing techniques. While model correctness is extremely important, the difficulty of use and non-scalability of most formal verification techniques when applied to large-scale design models renders them insufficient for most applications. We have also repeatedly seen that even the most complete test coverage fails to find many problems. In contrast, sophisticated model analysis techniques can be applied without human interaction to large-scale models. A discussion of the model analysis techniques and the model defects that they can detect is provided, along with some real-world examples of defects that have been caught.

1 Introduction

This paper summarizes several years of experience related to Model-Driven Engineering (MDE) in the context of developing large telecommunications applications. The experience covers the development of infrastructure components (e.g., cell site controllers) from the architecture and analysis phases though their field deployment. These projects used full automatic generation of C code from SDL and UML models and from protocol Domain Specific Languages (DSLs).

These models were developed by teams of between 10 and 30 engineers, and the size of the resulting code was from 300,000 to 600,000 lines of C code per application. The larger models contain over 300 signal definitions and a half-dozen state-machine-based processes (active classes). The system often has thousands of simultaneous active instances, each handling on the order of 100 different signals. Our code generation system has been used on dozens of such modelings projects. Through working on these projects and observing the types of model defects that are common, we have developed model analysis capabilities to catch these problems at the model level before code generation is even done.

Model correctness is fundamentally important to MDE. The central tenet of MDE—that the models are the basis for development—implies that the models must be correct. If they are not, then the principle of "garbage in, garbage out" applies. The effort spent in finding defects in the design phase when models

G. Engels et al. (Eds.): MoDELS 2007, LNCS 4735, pp. 166–180, 2007.
© Springer-Verlag Berlin Heidelberg 2007

are being created is paid back many-fold. The increased emphasis recently on network security, and the negative publicity generated when the security fails, provide additional reasons to exercise due diligence in finding defects.

Formal validation and verification are frequently cited as ways to ensure correctness. However, full formal verification is only possible on simple systems using simple properties, and even that requires some expertise in the underlying techniques (model checking, theorem proving, etc.). Testing is also used at both the modeling and implementation levels to uncover defects, but as is discussed below, this testing rarely uncovers more than the obvious problems with typical "sunny day" scenarios.

Model analysis provides an additional level of correctness checking to what can be realistically done with these verification techniques. Model analysis attempts to uncover model defects by looking for violations of the semantics of the underlying modeling language. While such analysis techniques are independent of what the model *should* do, they are quite fruitful at detecting situations where the model will do something that it *should not* do.

2   Model Development Environment

The "environment" in which the model is developed (i.e., the modeling domain, the semantics applied to close semantic variation points in the modeling language, the abstractions available in DSLs, etc.) provides a context in which to apply model analysis techniques. The domain in which the work described here was done is telecommunication systems. That is, the fundamental abstractions are protocol data units (PDUs) and communicating, asynchronous, finite state machines. Within Motorola, various languages such as ASN.1 are used for PDU descriptions, and the majority of MDE development is done using either SDL or UML plus the SDL Profile. The work in this paper is based on experience with the most commonly used modeling tools within Motorola, Telelogic's *SDL Suite* and *TAU*. However, virtually every commercial modeling tool is used within Motorola, and the discussions apply to these other tools as well.

The basic tool chain for MDE is shown in Figure 1. It is assumed that the modeling tool provides syntactic checking and some forms of simple static semantic checking, so there is no need for the model analysis tool to duplicate that capability. The model is first checked by the model analyzer for pathologies that can be found without "deep" analysis. These checks are done early to provide feedback to the user as quickly as possible and in a form that is most easily related to the model. The model is then passed to the code generator where, as a by-product of the sophisticated analysis required for code generation, additional pathologies can be uncovered. While the analysis here is more in-depth, it is also slower and more difficult to relate back to the original model. It is also in the code generation stage that defensive tactics can be inserted to account for potential problems (see Section 7).

The techniques used in the model analysis tools vary depending on whether they are done in the initial pass or as part of code generation. For initial

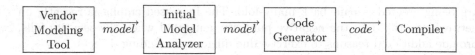

Fig. 1. Basic MDE Tool Chain Overview

analysis, the main technique is traversal of the model, applying rules that can be statically checked (e.g., the presence or absence of an element). Analysis done during code generation has access to the full power of the transformation engine. The major analytic algorithms are partial evaluation, constant folding, dead-code elimination, data-flow analysis, control-flow analysis, type propagation, def-use chaining, liveness analysis, alias analysis, side-effect analysis, and variable range propagation. All of these techniques are useful in finding different types of pathologies, and all of them are currently in use as part of the code generation system. For example, alias analysis is critical for enabling both performance and space optimizations, and as a by-product it can help identify uses of variables before they are defined.

2.1   Formal Verification of Models

Two techniques are commonly used to prove model correctness: model checking and automated theorem proving. Briefly, model checking exhaustively explores the state space of the model to verify that a property, usually written in temporal logic, holds in every state. Automated theorem proving is a technique for proving a mathematical statement (a *theorem*) about a model by automatically applying, possibly with some human direction, inductive or deductive steps to generate a proof of the truth of a theorem, usually stated in a higher-order logic.

There have been some tremendous advances and successes in these areas in recent years. For example, the VRS system [6] has found problems in requirements specifications for telecommunications components that had previously escaped formal inspections. The SLAM toolkit [2] statically analyzes sequential C code to check temporal safety properties of the interfaces, and has found several defects in Windows device drivers. The FLUID project [4] focuses on verifying non-local properties such as locking policies, type consistency, etc., in programs.

However, these techniques either have not been applied to large-scale design models in a language such as UML or SDL, or they are relatively specific in the types of pathologies they look for (or both). There are two main reasons for this. Firstly, the general techniques do not scale well to the size required, even when techniques such as supertrace exploration [5] are used to reduce the state space. Secondly, it is beyond the expertise of most developers to write the mathematically formal statements of "correctness" for general requirements. Formal verification can be useful in safety-critical systems where the cost of finding and removing even a single defect can be dwarfed by the cost of failure, but these systems are not typical, especially in the telecommunications domain.

2.2   System Testing

Our experience has shown that testing done on the models during the design phase is typically done to various levels of completeness, and that the more model testing that is done, the more defects that are found. However, this testing suffers from four main drawbacks.

Firstly, truly exhaustive testing is not possible. Even 100% coverage (of the paths, inputs, symbols, states, ...) does not mean that the model is being tested for all possible combinations of circumstances.

Secondly, test cases are often created by the same developers that are creating the model, and these test cases are based on the same requirements specification from which the model was created. At best, this ensures that the model will generate execution traces that correspond to the original Sequence Diagrams, but these traces are only a small fraction of the complete behavior of the system.

Thirdly, test cases and Sequence Diagrams tend to focus heavily on the correct behavior of the system. The exceptional behavior of the system, the so-called "rainy-day" scenarios, are usually added as an afterthought or provided to give a few examples of what should happen in certain undesirable scenarios.

Finally, the test cases tend not to exercise the system's boundary conditions, and the execution of a test case that encounters a defect may not trigger detectably bad behavior. For example, the implementation languages that are prevalent in telecommunications network components, C and C++, have little in the way of built-in protection against memory corruption. Having an off-by-one defect that causes an array to be indexed at −1 in C is not illegal, and the memory accessed may not be in use when the scenario is executed. However, the more loaded a system is, the more likely it is that the memory in question *will* be in use, and the more likely it is that the defect will cause a system fault. The most likely time for these faults to occur, then, is the worst possible time for them to occur in practice. We have seen this specific off-by-one defect bring down a cell site in a major urban center at peak system load, even though a test case exercised this situation.

We are not implying here that model testing is not useful. We have seen that having a comprehensive and automated regression test suite for a model lowers the overall development effort. When using this approach, product development groups have seen up to a 30x reduction in the time required to fix a defect, partly due to the fact that the defect is demonstrably fixed and that it very rarely introduces any new defects.

3   Model Analysis

Given the shortcomings of formal model verification and model testing, it is beneficial to apply additional techniques of determining the correctness of a model. This is especially true if these techniques are inexpensive and easy to apply and are unobtrusive to the development process. The technique discussed in the remainder of this paper is *model analysis*. Model analysis is the process of applying semantic rules that look for situations that either are semantically incorrect (e.g., they represent a type conflict) or are semantically suspect although not technically incorrect (e.g., silently discarding a received signal).

There are two assumptions being made here. Firstly, it is not relevant to talk about a model being semantically correct if it is not syntactically correct. Secondly, many violations of the static semantics are trivially easy to detect. For example, the declaration and initialization shown below are not semantically correct for the usual definition of `Integer`:

```
i : Integer = "this is not an integer";
```

Most UML and SDL modeling tools will detect syntax errors and simple semantic errors, and this aspect of model analysis will not be discussed further.

Model analysis requires the existence of sophisticated analysis tools, but the applied techniques tend to scale well. Most of these techniques are linear or better in the size of the model, *not* the size of its state space. There are some algorithms, such as forward analysis, that are almost quadratic in the depth of looping constructs in a given execution path, but this number is very small in practice. For those rare cases when it is not and the quadratic time begins to dominate, the time can still be limited in practice by truncating the iterations with very little effect on the results.

Many of the pathologies listed here are not specific to modeling and are familiar to compiler developers [1,3]. We present all of the pathologies listed here in the context of model analysis. It should also be noted that it is not sufficient to delay finding these problems until a compiler possibly detects them in the generated code. Firstly, some problems have no equivalent in the target language (e.g., non-deterministic signal sends). Secondly, compilers will generate messages related to the generated code, not to the model. Ideally, the modeler should never have to try to understand the generated code. Finally, the semantics of the modeling language and the target language may be different, and this is especially a problem when the target language semantics are looser than those of the modeling language. For example, C has different semantics than SDL for type compatibility, unions, and case-statement fall-through.

3.1   Model Pathologies Overview

Both static semantic errors and dynamic semantic errors can be detected through model analysis. Static semantic errors are those that can be detected through analysis of the structure and contents of the model without the need for symbolic execution. Dynamic semantic errors are those that cannot be caught through static evaluation. In practice, there are some cases of "dynamic" semantic errors that can be caught by static analysis. For example, detecting a divide-by-zero error cannot be done statically in the general case. However, constant folding and partial evaluation of constant expressions makes it possible to detect many of these errors at code generation time.

While one might be tempted to think that such "obvious" cases are rare, they are surprisingly common. We have seen this happen when a developer forgets to fill in a constant value that was set to a default value during initial development, and also when a developer changes one part of a model but does not anticipate all the consequences in another part of the model. Just as an experienced speaker of English will occasionally construct a grammatically incorrect sentence, even an experienced modeler will occasionally construct a model with semantic errors.

Another perspective of the capability that model analysis provides is that, even though it is a requirements-agnostic technique, it can find problems in the translation of requirements to design. Technically correct but semantically suspect model behaviors almost always indicate an incorrect translation of the requirements into the design model. For example, it is hard to imagine the requirement that corresponds to the sending of an internal signal that cannot possibly be received. As such, it is unlikely that this behavior was intended by the modeler. In practice, this type of problem can be caught during static semantic analysis and should be reported as a warning, whereas the semantic errors should be reported as defects.

When model analysis uncovers a problem or a potential problem with the model, reporting the issue back to the developer with sufficient detail to locate and understand the concern has proved to be generally sufficient. The model analysis needs to use the context of the problem to determine its severity and to show the implications of the problem. For example, consider what could happen when a violation is detected of an integer range (e.g., having a value of 6 in a context where the range is expected to be 1 to 5). In the context of a conditional comparison, it may or may not have any immediate consequences since the comparison will yield a value and execution will continue normally. In the context of a loop index check, it may lead to an infinite loop since it may result in the check part of the loop being optimized away. In the context of an array bounds, it is a violation of the underlying semantics of the data type and represents an incorrect capture of the requirements.

From experience, we have found that one additional step is helpful beyond a warning message to the user. It is an invaluable aid to debugging to put a marker in the code (e.g., an **assert** or a special comment) that indicates the suspicious condition that occurred and the potentially unintended consequences. For example, if a branch of a decision is being optimized away due to constant folding in the condition, then add a comment at that place indicating that the branch was optimized away and why. When the consequences of that code no longer being present manifests itself, it is at least immediately apparent why.

The pathologies described in the following sections are divided into three categories: structural and communications pathologies, domain pathologies, and realizability pathologies. Each will be discussed with examples.

4   Structural and Communications Pathologies

The structural and communications pathologies are those that depend on how the components of the model are put together and the communication paths that that structure implies.

4.1   Sending an Unreceivable Signal

Signals should be sent only to instances that exist. That is, it should always be true that when a signal is sent to an instance, either the instance must exist due to constraints in the model, or the logic of the model must ensure that if the receiving instance does not exist, the signal is not sent. From the point of view of clarity of purpose, the modeler should not rely on the semantics of the modeling language to silently ignore certain behavior. Otherwise, there will be no way to distinguish between intentional uses and unintentional ones. In practice, however, it can be very difficult to statically prove that a signal sent to a dynamic instance will be received. Typically, one must resort to the scheme described in Section 7.

4.2   Ignored Behavior

Behavior that is explicitly present in a model should rarely be ignored. For example, it is possible that the structure of a model allows an instance to send a signal to another instance even if the receiving instance does not explicitly handle the signal. Although the default behavior may be to silently discard the signal, this situation typically indicates behavior that was overlooked. It is usually quite straightforward to add an explicit discard of the signal to the behavior of the receiving instance, and it makes it clear to users of the model that that behavior is what was intended. The model analysis should check that every state in a state machine explicitly deals with every signal listed in the associated input ports.

4.3   Unused Elements

Any declaration of an element that is never used should be detected and warned about. Typical elements in this category include signals, signal lists, variables and attributes, operations and operators, classes, exported variables that are never imported, and external functions. These declarations are often an artifact of early development phases that were later not needed. Whether or not an unused element really indicates the presence of a problem does depend on context in some cases. For example, a local variable that is unused in an operator indicates an inconsistency, but a signal listed in a provided interface may simply indicate unused behavior in a reused component. However, it is still best to issue a warning because, to continue the previous example, the system may also use a signal in the component's required interface that is only generated when the unused provided signal is received, again indicating an inconsistency.

4.4   Statically Unreachable Behavior

While it is possible to have behavior that is in itself unreachable (e.g., a state with no transition into it), the more typical case involves behavior that is associated with paths of a branch point that cannot be executed. That is, static analysis can show that the possible values of a branch decision do not allow some branches ever to be taken. This includes loop termination expressions that can be shown to be always true (indicating that the loop will never execute) or always false (indicating an infinite loop). Behavior that can in reality never be executed but is included in a model typically indicates an undesired consequence of some related portion of the model. The more precise the reason can be given that the behavior is unreachable, the easier it is to determine the root cause of the problem.

4.5 Nondeterminism

Most modeling languages contain one or more nondeterministic features. One of the guiding principles of good design is that nondeterminism should be avoided since nondeterminism does not imply randomness or fairness. This distinction is extremely important, but is one that is often misunderstood by modelers. Nondeterminism means that there is no criteria for determining which choice will be made. A perfectly valid interpretation is that one choice is *always* made to the exclusion of the others. It will rarely, if ever, be the case that this is really the desired behavior.

Typical nondeterministic behavior that can be detected through model analysis is described here.

Nondeterministic signal sends: There is no issue if a signal send uses direct addressing to indicate the receiver of the signal (i.e., to a process ID in SDL or using the `receiver.signal` notation in UML). However, this generally limits the reusability of a component since it makes the behavior of the instance context dependent. Without using direct addressing, there are cases when a unique destination instance for a signal send cannot be determined, either because the signal is provided on multiple ports or because more than one instance that requires the signal is connected to the providing port. In either case, if the structure of the model does not enforce a single receiver, then the modeler should be warned that the behavior of the model is nondeterministic, with all possible receivers being listed as part of the warning. Note that this behavior is a semantic variation point in the UML specification, so UML tools are free to choose whatever behavior they want for this, which in itself is a problem for understandability.

Spontaneous transitions: Some modeling languages allow a transition out of a state in a state machine to happen without an associated event. For example, SDL has the concept of a spontaneous transition, which indicates a transition that can occur at any time, whether or not there is a relevant signal on the input port of the instance. There is no implied priority between normal transitions and spontaneous transitions.

Nondeterministic values: Some modeling languages allow an unspecified and arbitrary value of a data type to be indicated. For example, SDL allows the use of the keyword **any**. As before, nondeterministic does not mean random. If a random value is needed, an operation can be written to return one (for some suitable definition of random). If the value is truly not important but it must be set in the model, it is best for the modeler to pick a valid value and use it by documented convention.

4.6 Informal Text

Modeling languages often provide a way to indicate behavior that is not part of the semantics of the language itself. Instead, the text must be interpreted in the semantics of some other language such as C++. This construct is given various names such as *informal text* or *uninterpreted strings* and is distinct from comments, which do not specify behavior. While it could be argued that it is sometimes a necessity to drop into another language in a model, our experience has shown that there is almost always a way to specify the desired behavior within the modeling language itself. Informal text specifies behavior that is totally dependent on the context in which the model is executed and on the usually unspecified interplay between the semantics of the two languages.

4.7 Use Before Definition

A data value that is accessed before it is given a value yields undefined results, possibly even a value that is not legal. For example, assume that the model defines variables u and v to be Integers in the range of 1 to 10. If u is not given a value—that is, it is undefined—but it is assigned to v, there is no guarantee that the garbage in the memory location for u in an implementation will correspond to an integer value between 1 and 10. This discussion applies not only to scalar variables, but also to array elements, structure fields, etc.

5 Domain Pathologies

The domain pathologies are those that violate some inherent property of an underlying semantic domain such as arithmetic operations, integer ranges, data type properties, etc.

5.1 Underconstrained or Overconstrained Branch Points

Whenever a branch point (e.g., a static branch point in UML or a `decision` statement in SDL) exists in the model, the potential for a problem exists. The conditions covering the branches must cover *all* of the possible values of the associated decision, and multiple branches cannot cover the same value. For example, a branch point may cover all the values of an enumerated type. If a new value is later added to the type, this new case may not have been added to all branch points, and execution of the branch point leads to undefined behavior. This underconstrained case can be readily detected, though, by comparing the values of the decision's type with the union of all the conditions of the branches. The overconstrained case can be easily detected by intersecting the conditions on the branches. If the intersection is not empty, then there is a problem.

In one example we have seen, a type with three possible values (indicating the state of an environment actor) was used in a routine where only two of the three values were expected, and naturally this routine contained an underconstrained branch point. When the requirements of the model changed, the third value was passed to this routine and caused abnormal execution during system testing. If the third value was originally handled in the branch point by raising an error, the defect could have been traced to the incorrect routine without debugging.

The same discussion of overconstrained choices also holds for guard conditions on a state transition for the same input signal.

5.2   Real Number Equality

Modeling languages typically do not specify a precision for real (floating-point) numbers, but it should not be assumed by the modeler that it is infinite. For example, the comparison $(1.0/3.0) * 3.0 == 1.0$ may or may not be true in an underlying implementation. Uses of direct comparison operations for real numbers, including $<$, $>$, $<=$, etc., can be easily detected during model analysis.

5.3   Incompatible Subranges

Modeling languages allow data types to be based on a given type but with a restricted range of values. This is a very valuable feature in that it makes the intent more clear, it allows potentially infinite data types (such as Integers) to be implemented, and it enables additional model analysis checks. Anywhere that subranges are used, analysis can check that the actual values are within the defined range. Any attempt to assign a value out of the assigned range is an error, including for implicit assignments such as parameter values. Range checking can also detect situations that result in dead code such as is described above. In the context of iteration, these "fencepost errors" can result in an infinite loop due to a comparison to a constant which is always less than or greater than the index variable for the iteration.

One manifestation of this issue which we commonly see in models is a conditional statement which attempts to detect an exceptional situation and raise an error condition. These conditionals often compare a value with a defined range against a value which is outside of this range. These comparisons are never true, and the error recovery code can never be invoked. By definition, these errors will manifest themselves when the system is most heavily loaded, and failure to respond properly to the situation can create a catastrophic system failure.

5.4   Out-of-Bounds Indexes

One of the most common sources of field failures (i.e., system crashes) that we have seen involves either accessing or assigning to an element past the defined bounds of the range of an indexed type (array, string, etc.). As was previously stated, these errors can be extremely difficult to debug because the effects may not be local to the error. For example, if an array is defined to have 10 elements, then assigning a value to the "11th element" will corrupt memory.

A surprisingly large number of errors come from off-by-one mistakes in indexes, especially those caused by confusion on whether indexing starts at zero or one. In one example, we have observed a model where an array is indexed from the initial value 3, and a value of 0 was passed to a function for use as an index into this array. This situation led to heap corruption of a different array which happened to precede the indexed array in memory. Since the eventual application failure occurred long after the heap corruption, the modeler was unable to determine the cause of the fault through ordinary debugging techniques.

When an index expression is not a simple variable of an appropriately ranged type, the analysis should attempt to prove that the expression value *must* be in the correct range. If it cannot, it is critical to provide detailed feedback to the modeler. While not all index uses can be proved to be correct, it is better to have a false positive here than to have a false negative. When proof is impossible, the scheme described in Section 7 can be applied.

5.5 Dereferencing Null

Similar to out-of-bounds indexes, attempting to dereference (get the contents of) a Null pointer can cause runtime failures. The model analyzer should warn whenever a dereference operation occurs and the analyzer cannot prove that the value being dereferenced has been initialized and is not Null.

5.6 Invalid Union Access

Modeling languages often have the notion of a union of data types (e.g., the union property in UML and the choice data type in SDL). This type should not be used in the same way as a union is typically used in C; that is, as an implicit casting operation. The selection of the contained type should be explicit in the model. That is, one should not write a value of one type and read it back out into a variable of another type. Model analysis should try to determine that the type of the value being accessed matches the type of the value written. In other words, the union should be thought of as a *tagged union*, where the tag indicates the type of the latest value written to the union. A problem occurs with assignments that do not respect the tag. If this condition is not checked for and enforced, subtly unexpected results can occur. For example, consider the case where a union is composed of two structure types. If an assignment is made to a field of the first structure and a following assignment is made to a field of the second structure, the contents of the union as a whole are undefined. This can cause consequences ranging from a memory leak to heap corruption. This defect has been present in the majority of models we have seen.

5.7 Invalid Operations on a Type

Certain operations are invalid according to the constraints of the input types, such as division by zero. This concept—detecting obviously invalid operations on a given data type—can be extended to all data types. Model analysis can often detect when operations are inconsistent with the size of a string or other collection. For instance, we have seen models where a string is declared to hold at least some minimum number of elements, but values containing fewer elements than the minimum are used within the model.

6   Realizability Pathologies

The realizability pathologies are related to needing to create a design model that corresponds to limitations from the real world, such as how much memory is available in the system and what should be done when it is exhausted.

6.1   Unbounded Number of Instances

No implemented system can have an unbounded number of instances of any active class. It is better from a design perspective to limit the number of instances to some realizable number and then explicitly handle the case when the maximum is reached. If this is not done, the implemented system may fail in unexpected ways (e.g., exhausting available memory, reaching the limit of the number of simultaneous threads, etc.).

6.2   Infinite or Unrealizably Large Data Structures

It is possible to specify a data structure in a design model that cannot be implemented due to its sheer size. For example, one could specify an array whose keys (domain values) are Integers. However, even if the Integer values are restricted to 32 bits (and most modeling languages do not impose such explicit constraints), such an array could contain 2^{32} elements. As with unbounded process instances, it is better to specify a realistic maximum and then to explicitly handle the limit case. The definition of "overly large" is open to interpretation here, but the more obvious cases such as specifying `Integer` can be easily detected. As a representation of the requirements, it would be very unusual for a real system to have requirements that say that one must allow this large a number of elements.

7   When Model Analysis Is Not Sufficient

As has been mentioned in previous sections, it is not always possible to prove the absence of the above semantic issues. Two approaches can be taken in such a case. Both options are essentially equivalent in that they have the same end result. They differ only in their scalability.

The general scheme is to have safety code inserted in the resulting code at every point where the absence of an issue cannot be proved by the model analysis. While this technique will increase the size of the generated code, allowing incorrect and potentially catastrophic behavior to happen is not a valid trade-off. The two techniques to insert this safety code into the generated code are:

1. Insert safety code at the location of every potential issue prior to analysis. In those cases where this code was not necessary, model analysis will indicate that it can be safely removed again. That is, the failure branch can be proved to be dead code and can therefore be eliminated. Any safety code left after model analysis indicates a real issue, or at least one that cannot be shown to be not an issue.

2. Perform model analysis before inserting any safety code. At any point where a warning would be generated by the model analysis, safety code must be inserted.

In practice, the second option is somewhat more efficient in terms of the model analysis itself. However, depending on how integrated the code generation and the model analysis are, one approach may be more straightforward to implement than the other.

The following example illustrates the first scheme. In this assignment statement, assume that my_array is an array defined over a subrange of Integers, and that index is a variable declared to be of type Integer:

```
my_array[index] = 3;
```

The model analysis must prove that the value that index has at the time of the assignment is within the valid subrange used for indexing the array. The following safety code can be inserted around the array access:

```
if ((index >= LOWEST_INDEX) and (index <= HIGHEST_INDEX))
    my_array[index] = 3;
else
    error();
```

Appropriate values of LOWEST_INDEX and HIGHEST_INDEX can easily be obtained from the declaration of my_array. If the model analysis shows that the else branch is dead code, then the check can be optimized away again since there can be no indexing problem. The second scheme would simply flag the array access as suspect if the above condition could not be statically proved to evaluate to true, and the check could be inserted after the fact.

This general scheme, however, leaves an interesting open question: What should the definition of error() be? A context-independent answer would be that anything more ordered than memory corruption would be an improvement. From this point of view, the error functionality could simply log the problem and then shut the system down gracefully.

From the viewpoint of the system developer, however, perhaps a better answer is that the error functionality should depend on the context. For example, accessing an array value out of the bounds of the array may be a fault worthy of causing the system to shut down, but assigning to an integer variable a value out of its declared subrange may not. One may be tempted, therefore, to insert error indicators that differ based on the type of fault: index_error(), subrange_error(), etc.

In practice, the context of an error goes beyond the structure of a model to its purpose, so this scheme is also inadequate. For example, in a telecommunications network controller, dropping a single call is undesirable but acceptable; bringing down an entire cell site is not acceptable. Functionality such as is required by index_error() will need to differ based on where it is in a model. In an instance of an active class handling a single call, killing the instance may be appropriate.

In an instance of an active class that is the call handler for several thousand calls, more strenuous efforts at error recovery would probably need to be attempted.

8 Summary

The model analysis techniques described in this paper have been successfully deployed in an industrial context. Results of applying these analysis techniques to large models of telecommunications components indicate that a typical model will result in hundreds of potential problems being flagged. These issues are discovered *after* the model passes the analyzers from commercial modeling tools with no errors and no warnings.

The model analyzers have been applied to SDL and UML models from more than a dozen development teams, encompassing more than 50 production models. While it is difficult to give an exact measure of the size of these models, they correspond to a total of a few tens of millions of lines of generated C code. In analyzing these models, we have found that a typical model has four to six major semantic errors (e.g., an array access out of bounds). Once these errors have been fixed, a model will generally still have an average of about 1200 warnings, ranging from just over 100 to more than 4200 per model. About 75% of the total warning involve violations of range bounds (e.g., assigning 0 to a variable declared to range from 1 to 9). Many of these subrange warnings relate to assumed external constraints, such as when two protocols specify different value ranges, but the developer "knows" that values will be compatible due to system usage. This assumption, however, can lead to defects that are very difficult to diagnose when assumption change in subsequent versions of the system. The next biggest class of warnings, accounting for about 20% of the total, involves assuming that integers can be arbitrarily large, which, depending on the target platform, can lead to a system that is impossible to implement or unacceptibly slow. Surprisingly, the third most common type of warning, accounting for about 2% of the total, involves decision points (`if` statements or `switch` statements). Of these, 80% are under-constrained branch points, and 20% are statically computable Boolean conditions in `if` statements (which also implies dead code in most cases).

Approximately 10% to 40% of the uncovered issues correspond to problems that have caused a field failure, and virtually all of them have been discovered in deployed or soon-to-be-deployed telecommunications products. Many of the model pathologies discussed in this paper, and the techniques for finding them, have been driven by joint post-mortem analysis by the development team and the analysis team of actual field failures.

The importance of applying automated checks to uncover human error cannot be overestimated. In a large and complex model, even simple cut-and-paste errors can make it through formal inspections and rigorous testing. For example, the authors have seen a field failure that caused a cellular network control node to crash. The root cause of the problem was memory corruption caused by assigning to an array out of bounds. This access happened because the modeler cut-and-pasted similar code but forgot to change the name of the index variable in one place. This incorrect index variable had a largely overlapping subrange with the correct variable, and, due to the long and similar names of the variables, no one noticed it during inspections. Testing failed to trigger the case that caused the crash, but several weeks of sitting in a live network in a heavily used market did. The extra effort for a developer to read through and act on a detailed analysis report to fix this one warning before going to the field, even if this were several staff-days of effort, would have been far less costly than having the field failure. This is the reason that it is important to warn about situations that may not be what was intended (such as silently discarding a signal) and to appropriately change the model to address them.

The model analysis described in this paper has proved to be extremely valuable. These techniques can also be easily extended to apply local "good modeling practices", such as the use of named constants instead of literal values, naming conventions, etc. Because these model analysis techniques do not require specialized knowledge such as temporal logic, they can be effectively deployed across large organizations and incorporated into current development processes with a minimum of overhead.

References

1. Aho, A., Sethi, R., Ullman, J.: Compilers: Principles, Techniques, and Tools. Addison-Wesley, Reading, MA (1986)
2. Ball, T., Rajamani, S.: The SLAM Project: Debugging System Software via Static Analysis. In: Proceedings of POPL 2002, pp. 1–3 (2002)
3. Free Software Foundation, GNU Organization, GCC Online Documentation, http://gcc.gnu.org/onlinedocs/gcc-4.2.0/gcc/Warning-Options.html
4. Greenhouse, A., Halloran, T., Scherlis, W.: Observations on the Assured Evolution of Concurrent Java Programs. Science of Computer Programming 58(3), 384–411 (2005)
5. Holzmann, G.: Design and Validation of Computer Protocols. Prentice-Hall, Englewood Cliffs, NJ (1991)
6. Letichevsky, A., Kapitonova, J., Letichevsky Jr., A., Volkov, V., Baranov, S., Weigert, T.: Basic Protocols, Message Sequence Charts, and the Verification of Requirements Specifications. Computer Networks 49(5), 661–675 (2005)

Piecewise Modelling with State Subtypes

Friedrich Steimann and Thomas Kühne

Lehrgebiet Programmiersysteme
Fernuniversität in Hagen
D-58084 Hagen
steimann@acm.org
Fachgebiet Metamodellierung
Technische Universität Darmstadt
D-64289 Darmstadt
kuehne@informatik.tu-darmstadt.de

Abstract. Models addressing both structure and behaviour of a system are usually quite complex. Much of the complexity is caused by the necessity to distinguish between different cases, such as legal vs. illegal constellations of objects, typical vs. rare scenarios, and normal vs. exceptional flows of control. The result is an explosion of cases causing large and deeply nested case analyses. While those based on the kinds of objects involved can be tackled with standard dynamic dispatch, possibilities for differentiations based on the state of objects have not yet been considered for modelling. We show how the handling of class and state-induced distinctions can be unified under a common subtyping scheme, and how this scheme allows the simplification of models by splitting them into piecewise definitions. Using a running example, we demonstrate the potential of our approach and explain how it serves the consistent integration of static and dynamic specifications.

1 Introduction

In the age of model-driven development, models are required to specify both structure and behaviour of a system. Although there will always be the case for coarse abstractions omitting many particulars, a large class of models must address the specification of detail. Such models have to deal with considerable complexity introduced by the necessity to distinguish between many cases of alternative collaboration and control flow. Particular problems present cases of undefinedness, i.e., illegal constellations of objects or method invocations with which no reasonable behaviour can be associated. Excluding such cases through many explicit conditions disrupts the primary concern of a model, the depiction of what is right and what should be done. As a result, models are more difficult to write, to read, and maintain than they should be.

In this paper, we suggest to use subtyping and overloading of relations (where relations include associations, attributes, and operations) as a means for structuring a domain into defined and undefined cases, and to provide modular, piecewise definitions for the defined cases. Our approach shares some similarity with multi-dispatching known from programming languages, but significantly extends it with the possibility

G. Engels et al. (Eds.): MoDELS 2007, LNCS 4735, pp. 181–195, 2007.
© Springer-Verlag Berlin Heidelberg 2007

to take the dynamic state of objects into account. By doing so, our approach not only allows simpler models, it also improves the integration of static and dynamic specifications that were previously thought to be rather isolated.

In the remainder of this paper we first we show how the declaration of relations is used as an essential ingredient of type-level specifications, and how systematic overloading can be used to refine the information conveyed (Section 2). Following the concept of multi-dispatch, we show how attaching definitions to different (overloaded) branches of a declaration can eliminate the need for certain types of case distinctions (Section 3). We extend this idea to cover the state of objects, by introducing state subtypes and overloading relation declarations on these (Section 4). Furthermore, we show how state subtypes can be automatically derived from the statecharts associated with classes, and how in general statecharts can be integrated with structure and sequence diagrams. In Section 5, we elaborate process issues associated with our approach. We then present a discussion of the limitations of our approach and a comparison with related work (Section 6), and conclude with Section 7.

2 Modelling with Declarations

In object-oriented modelling, declarations are an accepted form of specification on the type level. For instance, an excerpt from a static structure (class) diagram such as

print

| Printer |————————————————| Document |

expresses that documents and printers can engage in a relation[1] named *print*. Its textual equivalent is the declaration of the signature of a relation:

$$print: Printer \times Document \tag{1}$$

Following the general understanding in programming (which suggests that the types in a declaration must be substitutable by all of their subtypes), such a declaration is usually (mis)interpreted as a statement of total definedness, i.e., it is assumed that printing is defined for all combinations of printers and documents. Looking at the problem domain more closely, however, one notices that there are such different entities as diagrams and texts, as well as line printers and plotters, and that texts are printed only on line printers, while diagrams are printed only on plotters. This refinement can be modelled by *overloading* the declaration of *print*, either graphically by

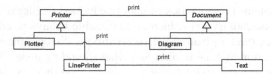

or textually by

$$print: Plotter \times Diagram$$
$$print: LinePrinter \times Text \tag{2}$$

---

[1] We deliberately speak of relations here and not of associations, since following [20] we take relations to cover associations, attributes, and methods.

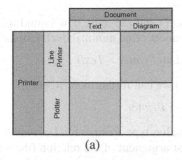 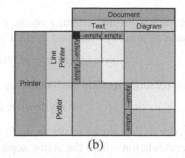

(a)   (b)

Fig. 1. (a) Definition holes (dark – or red – squares) induced by undefinedness of certain class combinations. (b) Additional holes imposed by dynamic conditions. The relation's domain is irregular and can only be declared piecewisely; besides, it cannot be declared using static classification (classes) alone.

However, there is no way to explicitly declare that

$$\neg print: LinePrinter \times Diagram \tag{3}$$

Instead, declaration (1) seems to warrant the printability of diagrams on line printers, a misunderstanding that results from the erroneous interpretation of *declarations* as *definitions*. In fact, in mathematics, a declaration such as (1) expresses that

$$print \subseteq Printer \times Document \tag{4}$$

i.e., that the extension of *print* is a subset of the Cartesian product of the domains of its places (which subset exactly typically being subject to further definition). The Cartesian product *Printer × Document* provides merely an upper bound, which is too high in our example since as mentioned above there are combinations of documents and printers that will *never* engage in the *print* relation. In fact, least upper bounds are usually non Cartesian, as illustrated by Fig. 1(a).

Moreover, in modelling and programming we have an additional temporal dimension, which means that the extension of *print* grows and shrinks with time. So how are the declarations (1) and (2) to be interpreted? Generally, we assume that more specific declarations (i.e., declarations of the same relation, but involving subtypes) are intended to overrule the more general ones. In fact, we go as far as requiring that only *minimal overloadings* of a declaration (i.e., overloadings such that there is no other overloading that involves only subtypes of the first) may have tuples. This allows us to define relations whose domain is not Cartesian, but rather a hypercube with (hypercubic) holes (*cf.* Fig. 1), with each tuple of the relation binding to a minimal overloading (not necessarily precisely one; *cf.* the discussion in Section 0). In Sections 0 and 0, we will attach specifications[2] to minimal overloadings and show how they represent *branches of a piecewise definition of the relation* within the model. One might be tempted to conclude that any tuple for which a minimal declaration exists is defined – however, as argued in Section 6.1 this is not necessarily the case.

---

[2] A more appropriate term would be "implementation", but this word is problematic in the context of modelling. "Definition" is another alternative; however, this would make our definition of definedness appear circular.

Returning to the meaning of the declarations (1) and (2): they now bound the extension of *print* by the union of the Cartesian products of its minimal declarations:

$$print \subseteq (Plotter \times Drawing) \cup (LinePrinter \times Text) \tag{5}$$

Since relations include attributes (see Footnote 1) we can furthermore write

$$myPrinter: Document \to Printer \tag{6}$$

in order to declare an attribute *myPrinter* with value type *Printer* for *Document* as a special relation where the arrow separates the last argument of the relation (the value) from the rest. Again, this declaration must not be interpreted as statement of total definedness, since (6) can be overloaded as in (2), e.g., with the branches

$$myPrinter: Diagram \to Plotter$$
$$myPrinter: Text \to LinePrinter \tag{7}$$

defining the range of the attribute as being dependent on its domain (*cf.* the discussion of dependent types in Section 0).

3 Attaching Definitions to Declarations

To illustrate how piecewise definitions based on overloaded declarations can simplify modelling, we extend our printing example by first adding a print manager that prints documents on printers. Second, we let texts consist of pages and let them be printed page by page, in contrast to diagrams, which are printed on a single page each. Last but not least, a printer can run out of paper in which case all printing attempts fail (i.e., are undefined), the only exception being the printing of an empty text (i.e., a text that has no pages). Later, we will be confronted with a niggling user who is dissatisfied with the undefinedness of the out-of-paper situation; fortunately, thanks to our piecewise definition approach pleasing him will turn out to be easy.

The static structure of our domain is modelled by the class diagram in Fig. 2. *PrintManager* is a singleton that acts as a façade to the printing module. Its sole purpose is to accept printing requests and forward them to the printer. *Printer* and *Document* are linked by an association *print*, which represents the same *print* relation as expressed by the operation *print(Document)* in class *Printer*, but shows additionally how it is overloaded. Since *Printer* and *Document* are both abstract, the *print* relation must recruit its elements (tuples) from its concrete subtypes. Following our binding rules from Section 2, missing combinations (e.g., *Text × Plotter*) are undefined.

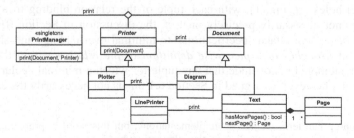

Fig. 2. Static structure of the printing example

The sequence of actions required to process a printing request is shown in the sequence diagram of Fig. 3. Although the overall behaviour is rather simple, the many case distinctions make the diagram appear complex. Note that the branches depend on the type of the arguments of the print relation (as expressed by the *instanceof* operator) and on the state of the individual involved objects (whether or not the printer is empty, whether or not the text has more pages). With growing detail in the modelled scenarios the number of special cases needed to be considered steadily increases, quickly leading to a combinatorial explosion. Since nested branches are known to be extremely error-prone (both in programming and in modelling), significant improvements can be expected from any modelling construct that does away with them.

This is where our overloaded declarations come into play. They allow us to split the sequence diagram of Fig. 3 into the two pieces shown in Fig. 4, one for each admissible combination of documents and printers. A message *print(doc, prn)* sent to a print manager *pm* is then bound to one of the two diagrams, or rejected as undefined. We thus have separated *defined* from *undefined* cases and provided *alternative definitions* depending on the *types* of arguments.

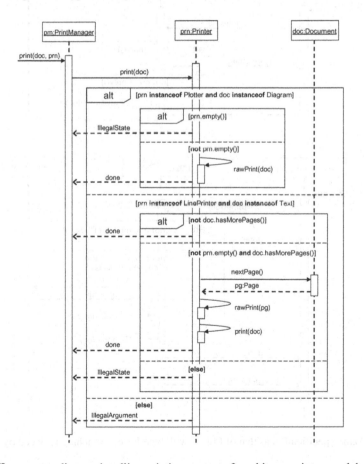

Fig. 3. UML sequence diagram handling printing requests for arbitrary printers and documents

4 Introduction of State Subtypes

The sequence diagrams of Fig. 4 still contains undesirable case analyses, but this time, the distinctions are induced by the *states* of the involved objects. If we could capture the states of the objects with corresponding (sub)types, we could use the same technique as before, namely overloading and piecewise definition, to further reduce the complexity of each sequence diagram. Assuming that we declare the state sub-types and the new minimal branches of *print* as shown in Fig. 5, we can extend Fig. 1(a) to 1(b) and replace the sequence diagrams of Fig. 4 with those of Fig. 6. Note that each object may only be in one state at a time; hence we have marked the classes that have state subtypes (*Plotter*, *LinePrinter*, and *Text*) as abstract.

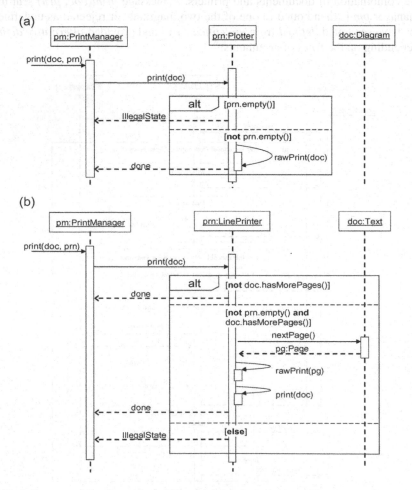

Fig. 4. Same specification as that of Fig. 2, with type-based branching replaced by binding

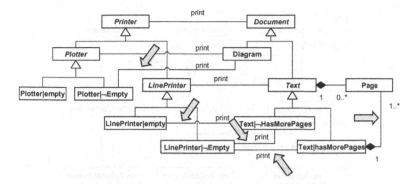

Fig. 5. Addition of state subtypes and corresponding overloadings (marked by arrows). Note the missing link from *Plotter|empty* to *Diagram*. Also, the multiplicity of the aggregation between *Text* and *Page* has been restricted for *Text|hasMorePages*, while *Text|¬hasMorePages* does not relate to *Page*.

If a state subtype does not engage in an overloading then this subtype contributes no elements to the relation. For instance, the absence of an overloading involving *Plotter|empty* and *Diagram* expresses that diagrams cannot be printed on empty plotters. Note that we use a vertical bar to create a state subtype's name from the name of the class it subtypes and the name of the state.

The diagram in Fig. 6 (c) raises an important issue. Because the state types of *doc* and *prn* may change after each printed page, the types specified in the lifelines heads may no longer be valid. However, subsequent relation tuples (i.e., method sends, including the recursive *print(prn, doc)*) will then *bind to branches based on the new state subtypes*. This allows us to elegantly model the printing of pages until either the printer is empty or all pages of the text have been printed, where all the control logic is implicit in the binding of message sends. Not inserting a new binding after possible state changes would require explicit tests and branching (in our example, a loop with explicit loop conditions), which are still possible but cumbersome.

Note that the change in state could be reflected by using a state invariant on the lifeline of the object, a feature of UML 2.0 [15]. As we will see below, possible state changes can be automatically derived from statecharts modelling the state transitions of objects.

4.1 Definition of State Subtypes

By definition, subtypes add to the intension of their supertypes: they add properties to and pose additional constraints on their elements. Hence, the extension of a subtype is always a subset of that of its supertypes, implying that relations that are only partially defined on supertypes may be totally defined on subtypes.

We define a *state subtype* as a subtype of a class that collects all objects of that class that are in a certain state. A state subtype adds to the intension of a class by restricting the range of attribute values, and by restricting the associations[3] and methods

---

[3]   Indeed, the fact that an object plays a certain role (sits in the place of a relationship) can be considered (part of) its state.

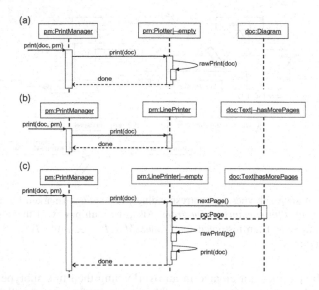

Fig. 6. Further simplification of behaviour specification made possible by the introduction of state subtypes. Note that Figs. 5 and 6 are equivalent to Figs. 2 and 3.

its objects can engage in. Although state subtypes need not generally be mutually exclusive, those that are (because they are generated from the same statechart; see below) provide a complete partition of the class's extension (*cf.* also Fig. 1(b)).

State subtypes allow piecewise total definitions of otherwise partial relations using subtyping and overloading, even if definedness depends on dynamic conditions. For instance, the relation *print* from above is totally defined for all objects of state subtype *Plotter|¬empty* (but not for all objects of class *Plotter*[4]). Like the subclasses from Fig. 2, state subtypes may serve to express definition holes by overloading relations.

4.2 Integration with Statecharts

One might argue that the static structure diagram of Fig. 5 has taken over much of the complexity removed from the sequence diagram of Fig. 3. Particularly larger examples may cause such structure diagrams to quickly become unwieldy. Fortunately, the latter can be automatically generated from much simpler diagrams.

The key to only specify a regular class diagram and obtain the additional information contained in Fig. 5 for free is to exploit supplementary statecharts. Fig. 7 shows

---

[4] At first glance, our notion of state subtypes seems to contradict the idea of subtyping, because objects of a (state-) subtype can no longer substitute for those of its supertype. However, we remind the reader that the (assumed) all-quantification of a declaration (cf. Section 2) is untenable anyway: a stack cannot be popped if it is empty, no matter what the declaration of the class promises. Since at any point in time every instance of a class is also an instance of one of its state subtypes, the notion of substitutability can only involve those properties that are independent of state.

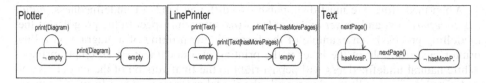

Fig. 7. Statecharts for classes *Plotter*, *LinePrinter*, and *Text*. States correspond to the state sub-types in Fig. 4, events correspond to the overloaded declarations in Fig. 4 and to the methods in Fig. 5; their absence determines which method is undefined for which state. Note that UML's semantics for statecharts is different: events for which no transitions are shown are ignored.

three simple statecharts, describing the behaviour of the objects of classes *Plotter*, *LinePrinter*, and *Text* respectively. Note that the states from each statechart partition the dynamic extension of the class it is associated with since each object of a class must be in exactly one state at a time. Furthermore, the events of Fig. 7 code the de-finedness of relations in each state: they correspond to the minimal overloaded decla-rations of Fig. 5 as well as to the method signatures found in Fig. 6. This allows us to *automatically* derive the state subtypes and overloadings of *print* in Fig. 5 from Fig. 7. In addition, the transitions of Fig. 7 specify which possible state changes need to be considered in a sequence diagram, e.g., that the state subtype of both a text and a line printer may change after printing each page (*cf.* section 4). Last but not least, the se-quence diagram specifies the interaction between objects, i.e., it binds the different statecharts together by showing which events in one statechart lead to which events in another (*print* of *LinePrinter* leads to *nextPage* of *Text*). It follows that state subtypes are natural pivotal points for the integration of static and dynamic specifications.

5   Process Issues

Model Evolution. Modelling is an iterative, incremental process, and the usability of modelling languages that do not support this development style is severely limited. Modelling with piecewise definitions based on subtypes and overloading supports in-cremental development rather well, as the following considerations suggest.

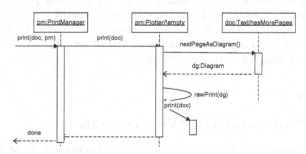

Fig. 8. Added behavior specification for printing texts on plotters. No interference with other specifications must be considered.

When modelling, we must distinguish two different kinds of undefinedness: *natural undefinedness* and *as-yet undefinedness* (*as-yet* with respect to the progress of the modelling process). An example of the former is the printing of a diagram on a line printer, and example of the latter is the printing of a document on an empty printer. While natural undefinedness will persist right to the final version of the model, as-yet undefinedness is usually removed while the model progresses: it is a form of "temporary omission". In terms of supporting model evolution, the question is how much re-arrangement of the model is required to remove such temporary omissions.

In the case of printing on an empty printer, the corresponding refactoring is trivial: all that needs to be added is a branch that lets the printer wait until paper is refilled; no interaction with other objects is required. Note how this added behaviour also fixes the undefinedness problem of a text having more pages than the printer has paper.

Also the decision that texts cannot be printed on plotters can be easily revoked. Fig. 8 shows the corresponding new collaboration which can be added without inter-fering with the rest of the model. These "low impact changes" are possible because we extend the definition of relations rather than that of single classes, thereby allow-ing *a form of refinement that is automatically coordinated among classes*. Our ap-proach thus fulfils the old promise of object-orientation, namely to be able to *refine models locally through subtyping*.

Tool Support. Just as programming today is unthinkable without programming environments, the quality of models is an increasing function of the quality of modelling tools used. However, this is only true if a modelling language allows tools to contribute to the quality of models. In particular, consistency and completeness checks should be supported; otherwise tools are reduced to mere drawing aids.

Fortunately, our framework offers a wealth of opportunities for tool support. Com-puting the coverage of minimal declarations of the domain spanned by the most gen-eral declaration of a relation shows potential definition holes, which can then be marked by the modeller as either natural or as as-yet undefined. Overlapping of minimal declarations can also be flagged: for instance, if default behaviour for all empty printers is added as suggested above, the resulting overlap with the specifica-tion for printing empty texts (*cf.* Fig. 1(b)) can be discovered. In fact, we envision a relationship browser that presents the definition of relations in a form similar to that of Fig. 1, collecting all piecewise definitions and allowing their quick access and edit-ing. Also, consistency of statecharts with associated structure and sequence diagrams can be automatically checked, and changes in one diagram may propagate directly to changes in the others. Last but no least, the state transitions of statecharts can be used to check whether possible state changes are adequately accounted for in the sequence diagrams. This may include the computation of sets of operations that are admissible in all possible post-states of a previous operation.

6   Discussion

6.1   Total and Partial Relations, and Error Propagation

At first glance it appears that our piecewise definition approach is capable of turning all partially defined relations into totally defined ones, by separating out the undefined cases. However, this need not always be the case: even though the minimal branch *print*: *LinePrinter*|¬*empty* × *Text*|*hasMorePages* appears to be OK, printing text on a non-empty line printer is in fact undefined if the text has more pages than the printer has sheets of paper (see Fig. 6). Unfortunately we cannot solve this problem by introducing a finite number of new state subtypes, since the number of pages is generally not limited. In fact, total definedness of a branch is granted if and only if

1. its specification does not rely on other branches, or
2. it involves only branches that are themselves totally defined.

In other words, the partiality of a branch automatically propagates to all specifications that depend on it. While this is a feature of our approach (because it frees the modeller from dealing with partiality explicitly), it is also a problem, because undefinedness can creep into seemingly innocuous parts of a model.

We address this problem by making definition holes explicit, for instance by adding exceptions to the signature of relations. These exceptions could be interpreted as the names of sets of tuples that are to be subtracted from the domain of a relation. For instance, we could write

$$print\colon LinePrinter|\neg empty \times Text|hasMorePages \setminus TooManyPages \qquad (8)$$

in order to denote that the set labelled *TooManyPages* must be subtracted from the domain of *print*. This set would contain all pairs of texts and printers of which the text has more pages than the printer has paper denotes; its specification could either be left implicit, or tied to the fact that the printing results in an empty printer with pages left to print. Relations that depend on *print* will inherit the exception, as for instance the ternary version of *print* that is associated with the print manager (see Fig. 3). We can unify this explicit form of error propagation with the implicit one (the inability to bind a tuple to a branch) since the latter can be automatically translated to explicit exceptions.

6.2 Open Problems

Although we could demonstrate some appealing properties, our approach certainly deserves further elaboration. For instance, in order to eliminate all explicit branches of a model, large numbers of small specifications will have to be provided. While this is considered good practice in object-oriented programming, it may lead to models that are difficult to trace (the negative impact dynamic binding may have on program understanding applies accordingly). Also, many of the opportunities for model integration suggested here depend on simple diagram languages. Given the complexity of UML's current statechart specification, it would be naïve to believe that integration with sequence diagrams can remain as simple as shown here. Another problem we did not touch on is that a single class can have several statecharts attached, and that the resulting state subtypes need not be unrelated, so that possible interdependencies between statecharts must be considered (for instance through attributes whose range is constrained by states of more than one statechart).

6.3 Possible Improvements

Many basic states such as *empty* occur over and over, as do the transitions linking them. Rather than specifying one statechart for each class separately, one could envision the definition of "abstract" statecharts that are "implemented" by various classes, each possibly adding variations. For instance, the statecharts for *LinePrinter* and *Plotter* in Fig. 7 are sufficiently similar to think about deriving them from a common generalization. However, a corresponding investigation is beyond the scope of this paper.

The notion of state subtypes and piecewise definition could also be extended to other diagram types, for instance activity diagrams. As the latter – together with sequence diagrams – are a popular means to formalize use cases, this immediately points out a path to piecewise definitions of use cases as well. Note that the "extends" relationship between use cases may already be considered as an existing way to piecewisely specify exceptional or alternative behaviour, clearly demonstrating the need for modularity in use case specifications.

Last but not least, an automatic translator of models into declarations of a statically typed language with (checked) exceptions (such as Java) could be devised. This translator may automatically add *illegal argument* and *illegal state* exceptions as shown in Fig. 3 to method signatures, and translate subtype information of overloaded declarations to preconditions firing these exceptions. For more sophisticly typed languages, the generation of dependent types should also prove to be a fairly simple exercise.

6.4 Related Work

Interestingly, the UML standard does not even attempt to specify binding rules: "The dispatching method by which a particular behaviour is associated with a given message depends on the higher-level formalism used and is not defined in the UML specification (i.e., it is a semantic variation point)." [15] It appears that specification of binding is deliberately left to the target programming language selected for the project. The consequences of such a policy have been discussed in [1]; we add that UML still lacks an explicit notion of overloading (cf. the discussion in [19]).

Avoiding case analyses caused by state-induced behaviour differentiation is also the goal of the *State* pattern [8]. Its application might be a viable alternative in the context of programming (e.g., with Java) but models should not contain explicit realization structures that are designed to address lacking language support. Applications of the *State* (addressing state-based dispatch) or the *Visitor* pattern (addressing multi-dispatch) would introduce realization structures to models which are not induced by an original to be described or a system to be specified.

Our approach has some similarity to order-sorted algebraic specifications [11], in particular the universal order-sorted algebras independently developed by Gogolla and others [10, 13]. This form of algebraic specification assumes that each operation symbol represents a single operation defined on a universe of individuals, with sorts corresponding to subsets of that universe. In this framework, overloaded operations whose domains overlap must be identical on the overlap (because they denote the same operation). This feature has particular appeal to us, since it saves the modeller from having to deal with such things as overriding and calls to *super*, programming constructs that are believed to increase the compactness of specifications, but really introduce a lot of problems. In fact, our requirement that specifications can only be attached to minimal declarations should be seen as a first step towards congruent relation specifications (as it can flag possible inconsistencies, *cf.* Section 0). Conditions required so that all tuples bind to one minimal declaration unambiguously (corresponding to the regularity of signatures simplifying the implementation of overloaded order-sorted algebras [10]) have not been formulated explicitly here, because they would have required a more formal exposition which we sacrificed in favour of a more readable description of our work.

Also in the context of algebraic specifications, Gogolla et al. have introduced the distinction between *unsafe* and *OK* functions, the former of which may lead to errors

[8]. Terms containing unsafe functions are themselves unsafe; in analogy to our relations that are specified in terms of other, partial relations. Interestingly, the partiality of operations and the propagation of errors remain hard problems for algebraic specifications [13]; given the analogies pointed out above, universal order-sorted algebras with exceptions can provide a nice formal semantics of our approach.

Shang shares our view that is inadequate to interpret declarations as statements generally all-quantified over their parameters [18]. For instance, a declaration *Animal.eat*(*Food*) (where *Animal* and *Food* are supertypes) should not be read as "all animals eat all food". He notes that what he calls *component types* (types of fields or of parameters to methods) are sometimes dependent on the type of the enclosing object (e.g., the specific kind of food is dependent on the specific kind of animal), and that declarations on (abstract) supertypes should explicitly express this dependency. Piecewise definitions of relations as suggested above cover the idea of dependent types, but are more general: in particular, they do not require a statement of which parameter type depends on which (which would be somewhat arbitrary in our printing example). Also, dependent types do not account for state-induced case analyses.

Castagna has argued the case for overloading by way of covariant redefinition and multiple dispatching in object-oriented programming [2]. In his λ&-calculus, he defines a message as a set of methods, where each method defines a branch of a function distinguished by the types of its parameters. Castagna's branches roughly correspond to our piecewise definitions, and his multi-dispatch to our binding rules. Castagna allows contravariance of method parameters to ensure substitutability for parameters on which no dispatching is desired; by contrast, we prefer to treat all parameters equally. Like Shang, Castagna does not consider dynamic typing.

Based on work by Ernst et al. [6], Millstein has extended Java with "predicate dispatch", i.e., a generalization of method dispatch that not only includes the types of parameters (as in multi-dispatch), but also potentially their values [12]. The emphasis in this work is on creating a modular static type system and the automatic detection of ambiguous method definitions. For this purpose, Millstein considers structural types, integers, and Boolean values; although dispatch on state subtypes could be emulated, such is not explicitly addressed.

Chambers introduced the concept of predicate classes (corresponding to our state subtypes) to the programming language CECIL and demonstrates their utility with a number of examples [3], thereby validating the concept in a programming language context. Our work differs from and goes beyond the work by Chambers by suggesting the applicability of state subtypes in the context of modelling and by using piecewise definitions for a number of diagram types ranging from sequence diagrams to use case diagrams. Working in a modelling context we can draw on the existence of statecharts associated to classes and *automatically* derive state subtypes from them, thus advancing the integration of static and dynamic diagram types. We furthermore show how to unify the treatment of class and state-based behaviour with undefined behaviour and hint at potential support by modelling tools.

Fickle$_{II}$ is a proposed extension to the static type system of object-oriented programming languages like Java that allows the type-safe, dynamic reclassification of aliased objects stored in temporary variables (including *this*) [5]. It uses state subtypes that extend so-called root types (classes) by adding and overriding members. The state type of an object can be changed at runtime, so that subsequent dispatches of same members on same variables may yield different results. Since different state subtypes of the same root can have different members, the contents of attributes that are not common to all state subtypes are dropped upon state change. In contrast, our state subtypes can only restrict the ranges of attributes and the applicability of operations.

Salzman and Aldrich have also suggested removing explicit branching by multi-dispatch based on the state of objects [17]. However, they use prototypes rather than state subtypes for this, and do not cater for undefined constellations as we do, but in-stead require that methods are provided for all possible parameter constellations, thus requiring explicit handling of undefinedness. Regarding natural "definition holes", their approach is not as expressive as dependent types, or our work.

Nierstrasz defines "regular types" by specifying (or at least approximating) their protocol through non-deterministic finite state machines [14]. Thanks to the existence of an equivalence test for this category of automata, he can check whether one type is a behavioural subtype of another type, with respect to their specified protocols. Although Niersrasz's work would be helpful for dealing with the inheritance of state-charts (cf. Section 0), unfortunately we cannot directly draw on it since we must as-sume more powerful automata types in general.

Similar to Nierstrasz, Paech and Rumpe specify rules for behavioural subtyping re-lationships but see their relevance predominantly in refining existing types in an in-cremental model development process [16]. Similar to our discussion of covariantly redefined attribute values in Section 0, Paech and Rumpe also constrain the attribute values of objects, depending on which state (of an automaton) the object is in.

DeLine and Fähndrich introduce type states to object-oriented programming in or-der to make statements about object states through the help of a modular, static type system [4]. They use type states in pre- and postconditions, for instance, to guarantee the error free execution of methods. As a result, state transitions are scattered over methods and – in contrast to our explicit state diagrams – the corresponding state transition graph is only implicitly defined. DeLine and Fähndrich thoroughly investi-gate state subtypes in the context of inheritance between classes, but do not consider the extension of dynamic binding to include the state (-types) of objects.

The type system of ALLOY [7] computes for every model two kinds of types, called bounding types and relevance types. Bounding types restrict the possible set of values of expressions from above, whereas relevance types approximate the sets of elements that make a difference in the evaluation of an expression in a given context; contrary to our approach, they are a derived, rather than a declared, property. Empty relevance types flag modelling errors (since no object can make a difference). Non-empty rele-vance types on the other hand do not indicate total definedness of an expression; in fact, ALLOY has no explicit relation declarations, and undefined expressions evaluate to the empty set (ALLOY has no exceptions). Interestingly, ALLOY also allows over-loading and interprets it as union of relations; in addition, it resolves all non-abstract supertypes to abstract supertypes with one additional, (implicit) subtype containing the remainder. Issues of openness (and modularity) of a model are not addressed.

7 Conclusion

While much work remains to be done, we believe to have shown that state subtypes can significantly reduce the complexity of models. In particular, in combination with overloading they allow the definition of models in a piecewise fashion, thus avoiding explicit case analyses. Hence state subtypes support incremental model development, allowing modellers to easily address temporary omissions by simply adding new local definitions, leaving the rest of the model unchanged. Our presented framework furthermore enables a unified treatment of special type-based, state-based, and undefined behaviour. Finally, state subtypes turn out to be natural pivotal points for the integration of static and dynamic specifications.

References

1. Beugnard, A.: Is MDA achievable without a proper definition of late binding? In: UML Workshop in Software Model Engineering (WiSME) (2002)
2. Castagna, G.: Covariance and contravariance: conflict without a cause. ACM TOPLAS 17(3), 431–447 (1995)
3. Chambers, C.: Predicate Classes. In: Nierstrasz, O. (ed.) ECOOP 1993. LNCS, vol. 707, pp. 268–296. Springer, Heidelberg (1993)
4. DeLine, R., Fähnrich, M.: Typestates for Objects. In: Odersky, M. (ed.) ECOOP 2004. LNCS, vol. 3086, pp. 465–490. Springer, Heidelberg (2004)
5. Drossopoulou, S., Damiani, F., Dezani-Ciancaglini, M., Giannini, P.: More dynamic object reclassification: Fickle$_{II}$. ACM Trans. Program. Lang. Syst. 24(2), 153–191 (2002)
6. Ernst, M., Kaplan, C., Chambers, C.: Predicate Dispatching: A Unified Theory of Dispatch. In: Jul, E. (ed.) ECOOP 1998. LNCS, vol. 1445, pp. 186–211. Springer, Heidelberg (1998)
7. Edwards, J., Jackson, D., Torlak, E.: A type system for object models. In: Proc. of the 12th ACM SIGSOFT Int. Symp. on Foundations of Software Engineering, pp. 189–199. ACM Press, New York (2004)
8. Gamma, E., Helm, R., Johnson, R., Vlissides, J.: Design Patterns: Elements of Reusable Object-Oriented Software. Addison-Wesley, Reading (1995)
9. Gogolla, M., Drosten, K., Lipeck, U. W., Ehrich, H.D.: Algebraic and Operational Semantics of Specifications Allowing Exceptions and Errors. Theor. Comput. Sci. 34, 289–313 (1984)
10. Goguen, J.A., Diaconescu, R.: An Oxford Survey of Order Sorted Algebra. Mathematical Structures in Computer Science 4(3), 363–392 (1994)
11. Meseguer, J., Goguen, J.A.: Order-sorted algebra solves the constructor-selector, multiple representation, and coercion problems. Information and Computation 103(1), 114–158 (1993)
12. Milstein, T.: Practical Predicate Dispatch. In: OOPSLA. Conference on Object-Oriented Programming Systems, Languages, and Applications, pp. 345–364 (2004)
13. Mosses, P.D.: The Use of Sorts in Algebraic Specifications. In: COMPASS/ADT, pp. 66–92 (1991)
14. Nierstrasz, O.: Regular types for active objects. In: OOPSLA. Conference on Object-Oriented Programming Systems, Languages, and Applications, pp. 1–15 (1993)
15. OMG Unified Modeling Language: Superstructure Version 2.0 (2005)
16. Paech, B., Rumpe, B.: A new concept of refinement used for behaviour modelling with automata. In: Naftalin, M., Bertran, M., Denvir, T. (eds.) FME 1994. LNCS, vol. 873, pp. 154–174. Springer, Heidelberg (1994)
17. Salzman, L., Aldrich, J.: Prototypes with Multiple Dispatch: An Expressive and Dynamic Object Model. In: ECOOP, pp. 312–336 (2005)
18. Shang, D.L.: Covariant deep subtyping reconsidered. SIGPLAN Notices 30(5), 21–28 (1995)
19. Steimann, F.: A radical revision of UML's role concept. In: Evans, A., Kent, S., Selic, B. (eds.) UML 2000. LNCS, vol. 1939, pp. 194–209. Springer, Heidelberg (2000)
20. Steimann, F., Kühne, T.: A radical reduction of UML's core semantics. In: Jézéquel, J.-M., Hussmann, H., Cook, S. (eds.) UML 2002. LNCS, vol. 2460, pp. 34–48. Springer, Heidelberg (2002)

Deriving Operation Contracts from UML Class Diagrams

Jordi Cabot[1] and Cristina Gómez[2]

[1] Estudis d'Informàtica, Multimedia i Telecomunicacions, Universitat Oberta de Catalunya
jcabot@uoc.edu
[2] Dept. de Llenguatges i Sistemes Informàtics, Universitat Politècnica de Catalunya
cristina@lsi.upc.edu

Abstract. Class diagrams must be complemented with a set of system operations that describes how users can modify and evolve the system state. To be useful, such a set must be complete (i.e. through these operations, users should be able to modify the population of all elements in the class diagram) and executable (i.e. for each operation, there must exist a system state over which the operation can be successfully applied). Manual specification of these operations is an error-prone and time-consuming activity. Therefore, the goal of this paper is to automatically provide a basic set of system operations that verify these two properties. Operations are drawn from the elements (classes, attributes, etc) of the class diagram and take into account the possible dependencies between the different change events (i.e. inserts/updates/deletes) that may be applied to them. Afterwards, the designer could reuse our proposal to build up more complex operations.

1 Introduction

The specification of an information system must include all relevant static and dynamic aspects of the domain [9]. The static aspects are collected in structural diagrams, class diagrams in the UML. Dynamic aspects are usually specified by means of a behavioral schema consisting of a set of system operations [11] (also known as domain events [14]) that the user may execute to query and/or modify the information modeled in the class diagram. A system operation consists of a non-empty set of basic modifications over the system state that is perceived by the user of the information system as a single change in the domain. We refer to these basic modifications as structural events. Each structural event, such as "create object", "update attribute" or "delete link", represents an elementary change to the elements of a class diagram.

Behavioral schemas must be complete [14] and executable [7]. A behavioral schema bs is complete when, through the system operations in bs, a user can apply all kinds of structural events to any modifiable element of the class diagram (i.e. given an element e of the class diagram and a possible structural event s over e, there is at least an operation in bs that includes s). It is executable, when, for each operation op, there exists at least an initial system state and a set of argument values that ensure a successful execution of op (an execution is successful when the new system state is consistent with the class diagram's integrity constraints). Incomplete behavior

G. Engels et al. (Eds.): MoDELS 2007, LNCS 4735, pp. 196–210, 2007.

schemas result in information systems that have parts that the user cannot modify since no available operations address their modification. Non-executable behavior schemas result in information systems with operations that can never be successfully executed.

For instance, given the simple example shown in Fig. 1.1, we must specify an operation to create new employees, an operation to delete employees and two operations to update the *name* and *salary* attributes. This behavior schema is complete since all the modifiable elements in the class diagram (*dateOfBirth* is marked as read only) can be created, updated and deleted through the execution of the system operations. Moreover, it is also executable. The deletion operation can be executed in all states with at least one employee instance. The creation and update operations can be applied provided that the argument corresponding to the new salary value is greater than 600, which is the only restriction imposed by the *ValidSalary* constraint.

```
┌─────────────────────────────────┐
│          Employee               │
├─────────────────────────────────┤
│ name: String                    │      context Employee ValidSalary inv:  self.salary>600
│ dateOfBirth: Date {readOnly}    │
│ salary: Money                   │
└─────────────────────────────────┘
```

Fig. 1.1. Example of a simple structural schema

For all non-trivial class diagrams the number of required system operations rapidly increases. Therefore, the specification of a complete and executable set of operations becomes an error-prone and time-consuming activity.

We believe that an automatic generation of behavior schemas from UML class diagrams would offer two main benefits. Firstly, it would guarantee the quality (in terms of completeness and executability) of the specified system operations. Secondly, the software development process would be sped up by avoiding a systematic definition of all operations. In this sense, given a class diagram, the main goal of our paper is to provide a method for the automatic generation of a basic behavior schema that satisfies the completeness and executability properties. We define our generated behavior schema as a basic one since we try to keep all defined operations as simple as possible. Operations are declaratively specified by means of OCL contracts. As far as we know, ours is the first approach to provide an automatic generation of a complete and executable behavior schema.

Although our basic schema suffices to cover most common operations appearing in class diagrams, designers may want to generate arbitrary complex operations. Such complex operations may be defined as a combination of our basic ones in order to guarantee their executability as well. Ideally, these more complex operations could also be (semi)automatically generated when additional diagrams (such as the use case diagram [21]) are considered but this is left as further work.

The rest of the paper is organized as follows. Section 2 introduces several preliminary concepts. Section 3 and 4 define the completeness and executability of a behavior schema, respectively. Section 5 presents our generation of a basic complete and executable behavior schema. A case study is shown in Section 6. Finally, Section 7 presents related work and section 8 puts forwards the conclusions and ideas for further research.

2 Preliminary Concepts

Class Diagrams. We represent a class diagram *CD* using the tuple:

$$CD=<CL, ATT, ASS, AC, GEN, IC>$$

where *CL*, *ATT*, *ASS*, *AC*, *GEN* and *IC* represent the set of classes, attributes, associations, association classes, generalizations and constraints of the class diagram *CD*, respectively. All elements in *CD* are assumed to be correct instances of the corresponding metaclasses of the UML metamodel. We assume that all associations are binary associations. N-ary associations can easily be expressed in terms of a set of binary ones plus additional constraints [2].

Structural events. The concrete number (and specification) of the system operations required by a class diagram depends on the exact types of structural events provided by the modeling language. The structural event types (and their effect) being considered in this paper are the following:

1. $iCl(x)$: inserts a new object (i.e. instance) x into class *Cl*. If *Cl* participates in a class taxonomy, x is inserted into all (direct or indirect) superclasses as well.
2. $dCl(x)$: deletes an existing object x from *Cl* and from all its direct and indirect superclasses and subclasses.
3. $uAt_iCl(x,v)$: sets v as the new value for the attribute At_i of object x (of class *Cl*).
4. $iAs(x_1:Cl_1,x_2:Cl_2)$: inserts a new link in *As* between objects x_1 of type Cl_1 and x_2 of type Cl_2.
5. $dAs(x_1:Cl_1,x_2:Cl_2)$: removes the link between objects x_1 and x_2 in *As*.
6. $gCl_cCl_p(x)$: generalizes an object x of a (child) subclass Cl_c to a (parent) superclass Cl_p.
7. $sCl_pCl_c(x)$: specializes an object x of a superclass Cl_p to Cl_c.

Creation/deletion of instances of association classes requires creating/deleting both the class and association facets of the association class instance with the corresponding events.

Our events are more basic than those proposed in the UML (see the list of *actions* in the UML metamodel [15]). This permits a more fine-grained reasoning. Nevertheless, we could easily define a correspondence between the two sets.

3 Completeness of a Behavior Schema

A behavior schema *bs* is complete when users are able to apply all kinds of changes to the modifiable elements of a class diagram *CD* through the execution of the operations in *bs*, that is, when for each modifiable element *e* in *CD* and each possible structural event *s* over *e*, there is at least one operation in *bs* that includes *s*.

Therefore, completeness is guaranteed if we first compute the set set_{ev} of structural events that may be applied over *CD* and then we ensure that each event *ev*, $ev \in set_{ev}$, is included in one of the system operations in *CD*.

In Section 3.1 we define the notion of modifiability for each kind of model element appearing in a class diagram. Then, in Section 3.2 we compute the set of structural events relevant to a given class diagram (i.e. the set of events that can be possibly executed over the diagram), taking into account the modifiability of each element in the diagram. To illustrate the process we use the class diagram shown in Fig. 3.1 as a running example.

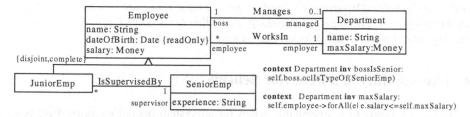

Fig. 3.1. Class diagram used as a running example

3.1 Modifiability of a Model Element

The modifiability of a model element (that is, the possibility of changing the value or population of that element) depends on the type of element and on the (metamodel) properties specified by the designer during its definition.

A class c is modifiable as long as c is not an abstract class (i.e. when its *isAbstract* property, defined in the *Class* metaclass evaluates to false) and c is not the supertype of a covering generalization set (covering is also known as complete). In a covering generalization set no instances of the supertype can be directly created, they can only be created when one of their subtypes is being instantiated.

An attribute a is modifiable when a is neither read only nor derived (i.e. *a.isReadOnly=false and a.isDerived=false*, where *isReadOnly* and *isDerived* are properties of the *Property* metaclass).

An association is modifiable when none of its member ends is read only or derived. An association class is modifiable when both its class facet and its association facet are modifiable.

Generalization sets are always modifiable.

All elements in the class diagram in Fig. 3.1 are modifiable except for the *dateOfBirth* attribute, which is marked as *readOnly*.

3.2 Computing the Relevant Structural Events for a Class Diagram

Given a class diagram $CD=<CL, ATT, ASS, AC, GEN, IC>$ the set of structural events that may be applied to CD are the following:

- iCl and dCl events for each modifiable class Cl in CD.
- iAs and dAs events for each modifiable association As in CD.
- An uAt_iCl event for each modifiable[1] attribute At_i of a class Cl.
- gCl_cCl_p and sCl_pCl_c events for each subclass Cl_c of a superclass Cl_p in a generalization set.

---

[1] Update events for non-modifiable attributes are only admitted just after the object has been created, as a way of initializing the attribute's value.

In the example of Fig. 3.1, *iJuniorEmp, dJuniorEmp* and *gJuniorEmpEmployee* events may be applied over *JuniorEmp*. Similarly, *iSeniorEmp, dSeniorEmp, gSeniorEmpEmployee* and *uExperienceSeniorEmp* may be applied over *SeniorEmp*. Relevant events for *Department* are *iDepartment, dDepartment, uNameDepartment and uMaxSalaryDepartment* and for *Employee* are *uNameEmployee, uSalaryEmployee, sEmployeeJuniorEmp* and *sEmployeeSeniorEmp*[2]. For *Manages* and *WorksIn*, insertion and deletion events may be applied.

4   Executability of a Behavior Schema

A behavior schema *bs* is executable when for all system operations in *bs* there is at least a system state and a set of arguments for the operation parameters that permit a successful execution of the operation. An operation succeeds when its execution evolves the initial system state to a new state that satisfies all integrity constraints.

Defining an operation as executable does not imply that every time the operation is executed the new system state will be consistent (this depends on the previous state and on the exact arguments passed as parameters for the operation). We just guarantee that it is at least possible to successfully execute it sometime. Otherwise, the operation is completely useless and should be removed.

Executability depends on the set of structural events that the operation applies over the system state. The basic idea is that some events require the presence of other events within the same operation in order to leave the data in a consistent state at the end of the operation execution. As an example, an operation *createDepartment* creating new instances of department (that is, an operation applying the *iDepartment* event) must be in charge of creating a new link in the *Manages* association relating the new department with its boss (*iManages* event). Otherwise, every time this operation is executed the minimum multiplicity of the *boss* role (see Fig. 3.1) becomes violated, and thus, the operation never succeeds.

Therefore, executability is guaranteed if, for each event *ev* included in the effect of a system operation *op,* all other events required by *ev* appear in *op* as well. A behavior schema is executable when all operations are executable.

Dependencies between structural events depend on the type of the event and on the integrity constraints of each particular class diagram. When the dependencies for an event *ev* are being computed, all we need to consider are the minimum multiplicity constraints for associations and attributes[3] and *disjoint* and *complete* constraints (either graphically represented or implicitly induced by textual OCL constraints[4]).

---

[2] *iEmployee* and *dEmployee* events may be applied over *Employee* only when the generalization set in which *Employee* participates as a supertype is defined as incomplete.

[3] Although it is also possible to define minimum multiplicities for the number of objects in a class, they are quite rare. We are therefore not going to consider them in our approach.

[4] Some textual OCL constraints may exactly correspond to minimum multiplicity, disjoint or complete constraints. Also, some may indirectly imply them (for instance, stating that navigating from an object of type *X* to the related *Y* objects we must find more than *N* objects satisfying condition *cond* implies that the minimum cardinality of the *Y* role in the navigated association must be at least *N*).

For other constraints, we can always find a combination of a system state and/or a set of arguments for which the execution of *ev* results in a consistent state. For instance, maximum multiplicity constraints are never violated when *ev* is applied to an empty system state. Constraints restricting the value of the attributes of an object may be satisfied when passing the appropriate arguments as parameters for the event. The same situation occurs with constraints restricting the relationship between an object and related objects. Therefore, all these constraints are ignored when computing the dependencies of *ev*, and thus, when determining the executability of operations including *ev*. As an example, the *maxSalary* constraint (Fig. 3.1) does not affect the executability of operations modifying departments, employees and the links between them. The creation of employees and departments is always successful. Updates of *Salary* and *MaxSalary* attributes may be successful when choosing the right values for the corresponding attributes. The creation of a new *WorksIn* link is successful when the state has at least a department and an employee (who is not already related to a department) that satisfies the *maxSalary* condition.

For the class diagram of Fig. 3.1, several dependencies between the relevant structural events are necessary. For instance, an *iJuniorEmp(x)* event requires the presence of events *uNameEmployee(x,name)*, *uDateOfBirthEmployee(x,date)* and *uSalaryEmployee(x,sal)* to initialize the values of its non-derived attributes. Otherwise, operations that do not include them will always violate the minimum '1' multiplicity of these attributes. Additionally, this event also requires the events *iIsSupervisedBy(x,y)* and *iWorksIn(x,z)* to avoid violating the minimum multiplicity of the *supervisor* and *employer* roles. The complete list of dependencies for this example can be found in Section 5.2.1.

5 Generation of a Complete and Executable Behavior Schema

In this section, we show how to automatically generate a complete and executable behavior schema for a given class diagram *CD*, according to the previous complete and executable properties. Our method has two main phases:

-   The assignment of all relevant events for *CD* to a set of new system operations (*completeness*)
-   The definition of the actual operation parameters and body in view of the dependencies of the assigned events (*executability*)

In our approach, system operations are assigned to an appropriate class of the class diagram. Other authors argue that it is better to first assign all operations to an artificial class called *System* [11]. The adaptation of our method to this case is straightforward.

Furthermore, operations can be specified in one of two ways: imperatively or declaratively [22]. In an imperative specification, the set of structural events that the operation applies during the operation execution are explicitly defined. In a declarative specification the designer defines a contract for each operation. The contract consists of a set of pre and postconditions. A precondition defines a set of conditions on the operation input and the system state that must hold when the operation is invoked. The postcondition states the set of conditions that must be satisfied by the system state at the end of the execution. In our approach, the imperative version of each operation can be directly deduced from the structural events we assign to the operation during the generation process. Therefore, we focus on the declarative version.

Note that our pre- and postconditions do not include the verification of the integrity constraints in *CD* (strict interpretation of operation contracts [18]) in order to avoid redundancies between the contracts and the constraints (this improves the quality of the resulting specifications, see [6]). Only those constraints that could potentially affect the executability property of the operations are considered (see the discussion presented in Section 4) and already tackled when reasoning on the dependencies between the structural events assigned to the operation.

5.1 Creating the Required System Operations

Assignment of relevant events for a class diagram *CD* into a set of system operation can be done in many different ways. Since we intend to create a basic behavior schema, our goal is to minimize the complexity of the generated operations. Roughly, we create a different operation in *CD* for each relevant structural event.

Given that set_{ev} is the set of relevant events for *CD* (as computed in section 3.2) the system operations our method generates are the following (operations are assigned to the appropriate class according to the *GRASP* patterns [11]):

- A class operation *Cl::Create* for each *iCl* event in set_{ev}
- A *Cl::Delete* operation for each *dCl* event in set_{ev}
- A $Cl_c::GeneralizeCl$ operation for each gCl_cCl_p event in set_{ev}
- A $Cl_p::SpecializeCl$ operation for each sCl_pCl_c event in set_{ev}
- A $Cl::UpdateAt_i$ operation for each uAt_iCl event in set_{ev}
- Two *P::CreateLinkAs* operations (one for each participant class *P*) for each *iAs* event in set_{ev}.
- Two *P::DeleteLinkAs* operations (one for each participant class *P*) for *dAs* events in set_{ev}

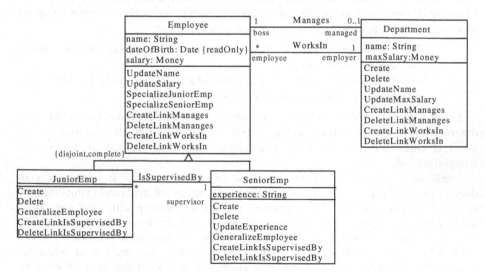

Fig. 5.1. Class diagram with the generated system operations

In UML, operations cannot be assigned to associations (except for association classes). Therefore, operations on associations are assigned to the participants of the association. For recursive associations we can use the name of the opposite role rather than the association name when creating the operation. To satisfy the completeness property, it would be enough to add the operations to one of the participants. However, on behalf of the usability of the generated behavior schema, we prefer to add the operations to both participants. When designing the specified system, designers may add navigability information to the diagram and remove the operations using non-navigable association ends.

Fig. 5.1 shows the running example of Fig. 3.1 once it has been extended to include the set of generated operations.

5.2 Completing the Operation Definition

Following on from the previous section we get a set of operations, each one attached to a class in *CD* and with one of the relevant structural events assigned to it.

To complete the operation definition we need to first determine whether the operation behavior must be extended with new events due to dependencies between them (Section 5.2.1). We will then be able to define the set of parameters for the operation (Section 5.2.2) and its final body (Section 5.2.3).

5.2.1 Computing the Dependencies

A simple dependency for a structural event *ev* is defined as a tuple *<direction, event>* where *event* is the name of the structural event required by *ev* and *direction* indicates whether that event should be executed before *ev* (symbol ←), after *ev* (symbol →) or if the exact position of *ev* is irrelevant (symbol ↑). In fact, the *direction* field is not strictly necessary (in the same way that the exact order of predicates in a postcondition is also irrelevant); as long as all dependent events are applied over the system state, the state at the end of the operation will be consistent. Nevertheless, the *direction* field helps to obtain a more clear and readable contract.

More complex dependencies are expressed as a sequence of simple ones joined with the logical *AND* and *OR* operators (for example, *ev* may require the existence of the events ev_1 and ev_2 or, alternatively, the existence of an event ev_3).

Computing the list of dependencies for an event *ev* is a recursive process. If *ev* requires an event ev_2 we must take into account also the dependencies of ev_2 and so on. Otherwise an operation including *ev* and ev_2 may be non-executable due to the dependencies of ev_2 not being satisfied within the operation. Therefore, to ensure executability, we must compute the *transitive closure* of the dependencies of *ev*. The *transitive closure* can be computed by means of recursively applying the following dependency rules over the new events added to the initially empty list of dependencies for an event *ev* until no more dependencies are added[5].

Note that, due to *OR* dependencies (stating that *ev* depends on an event ev_1 or, alternatively, on an event ev_2) we may obtain different alternative dependency lists at the end of the computation process. Each *OR* dependency is a splitting point. From that point, two new lists are created. The lists are initialized with the contents of the

---

[5] Termination is guaranteed except in the case of rare multiplicities combinations (as a cyclic sequence of exact one-to-one associations), which require the designer to take part in the process.

original one and then the process continues with each list separately. Each list generates a different operation specification.

The list of dependency rules is the following (in the rules $min(Cl,As)$ returns the minimum multiplicity of Cl in As, i.e. the minimum number of links in As in which all instances of Cl must participate, and $max(Cl,As)$ its maximum multiplicity).:

Rules for computing the dependencies for a structural event $iCl(x)$:
-   A dependency $dep_{At1} = <\rightarrow, uAt_iCl(x,v)>$ for each non-derived attribute At_i of Cl AND
-   A dependency $dep_{At2} = <\rightarrow, uAt_kCl_p(x,v)>$ for each non-derived attribute At_k of Cl_p where Cl_p is a direct or indirect superclass of Cl AND
-   A number of $min(Cl,As_j)$ dependencies $dep_{As1} = <\rightarrow, iAs_j(x,y)>$ for each non-derived association As_j where Cl has a mandatory participation $(min(Cl,As_j) >=1)$ AND
-   A number of $min(Cl_p,As_k)$ dependencies $dep_{As2} = <\rightarrow, iAs_k(x,y)>$ for each non-derived association As_s where Cl_p is a direct or indirect superclass of Cl and Cl_p has a mandatory participation in As_k.

Dependencies for a $dCl(x)$ event :
-   A number of $min(Cl,As_j)$ dependencies $dep_{As1} = <\rightarrow, dAs_j(x,y)>$ for each non-derived association As_j where Cl has a mandatory participation AND
-   A number of $min(Cl',As_k)$ dependencies $dep_{As2} = <\rightarrow, dAs_k(x,y)>$ for each non-derived association As_k where Cl' is a direct or indirect superclass or subclass of Cl and Cl' has a mandatory participation in As_k.

Dependencies for a $sCl_pCl_c(x)$ event:
-   A dependency $dep_{At} = <\rightarrow, uAt_iCl_c(x,v)>$ for each non-derived attribute At_i of Cl_c AND
-   $min(Cl_c, As_j)$ dependencies $dep_{As} = <\rightarrow, iAs_j(x,y)>$ for each non-derived association As_j where Cl_c has a mandatory participation AND
-   A dependency $dep_{Spec} = <\rightarrow, gCl_{c'}\cdot Cl_p(x)>$ for a $Cl_{c'}$ class such that $Cl_c \neq Cl_{c'}$ and that x is an instance of $Cl_{c'}$. This dependency only applies when the generalization set, for which Cl_p is the supertype, is disjoint and complete; in such a case, specialization of x to Cl_c forces the removal (generalization) of x from a different subtype $Cl_{c'}$ to satisfy the disjoint constraint. We know that x was instance of some $Cl_{c'}$ because the generalization set is complete.

Dependencies for a $gCl_cCl_p(x)$ event:
-   $min(Cl_c,As_j)$ dependencies $dep_{As} = <\rightarrow, dAs_j(x,y)>$ for each non-derived association As_j where Cl_c has a mandatory participation AND
-   A dependency $dep_{Gen} = <\rightarrow, sCl_pCl_{c'}(x)>$ such that $Cl_c \neq Cl_{c'}$. Again, this dependency only applies when the generalization set, for which Cl_p is the supertype, is disjoint and complete.

Dependencies for an $iAs(x{:}Cl_1,y{:}Cl_2)$ event when $min(Cl_1,As)=max(Cl_1,As)$ (the process must be repeated for Cl_2 when $min(Cl_2,As)=max(Cl_2,As)$):
-   A dependency $dep_{As} = <\uparrow, dAs(x,z)>$ such that $<x,z>$ is an existing link in As, if $min(Cl_2,As) \neq max(Cl_2,As)$ OR
-   A dependency $dep_{Ins} = <\leftarrow, iCl_1(x)>$ OR
-   A dependency $dep_{Spec} = <\leftarrow, sCl_pCl_1(x)>$ if Cl_1 has a supertype Cl_p

Dependencies for a $dAs(x{:}Cl_1,y{:}Cl_2)$ event when $min(Cl_1,As)=max(Cl_1,As)$ (the process must be repeated for Cl_2 when $min(Cl_2,As)=max(Cl_2,As)$):

-   A dependency $dep_{As} = <\uparrow, iAs(x,z)>$ if $min(Cl_2,As)\neq max(Cl_2,As)$ OR
-   A dependency $dep_{Ins} = <\leftarrow, dCl_1(x)>$ OR
-   A dependency $dep_{Gens} = <\leftarrow, gCl_1Cl_p(x)>$ if Cl_p is a supertype of Cl_1.

No dependencies are needed for uAt_iCl events since changes on attribute values do not violate cardinality, complete or disjoint constraints. Table 5.1 summarizes the result of the (recursive) application of these rules over the relevant structural events for the class diagram shown in Fig. 3.1.

Table 5.1. Dependencies for the relevant structural events in the class diagram in Fig 3.1

Structural event	Required events (dependencies)
iJuniorEmp(x)	uNameEmployee(x,v_{name}) *AND* uDateOfBirthEmployee(x,v_{date}) *AND* uSalaryEmployee(x,v_{sal}) *AND* iIsSupervisedBy(x,y) *AND* iWorksIn(x,z)
dJuniorEmp(x)	dIsSupervisedBy(x,y) *AND* dWorksIn(x,z)
sEmployeeJuniorEmp(x)	iIsSupervisedBy(x,y) *AND* gSeniorEmpEmployee(x)
sEmployeeSeniorEmp(x)	uExperienceSeniorEmp(x,exp) *AND* gJuniorEmpEmployee(x) *AND* dIsSupervisedBy(x,y)
gJuniorEmpEmployee(x)	sEmployeeSeniorEmp(x) *AND* dIsSupervisedBy(x,y) *AND* uExperienceSeniorEmp(x,exp)
iSeniorEmp(x)	uNameEmployee(x,name) AND uDateOfBirthEmployee(x,date) AND uSalaryEmployee(x,sal) AND uExperienceSeniorEmp(x,exp) AND iWorksIn(x,z)
dSeniorEmp(x)	dWorksIn(x,z)
gSeniorEmpEmployee(x)	sEmployeeJuniorEmp(x) AND iIsSupervisedBy(x,y)
iDepartment(x)	uNameDepartment(x,name) AND uMaxSalaryDepartment(x,maxSal) AND iManages(x,y)
dDepartment(x)	dManages(x,y)
iManages(x,y)	dManages(x:Department,z:Employee) OR iDepartment(x)
dManages(x,y)	iManages(x:Department,z:Employee) OR dDepartment(x)
iWorksIn(x,y)	dWorksIn(z:Department,y:Employee) OR iEmployee(y)
dWorksIn(x,y)	iWorksIn(z:Department,y:Employee) OR dEmployee(y)
iIsSupervisedBy(x,y)	dIsSupervisedBy(z:SeniorEmp,y:JuniorEmp) OR iJuniorEmp(y) OR sJuniorEmp(y)
dIsSupervisedBy(x,y)	iIsSupervisedBy(z:SeniorEmp,y:JuniorEmp) OR dJuniorEmp(y) OR gJuniorEmpEmployee(y)

5.2.2 Defining the Operation Signature

The signature of an operation *op* depends on the list $list_{ev}$ of structural events the operation consist of (computed as shown in the previous section) and the class where *op* is attached.

Each event $ev \in list_{ev}$ may require the addition of new parameters in the signature. The basic idea is that every variable that appears as a parameter of *ev* must also appear as a parameter (of the same type) in the operation. Four exceptions apply:

1.   Variables for *iCl* events are not parameters of the operation. These new objects are created *during* the operation execution.
2.   A parameter variable that has already appeared in a previous event does not generate a new operation parameter (i.e. if an operation consists of two events $iAsX(x_1,x_2)$ and $iAsY(x_1,x_3)$ only three parameters x_1, x_2 and x_3 are defined).

3. We use the implicit parameter *self* as a replacement for one of the parameters whose type is the class to which the operation is attached (i.e. if an operation defined in a class *Cl* has the event $uAt_iCl(x,v)$ only a parameter for *v* is generated; the implicit *self* parameter is used whenever *x* appears).

4. Variables for *dAs* events not included in a *DeleteLinkAs* or a *CreateLinkAs* operation are not parameters of the operation. Those deletions are required by *dCl* or gCl_cCl_p events. In those cases, the link/s to be deleted are the ones in which the *self* parameter participates, and thus, they can be determined automatically.

For instance, the operation *JuniorEmp::Create* of Fig 5.1 consists of the $iJuniorEmp(x)$ event and all its dependencies defined in Table 5.1 $(uNameEmployee(x,v_{name})$, $uDateOfBirthEmployee(x,v_{date})$, $uSalaryEmployee(x,v_{sal})$, $iIsSupervisedBy(x,y)$, $iWorksIn(x,z))$. From this list of events we may determine the signature of the *Create* operation as follows:

$Create(v_{name}{:}String, v_{date}{:}Date, v_{sal}{:}Money, y{:} SeniorEmp, z{:}Department)$.

Similarly, the signature of *Department::Delete* is simply *Delete()*. This signature is calculated from the list of events for this operation *(dDepartment(x), dManages(x,y))*. In accordance with the rules above, none of the event variables must be added as an explicit parameter of this operation.

5.2.3 Defining the Operation Body

In an imperative specification of the operation effect, the operation body is simply defined as the ordered list of structural events computed for the operation as shown in the previous sections. However, in a declarative specification we must transform the list of structural events into an OCL contract such that the application of the events over a state satisfying the contract preconditions evolves this initial state into a new state that satisfies the contract postconditions.

As discussed in the introduction of Section 5, our operations do not explicitly include the integrity checking of the class diagram's constraints. Therefore, our operations do not include preconditions and the postconditions refer solely to the operation's own behavior. Constraints that may affect the executability property of the operation will already have been considered when computing the dependencies.

The initial postcondition is obtained by means of translating each single event into an equivalent boolean condition and concatenating the different conditions with *AND* operators (this translation is not unique, see [3]). In the following we provide a possible boolean condition for each event.

1. $iCl(x)$: *x.oclIsNew() and x.oclIsTypeOf(Cl)*
2. $dCl(x)$:*OclAny::allInstances()=OclAny::allInstances()@pre->excluding(x)*
3. $uAt_iCl(x,v)$: *x.At_i=v*
4. $iAs(x_1,x_2)$: $x_1.r_2$->*includes(x_2)* (r_2 is the role corresponding to x_2 in As)
5. $dAs(x_1,x_2)$: $x_1.r_2$->*excludes(x_2)* (r_2 is the role corresponding to x_2 in As)
6. $gCl_cCl_p(x)$: *x.oclIsTypeOf(Cl_p)*
7. $sCl_pCl_c(x)$: *x.oclIsTypeOf(Cl_c)*

Note that *OclAny* is the supertype of all types in the UML class diagram [16]. Using *OclAny* instead of *Cl* in the definition of the *dCl(x)* event condition guarantees that the object *x* is completely removed from the system (and that it does not remain, for example, as an instance of a supertype of *Cl*).

The resulting postcondition may need to be refined depending on the combination of translated structural events. For instance, if several sCl_pCl_c events are applied over an instance *x*, only the translation for the event over the more specific class is necessary. Translation for events $dAs(x_1,x_2)$ can be discarded when $dCl(x_1)$ and/or $dCl'(x_2)$ events also appear (usually, the deletion of links is implicitly assumed).

As an example, we provide the contract for the operation *JuniorEmp::Create* and for the operation *Department::Delete*.

context JuniorEmp::Create(v_{name}:String, v_{date}:Date, v_{sal}:Money,y:SeniorEmp, z:Department)
post: x.oclIsNew() and x.oclIsTypeOf(JuniorEmp) and x.name=v_{name} and x.dateOfBirth=v_{date} and x.salary=v_{sal} and x.supervisor->includes(y) and x.employer->includes(z)

context Department::Delete()
post: OclAny::allInstances()=OclAny::allInstances()@pre->excluding(self)

6 Case Study

To show the benefits of our proposal, in this section we compare the behavior schema for a real-life application when it has been generated by our method with the behavior schema originally specified by the designer by hand for the same application.

In particular, we analyze a system for a Conference Management Application as specified in [19]. This system provides functionalities to support paper submissions, assignment of papers to reviewers and the evaluation process. The class diagram consists of 13 classes, 13 binary associations, 2 non-covering generalization sets and several constraints. The proposed behavior schema includes 29 operations.

Our method is able to completely generate 13 of the 29 operations (6 creation operations, 3 deletion operations and 4 update operations). Seven additional operations (each one assigning one or more constant values to attributes of the class diagram) can be directly mapped to our generated system operations by passing these constant values as parameters of our *UpdateAt$_i$* operations. The rest of the system operations, 9 out of 29, can only be partially generated by our method. This means that the designer must manually complete their specification. Mainly, the difference is that in the original schema these nine operations include some ad hoc if-else conditions that restrict the applicability of the operations depending on the system state. Clearly, it is not possible to automatically generate these conditions.

From the results presented above, we see that the application of our proposal helps designers by reducing by 69% (20 of 29) the number of operations to be defined and by providing at least an initial contract specification for the remaining ones.

Moreover, our approach generates several operations that did not appear in the manually specified schema (for instance, all *GeneralizeCl* and *SpecializeCl* and some *UpdateAt$_i$* operations). Designers could use this information to detect whether some required operations must be added to the class diagram or if the specification of the class diagram is incomplete (for instance, attributes are not marked as *readOnly* or *derived*, completeness and/or disjointness of generalizations sets is not defined, etc).

7 Related Research

As far as we know, ours is the first approach to study the application of the completeness and executability properties to the automatic generation of a basic behavior schema.

[10] partially determines the set of possible structural events to be applied to a class diagram (generalizations are not considered). However, in this approach system operations must be manually defined as a combination of a set of structural events. Therefore, the completenesss and executability of these operations must be guaranteed by designers. In this approach, operations are specified using the formal notation B.

[8] derives a set of basic operations (similar to our concept of structural events) and a set of elementary operations from an EER diagram. These latter operations are not necessarily executable since cardinality constraints are not considered in any case.

Alternatively, other approaches try to generate system operations from the information provided in different diagrams, such as the use case diagram. For instance, [21] presents a method for generating system operations from use cases specifications. Nevertheless, this method is not automatic and completeness and executability properties of the generated behavior schema are not analyzed.

The idea of dependencies between structural events is not new. This problem has been addressed in the (deductive) database field as part of the more general problem of integrity maintenance at compile-time (see [20], [12], [17]). In those cases, the goal was similar: to extend a (predefined) given transaction/operation with additional events to always ensure its successful execution. However, their expressivity regarding the definition of the structural diagram and the set of admitted structural event types is more restricted than in our method.

Regarding OCL contracts, [1] provides some patterns to help designers in their manual definition but automatic generation of contracts is not studied.

Some IDEs (as [5] or [13]) are able to automatically generate basic getter/setter and creator methods for classes. However, the methods do not take into account the possible dependencies among the included events.

8 Conclusions and Further Research

The complete definition of the behavior schema is one of the most important tasks in the analysis and design stages of an information system. The method presented in this paper facilitates this task by automatically generating an initial set of system operations. Operations are drawn from the structure of the class diagram.

The executability and completeness properties of this set of operations guarantee the quality of the behavior schema. Designers are free to directly use our operations (avoiding the manual definition of the behavior schema) or to reuse our method in order to create more complex operations (while maintaining the previous properties).

We believe our method is useful even when the designer is not interested in a complete and automatic generation of the behavior schema. If integrated in an OCL editor, our method could assist the designer during the definition of OCL contracts by means of *suggesting* additional predicates to complete the postconditions. These suggestions would be provided based on our dependencies computation.

As a further research, we would like to study how we can reuse the information of use cases (as in [21]) and state diagrams to automatically derive more complex system operations. We are also interested in studying how to integrate the efficient verification of all constraints that may be violated by the operation execution (the relevant constraints can be determined with [4]) into the preconditions of our generated operation contracts so that a successful execution of the operation is *always* guaranteed (providing that the precondition is satisfied). Additionally, we plan to apply the completeness and executability properties to the verification of existing behavior schemas.

Acknowledgments

We would like to thank the people of the GMC group and the anonymous reviewers for their many useful comments in the preparation of this paper. This work has been partially supported by the Ministerio de Ciencia y Tecnología under project TIN2005-06053.

References

1. Ackermann, J., Turowski, K.: A Library of OCL Specification Patterns for Behavioral Specification of Software Components. In: Dubois, E., Pohl, K. (eds.) CAiSE 2006. LNCS, vol. 4001, pp. 255–269. Springer, Heidelberg (2006)
2. Andrew, J., Mcallister, D.S.: An approach for decomposing N-ary data relationships. Software: Practice and Experience 28, 125–154 (1998)
3. Cabot, J.: From Declarative to Imperative UML/OCL Operation Specifications. In: ER 2007. LNCS, Springer, Heidelberg (to appear, 2007)
4. Cabot, J., Teniente, E.: Determining the Structural Events that May Violate an Integrity Constraint. In: Baar, T., Strohmeier, A., Moreira, A., Mellor, S.J. (eds.) UML 2004. LNCS, vol. 3273, pp. 173–187. Springer, Heidelberg (2004)
5. CincomSmalltalk. VisualWorks, http://www.cincomsmalltalk.com/
6. Costal, D., Sancho, M.-R., Teniente, E.: Understanding Redundancy in UML Models for Object-Oriented Analysis. In: Pidduck, A.B., Mylopoulos, J., Woo, C.C., Ozsu, M.T. (eds.) CAiSE 2002. LNCS, vol. 2348, pp. 659–674. Springer, Heidelberg (2002)
7. Costal, D., Teniente, E., Urpí, T., Farré, C.: Handling Conceptual Model Validation by Planning. In: Constantopoulos, P., Vassiliou, Y., Mylopoulos, J. (eds.) CAiSE 1996. LNCS, vol. 1080, pp. 255–271. Springer, Heidelberg (1996)
8. Engels, G., Gogolla, M., Hohenstein, U., Hüllmann, K., Löhr-Richter, P., Saake, G., Ehrich, H.-D.: Conceptual Modelling of Database Applications Using an Extended ER Model. Data & Knowledge Engineering 9, 157–204 (1992)
9. ISO/TC97/SC5/WG3: Concepts and Terminology for the Conceptual Schema and Information Base. ISO (1982)

10. Laleau, R., Polack, F.: Specification of Integrity-Preserving Operations in Information Systems by Using a Formal UML-based Language. Information and Software Technology 43, 693–704 (2001)
11. Larman, C.: Applying UML and Patterns: An Introduction to Object-Oriented Analysis and Design and the Unified Process, 2nd edn. Prentice-Hall, Englewood Cliffs (2001)
12. Link, S.: Consistency Enforcement in Databases. In: Bertossi, L., Katona, G.O.H., Schewe, K.-D., Thalheim, B. (eds.) Semantics in Databases. LNCS, vol. 2582, pp. 139–159. Springer, Heidelberg (2003)
13. Microsoft. Visual Studio 2008, http://msdn2.microsoft.com/en-us/vstudio/default.aspx
14. Olivé, A.: Conceptual Modeling of Information Systems. Springer, Heidelberg (2007)
15. OMG: UML 2.0 Superstructure Specification. OMG Adopted Specification (ptc/03-08-02) (2003)
16. OMG: UML 2.0 OCL Specification. OMG Adopted Specification (ptc/03-10-14) (2003)
17. Pastor, J.A., Olivé, A.: Supporting Transaction Design in Conceptual Modelling of Information Systems. In: Iivari, J., Rossi, M., Lyytinen, K. (eds.) CAiSE 1995. LNCS, vol. 932, pp. 40–53. Springer, Heidelberg (1995)
18. Queralt, A., Teniente, E.: Specifying the Semantics of Operation Contracts in Conceptual Modeling. Journal on Data Semantics 7, 33–56 (2006)
19. Raventós, R.: A conceptual schema for a conference management application. UPC, LSI Technical Report, LSI-05-1-R (2005)
20. Schewe, K.-D., Thalheim, B.: Towards a theory of consistency enforcement. Acta Informatica 36, 97–141 (1999)
21. Sendall, S., Strohmeier, A.: From use cases to system operation specifications. In: Evans, A., Kent, S., Selic, B. (eds.) UML 2000. LNCS, vol. 1939, Springer, Heidelberg (2000)
22. Wieringa, R.: A survey of structured and object-oriented software specification methods and techniques. ACM Computing Surveys 30, 459–527 (1998)

Finding the Pattern You Need: The Design Pattern Intent Ontology

Holger Kampffmeyer and Steffen Zschaler

Technische Universität Dresden, Germany
Holger.Kampffmeyer@googlemail.com,
Steffen.Zschaler@tu-dresden.de

Abstract. Since the seminal book by the Gang of Four, design patterns have proven an important tool in software development. Over time, more and more patterns have been discovered and developed. The sheer amount of patterns available makes it hard to find patterns useful for solving a specific design problem. Hence, tools supporting searching and finding design patterns appropriate to a certain problem are required. To develop such tooling, design patterns must be described formally such that they can be queried by the problem to be solved. Current approaches to formalising design patterns focus on the solution structure of the pattern rather than on the problems solved. In this paper, we present a formalisation of the intent of the 23 patterns from the Gang-of-Four book. Based on this formalisation we have developed a Design Pattern Wizard that proposes applicable design patterns based on a description of a design problem.

1 Introduction

Since Gamma, Helm, Johnson, and Vlissides (the so-called Gang of Four (GoF)) published their seminal book [14], design patterns have proven a useful tool in software development. A design pattern encapsulates a solution for a recurring design problem in template form, ready to be applied to new instances of the problem. It, thus, is a form of encoding and transferring design knowledge between projects and developers.

Because design patterns are so useful, lots of them have been discovered or developed and documented since the publication of the GoF book. Current design patterns have appeared in specific application domains (J2EE patterns [5,6,11], User-Interface patterns [20]), as language-dependent patterns (also called idioms), as patterns at different abstraction levels (analysis patterns [10], architectural patterns [11,12]), or simply as large collections of design patterns in pattern catalogues [7,25]. Even though the GoF book only contains 23 design patterns, the authors state that "it might be hard to find the one (design pattern) that addresses a particular design problem especially if the catalogue is new and unfamiliar to you." [14]. The sheer number of design patterns available today impedes effective reuse of design patterns, because it is very difficult to find the

G. Engels et al. (Eds.): MoDELS 2007, LNCS 4735, pp. 211–225, 2007.

right design pattern for a given design problem. This is especially true for inexperienced developers who do not yet know a large number of design patterns by heart. To deal with the large number of patterns effectively, software developers require tool support for finding design patterns that can solve a certain design problem. This paper is a step towards such tooling.

For this, we require a description of design patterns to be available in a machine-readable format. This description must contain a formal specification of the design patterns. It must be constructed in such a way as to allow querying based on the design problem to be solved.

Existing approaches to formalising design patterns generally cover only the formal description of the *solution structure* of design patterns. While the *structure* of a design pattern explains *how* it is applied in software design, it does not explain *when* to apply a design pattern for a given design problem. Only the *intent* section of a design pattern description explains the purpose of a design pattern. To the best of our knowledge, no work exists trying to formalise the intent of design patterns. However, software tools based on such a formalisation could enable users to query for a design pattern by giving a description of their design problem based on terminology defined in the specification. Ontologies are one way of expressing such a formalisation, because they directly support the creation and querying of such knowledge bases.

The main contribution of this paper is, therefore, a Design Pattern Intent Ontology (DPIO); that is, an extensible knowledge base of design patterns (in our case the 23 GoF patterns) classified by their *intent*.

The remainder of this paper is structured as follows: The following two sections give a short introduction to design patterns and to ontologies. Section 4, the main section of the paper, presents the DPIO. In Sect. 5 the ontology is evaluated by checking that certain competency questions (sample queries) can be formalised and answered based on the ontology. Section 6 discusses the design pattern wizard developed on top of the DPIO. The paper closes with a discussion of related work (Sect. 7) and a conclusion (Sect. 8).

2 Design Patterns

A design pattern is "*a solution to a problem in a context*" [13]. It is a way to achieve reusability in software design. Design patterns first emerged in the context of architecture and town building [4]. However, the idea of reusing design by applying patterns to recurring design problems has been ported to object-oriented software design in the GoF book *Design Patterns: Elements of Reusable Object-Oriented Software* [14].

In the GoF book, design pattern descriptions are structured into the following parts: *pattern name and classification, intent, motivation (forces), applicability, structure, participants, collaboration, consequences, implementation, sample code, known uses,* and *related patterns*. Formalisations of design patterns typically focus on the structure of the solution proposed in the pattern (for example, [17]). This does not, however, uniquely characterise a design pattern. Consider, for

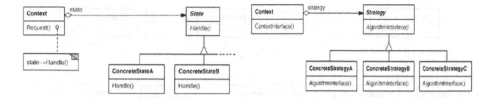

Fig. 1. The structure of the (a) *State* and (b) *Strategy* patterns. Copied from [14].

example, the patterns *State* and *Strategy* (cf. Fig. **??**). Their structure is more or less identical. However, their intent is not. The intent of *State* is given in [14] as

"Allow an object to alter its behavior when its internal state changes. The object will appear to change its class." [14]

In contrast, the intent of *Strategy* is

"Define a family of algorithms, encapsulate each one, and make them interchangeable. Strategy lets the algorithm vary independently from clients that use it." [14]

The intent of a design pattern is the first section a developer reads when trying to understand whether a design pattern is a solution to the developer's current problem. Hence, this is the section that should form the basis of a formalisation of design patterns that can help developers find the pattern they need.

3 Ontologies

"An ontology is an explicit specification of a conceptualisation" [16]. Ontologies were developed by the Artificial Intelligence community to support the sharing and common understanding of domain knowledge. Every ontology consists of a hierarchy of `classes`, `properties` (attached to the classes and used to model relationships between them), and `individuals` (instances of classes).

Ontologies are suitable means for formalising the intent of design patterns, because they allow to encode domain knowledge in a simplified abstract way and enable queries to be evaluated against a knowledge base defined by an ontology. For this reason, in this paper, we present an ontology-based formalisation of the intent of design patterns, thus defining a machine-readable, queryable catalogue of design patterns. We use OWL, the Web Ontology Language [26], as the formalisation language. We have chosen OWL, because it is a W3C recommendation (that is, an accepted standard) and because its good tool- and framework-support (see [2,3]) allows easy extension of the ontology and development of tools using the ontology as a knowledge base.

4 The Design Pattern Intent Ontology

The aim of the Design Pattern Intent Ontology (DPIO) is to support developers in choosing a design pattern for a given design problem. That is, the domain

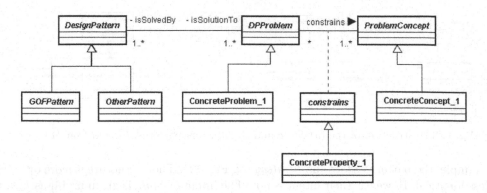

Fig. 2. The parent classes of the three hierarchies *Design Pattern*, *Design Problem* and *Problem Concept*

of the ontology is the area of software development. In this paper, we constrain the scope of the ontology to cover only the design patterns defined in the GoF book.[1] Thus, the ontology should provide the terms and concepts the GoF book uses to describe software design and design patterns.

The scope of the ontology is restricted to the intent and the application of design patterns in software design. The ontology must be elaborate enough to enable the querying for solutions to design problems. However, it is not intended to describe the *structure* of a design pattern. There is other work that is formalising these aspects of design patterns (see Sect. 7). For the scope of this work, the formalisation of the structure of design patterns does not give any additional benefits.

Competency questions are a way of determining the scope of an ontology [24]. They are the kind of questions the ontology should be able to answer. Here are some possible competency questions for the DPIO:

- Which design patterns are contained in the ontology?
- Which concepts are contained in the ontology that can be used to model a design problem?
- Which design pattern is a solution to the problem of varying an algorithm?
- Which design pattern is a solution to the problem of objectifying state?

In Sect. 5 we show how a formalised representation of the competency questions can be used to evaluate the ontology.

In designing the structure of the DPIO, we need to take into account the possible relations between design patterns and the problems they solve: One design pattern can be the solution to more than one design problem. But one design problem can also be solved by more than one design pattern. For example, both the *Prototype* pattern and the *Builder* pattern are concerned with object creation. Consequently, there exists an n:m relationship between design patterns and design problems. The solution to model this n:m relationship is

---

[1] This is for reasons of associated effort only. The basic structure of the DPIO has been designed to be extensible to arbitrary design patterns.

```
1   Class: StrategyDesignPattern
2
3   SubClassOf: GOFPattern
4
5      and isSolutionTo some
6          AlgorithmDecoupling
7      and isSolutionTo some
8          AlgorithmSelection
9      and isSolutionTo some
10         AlgorithmVariation
11
12   ...
```

Listing 1.1. Definition of the design pattern class `StrategyDesignPattern`

by defining a set of *design problems* and relating these design problems to design patterns. Therefore, a design pattern is a *solution* to one or more design problems. We furthermore describe a design problem by using *problem concept* terms. A design problem is *constrained* by problem concepts. Figure **??** gives a graphical overview of the core structure of the DPIO. The relations between `DesignPattern`, `DesignProblem` and `ProblemConcept` classes are depicted using UML-notations.[2] UML classes symbolise OWL classes, UML associations symbolise OWL object properties. The association between `DesignPattern` and `DesignProblem` indicates an object property `isSolutionTo` that relates `Design-Pattern` classes to `DesignProblem` classes. The property `isSolvedBy` is an inverse property of `isSolutionTo`. The association class `constrains` indicates an OWL object property that can be further specialised by subproperties. Both `DesignPattern`, `DesignProblem` and `ProblemConcept` classes are the root classes of subclass hierarchies specialising the root concepts.

To discuss the structure of the ontology in more detail, we look at some example definitions. Listing 1.1 is an example of a subclass from the `DesignPattern` hierarchy and shows the definition of the class `StrategyDesignPattern` in Manchester OWL syntax [18]. It states that an individual is a `StrategyDesign-Pattern` if it is a subclass of `GOFPattern` (Line 3) and is a solution to certain problems as per the intent of the *Strategy* design pattern: "Define a family of algorithms, encapsulate each one, and make them interchangeable. Strategy lets the algorithm vary independently from clients that use it. [. . .] Strategies provide a way to configure a class with one of many behaviors [they can be used when] we need different variants of an algorithm" [14]. The *design problems* formalising these aspects are `AlgorithmDecoupling` (Line 6), `AlgorithmSelection` (Line 8), and `AlgorithmVariation` (Line 10). Together they define a set of design problem facets with the need to decouple, select and vary algorithms.

---

[2] There seems to be no commonly agreed visual representation of ontologies yet. We have chosen UML class diagrams because they are easy to understand.

Consequently, the *Strategy* design pattern is a solution to these design problems and connected to them via the object property `isSolutionTo`.

```
1  Class: AlgorithmDecoupling
2
3  SubClassOf: DecouplingProblem
4
5  and decouples some
6        Algorithm
```

Listing 1.2. Definition of the design problem class `AlgorithmDecoupling`

Listing 1.2 is an example of a subclass from the `DPProblem` hierarchy and shows the definition of the class `AlgorithmDecoupling`. For an individual to be of class `AlgorithmDecoupling` it is necessary to be a member of the anonymous class of things that is linked to at least one member of class `Algorithm` via the object property `decouples` (Line 5–6). The property `decouples` is thereby an example of a subproperty of the `constrains` hierarchy. An `AlgorithmDecoupling` design problem is therefore simply concerned with the decoupling of algorithms.

The examples discussed so far (Listing 1.1 and 1.2) have been modelled using vocabulary defined in the DPIO. Tables 1 and 2 show an excerpt of the terms and concepts we have defined in the DPIO[3]. Table 1 shows the design problem hierarchy. An abstract concept `DPProblem` is defined that is the root node for more specific problems. The inheritance relationship is represented by a ∟ symbol and an indentation. The description column gives a short comment on the intent of each problem concept. For example, a `Problem` "is the abstract base class and the root node of the problem hierarchy". A `DPProblem` is a `Problem` that can be solved by design pattern solutions. Further specialisations are `ControlProblem`, `AlgorithmSelection`, `DecouplingProblem` etc. The top level hierarchy is based on the Tichy catalogue [28], an initial classification of design-pattern intents. However, we extend and restructure Tichy's hierarchy to allow a more detailed modelling of the problem domain.

Similar to classes, OWL properties also can be specialised by sub-properties. The OWL object property `constrains`, modelled as a UML association class, is the root property of more specialised object properties. Table 2 lists an excerpt of modelled OWL object properties which are used to describe design problems.

The aforementioned framework structure of `DesignPattern`, `DesignProblem`, `constrains`-properties and `ProblemConcept` is what allows developers to formulate queries without knowing the design patterns in the knowledge base. To do so, developers use the terminology provided in the `ProblemConcept` hierarchy and the `constrains`-properties to model their design problem. Standard ontology reasoning can then be used to determine the design patterns solving such

---

[3] The whole ontology can be downloaded from
http://www.holger-kampffmeyer.de/DesignpatternsIntentOntology.owl.

Table 1. The design problem taxonomy

Design Problem	Description
Problem	Abstract base class and root of the hierarchy.
∟ DPProblem	A problem from the domain of design patterns.
∟ ControlProblem	controls execution and method selection.
∟ ...	...
∟ AlgorithmSelection	controls algorithm selection.
∟ ...	...
∟ DecouplingProblem	divides a software system into independent parts. The parts can be built, changed, replaced.
∟ AlgorithmDecoupling	decouples algorithms from the rest of the system.
∟ ...	...
∟ VariantManagementProblem	treats different objects uniformly.
∟ ...	...
∟ AlgorithmVariation	varies an algorithm.
∟ ...	...

```
1  (retrieve (?solution )
2     (and (?solution |GOFPattern|)
3        (?solution ?designproblem |isSolutionTo|)
4        (?designproblem ?problemConstraint |<objectproperty>|)
5        (?problemConstraint |<someclass>|)
6  ))
```

Listing 1.3. Template-structure of a nRQL query to retrieve all design patterns that are a solution to some design problem

problems. The structure of a query asking for those design patterns solving a specific problem can look like Listing 1.3, showing the structure of a query in nRQL syntax [19]. In nRQL, a query is composed of a preliminary *command*, followed by the query *head* and *body* put in parentheses, respectively. The head of the query contains variables that are bound to the result set of the query. The body contains constraints of a query, similar to those of SQL where-clauses. It consists of one or more query atoms. Query atoms can be combined to complex queries by using logical operators such as AND, OR, and NOT. The query in Listing 1.3 retrieves all design patterns being a solution to a specified design problem. The placeholders <objectproperty> and <someclass> need to be replaced by concrete subproperties of **constrains** and subclasses of **ProblemConcept**, respectively. For example, <objectproperty> could be replaced by **distributes** and <someclass> could be replaced by **Behavior** to retrieve all *behavioral* design patterns. The Design Pattern Wizard we introduce in Section 6 generates queries with a similar structure to the query shown in Listing 1.3.

Table 2. The OWL object properties used to model design problems

Object Property	Description
constrains	binds a problem concept to a design problem, super property for all other properties.
∟ varies	allows diversity.
∟ ...	...
∟ controls	handles something.
∟ ...	...
∟ handles	equivalent property to "controls".
∟ ...	...
∟ selects	chooses some behavior.
∟ ...	...
∟ decouples	loosens the coupling of objects.
∟ ...	...

```
1  (concept-descendants |DesignPattern|)
```

Listing 1.4. nRQL query to retrieve all design patterns defined in the ontology

5   Evaluation

There are different possibilities for evaluating an ontology such as the DPIO. The final proof of the concepts can, of course, only be found through a controlled experiment in which developers are asked to use the ontology to solve certain design problems. Such an experiment would show if the ontology achieves our underlying goal of providing a formalisation of design patterns more appropriate to the problem of finding the design pattern one needs to solve a specific problem in software design. Performing such experiments is costly and time consuming. For this reason, we have not done so yet. However, we acknowledge the importance of such work and propose to perform it in future research.

A different approach to evaluating our ontology is to check whether it can answer the kinds of questions that are likely to be asked of it. To this end, we have developed a catalogue of *competency questions*. Here, we translate the competency questions into a formal version we can use to query the DPIO. In a second step, we test the ontology by verifying that the result set of the queries correspond to the intended meaning of the ontology. We use the nRQL query language [19] for formalising the competency questions.

Which design patterns are contained in the ontology? Listing 1.4 shows the formal representation of this competency question. It retrieves all children of the concept **DesignPattern**. The result set of the query contains the 23 GoF patterns modelled in the ontology.

Which concepts are contained in the ontology that can be used to model a design problem? Listing 1.5 shows the formal representation of this competency

```
1  (concept-descendants |ProblemConcept|)
```

Listing 1.5. nRQL query to retrieve all problem concepts defined in the ontology

```
1  (retrieve (?x )
2    (and (?x |GOFPattern|)
3    (?x ?p |isSolutionTo|)
4    (?p ?a |varies|)
5    (?a |Algorithm|)
6  ))
```

Listing 1.6. nRQL query to retrieve all design patterns that are a solution to varying an algorithm

question. Concepts that are intended to model a design problem are subclasses of the class **ProblemConcept**. Consequently, the nRQL query retrieves all children of **ProblemConcept**.

Which design pattern is a solution to the problem of varying an algorithm? Listing 1.6 shows the formal representation of this competency question. The query asks for those subclasses of **GOFPattern** that are linked to a design problem via the object property **isSolutionTo**. It furthermore asks for only those design problems that have an object property **varies** that relates the concept **Algorithm**. The query results in the Template Method and the Strategy design patterns.

Which design pattern is a solution to the problem of objectifying state? Listing 1.7 shows the formal representation of this competency question. The query asks for those subclasses of **GOFPattern** that are linked to a design problem via the object property **isSolutionTo**. It furthermore asks for only those design problems that have an object property **objectifies** that relates the concept **State**. The result set for this query contains the Memento and the State design patterns.

```
1    (and (?x |GOFPattern|)
2    (?x ?p |isSolutionTo|)
3    (?p ?a |objectifies|)
4    (?a |State|)
5  ))
```

Listing 1.7. nRQL query to retrieve all design patterns that are a solution to objectifying state

6   The Design Pattern Wizard

As we have outlined in the introduction, an ontology can be used in tools as a knowledge base. The Design Pattern Intent Ontology contains the vocabulary for describing the intent of design patterns. In order to extract knowledge from the ontology, the user has to execute queries on it. However, the construction of these queries can be quite complicated. This reduces the usability of ontologies to domain experts only. The Design Pattern Wizard serves as a front-end for generating well-defined queries. It allows design problems to be described visually and suggests a set of matching design patterns for a given design problem. It, furthermore, provides inexperienced users with vocabulary they can use to define design problems. The Design Pattern Wizard can be obtained from the first author at email request.

The Design Pattern Wizard is a prototype and proof of concept. It shows the applicability of ontologies for tool support in the area of software design. The Design Pattern Wizard has been implemented as an Eclipse RCP (Rich Client Platform) application [1]. The RCP architecture allows the developer to configure an application to either be integrated as a plug-in into the Eclipse platform or to be deployed as a stand-alone application.

Figure 3 shows the problem description window when the Design Pattern Wizard is started. The main part of the dialogue is filled by the problem description table. It consists of a predicate constraint column and an object (concept) constraint column. Each row represents a statement constraining the design problem a design pattern should solve. The first column consists of check boxes used to select rows for editing or deletion. Below the constraint table resides a button for adding constraint rows to the table and a button for deleting a selected row. In the bottom right corner of the dialogue the button for retrieving a design pattern suggestion based on the design problem description is located.

Clicking on the *predicate constraint* in one of the rows allows to open the *Predicate Constraint Dialog* shown in Fig. 4 a). A tree representation of all *predicates* modelled in the ontology is presented. The user can choose a predicate and the

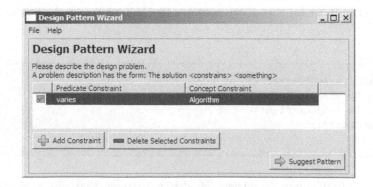

Fig. 3. Screenshot of the Design Pattern Wizard

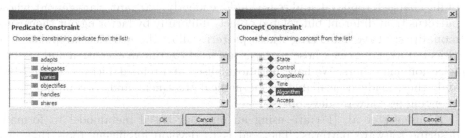

(a) constraining problem predicates   (b) constraining problem concepts

Fig. 4. Screenshot of the dialogues constraining problem predicates and problem concepts

chosen predicate value will be set in the constraint row of the wizard. A similar dialogue is opened when the user wants to constrain the *concept constraint* part of a problem constraint. The *Concept Constraint Dialog* (Fig. 4 b)) allows the user to select a concept that further constrains a predicate. The selected concept is then set in the constraint cell.

Figure 5 shows the *Result Dialog* that is opened when the user hits the *Suggest Design Pattern* button. It presents all suitable design patterns matching the modelled design problem in a simple table. The first column shows the name of the design pattern while the second shows a description of the pattern.

Suitable Design Pattern(s)

The following Design Patterns solve the given problem description!

Design Pattern	Description
Strategy	Define a family of algorithms, encapsulate each one, and make them interchangeable. Strategy let...
Template	Define the skeleton of an algorithm in an operation, deferring some steps to subclasses. Template ...

Fig. 5. Screenshot of the result window suggesting suitable design patterns

7   Related Work and Discussion

Our survey of current work in design pattern formalisation indicates that no other work before has tried to classify design patterns according to their intent by using ontologies.

The formalisation of design problems in this paper is based on the work of Tichy [28], who developed a catalogue (but no formal ontology) listing over 100 design patterns. In his classification, he concentrates on the problems patterns solve. Among Tichy's design problem categories are `decoupling` which refers to dividing a software system into independent parts such that the parts can be built, changed, replaced, and reused independently; `variant management` whose patterns treat different but related objects uniformly by factoring out their commonalities; `state handling` whose patterns allow the generic manipulation of object state; `control` whose patterns are used to control execution and method selection and others. We have both formalised and refined Tichy's classification and, thus, made it available to mechanical treatment and to computer-aided querying by software developers.

Henninger et al. [17] are using an ontology-based metamodel to formally describe software patterns. The goal is to develop intelligent tools that provide a computational basis utilising software patterns. As a use case they mention usability patterns. They base their core metamodel on properties such as `hasProblem, hasSolution, hasContext, hasRationale, hasForces`, properties that are developed from the original structure of the pattern description of [4]. They extend the core metamodel with properties that describe the relationships between patterns, such as `uses, requires, alternative`, and `conflictsWith`. The ontology they have developed does not so much concentrate on the support of selecting a suitable pattern for a given problem, but rather on the relationships between patterns. Furthermore, Henninger et al. concentrate on the domain of usability patterns of web sites, but not on GoF patterns or design patterns in general. The main goal of their work is to define a shareable vocabulary in the domain of usability patterns. Furthermore, they do not describe how to query their ontology, nor do they describe how to model design problems in an ontology. A very similar approach to [17] is [23]. Here, ontologies are used to formalise hypermedia and web design patterns. The vocabulary is almost identical to [17] but includes additional concepts such as `PatternComponent, Category, Problem` and `Solution`. The scope of their work is the support of and an integration into hypermedia design tools.

Pereira de Medeiros et al. [8] have developed the Kuaba Ontololgy, an ontology and design vocabulary to describe the design rationale of software design. The goal is to make explicit the decisions and justifications that have led to a design. The formal description of the design rationale enables reuse at a high abstraction level. Important reasoning elements of the ontology are `Question, Idea` and `Argument`. The vocabulary defined in the Kuaba Ontology helps in the decision-making process of software design. Their work is not particularly focused on design patterns, but more on software design in general.

The work of Dietrich et al. [9] uses OWL to formally describe design patterns. However, only the structure of design patterns is considered, not the intent or applicability. They use their ontology to detect patterns in software artefacts, not to help in the selection process of a design pattern. Other approaches that use formal languages to describe the structure of design patterns include [21] who developed a design pattern modelling language DPML, and [22,27]. What is common to all this work is the sole concentration on the structural aspects of design patterns and the omission of the intent of design patterns.

8 Conclusion and Further Work

In this paper, we have presented a novel approach to formalising design patterns, based on their intent. We have proposed the Design Pattern Intent Ontology providing terminology for formulating intents and classifying the 23 GoF design patterns by their intent. To the best of our knowledge, this is the first formalisation of design patterns based on their intent. This ontology forms the basis for the Design Pattern Wizard, a tool supporting software developers in finding the right design pattern(s) for a given design problem.

The work presented in this paper enables software developers to efficiently find design patterns applicable for their design problems. It is, thus, a measure countering the effect of the ever increasing number of design patterns available in various design pattern catalogues. The hierarchical structure of the ontology allows developers to provide incomplete descriptions of their design problems and still receive valuable responses. A developer with only a rough picture of her design problem can simply choose predicates from the top parts of the hierarchies. Such a query results in the retrieval of all design patterns matching this rough problem description. Iteratively, the developer can describe her design problem more precisely, based on the results of the former queries. The basic structure of `DesignPattern`, `DesignProblem`, and `ProblemConcept` is valid for all design patterns. Therefore, the ontology can easily be extended to cover other design patterns beyond the GoF book. To this end, it may be necessary to add new vocabulary to the `DesignProblem` and `ProblemConcept` hierarchies. Testing the ontology with other catalogues of design patterns remains for future work.

So far, the Design Pattern Wizard is a stand-alone application. In the future, this should be integrated with CASE tools to allow direct integration of a pattern found into a design under development. Developers could select a design and add a design pattern to it, using the Design Pattern Wizard to select the pattern. For this to work, we need to study ways of selecting and manipulating the parts of a model where a design pattern should be added, in addition to the formalisation of design-pattern intent.

Another interesting question is how we can help developers understand the problem they are trying to solve. We believe, developers often know that there are flaws in their design, but cannot immediately understand the source of the problem. Thus, they will experiment with different design choices, which with good designers will eventually lead to a better understanding of the problem, and, thence, to a better design. An interesting question is if this problem-finding process can be supported by CASE tools observing the different experiments of a designer and using the DPIO to suggest design patterns that might be helpful.[4]

Additionally, the basic structure of the DPIO should also be applicable to patterns that are not strictly design patterns. For example, [15] have proposed an approach to graphically organise, analyse and refine non-functional requirements for the structuring, understanding, and applying of design patterns during design. We plan to study how the DPIO can be applied in this context.

---

[4] Thanks to Mirko Seifert for this suggestion.

References

1. Eclipse – Rich Client Platform (2006),
   http://www.eclipse.org/home/categories/rcp.php
2. Jena – a semantic web framework for Java (2006),
   http://jena.sourceforge.net/
3. Protégé ontology editor and knowledge acquisition system (2006),
   http://protege.stanford.edu
4. Alexander, C., Ishikawa, S., Silverstein, M.: A Pattern Language. Center for Environmental Structure Series, vol. 2. Oxford University Press, New York, NY (1977)
5. Alur, D., Crupi, J., Malks, D.: Core J2EE Patterns: Best Practices and Design Strategies. Pearson Education (2001), Patterns catalog available at
   http://java.sun.com/blueprints/corej2eepatterns/index.html
6. Alur, D., Crupi, J., Malks, D.: Sun Java center – J2EE patterns (March 2001),
   http://java.sun.com/developer/technicalArticles/J2EE/patterns/
7. PatternShare Community. Patternshare community (2006),
   http://patternshare.org/
8. de Medeiros, A.P., Schwabe, D., Feijó, B.: A design rationale representation for model-based designs in software engineering. In: Belo, O., Eder, J., e Cunha, J.F., Pastor, O. (eds.) CAiSE Short Paper Proceedings, CEUR Workshop Proceedings. CEUR-WS.org. vol. 161 (2005), http://www.ceurws.org/Vol-161/FORUM_27.pdf
9. Dietrich, J., Elgar, C.: A formal description of design patterns using OWL. In: Australian Software Engineering Conference (ASWEC'05), pp. 243–250. IEEE Computer Society, Los Alamitos (2005), http://doi.ieeecomputersociety.org/10.1109/ASWEC.2005.6
10. Fowler, M.: Analysis Patterns: Reusable Object Models. The Addison-Wesley Object Technology Series. Addison-Wesley Professional, Reading (1996)
11. Fowler, M.: Patterns of Enterprise Application Architecture. The Addison-Wesley Signature Series. Addison Wesley, Reading (2003)
12. Fowler, M.: Patterns in enterprise software (2005),
    http://www.martinfowler.com/articles/enterprisePatterns.html
13. Fowler, M.: Writing software patterns (August 2006),
    http://martinfowler.com/articles/writingPatterns.html
14. Gamma, E., Helm, R., Johnson, R., Vlissides, J.: Design Patterns: Elements of Reusable Object-Oriented Software. Addison-Wesley Professional Computing Series. Addison-Wesley Publishing Company, New York, NY (1995)
15. Gross, D., Yu, E.S.K.: From non-functional requirements to design through patterns. Requirements Engineering 6(1), 18–36 (2001),
    http://citeseer.ist.psu.edu/gross00from.html
16. Gruber, T.R.: A translation approach to portable ontology specifications. Knowledge Acquisition 6(2), 199–221 (1993),
    http://portal.acm.org/citation.cfm?id=173747
17. Henninger, S., Padmapriya, A.: An Ontology-Based Metamodel for Software Patterns. In: Seke2006. 18th Int. Conf. on Software Engineering and Knowledge Engineering, San Francisco (July 2006) (to be presented),
    http://cse.unl.edu/~scotth/papers/SEKE06-TechReport.pdf
18. Horridge, M., Drummond, N., Goodwin, J., Rector, A., Stevens, R., Wang, H.H: The Manchester OWL syntax (2005), http://owl-workshop.man.ac.uk/acceptedLong/submission_9.pdf

19. Racer Systems GmbH & Co. KG. Racerpro users guide version 1.9 (2005), http://www.racer-systems.com/products/racerpro/manual.phtml
20. Mahemoff, M., Johnston, L.: Principles for a usability-oriented pattern language (1998), http://citeseer.ist.psu.edu/article/mahemoff98principles.html
21. Mapelsden, D., Hosking, J., Grundy, J.: Design pattern modelling and instantiation using DPML. In: CRPIT '02. Proceedings of the Fortieth International Conference on Tools Pacific, pp. 3–11. Australian Computer Society, Inc., Darlinghurst, Australia, Australia (2002)
22. Mikkonen, T.: Formalizing design patterns. In: ICSE '98. Proceedings of the 20th international conference on Software engineering, pp. 115–124. IEEE Computer Society, Washington, DC, USA (1998)
23. Montero, S., Diaz, P., Aedo, I.: Formalization of web design patterns using ontologies (2005)
24. Noy, N.F., McGuinness, D.L.: Ontology development 101: A guide to creating your first ontology. Technical Report KSL-01-05, Knowledge Systems Laboratory, Stanford University, Stanford, CA, 94305, USA (March 2001)
25. Portland Pattern Repository. Portland pattern repository (2006), http://c2.com/ppr/
26. Smith, M.K., Welty, C., McGuinness, D.L.: OWL web ontology language guide (2004), http://www.w3.org/TR/owl-guide
27. Taibi, T., Ngo, D.C.L.: Formal specification of design patterns - a balanced approach. Journal of Object Technology 2(4), 127–140 (2003), http://www.jot.fm/issues/issue_2003_07/article4.pdf
28. Tichy, W.F.: A catalogue of general-purpose software design patterns. In: TOOLS '97. Proceedings of the Tools-23: Technology of Object-Oriented Languages and Systems, IEEE Computer Society, Washington, DC, USA (1997)

Model-Driven Approach for Managing
Human Interface Design Life Cycle

Noi Sukaviriya[1], Vibha Sinha[2], Thejaswini Ramachandra[3], and Senthil Mani[2]

[1] IBM TJ Watson Research Center,
19 Skyline Drive, Hawthorne, NY 10532, USA
[2] IBM India Research Lab,
ISID Campus, Vasant Kunj, New Delhi 110070, India
[3] IBM India Software Group,
Embassy Links, Indiranagar-Koramangala, Intermediate Ring Road, Bangalore 560071, India
noi@us.ibm.com, {vibha.sinha,thejaswini,sentmani}@in.ibm.com

Abstract. Designing a large application user interface is an iterative process. Commonly used tools lack models to support this iterative process. Research on model-driven UI design has over the years focused on modeling UI at a higher level of abstraction but lacked support during in the iteration process. This paper briefly presents the context of our research – transforming a business model into a base UI model for further customization. Specifically, we present a feature that helps reflect changes from the business model in the user interface design tool. We designed it so that the human designers can choose to react to these changes as they see appropriate. The technique is one of our attempts to apply the model-drive approach to better support design iteration through requirement changes.

Keywords: User Interface Model, Human Computer Interaction Model, User-Centered Design, User Interface Modeling Tools, Model Transformations, Model Engineering Methodologies.

1 Introduction

For large enterprise solutions, designing human interactions is a complex process. Products of this design process – solution storyboards, low-fidelity UI mock-ups, and high-fidelity designs – are the bridge between business requirements and solution development. Business stakeholders express requirements more explicitly as a solution design become more concrete through these design products. Users give better feedback when they can see a design in early prototyping phases. Keeping user interface design products in sync with requirement changes is very tedious via today's popular tools such as Adobe products, Macromedia family products, or Microsoft PowerPoint. A small change in requirement could mean updating many static design pages embedded in multiple documents and websites. One can easily imagine the multitude of this task.

As the business world is becoming more competitive, the overall design process is on a constant pressure to achieve more with less. Quality often goes down with the

G. Engels et al. (Eds.): MoDELS 2007, LNCS 4735, pp. 226–240, 2007.

time pressure. When there is less time to propagate requirement changes to design, it often leads to requirement compromises. While some user-centered design activities such as user interviews cannot take shortcuts, design tools can be improved to accelerate the design maintenance process. When a high-level design model is used to generate UI code, inefficiency from integrity loss from design to development can be reduced.

Our team focused on model-driven business transformation research, specifically on deriving methodologies that enable better and automated connections from business process modeling to the underlying IT solution [1,2]. Through several business transformation engagements, we have experienced the enterprise-level solution design process and gained valuable insights into the iteration bottleneck in the large enterprise UI design process. Our goal is to better the solution design process to meet business requirements and keep up with high-quality UI. The research tools reported in this paper is part of how we enable designers to conform and connect UI design with business domain model elements, and to help maintain design integrity through automatic generation. Our work primarily focuses on the UI design process and not started on UI design management. The latter is specifically reported in this paper.

Since our research is driven by case studies from real business engagements, we will use an example from a recent engagement to drive the discussion in this paper.

2 Related Work

Model-driven user interface research has been around as several references to early efforts can be found in [3]. According to Myers, Hudson & Pausch in [3], the motivation behind model-based UI design tools was to enable programmers who, without user interface design experience, could implement only the functionality and let user interfaces be automatically generated. Early efforts focused on exactly that by explicitly representing components that makes up the user interfaces from user tasks to interaction techniques [4,5,6,7]. Nowadays design skills are abundant but tools that could help these designers manager their jobs, not eliminating them, are seriously lacking.

Automatic generation of user interfaces dated back to the early 90's. In [8], a comprehensive notation for expressing the quality of data attributes was used. This work also attempted to lay out the whole page detail, a task that is quite easy for a designer to do but tremendously hard for a machine. Related work in [9] was at a much smaller and practical scale, automatically generating for only selected attributes, that inspired the automatic generation in our work. Venderdonkt in [7] does semi-automatic generations through Automatic Interaction Objects structure.

Generating user interfaces from task-based models was done in a series of publications by Paterno in [4,5,10]. His approach is quite similar to ours while our work use business tasks, which tend to be much higher level, as a starting point. Our work uses the business task model as the input into the design process. One of the major embarkations on using UML to represent high-level abstraction of user interface was reported in [11.] In this work, Nunes and Cunha put the interaction design methodology of separating the information from the design and the application logic into perspectives through a set of corresponding UML profile elements. Nunes' later work with Campos [12] attempted to provide User-Center Design (UCD) support, a goal similar to ours, though their work focused on reusable interaction patterns.

With the prevalence of XML, we have seen more of modeling languages that drive user interface creation such as XForms, Laszlo, XIML, though these models are fairly page-based low-level models without much inter-connections between various parts of the design or to design intentions. We have not found other related work that attempts to understand the role of user-centered design in the context of business process modeling.

3   Approach for User Interface Life Cycle Support

Our approach begins with business modeling, a starting point of many IT business engagements in consulting practices today. A business model expresses a business flow by means of sequences of business tasks that must be performed, user roles or an organization that performs each task, and the information which flows through tasks. Our design methodology harvests information about "people" and their work context associated with business tasks from the business model. We collect information artifacts that are passed through people as inputs to and output from business tasks. We then churn the collected information into views that focus on the "users."

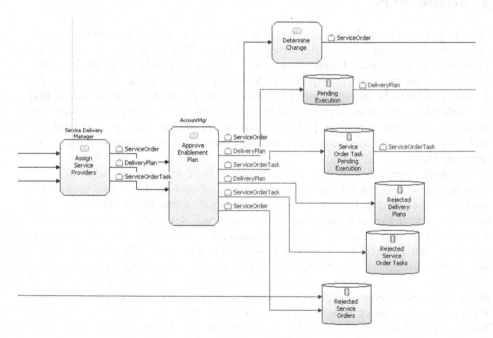

Fig. 1. A snippet of a service order management process

Figure 1 illustrates a snippet of a service order management process drawn in a particular software product called IBM Websphere Business Modeler (WBM). It shows the "Approve Enablement Plan" task of the account manager user role. In this paper, we will discuss various elements pertaining to a page design for this task. In this task, the account manager views the content of a "Service Order" with a list of "Delivery Plan" and "Service Order Tasks" to be performed by various service providers. The service providers were assigned by a "Service Delivery Manager" in the previous task. When the designers discuss with business analysts and subject matter experts (SMEs) about this task, they have the context that the account manager will either approve or reject the order based on the validity and plausibility of the combined plan. We want to make it clear that the designer's awareness of the meaning of this diagram is much deeper than what the diagram is capable of expressing; hence a business model like this is not the end-all be-all source of information for interaction designers.

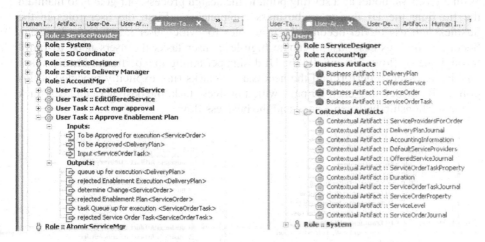

Fig. 2a. The "User – Task" view   **Fig. 2b.** The "User – Artifact" view

To begin the human interface modeling, the business model with diagrams as such is transformed into a human-centric perspective in our Model-Driven Human Interaction (MDHI) tool. In this particular implementation, the WBM model in its custom format is converted into the human interaction model in UML2. The transformation turns the business model around to show "what each user role" does in the "User – Task" view (Figure 2a) and what kind of information altogether each role has to manage in the "User – Artifact" view (Figure 2b). Information that flows through user tasks is turned into either "business" artifacts or "contextual" artifacts in this view. Contextual artifacts are information that are inputs to single tasks while business artifacts are passed around among tasks. We use the artifact notions to inform designers of the significance of the information in the context of the business model.

Please note that though our work is specific to the IBM WBM business model, the transformation from the business domain to user-centered design domain harvests generic business information such as user roles, tasks, task inputs and outputs, as well as task flow in the business context. In other words, we connect the business domain concepts to user-centered design concept through shared vocabulary. The transformation concept of this work is transferable though the specific code will not be generically usable in other business modeling tools.

It is the user interface model that is gradually built up by the designers in the "User – Design" view (for example shown in Figure 3a) that, at the end of the design process, captures the structure of the pages and UI components that can generate an executable interface. Since the focus of this paper is not the overview of the modeling environment, readers can refer to [13] for more information about the views and the transformation. We will only touch upon particular aspects of this environment that support the technical discussions in the following 3 sections.

3.1   Remaining in Scope with Business Design

With a business model as a starting point to the design process, our goal is to maintain the tie from the UI design to business components. First, the user roles in the business process model become the user roles for which user interface is designed. Secondly, the business tasks become high-level user tasks the users need to fulfill through the UI flow. Essentially the UI design pertaining to a particular business task enables the user to perform all the necessary subtasks (not explicitly modeled at this point.) By associating each UI page with a business task, the designer can keep track of the relationship between a page and the business flow.

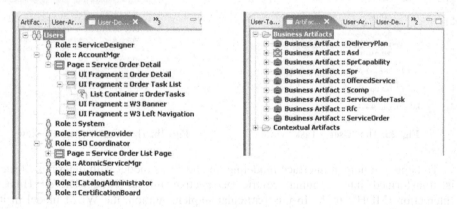

Fig. 3a. The User-Design view showing an artifact has been removed

Fig. 3b. The "Artifact – User" view showing a user role has been removed

Typically, the business model and the user interface design are iterated for a period of time before they stabilize. The ties between elements from the business model to the elements added as part of the human interaction design process help the designer keep track of the business connections, hence help maintain business integrity. However, the ties are not designed to be too rigid as to hinder the design process. For example, we cannot quite assume removal of elements in the business model should automatically eradicate human interaction elements related to them. When a design has started for some time, there is quite a bit of human investment into the design efforts. Design of pages associated to a single user or a single artifact that may have been removed from the business model could be reused in other parts of the solution or archived for future use. In MDHI, we only inform the designer of the changes from the business domain. It is up to designers to evaluate the consequence of the changes as human are much better at this kind of decisions. It is up to them to remove these elements from their design space. Figure 3a shows an example from the "User – Design" view when a user role is removed form the business model. Figure 3b shows the "Artifact – User" view after a business artifact has been removed. We currently handle only changes in user roles, user tasks, artifacts and artifact attributes. We do not deal with changes in task sequence at this point.

3.2 Coping with Iterations

Changing in business requirements happens continuously during design evolution, regardless of how careful one might be from the beginning. From our experience, once an initial cut of a user interface design is presented, it gives business analysts and subject matter experts more concrete handles which they can discuss in further detail. They then can recognize shortcomings of the process or wrong attempts to pull unrelated actions together. By going from a high-level design to a concrete design, business analysts can visualize and actually say "you need to add total amount to this order detail view" or "this task needs to happen first." Repeatedly, requirements become more detailed; common understandings of business solution are confirmed in finer details.

The traditional process of using drawing tools such as Adobe Photoshop, Adobe Illustrator or Microsoft PowerPoint to produce design snapshots has at least two drawbacks when it comes to dealing with requirement changes. First, little changes such as adding or dropping a field to a set of information could result in changes in many mock-up pages, making it rather tedious and time consuming to keep up with requirement changes. The other drawback is lack of a visible structure that could help designers locate change locality. A particular screen snapshot may be identified for changes, but the designers would have to search for the source of drawings and the right drawing layers to pinpoint where corrections must be made.

One of the modeling concepts we use to link design elements to the business model, in particular information artifacts, is "data group." A data group is a subset of data elements in an artifact that appears visually together on a UI page. From our example, a service order has many attributes (which can be seen in Figure 5,) some of which are key to provide a summary of a service order. Figure 4 schematically illustrates the "data group" concept that allows a subset of attributes to be selected from the full list of artifact attributes. A data group is then related later on to a UI element on a page. A data group can be reused multiple times on different pages. Data groups can overlap. In other words, the same attributes can be included in multiple data groups as the designer see appropriate. We also support data groups that can span across artifacts but it must be used with caution. The intention is not to encourage careless information mash-up but to provide the users with good proximity of information they need for a task, even if it means crossing boundary of artifacts.

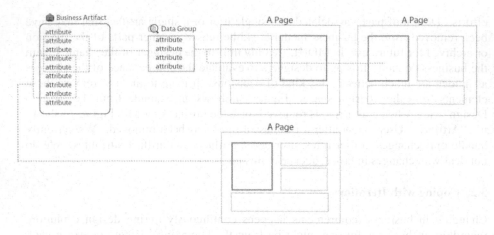

Fig. 4. Illustrating the relationship of a data group, and artifact, and a display

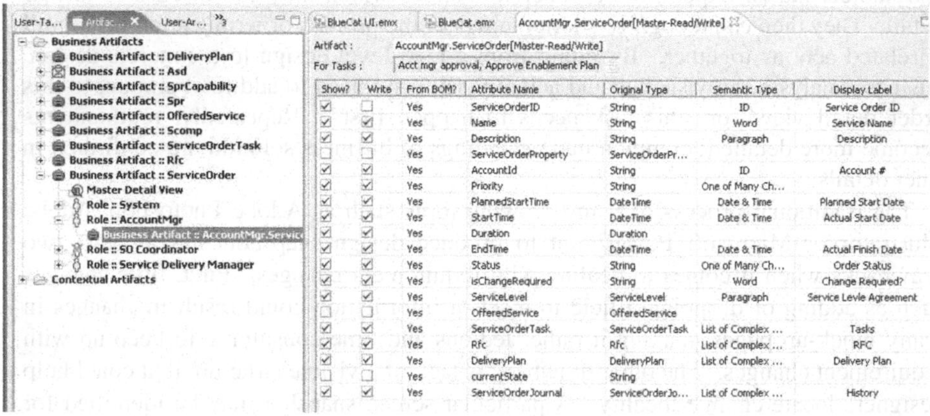

Fig. 5. A full set of attributes associated with the service order business artifact

Figure 5 above shows the full list of attributes of the service order artifact and Figure 6 on the following page shows a data group that is created as a subset. Both figures display the "Artifact – User" view which supports the views of artifact attributes and their types. Notice that this view is also used to define UI semantic types and display labels of each attribute. UI semantic types are the modeling concept we use to associate a UI hint to each attribute. MDHI uses semantic types and display labels to generate UI elements when a data group is associated to a UI element, with automatic generation capability, on a page. Semantic types should be defined globally across user roles in this view for each artifact so the same types hold true through out the design scope. Display labels can be defined globally, per user role, and per page. When a data group is created, semantic types and display labels are inherited from parent definition.

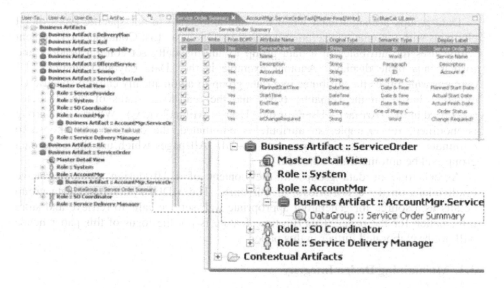

Fig. 6. Service Order Summary Data Group Showing a Subset of Service Order Attributes

Now let us demonstrate how a small change is handled in MDHI. Let's say the attribute priority is no longer needed. Business executives decide there is no need to define service order priority. The attribute is removed from the business model. Figure 7 illustrates the indication in the "Artifact – User" view that the priority attribute has been removed from the business model. It is up to the designer to remove this attribute from the master view of the service order artifact, which will be reflected in all child data groups. Or if the designer remains skeptical of such a change, she may choose to only remove the priority attribute from a particular data group while continuing with her design.

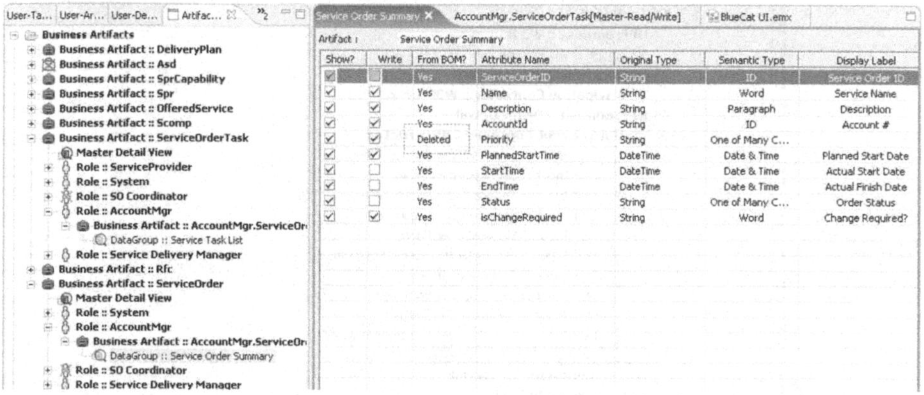

Fig. 7. The "Service Order Summary" data group shows the priority attribute deleted removed from the business model

Data groups are significant when they are associated to our specific types of UI elements with automatic generation capability. One is a "List Container" UI element which uses the attributes in a data group to automatically set column headers and format in a table display. Another example is the "Data Layout Container" UI element which uses the attributes in a data group to automatically call out specific UI elements based on the semantic types and the read/write status of the attributes (specified in the "Artifact – User" or the "User – Artifact" view.) When a data group is modified, for example, an attribute is eliminated, the designer can choose to eliminate the attributes and re-generate the UI. All pages which use the same data group will be automatically adjusted.

Associations of data groups to the content of a UI design page also have implications on the data retrieval at run time. The MDHI system uses data group details as a source for generating appropriate web service calls to retrieve and store information for the page. This run-time feature is not the focus of this paper hence will not be elaborated further in this paper.

3.3 Maintaining Design Integrity

One of the major problems in delivering a user interface from design to implementation is the lost of integrity. We have seen repeatedly high fidelity designs losing element placements and precision, which often means losing the semantic clusters of the application semantics. While a WSIWYG editor is a better solution for this aspect, we have developed a model that would be appropriate for such a tool that also works with our framework. Our unique contribution is in how we instrument the meta-model to provide a design hierarchy that structures the design per user role. This makes it simpler to manage a large design space.

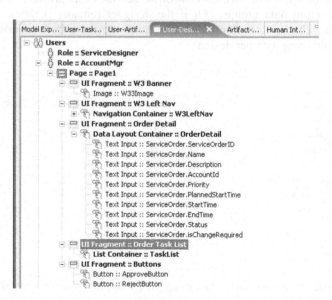

Fig. 8. The "User – Design" view showing a model structure of the "Service Order Detail" page that shows the summary of the service order and its related service order tasks

Figure 8 shows the "User – Design" view where pages designed for each user role are organized and structured. Notice a structure of "Page1" designed for the "account manager" user role. The "Service Order Summary" data group in Figure 6 is associated to the "OrderDetail" data layout container; its content shown in the list below is automatically generated by MDHI when the designer clicks on "Auto Gen" button in data layout editor (not shown in the paper.) This portion of the UI can be seen in the top half of the generated UI in Figure 9. Another data group for the service order task artifact is associated with the list container, which generates a table of service tasks shown in the bottom half of Figure 9. The "Approve" and "Reject" buttons, a left navigation, and a low-fidelity banner are also defined for this page.

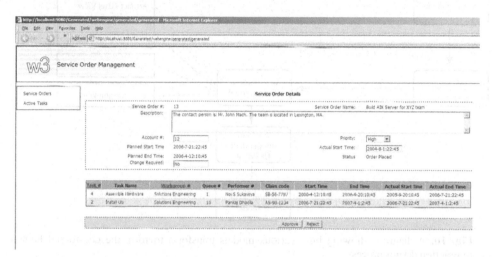

Fig. 9. The page that is automatically generated from the model description in Figure 8

The precision of UI elements are stated in the MDHI design model as a function of the designer-defined grid. For example, the screen in Figure 9 is designed with a grid size of 16 by 9, where the width of 16 units occupies 100% of the screen width and the height of 9 units occupies 100% of the screen height (minus what is lost from the browser's overhead.) The precision control is not accurate to the pixel levels as the underlying web programming model doesn't allow such precision. Even so we are still refining our precision control and will continue to improve in this respect. Currently the precision and the integrity of the design are reasonably supported.

4 Models

In this section, we will discuss part of our human interaction model that supports the design process with the selected features mentioned in this paper. Due to space limitation, we cannot elaborate on all aspects of the model.

Figure 10 shows the schematic view of various models in our research tool. To enable parallel yet collaborative iterations of the business model and the human interaction design, we instrument the MDHI model as an extension to the core translation of the business model. First, the business model is transformed into a "core model." Unique identifications of business model elements are maintained here and the core model functions as a reflection of the current state of the business model. The UI extension model maintains the references, which needs to be established once when the UI extension model is first created, to the core model elements. Any changes in the business model that follow are updated in the extension model when MDHI views open. New references are new elements and missing references are shown as eliminated elements.

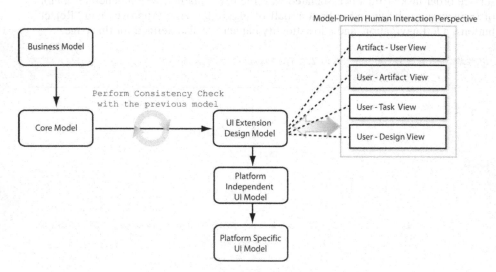

Fig. 10. A diagram showing how various models transform to form the end-to-end human interaction design process

The UI design model captures both the structure of the screen designs, the human perspective interpretation of the business model, as well as the relationships from the UI design to business model elements. To generate an executable UI, the screen content portion of the model is used to generate a more efficient platform independent model which then is used to generate a platform specific model that enables a running user interface. The platform independent model is currently captured in XML. The next 3 sections highlight selected parts of the MDHI model pertaining to this paper.

4.1 Design Model

The design model is centered on the user role and how each user role has references to various aspects of the business and design meta-structures. Information pertaining to the human interaction design process such as user tasks, task inputs and outputs, the types of information the user needs to juggle in a day (list of artifact types in the "User – Artifact" view) and the scope of the user functionality (list of business tasks as user tasks in the "User – Task" view) all point to the user role. The central connection to the user role is one key dimension in making and exposing the user-centric perspective of our design model.

Another dimension of our design model comes in the support for the design hierarchy and the inheritance of semantically enhanced UI information in order to help speed up the design process. The data group model to be discussed in Section 4.2 is an example of how we support sharing information among multiple design components. Another example is in the use of the pseudo "users" user role. A page designed for this user role can be used across the design space. Figure 11 portrays most of the stereotypes define in the Human Interaction UML2 profile. The meta-classes from which MDHI stereotypes extend are shown in parentheses.

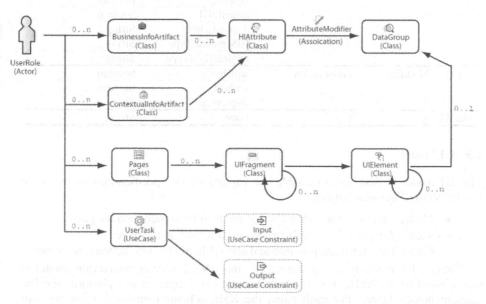

Fig. 11. A meta-model diagram of the human interaction design stereotypes. In this diagram, the user role is centric to aspects that are considered in the human interaction design.

4.2 Group Model

Two modeling elements, the HIAttribute and AttributeModifier stereotypes, are the key elements that allow artifact attributes to be shared yet customizable for various design contexts. Table 1 shows the attributes of these 2 stereotypes.

Notice that while business artifacts, contextual artifacts, and data groups appear to the designers to have the most information in the MDHI various views, the information load is internally on instances of the HIAttribute and AttributeModifier stereotypes. Via instances of these two stereotypes, the system maintains and manages the consistency of information among them.

In the current implementation, we assume artifact attributes come directly from the business model. In practice, we find that this level of detail is often not available unless it is modeled by someone with a more technical background. The UI extension model in a way is insensitive to the data source. We foresee that, in the future, the data model source could be other data modeling tools while the business model still gives the essence of how the information is aggregated in the business flow. Our approach remains in tact with close ties to the business model, though detailed data could come from elsewhere.

Table 1. Attributes in the HIAttribute, AttributeModifier, and DataGroup stereotypes

Stereotype	Base Meta-Class	Attribute	Type
BusinessInfoArtifact	Class	userFriendlyLabel	String
ContextualInfoArtifact	Class	userFriendlyLabel	
HIAttribute	Class	writability	boolean
		visibility	boolean
		addedInHI	boolean
		displayLabel	String
		HISemanticType	String
		originalDataType	String
AttributeModifier	Association	writability	boolean
		visibility	boolean
		displayLabel	String
DataGroup	Class	name	String

4.3 UI Model

The UI model is designed to enable the capture of UI specifications in the usual Model-View-Controller pattern.

- Model – the set of information that needs to be rendered on the page
- View – how this set of information is presented on the page
- Controller – what are possible and allowable human interactions on the page

When a UI is ready to be generated, the UML2 human interaction model is transformed to an XML model which becomes the input to the platform specific implementation layer. For each page, the XML schema contains 3 main types of information:

- Page Content – describes the page layout and the placement of UI elements. This content is specified in the "User - Design" view and constitutes the "view" part of the MVC pattern. Currently, we decorate style information on appropriate elements in the model here. This will change in the future when MDHI views are enhanced.
- Page Bindings – list the bindings from UI elements to data elements. This information is specified in data groups and their mappings to UI elements, which altogether translates to the "model" part of the MVC pattern.
- Page Interactions – defined actions and page transitions on the page. This information is captured in "user task" view (not discussed in the paper) and makes up the "controller" part of the MVC pattern.

Figure 12 shows snippets of the XML model for the service order details page (Page1) example shown in Figure 9.

Finally, the XML UI model specified as above is transformed into a platform specific IBM "Websphere Portlet Factory" model (a commercial software product) which in turn generates the corresponding JSPs and backend java beans. We follow the loosely coupled consumer-provider pattern to link the generated UI to the backend system that provides the data. The page content is used to generate the consumer code. The page bindings and interactions define the provider. The generated provider mainly consists of stubs to connect to the backend which can provide actual data and backend functions through web services, simple java classes, or RMI. In absence of real connections to the backend system, the provider gives dummy data for different UI elements. The dummy data is generated based on the semantic types associated with the attributes in the artifact model. We use a test database which contains sample data for different semantic types. The test data contains multiple values for each semantic types and picks up data randomly from each value set. For lists, we generate pre-specified number of records. To ensure uniqueness between values picked for successive records in a list, we remove the value which has been already picked in prior runs for list records from the candidate value-set.

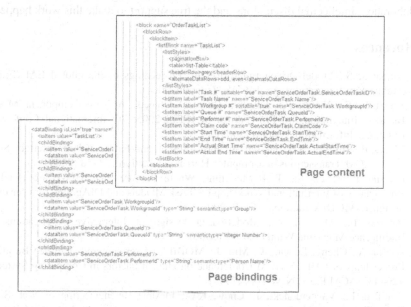

Fig. 12. Page binding and page content XML snippets of the UI page shown in Figure 9

During initial phases of the project, when the system requirements and the design are still evolving, the UI model is used primarily to generate screen mockups and UI prototypes with dummy data. During later stages in the project, the same model can be crafted to generate the high-fidelity UI and backend connections to the actual data sources and process instances. This would require inputs from the designers for the style information and input from the developers on what information is passed from one page to another or from the generated UI to the backend for different events.

5 Summary

In this paper, we presented our general model-driven UI research approach. While have a full model to capture UI design for automatic code generation, our research focus is on connecting the UI model to the business model and requirements. We used the business model to bootstrap the UI design process through sharing and connecting common domain elements. Specifically presented in this paper is a rather new work we are embarking on – supporting design iterations while business model changes. We presented how changes in business model are reflected in the UI design model and the particular modeling components to maintain and support the changing information.   Our endeavor will continue to support more automation paralleling business model changes with changes in UI design work products.

Acknowledgements. We would like to acknowledge the rest of the "Model-Driven Business Transformation Team" and its extended members – Santhosh Kumaran, Prabir Nandi, Pankaj Dhoolia, Markus Stolze, Terry Heath, Anil Nigam, Susan Spraragen, Frederick Wu, Florian Pinel, Prasun Dawan, and Jun-Jang Jeng – for their continual collaboration, intellectual discussions and the true support to make this work happen.

References

1. Kumaran, S.: Model-Driven Enterprise. In: Proceedings of the Global EAI (Enterprise Architecture Integration), pp. 166–180 (summit. 2004)
2. Kumaran, S., Nandi, P.: Adaptive Business Objects: A New Component Model for Business Integration. In: Proceedings of ICEIS 2005: 7th International Conference on Enterprise Information Systems (2005)
3. Myers, B., Hudson, S.E., Pausch, R.: Past, Present, and Future of User Interface Software Tools. ACM Transactions on Computer-Human Interaction 7, 3–28 (2000)
4. Paterno, F.: Tools for Task Modeling: Where We are, Where We are Headed. In: Proceedings of International Workshop on TAsk MOdels and DIAgrams for user interface design TAMODIA 2002, pp. 10–17 (2002)
5. Paterno, F., Mancini, C.: Model-based Design of Interactive Applications. ACM Intelligence Magazine Winter, 26–37 (2000)
6. Puerta, A., Cheng, E., Ou, T., Min, J.: MOBILE : User-centered Interface Building. In: Proceedings of CHI 99: ACM Conference on Human Factors in Computing Systems, pp. 426–433. ACM Press, New York (1999)
7. Bouillon, L., Vanderdonckt, J., Chow, K.C.: Flexible Re-engineering of Web Sties. In: Proceedings of ACM Conference on Intelligent Interfaces, pp. 132–139. ACM Press, New York (2004)
8. Kim, W.C., Foley, J.D.: Providing High-level Control and Expert Assistance in the User Interface Presentation Design. In: Proceedings of CHI 93: ACM Conference on Human Factors in Computing Systems, pp. 430–437. ACM Press, New York (1993)
9. de Baar, D., Foley, J.D., Mullet, K.E.: Coupling Application Design and User Interface Design. In: Proceedings of CHI 92: ACM Conference on Human Factors in Computing Systems, pp. 259–266. ACM Press, New York (1992)
10. Paterno, F., Santoro, C.: One Model, many Interfaces. In: CADUI 2002. Proceedings of 4th International Conference on Computer-Aided Design of user Interfaces, pp. 143–154 (2002)
11. Nunes, N.J., Cunha, J.F.: Towards a UML Profile for Interaction Design: the Wisdom Approach. In: Evans, A., Kent, S., Selic, B. (eds.) UML 2000. LNCS, vol. 1939, pp. 101–116. Springer, Heidelberg (2000)
12. Campos, P.F., Nunes, N.J.: CanonSketch: a User-Centered Tool for Canonical Abstract Prototyping. In: Proceedings of DSV-IS 2004: International Workshop on Design, Specification and Verification of Interactive System (2005)
13. Sukaviriya, N., Sinha, V., Ramachandra, T., Mani, S., Stolze, M.: User-Centered Design & Business Process Modeling: Cross Road in Rapid Prototyping Tools. In: INTERACT 2007. IFIP Conference on Human Computer Interactions (to appear, 2007)

Integrating Heterogeneous Tools into Model-Centric Development of Interactive Applications

Andreas Pleuss, Arnd Vitzthum, and Heinrich Hussmann

Department of Computer Science, University of Munich
Munich, Germany
{pleuss,vitzthum,hussmann}@cip.ifi.lmu.de
http://www.medien.ifi.lmu.de

Abstract. The development of successful interactive applications often requires high efforts in creative design tasks to build high quality user interfaces. Such creative development tasks – such as user interface design or design of specific features like 3D objects – are usually performed using different tools optimized for the respective task. For example, in early development stages, tools like *Photoshop* or *Flash* are established for creating user interface prototypes. 3D graphics is usually developed using 3D authoring tools.

In this paper we propose a general approach to integrate heterogeneous tools into model-centric development. Thereby, the models act as central hub between different specific tools and development steps. This enables excellent support for creative design while using concepts from model driven engineering, such as explicit metamodels and transformations, to facilitate a well-coordinated development and ensure consistency of the resulting overall application. We illustrate this approach by concrete examples from different application domains.

1 Introduction

From an engineering point of view, the approach of model-driven development is very appropriate for the production of high-quality software. However, a key problem in using the state-of-the-art modelling languages (like UML), is that the audience which can understand the models is severely restricted to people being trained in using high abstraction levels. This in practice mostly means engineers or scientists, may they be part of the development team or work on the customer side. It is already difficult for domain specialists without computer science background to deal with abstract models, and for most end users it is completely impossible. This observation is in unpleasant contrast to the fact that the end users are the target group for which the whole design effort is directed. The overall acceptance of software systems by its end users is often determined by properties of its user interface [1,2], so it makes good sense to adopt a user-centred design process. Such a process typically uses very early prototypes and interface mock-ups to obtain feedback from end users. These

G. Engels et al. (Eds.): MoDELS 2007, LNCS 4735, pp. 241–255, 2007.

steps produce artefacts which are of a completely different nature compared to the abstract (e.g. UML-based) engineering models. Moreover, there is a rapidly growing species of applications where the key features are located rather in the user interface than in the background logic. Examples are all kinds of product presentations on the Web or multimedia applications in entertainment sector. These applications cannot be developed without taking care of the user interface and its graphical design from the beginning. So it can be stated that development projects face a clash of cultures between engineering people taking the abstract view and interface designers and graphical artists working creatively and taking a very concrete view of the system.

This paper presents thoughts and technologies which may in the long run help to bridge the gap between these cultures. We concentrate here on possibilities to conceptually integrate the tools which are used in the development process. In tool usage, the clash of cultures is apparent as well. Engineers typically use CASE tools (like *MagicDraw, EclipseUML, Rational Rose, Poseidon*) which are not usable in any way for a typical graphical designer, for instance. On the other hand, a user interface designer has a number of very advanced tools available, e.g. drawing tools like *Adobe Illustrator*, image processing tools like *Adobe PhotoShop* or animation workbenches like *Adobe Flash*, or even 3D graphics tools like *3D studio max*. This kind of tools at the first look is completely incompatible with the CASE tools. However, we will point out in this paper that there are ways how such a landscape of different tools can work together in an orchestrated way, to produce a common vision of the system under development.

The ideas presented here are partially motivated by the results of a workshop on Model-Driven Development of Advanced User Interface [3] which was held at the UML 2005 conference. There, an agreement among people with different perspectives on user interface development emerged that the role of an abstract model (based on an appropriate metamodel) can be to integrate the heterogeneous artefacts produced with tools of different application context.

The paper is based on work by the authors on extending UML for the specification of multimedia and 3D user interfaces, which provides already a step from abstract models towards more concrete representations at the user interface. There exists already some work on generating (multimedia) interfaces, or better skeletons for them, from such an abstract description. In this paper we will take the opposite view and will discuss how a concrete interface prototype or mock-up can be abstracted onto a level where it fits to the abstract model. It is obvious that a long-term goal will then be a seamless transition across abstraction levels (and across tools). In order to avoid analysing completely unstructured artefacts, we make the assumption that it will be possible to agree on a number of structure and naming conventions in a development project, for all people, including user interface designers and artists. So, the vision is that the user interface specialists work with the tools they are used to (PhotoShop, Flash), but obtain a number of conventions which make it easier to integrate their work with abstract models. We are confident that designers are willing to adhere to such conventions, since this is already daily routine at many places. A simple example are the naming conventions in Flash ActionScript code (where a movie clip object for instance should have a name ending in "_mc"), which are well accepted in the community. We propose below more "invasive" conventions which also give guidelines on how to structure a user interface prototype or mock-up. However, we believe that such guidelines will be used like design patterns, in giving people more advice and confidence when solving a problem, so the acceptance problem will be small. It is obvious that the fact of being able to use familiar tools will contribute significantly to bring abstract models into the world of interface designers.

The paper is structured as follows. Section 2 gives a general discussion of the relationship between models and tools for creative design. Sections 3 through 5 give concrete examples for the integration of models with various state-of-the art interface creation tools. Section 6 briefly discusses related work.

2   Models as Central Hub

As explained in section 1, there is a need for enabling the usage of tools which support creative design. The goal is to better integrate different experts with background in creative design into the development process. This consideration affects not only the final implementation of the system, but rather all steps during the development process to ensure the usability of the system and the required quality of the user interface.

Models are an excellent vehicle for integrating different stakeholders and different views on the system during the whole development process. Thus, in our vision, models are also used to integrate the different tools and the resulting products. Thereby the concepts from model-driven development, like explicit transformations, are applied for computer-supported transitions between tools and artefacts. This ensures consistency between the artefacts produced by heterogeneous tools and furthermore reduces effort as subsequent steps can start directly from earlier results instead of taking them over manually.

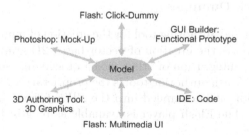

Fig. 1. Models as central hub in the development integrating different specific development steps and tools

Figure 1 visualizes this idea on models acting as such an "central hub". The upper arrows contain examples for earlier development phases where prototypes play a central role in interactive systems development. For example, Photoshop mock-ups (as described in section 3.2) can be used to select first ideas about the system to be developed. When this step is finished, transformations are used to transmit the significant abstract information from the mock-ups into the model where it can be used for further development steps, like creating corresponding Flash click-dummies (see section 3.1) for gaining more specific user feedback. During this step, additional abstract information about the system is added which should again be kept in a central place, i.e. in the model. Thus, it is important to allow transitions into both directions: extraction of relevant abstract information from the tools (kind of "reverse engineering") and generation of artefacts for the desired tools based on the existing model information.

The lower part of figure 1 shows examples for later development steps, such as implementation of a final release. Here, the kinds of tools are more diverse and depend on the application domain and the target platforms. Models can be used to distribute the final implementation on different tools optimized for realizing different aspects of the system. For example, in multimedia development with Flash, it is a common practice to develop the code for system's application logic within an external programming IDE, instead of using the Flash authoring tool's built-in code editor.

In the following sections we describe four of the mentioned examples more in detail and present how the transitions between models and tools can be realized.

3   From Prototypes to Models

An important part of interactive systems development is the usage of prototypes. Standard tools for user interface designers and people with background in creative design are Adobe Photoshop and Adobe Flash. In the following we show as an example how the information from prototypes created with these two tools can be extracted into models.

3.1   Flash Click-Dummies

Flash is a professional authoring tool for the creation of multimedia applications. It especially supports the creation of vector-based 2D graphics and animations and provides easy integration of other media objects like sound and video. The file format for the Flash authoring tool is the proprietary *FLA* format. To execute a Flash application, it is compiled into the *SWF* format which is executed by the Flash player. The Flash player is available as plug-in for all common web-browsers.

Due to the possibility to create complex and individual visual user interfaces very quickly, Flash is often used as a tool for creating prototypes. In this section we focus on so-called click-dummies, i.e. (horizontal) prototypes which show a broad range of user interface screens of a visual application without any underlying functionality. Sometimes some basic application functionality is simulated through predefined visualizations. Therefore, and for the navigation between the scenes, some user interface elements are (partially) enabled. This provides the user a good impression of the overall look and feel of the system and gives an idea of the intended task flow and overall functionality.

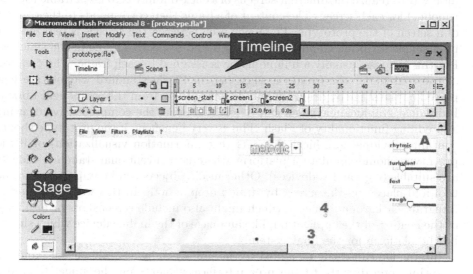

Fig. 2. Screenshot of the Flash authoring tool (reduced to the most important elements.) On the stage, a part of a UI prototype is visible including individual graphics and standard widgets.

Using the Flash Authoring Tool for Creation of Click-Dummies. Figure 2 shows a screenshot of the Flash authoring tool. The tool is timeline-based, i.e. it uses the metaphor of a timeline to visualize the temporal aspect of Flash applications (important e.g. to support easy creation of animations). A timeline (upper window in fig. 2) contains a sequence of frames. Each frame can contain individual content, i.e. graphics, sound, video, text, or a Flash component such as a standard user interface widget (buttons, list boxes, text fields, etc.). In the authoring tool, the content of the currently selected frame is shown on the stage (centre window in fig. 2). The toolbar on the left hand side in figure 2 provides tools to create or modify the content on the stage, such as creating or editing 2D graphics or text on the stage. During the execution of a Flash application in the Flash player, the content of the frames is displayed. By default, the frames are shown successively according to their order on the timeline. However, it is also possible to stop the play-back of the timeline, which means that the content of the current frame remains on the stage (while it is still possible to interact with the elements in the displayed frame). Starting and stopping play-back of the timeline as well as jumping to a specific frame can be controlled by scripting code. The scripting language provided in Flash is *ActionScript*, which is a complete programming language with object-oriented features. ActionScript code can either be embedded within Flash documents directly, or can be specified in external class files (analogous to class files in Java).

The Flash authoring tool enables very easy creation of click-dummies. A common way to realize the different screens of a click-dummy is to use a frame (on the timeline) for each screen. The content of a frame then corresponds to the content of a screen of the prototype. The navigation between the screens is realized just by a simple scripting command to jump to the respective frame in the timeline. For this purpose, usually a unique name is assigned to each frame. The creation of the content for the frames is very easy in Flash: standard widgets are just placed on the stage by dragging and dropping them from the component library. Individual graphical elements can be created using the comprehensive drawing and editing functionality of Flash. This is very beneficial if the prototype should contain individual graphical elements (e.g. information visualization) or if the prototype should simulate a platform where no conventional standard widgets are suitable (e.g. on TV devices). Other media objects can be imported into the tool and placed on the screen by drag'n drop as well. In this way, it is easy to integrate background images, which might also include screenshots or an image of the context of the application, like an image of the mobile device which should be the platform for the final system.

Interaction can easily be added to the prototype by assigning event handling scripting code directly to the user interface elements on the stage. It is also easily possible to assign scripting code to a specific region within an image, which is useful for screenshots part of the prototype (e.g. it is possible to simulate interaction with a button on a mobile phone using an image of the mobile phone).

Extracting information from Flash into a Flash model. A general problem when extracting information from a Flash document is the proprietary file format of Flash documents. We aim to deal with the FLA files, as they contain more information than the complied SWF files. To solve this problem we use the extension mechanism which exists for the Flash authoring tool. The extensions must be specified in *JSFL*, i.e. JavaScript using the JSFL library to access all elements in the authoring tool in terms of a kind of "document object model", similar to DOM in web-browsers. JSFL allows creating, accessing and editing of Flash documents just like in the authoring tool. We have specified a JSFL script which walks through an arbitrary Flash document, accesses its content, and outputs it as a Flash model. For the Flash model, we use the Flash metamodel we proposed in [4]. The Flash metamodel is implemented with the *Eclipse Modelling Framework* (*EMF*); the Flash models resulting from our JSFL plug-in are thus saved as an EMF-compliant XML file. Consequently, the resulting Flash model can be further processed using the conventional technologies from model driven engineering without any knowledge on JSFL or on the Flash authoring tool itself.

Example Transformation from Flash Model to Abstract User Interface Model. A common concept to model user interfaces in an abstract way are abstract user interface models as used in the user interface modelling community (see e.g. [5]). In other domains, like web-engineering, similar concepts are used. In an abstract user interface model the user interface consists of *presentation units* (abstraction of screens) which contain *abstract user interface elements*. Possible abstract user interface elements are *input elements* which allow the user to input information into the system (such as a text input), *output elements* which present some information to the user (such as a text label or an image), and *action components* which allow the user to invoke an action of the system without additional data input (such as a button). *UI containers* are used to structure the UI elements. (These elements are often further subdivided into more specific subtypes, like *selection element*, *notification element*, etc., depending on the specific modelling approach.)

For abstract user interface models as described above, the transformation can be defined as follows: widgets in the prototype are mapped to the corresponding abstract user interface elements (text input to input component, button to action component, etc.). Videos and sounds are mapped to output components. 2D graphic objects are mapped to input components when they have assigned an event listener and otherwise to output components. Images are mapped to UI containers, as images might be screenshots which represent several UI components (the content of the UI container then can not be derived automatically from the model). Frames on the timeline are mapped to presentation units. The transitions between the presentation units can be derived by searching for script commands in the ActionScript code which specify the jumps to another frame.

To ensure the success of the transformation, the designer is asked to comply to some conventions: first, the prototype must follow the explained timeline-based structure, instead of e.g. using external ActionScript code for the different screens (which is usually not useful for quick creation of prototypes, anyway). Second, the names for the frames representing screens have to follow a naming convention (e.g. a prefix "screen") to distinguish them from other frames in the document.

3.2   Photoshop Mock-Ups

Photoshop is an image editing software which can be used for the very fast creation of user interface mock-ups, i.e. pictures of the user interface to present possible ideas about the user interface to the customer or the target user group. Based on the mock-ups the most promising approaches are selected and can then be further refined using more advanced prototypes e.g. created with Flash as described in section 3.1. The main advantage of using bitmaps (instead of e.g. GUI builders) is that any new and arbitrarily complex user interface can be visualized in very short time. The mock-ups are usually composed of different image pieces which are arranged and manipulated to create the desired new user interface. For standard widgets or user interface elements similar to already existing elements (like a dialogue window, a menu bar, or icons), the designer simply takes a screenshot of the desired element, cuts it, and integrates it in the mock-up. More individual elements, such as complex information visualization, are drawn manually or put together from any other existing image snippets. Due to the unlimited flexibility of an image editing tool, any visual user interface can be composed in very short time, including user interfaces for new devices. If required, an image of the device (e.g. a mobile device) can also be part of the mock-up: the actual (software) user interface is then placed on the "screen" location within the image of the device which provides the user a good idea how the application will look like on the final device.

After the usage of the mock-ups is finished and some user interfaces have been identified as promising, it is useful to capture the basic information from the mock-ups. As described in section 2, models in their role as "central hub" are an ideal instrument for this purpose. The most relevant information from the mock-ups are the user interface elements required for the user interface. Therefore, we provide a transition from the mock-ups to abstract user interface models as described in section 3.1. The identification of the type of user interface element requires some conventions for the designer: usually, each piece of image in a Photoshop document is put on a different layer (because otherwise it would not be possible to move it later within the whole image). By convention, the designer specifies the type and the name of the user interface element as the layer name (e.g. "address_input" for an input component intended to input an address). Complex pieces which represent multiple user interface elements are specified as UI container.

Technically, the transformation is performed by executing the built-in command "Save Layers..." in Photoshop which causes all layers to be saved on disk in separate files. The resulting file names then correspond to the layer names which contain by convention the type and the name of the user interface element. A simple Java application then collects the file names and creates a corresponding abstract user interface model. The model is used as base for further development steps, which may include, for instance, generating skeletons for a Flash click-dummy according to the mock-ups.

4 MML: From Models to Multimedia Authoring Tools

In the foregoing sections we discussed by the example of prototypes how in early development phases abstract model information can partially be derived from different external tools which support informal, creative design. In the following two sections we briefly present two existing examples showing how professional existing tools can be integrated into model-centric approaches to support creative tasks during the final implementation of the system.

In this section we show how to integrate the Flash authoring tool for the final design of user interfaces and media objects within a model-driven development approach for multimedia applications. By the term "multimedia applications" we refer to applications which provide a individual, complex and interactive user interface using different kinds of media objects, like graphics, animation, sound and video. Classic examples are training and simulation applications, newer examples are infotainment systems in cars or home entertainment applications. The modelling language used here is called *Multimedia Modeling Language* (*MML*) [6,7], a platform-independent language for model-driven development of multimedia applications. Motivation for the language is the lack of a structured development process, claimed by many publications in this area. A specific problem of these applications are the different developer groups involved in the development process, as interactive multimedia applications require three different kinds of design: *software design* as current applications often include complex application logic, *user interface design* as the user interface is one of the core issues of the application and should often be individual and provide a high degree of usability, and *media design* which often requires large effort and specific expert knowledge. A design phase using models ensures the coordination between the different developer groups and their results.

MML reuses concepts from UML, user interface modelling, and existing approaches for modelling of multimedia applications. It provides four kinds of models: the *structural model* specifies the structure of the application logic in terms of an UML class diagram. In addition, the model allows specifying media objects which are part of the system. The *scene model* mainly describes the presentation units (in multimedia context referred to as *scenes*) and the transitions between them. The *abstract user interface model* specifies the user interface for each scene in terms of abstract user interface elements similar as described in section 3.1. The abstract user interface elements are associated with classes or class properties from the structural model, e.g. an output component can represent a class attribute. In addition, media objects can be integrated into the user interface. Finally, the *interaction model* shows the main control flow in each scene based on events received from the abstract user interface elements or caused by media objects.

MML is intended to allow transformations into code skeletons for any platform which supports multimedia user interfaces. However, particularly interesting is the idea to generate code skeletons for multimedia authoring tools like Flash. As described in [4], we perform this in several steps. The MML models are transformed into Flash models. For this purpose we have specified a Flash metamodel and a transformation. The Flash models are transformed into code skeletons: Flash documents (FLA files, generated using JSFL analoguous to section 3.1) which can be directly loaded into the Flash authoring tool and ActionScript code.

For the application logic we use external ActionScript class files. Thus, the ActionScript code can be edited as any other object-oriented code e.g. within an IDE like Eclipse using Eclipse plug-ins for ActionScript support. The relationships between the elements in the Flash document and the ActionScript code are generated using JSFL, which ensures consistency with the model. The ActionScript code contains the properties from the class diagram including attributes and operation signatures. The operations have to be filled out by conventional programming in order to realize the application logic and to control the user interface objects and media objects they are associated with.

The generated Flash documents implement the overall structure of the application and contain the required standard widgets and placeholders for the media objects. In order to complete the application, the developer has to fill out the placeholders by selecting them on the stage and creating content inside them using all available functionality of the authoring tool in conventional way. Furthermore she arranges the filled-out placeholders and the generated widgets on the stage to create the final user interface layout. Of course, she is free to add any adornments, change the size and the appearance of the generated elements, or their type. As the navigation between the screens has been generated as well, it is possible to run and test the application immediately after generation.

The described transformation has been implemented using the *Atlas Transformation Language* (ATL) and JSFL realizing the concepts described above (currently not including MML interaction models). First tests with MML have been run in several student projects. MML is currently not supported by its own visual modelling editor, but as a temporary solution we provide a plug-in for the UML tool MagicDraw supporting all types of MML models.

5   SSIML: From Models to 3D Authoring Tools

Just as multimedia system development, the creation of interactive 3D applications is an interdisciplinary activity. Different developer groups such as 3D content creators and programmers are involved in the development process. These groups use a variety of tools, such as 3D authoring tools and integrated development environments. Furthermore, a 3D user interface displaying 3D contents (represented by a so-called *3D scene* which contains 3D objects) is often tightly coupled with the base application containing the application logic. Thus, these software components must seamlessly integrate with each other. Therefore concepts and tools are needed to support the collaboration of developers.

The visual modeling language *SSIML (Scene Structure and Integration Modelling Language)* [8], which was developed by our research group, addresses this challenge. SSIML has the goal to facilitate the integration of 3D user interfaces into a broad spectrum of applications such as 3D product presentations and product configurators, virtual galleries, interactive 3D-manuals, 3D guides, Virtual Reality (VR) applications and even Augmented Reality (AR)-applications.

The Visual Modeling Language SSIML and its Extensions. Unlike most conventional 3D development approaches, SSIML supports a model-based design prior to implementation. The elements of SSIML were specified in a MOF-conform metamodel which was also mapped to a UML profile in order to allow an easy integration of SSIML into existing UML tools.

SSIML allows a semi-formal and platform-independent specification of important 3D application parts. In particular, it offers different possibilities to interconnect complex-structured 3D contents (represented by a 3D scene) and application logic. One possibility for such an interconnection is the *inter-relation model* described below in this section. Thereby SSIML abstracts from implementation-level details such as concrete geometries of 3D objects. SSIML provides a comprehensible and compact graphical notation which was designed to be easily understood by the different persons involved in the development process. Moreover, the visual SSIML models can not only serve as communication aid for the developers, but can also be used for documentation purposes. In order to reduce implementation errors and to save implementation time, platform-specific code skeletons can be generated automatically from the models.

The basic SSIML language comprises two model types: the *scene model* and the *interrelation model*. The *scene model* allows describing the 3D scene structure (including all scene objects) and therewith the structure of the *3D user interface* in a scene graph-oriented manner. A scene graph is a directed acyclic graph (DAG) representing the transformation hierarchy of the objects contained in the 3D scene. A scene object in SSIML is characterized by a *unique name* (e.g. table1), a content type identifier (e.g. Table), the *type of the model element* representing the 3D object (e.g. group, 'atomic' object), its *attributes* (e.g. a transformation attribute) and - where applicable - by its *children* objects (e.g. a group tables could have two children: table1 and table2). In order to manage complex scene structures it is also possible to encapsulate whole subgraphs of a SSIML scene model in specialized nodes. Due to the high level of abstraction, 2D-scenes instead of 3D-scenes can be described using SSIML, even if we focus especially on 3D scenes.

The *interrelation model* comprises the *scene model*, *application components* represented by UML classes and *interrelations* between classes and scene elements. For example, in the interrelation model one can specify that instances of a certain class are able to modify the transformation value of a specific 3D object (the transformation value represents the orientation, position and scale of an object in the three-dimensional space).

Based on the core elements of SSIML we have defined several SSIML extensions, e.g. for the description of 3D object behavior [9] and task-dependent information presentation [10]. Partial results of an informal evaluation of an extended SSIML version (SSIML for Augmented Reality - SSIML/AR) were presented in [10].

Model-Centric Development with SSIML. Along with SSIML, we propose a sketch of a suitable development process. In this process three main developer roles are distinguished: *software designer*, *3D developer* and *programmer*. The *software designer* creates the visual SSIML models. The design can be discussed with other development team members and can also be changed if necessary. Afterwards, platform-specific code is generated from the models. The scene model is translated into a so-called *3D template* which is encoded e.g. in a 3D markup language such as VRML or X3D. The *3D developer* is then able to import the 3D template into a 3D authoring environment and to enrich it with 3D contents such as concrete object geometries. The interrelations in the interrelation model are transformed into program code, e.g. Java or C++ code. The *programmer* fills out the gaps in the generated program code.

Since different code components in different languages (e.g. 3D markup language code, Java code) originate from the same SSIML model, the model can be seen as the central element in the SSIML development process. It represents a kind of *contract* between the different developer groups and therewith also ensures the *consistency* between (generated) code components.

Ideally, from a model-centric point of view and as mentioned in section 1, modifications of the code should be also reflected by changes in the SSIML models and vice versa, while the consistency of the changed system would be checked automatically. It should be even feasible to integrate existing code components such as an existing 3D scene into a SSIML model. Unfortunately, the present SSIML tools (see below) only support forward engineering. However, at the end of this section we briefly sketch a possibility how an existing 3D scene can be translated back into a SSIML scene model.

Present Tool Support. As mentioned above, we have defined a UML-profile based on the SSIML metamodel in order to enable the integration of SSIML into existing UML tools. We have integrated SSIML into the well-known UML tool *MagicDraw* , to name an example. In addition, this allows us encoding a SSIML model in the XML Metadata Interchange format (XMI), which is supported by a variety of UML tools. The XMI-format provides a suitable basis for the automatic translation of SSIML models into platform-specific code using XSLT stylesheets. More precisely, we have chosen X3D and VRML as target formats for automatically generated 3D code skeletons and Java as target language for the XSLT-based generation of program code.

From the interrelations specified in a SSIML interrelation model we generate Java 'glue code' in order to enable a seamless integration of the 3D scene into the corresponding overall application.

Deriving SSIML Models from Existing 3D Scenes. As discussed in section 1, in some cases it might be useful to derive a model from an existing user interface prototype. In the context of SSIML this means that a SSIML model can be extracted from an already existing 3D scene which was created by a 3D content developer using a 3D authoring tool. As a first approach which actually takes into account an existing 3D scene, we have developed a validation tool which compares two 3D scenes in terms of their structure (object hierarchy, object types) and the naming of the contained 3D objects. After generating a 3D template from a SSIML scene specification, the tool allows checking if a present scene conforms to the generated template and therewith conforms to the SSIML model. The tool reports structural differences between the two 3D scenes. However, this approach still requires an existing abstract specification of the scene in a SSIML model.

The next step would be to completely generate a SSIML model from a 3D scene which was already created with a 3D authoring tool. A suitable basis for such a transformation is the X3D-format. X3D is the successor of VRML. X3D and VRML are standardized formats, integrate with different 3D platforms which provide VRML/X3D scene loaders (e.g. Java3D), can be imported and exported by many 3D authoring tools and are supported by several 3D format converters. Since X3D is an XML-based format, it is possible to translate an X3D scene into an XMI-encoded SSIML model using XSLT. Like SSIML, X3D uses a scene graph-oriented structure. Thus, the mapping between X3D and SSIML is quite straightforward. Names and types of SSIML elements can be derived from the names and types of corresponding elements of the X3D scene. A transformation of an X3D scene into a SSIML model should also preserve property values of scene objects (e.g. transformation and color values). Although these values are not visible in the SSIML diagram editor, they could be later incorporated again in a retranslation of an (adapted) SSIML model into X3D code or another 3D format.

6 Related Work

Important related work are the concepts from the user interface modelling community. While many approaches in this field are restricted to the usage of modelling tools or markup languages without specific tool support, some approaches basically allow the usage of visual user interface design tools. UsiXML [11] allows transformations between the different levels of abstraction which could help to extract abstract information from a concrete user interface model. [12,13] provide reverse engineering for HTML-based user interfaces. User interface modelling tools like CanonSketch [14] integrate different views on the user interface models, including a concrete HTML view. Similarly, in domain specific tools for model-driven development, such as [15], GUI builders are integrated to visually specify the application's user interface. However, concerning our goals, all approaches have in common that they are restricted to standard elements like HTML widgets.

Similar restrictions exist (naturally) in the area of model-driven development of web-application (web-engineering). However, it is worth to note that some approaches in this area (e.g. [16,17]) claim that the integration of existing tools for visual, creative user interface design is mandatory for the success of the development approach. As a consequence, they enable the usage of external web-design tools for the specification of the user interface design.

In the field of 3D application development (e.g. Virtual Reality and Augmented Reality development) model-driven methods are rarely applied, although e.g. Smith et al. [18] underline the need of such approaches in the context of VR development. Nevertheless, some modelling approaches exist which allow the generation of platform-specific code (e.g. for interaction techniques [19] or presentation flows [20]) and are often directed towards a better support of developers without deeper experiences in programming such as designers [21]. However, all these approaches focus mainly on forward engineering and don't consider a retranslation of an implementation level specification into a more abstract representation. Furthermore, unlike in SSIML (see previous section), 3D contents have to be created independently from the modelling approaches and must be connected manually (by programming) with the code generated from the models.

In the domain of multimedia applications, [22] presents a model-based approach for the development of training applications using the authoring tool Adobe Director. However, the transition between the models and the implementation in the tool has to be performed manually.

In summary, several tendencies towards a better integration of tools for creative design and model-centric development exist, but there is clearly a lack of a more consequent and explicit integration of them, especially for earlier development steps like prototyping.

7   Conclusion and Outlook

In this paper, we have discussed a number of quite heterogeneous approaches, without providing much detail on each individual approach. The common theme among all the proposed tools is that there is an abstract software model on one side and an artefact of a concrete state-of-the-art commercial tool for user interface design on the other side. Both directions of transitions have been described: Creating artefact skeletons from the abstract model (forward engineering) as well as creating abstract views from artefacts (reverse engineering). Like in the traditional transitions between models and program code, the ideal world would be a "round trip" engineering between the levels. However, in the case of user interface design, it is also common to deal with many tool technologies at once. So a development may start from a user interface design (made in Photoshop), then derive an abstract interface model from it, create a skeleton for a Flash click-prototype from it, refine the abstract model based on some user studies, and finally make a transition towards Java code skeletons for a final implementation. The abstract model helps for establishing much more formal and traceable links between various design artefacts than in todays practice. Moreover, using the model as a central hub for transformations may in the long run help to reduce the (nowadays common) thinking in terms of one single design platform only, and therefore helps to introduce a more abstract way of thinking in projects where graphical designers and software designers work together.

Of course, the prototypes for transformators which were mentioned above are far from being exhaustive and not yet well integrated. Nevertheless, they show that this is a road which is viable, and we will further elaborate the "model as design hub" idea in future work. The long-term vision is that import and export functions to and from abstract models will become standard features for any kind of design tool used in the development process.

References

1. Davis, F.D.: Perceived usefulness, perceived ease of use, and user acceptance of information technology. MIS Quarterly 13 (1989)
2. Cooper, A., Reimann, R.M.: About Face 2.0: The Essentials of Interaction Design. Wiley, Chichester (2003)
3. Pleuß, A., van den Bergh, J., Sauer, S., Hußmann, H.: Workshop report: Model driven development of advanced user interfaces (mddaui). In: MoDELS Satellite Events. LNCS, vol. 3844, Springer, Heidelberg (2005)
4. Pleuß, A., Hußmann, H.: Integrating authoring tools into model-driven development of multimedia applications. In: HCII'07, Springer, Heidelberg (to appear, 2007)

5. Calvary, G., Coutaz, J., Thevenin, D., Limbourg, Q., Souchon, N., Bouillon, L., Florins, M., Vanderdonckt, J.: Plasticity of user interfaces: A revised reference framework. In: TAMODIA'02 Proc. INFOREC (2002)
6. Pleuß, A.: Modeling the User Interface of Multimedia Applications. In: Briand, L.C., Williams, C. (eds.) MoDELS 2005. LNCS, vol. 3713, Springer, Heidelberg (2005)
7. Pleuß, A.: MML: A Modeling Language for Interactive Multimedia Applications. In: ISM'05 Proc., IEEE, Los Alamitos (2005)
8. Vitzthum, A., Pleuss, A.: SSIML: Designing structure and application integration of 3d scenes. In: Web3D'05 Proc., ACM Press, New York (2005)
9. Vitzthum, A.: SSIML/behaviour: Designing behaviour and animation of graphical objects in virtual reality and multimedia applications. In: ISM'05 Proc., IEEE, Los Alamitos (2005)
10. Vitzthum, A.: SSIML/AR: A visual language for the abstract specification of augmented reality user interfaces. In: Proc. of 3DUI'06, IEEE, Los Alamitos (2006)
11. Limbourg, Q., Vanderdonckt, J., Michotte, B., Bouillon, L., López-Jaquero, V.: Usixml: A language supporting multi-path development of user interfaces. In: Bastide, R., Palanque, P., Roth, J. (eds.) EHCI-DSVIS 2004. LNCS, vol. 3425, Springer, Heidelberg (2005)
12. Bouillon, L., Limbourg, Q., Vanderdonckt, J., Michotte, B.: Reverse engineering of web pages based on derivations and transformations. In: LA-Web'2005. Proc.of 3rd Latin American Web Congress, IEEE, Los Alamitos (2005)
13. Paganelli, L., Paternò, F.: A tool for creating design models from web site code. International Journal of Software Engineering and Knowledge Engineering 13 (2003)
14. Campos, P.F., Nunes, N.J.: Canonsketch: A user-centered tool for canonical abstract prototyping. In: Bastide, R., Palanque, P., Roth, J. (eds.) EHCI-DSVIS 2004. LNCS, vol. 3425, Springer, Heidelberg (2005)
15. Tangible Architect (2007), http://www.tangiblearchitect.com
16. Hennicker, R., Koch, N.: Modeling the User Interface of Web Applications with UML. In: Practical UML-Based Rigorous Development Methods, Workshop of the pUML-Group held together with the UML2001, GI (2001)
17. Ceri, S., Fraternali, P., Bongio, A.: Web modeling language (webml): a modeling language for designing web sites. Computer Networks 33 (2000)
18. Smith, S.P., Duke, D.J., Willans, J.S.: Designing world objects for usable virtual environments. In: Workshop on Design, Specification and Verification of Interactive Systems 2000, Limerick (2000)
19. Willans, J.S., Harrison, M.D.: A toolset supported approach for designing and testing virtual environment interaction techniques. International Journal of Human-Computer Studies 55 (2001)
20. Ledermann, F.: An authoring framework for augmented reality presentations. Master's thesis, Vienna Technical University (2004)
21. Dubois, E., da Silva, P.P., Gray, P.D.: Notational support for the design of augmented reality systems. In: Forbrig, P., Limbourg, Q., Urban, B., Vanderdonckt, J. (eds.) DSV-IS 2002. LNCS, vol. 2545, Springer, Heidelberg (2002)
22. Depke, R., Engels, G., Mehner, K., Sauer, S., Wagner, A.: Ein Vorgehensmodell für die Multimedia-Entwicklung mit Autorensystemen. Informatik: Forschung und Entwicklung (1999)

A Business-Process-Driven Approach for Generating E-Commerce User Interfaces

Xulin Zhao, Ying Zou, Jen Hawkins, and Bhadri Madapusi

Electrical and Computer Engineering Depart
Queen's University
Kingston, Ontario, Canada
4xz5@qlink.queensu.ca, ying.zou@queensu.ca
Jen Hawkins, Bhadri Madapusi
IBM Toronto Laboratory
IBM Canada Ltd.Markham, Ontario, Canada
{jlhawkin, bhadrim} @ca.ibm.com

Abstract. A business process contains a set of interdependent activities that describe operations provided by an organization. E-commerce applications are designed to automate business processes. A business process specification (i.e., a workflow) is defined by a business analyst from the viewpoint of the end-users. The process encapsulates the knowledge related to the natural work rhythms that a business user would follow when using an e-commerce application. In this paper, we analyze the information embedded in business process specifications, and infer the functional and usability requirements. We use the inferred information in a model-driven approach to automatically generate user interfaces (UIs) from a business process specification through a set of transformations. To improve the usability of UIs for the e-commerce applications, each transformation is guided by usability principles.

Keywords: Business process, User interface generation, Usability, Model driven engineering, Task model, Dialog model, Presentation model.

1 Introduction

A business process is a set of linked activities that need to be carried out to achieve business goals for an organization. These activities are called tasks and may be performed by multiple roles. For example, *"Find product"* and *"Validate credit card"* are two tasks in an online shopping business process and are performed by the role *"Customer sales representative"* (CSR) and the role *"Bank clerk"*, respectively. E-commerce applications are designed to automate business processes in an organization [1]. In today's fast-changing business environments, marketing strategies and technical innovations drive the evolution of business processes. To maintain their competitive edge, organizations must keep on updating the business logic and UIs of e-commerce applications in order to reflect new business initiatives. However, the updates and evolution of UIs are often performed without referencing the underlying business processes. Therefore, the UIs of e-commerce applications and the business processes rarely evolve consistently.

G. Engels et al. (Eds.): MoDELS 2007, LNCS 4735, pp. 256–270, 2007.

The usability of UIs of e-commerce applications is crucial for the survival of a business and its success. The key to attract customers is e-commerce applications with high quality content, ease of use, quick response and frequent updates. However, studies [2] show that the majority of the UIs of e-commerce applications suffer from usability problems. For example, applications often provide more functionalities than necessary. It is not obvious for users to navigate through a data centered UI in order to fulfill business tasks. Approaches have been proposed for automating the design and implementation of UIs [3, 5]. However, these approaches create the designs of the UIs from the perspectives of the software development domain. Without the participation of business users, the UI might not satisfy the usage patterns of business users. Therefore, it is crucial to provide an approach that can directly involve the business user in the design of UIs in order to ensure the usability of these interfaces.

In this work, we propose a framework that can automatically generate UIs from business processes using a sequence of model transformations. More specifically, we analyze business processes to obtain the functional requirements specified by the business analysts. An initial task model is generated to reflect the functionalities in the UI. We extract contextual information from business process specifications to provide supports for users. Examples of contextual information are the order of task execution, the data flow from one task to another, and the pre-conditions and post-conditions for task execution. To generate UIs with different quality goals (e.g., high learnability and high efficiency), we apply usability practices and UI design principles to guide the transformations of business processes into UI design models. For example, UIs with high learnability can provide more guidance for the users. Developers can pick their preferred UI based on their quality requirements.

In the following sections, we introduce the details of our approach. Section 2 describes other researchers' work related to our approach. Section 3 introduces our model-driven UI generation framework. Section 4 presents models and their transformations. Section 5 illustrates our approach using case studies. Section 6 presents our conclusion and future work.

2 Related Work

To improve the efficiency in software development, approaches for automatic UI generation were proposed to separate UIs from business logic [5]. Model based UI development typically creates a series of declarative models, such as models of user tasks, dialogs and presentations [6]. User task models are used to elicit the requirements of functionalities. Derived from the identified user tasks, dialog models describe the interactions between the user and the UI. Furthermore, presentation models are created by assigning abstracted graphic user interface (GUI) widgets to dialog models. Finally, UIs are implemented by mapping presentation models to specific UI implementation technologies, such as HTML pages or Java™ Swing components. Proposed by Elkoutbi et al. [3], UML collaboration diagrams annotated with UI information can be transformed into UI prototypes. In our work, we provide an approach that generates UIs from business processes modeled by business analysts.

Domain-specific languages (DSL) provide high-level languages for code generation in certain problem domains [8]. For example, business process specifications, such as BPEL (Business Process Execution Language) are widely used to integrate Web services using the standards of the service-oriented architecture [9]. In our research, we infer the information embedded in business process specifications to generate UIs for applications.

3 Model-Based UI Generation Framework

To reduce the development and maintenance efforts of a UI and its corresponding business process, our UI generation framework uses model transformation techniques that analyze the content of business processes (e.g., tasks, data, and control flows) and produce the source code of an application. Figure 1 illustrates the overall steps in our UI generation framework. However, business processes usually do not contain enough information to automatically generate code. For example, a business process does not describe the threading model or the database schemas to be used in the implementation. We generate intermediate models with increasing details towards the final UI code. The intermediate models include task models, dialog models, and presentation models. Task models recognize the structural and temporal information and describe how roles execute various tasks defined in business processes.

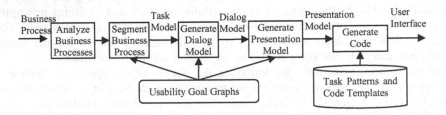

Fig. 1. UI generation framework

   To derive such functional requirements, we must analyze business processes and examine task annotations that describe the functionality of a role (i.e., the person who performs a task). The challenge lies in the fact that the granularity of tasks, expressed in business processes, varies considerably. For example, primitive tasks, such as *"Create an order"* and *"Add a new category"* represent the lowest level of detail. Other tasks may convey functional requirements that can be composed by a sequence of primitive tasks. Instead of considering every task as delivering individual functional requirements, we analyze data dependencies among tasks and data coupling/cohesion in the business processes to group tasks into functional segments. We propose a set of segmentation rules that can group relevant tasks and data into meaningful functional segments. A task model consists of a collection of functional segments. Furthermore, we map one segment into one dialog that specifies the interaction between users and applications.

   A dialog model defines the flows (i.e., transitions) of UI windows. We distinguish two types of windows in a dialog model: task windows and support windows. A task window maps to a segment in the task model. A task window allows a user to interact with an application in order to accomplish the tasks in a segment. To improve the user experience, we need to provide more precise information (i.e., input data or output data) about a task. In this way, a business user can understand the functionality of a task, verify the correctness of their operations once the task is carried out, and easily access auxiliary information for the fulfillment of a task. For example, a picture is a required input to the *"Add product attributes"* task, and a picture of a product may be added into a product catalog after performing this task. A user who is performing the *"Add product attributes"* task would find it useful to be able to preview a picture of the product before inserting it into a product catalog. It may be valuable for the user to be able to verify the product description after the product description is created in a catalog. Support windows provide auxiliary data operations inferred from business processes and assist users performing tasks in the task windows. Window transitions are generated based on the control flows between segments in a task model.

In the presentation model, a dialog is associated with a set of abstract widgets and implements task windows and support windows. An abstract widget describes the generic properties (e.g., title, size, and alignment) of a concrete widget in a specific implementation platform. General speaking, a task can be realized by a set of abstract widgets. For example, a *"Find product"* task can be implemented by a text field that allows a user to enter searching criteria, and a table that displays the search result. To improve the usability of the generated UI, each task or group of tasks is associated with a UI design pattern (e.g., find pattern, output display pattern, and shopping cart pattern). The layout of each window is guided by various UI best practices.

4 Models and Their Transformations

In this section, we introduce the background knowledge about our model-based UI generation. We illustrate the concepts of task models, dialog models, presentation models and transformation rules using examples.

4.1 Analysis of Business Processes

Our UI generation approach starts by analyzing business processes in order to extract functional requirements and usability information from them. As illustrated in Figure 2, we model a business process in terms of tasks and connectors. Tasks (e.g., *"Find product"*) describe activities that a user needs to perform in order to achieve a business goal. Connectors define the control flow and data flow of a business process. As depicted in Figure 4, an example business process is composed of 9 tasks and 8 connectors. More specifically, a task has a name, a role, zero or more input connectors, and zero or more output connectors. The role specifies who is allowed to perform the task. Input connectors and output connectors describe the input data and the output data exchanged between tasks. A connector links a source task with a target task along with data flows and conditions. For example, as shown in Figure 4, a connector, < *"Add to cart"*, *"Enter quantity"*, *Shopping cart, null*> links its source task *"Add to cart"* and its target task *"Enter quantity"* along with a data item *Shopping cart* that is created by the *"Add to cart"* task and serves as an input to the *"Enter quantity"* task. No data flow conditions are specified in this example. There are three types of control structures in business processes: sequence, decision, and loop. Tasks in a sequence structure will be performed continuously. A decision structure defines more than one possible execution path. Each path has a precondition. Once the precondition of a path is satisfied, the tasks in the path are executed. Tasks in a loop structure will be performed multiple times. A loop may have a condition to decide when it stops. Moreover, a data item is composed of a set of data attributes, such as name and type. The data attributes in a business process is specified in an information model that defines the relations among the data items in business processes. Each data item is considered as a class in the information model. The attributes of the data item are interpreted as fields in the corresponding class. A data attribute can be a primitive type (e.g., a string and an integer) or a reference to other data item. For example, the data attribute *billingAddress* in *Payment* data item is an instance of a type *Address*. Moreover, a data attribute may have a default value.

```
BP = <Task+, Connector* >
Task = <Name, Role, InputConnector*, OutputConnector*>
Connector = <SourceTask, TargetTask, [DataItem], [Condition]>
DataItem = < DataAttr+>
DataAttr = < Name, Type, DefaultValue >
+ means that the occurrence of an element in a tuple is one or more
* means that the occurrence of an element in a tuple is zero or more
[ ] means the this element is optional
```

Fig. 2. The definition of business processes

4.2 Task Model

In the design of UIs, limiting the number of widgets in a window can improve the efficiency of the interactions between users and computers because users can concentrate on performing one task at a time. Elkoutbi et al. suggest a threshold of 20 widgets in a window [3]. As aforementioned, a task can be implemented by a group of widgets. Our aim is to identify the relevant tasks that need to be performed together in the same window, and to limit the number of widgets in a window. We propose several heuristic rules that identify the relevance between tasks and segment tasks by

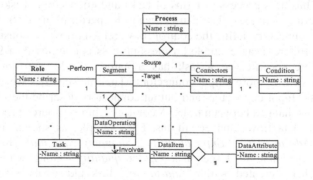

Fig. 3. The meta-model of the task model

analyzing control and data flows in the business processes. Intuitively, grouping relevant tasks divides a business process into a set of smaller segments of task groups. Each segment is mapped to a window in the generated UI. The segments are linked by the connectors between segments in a business process. The meta-model of task models is depicted in Figure 3. A business process consists of a set of *Segments* and *Connectors*. Each Segment encapsulates tasks and data items. In the following subsection, we discuss our heuristics for dividing a business process.

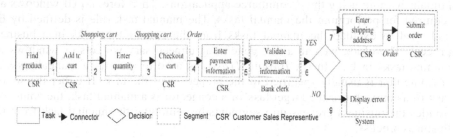

Fig. 4. An example business process

4.2.1 Role Rule

$$B_1=\{c|c.SourceTask.Role \neq c.TargetTask.Role\}$$

where c is a connector, $c.SourceTask$ and $c.TargetTask$ describe c's source task and c's target task respectively (Figure 2). $c.SourceTask.Role$ and $c.TargetTask.Role$ refer to the roles of these two tasks, respectively.

We want to provide a personalized UI for different roles in a business process. A personalized UI provides the necessary functional features and UI widgets with respect to individual roles. A user can focus on their own work and easily select the widgets from fewer widget sets in the UI. We derive a role rule that divides a business process into a few groups as specified in B_1. Each group of tasks is performed by the same role. We identify a set of connectors that link the source tasks of connectors and target tasks of the connectors, performed by different roles. Furthermore, connectors in B_1 divide a business process into *Segments*. Each *Segment* contains the task and related data items performed by one role. For example, as illustrated in Figure 4, the business process is segmented by connectors $\{5, 6, 7, 9\}$ by applying the role rule.

4.2.2 Primitive Task Rule

$$B_2=\{c|c.SourceTask.IsPrimitiveTask()\}$$

where c is a connector, *IsPrimitiveTask()* can identify whether a task is primitive or not

Primitive tasks, such as *"Add to cart"* and *"Submit order"*, describe the lowest level of details in a business process. Such tasks are usually implemented as button widgets and used in combination with other tasks that share the same UI window. For example, when a product is found, the product can be selected from a list of products, and added to a shopping cart. To improve the efficiency of the UI, the *"Add to cart"* task is often combined with the *"Find product"* task and implemented in one window, as illustrated in Figure 4. We identify primitive tasks using naming conventions. For example, we look for tasks with a name containing *"submit"*, *"save"*, *"add"*, and *"delete"*. A primitive task alone does not produce a segment in a business process.

4.2.3 Manual Task Rule

$$B_3=\{c|c.SourceTask.IsManualTask() \oplus c.TargetTask.IsManualTask()\}$$

where c is a connector, *IsManualTask()* checks whether a task is a manual task or not

Manual tasks, such as sending surface mail, are manually accomplished by humans without using the e-commerce applications. Therefore, no UI windows are required for performing the manual tasks. The manual task rule is defined by B_3, and used to exclude the manual tasks from the rest of the tasks in a business process. Therefore, we can avoid generating unnecessary widgets or windows for manual tasks in UIs. In B_3, *IsManualTask()* is used to identify a manual task, an attribute that is usually specified in the properties of the task. If either the source task of a connector or the target task of a connector is a manual task, the connector divides the business process into segments by removing the manual task from the business process.

4.2.4 Optional Task Rule

$$B_4=\{c|c.SourceTask.IsOptionalTask() \oplus c.TargetTask.IsOptionalTask()\}$$

where c is a connector, *IsOptionalTask()* identify whether a task is a optional task or not

To improve their work efficiency, users would rather click fewer buttons. In this case, the developers may choose to provide default values or settings for UI widgets, such as combo boxes and text boxes. We use the default values specified in the attributes of a data item, which is defined as an output of a task, to provide the default values for the widgets that are used to implement the task. Such a task is considered an optional task to perform only if the user would like to change the default values. For example, the payment by credit card is defined as the default payment method. The task *"Select payment method"* can be skipped. Optional tasks separate the business processes into segments. This rule can improve efficiency because users perform optional tasks only when they cannot use default values. We use the equation B_4 to identify connectors that link to optional tasks. If a task is an optional task, then *IsOptionalTask()* returns true. If either the source task or the target task of a connector is an optional task, such a connector breaks the business process into segments.

4.2.5 Branch Rule

$$B_5=\{c|NumberofInputConnectors(c.SourceTask)>1$$
$$\vee NumberOfOutputConnectors(c.TargetTask)>1\}$$

where c is a connector, *NumberofInputConnectors ()* and *NumberOfOutputConnectors ()* count the number of input connectors and the number of output connectors of a task.

If a task has more than one outgoing connector, this structure indicates alternative branches or parallel paths. We include all the tasks in one branch as a segment. For example, the business process in Figure 4 is divided into three segments using the branch rule by connectors $\{6, 7, 9\}$.

4.2.6 Data Sharing Rule

$$B6=\{c|(c.SourceTask.InputConnectors{\neq}null \wedge$$
$$ic{\in} c.SourceTask.InputConnectors()) \wedge ic.DataItem() {\neq} c.DataItem()\}$$

where c is a connector, ic is one of the input connectors of the source task of connector $c;$ $c.DataItem()$ returns the data item associated with connector c.

If a sequence of tasks operates on the same data item, these tasks share the same data information. In the UI design, it would be inefficient if a user had to enter the same information multiple times when performing different tasks. This data sharing rule improves the efficiency of the UI design by grouping the tasks and their shared data item in a segment. As specified in B_6, if the data item of the connector c and the data item of the connector ic are not equivalent, c is a connector that divides the business process into different segments. For example as shown in Figure 4, two segments are identified using the data sharing rule. The *"Enter quantity"* task and the *"Checkout cart"* task are included in the same segment since both tasks share the data item, *Shopping cart*. The *"Enter payment information"* task and the *"Submit order"* task share the data item, *"Order"* and are grouped into one segment.

4.3 Application of Rules

Each rule has preconditions that specify the context for applying the rule. For example, the branch rule would not be applied for sequentially ordered tasks. In the process of dividing a business process into a collection of segments in a task model, we identify an applicable set of rules using the preconditions of each rule. The role rule, data sharing rule, branch rule, manual task rule, and optional task are independent from each other. Therefore, the result of the segments is independent from the order of applying these rules. The primitive tasks, such as the *"Add to cart"* and *"Submit order"* tasks are not associated with a separate window. As the result, the primitive task rule identifies primitive tasks, and merges the identified primitive task with the prior task. In this case, the primitive task and their prior task are treated as one merged task. Therefore, we apply the primitive task rule before any other rules. Figure 4 illustrates a segmentation example. The primitive task rule is applied before any other rules and it identifies the connector set B_2. Moreover, the role rule, branch rule, and data sharing rule are applied independently in any orders. These will identify the connector set $(B_1 \cup B_5 \cup B_6)$. At last, we derive six connectors that divide the business process into segments (i.e., $B_2 \cup (B_1 \cup B_5 \cup B_6) = \{2, 4, 5, 6, 7, 9\}$).

4.4 Dialog Model

The dialog model is composed of a set of linked windows, as illustrated in Figure 5. Each window is generated from a segment in the task model. Window transitions are generated from connectors in the task model. In the dialog model, task windows provide widgets that allow users to interact with the application in order to fulfill tasks following predefined orders as specified in the business processes. The Support windows provide contextual support that assists users in performing related tasks. A task window contains task and data operations. A support window can only include data operations inferred from data items. An example of data operations can be viewing the product specification before performing the "*Add to cart*" task. To prevent users from making errors when performing tasks, a window (i.e., editor) can be only editable when a task requires input from the user. Otherwise, we set the window as a viewer that is used for displaying information.

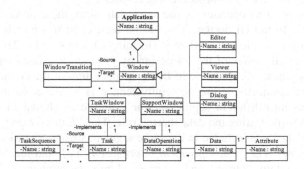

Fig. 5. The meta-model of the dialog model

Our rules for transforming task models to dialog models are inspired by [14, 15], which derive navigation structures and window structures from task structures. Our window transactions are directly derived from the connectors in the task models. To generate window structures, we have designed the following three rules to transform segments to windows.

Single Window Rule

$$S \overset{T}{\Rightarrow} W,$$ where S is a segment, W is a window and $\overset{T}{\Rightarrow}$ is a transformation

We transform all tasks and data operations in one segment into one window. This rule results in a UI that is ideal for expert users, since the users can complete tasks using fewer clicks. But the UI might be difficult for novice users, who have to learn and remember the functionality and order of use of all the widgets in one window.

Data Window Rule

$$S \overset{T}{\Rightarrow} W_T + \sum_{i}^{N} W_{Di},$$ where S is a segment, W_T is the task window for the

segment S, W_{Di} is the window of the data item i, and N is the number of data items of a segment

This rule transforms all tasks in one segment in a task window and the data operations related to one data item into a support window. In this case, support windows can be reused in many other segments that contain data operations for the same data item. This rule permits us to reduce the number of tasks and data operations implemented in one window. Moreover, users' knowledge about a support window can be reused in other contexts when the same data items recur.

Functional Window Rule

$$S \overset{T}{\Rightarrow} W_T + W_S,$$

where S is a segment, W_T is the task window for the segment S, and W_S is the support window for the segment S.

This rule transforms all tasks in a segment into a window and all data operations in the segment into a support window. This rule decreases the number of tasks and data operations in a window. The generated UI is easier to learn than the one generated using the single window rule. However, it may be more difficult to learn than the one generated using the data window rule. This rule is a trade-off between the single window rule and the data window rule.

4.5 Presentation Model

The presentation model transforms the dialog model into a collection of platform-independent abstract interface objects. As illustrated in Figure 6, a window in the presentation model is associated with a collection of abstract interface objects, such as input fields, output fields, selectors and action invokers. The generation of the presentation model is accomplished by matching tasks with task patterns, which are collections of widely used operations in the design of UIs, such as find operation, display contents operation. Two well-known collections of UI task patterns are summarized in [10, 11]. For example, Figure 7 shows the structure of a search task pattern, adapted from [12, 13]. Specifically, a search task can be decomposed into two sub-tasks including entering search criteria and displaying search results. Furthermore, an input field (e.g., text field) and a search action invoker (e.g., button) allow a user to enter search criteria. Similarly, an output field (e.g., a table) and search action invoker (e.g., button) display the returned result. We use naming similarities in order to match tasks in the business processes with task patterns used in the UI designs. The screenshot for our generated UI after performing the search task pattern on the *"Find product"* task is shown on the right side of Figure 7.

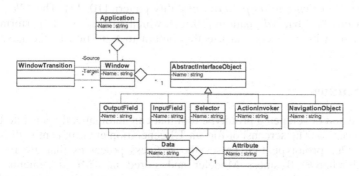

Fig. 6. The meta-model of the presentation model

To transform a platform-independent presentation model into a platform-specific concrete UI implementation, a set of platform-specific code templates are used to implement abstract interface objects. The selection of a concrete widget is determined by the tasks and data items specified in the business process. For example, when a task generates output data, the window that implements this task is set to be an editor, for the reason that a user can modify the intermediate result when performing the task. When a data item, such as a *Picture* of a *Product*, is used as an input to a task, the content of this data item can be previewed using a viewer assuming that we can't modify the content of the input data item. We map the types of the data attributes of a data item with different widgets that are used to display the context of a data item. For example, the data item, *Product*, contains string typed attributes, such as *Product ID*, *Product name,* and *Product type*. The string typed data attributes are interpreted into text fields, as illustrated in Figure 7. A data item that can appear multiple times (e.g., an array of resulting products), is converted into a table, as shown in Figure 7.

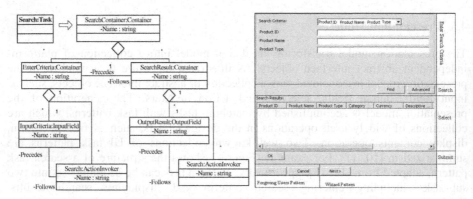

Fig. 7. The structure of search task pattern and an example implementation

Moreover, we incorporate several UI design principles during the generation of the UI. For example as shown in Figure 7, three buttons (i.e., *"Undo", "Cancel",* and *"Next"*) at the bottom are generated to improve the usability of this window. The *"Undo"* button and the *"Cancel"* button allow user to correct their mistakes by adapting the *"Forgiving users pattern"* usability pattern [10, 11]. The *"Next"* button is created using the *"Wizard"* pattern [10, 11], which guides users to perform tasks in a step-by-step fashion. Users can use this button to transition to the next page to perform the next task.

5   Case Studies

We designed and developed a prototype tool that can automatically generate UIs from business processes. The screenshot for our UI generation environment is illustrated in Figure 8.a. Our prototype tool parses the business processes that are modeled in IBM® WebSphere® Business Modeler and stored in XML documents. For the business process modeled using other modelers, a separate parser needs to be developed in order to handle a particular business process specification language. In our UI generation environment, multiple UIs can be generated by applying different UI design principles on the task models, dialog models, and presentation models. Eventually, we allow developers to specify the code templates for implementing abstract UI objects defined in the presentation model.

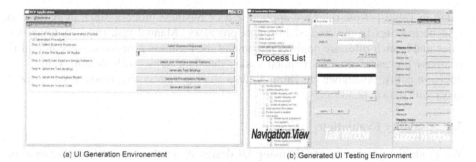

(a) UI Generation Environement   (b) Generated UI Testing Environment

Fig. 8. Screenshots of our prototype UI generation tool

Moreover, our generated UIs can be previewed in a test environment, as illustrated in Figure 8.b. The process list view allows developers to access the UI generated from the selected business process. Once a UI for a business process is selected, the structure of the business process is displayed in the navigation view. The task window of the initial task is launched in the area of the task window. The data operations of the task are initiated in the area of the support window. In this screenshot, we applied the data window rule to generate separate windows for each data items and their corresponding data operations.

In the following subsections, we report on the following studies:

1) Evaluate the usefulness of the business process segmentation rules. We track the frequency of the application of each rule in the process of generating task models from business processes.

2) Evaluate the usability of the generated UI.

5.1 Application of Rules

We performed an experiment to count the frequency of the rule application during model transformations. We selected 10 business processes from an existing call center application and counted the frequency of the application of each rule. The results are

Table 1. Frequency of rule application

Business process	Role rule	Manual task rule	Optional task rule	Branch rule	Data sharing rule
1	1	1	0	0	3
2	1	0	0	0	2
3	1	0	0	1	6
4	1	1	3	2	11
5	1	1	3	1	5
6	1	1	2	3	5
7	1	0	0	1	1
8	1	1	0	0	2
9	1	1	0	0	4
10	1	1	0	0	4

listed in Table 1. Role rules are applied in each business process, since more than one role participates in the studied business processes. Manual task rules, optional task rules and branch rules are applied in some of the business processes, since not all the business processes contain manual tasks, the data items with default values, and decision paths. As shown in Table 1, data sharing rule is frequently used to merge tasks with data dependencies.

5.2 Usability Evaluation

We guide the generation of UIs using UI design principles. To evaluate the usability of the generated UI, we use a standard questionnaire proposed by Chin, Nielsen, and Lewis et al. [7], as illustrated in Figure 9.

	1	2	3	4	5	NA
1. Visibility of system status	bad $\subset$	$\subset$	$\subset$	$\subset$	$\subset$ good	$\subset$
2. Match between system and the real world	bad $\subset$	$\subset$	$\subset$	$\subset$	$\subset$ good	$\subset$
3. User control and freedom	bad $\subset$	$\subset$	$\subset$	$\subset$	$\subset$ good	$\subset$
4. Consistency	bad $\subset$	$\subset$	$\subset$	$\subset$	$\subset$ good	$\subset$
5. Prevent Errors	bad $\subset$	$\subset$	$\subset$	$\subset$	$\subset$ good	$\subset$
6. Recognition rather than recall	bad $\subset$	$\subset$	$\subset$	$\subset$	$\subset$ good	$\subset$
7. Flexibility and efficiency of use	bad $\subset$	$\subset$	$\subset$	$\subset$	$\subset$ good	$\subset$
8. Aesthetic and minimalist design	bad $\subset$	$\subset$	$\subset$	$\subset$	$\subset$ good	$\subset$
9. Help users recognize, diagnose, and recover from errors	bad $\subset$	$\subset$	$\subset$	$\subset$	$\subset$ good	$\subset$
10. Help and Documentation	bad $\subset$	$\subset$	$\subset$	$\subset$	$\subset$ good	$\subset$
	1	2	3	4	5	NA

Fig. 9. The Nielsen's ten usability heuristic questionnaire [7]

We generate two versions of the UI following different UI design principles. One version has high learnability and the other one has high efficiency. For example, high learnability can be achieved by giving more guidance to users. We generate navigation buttons for each window. High efficiency is achieved by grouping the widgets for accomplishing the tasks in one segment into one window. We invite 5 usability researchers to evaluate these generated UIs. The evaluators walked twice through the two versions of UIs of the 10 business processes. Then they evaluated the

Table 2. Usability evaluation result

Question	E1	E2	E3	E4	E5
Q1	5	5	5	5	5
Q2	5	5	5	5	5
Q3	4	5	5	4	5
Q4	5	5	5	5	5
Q5	5	4	4	5	4
Q6	5	5	5	5	5
Q7	5	5	5	5	5
Q8	5	5	4	5	5
Q9	5	4	5	4	5
Q10	NA	NA	NA	NA	NA

generated UIs using the questionnaire shown in Figure 9. Each question is scored from 1(bad) to 5(good). If the feature corresponding to a particular question is not available, the score is marked NA. The evaluation results are listed in Table 2. All scores for Q10 are NA, for the reason that we generate UIs without considering context-sensitive help and documentation. Other questions are marked as good or above average. Hence, we believe that the usability of the generated UI is acceptable for the professional usability researchers.

6 Conclusion and Future Work

In this paper, we present our approach for automatically generating UIs from business processes. Business processes are used as business requirement models that capture business knowledge. The task models are directly derived from business processes. The UIs are generated using two other intermediate models: the dialog models and the presentation models with increasing levels of details. To ensure that the generated UIs are easy to use and learn, the transformations between models are guided by UI design principles and task patterns. As a result, the generated UI has strong usability supports such as consistent look and feel, and transition guidance. Moreover, any changes to business processes can be automatically propagated to the UIs by regenerating the code using our prototype tool. So, it is easy for developers to integrate their feedback to the generation process and to fine-tune their design.

In the future, we plan to extend our prototype tool to allow users to integrate their own task patterns and UI design principles into the generation process so that they can reuse their existing knowledge to generate UI following their existing style. We will improve the business process analysis technique by identifying additional contextual information from business processes.

Acknowledgement

This research is sponsored by IBM Canada, Centers for Advanced Studies (CAS), and National Sciences and Engineering Council (NSERC). We would like to thank Mr. Qi Zhang at the University of Waterloo and Mr. Tack Tong at the IBM Canada Toronto Laboratory for their suggestions on this work.

Trademarks

Disclaimer

References

1. Zhang, Q., Chen, R., Zou, Y.: Reengineering User Interfaces of E-Commerce Applications Using Business Processes. In: ICSM, pp. 428–437 (2006)
2. Guruge, A.: Corporate Portals Empowered with XML and Web Services. Elsevier Science, USA (2003)
3. Elkoutbi, M., Keller, R.K.: User Interface Prototyping Based on UML Scenarios and High-Level Petri Nets. In: Nielsen, M., Simpson, D. (eds.) ICATPN 2000. LNCS, vol. 1825, pp. 166–186. Springer, Heidelberg (2000)
4. Schmidt, D.C.: Guest Editor's Introduction: Model-Driven Engineering. IEEE Computer 39(2) (2006)
5. Nichols, J., Faulring, A.: Automatic Interface Generation and Future User Interface Tools. In: CHI. Proceedings of the Workshop on the Future of User Interface Design Tools (2005)
6. Puerta, A.R.: A Model-Based Interface Development Environment. IEEE Software 14(4), 41–47 (1997)
7. Nielsen, J.: Nielsen's Heuristic Evaluation. In: Usability Engineering. ch. 5, p. 115. Academic Press, San Diego (1993)
8. Zdun, U.: Concepts for model-driven design and evolution of domain-specific languages. In: OOPSLA. Proceedings of the International Workshop on Software Factories, USA, pp. 1–6 (2005)
9. Leymann, F., Roller, D., Schmidt, M.-T.: Web services and business process management. IBM Systems Journal 41(2) (2002)
10. van Welie. M.: Patterns in Interaction Design, http://www.welie.com/
11. Tidwell, J.: Designing User interfaces, http://designinginterfaces.com/
12. Sinnig, D., Gaffar, A., Reichart, D., Forbrig, P., Seffah, A.: Patterns in Model-Based Engineering. In: Proceedings of CADUI, pp. 197–210 (2004)
13. Paternò, F.: Model-Based Design and Evaluation of Interactive Application. Springer, Heidelberg (1999)
14. Zanden, B.T.V., Myers, B.A.: Automatic, Look-and-Feel Independent Dialog Creation for Graphical User Inter-faces. In: Proceedings of SIGCHI, pp. 27–34 (1990)
15. Vanderdonckt, J.: Knowledge-Based Systems for Automated User Interface Generation: the TRIDENT Experience. In: Proceedings of the CHI Workshop on Knowledge-Based Support for the User Interface Design Process (1995)

Enhancing UML Extensions with Operational Semantics
Behaviored Profiles with Templates

Arnaud Cuccuru, Chokri Mraidha, François Terrier, and Sébastien Gérard

CEA, LIST, Gif-sur-Yvette, F-91191, France
{arnaud.cuccuru, chokri.mraidha, francois.terrier,
sebastien.gerard}@cea.fr

Abstract. The objective of the ongoing OMG standard about a foundational UML subset semantics (fUML) is twofold: providing operational semantics for a UML subset, and ease unambiguous and automatic model exploitations. Its impact could however be limited if usual UML profiling practices do not evolve. Profiles are the traditional way to specialize UML semantics and handle semantic variation points. However, they are usually defined in a way that only informally addresses the semantic issue, potentially limiting the benefits that fUML could bring in UML based methodologies. UML profiling practices must evolve: we propose to explicitly encapsulate operational semantics into stereotype operations, and provide a way to intuitively handle semantic variation points through template parameters. We illustrate the usage of these mechanisms and demonstrate their potential benefits. We also show that no UML metamodel modifications are required to support them, so that their implementation in L3-compliant UML tools is straightforward[1].

1 Introduction

Since the earliest days of the Unified Modeling Language and right up to its current version (2.1.1), the most fundamental criticism made by the "models" community has been its lack of semantics. The informal nature of UML's semantic description (i.e. natural language) inevitably generates ambiguities, often making it difficult to agree on the meaning of a given model. These ambiguities involuntarily leave the door open to multiple and potentially contradictory interpretations of a same model. In [6], Steve Cook thus talks about the "cognitive semantics" of UML, as opposed to the "objectivist semantics" of a well-defined programming or modeling language. This drawback limits the impact that "pure" UML could have on model-driven engineering, where (semi-)automatic model exploitations

---

[1] This work has been performed in the context of the Usine Logicielle project (www.usine-logicielle.org) of the System@tic Paris Region Cluster. This project is partially funded by the "Direction Générale des Entreprises of the French administration", the "Conseil Régional d'Île de France", the "Conseil Général des Yvelines", the "Conseil Général de l'Essonne" and the "Conseil Général des Hauts de Seine".

G. Engels et al. (Eds.): MoDELS 2007, LNCS 4735, pp. 271–285, 2007.

(and therefore automatic and unambiguous interpretations of the underlying semantics) are required (the most obvious example of this is code generation).

UML profiles are usually used to overcome these undesirable side-effects. A profile basically provides users with a specialization of the UML 2 metamodel, through stereotypes and constraints definitions, thus enabling to clarify and/or specialize its semantics. Often associated with dedicated methodologies, profiles have proven useful for many modeling domains. The Accord/UML methodology [13][14] we proposed in previous works is a good example of UML profiling efficiency, namely for real-time systems engineering. More generally, for a given profiled model, a given stereotyped model element can be "recognized" by the tool chain associated with the methodology, and interpreted in a proper way. OCL constraints associated with the definition of profile elements are typically used to guarantee that the underlying model repository will be well-formed and interpretable for the tool chain. A similar usage of UML profiles is of course also recommended to fix the explicitly identified semantic points for which potential interpretations may vary (i.e. fix the "Semantic Variation Points").

More pragmatically, UML profiling mechanisms have achieved their primary objective: providing UML users with an intuitive means for customizing UML, from both syntactic and semantic terms. However, as in the case of the UML 2 metamodel, semantic descriptions associated with a profile remain "cognitive" (i.e. expressed in natural language), and are not an explicit part of the profile definition: "Automatic semantic interpretation" (as mentioned above) must therefore be "hard-coded" in one way or another in the tool chain. At a time when the OMG officially recognizes a need for formalizing UML 2 semantics (see its RFP of August 2005 [9], on the Semantics of a Foundational Subset for Executable UML Models), we are convinced that the role of profiles in UML-based methodologies should be enhanced. The offshoot of this RFP (which is further described in section 2.3) should be an executable semantics description (in operational style) for a UML subset. UML profiles (traditionally used to define specialized UML semantics) would thus directly and explicitly influence operational semantics: UML profiles should explicitly encapsulate such operational semantic description and provide users with an explicit means for varying semantics where needed (i.e. for identifying and/or fixing semantic variation points).

In section 2, we first clarify the meaning of "semantics", and provide an overview of existing works on formalizing UML semantics. This focuses specifically on the OMG standard being developed for UML semantics and motivates our own "operational profiles" approach. To better explain how traditional use of profiling mechanisms can be enhanced, section 3 gives a recap of UML profiling basics, by defining a simple profile for specializing UML state machines. In the rest of the paper, we use this basic example to explain and illustrate how the role of UML profiles could be enhanced. In section 4, we elaborate on the basic profile by first showing how operational semantics can be encapsulated in stereotype definitions, then how template parameters help to explicitly handle semantic variation points. We also show why and how these original mechanisms are easily supported by our open source UML 2 tool, Papyrus[2]. We conclude with a discussion of relevant future works and their potential implications for UML-based model-driven approaches.

2   UML and Semantics

There has been much previous work on formalizing UML semantics. Before describing the approaches most frequently used to address this issue, we propose to clarify the definition of the much debased word "semantics". This is followed by a thorough description of what is probably one of the most promising documents about the subject: the future OMG standard on "Semantics of a Foundational Subset for Executable UML Models".

2.1   The Semantics of Semantics

As stated by D.Harel and B.Rumpe in [5], a language is defined by its syntax, itself defined by a lexicon and a grammar (or, in our context, a metamodel), and its semantics, i.e. the meanings associated with each of its syntactic elements. The authors explain that defining the semantics of a language is like defining a mapping between the identified syntactic elements and a given semantic domain (the set of natural numbers $\mathbb{N}$, and their underlying properties, is an example of a semantic domain). A language can be considered well-defined if its semantic mapping M leads from a clear and expressive syntax L to a well-defined and well-understood semantic domain S (concisely denoted as $M : L \to S$ in their article). Whatever the concrete formalism used to do so, a description of programming or modeling language semantics most often uses one the three following "semantic description styles" [4]:

- **Denotational:** Semantic mapping provides a link between language constructs and mathematical models. For a given language being defined, a conventional (and very simple) example of such mapping would be linking the keyword "nat" to the set of natural numbers $\mathbb{N}$. In the same way, the keyword "+" could be mapped to the addition function on natural numbers.
- **Axiomatic:** Semantic mapping links language constructs and a set of logical rules. A typical example is association of the keyword "function" with a rule stating that "applying equal functions to equal arguments produces equal outputs".
- **Operational:** Semantic mapping links language constructs with rules for their execution on an abstract machine. This style could also be qualified "algorithmic", in the sense that it usually relies on a sequence of "instructions" on the abstract machine. The word "operational" is understood intuitively in connection with the semantics of dynamic/executable parts of a language. However, the same style could also be used to describe the non executable, or static, parts of a language. Extensively used for the definition

---

[2] http://www.papyrusuml.org

of the UML metamodel, the "PackageMerge" relationship (which is a purely structural construct) is a typical case in which the same description mechanisms could be applied. In the UML Infrastructure [10], the Merge concept is in fact described using an informal algorithmic description (through a set of 10 algorithmic statements).

According to these very general definitions, UML does not currently fall into the category of well-defined languages, since the semantic domain and semantic mapping are described only through an informal natural language description associated with the metamodel. The next section of this article gives a brief overview of UML formalization approaches commonly used to overcome these limitations.

2.2 Related Works

In his article on improving UML definition [2], G.O'Keefe gives (along with a set of general recommendations) an interesting and illustrative survey of work related to formalizing UML (examples include [8] and [12]). All the work identified by O'Keefe basically shares one principle: linking parts of the UML metamodel to formal languages such as Z, Object-Z, or CSP, thus providing a strong mathematical basis for the targeted UML subsets.

However, the traditional UML user usually does not have an "academic profile". This assertion, which is not meant to be pejorative, refers to the fact that formalizing an UML subset in Z language will probably not help the traditional UML user better understand state machine or sequence diagram semantics. It is not a criticism of the fundamental value or quality of the work mentioned in the previous paragraph. We simply think that the answers given so far are not suited to the typical UML user, who is looking for an unambiguous but intuitive semantic description, and a degree of formalism that affords reliable automatic model exploitations. G.O'Keefe has quipped that "Too much of the work on UML semantics looks like a technical answer which is glad to have found a good practical question". We have reason to believe that the future OMG standard "Semantics of a Foundational UML Subset for Executable UML Models" will provide the practical answers hoped for by most of the UML users. The following section tells the reader why.

2.3 Semantics of a Foundational UML Subset

The RFP of august 2005 on "Semantics of a Foundational UML Subset for Executable UML Models" RFP [9] requests proposals of a definition for a "computationally complete and compact subset of UML 2". Here, "computationally complete" means that the elements contained in the identified subset (usually called fUML, for "foundational UML"') should have unambiguous execution/interpretation semantics, so that any compliant tool chain can fully and automatically interpret a model specified using that subset. For example, fUML will identify and precisely define a minimal set of actions and activities that could be combined to propose semantic definitions of higher level modeling constructs, such as state machines. The RFP states (at the end of §6.2) that "the foundational subset defines a basic virtual machine for the Unified Modeling Language, and the specific abstractions supported thereon...". In other words, fUML elements are given an unambiguous interpretation for that virtual machine.

Using a combination of fUML elements to specify semantics of higher level UML constructs (such as state machines) clearly boils down to defining operational semantics for these constructs (see section 2.1). From a traditional UML user's point of view, this is basically enough to understand the exact meaning of a UML model: on the one hand, the informal semantic description associated with the UML metamodel can intuitively lead to a general understanding of the model; on the other hand, the operational semantic description of the foundational UML subset elements (and their combination for defining higher level modeling constructs) provides unambiguous clarification where needed, and are also available for automatic model exploitations such as code generation or interactive model debugging. In fact, we have something that is almost equivalent to a well-documented Java program, where generated Java doc provides the overall intuition, and Java statements the precise meaning.

Note that a subset of fUML, usually called bUML (for base UML), will be identified and formalized in a way that "breaks circularity" (this subset will be mathematically founded, probably on a denotational style. This is not clearly stated in the RFP, or indirectly in requirement 6.5.2.4: "To avoid circularity, the base semantics shall not itself depend on the semantic definition provided by the execution model."). Beyond the inherent elegance of strong mathematical formalization, this kind of information is potentially useful for formal model-based activities such as model compiling or optimization, where, for example, a strong definition of "model equivalence" properties (based on the denotational semantics of bUML) may be necessary for suitable optimizations, in early stages of the development process.

Beyond the exact content of fUML (which is supposed to be delivered before the end of 2007), the main point to keep in mind is that semantics of high level modeling constructs will be specified through a combination of lower level fUML elements, using an operational semantics style. For consistency purposes, and in order to maximize the potential benefits of fUML, we believe that UML profiling, which is the main current practice for UML specializations/extensions, should enable "objectivist" semantic description of UML semantics specializations (as opposed to the "cognitive" semantics we mentioned in the introduction to this article). According to the RFP quoted above, the most natural way to do so would be to in some way or another encapsulate an operational semantic description within the profile definition, using elements provided by fUML (or even higher level elements specified through a combination of fUML elements) to express it. A means for explicitly identifying and fixing semantic variation points and their related operational semantics should also be provided. The next section of this article proposes the definition of a "naïve" UML profile, illustrating common profile definition practices. This basic profile definition is then enhanced (in subsequent later sections of this article), with emphasis on the added value of our proposals.

3   UML Profiling Basics

To illustrate the way that UML profiles are traditionally defined, let us describe (in a cognitive way) a simple profile for specializing UML 2 state machine semantics. In UML, state machines are mainly used to describe the behavior associated with instances of active classes. A state machine owns one or more regions, which in turn own vertices (pseudo states and states) and transitions that relate vertices to each other. Transitions may be guarded by a constraint, and fired according to certain triggers referencing a firing event. Firing a transition results in the execution of the potentially associated effect behavior, and an evolution of the region's current state from the source vertex to the target vertex of the fired transition.

At execution time, the state machine accesses an event pool that is managed by the context object owning this state machine. The context object interacts with its environment, and updates the content of the event pool according to certain event occurrences (reception of an operation call, a signal, etc.). The state machine takes, dispatches and processes events from the pool only once the previous event occurrence has been fully processed (i.e. Run-to-completion semantics). On the basis of such simple dynamic semantics, several semantic variation points are identified in the UML 2 Superstructure. A semantic variation point can be basically defined as a particular aspect of a language where semantic interpretation can vary according to the domain where the language is used. Given the limited size of this article, the profile we are about to describe will address only two of them: event selection and transition selection policies.

According to the current state of the state machine and the set of relevant events contained in the pool (i.e. events that can trigger a transition from the current state), the event selection policy determines an event dequeuing order, and leaves open the possibility of modeling different priority-based schemes. In the following sections, we consider simple LIFO and FIFO policies as possible concrete semantics for the event selection policy. When the "highest priority event" has been selected according to a concrete event selection policy, the transition selection policy determines the transition to be fired where the same event may trigger several transitions. In the following paragraphs, we consider a Random policy and a Stochastic policy as concrete semantics for the transition selection policy.

In fig.1, we define a profile that addresses the two previously identified semantic variation points for the UML state machines. We implicitly consider other semantic aspects to be handled by the tool chain associated with this profile. This simple profile contains the definition of the *UnambiguousStateMachine* stereotype, which applies to the StateMachine metaclass. This stereotype embeds two tagged values: *eventPolicy* and *transitionPolicy*, typed by *EventSelectionPolicy* and *TransitionSelectionPolicy* enumerations respectively. These

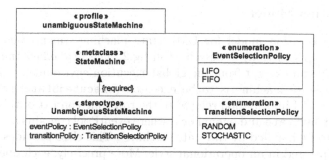

Fig. 1. Simple Profile Example

enumerations contain identifiers for the possible interpretations associated with the semantic variation points identified in the source state machine metamodel. Note that they could be extended to account for other policies.

Applying this profile to a given UML model can make this model unambiguously and automatically interpretable for the associated tool chain. However, this automatic interpretation must be hard-coded in some way or another in the tool chain. Moreover, the semantic description of the profile remains "cognitive" (i.e. with an associated natural language description). The next section demonstrates how use of UML profiling mechanisms can be enhanced to overcome these limitations.

4   Behaviored and Templated Profiles

To better demonstrate how the definition of the basic profile proposed in the previous section can be enhanced, let us consider the following hypothesis: The UML tool used for the current design proposes the default operational semantics for state machine interpretations, expressed using fUML (or any formalism respecting the fUML approach described in section 2.3). In section 4.1, we first describe this default operational semantics, providing what we will call the "execution model". We also show how this default execution model could be specialized to take into account the basic profile of fig.1 and its connoted semantic specializations. In section 4.2, we propose a L2-compliant enhancement of the basic profile (where L2 is the second UML compliance level, as it is described in the superstructure [11], p.5) making semantic specializations explicit and objectivist, through the definition of operations on stereotypes, and encapsulation of operational semantics into these operations. In 4.3, we propose a L3-compliant enhanced version of the profile defined in section 4.2, where semantic variation points are: *(a)* explicitly identified through template parameter definitions, *(b)* described using the strategy design pattern, and *(c)* fixed through bindings of the formal parameters to actual strategy classes. Section 4.4 finally explains how and why the profile enhancements we propose are actually legal according to UML compliance levels.

4.1 Execution Model

The default execution model for UML state machines (mentioned in the introduction to this section) is described in fig.2. The `StateMachineExecution` class represents an excerpt from the global execution environment[3]. It basically knows which state machine is to be executed (`executedStateMachine` role), and knows a set of information related to the run-time context of this execution (namely `currentState` and `eventPool`, where eventPool contains the set of run-time events to be handled by the state machine). This class provides executable semantics expressed in an operational style. More precisely, each operation embeds a behavior, and these operations are properly called by the encapsulating execution environment. The `step()` operation provides a global scheme for the execution of one step of the state machine. The operational semantic description provided in fig.2 is expressed using a Java-like language as its concrete syntax, but the description could have been based on an activity (where each action has a corresponding fUML definition, or a combination of fUML elements). The Java-like syntax is only used for conciseness.

The behavior associated with the step operation first extracts from the event pool the events that are currently relevant (i.e. events that may fire a transition according to the current state). This is done through a call to the selectRelevantEvents() operation. From this set of relevant events, the event with the highest priority is then selected, through a call to the selectEventToHandle() operation. This step corresponds to the "event selection policy" semantic variation point we have identified in the previous section. A default implementation (i.e. the default semantics) is provided by the tool for this operation. Then, from the highest priority selected event, the transition to fire is determined through a call to the selectTransitionToFire() operation. This step is related to the previously identified "transition selection policy" semantic variation point. Finally, the selected transition is fired through a call to the fireTransition() operation.

To credit the connoted semantic specializations introduced in the profile described in fig.1, the selectEventToHandle() and selectTransitionToFire() operations of StateMachineExecution need to be specialized. This could be done by defining a SpecializedStateMachineExecution class, with a generalization relationship between this class and StateMachineExecution. A constraint would be defined on that class to ensure that the `executedStateMachine` is an instance of an «UnambiguousStateMachine» stereotyped StateMachine. Then the behavior associated with the selectEventToHandle() and selectTransitionToFire() operations of the SpecializedStateMachineExecution class would be redefined to account for the eventSelectionPolicy and transitionSelectionPolicy tagged values. A similar methodology could be applied for each new stereotype definition, each

---

[3] This architectural pattern (one *Execution* class, embedding operational semantics, for each metaclass that may have a run-time and executable manifestation) seems to be the one used for the definition of fUML, as it is described in this official and publicly available presentation: *http://www.omg.org/docs/ad/06-06-16.pdf*. The slides are extracted from the initial submission, which is unfortunately not publicly available.

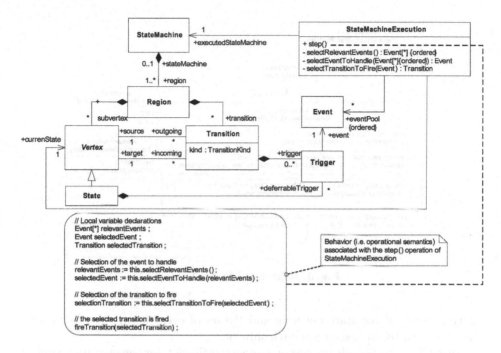

Fig. 2. State Machines Default Execution Model

time involving the definition of a new specialization of StateMachineExecution. This methodology can be improved. The next section describes a L2-compliant enhanced version of the profile where operational semantics is made explicit and objectivist, and shows how the default execution model previously described can be made more generic.

4.2   L2-Compliant Profile Enhancement

The profile described in fig.3 is an enhanced version of the "unambiguousState-Machine" profile shown in fig.1. The «UnambiguousStateMachine» stereotype now owns two (meta-)operations: selectEventToHandle() and selectTransition-ToFire(). Each operation encapsulates a behavioral description, thus providing specialized operational semantics for these particular aspects of state machine meaning (with semantics of course expressed using fUML).

These operations are intended to be called by the execution environment. Indeed, in fig.4, the default StateMachineExecution class has been specialized by the definition of GenericStateMachineExecution. In GenericStateMachine-Execution, selectEventToHandle() and selectTransitionToFire() operations are overridden so that each behavior actually provides user-defined operational se-mantics, through explicit delegation to the selectEventToHandle() and select-TransitionToFire() operations of the «UnambiguousStateMachine» stereotype. The run-time information required for execution of these operations (i.e. the

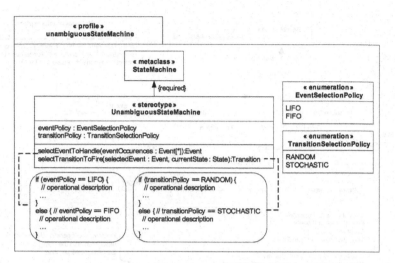

Fig. 3. Behaviored Profile Example

current state of the state machine and the set of event occurences to consider) are provided by the execution environment.

In addition to the advantages of a generic definition for the execution environment (we will not in fact further modify its definition until the end of this article), the main benefit to be had from this approach is that semantic aspects of interest to the user (in this case event selection policy and transition selection policy) have now become an explicit part of the profile definition, through an explicit operational semantic description: From the first version shown in fig.1 to the one in fig.3, we progressed from a "cognitive" semantic description to an "objectivist" semantic description, automatically interpretable by the execution environment. A very perceptive reader might object that the objectivist semantic we are talking about was already part of the previously introduced SpecializedStateMachineExecution class. We would reply that the traditional way to specialize UML semantics is to define UML profiles. From our point of view, the most natural place to describe the semantic specializations is therefore the profile itself. The next section describes another possible enhancement for the profile definition, providing a L3-compliant elegant and intuitive solution for handling semantic variation points, through template parameters and strategy classes.

4.3   L3-Compliant Profile Enhancement

In [1], we proposed a technique for explicitly denoting semantic variation points in Domain-Specific Modeling Languages (DSML) metamodels. This approach is based on the definition of template parameters and the use of the strategy design pattern [3]. In the strategy design pattern, a "strategy class" embeds only one operation and is used as a server. A client class then embeds several operations, in which the behavior of each operation is expressed as a call for a strategy class operation.

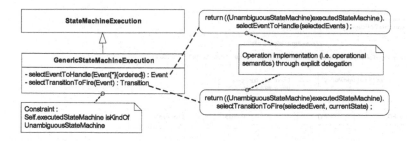

Fig. 4. Execution Model Refinement

In our proposal, semantic variation points are encapsulated in strategy classes, the concrete type of which is left open as a template parameter in the meta-model. Binding such a template parameter creates a new specialization of the language, thus fixing the semantic variation point. The profile enhancement proposed in fig.5 is an almost direct application of this approach. The powerful templatization mechanisms provided by UML[4] are applied to define this enhanced profile.

The `concretePolicies` model library contains strategy classes. *Event-SelectionPolicy* and *TransitionSelectionPolicy* are abstract classes representing abstract strategies (i.e. strategies having no behavior associated with selectEvent-ToHandle() and selectTransitionToFire() operations) and concrete policies (i.e. with an actual behavior description) are defined in the library through definition of new specialized classes: FIFO, LIFO, Random and Stochastic (other policies could be defined in the same way). The template signature associated with the unambiguousStateMachine profile owns two formal template parameters, exposing the following ParameterableElements[5]: ESP, a class that must comply with the abstract *EventSelectionPolicy* strategy class (i.e. *EventSelectionPolicy* is the constraining classifier of the corresponding template parameter), and TSP, which must comply with *TransitionSelectionPolicy*. The «UnambiguousStateMachine» stereotype owns two properties: eventSelectionPolicy and transitionSelectionPolicy, of type ESP and TSP respectively (i.e. the types exposed as formal template parameters by the owning profile). The behavior associated with the selectEvent-ToHandle() and selectTransitionToFire() operations of the «UnambiguousState-Machine» stereotype are now specified as explicit delegations, through the following operation calls (parameters are not written for conciseness): eventSelectionPolicy. selectEventToHandle() and transitionSelectionPolicy. selectTransitionToFire() respectively.

From a functional point of view, this solution is equivalent to the one proposed in fig.3. However, from a cognitive point of view ("cognitive" is used here in a non-pejorative way), semantic variation points have been clearly and explicitly identified through template parameter definitions on the owning profile.

---

[4] Superstructure [11], chapter 17.5.

[5] ParameterableElement is a metaclass introduced in the UML metamodel subset about templates. It represents an element that can be exposed as a formal template parameter, and passed as an actual value in the context of a template binding.

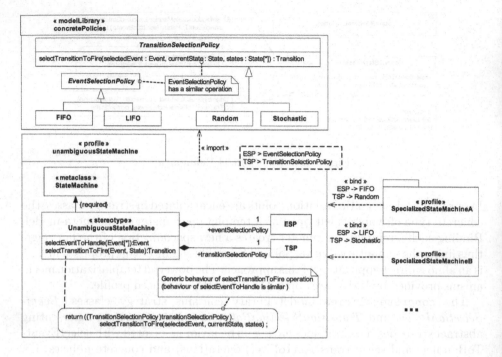

Fig. 5. Templated profile example

Users of the profile now have a clear indication of how they can make state machine semantics vary. In fig.5, this is done by defining new specialized profiles (SpecializedStateMachineA and B), where formal parameters of the unambiguousStateMachine profile are actually bound to concrete policies from the model library (through the «bind» relationships). Each profile thus provides a variant for state machine semantics, and these variants can be unambiguously and automatically interpreted by the execution environment. Note that the example we propose defines formal parameters for the profile, but these parameters could have been specified for the stereotype itself in a similar way. In the latter case, new specialized stereotypes could also be defined through binding relationships. The next sections now explains how and why these profile enhancements are legal according to the various UML compliance levels.

4.4   Profile Enhancements and UML Compliance Levels

Our proposed use of UML profiles in the previous sections may seem strange. We have not in fact seen similar uses in the relevant literature; nor are any mentioned in the Superstructure itself. However, the uses we are promoting are legal according to UML specification. Fig.6 is based on figures 2.2, 2.3 and 2.4 from the UML superstructure [11]. The packages on the bottom of the diagram represent the various UML compliance levels, and those on the top provide a particular focus that helps us to argue our point. The various merge relationships illustrated in fig.6 justify our affirmation, and hence the profiles we defined in fig.3 and fig.5 are actually legal. The interpretation we are about to describe is based on a similar case proposed in the UML 2 superstructure to illustrate the use and the meaning of package merges. This example is illustrated in figures 7.68 and 7.69 of the UML 2 superstructure.

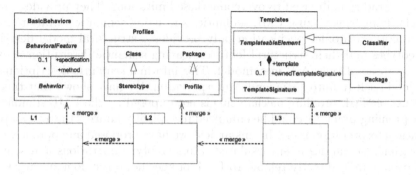

Fig. 6. UML Compliance Levels (Excerpt and Focus)

At the L1 level, the BasicBehaviors package is merged, so that operations of classes can now encapsulate behaviors (Operation is indeed a subtype of BehavioralFeature). At the L2 level, the Profiles package is merged, so that Stereotype can have operations that encapsulate behaviors (because Stereotype is a subclass of Class, and Class has this capability from L1 level). Thus we conclude that the L2 compliance level is enough to support profile definitions similar to the one proposed in fig.3 (where behaviors are encapsulated in the selectEventToHandle() and selectTransitionToFire() operations of the «UnambiguousStateMachine» stereotype). Then, at the L3 level, the Templates package is merged, so that classifiers and packages defined with an L3-compliant tool can own a template signature (they are indeed subtypes of the abstract TemplateableElement metaclass, that is to say an element that can own a TemplateSignature). As a consequence, profiles and stereotypes specified with an L3-compliant tool can also own a template signature (because Profile is a subclass of Package, and Stereotype is a subclass of Class, which is a subclass of Classifier). We therefore also conclude that the profile we proposed in fig.5 is legal, and requires the use of a L3-compliant tool.

Pragmatically, the designers of the Eclipse UML 2 plugin[6] (the de-facto standard implementation of the UML 2 metamodel) came to the same conclusion. The plugin indeed provides an L3-compliant implementation of UML [7], and hence it directly supports the encapsulation of behaviors into stereotype operations and the definition of template signatures for both stereotypes and profiles. This is all what we need to support the description part of the mechanisms proposed in section 4. Implementation of the visual editing part of these mechanisms in our open source UML 2 tool, Papyrus, is then quite straightforward. We discuss this point again in the conclusion.

---

[6] http://www.eclipse.org

5 Conclusion

The informal nature of UML semantic description leaves the door open to multiple and possibly contradictory interpretations. This strongly limits its adoption for constrained domains, such as real-time embedded systems, while the demand and potential benefits of using such a widespread standard are strong. UML Profiles are traditionally used to overcome these limitations. They provide a means for clarifying/specializing UML semantics and fixing semantic variation points. In practice, a profile specification only provides a structural model describing stereotypes and their structural features, leaving semantic description as informal as those of the UML metamodel. This informal semantic description does not lend itself to automatic model interpretation. We have shown that this limited use of UML profiling mechanisms is not imposed by the UML standard, so that profiling practices could be enhanced to explicitly address the automatic semantic interpretation issue. In this article, we have proposed an approach where operational semantics is encapsulated into stereotype operations and semantic variation points (directly related to UML or specific to the domain targeted by the profile) are explicitly handled through template parameters and strategy classes. We have demonstrated that this approach is UML compliant, and described how it could explicitly and intuitively influence the default semantics of UML models, where this default semantics could be expressed with fUML (the future OMG standard about the executable semantics of a foundational UML subset), or any equivalent formalism. Papyrus, our open source UML tool, is based on the Eclipse UML2 plugin. Therefore, the 1.6 version of Papyrus supports the description part of the mechanisms we propose.

Most of our current work now concerns the application of the enhanced profiling mechanisms we propose to our Accord/UML approach. Accord/UML is a UML-based methodology for real-time systems development. It basically relies on: *(a)* a UML profile (implicitly specializing UML semantics about state machines and active classes, and providing time related annotations), *(b)* a refinement process (involving model transformations and code generation), *(c)* an execution framework (encapsulating the semantic interpretation of profiled UML models). We plan to make Accord/UML evolve from this hard-coded semantic interpretation to a more flexible and high level semantic description directly integrated into the profile. The first step of this evolution is to provide users with a means for varying certain hard-coded aspects of the refinement process, such as the definition of the event selection policy, while the "global" refinement process remains hard-coded. This could be done through template parameters, strategy classes and stereotype operations (i.e. something similar to the templated profile provided as an example in this article), and code generation from these particular semantic aspects (described in operational style within behaviors encapsulated into strategy classes and stereotype operations). This goal is independent from fUML, and it will be achieved before the end of the year. Concerning long-term perspectives, we would like to integrate in the profile a complete semantic interpretation (i.e. not limited to semantic variation points), and explicit links with UML default operational semantics (default operational semantics should be partially provided by fUML). This would enable interactive model level debugging or model level simulation from a profiled UML model. The main difficulty to handle concerns the way to address concurrency and time-related aspects of the semantics. One possible way we are exploring is to define a semantic foundation for time-related concepts of MARTE[7] (Modeling and Analysis of Real Time and Embedded systems). Then, the time management mechanisms provided by MARTE (logical clocks, chronometric clocks, etc.) could be used unambiguously to integrate timed semantics within the Accord/UML profile definition.

References

1. Cuccuru, A., Mraidha, C., Terrier, F., Gérard, S.: Templateable Metomodels for Semantic Variation Points. In: ECMDA-FA. European Conference on Model Driven Architecture - Foundations and Applications, Haifa, Israel (2007)
2. O'Keefe, G.: Improving the Definition of UML. In: Nierstrasz, O., Whittle, J., Harel, D., Reggio, G. (eds.) MoDELS 2006. LNCS, vol. 4199, Springer, Heidelberg (2006)
3. Gamma, E., Helm, R., Johnson, R., Vlissides, J.: Design patterns: Abstraction and reuse of object-oriented design. In: Nierstrasz, O. (ed.) ECOOP 1993. LNCS, vol. 707, pp. 406–431. Springer, Heidelberg (1993)
4. Gunter, C.: Semantics of Programming Languages: Structures and Techniques (Foundations of Computing). MIT Press, Cambridge (1992)
5. Harel, D., Rumpe, B.: Meaningful Modeling: What's the Semantics of "Semantics"? Computer 37(10), 64–72 (2004)
6. Henderson-Sellers, B.: UML - the Good, the Bad or the Ugly? Perspectives from a panel of experts. Software and System Modeling 4(1), 4–13 (2005)
7. Hussey, K.: What do YOU want UML to be? (Slides). In: EclipseCON, Santa Clara, CA, USA (2007)
8. Kim, S.-K., Burger, D., Carrington, D.A.: An MDA Approach towards Integrating Formal and Informal Modeling Languages. In: Fitzgerald, J.A., Hayes, I.J., Tarlecki, A. (eds.) FM 2005. LNCS, vol. 3582, Springer, Heidelberg (2005)
9. OMG. Semantics of a Foundational Subset for Executable UML Models RFP (2005)
10. OMG. Unified Modeling Language: Infrastructure (2005)
11. OMG. Unified Modeling Language: Superstructure version 2.1.1 (2007)
12. Rasch, H., Wehrheim, H.: Checking Consistency in UML Diagrams: Classes and State Machines. In: Najm, E., Nestmann, U., Stevens, P. (eds.) FMOODS 2003. LNCS, vol. 2884, Springer, Heidelberg (2003)
13. Tanguy, Y., Gérard, S., Radermacher, A., Terrier, F.: Model Driven Engineering for Real Time Embedded Systems. In: ERTS. 3rd European Congress Embedded Real Time Software, Toulouse (France) (January 2006)
14. Tessier, P., Gérard, S., Mraidha, C., Terrier, F., Geib, J.-M.: A Component-Based Methodology for Embedded System Prototyping. In: 14th IEEE International Workshop on Rapid Systems Prototyping, San Diego, CA, USA, IEEE Computer Society Press, Los Alamitos (2003)

---

[7] Revised submission is available at *http://www.promarte.org*

Integrated Definition of Abstract and Concrete Syntax for Textual Languages

Holger Krahn, Bernhard Rumpe, and Steven Völkel

Institute for Software Systems Engineering
Technische Universität Braunschweig, Braunschweig, Germany
http://www.sse-tubs.de

Abstract. An understandable concrete syntax and a comprehensible abstract syntax are two central aspects of defining a modeling language. Both representations of a language significantly overlap in their structure and also information, but may also differ in parts of the information. To avoid discrepancies and problems while handling the language, concrete and abstract syntax need to be consistently defined. This will become an even bigger problem, when domain specific languages will become used to a larger extent. In this paper we present an extended grammar format that avoids redundancy between concrete and abstract syntax by allowing an integrated definition of both for textual modeling languages. For an amendment of the usability of the abstract syntax it furthermore integrates meta-modeling concepts like associations and inheritance into a well-understood grammar-based approach. This forms a sound foundation for an extensible grammar and therefore language definition.

1 Introduction

The definition of a language involves various kinds of activities. Usually a concrete and an abstract syntax is developed first, and then a semantics is designed to define the meaning of the language [9]. These activities are complemented by developing a type system, priorities for operators, naming systems etc. if appropriate. Especially the definition of concrete and abstract syntax show a significant redundancy, because domain concepts are reflected in both artifacts. This leads to duplications which are a constant source of problems in an iterative agile development of modeling languages. Despite general problems that occur when two documents are used, like inconsistency checking and supporting adequate ways to resolve them, domain concepts in the abstract syntax may be expressed on different ways in the concrete syntax. On the contrary the abstract syntax may contain elements that cannot be expressed in the concrete syntax of the language. These potential problems unnecessarily complicate an efficient development and evolution of languages, and therefore, an integrated development of both artifacts is highly desirable.

Meta-modeling is a popular method to define the abstract syntax of languages. It simplifies the language development by allowing the designers to directly map the classes of a domain analysis [2] to classes in the meta-model, associations

G. Engels et al. (Eds.): MoDELS 2007, LNCS 4735, pp. 286–300, 2007.

and inheritance are directly part of the language definition. On the other side, grammar-based language definitions yield trees with single root-objects. Associations between leaves and an inheritance-based substitutability are not commonly existent in grammars.

There are several approaches to derive a grammar and thus, a textual representation from a given metamodel. These approaches often lack flexibility in defining an arbitrary concrete syntax, it may even happen that the desired concrete syntax must be adapted in order to get an automatic mapping between abstract and concrete syntax (e.g., [11]). This stands in contrast to a basic design principle for DSLs [19] that already existing notations of the domain shall be used unaltered. Beyond that, one main argument for defining concrete syntax and abstract syntax separatly is that more than a one concrete syntax can be used with a single abstract syntax. We argue, when dealing with DSLs this is a minor aspect because usually in a single domain no two notations are used that have the same expressiveness and therefore apply to same abstract syntax. However, we suggest the use of two similar abstract syntaxes and a (simple) model transformation for these rare cases.

The concrete syntax of a language is either texual or graphical. The graphical concrete syntax is often defined by the structure of the abstract syntax and a set of graphical representations for classes and associations in the abstract syntax (e.g. [7]). Especially for languages that do not have an adequate graphical representation, a textual syntax is used which is usually described by a context-free grammar. Parser-generators, e.g. Antlr [14] or SableCC [5] can be used to generate language recognition tools from this form of language definition. Since we aim at textual concrete syntaxes, we concentrate in this paper on the second approach.

The MontiCore framework [8] can be used for the agile development of textual languages, in particular domain-specific modeling languages (DSMLs). In order to reduce the abovementioned redundancy, one of the main design goals of the underlying grammar format was to provide a single language for specifying concrete as well as abstract syntax in a single definition. Associations, compositions, and inheritance as known from meta-modeling can directly be specified in this format. Such a language definition can be mapped to an object-oriented programming language where the each production is mapped to a class with strongly typed attributes. A parser is generated to create instances of the abstract syntax from a textual representation.

Despite these main design goals, we decided to develop the MontiCore grammar format in such a way, that the recognition power of the resulting parser is only limited by the underlying parser generator, namely Antlr [14], which is a predicated-LL(k) parser generator. Thus, it can not only be used for the development of domain specific modeling languages but for general-purpose languages like variants of Java or C++. The concrete syntax of the grammar format is similar to the input format of common parser-generators. Therefore, users that have already worked with such tools shall easily become familiar with it. The context-free grammars can be extended with meta-modeling concepts like associations and inheritance to define the abstract syntax of the modeling language.

The rest of the paper is structured as follows. Section 2 describes the syntax of the MontiCore grammar format and its semantics in form of the resulting concrete and abstract syntax of the defined modeling language. Section 3 describes an example that illustrates the clarity of the specification in the MontiCore grammar format. Section 4 relates our approach to others whereas Section 5 concludes the paper and outlines future work.

2   The MontiCore Grammar Format

The MontiCore grammar format specifies a context free grammar with productions that contain nonterminals (reference to other rules) and terminals. Terminals (also called identifiers) are specified by regular expressions. To simplify the development of DSLs, the identifer IDENT and STRING are predefined to parse names and strings. Identifiers are usually handled as strings, but more complex identifers are possible by giving a function defined in the programming language Java that maps a string to an arbitrary data type. Default functions exist for primitive data types like floats and integers. Examples are given in Figure 1. In line 2 the simple identifier IDENT is specified which is mapped to a String in the abstract syntax. The identifier NUMBER in line 5 is mapped to an integer in the abstract syntax whereat the default mapping is used. In line 8 the identifier CARDINALITY is mapped to an int. The transformation is specified in Java (line 10 and 11) and the unbounded cardinality is expressed as the value -1.

```
                        ─────── MontiCore-Grammar ───────
 1  // Simple name
 2  ident IDENT ('a'..'z'|'A'..'Z')+ ;
 3
 4  // Numbers (using default transformation)
 5  ident NUMBER ('0'..'9')+ : int;
 6
 7  // Cardinality (STAR = -1)
 8  ident CARDINALITY ('0'..'9')+ | '*' :
 9    x -> int {
10      if (x.equals("*")) return -1;
11      else return Integer.parseInt(x);
12    };
```

Fig. 1. Definition of identifiers in MontiCore

The definition of a production in the grammar leads to a class with the same name in the abstract syntax. The nonterminals and identifiers on the right hand side of a rule can explicitly be given a name that maps to attribute names. For unnamed elements we derive default names from the name of the nonterminal. The identifiers form the attributes of the class whereas the nonterminals lead to composition relationships between classes in the abstract syntax. The type of attributes in the abstract syntax is inferred automatically, the types of identifieres are handled as described before, attributes which form a composition relationship are typed with the class of the target rule. Thus, the attribute **name** of the rule **Client** in line 7 of Figure 2 results in a string attribute in the corresponding class of the abstract syntax, whereas **Address** in line 8 results in an attribute of the type **Address**. Additionally, the structure of a production is analyzed to determine the cardinality of the attributes and compositions. Doing so, attributes that occur more than once are realized as lists of the corresponding data type. This approach allows to specify constant separated lists without an extra construct in the grammar format. The term **a:X ("," a:X)*** can be used on the right hand side of a grammar rule and desribes a comma-separated list of the non-terminal **X**. It results in a composition named **a** with unbounded cardinality that contains all occurrences of **X**. Therefore, terminals and identifiers with the same name contribute to the same attribute or composition.

Figure 2 shows an excerpt of a MontiCore grammar. The class section shows a UML class diagram of the abstract syntax that is created from the productions. In the MontiCore framework this class diagram is mapped to a Java implementation where the production names are used as class names. All attributes are realized as private fields and public get- and set-methods. The composition relationships are realized in the same way as attributes and contribute to the constructor of the class. All classes support a variant of the Visitor pattern [6] to traverse the abstract syntax along the composition relationships.

In addition to the already explained nonterminals and identifiers, constant terminal symbols like keywords can be added to the concrete syntax of the language. These elements are not directly reflected in the abstract syntax if they are unnamed. Note that in contrast to many parser generators and languages like TCS [11] there is no specific need for distinguishing between keywords like "public" and special symbols like ",". To further simplify the development of a modeling language we generate the lexer automatically from the grammar. By this strategy the technical details like the distinction between parser and lexer (necessary for the parser generator) are effectively hidden from the language developer.

As explained above, keywords are not directly reflected in the abstract syntax as attributes, but may influence the creation of the AST by distinguishing productions with the same attributes from each other. The situation is different for reserved words that determine certain properties of domain concepts. An example is shown in Figure 3 where the reserved word *premiumclient* determines the value of an attribute of the domain concept client. The grammar format uses constants (inside brackets) to express this fact. Single value constants are translated to booleans whereas multi-value constants are mapped to enumerations.

The languages defined through the grammar in Figure 2 and the substituted grammar in Figure 3 are equal. But their abstract syntax is quite different. The concrete syntax poses the invariant that clients cannot have a discount whereas premium clients do have one. This invariant is not visible in the abstract syntax

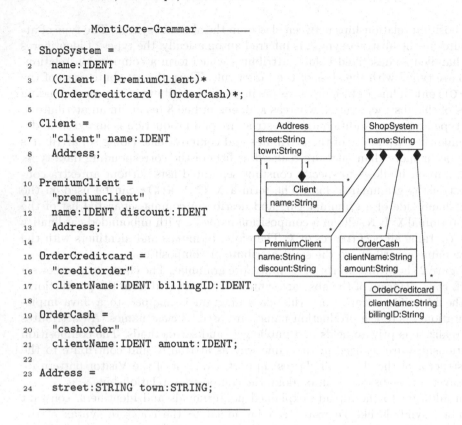

```
          MontiCore-Grammar
1 ShopSystem =
2   name:IDENT
3   (Client | PremiumClient)*
4   (OrderCreditcard | OrderCash)*;
5
6 Client =
7   "client" name:IDENT
8   Address;
9
10 PremiumClient =
11   "premiumclient"
12   name:IDENT discount:IDENT
13   Address;
14
15 OrderCreditcard =
16   "creditorder"
17   clientName:IDENT billingID:IDENT
18
19 OrderCash =
20   "cashorder"
21   clientName:IDENT amount:IDENT;
22
23 Address =
24   street:STRING town:STRING;
```

Fig. 2. Definition of productions in MontiCore

```
          MontiCore-Grammar
1 Client =
2   premium:["premiumclient"]
3     name:IDENT discount:IDENT
4   | "client" name:IDENT;
```

Client
premium:Boolean
name:String
discount:String

Fig. 3. Use of constants

from Figure 3. On the other hand, the abstract syntax resulting from Figure 2 doesn't reflect the similarities between Client and PremiumClient resp. Order-Cash and OrderCreditcard. This example motivates the use of more advanced features of the MontiCore grammar format to represent the invariants and similarities directly in the grammar. Despite very general constraint definitions (like OCL), inheritance allows us to deal with similarities and associations with connections between related nodes of the AST.

2.1   Interfaces and Inheritance Between Nonterminals

The abstract syntax shown in Figure 2 raises the question how an interface
Order that both classes OrderCreditcard and OrderCash implement can be
added to the abstract syntax and how it can be expressed that premium clients
are special clients. For this purpose we decided to extend the grammar format
by allowing to express an inheritance relationship and to define interfaces which
can be implemented by nontermianls.

The inheritance relationship between two productions is expressed by including the rule name of the super-production after the production name of the sub-
production using the keyword extends (Figure 4, left, line 16). This inheritance
of rules is directly reflected in the abstract syntax as an object-oriented inheritance. In the parser an additional alternative is added to the super-production.
This concept is motivated by the definition of object-oriented inheritance where
each occurrence of a superclass can be substituted by a subclass. The EBNF section in Figure 4, right, shows a representation with equivalent concrete syntax
to explain the mapping of the MontiCore grammar format to the input format
of a parser generator.

The grammar on the right hand side in Figure 4 defines the same concrete
syntax as the one on the left, but has additional Order and Client rules. However, we have decided to use an OO style of inheritance instead of the traditional
grammar style to get more flexibility in extending languages. In the left grammar,
Client need not be changed when extending the language with PremiumClients.
This is a significant benefit that we will further explore when defining operators
on the language.

Due to experience in designing languages with this grammar format, we decided to decouple the concrete syntax of the both productions (sub- and super-
production). This allows the language designer to decide freely on the concrete
syntax and minimizes non-determinisms in the grammar.

This form of inheritance also allows the definition of superinterfaces using
the keyword implements (Figure 4, left, line 5 and 9). Interfaces can be used
as normal nonterminals on the right hand side of any production. By default
a interface does not contain any attributes. We decided against an automatic
strategy where all common attributes of known subclasses are taken, as the
interface may be a good place for future extensions of the defined language
which only use a subset of all available attributes.

Additional attributes may be added to interfaces and classes by using the ast
section in a grammar (Figure 4, left, line 25). This block uses the same syntax
as inside the production but only produces attributes in the abstract syntax
and does not interfere with the concrete syntax. The attributes of interfaces are
realized as get- and set-methods in the Java implementation. Figure 4 illustrates
the inheritance capabilities of the grammar format extending the example. The
EBNF section shows the equivalent EBNF syntax used for parsing and the resulting abstract syntax can be found in Figure 5.

```
——————— MontiCore-Grammar ———————          ——————— EBNF ———————

1  ShopSystem =                              1  ShopSystem ::=
2    name:IDENT                              2    IDENT Address
3    Client* Order*;                         3    Client* Order*
4                                            4
5  OrderCreditcard implements Order =        5  OrderCreditcard ::=
6    "creditorder"                           6    "creditorder"
7    clientName:IDENT billingID:IDENT;       7    IDENT IDENT
8                                            8
9  OrderCash implements Order =              9  OrderCash ::=
10   "cashorder"                             10   "cashorder"
11   clientName:IDENT amount:IDENT;          11   IDENT IDENT
12                                           12
13 Client =                                  13 Client ::=
14   "client" name:IDENT Address;            14   "client" IDENT Address
15                                           15   | PremiumClient ;
16 PremiumClient extends Client =            16
17   "premiumclient"                         17 PremiumClient ::=
18   name:IDENT discount:IDENT;              18   "premiumclient"
19                                           19   IDENT IDENT
20 Address =                                 20
21   street:STRING town:STRING;              21 Address ::=
22                                           22   STRING STRING
23 interface Order;                          23
24                                           24 Order ::=
25 ast Order =                               25   OrderCredit | OrderCash
26   clientName:IDENT;                       26
```

Fig. 4. Inheritance and use of interfaces

2.2 Associations Between Nonterminals

The attributes `name` in `Client` and `clientName` in `Order` (see Figure 5) are
obviously semantically connected. The invariant that an order may only use
client names that exists cannot be expressed in a context-free grammar format.
When designing a meta-model this relation is usually expressed by an association where an order references a client as the ordering person. Therefore, the
extended context-free grammar allows to add associations and mimic typical
meta-modeling techniques in grammars. The result of this extension is an arbitrary graph with an embedded spanning tree that results from the original
grammar.

The block `association` allows to specify non-compositional associations between rules which enables the navigation between objects in the abstract syntax.
This concept allows to specify a uni-directed navigation from one object to a
specified number of other objects. In addition, an opposite association can be
specified that reverses the first association. An example for an association can
be found in Figure 6 (line 17-21) where the association OrderingClient connects

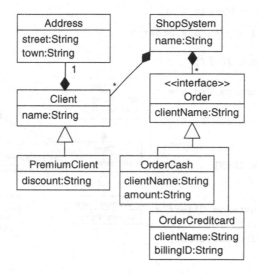

Fig. 5. Abstract syntax of the language defined in Figure 4

one Order object with a single Client. The reverse association is named Order (the name is automatically derived from the target) which connects one Client object with an unbounded number of Order objects. This form is very similar to the associations in EMF [1].

The main challenging question for associations in a unified format for concrete and abstract syntax is not the specification but to automatically establish the links between associated objects from a parseable textual input. Grammar-based systems usually parse the linear character stream and represent it in a tree-based structure that has the same structure as the grammar. Then symbol tables are used to navigate between nodes in the AST that are not directly connected. The desired target of navigation is determined by identifiers in source and target nodes and a name resolution algorithm.

Due to the simple nature of many languages that lack namespaces, simple resolution mechanisms like file-wide unique identifiers can often be used for an establishment of associations. This of course does not always work. E.g., languages like Java and many UML-sublanguages do have a more sophisticated namespace concept.

Therefore, we decided to use a twofold strategy: First, we generate interfaces that contain methods induced by the association to navigate between the AST-objects. The resulting classes of the abstract syntax allow the access of associations in the same way as attributes and compositions are accessed. Second, we generate implementations for simple resolving problems like file-wide flat simple or hierarchical namespaces. As an alternative, the DSL developer can program his own resolving algorithms in the second step if needed.

Figure 6 extends the example from Figure 5 by adding an association specification. The association `orderingClient` connects each `Order` to a single Client

——— MontiCore-Grammar ———

```
1  OrderCreditcard implements Order =
2    "creditorder"
3    iD:IDENT amount:IDENT;
4
5  OrderCash implements Order =
6    "cashorder"
7    iD:IDENT amount:IDENT;
8
9  Client =
10   "client" name:IDENT
11   Address;
12
13 PremiumClient extends Client =
14   "premiumclient"
15   name:IDENT discount:IDENT;
16
17 association {
18   Order.orderingClient
19   * <-> 1
20   Client
21 }
```

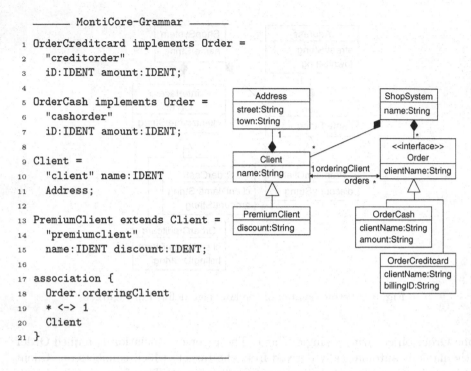

Fig. 6. Specification of associations

(as specified by 1). Order is the inverse association from Client to Order with unbound cardinality (as specified by *). In addition to the shown cardinalities, ranges like 3..4 are possible values.

Figure 7 sketches a Java implementation of the class diagram from Figure 6 with the most important methods. A Binding-Interface is generated for each

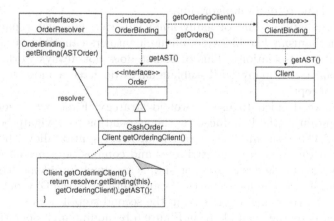

Fig. 7. Java implementation of an association

interface or class that is involved in an association as either source or target. This interface contains the relevant methods for the navigation between different nodes. In addition a Resolver is generated for each class or interface which allows the resolving of a Binding-object from an AST-object.

Note that these interfaces are generated to simplify the use of associations for a DSL. When simple resolving algorithms are appropriate, MontiCore can generate both Binding-implementations and a single Resolver-implementation that resolves all objects automatically. The complexity of multiple classes with different responsibilities is hidden from the user of the abstract syntax, e.g. a programmer of a code generator for the developed language. He simply uses the get- and set-methods like the shown getOrderingClient() method that returns the client object which is referred from this order.

3   A Demonstrating Example

To demonstrate the usability of our approach to specify a modeling language we use a simplified version of finite hierarchical automata as shown in Figure 8.

---------------------------------- Automaton ----------------------------------

```
1  automaton PingPong {
2      state <<initial>> NoGame;
3      NoGame - startGame > InPlay;
4      InPlay - ["doStopGame()"] stopGame > NoGame;
5      state InPlay {
6          state <<initial>> Ping;
7          state Pong;
8      }
9      Ping - returnBall > Pong;
10     Pong - returnBall > Ping;
11  }
```

Fig. 8. Example for finite hierarchical automata

An automaton has a name and consists of several states and transitions. States in turn may be initial or final and may have substates. Transitions have a source and a target state, an event models the condition for the transition. In order to demonstrate a possible field of application for grammar rule inheritance, a transition may have an action which will be executed when a transition was performed. Figure 9 shows a first version of the MontiCore grammar.

In our example transitions refer to states as source and target which will be identified by their name. Thus, the generated class Transition contains string attributes from and to containing the names of these states. This is ineffective because a direct navigation from transitions to their source or target is not possible. Furthermore, there is no information in states about their ingoing and

———————————————— MontiCore-Grammar ————————————————

```
1  package mc.languages.automaton;
2
3  grammar Automaton {
4
5    Automaton = !"automaton" name:IDENT "{"
6      ( State | Transition )*
7    "}";
8
9    State =
10     !"state" name:IDENT
11     ( "<<" initial:[!"initial"] ">>" | "<<" final:[!"final"] ">>" )*
12     ( "{" State* "}" | ";" ) ;
13
14   Transition = from:IDENT "-" event:IDENT ">" to:IDENT ";";
15
16   TransitionWithAction extends Transition =
17     from:IDENT "[" action:STRING "]"  "-" event:IDENT ">" to:IDENT ";";
18 }
```

Fig. 9. Basic definition of finite hierarchical automata in MontiCore

———————————————— MontiCore-Grammar ————————————————

```
1  association {
2    Transition.fromState * <-> 1 State.outgoingTransitions * ;
3    Transition.toState * <-> 1 State.incomingTransitions * ;
4  }
```

Fig. 10. Definition of non-compositional associations between states and transitions

outgoing transitions. The abovementioned concept association can be used to solve this. An appropriate association extending Figure 9 is defined in Figure 10.

This definition leads to attributes which refer to states and transitions directly. They have to be filled by an appropriate resolve mechanism depending on the underlying naming system. As described in Section 2.2, MontiCore supports different kinds of resolve mechanisms, in this example we use the simplest version, namely file-wide unique identifiers which is defined by the concept simplereference. Therefore, states must have a unique name within the automaton and transitions use that name in order to reference these states. Beyond that, we have to specify that our concept for simple references should be used in order to resolve the associations defined in Figure 10. Therefore, the code of Figure 11 has to be added to our automaton grammar (Figure 9).

———————————————— MontiCore-Grammar ————————————————

```
1   concept simplereference {
2     fromState: Transition.from -> State.name;
3     toState: Transition.to -> State.name;
4   }
```

Fig. 11. Using concept simplereference to resolve source and target of transitions

Given both, the definition of simple references and the concept association, the MontiCore framework ensures the following constraints.

1. Each transition refers exactly one state as source and target, respectively.
2. Each referenced state has a unique name within all referenced states.
3. The method getIncomingTransitions() of a state returns all transitions which refer to that state as target. Therefore, `incomingTransitions` is the opposite association of ToState.
4. The method getOutgoingTransitions() of a state returns all transitions which refer to that state as source. Therefore, `outgoingTransitions` is the opposite association of FromState.

Another useful feature we want to present in this example is the concept classgen. It can be used in order to add attributes as well as methods into the abstract syntax. Therefore, it has to be defined which class should be extended and what should be added to that class. Again, the example shown in Figure 12 can be added to the basic grammar (Figure 9). It adds the boolean method isDirectlyReachable to the abstract syntax class State which calculates if one state is directly reachable from this state.

———————————————— MontiCore-Grammar ————————————————

```
1   ast State =
2     method public boolean isDirectlyReachable(State target) {
3       for (Transition t: getOutgoingTransitions()){
4         if (t.getToState().equals(target)){
5           return true;
6         }
7       }
8       return false;
9   };
```

Fig. 12. Using an ast block to add methods to states

Summarizing, we have developed a grammar for finite hierarchical automata in a few lines of code. Non-compositional bi-directional associations are supported and filled automatically by a simple naming system which ensures correct cardinalities. An example of an additional method defined in Java enhances usability as well as it prevents editing the generated code.

4   Related Work

We are currently not aware of a language that allows specifying both a textual concrete syntax and an abstract syntax with (non-compositional) associations in a coherent and concise format. Grammar-based approaches usually lack a strongly typed internal representation (for exceptions see below) and the existing model-based approaches use two forms of description, a meta-model for the abstract syntax and a specific notation for the concrete syntax.

In [15] a phylum/operator-notation is used to describe the abstract syntax of a language. The notation of alternate phylums achieves similar results as the object-oriented inheritance we use, although our tied coupling of the abstract syntax to a programming language allows the direct use of the inheritance to simplify the implementation of algorithms working on the abstract syntax.

SableCC [5] is a parser-generator that generates strictly-typed abstract syntax trees and tree-walkers. The grammar format contains actions to influence the automatic derivation of the AST. In contrast to MontiCore, SableCC does not aim to include associations in its AST.

In [18] an algorithm is presented that derives an (strongly typed) abstract syntax from a WBNF grammar (an BNF variant). The main difference in the derivation to our approach is the use of an explicit notation for lists that are separated by constants and that nonterminals with same name do not contribute to the same attribute in the abstract syntax.

The Grammar Deployment Kit (GDK) [12] consists of several components to support the development of grammars and language processing tools. The internal grammar format can be transformed into inputs of different parser generators, such as btyacc [3], Antlr [14] or SDF [10]. Furthermore, it provides possibilities for grammar adaption, like renaming of rules or adding alternatives. In opposition to our approach it does not support extended concepts like inheritance or associations.

In [4] and [13] the Textual Concrete Syntax Specification Language (TCSSL) is described that allows the description of a textual concrete syntax for a given abstract syntax in form of a meta-model. TCSSL describes a bidirectional mapping between models and their textual representation. The authors describe tool support to transform a textual representation to a model and back again.

In [11] a DSL named TCS (Textual Concrete Syntax) is described that specifies the textual concrete syntax for an abstract syntax given as a meta-model. The described tool support is similar to the one we used for the MontiCore framework and the name resolution mechanisms are the same that we generate automatically from the grammar format. In contrast to our approach, two descriptions for abstract and concrete syntax are needed.

5   Conclusion

This work presents a new approach where an extended grammar format is used to specify both, abstract and concrete syntax of a modeling language. By using a single format it avoids general problems that occur when abstract and concrete syntax are described by two different languages like inconsistency checking and resolving by construction.

As special concepts, we added the possibility to define associations between AST nodes based on name references and we allow inheritance of grammar rules that does not affect the super-rule at all. This paves the way for extensible languages.

We also implemented a prototypical framework called MontiCore that is based on an established parser-generator. It is able to parse textual syntax and generates the model representation as both Java and EMF classes. The prototype is able to parse multiple language definitions like UML/P [17,16] and a complete Java 5 grammar. In addition the system is bootstrapped and currently about 75% of the code is generated from several DSLs.

In future we especially want to explore which resolution mechanisms can be used to create links between objects (that conform to the specified associations). The mechanisms for resolving imports in models and inheritance seem to be promising candidates for generalization.

We mainly treated the transformation from concrete syntax to abstract syntax representation in this paper. The opposite transformation, where a model is transformed to a concrete (textual) representation could be realized in different ways: Hand coded java code, template engines etc. In the future we like to explore which additional information has to be included in the grammar format to allow the automatic generation of concrete syntax representations.

The current implementation of the parser generation in the MontiCore framework is based on Antlr version 2.x. The new version 3 simplifies the creation of grammars by automatically calculating the necessary syntactic predicates for alternatives where the linearized lookahead algorithm predicts false results. This will reduce the number of required syntactic predicates and simplify the development of readable grammars in MontiCore.

Acknowledgement. The work presented in this paper is undertaken as a part of the MODELPLEX project. MODELPLEX is a project co-funded by the European Commission under the "Information Society Technologies" Sixth Framework Programme (2002-2006). Information included in this document reflects only the authors' views. The European Community is not liable for any use that may be made of the information contained herein.

References

1. Budinsky, F., Steinberg, D., Merks, E., Ellersick, R., Grose, T.J.: Eclipse Modeling Framework. Addison-Wesley, Reading (2003)
2. Czarnecki, K., Eisenecker, U.W.: Generative Programming: Methods, Tools, and Applications. Addison-Wesley, Reading (2000)

3. Dodd, C., Maslov, V.: BTYACC – backtracking YACC. Siber Systems (2006), http://www.siber.com/btyacc/

4. Fondement, F., Schnekenburger, R., Gerard, S., Muller, P.-A.: Metamodel-aware textual concrete syntax specification. Technical report, LGL-REPORT-2006-005 Swiss Federal Institute of Technology in Lausanne, Switzerland (2006)

5. Gagnon, E., Hendren, L.: SableCC – An Object-Oriented Compiler Framework. In: Puigjaner, R., Savino, N.N., Serra, B. (eds.) Tools 1998. LNCS, vol. 1469, Springer, Heidelberg (1998)

6. Gamma, E., Helm, R., Johnson, R., Vlissides, J.: Design Patterns: Elements of Reusable Object-Oriented Software. Addison-Wesley, Reading (1995)

7. Graphical Modeling Framework (GMF, Eclipse technology subproject), http://www.eclipse.org/gmf/

8. Grönniger, H., Krahn, H., Rumpe, B., Schindler, M., Völkel, S.: MontiCore 1.0 - Ein Framework zur Erstellung und Verarbeitung domänenspezifischer Sprachen. Technical Report Informatik-Bericht 2006-04, Software Systems Engineering Institute, Braunschweig University of Technology (2006)

9. Harel, D., Rumpe, B.: Meaningful modeling: What's the semantics of "semantics"? Computer 37(10), 64–72 (2004)

10. Heering, J., Hendriks, P.R.H., Klint, P., Rekers, J.: The syntax definition formalism SDF–Reference Manual—. 24(11), 43–75 (1989)

11. Jouault, F., Bezivin, J., Kurtev, I.: TCS: a DSL for the Specification of Textual Concrete Syntaxes in Model Engineering. In: Proceedings of GPCE '06 (2006)

12. Kort, J., Lämmel, R., Verhoef, C.: The grammar deployment kit. In: van den Brand, M., Lämmel, R. (eds.) ENTCS, vol. 65, Elsevier Science Publishers, Amsterdam (2002)

13. Muller, P.-A., Fleurey, F., Fondement, F., Hassenforder, M., Schneckenburger, R., Gérard, S., Jézéquel, J.-M.: Model-driven analysis and synthesis of concrete syntax. In: Nierstrasz, O., Whittle, J., Harel, D., Reggio, G. (eds.) MoDELS 2006. LNCS, vol. 4199, pp. 98–110. Springer, Heidelberg (2006)

14. Parr, T., Quong, R.: ANTLR: A Predicated-LL(k) parser generator. Journal of Software Practice and Experience 25(7), 789–810 (1995)

15. Reps, T., Teitelbaum, T.: The synthesizer generator. In: Proceedings of the first ACM SIGSOFT/SIGPLAN software engineering symposium on Practical software development environments, pp. 42–48. ACM Press, New York (1984)

16. Rumpe, B.: Agile Modellierung mit UML: Codegenerierung, Testfälle, Refactoring. Springer, Berlin (2004)

17. Rumpe, B.: Modellierung mit UML. Springer, Berlin (2004)

18. Wile, D.: Abstract syntax from concrete syntax. In: ICSE '97. Proceedings of the 19th international conference on Software engineering, New York, NY, USA, pp. 472–480 (1997)

19. Wile, D.: Lessons learned from real DSL experiments. Science of Computer Programming 51(3), 265–290 (2004)

Architectural Aspects in UML

Jon Oldevik[1,2] and Øystein Haugen[2,1]

[1] University of Oslo, Department of Informatics, Oslo, Norway
[2] SINTEF Information and Communication Technology, Oslo, Norway
jonold@ifi.uio.no, oystein.haugen@sintef.no

Abstract. Architecture descriptions are important for reasoning about system properties in order to make the right architectural decisions for building systems with adequate quality. Modularising concerns at the architecture description level may ease system configurability and cater for variations in architectural requirements. We devise a technique for modularising and composing complex architectural connectors described in UML using structured classes. We define a binding language with lexical and graphical syntax to support the composition. Finally, we discuss the relationship with standard UML constructs.

1 Introduction

Architecture represents essential building blocks of software. The quality of the architecture correlates strongly with the performance of the system as a whole. The importance of architecture in software development has long been recognised by the software engineering community, as seen in books (e.g. [1]), articles ([2,3,4]), and standards (IEEE 1471[5]).

To support the need for architecture descriptions, many different architecture description languages (ADLs) have evolved over the years, such as Architecture Analysis and Design Language (AADL)[6], Koala[7], and ACME [8]. Some of them targets specific domains of architecture, others are more general-purpose languages. Being the de-facto standard for software modelling, UML is also being used for architecture description. UML, however, has limitations both regarding architectural descriptions and architectural composition. Specifically, the connector concept in UML has been argued to be too simple to represent connections between components in architectural descriptions (architectural connectors) [9,10].

As system complexity and size increases, the need and benefit of modularising and combining architectural features become evident. Architectural features that cross-cut the architecture could be separated out for simplicity and comprehension. Architectural features that are not necessarily cross-cutting, but variable in an architecture configuration, could also be modularised to enable flexible composition with a common (product line) architecture. The standard mechanisms in UML cannot provide this out of the box.

We address how complex architectural connectors can be described in UML, and how these provide architectural variance when they are composed with base architectures. We call the approach *ArchSpect*. We devise a binding language that is used to describe the composition between archspects and base architectures. Finally, we discuss how standard UML mechanisms relate to the proposed approach. Initial ideas on this topic were presented in [11].

G. Engels et al. (Eds.): MoDELS 2007, LNCS 4735, pp. 301–315, 2007.
© Springer-Verlag Berlin Heidelberg 2007

In the following, we detail our approach for architectural description and composition in UML. In Section 2, we present a motivating example. In Section 3, we describe the main ideas of archspects. In Section 4, we compare our approach with standard UML mechanisms. In Section 5 we describe related work, and in Section 6 we conclude.

2   A Motivating Example

We will use an example called *ICU (I See You)* to motivate the ArchSpect approach. The ICU-System is a buddy positioning application based on mobile messaging (SMS), where users can register, manage their buddies, and perform positioning services on those buddies. Mobile terminals are used to interact with the system.

The system is described completely in terms of UML models and it represents a fully executable system. UML structured classes describe the system architecture while UML state machines describe its behaviour. Our focus in this paper is the structural parts of the system.

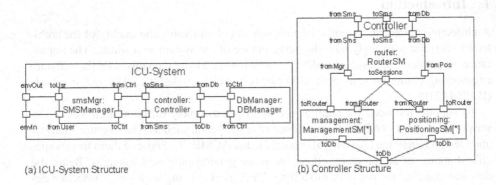

(a) ICU-System Structure   (b) Controller Structure

Fig. 1. The ICU-System Architecture

The ICU-System is a state-machine-based, reactive system with asynchronous messaging between system objects. The system architecture is defined by structured classes with parts, ports, and connectors. The main system architecture (Fig 1a) consists of three parts: the *SMSManager* receives and sends SMS messages from and to the user terminals. The *Controller* does the further handling of the messages and the *DBManager* handles persistent storage of data. Fig 1b details the structure of the Controller class.

Extending the System. At some point, we get new requirements to our system. In this case, the requirement is to add access restrictions to the system, which we choose to implement using an access control component. We also want to keep our current system without this new feature, since it is the common core of the *ICU* software product line. Fig 2a shows the desired new architecture. An access control part intercepts incoming messages to authorise users and then to forward or reject them.

The new architecture could be made in UML by modifying the original system directly. This would leave us without the original system intact and prevent other variant architectures. We could also try to specialise the original system and redefine the necessary parts. The desired result cannot be achieved by redefinition in UML. Simply copying the original system and modifying it would lead to unwanted duplication and loss of traceability to the original system.

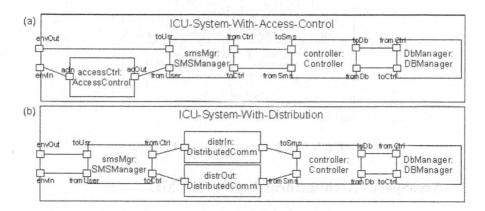

Fig. 2. System with Access Control

Another requirement to our original system appears independently of the previous one. A customer requests that the SMSManager and the Controller must execute on different machines. This is solved by introducing distribution layers between the two parts. Fig 2b shows our new architecture variant, imposing two new parts to our architecture: one handling local-to-remote communication from the SMSManager to the Controller and another handling local-to-remote communication the opposite direction.

Fig 3 illustrates a third variant, a modification to the Controller structure to manage transactions on user requests. In the original Controller (Fig 1b), the router figures out the destination for requests and instantiate the appropriate type for forwarding. When the task is finished, a response is sent back to the router and then to the SMS-Manager. In Fig 3, transaction managers have been inserted between the router and its communicating parts. The transaction managers are typed by a state machine *TransactionSM* that is responsible for initiating and committing transactions. In this refinement, two connectors are replaced with one part handling the communication on both connectors.

The new models illustrated by Figs 2 and 3 are refinements of the original architecture. We cannot obtain these in UML while preserving the original architecture and avoiding model duplication. Requirements may also demand more cross-cutting modifications, e.g. that all communication between components should be a specific protocol.

In order to cope with such architectural variance, we devise ArchSpect, a technique for describing UML architectural connector aspects and their binding to base architectures. ArchSpect is detailed in Section 3.

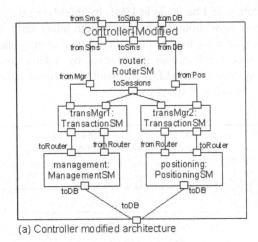

(a) Controller modified architecture

Fig. 3. Adding Transactions to the Controller

3   ArchSpect - Architectural Aspects in UML

An archspect is a UML-based architectural aspect, which can be composed with and possibly cross-cut base architectures. It represents an architectural connector that refine connectors in base architecture with more complex structures.

An archspect is represented by a UML structured class with parts, ports and connectors. No specific UML extensions are used for describing it; it is syntactically like any other structured class. In order to be useful in a composition, however, an archspect should be designed according to constraints that restrict how it can be composed with other models. We use the ports owned by the archspect class (formal ports) to represent elements that should be bound in a composition. The internal structure of the class is fixed and will be instantiated during a composition. We restrict this further by only allowing pairs of formal ports that may represent end points of a connector. These restrictions are not explicit in the archspect itself, but a property of the binding language. The binding language requires the input archspect to provide formal ports that can be bound to connectors. With ArchSpect, we provide a flexible way of refining connectors in multiple locations in a base architecture.

An archspect is composed with a base architecture using a binding specification. Fig 4 illustrates the approach. The base model (a) and the AccessControl archspect (b) are combined using a binding (c). The binding specifies how the unbound elements are bound to the base model. A composed model (d) is the result.

The specification of the archspect is done in UML, as illustrated in Fig 4b. It is defined by a class, which in this case have two formal ports. These are bound to the connector identified by the ports *envUsr* and *fromUsr* in the binding. Fig 5 shows the *Distribution Archspect*. It has two formal ports, *localPort* and *remotePort*, which represents unbound elements of the archspect. The *DistributedComm* class implements the distributed communication and is decomposed in proxy and skeleton parts.

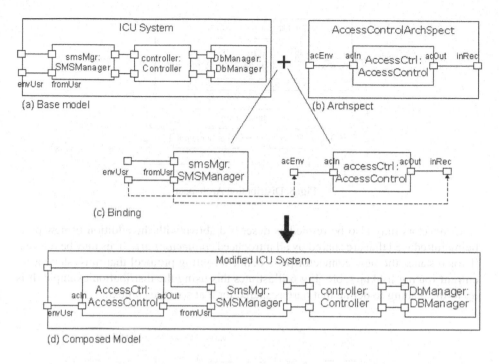

Fig. 4. ArchSpect Illustration

The bindings are specified separately from the architectural models in a binding language that establishes the relationships between the architectures. Section 3.3 details the binding language.

3.1 Architectural Variance with ArchSpect

Many architectural specialisations can be handled by standard UML mechanisms such as specialisation and redefinition. We address a class of refinement that the standard mechanisms cannot cope with; the refinement of connectors into more complex structures. The examples in Figs 2 and 3 are examples of this. The refinement of connectors has implications also for the ends to which connectors attach, the ports, the parts, and their types.

Single Connector Replacement. A single connector can be replaced by a structure such that inner connectors of that structure attach to the connection ends of the original connector. This structure may be a single part or several connected parts. This is illustrated by Fig 2. The types of the new parts may contain additional structure or behaviour. We refer to these newly introduced parts as *connector parts*.

Replacement of Multiple Connectors Between two Parts. Two (or more) connectors between two parts are replaced by a structure with twice as many connectors that must attach to the connection ends of the original connectors. This is illustrated by Fig 3.

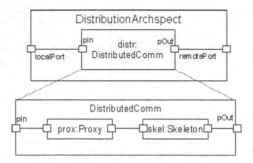

Fig. 5. Distribution Archspect

Connectors may also be refined as described above with the addition of new parts being introduced that are shared by all introduced *connector parts*. This may be relevant if for instance the new connector parts define a routing protocol that must share a re-entrant routing lookup table. Fig 6 illustrates this using the distribution example. It is up to the binding to define that part as shared for that specific context.

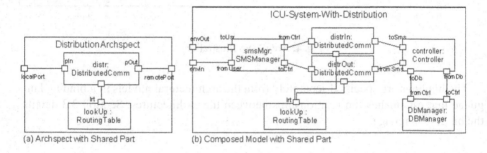

Fig. 6. Example - Shared Part in the Distribution Example

Connector replacement may cross-cut several connectors in the base architecture, as defined by the binding. If this is the case, several occurrences of the archspect will be populated in the composed model, corresponding to the connectors matched by the binding.

The Connector Binding. Architectural connectors are the main focus of ArchSpect. In order to refine a connector with more complexity, we need to refer to the endpoints of the connector, which in UML terms are ConnectorEnds. In our context, these are ports or parts owned by an archspect class in the UML specification. In ArchSpect, only formal ports represent unbound elements and are subject to binding in a composition.

Unbound ports are bound pairwise to connectors in the base architecture. Each base connector has two endpoints that are targets of the binding. The end points for the binding are the ConnectorEnds of that connector, which may be either ports or parts. (Although a connector in UML formally may have N-ary connectors, we make a restriction to binary ones.) The binding to a connector identifies the connector and its connector ends and establishes the relation between the archspect formal ports and the connector ends in the base architecture.

Type Implications. The binding of ports to the base architecture intuitively has impact on the typing of the ports and parts of the final system. The binding of a port to another leads to augmenting the provided and required interfaces. Eventually, the owning classifier must provide an implementation of those interfaces. A port added to a part represents a corresponding property on the owning classifier.

Ports originating from the archspect may also be subject to change in order to maintain a coherent architecture. If the connector ends targeted by an archspect binding are ports that provide specific interfaces, these must be propagated to the ports of the archspect occurrence replacing their role in the composition. Fig 7 illustrates this with the access control example.

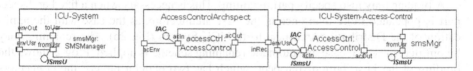

Fig. 7. Interface Propagation

3.2 Asymmetrical vs. Symmetrical Concerns

In our approach, we define archspects using standard structured classes from UML. However, in the bindings, we require an archspect to be on a certain form: a class with inner structure and a set of formal ports that serve as parameters to the binding. A model without any formal ports is thus meaningless in the role of an archspect. Hence, with respect to the binding and the binding language, archspects are asymmetrical.

Since the archspect representations are just like any structured class they have the same capabilities and constraints as any other UML class. They may for example be specialised. This also means that archspects may be applied to other archspects in compositions, which implies element symmetry[12]. Furthermore, the specification of relationships between archspects and a base architecture is done separately from both in the binding specification, which implies relationship symmetry.

The separation between base architecture and archspect is important as they play different roles in a composition. The base architecture is refined to a new structure that contains more detail. The archspect contributes to that detailing in populating itself in the new structure.

3.3 Binding Language

The binding language specifies how archspects are composed with base architectures by relating unbound elements to base architecture elements. In that sense, it resembles UML CollaborationUse, which binds roles to elements in a classifier. The relationship with UML collaborations is discussed in Section 4.2.

A binding uses name references to specify specific unbound elements (ports) of an archspect. These are the actual parameters to the binding. A blend of name references and queries are used to identify target elements in the base architecture. A binding of the AccessControlArchSpect (Fig 4b) is shown in Prog 1. It simply binds the two ports *acEnv* and *inRec* from the archspect to the ports *envUsr* and *fromUsr*, respectively.

```
binding acBinding (AccessControlArchSpect ac, ICU-System base) {
    bind (acEnv, inRec)
        envUsr, fromUsr;
}
```

Prog 1. Binding the Access Control Archspect

A binding may also specify part renaming. This is necessary when a fixed archspect part is populated more than once in the base architecture, as the case with the application of the DistributionArchspect to the ICU-System (Fig 2b). The renaming may be omitted, in which case an automatic renaming scheme is applied. Prog 2 shows the binding of the DistributionArchspect, which has two explicit targets, the connectors between port pairs *(toCtrl, fromSms)* and *(toSms, fromCtrl)*.

```
binding dpaBinding (DistributionArchspect distr, ICU-System base) {
    bind (localPort, remotePort, "distrProxy"){
        Connector(toCtrl, fromSms), "distrIn";
        Connector(toSms, fromCtrl), "distrOut";
    }
}
```

Prog 2. Binding the Distribution Archspect

Variables and Quantification. The bindings of port pairs to connectors in the base architecture are either specified by direct matching with element references (names) or by variables representing queries that quantify sets of target elements. Variables are declared using an @ sign before a name and the queries are constrained by connectivity, typing, or naming.

In Fig 3 we illustrated a refinement of the controller structure where transaction parts intercept communication between the router and the service providers of the system. Fig 8 defines the archspect representing that refinement and Prog 3 defines the binding for that particular composition. The *TransactionMgmtArchspect* defines four unbound ports(a,b,c, and d), each associated with a connector to the transaction manager.

The binding (Prog 3) specifies the binding of the four unbound ports and the 'transMgr' part name. The ports are bound pairwise based on their connectivity. It binds port *a* to the *toSessions* port and port *b* to any port *p* that is connected with the *toSessions*

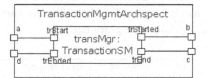

Fig. 8. Transaction Archspect

port. Furthermore, it binds ports *c* and *d* to any ports *x* and *y* that are connected, where the owner of *y* is the *router* and *y* is a different port than *toSessions*. The binding of the name 'transMgr' is omitted, which then defaults to an automatic renaming scheme.

```
binding transBinding (TransactionMgmtArchspect trans, ICU-System base) {
    bind ((a, b), (c, d), "transMgr"){
        Connector(toSessions, Port @p)), Connector((Port @x, Port @y)  |
              y.owner.name = "router" && y!=toSessions;
    }
}
```

Prog 3. Binding the Transaction Archspect

The *toSessions* reference in the binding in Prog 3 is an example of a named reference and the @*p* variable an example of a variable that quantifies over ports in the base model. A binding specification may also query elements that are not direct targets for a binding (free variables), but used as parameters to constrain the binding. These are declared as other variables, as a prefix to the constraint.

As seen by the examples, a binding may target specific connectors by name references. It may also specify more general queries that select sets of elements in the base architecture. A query may search for connectors adhering to specified constraints, which for instance could be the types of its connector ends. The constraint is defined by an ocl-like expression, extended with some helper operation related to parts, ports, and connectors (such as *isConnected(p1,p2)*, *hasType(t)*, *provides(i)*).

Binding Language Details. A binding is declared with the *binding* keyword followed by a name and a set of input model parameters (one or more archspect and a base model). The input model parameters refer to classes that define the archspect and the base architecture. The binding contains a set of *binds*. Each *bind* refers to the unbound elements - the ports - of the archspect and contains a body with one or more *binding queries*.

A binding may also specify an input constraint specifying an archspect element to be shared in the context of that binding, i.e. to allow multiple occurrences of the archspect to share a part in the composed system.

Each binding query defines a corresponding number of query elements to which the unbound elements are bound. It selects the connectors to which the parameter ports should be bound. This is done by direct reference to the connector or its ends, or by variables that represent selection of element sets in the base architecture. The result is limited by additional constraints related to the connector variables.

Prog 4 shows the binding language grammar.

```
binding               = 'binding' name '(' inputArchspect+ ',' inputBase '){'
                              inputConstraint?
                              bind+
                          '}'
inputConstraint       = 'shared' name (',' name)+ ';'
bind                  = 'bind' '(' portPair
                          (',' portPair)* (',' renameSource)* ')'
                          '{' (bindingQuery ';')+ '}'
portPair              = '(' portPair ')' | portRef ',' portRef
bindingQuery          = queryConnector (',' queryConnector)* (',' renameTarget)*
                          ('|' (freeVariable ',')* elementConstraint)? ';'
queryConnector        = 'Connector' '(' connectorEnd ',' connectorEnd')' |
                          'Connector' name
connectorEnd          = name | variableElement
variableElement       = (type)? '@'name
freeVariable          = variableElement
elementConstraint     = oclExpression
portRef               = name
inputArchspect        = modelName name
inputBase             = modelName name
renameSource          = '"' name '"'
renameTarget          = '"' name '"' | 'auto'
type                  = 'Port' | 'Part'
modelName             = name
name                  = <valid name>
```

Prog 4. Binding Language Grammar

Graphical Binding with UML constructs. A graphical, UML-based approach to binding specifications is an alternative to the lexical one. A benefit with this is improved visualisation of the composition, which may be easier to understand for the user. The user also gets the benefit of a similar notation for bindings as for architecture descriptions. Explicit links to the archspect and base model and quasi-UML conventions can be considered drawbacks.

In the UML-based binding notation, we use parts, ports and connectors to specify binding patterns. Dependencies to from the binding pattern to the archspect specify the bindings. Fig 9 illustrates bindings in UML.

The purpose of the binding pattern is to mimic the semantics defined by our lexical binding language. Fig 9a shows the elements of a binding pattern. Open ('?') names on parts, ports, or types signifies quantifications that can match anything in the base architecture. Names of parts and ports are matched exactly if specified.

Types of ports may be specified as a list of alternative or mandatory types. For parts, only list of alternatives can be specified. Alternatives are described as a comma-separated list of type names *(portA: IntA, intB)*. Mandatory types are separated by ampersands *(portA:intA & intB)*.

The main advantage with this graphical notation over UML CollaborationUse is the capability of quantification and thereby specifying cross-cutting bindings more compactly.

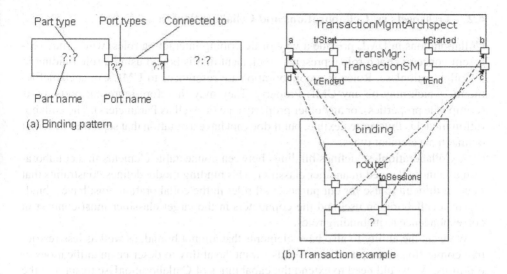

Fig. 9. Binding Using UML Notation

4   Comparing With UML Mechanisms

4.1   Archspects vs. Specialisation

Specialisation and redefinition of virtual elements open many configuration possibilities. A class can be specialised, and its contained parts, ports, and connectors can be redefined. However, the architectural extensions made in the example are not legal specialisations in UML, because they require changes to existing elements (connectors) that are not legal redefinitions in UML. A redefinition of a port, part, or connector can only be done with another, type compatible port, part, or connector, respectively.

In the examples, the changes in the composed system architecture are not compatible with specialisation. Another option with specialisation is to redefine the existing parts to contain the added complexity, such that for example, a redefined SMSManager contains the access control aspect. Redefined parts, ports and connectors must be type/interface-compatible with the redefined ones. The redefinition approach requires a full specification of the "new" system and its wiring. Concerns can still be described as separate components (aspects), but the benefits of quantification and cross-cutting are lost, and the architecture is different from our composed result.

Another possibility with specialisation is to take an approach closer to product line design, where possible future "aspect" extensions already are identified in the architecture as "virtual parts" that can be overridden in specialisations. This, however, is in many cases difficult as future requirements can be hard to predict.

We see that the intent provided by an aspect-oriented approach is a different one than that captured by inheritance and virtuality. In this case, the transformation defining the composed model from the archspects and the base model can be viewed as a set of connector refinements of the base model.

4.2   Archspects vs. Collaborations and CollaborationUse

Collaborations in UML provide a way of describing interacting roles, which may represent concerns. Each role represents an element that is bound using a role binding in a CollaborationUse. Roles in a collaboration are references to UML ConnectableElements, which may be any UML Property. They may therefore represent ports, parts (composite properties), or any other property type (as well as Parameters). The collaboration model is thus quite flexible, but it does not have a notation that separates different connectable element types.

A CollaborationUse defines bindings between connectable elements in a collaboration and those defined in another classifier. This binding model defines constraints that make it difficult to use for our purpose: all roles in the collaboration must have a binding in a collaboration use, and the connectors in the target classifier must connect in correspondence to its binding roles.

We want the ability to also have elements that are not bound, as well as less restrictive connection correspondence. We also want the ability to describe quantifications of bindings. We would need to extend the capabilities of CollaborationUse to support the semantics of the archspect binding language.

4.3   Archspects and Virtual Connectors

Connectors in UML can be virtual and may then be redefined by specialising classes. This is also true for ports and parts. A connector can only be redefined by another connector, and the redefining connector and its ends must be associated with compatible types. Consequently, it is not possible to redefine a connector into a more complex structure. A certain degree of connector modification can be achieved by using stereotypes to signify different kind of connectors, but this will not help in advancing the architecture.

The extensions illustrated by the examples in this paper (such as the distributed proxy example) represent a kind of connector refinement, where a connector is redefined by a part with two connectors. Such a part may represent further decomposition defined by its type, which may contain more complex structure or behaviour.

A UML connector is typed by an association, which is either defined by the user or inferred. If that association is an association class, the connector type may be an instance of a more complex structure defined by that class. This is a way of reasoning about the complexity of a connector. It also provides a possible means of refining the connector structure by specialising the system and redefine the connector with a specialised version of the original connector association class. The type of the connector, however, depends on the types of the corresponding connector ends; as such, it cannot be easily reused by connectors with different end types.

5   Related Work

There are many approaches to aspect-based architectural design with and without UML, which are surveyed in an AOSD Europe report [13]. Krechetov et al [14] integrate what they see as the key architectural design approaches that use UML. The resulting language uses stereotypes to represent aspects with cross-cutting interfaces and dependencies. They do not address the decomposition of UML structured classes with parts, ports and connectors, which is the main focus in ArchSpect.

In [15], Pinto et al present the language *DAOP-ADL* and the *Component Aspect Model (CAM)*, which combines architecture description and aspects. They provide a formalism to specify component architectures, aspects, and composition rules for these descriptions. XML is used to describe these, and a mapping to UML is provided as a profile. In contrast to ArchSpect, this is specific to the DAOP platform, it operates on component interfaces related to messages/events, and does not consider internal structure or connectors.

In [16], Bouckle et al describe an approach for composing cross-cutting concerns by introducing three composition operators in xADL[17]: substructure, unification, and mapping. Substructure composition is used to describe independent substructures that decompose higher level components. This resembles decomposition with structured classes in UML. Mappings define correspondences between interfaces in two architectural structures. This is similar to the bindings defined in this paper. The unification operator unifies interfaces in mapping operations.

In [18], Garcia et al describe an aspect-oriented extension to the ACME architecture description language[8], which introduces *aspectual connectors* that modularise cross-cutting interactions between components. An aspectual connector identifies a base role, a cross-cutting role, and a glue that specifies behaviour ordering of compositions (before/after/around). The mapping of roles are done with ACME attachments with quantifiers. The binding of ArchSpect ports resembles the binding of roles in AspectualACME. ArchSpect focuses on connector binding to refine connectors, not only ports. The specification of composition ordering is not supported by ArchSpect. A mapping of AspectualACME to UML introducing profile extensions for aspectual connectors is presented in Sande et al [19].

In [20], Haugen and Møller-Pedersen describe how architectural configurations can be done by combining inheritance with subsetting and constraints. The approach is complementary to the one described here, as it does not address connector refinement as part of configurations.

In [9], Perez-Martinez et al evaluate the usage of UML1.4 and UML2 for representing architecture connectors in the context of some specific architectural styles. One of the observations is that the connector concept in UML2 is not powerful enough to represent architectural connectors. The workshop 'Software Architecture Description and UML' in 2004 [10] discussed and presented various facets of UML used for architecture description. Roh et al [21] describe a UML2 profile for architecture modelling where a number of extensions are made to support architecture modelling, such as providing a new type of connector based on collaborations.

6   Conclusion

Architecture descriptions are important for understanding architecture and being able to reason about their properties. UML provides a standard notation for representing software system abstractions. We have shown how UML structures can be used to describe architectural concerns that can modularise complex system specifications, and how these concerns can be used in composition towards more complex system architectures.

We have described ArchSpect, a technique for describing architectural concerns specifically related to refinement of architectural connectors. It allows UML architectural concerns to cut across other architectures. We have defined a binding language that provides the means of binding architectural concerns to base architectures and proposed a lexical and UML-based notation. Our approach provides refinements that goes beyond those available in vanilla UML in terms of extension and redefinitions. We discussed the relationship with different UML constructs and how these are difficult to use for representing refinement of architectural connectors.

Our main contribution is a way of specifying and composing architectural concerns that specifically pertain to the connectors between components and parts, which improve how UML architectural descriptions can be modularised and composed.

Acknowledgement. This work has been carried out in the context of the SWAT project (Semantics-preserving Weaving - Advancing the Technology), funded by the Norwegian Research Council (project number 167172/V30).

References

1. Bass, L., Clements, P., Kazman, R.: Software Architecture in Practice. Addison-Wesley Longman Publishing Co., Boston, MA, USA (1998)
2. Garlan, D., Shaw, M.: An Introduction to Software Architecture. In: Ambriola, V., Tortora, G. (eds.) Advances in Software Engineering and Knowledge Engineering, pp. 1–39. World Scientific Publishing Company, Singapore (1993)
3. France, R., Ray, I., Georg, G., Ghosh, S.: Aspect-oriented Approach to Early Design Modelling. In: IEE Proceedings - Software (2004)
4. Baniassad, E., Clements, P., Araujo, J., Moreira, A., Rashid, A., Tekinerdogan, B.: Discovering Early Aspects. Software, IEEE (1), 61–70 (2006)
5. IEEE: IEEE Std 1471:2000 Recommended Practice For Architectural Description Of Software-Intensive Systems. Technical report, IEEE (2000)
6. Society of Automotive Engineers (SAE), SAE Architecture Analysis and Design Language (AADL). Technical report (2006)
7. van Ommering, R., van der Linden, F., Kramer, J., Magee, J.: The Koala Component Model for Consumer Electronics Software. Computer (3) (2000) 78–85
8. Garlan, D., Monroe, R., Wile, D.: Acme: An Architecture Description Interchange Language. In: CASCON '97. Proceedings of the 1997 conference of the Centre for Advanced Studies on Collaborative research, p. 7. IBM Press (1997)
9. Perez-Martines, J.E., Sierra-Alonso, A.: UML 1.4 Versus UML 2.0 as Languages to Describe Software Architectures. In: Oquendo, F., Warboys, B.C., Morrison, R. (eds.) EWSA 2004. LNCS, vol. 3047, pp. 88–102. Springer, Heidelberg (2004)
10. Avgeriou, P., Guelfi, N., Medvidovic, N.: Software Architecture Description and UML. In: Nunes, N.J., Selic, B., Rodrigues da Silva, A., Toval Alvarez, A. (eds.) UML Modeling Languages and Applications. LNCS, vol. 3297, pp. 23–32. Springer, Heidelberg (2005)
11. Oldevik, J., Haugen, Ø.: Architectural Aspects in UML. In: AOSD 2007. First Workshop on Aspects in Architectural Description (2007)

12. Harrison, W., Ossher, H., Tarr, P.: Asymmetrically vs. Symmetrically Organized Paradigms for Software Composition. Technical Report RC22685 (W0212-147), IBM (2002)
13. Chitchyan, R., Rashid, A., Sawyer, P., Garcia, A., Alarcon, M.P., Bakker, J., Tekinerdogan, B., Clarke, S., Jackson, A.: Survey of Analysis and Design Approaches, AOSD Europe. Technical report (2005)
14. Krechetov, I., Tekinerdogan, B., Garcia, A., Chavez, C., Kulesza, U.: Towards an Integrated Aspect-Oriented Modeling Approach for Software Architecture Design. In: AOSD 2006. 8th International Workshop on Aspect-Oriented Modeling, Bonn, Germany (2006)
15. Pinto, M., Fuentes, L., Troya, J.M.: A Dynamic Component and Aspect-Oriented Platform. The Computer Journal 4, 401–420 (2005)
16. Bouckle, N., Garcia, A., Holvoet, T.: Composing Architectural Crosscutting Structures in xADL. In: AOSD 2007. 10th Workshop on Early Aspects, Springer, Heidelberg (2007)
17. Dashofy, E.M., van der Hoek, A., Taylor, R.N.: A Comprehensive Approach for the Development of Modular Software Architecture Description Languages. ACM Trans. Softw. Eng. Methodol. 2, 199–245 (2005)
18. Garcia, A., Chavez, C., Batista, T., Sant'anna, C., Kulesza, U., Rashid, A., Lucena, C.: On the Modular Representation of Architectural Aspects. Springer, Heidelberg (2006)
19. Sande, M., Choren, R., Chavez, C.: Mapping AspectualACME into UML 2.0. In: AOM. 9th Aspect-Oriented Modelling Workshop (2006)
20. Haugen, Ø., Møller-Pedersen, B.: Configurations by UML. In: Gruhn, V., Oquendo, F. (eds.) EWSA 2006. LNCS, vol. 4344, Springer, Heidelberg (2006)
21. Roh, S., Kim, K., Jeon, T.: Architecture Modeling Language based on UML2.0. In: APSEC'04. Proceedings of the 11th Asia-Pacific Software Engineering Conference, pp. 663–669. IEEE Computer Society, Washington, DC, USA (2004)

Domain Specific Modeling Methodology for Reconfigurable Networked Systems

Gabor Batori[1], Zoltan Theisz[2], and Domonkos Asztalos[1]

[1] Software Engineering Group, Ericsson Hungary Ltd.
[2] Network Management Research Centre, Ericsson Ireland Ltd.
{gabor.batori,zoltan.a.theisz,domonkos.asztalos}@ericsson.com

Abstract. Our empirical study shows that reconfigurable networked systems executing software components deployed on interconnected heterogeneous hardware nodes highly benefit from an effective framework and a normative design methodology relying on domain specific models supporting both application and platform domains. The approach is based on building metamodels for various expert domains to enable precise knowledge codification in the form of interpretable and analyzable information frameworks. The core platform architecture is characterized by interlinked on-the-fly reconfigurable communicating components whose behavior is specified by finite state machine model of computation. The proposed methodology covers the whole development and operation life cycle.

1 Introduction

Effective high-quality software development supporting distributed reconfigurable networked systems is an actively researched topic in the domains of ad-hoc and sensor networks. One of the ambitions of the Reconfigurable Ubiquitous Network Embedded Systems (RUNES) [1] IST project has been to unburden the task of the application developers by providing a distributed component-based platform architecture on top of a heterogeneous network of computational nodes and by establishing a model-based software development methodology and a corresponding model-based development framework. The heterogeneity of the nodes are hidden from the applications. The reflective components are linked together by their interfaces, communicate by message sending and store their meta-data in a distributed database. Each computational node incorporates a Component Run-Time Kernel (CRTK), which provides the basic middleware APIs of component management. The corresponding model-based software development framework deals with both the application and the platform domains. The framework supports domain experts in the creation of formal metamodels and enables refinement steps in multiple iterations. The metamodels describe the entities, relationships and properties of the domains; the models represent the particular domain knowledge. The goal of the approach is to facilitate the creation of executable models through multi-stage model transformations until platform compatible executable code has been produced. The platform metamodel features deployment aspects to assist software architects in mapping the application components onto the available computational resources in a fully distributed environment. On one hand, the framework takes

G. Engels et al. (Eds.): MoDELS 2007, LNCS 4735, pp. 316–330, 2007.

advantage of Model-Integrated Computing [2], that is, the metamodels and the instantiated models support precise knowledge codification in the form of interpretable and analyzable information. On the other hand, it applies important elements of Service Oriented Architecture [3] for the behavior specification of the application. The core part of the generated application consists of interconnected communicating components controlled by the Finite State Machine (FSM) model of computation.

We have developed ErlCOM [4,5], a robust prototype RUNES platform implemented in Erlang [6], to illustrate effective code generation and component deployment on top of the Generic Modeling Environment (GME) [7]. Furthermore, we have implemented a MIC-based Deployment Tool demonstrating both initial application deployment and dynamic run-time reconfiguration of component systems. Our technology supports the whole development cycle of distributed applications for heterogeneous HW/OS architectures via model interpreters and graph transformators. Our prototype performed successfully at the IST2006 Conference [8] and vividly demonstrated the feasibility of the methodology and the related technology. In the remainder of the paper the elements of our model-based methodology, the development framework and the execution platform will be demonstrated by the fire detection service of the Tunnel Control Application; more detailed information is available in our RUNES technical report [9].

In Section 2, an overview of the design process is given, then the scenario analysis will be introduced in Section 3. Section 4 examines interaction modeling using the fire detection service to disseminate the approach and to show some important technical points in the case of a small real-life example. In Section 5, platform modeling and implementation details are presented. Then, Section 6 shows the connection between platform and interaction modeling. In Section 7, our model based deployment and reconfiguration facilities are introduced.

2   Process

Any kind of professional software development is usually accompanied by some development process which safeguards the industrial scale applicability of the technology. There are many widely applied development approaches in the model based software arena like Rational Unified Process or Microsoft's Software Factory. Our process has been significantly influenced by the existing methods; however, the ambition of our work aimed at covering all the stages of application development based on generative metamodeling technologies. The overview of the process stages are depicted in Figure 1.

Figure 1 is layered into five stages, namely, the Scenario (Section 3), the Application Modeling (Section 4 and 6), the Platform (Section 5), the Code Repository and the Running System (Section 7). The arrows connecting the artifacts of the various stages are labelled by sequence numbers regarding the process activities. This section only introduces the steps of the process and concentrates on their interwork; the technical details of the relevant steps are revisited in later sections of the paper.

The Scenario contains the set of scenario descriptions establishing the scope of the application domain. Our experience from the RUNES project has proved that applications can only be successfully developed if the scenarios are detailed enough to enable

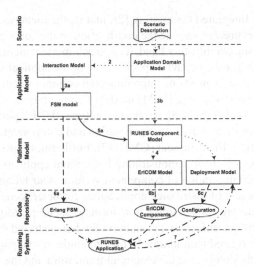

Fig. 1. Platform Independent Behavior for Fire Detector

non-trivial application modeling and quality analysis. Since any realistic distributed application involves intense interaction among application components both structural and interaction (Section 4) modeling are equally important. The Application Domain Model must be created to cover the scenario in such a way that all the use case details must be taken adequately into account and the stakeholders' roles have to be discovered. The roles make up the basic elements of the interaction model, hence, the dynamicity of the use cases must be translated into corresponding Message Sequence Charts (MSC) among the participating roles. The Application Domain Model and the Interaction Model should be detailed enough so that quality investigations could be carried out to check the feasibility of the design. Obviously, modeling is not a one shot activity, therefore, iterative scenario-modeling-to-quality-evaluation cycles are fairly probable. Each cycle involves creative decision making, so both arrow 1 and 2 in Figure 1 are dotted showing that the activity is carried out manually.

The Interaction Model is transformed into FSM Model (Section 4) and then a further translation step turns it into RUNES Component Model (Section 5). The solid arrow indicates that the translation is executed by graph transformations. The Application Domain Model usually requires creative refinements and only in rare cases allows totally automatically translation into the RUNES Component Model, which is indicated by the dotted arrow.

The Platform Model stage has been conceived to support total semantics elaboration, that is, the RUNES Component Model is extended by the semantics of the platform, components and the FSMs applying any high level process-oriented - preferably functional - language. The step involves some manual coding to produce executable specification of the whole application. During the evaluation of the process, we used ErlCOM (Section 5.3) as the platform and Erlang as its accompanying language of semantics embedment; therefore, the resulting model is called 'ErlCOM Model' in the paper.

The final application model must take into consideration the distributed nature of the application, hence, the so-called Deployment Model is established (Section 7). It fully specifies the component allocation of the application towards the available resources. The Deployment Model is an instance (dotted arrow without label) of the RUNES Component Model as it represents one particular deployment configuration of the application.

The Code Repository is the stage where the source code is stored. The code production is fully automatic indicated by the dashed lines. The Deployment Model is translated into run-time configuration information (initial configuration, run-time distributed database etc.) which is, finally, deployed onto the available platform nodes. The run-time changes of the component configuration are managed by the Deployment Tool and the Deployment Model is updated accordingly as indicated by arrow 7.

This process incorporates a set of meta-programmable MIC tools; the most important to be mentioned are the Generic Model Environment (GME) for metamodeling [7] and the Graph Rewriting and Transformation (GReAT) tool for graph transformation [10] purposes. The process is totally application agnostic, therefore, it is suitable for the development of any possible type of distributed applications. Our experiment and the success of the RUNES demo at the IST2006 Conference have clearly shown the viability of the process and the underlying modeling technology.

3 Scenario Modeling

In order to illustrate certain technical concepts of the integration of the various pieces of modeling work and code generators the RUNES project decided to choose the Fire in the Road Tunnel scenario [11]. The overview of the scenario story is the following:

"At the beginning of our story traffic is flowing normally in the road tunnel. Tunnel fires can be detected by the wired system that is part of the tunnel infrastructure. The fire sensors do, however, have the capability to operate wirelessly if required. An accident within the road tunnel has resulted in a fire. The fire is detected and is reported back to the Tunnel Control Room. The emergency services are summoned manually by Tunnel Control Room personnel. As a result of the fire the wired infrastructure is damaged and the link is lost between fire detection nodes. Using wireless communication, information from the fire detection nodes is still delivered to the Tunnel Control Room seamlessly. The first response team arrive from the fire brigade. Four firemen are sent into the tunnel on foot. As the firemen move towards the fire the sensors reporting periodic data on external temperatures detect a rise in temperature and respond by increasing the frequency of reporting so that the Emergency Control can assess the danger to the fire fighters. The fire becomes more severe. A node is lost..."

The methodology begins with the analysis of the scenario description, that is, the creation of an application domain specific metamodel. In this case, the metamodel contains the concepts and the relationships relating to the networked embedded system of the tunnel domain. The domain concepts of the scenario are e.g. Tunnel Control Room, Alarm, Sensor nodes, Filters etc. Based on the metamodel different configurations (a.k.a. application scenarios) of the domain can be built.

The goal of the application in this scenario is to monitor the state of the various tunnel networks connected to the Tunnel Control Room. The sensors in different sectors of the tunnel send measurements towards the control room. The control room filters the measurements and notifies the emergency units via a jurisdiction network when a fire situation has been detected. The crucial point of the scenario is that the application server in the control room is able to detect fire situations in an ever changing environment. The connection to the sensor network of the tunnel in disaster case is unreliable therefore some measurements can be lost. Furthermore, the network can reconfigure itself to be able to adapt to the changing environment. Since the reconfiguration of the network sometimes takes for a while the application has to handle delayed measurements due to the reconfiguration. The fire detection service is implemented by the Fire Detector (FD) unit of the application, which will be used as our example in the sequel.

4   Interaction Modeling

Large-scale networked systems can be efficiently comprehended as a large number of interacting services. By combining the various services an entity is involved in the complete behavior specification for that entity is obtained. Therefore, the service concept is effectively based on the interaction patterns among distributed entities. The notion of a role describes the contribution of the entity as it plays this role to a given interaction pattern. The main idea of service-oriented development presumes that when developing one service the designer may not necessarily have complete information about all the other services that might co-exist in the system. The mapping of the service specification onto a set of components is well established and follows the methodology [3] which advocates the use of state machine synthesis algorithms so that the scenarios can be quickly simulated and/or validated (see Figure 2). The generated state machines possess the intended dynamic behavior of the system that can be automatically incorporated into the architectural design.

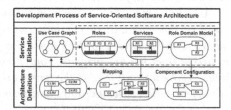

Fig. 2. Service-based development

The state machine generation is carried out by the tool called M2Code [3]. A centerpiece of its capability is the (semi)automatic derivation of state machines from the interaction patterns specified via Message Sequence Charts (MSC). Two types of MSCs are defined; the basic MSCs and the high level MSCs (HMSC). A basic MSC consists of a set of axes, each labelled with the name of a role. An axis represents a certain segment of the behavior displayed by its corresponding role. An HMSC is a graph whose nodes are references to other (H)MSCs. The semantics of an HMSC is obtained by following the paths through the graph and composing the interaction patterns owned by the nodes along the way. The output of M2Code is a role domain model (Figure 4) together with one Finite State Machine (FSM) for each role defined in the domain model; the FSM implements the respective role's contribution to the services it is associated with.

4.1 Behavior Specification

Interaction modeling focuses on the specification and elaboration of the dynamic behavior of the application. The interactions are defined manually based on the application scenario via Message Sequence Charts (MSC) in M2Code. The Fire Detector service is specified by the following five MSCs (Figure 3):

- MSC A: In **Normal** state the detector receives all the measurement events from the sensor.
- MSC B: If in **Normal** state no measurement event has been arrived in time from the sensor (due to loss of connection) **Fire** state is activated.
- MSC C: If in **Fire** state a new measurement event is received it indicates that the connection to the sensor has been re-established, therefore, **Normal** state is re-set.
- MSC D: If in **Fire** state no measurement event has been arrived in time from the sensor (due to destroyed sensor) **Dead** state is activated.
- MSC E: In **Dead** state any measurement that might have been received is considered faulty since the sensor must have already been destroyed by the fire.

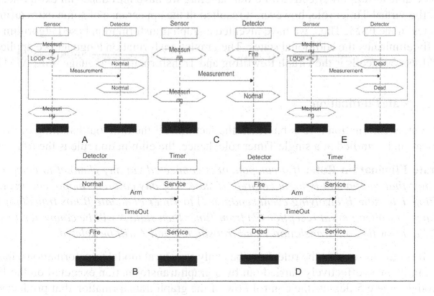

Fig. 3. Message Sequence Charts of Fire Detector

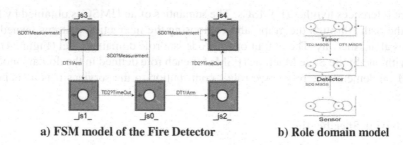

a) FSM model of the Fire Detector   b) Role domain model

Fig. 4. M2Code generated models imported into GME

The generated FSM is shown in Figure 4a. It can be seen that the state names are automatically generated; therefore, they must be relabelled to establish their semantics. The relabelling is a highly creative task since the model designer should be able to understand the correspondence between the generated FSMs' states and the states of the MSCs. Moreover, the correspondence must be thoroughly understood, otherwise, semantics anchoring is unfeasible and any further semantics elaboration into implementation code is fairly questionable. In general, the roles of the FSMs are later mapped onto the architectural model of the application. The mapping might happen in various manners depending on the needs of the application. In the particular case of the Fire Detector we applied the simple one-role-to-one-component style.

Although M2Code provides essential support for distributed application development currently it does not support time-out handling out-of-the-box. Therefore, in order to be able to support effective time-out handling we have introduced an extra role, that is, the virtual Timer role, however, it resulted in the appearance of some extra artificial states in the FSM. Thus, we have invented a graph transformation based algorithm that fully eliminates the additional states. The graph transformation language we applied to address our needs is the Graph Rewriting and Transformation language (GReAT) [10].

4.2   State Elimination

Our state elimination rule is based on the fact that all the time-out handling of the application is bundled in a single Timer role, hence, the elimination rule is the following:

State Elimination Rule: *If a component is connected via any channel to component Timer that component is to be colored. If state A of any colored component has transition T to state B involving event sending (!) to Timer and state B has transition Z to state C involving event reception (?) from Timer then state A will be connected to state C via a new timeout translation Y and transition T and Z will be deleted.*

It is easy to see that the rule describes only structural model transformations, that is, it can be most effectively carried out by a graph transformation executed on the FSM model. Figure 5 depicts the control flow of the graph transformation that produces the reduced FSM. The rules describe the sequence of pattern matchings applied to the incoming subgraphs of the FSM model. The packets enter via the blue ingresses and leave through the red egresses of the boxes. The boxes are connected together according to

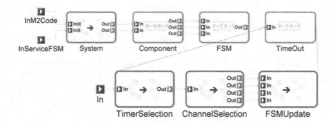

Fig. 5. Control Flow of Graph Transformation for State Elimination

the logic flow of the process. The pattern matching rules are located inside the boxes. Figure 6 depicts the first part of the elimination rule, which is located in box ChannelSelection. After the successful application of "State Elimination Rule" the unnecessary transitions (transition T and Z) and state (state B) are deleted (red rectangles) and a new transition (translation Y) is partially created (blue rectangle). The conditions of the "State Elimination Rule" are only fulfilled by the simultaneously successful pattern matching and the related satisfied conditions (TimeOutEvent). Figure 7 shows the second part of the rule (located in FSMUpdate) when the creation of the new transition (translation Y) is completed and the default value of the timeout is set.

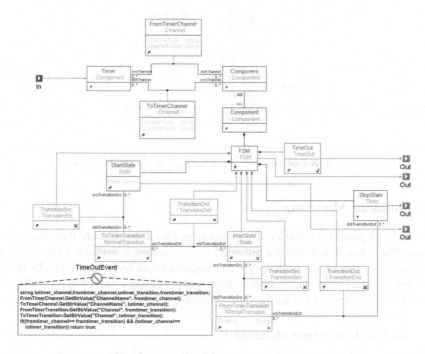

Fig. 6. First part of State Elimination

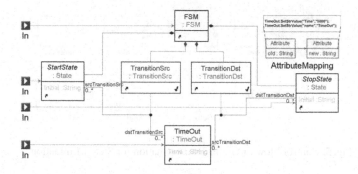

Fig. 7. Second part of State Elimination

The generated FSMs are stored in the GME database and they represent the Platform Independent Behavior (PIB) of the to-be-synthesized system. The resulted FSM of the Fire Detector is shown in Figure 8. It is important to emphasize that the states have already been named according to their semantics, however, the details of the semantics have not been elaborated yet so that the semantic details of the behavior could be kept totally platform independent.

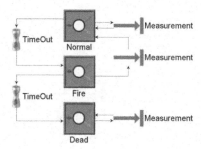

Fig. 8. Platform Independent Behavior for Fire Detector

5   Platform Modeling

The RUNES platform consists of a component-based middleware that decouples and encapsulates the functionalities provided by its various constituents via well-defined interfaces. Moreover, the middleware reaches down into layers that typically belong to the network and the operating system, therefore providing a unified approach to configuration, deployment and reconfiguration at multiple levels of abstraction.

5.1   RUNES Component Metamodel

An outline of the component metamodel is illustrated in Figure 9. Components are encapsulated units of functionality and deployment that interact with each other exclusively via their interfaces and receptacles. Interfaces are defined by a set of related

operation signatures and associated data types. Components can support multiple interfaces; this is useful in embodying separations of concern (e.g. between base functionality and component management). Receptacles are "required" interfaces to explicitly
state the dependencies among the components. Deploying a component into a capsule
its receptacles totally determine which other components must be present to satisfy
the dependencies. Capsules are platform containers providing access to run-time APIs.
Bindings ensure consistent connection setup between an interface and a receptacle. The
component model itself is complemented by two further architecture elements: component frameworks and reflective extensions. Component frameworks (CF) are groupings
of components with constraint guarantee to allow only meaningful component configurations. All entities of the metamodel (Component, Capsule, Interface, Receptacle,
Binding, Component Framework) can store arbitrary <key,value> attributes, which establish a reflective layer facilitating global discover at run-time. Component interactions
can be intercepted at the bindings by pre-actions and post-actions to enable additional
processing on the level of individual messages. An important aspect of the platform
is that both the concurrent activities represented by the components and the individual
interactions represented by the bindings can be reified and reasoned on.

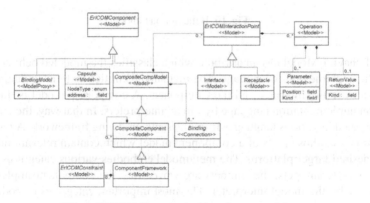

Fig. 9. RUNES Component Kernel metamodel

5.2 RUNES Semantics Metamodel

There are many ways of defining action semantics on model level. OMG (Object Management Group) has proposed to define a common, high level "Action Semantics" in
UML, which is capable to embody all kinds of actions. The main purpose of the Action
Semantics is to provide platform independent description of the actions from which language specific implementation can be generated. Unfortunately, the syntax of the Action
Semantics Language (ASL) and the additional extensions of proprietary ASLs are not
standardized, hence, all vendors have their own non-standardized syntactical representation of the UML Action Semantics. Moreover, strong abstraction and modeling skills
are needed to use the proprietary ASLs correctly, which might even be harder to find
than good coding skills and this usually restricts broader applicability in large corpo-
rations. In the RUNES project we have experimented with another way for describing

behavior. We have restricted our investigation only to the definition of behavior in component platform. Since platforms must support heterogeneity on software and hardware level so the modeling framework also has to cater for this diversity. The cornerstone of the component behavior description metamodel is an abstract model (see Figure 10), namely, the Behavior Model.

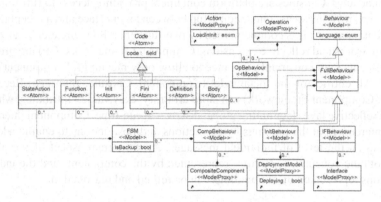

Fig. 10. Behavior metamodel

The Behavior Model has an attribute which classifies the modeled behavior according to the implementation language. The metamodel entities that may contain behavior description are the Interface and the Component. A component model is translated to the target implementation language by model interpreters. In that way, the components can be created in various languages from the same modeling framework. A model interpreter processes those parts of a component model which contain relevant information for the desired target platform. The metamodel embodies various categories of implementation code snippets; the snippets are woven into the component implementation accordingly by the model interpreter. The most important categories of code snippets are:

- Init - Initialization code for a component, an interface or the system.
- Body - Implementation of an interface operation. The signature of the operation is defined in the model and generated by the interpreter.
- StateAction - Specifies the semantics of a state. This is the connection point to the generated FSM Model. (see Section 6.2)

5.3 ErlCOM

The RUNES Semantics Metamodel relies on the RUNES Component metamodel and the heterogeneous implementations of the corresponding Run-Time Kernel (CRTK) to enable rapid application development by leveraging model based code generation. The Erlang implementation of the RUNES CRTK is called ErlCOM [4,5], which takes into account the beneficial aspects of Erlang, that is, concurrency, fault-tolerance, robustness, functional programming style and super-efficient message passing. Erlang is also used for semantics specification taking advantage of its functional style to eliminate side-effects and to benefit from complex pattern matching and higher order functions. The most important characteristics of ErlCOM are as follows:

- ErlCOM supports concurrently executing components that can be dynamically created, loaded, updated, unloaded and destroyed.
- The components can migrate from capsule to capsule by reconfiguring themselves in response to external events.
- The design and management of ErlCOM is built on top of the GME innovative modeling environment and GUI.

6   Application Modeling

6.1   Metamodel Mapping to ErlCOM

Section 4 described in details the method how the Platform Independent Behavioral (PIB) model is produced for distributed reactive systems from the application scenario specifications. The PIB model does not depend on the platform concepts of the RUNES metamodel (Secton 5). Nevertheless, since the component architecture represents the platform specific modeling (PSM) layer, thus, following OMG's MDA terminology the platform independent model (PIM) must be translated into the platform specific one.

Although the structural parts of the synthesized application model seem to be perfect candidates for graph transformations, in the case of complex real-life applications only some parts of the model can be handled accordingly. Fortunately, the behavior parts of any synthesized application model are available in the form of Platform Independent Behavior FSMs. The states of the FSMs are labelled, but the semantics of the transitions and the states have not been specified, yet. First, the transitions of the FSMs are mapped into StateAction atoms by graph transformation. Then, the FSMs' state structure is mapped onto ErlCOM. We have taken explicit advantage of Erlang's FSM design pattern (timeout handling included), hence, a GME interpreter automatically produces the necessary source code.

6.2   Component Mapping

The structural and behavioral mapping is followed by the component mapping. The FSM StateActions are automatically generated from the PIB FSMs, however, the other component functionalities must be elaborated manually according to the needed application semantics. When the complete synthesized platform specific application model (ErlCOM Model) is ready it is automatically translated into ErlCOM compliant Erlang source code. The translator is a complex GME interpreter which checks the model consistency and generates optimized Erlang code. It is important to emphasize here that as far as the RUNES metamodel and the ErlCOM API do not change the translator safeguards that good quality source code is produced from analyzable, semantically anchored, synthesized, reusable domain specific models.

7 Deployment Architecture

The complete synthesized platform specific application model contains both the structural configuration and the behavioral semantics of all the constituent components including their interconnecting bindings and the component framework constraints. That model represents the functional view of the application, however, it does not specify how the application is distributed on the available hardware nodes and how it starts. Therefore, the distributed configuration must be modeled, too. Although the deployed component configuration, in theory, is an instance of the complete synthesized platform specific application GME, unfortunately, only allows two layer model-metamodel relationships, so multi-layer model-metamodel relationships (three layer in our case) are only realizable with the help of References as it is shown in Figure 11. The model, which contains the complete synthesized platform specific application model and the initial configuration information of the components is called the total synthesized platform specific distributed application model.

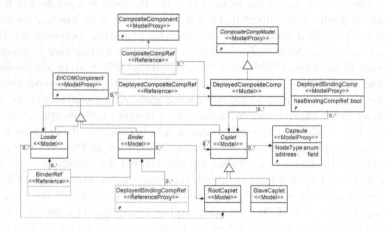

Fig. 11. Deployment configuration metamodel

From the point of view of model based development, the most important element of the deployment is the Deployment Tool, which establishes a soft real-time synchronization loop between the GME model repository and the running application. The schematics of Deployment Tool based reconfigurability is shown in Figure 12. The Deployment Tool analyzes the initial component configuration of the total synthesized platform specific application model and creates the ErlCOM elements by relying on the ErlCOM API. After the initial deployment has been completed the application starts running and ErlCOM's CRTK constantly observes any changes of the component configuration and sends notifications to the Deployment Tool accordingly. The Deployment Tool visualizes the actual component configuration of the running system by updating the total synthesized platform specific RUNES application model in GME.

The current implementation of the Deployment Tool is based on a GME Add-on running in its own thread, connects to a UDP/IP socket in order to be able to receive notifications and it relies on ErlCOM's CRTK for component management. ErlCOM's innovative architecture supports the following modes of reconfigurability:

- **Re-active component reconfiguration:** The application's control logic decides how to reconfigure the currently deployed component configuration in order to adapt it to dynamically changing environmental factors. GME only tracks the changes; the control intelligence lies inside the application code. The decision-making is based both on the component configuration graph and the current execution state of the application.
- **Pro-active component configuration:** GME continuously evaluates the actual component configuration of the deployed application and decides when and how changes should be carried out. The intelligence is incorporated either in one of the plug-ins of GME or in any legacy tool connected to GME via a versatile XML/XSLT importer facility. The decision-making is only based on the component configuration graph of the application.
- **Component behavior change:** The previous two reconfiguration types take effect only on the component configuration granularity, but the functionality of the component is kept unchanged. However, GME knows about the total synthesized platform specific model database, therefore, the programmer or any intelligent plug-in can modify the semantics of any of the components even dynamically. Via model-driven code generators, automatic deployment and the dynamic code change feature of Erlang the functionality of the relevant parts of the application can be changed on-the-fly without ever touching the current component configuration graph. Both the intelligence and the decision-making lie in this case inside GME.

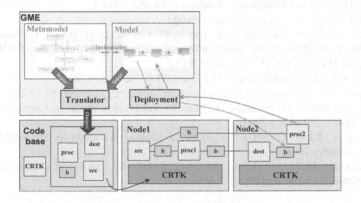

Fig. 12. Deployment Tool based reconfigurability of running component application

8   Conclusion

The paper proposed a software development process for reconfigurable networked systems. The evaluation of the software development process and the related technology enabled us to gather enough experiences to be able to summarize the most important advantages of the technique. The observed advantages are numerous; however, the most important one is that models and metamodel provide analyzable formal specifications, therefore, early validation and verification are possible semi-automatically. Moreover, applying the MIC philosophy available legacy system components can be easily integrated into the application and/or platform through the meta-programable development environment.

All in all, the described process and modeling technology seem to be highly beneficial in practical distributed application development in reconfigurable networked systems. The only major remaining challenge is related to the formal validation and verification of the transformations since the currently introduced technology works correctly only in the hands of talented and crafted software designers. Thus, having the framework put in place its accurate applicability will be the focus of our further research efforts in the future.

References

1. RUNES IST Project, http://www.ist-runes.org/
2. Karsai, G., Sztipanovits, J., Ledeczi, A., Bapty, T.: Model-integrated development of embedded software. Proceedings of the IEEE 91, 145–164 (2003)
3. Krüger, I.H., Mathew, R.: Component synthesis from service specifications. In: Leue, S., Systä, T.J. (eds.) Scenarios: Models, Transformations and Tools. LNCS, vol. 3466, pp. 255–277. Springer, Heidelberg (2005)
4. Batori, G., Theisz, Z., Asztalos, D.: Robust reconfigurable erlang component system. In: 11th Erlang User Conference, Stockholm, Sweden (2005)
5. Batori, G., Theisz, Z., Asztalos, D.: Configuration aware distributed system design in erlang. In: 12th Erlang User Conference, Stockholm, Sweden (2006)
6. Erlang/OTP, http://www.erlang.org/
7. Ledeczi, A., Maroti, M., Bakay, A., Karsai, G., Garrett, J., Thomason, C., Nordstrom, G., Sprinkle, J., Volgyesi, P.: The generic modeling environment. In: Proceedings of WISP'2001, Budapest, Hungary, pp. 255–277 (May 2001)
8. Information Society Technologies IST Event 2006 (November 2006), http://ec.europa.eu/information_society/istevent/2006/
9. Batori, G., Theisz, Z., Asztalos, D.: Platform independent model and repository. RUNES Deliverable 2.3 (January 2007)
10. Agrawal, A., Karsai, G., Shi, F.: Graph transformations on domain-specific models. ISIS-03-403 (November 2003)
11. Hailes, S., Hanna, L., Asztalos, D., Batori, G.: Small scale deployment specification. RUNES Deliverable 7.3.1 (January 2006)

A Modelling Method for
Rigorous and Automated Design of
Large-Scale Industrial Systems

Kari Leppänen, Sari Leppänen, and Markku Turunen

Nokia Research Center,
P.O. Box 407, 00045 NOKIA, Finland
{kari.j.leppanen, sari.leppanen, markku.turunen}@nokia.com
http://www.nokia.com/research

Abstract. Compositional architecture-driven and model-based system
design holds huge potential to increase design efficiency and improve design quality for large-scale industrial systems. Transition to such design
paradigm is hampered by the lack of domain-specific methods and tools
that give adequate support for both behavioral and structural modeling
and development automation. This paper introduces an enhancement
to Lyra, a rigorous service-oriented modeling method for the design of
communicating distributed systems that brings process algebraic thinking into industrial system specification with particular focus on behavior. This enhancement offers a sound basis for implementing the ideas of
MDA in automation of system design, functional verification and conformance testing. The Lyra method and its enhancement are exemplified
using UML2 to model a critical and complex part of the mobile WiMAX
wireless system.

Keywords: model-based system design, MDA, UML2, design automation, formal methods.

1 Introduction

The trend in industrial product development is towards ever more complex functionalities that are often distributed between multiple processing nodes. Traditional product creation paradigm puts a lot of emphasis on fast implementation
of the needed functionalities, with little consideration how to maintain overall
consistency over the component implementations. This has led to problems such
as time and cost overruns because of prolonged testing and debugging periods,
low quality due to difficulty of removing errors from complex implementations,
difficulty in making changes and new configurations, lack of reuse, and difficulty
in tracking product requirements. As an alternative to this implementation-
driven mode of development, architecture-driven product development based on
modeling offers a promising way to tackle these problems. This is partly due
to the fact that in model-based development the testing and verification can

G. Engels et al. (Eds.): MoDELS 2007, LNCS 4735, pp. 331–345, 2007.

cover the whole product creation cycle and not be limited to just the last hectic phase, and partly to the possibility to have an unbroken chain of rigorous machine-readable descriptions of the system from the first requirements to final implementation.

The chain of models represent system specifications at various levels of abstraction. To establish strong linkage between these specifications, and ultimately between the specifications and final implementations, they should include rigorous and verifiable definitions of system behavior for critical control points. Natural such control points are the system interfaces. They allow ensurance of conformance with respect to externally observable behavior of the system using various verification and testing methods.

The advance of modeling languages and methods opens up new possibilities for the deployment of model-based development and formal methods in industry. *The Lyra method*[3] is a domain-specific design method that incorporates the ideas of model-based development and process algebraic specification into existing practices of industrial system development. Lyra has been tested and streamlined in several industrial modeling pilots, which have shown the large potential of modeling and formal methods in industrial settings. These pilots have also shown that strong automation is both possible and crucial in industrial environment. Automation allows hiding of the unnecessary details in modeling or the underlying formal framework, and remarkably speeds up the process of creating executable and verifiable specifications. It was mostly this ambition towards automation and enhanced industrial applicability that drove us to enhance Lyra. In the following we will present the enhanced Lyra method and demonstrate the use of it with the WiMAX mobile broadband system.

This paper is structured as follows: Section 2 describes the Lyra design flow for specification of behavior. Section 3 introduces the additions to Lyra for rigorous structural definitions. An example case, using the enhanced method in the design of the WiMAX system, is described in section 4. Finally, the learnings and identified future work items are covered in Section 5.

2 Lyra Design Flow for the Specification of Behavior

The Lyra method[3] is a service-oriented rigorous design method that incorporates the ideas of model-based development and process algebraic specification (see e.g. [1,2]) into existing practices of industrial system development. Lyra is especially suitable for designing distributed reactive systems with asynchronous communication, like telecommunication systems and protocols. The design flow of Lyra focuses on the definition of system behavior and it is based on the concepts of decomposition, step-wise refinement and preservation of the externally observable behavior. The system behavior is modularized and organised into hierarchical layers according to the type of behavior. The main categories of behavior are internal and externally observable behavior. The internal behavior types include the control logic and internal computation. The externally observable behavior is further categorized according to the related communication and interface types (provided/used service interface). Strict categorization and labeling of behavior aims at facilitating automated refinement and compositional verification and testing of industrial-scale systems with enormous size and complexity. Embedding formal methods into industrial development process, as an inherent and invisible part of it, has been a main driver for the development of Lyra.

The Lyra method can be considered as an implementation of *MDA (Model Driven Architecture)*[15]. It refines the process framework set by *UP (The Unified Process)*[11] and *RUP (Rational Unified Process)*[10] into a domain-specific, and even company-specific, method. It adopts relevant ideas from the earlier approaches in the domain of distributed communicating systems, like *SDL design method*[6] and *ROOM (Real-time Object-Oriented Modeling)*[7,8]. All this, and the process algebraic specification approach, are combined with the existing industrial design practices, and consolidated into a harmonized rigorous design flow. The method is language independent, but used mostly with the UML2[16]. The subset of UML2 modeling concepts used and extended by Lyra has been defined in a Lyra UML2 profile, see e.g. [4]. In this work we have used *Telelogic Tau G2*[14] as the modeling tool.

This section describes the Lyra design flow, introduced in the original Lyra method[3]. It consists of four main phases, each of which focuses on either definition or refinement of a certain aspect in the system behavioral specification. In Lyra, all system behavior is encapsulated in services that are building blocks of either systems or larger services. Construction of systems from these building blocks will be described later. The overall picture of the Lyra design flow is presented in Figure 1.

2.1 Service Specification

The valid behavior of a service as observed by the users of that service is specified in the Service Specification phase. This specification for externally observable behavior consists of both static interface definitions and dynamic behavioral specifications. First, *Provided Service Access Points (PSAPs)* are created for user-service communication, usually one for each service. Next, logical interfaces are defined for communication with various user types and attached to the relevant PSAPs of the service. The interface definitions encapsulate the service messages and their parameters used for the communication between the user and the service. Finally, the valid dynamic behavior of the service as observed by the users (i.e. the PSAP interface behavior) is defined by a single *PSAP Communication* state machine. If independent state machines would be needed for different users, the service should be broken into multiple services in such a way that the PSAP behavior of each can be specified by a single state machine. State machines are used as the primary behavioral specification as they unambiguously and rigorously specify the valid behavior of the service.

Fig. 2 shows an example of a PSAP Communication state machine. State transitions (arrows) are tied to triggers, usually receptions of service requests,

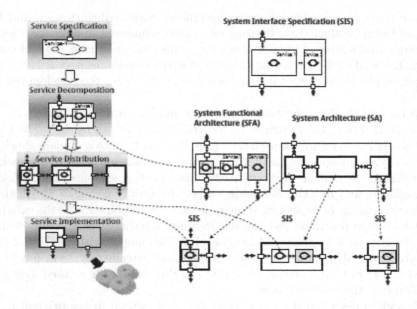

Fig. 1. Lyra design flow: behavioral specification flow (left) and structural specifications (right)

such as *RNG_REQ()*. Transitions can also cause actions, like sending responses (e.g. *RNG_RSP*) back to the user. Typically. the states in a PSAP Communication state machine are composite states (in UML2 terms), i.e. contain substate machines. Hierarchical state machines allow separation of the externally observable behavior from the internal behavior. As we do not specify the internal behavior in Service Specification phase, we replace these parts of the behavior with indeterministic substate machines. This allows already the Service Specification model to be executable. Stochastic models can be used to emulate the anticipated internal behavior for early simulations and analysis.

2.2 Service Decomposition

In Service Decomposition phase we specify how the externally observable behavior of a service is realized by its internal behavior. The internal behavior specification consists of *Execution Control* and *Internal Computation* state machines. Execution Control state machine specifies the order and logic of running internal functionalities of a service (service components). Internal Computation state machine encapsulates the internal algorithms and calculations. One more layer of behavioral hierarchy is still needed in those services that use other services. That is defined by *USAP (Used Service Access Point) Communication* state machines that specify how the service communicates with the external service providers. Both Internal Computation and USAP Communication are triggered by the Execution Control state machine. The most important driver

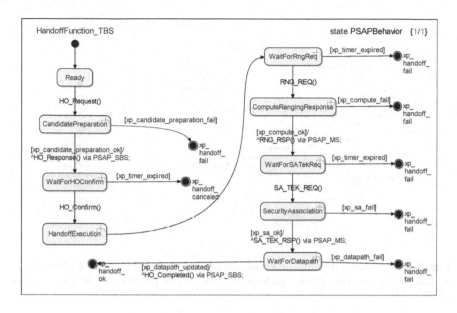

Fig. 2. An example of a PSAP Communication state machine

for such strong layering of behavior has been the intention to automate behavioral verification, as mentioned earlier.

The first step in defining the internal behavior is the decomposition of the service into smaller units, or *service components*. Decomposition is described in a *Functional Decomposition* diagram, which is a stereotyped use case diagram in UML2 (see Fig. 3.a). That shows the service components that the service consists of as well as their mutual relation. As an example, in Fig. 3.a the service component *HandoffExecution* itself consists of three components: *DatapathRegistration*, *RetrieveKeys* and *UplinkMap*. The Execution Control state machine for *HandoffExecution* (Fig. 3.b) has the same service components and specifies the logic and order how they are triggered. The decomposition of the service into smaller components continues recursively until the required level of atomicity is reached. For example, if the service will be distributed into multiple nodes (subsystems), there is no need to decompose further service components that will stay intact in all foreseeable distribution configurations.

2.3   Service Distribution

The Service Decomposition model defines the functionality (behavior) of a service by breaking it into service components and specifying how they are triggered. How the service is distributed into different logical or physical entities (subsystems) has been of no concern. The purpose of Service Distribution phase is to define this distribution. It requires that Service Decomposition for the given service exists and that the System Architecture (section 3.3) has been defined for

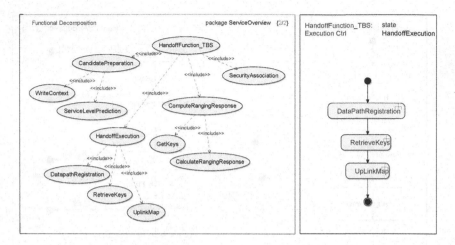

Fig. 3. a. Functional decomposition diagram of the *HandoffFunction_TBS* service (left). b. Execution Control state machine for the *HandoffExecution* service component (right).

the system where the service resides. Service Distribution consists of the following steps: creation of a new service for each subsystem where the original service will distribute into, assigning the service components into these new services, definition of the peer interfaces for the communication between the distributed services, and finally generation of the state machines to specify the behavior for peer communication (following the PSAP/USAP pattern introduced earlier). The new set of services is specified in such a way that the composition of that set is behaviorally equivalent to the original service specification.

2.4   Service Implementation

The previous Lyra steps have produced virtual specifications for the service interfaces. That is, all the external communication of the services is based on the specified real behavior but virtual representations of the signals. The purpose of the Service Implementation phase is to map the virtual signal representations into the real communication mechanisms used in the implementation (or specifications thereof). In its simplest form, this involves a signal mapping using an adapter element in the model. A more complicated situation arises often in communication systems when the virtual communication between peer entities needs to be routed through the underlying communication layers. Also then an adapter (or peer proxy) that maps the virtual signal representations into a form understandable by the underlying communication layer can be made. Such adapters open up good opportunities for automated testing of implementations against their (virtual) specifications, or even automatic generation of code from detailed specification models. The generation of such adapters is itself a process that can be mostly automated. This makes the mapping of specification models into different implementations, and thus realization of virtualization, efficient.

2.5   Automation of Service Design Flow

We have developed a set of wizards to automate the routine operations in creating Lyra models. When initiating a new phase in the service design flow, the *New Service Specification* wizard creates a complete model structure for a new service. The *New Service Decomposition*, *New Service Distribution* and *New Service Implementation* wizards create a transition path from an earlier development phase by creating a working area with a new set of refined model elements for the new phase. The refined model elements are generated according to patterns defined by the Lyra design flow, and by using the definitions (rules, constraints) described in the Lyra profile (see e.g. [5]). This allows refinement of a service or service component without modifications to the earlier, more abstract specifications. It also provides means for preserving the consistency between the Lyra phases, particularly when enhanced with the formal verification approaches, such as automated refinement producing models correct-by-construction or automated consistency checking at the end of the specification phase.

The wizards at the lower levels of hierarchy correspond to the design tasks inside the main Lyra phases. They create or update smaller parts of the phase-specific model. For example, the *New Execution Control* and *New USAP Communication* wizards create new hierarchical state machines for the behavior.

3   Lyra Enhancement for the Specification of System Structure

The Lyra design flow was primarily developed for rigorous and systematic modeling of system behavior. However, an equally systematic and rigorous approach for the description of system structure is required for full-scale design automation and efficient reuse. The systematically encapsulated behavioral components are easy to reuse in different systems and system configurations. Similarly, systems or parts of systems should be reusable.

We have developed an enhancement to Lyra (Lyra 2.0) that provides a consistent model structure both regarding the system behavior and the system structure. Adhering to a stable and systematic model structure has major advantages: it allows automation of tools to help generate models faster as well as development of view and document generators that work similarly on any model. Also, post-processing of the models for code generation as well as generation of analysis models and testware can be easily automated.

Lyra 2.0 preserves the basic requirements and goals of Lyra for the specification of systems. First, recursive top-down system specification has to be possible. Second, compositional mode of development and efficient reuse, where systems can be built from existing system or service definitions, has to be supported. Third, the enhancement has to allow scalability for truly large systems being developed by a multitude of teams in parallel. Lastly, it should be possible to separate three different kind of specifications for a system: external behavior, internal behavior and internal structure. In Lyra 2.0, these specifications are, in the same order, *System Interface Specification*, *System Functional Architecture* and *System Architecture*. In the following, the term *system component* is used to refer to the specification of any structural part of a system. The system components always appear similar regardless of whether they are at the top of the overall system hierarchy or at the very bottom.

3.1   System Interface Specification

The externally observable behavior of a system component is specified by its System Interface Specification (SIS). This specification is necessary for integrating a system component with the other system components and with its environment. SIS is usually given as input to the team responsible for the further development of this system component.

The behavioral components resulting from all phases of the Lyra design flow are used as building blocks when defining systems and system components. For example, some system components may have only a Service Specfication level of description, while the other ones may be results of Service Decomposition and Service Distribution phases. The systems may include also components, which have not been implemented according to the Lyra design flow, like COTS (Commercial-off-the-shelf) or legacy components and open source software. To include also such components into the overall system model, they are encapsulated as system components with SIS. Possibility to use a mixture of system components at different levels of abstraction allows early simulations and early start for systems pre-integration in a truly compositional development mode.

The SIS of each system component includes the specification of behavior for all its interfaces (which are attached to the system ports in UML2). To this end, SIS is defined as a composition of services that are appropriately connected to the system ports. A system interface specification is the composition of the interface specifications of all the services that communicate through that system interface. The systems ports consist of PSAP and USAP ports of the included servces, but the system ports do not have such categorization.

3.2   System Functional Architecture

System Functional Architecture (SFA) presents an encapsulation and view for the system as a composition of its services and internal functionalities (defined also as services in the Lyra design flow). Composite behavior of the included service components and internal functional components defines the emergent behavior of a system component.

SIS is the starting point for the definition of the SFA. Since the SIS is a more abstract specification for the system, it allows several SFAs for implementation. The consistency between the SIS and SFA has to be preserved with respect to the alphabet and externally observable behavior.

3.3 System Architecture

The third definition for a system component is needed when the system component itself consists of (sub)systems. System Architecture (SA) defines formally how the system is composed of its subsystems and how the subsystems are interfaced. Whereas SIS and SFA are compositions of service components, SA is a composition of system components. It is the formal SA that allows recursion in system design and ties together all system specifications at different levels of abstraction.

3.4 Recursive System Design Process

A typical cycle in recursive system design process (see Fig. 1) begins by the team responsible for system X development take the SIS of that system component as input. Unless the system is at the root of system hierarchy, i.e. not a subsystem of any larger system, the SIS has been generated as the output of the previous cycle. For the top-most system component, creation of the SIS is the very first step.

Next, the system X team generates the SFA for their system component. This typically is derived from the SIS by refining the services further and possibly defining a set of new services that are internal to the system component. If the services in SIS are found as an inadequate starting point, a new set of services can be defined by the team as long as consistency with the SIS external behavior is maintained. This may be, for example, due to need for using legacy or open source softwware as part of the system implementation. Note that the team will not be able to unilaterally change the SIS as that is the key for successful interfacing with the surrounding system components.

Parallel to the SFA development, also the System Architecture is developed. It is possible to have multiple SA variants that correspond e.g. to differing product configurations. The final step in the recursive system design cycle is executing the Service Distribution phase of the Lyra design flow, where the SFA services are distributed into the SA subsystems. The outcome of this is the SIS for each new system component. The SA generated this way is the integration model for system X, i.e., it specifies how the subsystems of X operate together.

Note that this description applies to a pure top-down system development flow for the sake of clarity and simplification. In practice, system design is a mixture of top-down and compositional mode, due to e.g. existing legacy. Also, the development of the SIS, SFA and SA, as well as different layers of architectural abstraction (e.g. systems and implementation architectures), always overlap in practical work.

3.5 Automation of System Design Flow

The implementation of wizards for the system component design flow follows the same approach used for the service design flow. There are top-level wizards that participate in the initiation of a new phase in the system component design flow.

The *New System Interface Specification*, *New System Functional Architecture* and *New System Architecture* wizards prepare transitions paths from one design phase to another. These are the largest wizards, creating several UML packages and classes for the Lyra concepts.

A second set of wizards create or update a smaller portion of a model within one development phase. The *New System User* and *New System Interface* wizards specify the system SAPs and interfaces for users. Finally the *New System Service* wizard specifies system services, thus making a link between the service and system component design flows. In order to provide visibility for the Lyra concepts, the Tau tool was extended with Lyra specific views using view generators. They show model elements grouped according to the relationships between the Lyra concepts. The *Service View* shows the specified service components and the outcomes of the refinement phases. The *System View* shows all the three possible specifications (SIS, SFA and SA) for the system components. Finally, the *System Tree* view shows the system structure, i.e. the hierarchical decomposition of the system into (sub)systems, in the form of a tree. In principle, each system designer could customize his/her own view generators that would work on all Lyra 2.0 models.

4   WiMAX Design Example

This example is part of a larger case study that was done to demonstrate the suitability of the enhanced Lyra for the functional specification of a real industrial distributed system. Also the applicability of a service-oriented approach to quite complex control functionalities was of interest. The purpose was to make an executable specification for product implementation purposes that could also be validated functionally by way of simulations. This exercise was part of a real system specification effort in a product program, and was done in close cooperation with the system architecture and specification team.

The target system for the study was a mobile version of WiMAX [17]. WiMAX is a broadband wireless communication system, that originally was developed to replace the last-mile wired broadband connection to Internet. The logical architecture of mobile WiMAX contains the following nodes: mobile station, base station, gateway, home agent and some other nodes that are of no interest to the case study because of their functionality.

To limit the scope of the case study, only the functionalities related to mobility, and especially those functionalities needed for handover, were modeled. Handover means here the sequence of events that the network has to master to change the serving base station of a given WiMAX terminal to another one. As one the most critical and complex functionalities of any mobile network, it meets well the goals set for this case study. To limit the scope of this example, only the functionality in the mobile station and the two base stations involved in the handover is addressed, omitting a lot of details. The case study was done using Telelogic Tau G2 UML2 tool and the Lyra profile developed in Nokia Research Center.

We dubbed the system to be specified (modeled) *WimaxMobility*. It is not a physical component of WiMAX, but can be considered a logical subsystem (encapsulation of functionality) that is distributed in many nodes of the complete WiMAX system. The first step would be to create the System Interface Specification (SIS) for *WimaxMobility*. However, somewhat atypically, *WimaxMobility* is a closed system with no communication with the external world and therefore cannot have a SIS. We will start by defining the Functional Architecture for *WimaxMobility*. Creating of a SIS will be illustrated for one of the subsystems of *WimaxMobility*.

4.1   Functional Architecture for *WimaxMobility*

The WiMAX standard specification [17] defines three primary mobility functions. *HandoffFunction* has the overall control of the handover process. *Context-Function* manages the context information in the network and, for example, moves the context information from source to target base station. *Datapath-Function* modifies the required routing tables keeping the data flows to and from the mobile node correctly routed as it moves in the network. *RRM* refers to Radio Resource Management, and its purpose in this limited example is to make a decision whether the target base station is able to accept a new mobile station to be served.

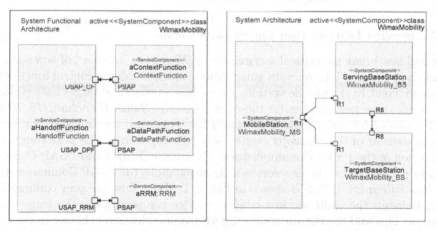

Fig. 4. *WimaxMobility* system. a) Functional Architecture (left). b) System Architecture (right).

The Functional Architecture of *WimaxMobility* is shown in Fig. 4.a (a composite structure diagram in UML2). HandoffFunction recognizes the need for handover, makes the decision to start it, controls its execution and uses three other services for certain handover tasks. It is clearly the most critical service of *WimaxMobility* and one that has to be distributed both to mobile station and to base station. The three others will only be visible in the base station, and at this stage of system development a definition at the level of Service Specification is adequate for them. As the *DataPathFunction* was available as an legacy implementation, there was no need to refine its definition beyond the Service Specification level at all. Instead of modeling existing legacy, another possibility would have been to encapsulate the implementation as a model component.

4.2   System Architecture for *WimaxMobility*

The logical nodes of the WiMAX network can be considered as the platform on which we have to implement the *WimaxMobility* system. With this in mind, we break the *WimaxMobility* into two subsystems *WimaxMobility_MS* and *Wimax-Mobility_BS*, for the mobile station and base station, respectively. Fig. 4.b shows the System Architecture for *WimaxMobility*. Note that it has one instantiation of *WimaxMobility_BS* for the source and a second for the target base station. As these are rather base station roles that change in the handover, defining a separate subsystem for both target and source base station would make little sense.

Note that the System Architecture so defined can not yet be used for system simulation, since at this point the behavior of the subsystems is not defined by way of services. Defining these is our next job.

4.3   Service Distribution for *HandoffFunction*

We will now break the critical service, *HandoffFunction*, into a set of new services that can be placed into relevant subsystems. As the handover control functionality is different in the mobile station, source base station and target base station, we define three new services for these, respectively: *HandoffFunction_MS*, *HandoffFunction_SBS* and *HandoffFunction_TBS*. This begins with the definition of the placement of the *HandoffFunction* service components. The next step is the definition of the PSAP Communication, Execution Control and USAP Communication behavior for the new services. As an example, the PSAP Communication of *HandoffFunction_TBS* is shown in Fig. 2. It shows how the peer communication between the child services triggers service components in the target base station. This and the corresponding USAP Communication behavior in the triggering service define a part of the peer communication protocol between the source and target base station.

4.4   System Interface Specification for *WimaxMobility_BS*

From now on, the two subsystems are considered full-fledged systems. Next we show how to construct the SIS for the *WimaxMobility_BS* (Fig. 5). As this system can take the role of both source and target base station, its SIS contains the corresponding distributed versions of the *HandoffFunction* service. Note that only these two services are directly connected to the system ports of *Wimax-Mobility_BS*. The three other services, *DataPathFunction*, *ContextFunction* and *RRM* are included in the SIS, because the preceding definition of *HandoffFunction* (see Fig. 4.a) already used them, and their omission would have required abstracting away already defined behavior.

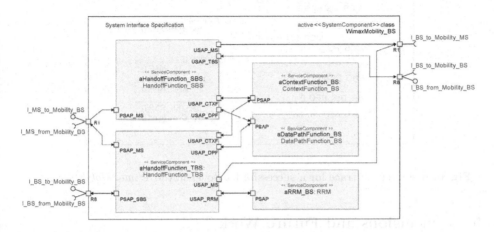

Fig. 5. System Interface Specification for the *WimaxMobility_BS* system

Fig. 5 also indicates how, for example, the R1 interface specification between a mobile station and a base station is composed. First, it is the composition of, at least, the three system interfaces of *WimaxMobility_BS*: *I_MS_to_Mobility_BS*, *I_MS_from_Mobility_BS* and *I_BS_to_Mobility_MS*, which are further compositions of the corresponding service interfaces. As an example, interface *I_MS_to_Mobility_BS* is a composition of two service interfaces, *I_MS_to_HandoffFunction_SBS* and *I_MS_to_HandoffFunction_TBS*. As these contain both interface alphabet and behavior, the logical system interface R1 is fully specified compositionally. Of course, full R1 specification requires the composition of all the systems that use R1, not only *WimaxMobility*. The model elements that are required to form the interface compositions can be automatically generated based on the information in the relevant System Interface Specification(s).

4.5   System Integration

Once an executable SIS has been defined for each subsystem, the System Architecture of *WimaxMobility* can be made executable with references to the subsystem SISs. The SA can then be used to validate the subsystem integration. As an example, a simulation trace in Fig. 6 shows the communication between the subsystem instances for a fully successful handover execution. Since the behavior of each subsystem if formally defined based on process algebra, also more sophisticated methods could be used to validate correct system behavior.

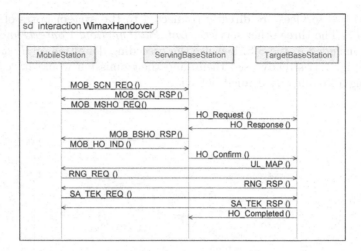

Fig. 6. A simulation trace for a successful handover in the *WimaxMobility* system

5 Conclusions and Future Work

We have introduced an enhancement to Lyra, a rigorous service-oriented modeling method for the design of communicating distributed systems. The driver for the enhancement has been automation of the industrial strength development process. Automation is necessary to get a wider acceptance for model-driven development and formal methods in industrial settings. In this paper we have also illustrated the applicability of model-based design, incorporating formal approaches, to the design of industrial systems with the design of critical control functionality in the WiMAX mobile wireless system.

Even though many powerful ideas of automation have not yet been implemented, already the first experiments with automation wizards have shown significant speed-up in the specification of complex systems. The first wizards take care of many of the mundane tasks related to creation of new services and system components, or building simulators, and allow the modeler to focus on the system structure and behavior, rather than the details of the model. The lack of such automation has been a significant obstacle in industrial projects up to date.

In future, a lot of work remains in further automation. As an example, the definition of service components and their relations during the Service Decomposition and Service Distribution phases takes a lot of manual labor, but have excellent potential for automation: the Execution Control state machines, which orchestrate the parallel execution of the service components, can be automatically created based on the information in the Functional Decomposition diagram, instead of specifying such state machines manually.

Another important area of automation is the consistency checking and functional verification of models, where good progress has been done e.g. in the

RODIN project [9] (see e.g. [5,12,13]). The project has used the Lyra design flow and the Lyra profile as a basis for academic research and development of automated refinement, fault-tolerance and plug-in verification tools. Finally, automatic checking of implementation conformance is an important current and future work item to close the circle for fully architecture-driven product development.

References

1. Hoare, C.A.R.: Communicating Sequential Processes. Prentice Hall, Englewood Cliffs (1985)
2. Milner, R.: Communication and Concurrency. Prentice Hall, Englewood Cliffs (1989)
3. Leppänen, S.: The Lyra Design Method. Technical report, Technical University of Tampere (2005) ISBN 952-15-1464-7, ISSN 1459-417X
4. Leppänen, S., Ilic, D., Malik, Q., Systä, T., Troubitsyna, E.: Specifying UML Profile for Distributed Communicating Systems and Communication Protocols. In: Proceedings of the Workshop on Consistency in Model Driven Engineering (2005)
5. Ilic, D., Troubitsyna, E., Laibinis, L., Leppänen, S.: Formal Verification of Consistency in Model-Driven Development of Distributed Communicating Systems and Communication Protocols, Technical Report (enhanced version submitted into a journal), Turku Centre for Computer Science (2006) ISBN 952-12-1749-9
6. Ellsberger, J., Hogrefe, D., Sarma, A.: SDL Formal Object-Oriented Language for Communicating Systems. Prentice Hall, Englewood Cliffs (1997)
7. Selic, B.: Tutorial: real-time object-oriented modeling (ROOM). In: Real-Time Technology and Applications Symposium, IEEE, Los Alamitos (1996)
8. Selic, B., Gullekson, G., McGee, J., Engelberg, I.: ROOM: an object-oriented methodology for developing real-time systems. In: Fifth International Workshop on Computer-Aided Software Engineering (1992)
9. EU project IST 511599 RODIN: Rigorous Open Development Environment for Complex Systems (2007), http://rodin.cs.ncl.ac.uk/
10. Kruchten, P.: The Rational Unified Process: An Introduction. Addison-Wesley, Reading (2000)
11. Jacobson, I., Booch, G., Rumbaugh, J.: The Unified Software Development Process: An Introduction. Addison-Wesley, Reading (1999)
12. Laibinis, L., Troubitsyna, E., Leppänen, S., Lilius, J., Malik, Q.: Formal Model-Driven Development of Communicating Systems. In: Lau, K.-K., Banach, R. (eds.) ICFEM 2005. LNCS, vol. 3785, Springer, Heidelberg (2005)
13. Laibinis, L., Troubitsyna, E., Leppänen, S., Lilius, J., Malik, Q.: Formal Service-Oriented Development of Fault Tolerant Communicating Systems. In: Butler, M., Jones, C., Romanovsky, A., Troubitsyna, E. (eds.) Rigorous Development of Complex Fault-Tolerant Systems. LNCS, vol. 4157, pp. 187–261. Springer, Heidelberg (2006)
14. Telelogic TAU G2 documentation (2007), http://www.telelogic.com/
15. Model Driven Architecture (2007), http://www.omg.org/mda/
16. UML 2.0: Superstructure, version 2.0. Object Management Group (2005)
17. WiMAX End-to-End Network Systems Architecture (Stage 3: Detailed Protocols and Procedures). WiMAX Forum (2006)

Relating Navigation and Request Routing Models in Web Applications

Minmin Han[1] and Christine Hofmeister[2]

[1] Amazon.com
[2] Lehigh University
minmin@amazon.com, crh5@lehigh.edu

Abstract. A navigation model describes the possible sequences of web pages a user can visit, and a request routing model describes how server side components handle each request. Earlier we developed formal models and analysis operations for such models. While each is useful independently, their utility is greatly improved by relating the models, which is the contribution described in this paper. We describe mappings between the models, and show that the mappings preserve navigation behavior and are bijective, thus supporting traceability and allowing the models to be used in round-trip engineering. With these mappings built into our Model Helper tool, it is now possible to automatically determine whether a Request Routing model conforms to the navigation design, and to automatically generate a Request Routing model from a navigation model. Finally, we describe one of a number of case studies where we used Model Helper in a round-trip engineering scenario.

1 Introduction

The *navigation* of a web application is the possible sequences of web pages a user can visit. For simple cases, the next page displayed depends only on which button or link the user selects in the current page. But web applications often use *adaptive navigation*, where the next page may also depend on: the user's mode, for example whether they are a customer or an administrator (mode-adaptive navigation); what pages the user has visited previously (history-sensitive navigation). Adaptive navigation is more common for applications with dynamic page content (sometimes called "dynamic web applications," but can also be used with static web pages.

Navigation is a very important part of a web application, but it is usually described informally in a diagram such as Fig. 1. Boxes are web pages, and arrows are navigation links between pages, with text describing the circumstances under which the navigation link is taken. From a diagram like this we can see that when a user is at the Home page, is logged on, and wishes to view their account, they next see the Account page. But it is problematic to determine whether the user must always be logged on in order to view the Account page. The link from the Home page explicitly states this, but the link from the Order page does not, so we must examine all links that reach the Order page. Clearly, as the number of pages and links grows, this becomes problematic with an informal diagram.

G. Engels et al. (Eds.): MoDELS 2007, LNCS 4735, pp. 346–359, 2007.

On the other hand, some web engineering approaches include formal navigation models, but these other approaches do not explicitly model adaptive navigation ([1] [4] [12]), thus cannot support automated analysis to answer questions like the one posed above.

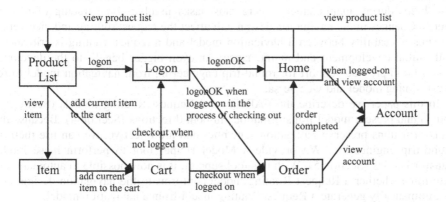

Fig. 1. Typical Description of Navigation

We earlier provided this with FARNav (a Formal Approach for Rich Navigation) [7]. A FARNav model uses statecharts to precisely describe navigation, including adaptive navigation. We use an existing general-purpose tool (SMV model checker) to analyze the model [3]. To use this tool, we translate the FARNav model into the input format needed, then write rules describing the desired properties of the model. The tool determines which rules hold and which are violated.

A navigation model describes how the application should behave, so it is the developer's responsibility to provide an implementation that conforms to the navigation model. With a platform such as J2EE, each navigation link first results in a request received by the server. The web application processes the request using a set of server components, then prepares a response page that is sent back to the user, which completes the navigation link. Thus for each navigation link the developer must determine which server components process it, in what order, and which response page will be returned. We call this a *request route*, the sequence of components that handle a request originating from a web page. The *request routing* is the union of all possible request routes from all web pages, so it provides implementation information for an application's navigation.

Unfortunately, identifying request routes with a platform such as J2EE is difficult. Each component along the request route participates in the request routing by checking and/or setting the associated URI, so this code is scattered across a number of components. In addition, although the request handling is in effect a pipeline, communication between server components is indirect, with web.xml mapping a request between components using its URI. This makes the tracing of request routes a complicated and error-prone task.

To support the developer in understanding request routing, we earlier developed a formal Request Routing model to explicitly represent the request routing of a web application, and we provide operations for analyzing the request routing [6]. The model and operations are specified with Z, and Jaza is used to read in a model and perform operations on it. No prior approach explicitly represents and/or supports the tracing of request routes.

The ability to see and analyze a request routing model can help the developer determine whether the implementation conforms to the desired navigation behavior. It also helps during maintenance, where most tasks involve some tracing of request routes. Changes in the navigation design will affect the request routing and vice versa.

Thus traceability between a navigation model and a request routing is critical for both initial development and maintenance. An even bigger benefit to the developer would be to provide support for round-trip engineering from navigation model to request routing model and vice versa.

In this paper we describe the FARNav and Request Routing models (Sections 2 and 3) and present model transformations in both directions (Section 4). Because the transformations preserve navigation semantics and are bijective, we can use them in round-trip engineering. We provide a Model Helper tool to perform these model transformations (Section 5). With this tool support, it is now possible to automatically determine whether a Request Routing model conforms to the navigation design, and to automatically generate a Request Routing model from a navigation model.

2   The FARNav Navigation Model

FARNav uses Statecharts with parallel state machines to model navigation. The main state machine is the Page Navigation state machine, which contains one state per web page, with navigation links represented by transitions between these pages. The other state machines are mode state machines, one for each mode, and one for each case of history-sensitive navigation. For the example in Fig. 1, the FARNav model consist of three state machines: Page Navigation, Logon Mode, and Logon NextPage (Fig. 2).

Page Navigation transitions may use information about the state of the modes as a guard, and may send events to zero or more mode state machines in order to change their states. Thus the mode state machines control which Page Navigation transition fires, and these transitions in turn cause a mode to change state. The format for a transition label is 'event[guard]/action':

- event: a user action; format is 'eTag.eRes', where eTag is the user action and eRes is the result of processing this action (e.g. logon.success)
- guard: navigation link applies only for these mode values; format is 'in gs_0, in gs_1,…, in gs_i' where gs is a state in a mode state machine.
- action: navigation link causes the mode in a parallel substate to change its value; format is '$a_0, a_1, …, a_j$' where a is a transition in a mode state machine.

For a mode state machine representing mode-adaptive navigation, there is one state for each possible value of the mode (Logon Mode in Fig. 2). For history-sensitive navigation, the state machine starts in a neutral state where previously visited pages are not relevant. There is an additional state for each case where a previously visited page must be remembered (Logon NextPage in Fig. 2). Transitions in mode state machines are usually triggered with events fired in the Page Navigation state machine, or with timing events such as logon timeout. All state names must be unique, across all state machines in the model.

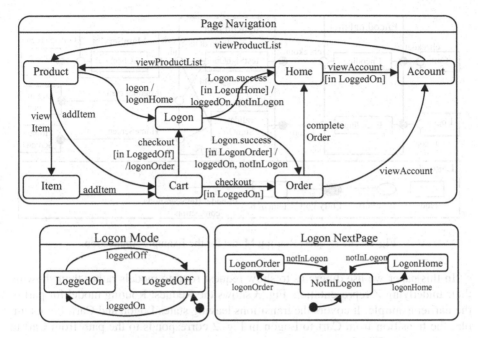

Fig. 2. FARNav Model for the Example Application

Although a navigation model such as that in Fig. 2 precisely describes the navigation, it can still be difficult to check properties of the navigation. For example, it is still not obvious that the Account page cannot be reached unless the user is logged on. So we formulate this property as a rule and use the tools to automatically determine whether the navigation model violates this rule.

3 The Request Routing Model

The Request Routing model describes a set of nodes that represent server components, which can be web pages, filters, or servlets. Each node has associated entry and exit ports. Each port has a URI, so an entry port on a node indicates that the node receives requests with the URI of the entry port, and an exit port on a node indicates that the node forwards requests with the URI of the exit port. Using URIs to route a request through server-side components is a convention of J2EE web applications.

An exit port of one node is connected to an entry port of another node when their URIs are the same (a between-node connection). Within a node, an entry port can be connected to an exit port (an in-node connection), and the in-node connection may have an associated function that processes requests arriving with the URI of the entry port. A function can be associated with multiple in-node connections of a node, and in this case each in-node connection has a different return value and exit port, or has a different entry port.

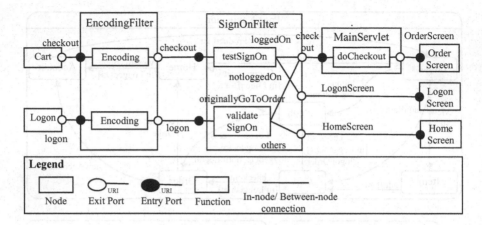

Fig. 3. The Request Routing Model for the Example Application

In this paper we use diagrams to show request routing models rather than showing their underlying Z representation. Fig. 3 shows the Request Routing model for part of the earlier example. It covers the transitions leaving states Logon and Cart. For example, the transition from Cart to Logon in Fig. 2 corresponds to the path from Cart to LogonScreen in Fig. 3.

Now we can more precisely define a request route to be a path through a sequence of ports (ExitP0, EntryP1, ExitP1, EntryP2, ExitP2, …, EntryPi, ExitPi, … EntryPn-1, ExitPn-1, EntryPn) where:

- EntryP$_i$ ($0 < i <= n$) are entry ports;
- ExitP$_i$ ($0 <= i < n$) are exit ports;
- ExitP$_0$ is attached to a request web page node;
- EntryP$_n$ is attached to a response web page node;
- EntryP$_i$ and ExitP$_i$ ($0<i<n$) are attached to Node$_i$ and are in-node connected;
- The in-node connection between EntryP$_i$ and ExitP$_i$ ($0<i<n$) may have an associated function (Func$_i$) and return value (FuncReturn$_i$).
- ExitP$_i$ and EntryP$_{i+1}$ ($0 <= i < n$) are attached to different nodes, and are between-node connected, thus the URI of both is URI$_i$.

If two request routes go through exactly the same set of ports, they must be the same request route, so the sequence of ports uniquely identifies a request route.

4   Relating the Models

The FARNav and Request Routing models both describe navigation behavior, so they are related. A transition between two states in the Page Navigation state machine of the FARNav model corresponds to a request route in the request routing model.

We wish not just to establish traceability between the models but to support round-trip engineering. With round-trip engineering the task of keeping the models synchronized is greatly simplified: if a developer changes the FARNav model, then the Request Routing model is generated or adjusted accordingly. Similarly, if the developer changes the Request Routing model, the FARNav model can be extracted from the Request Routing model.

We provide mappings to build a Request Routing model from a FARNav model (FARtoRR) and vice versa (RRtoFAR). We then show that the mappings are bijective and that they preserve the navigation-related behavior. Both of these properties are necessary for the mappings to be used in round-trip engineering. We do this first for the core part of the mappings, then extend the mappings to handle transformation operations on a Request Routing model.

Property 1: Correctness. The mapping result (a FARNav model or a Request Routing model) is semantically equivalent to the mapping source (a Request Routing model or a FARNav model) for the navigation-related features.

Property 2: Bijection. Let N be a FARNav model and R be a Request Routing model, $(\forall N, \text{RRtoFAR}(\text{FARtoRR}(N)) = N) \land (\forall R, \text{FARtoRR}(\text{RRtoFAR}(R)) = R))$.

4.1 The Core Mappings

The general idea of the core mapping from navigation to request routing is as follows. Fig. 4 shows an example of the request routing model that would be generated from the navigation model in Fig. 2. After leaving the request web page node, each request route will go through a ProcessRequest node. This is where the developer can provide code to evaluate the request, so a function is assigned to each request route. The function may have return values that split requests along two or more routes. An example is a login function that returns either 'success' or 'failure'.

At the other end of the request routes, the last node before the response page node is a PrepareResponse node. Again a function is assigned so that a developer can provide code to prepare the response page.

The nodes between ProcessRequest and Prepare-Response are created to handle the adaptive navigation. They check mode values and route a request accordingly, then set new mode values as required.

To describe the core mappings we represent the models in a tabular format. For a FARNav model, each row represents a Page Navigation transition and describes its transit-from state, event, guard, action and transit-to state (Tables 1 and 4). For a Request Routing model, we use one request route per table. Each row in the table describes how the request route travels through a node. So between-node connections are between adjacent rows (Tables 2 and 3).

CoreFARtoRR

Next we describe how the transition in Table 1 is mapped to the rows in Table 2.

Row 1: The transit-from state pageA is mapped to a node named pageA, with sequence number "1". This node has an exit port ExitP_0 with URI pageA_eTag.

Row 2: This row uses the ProcessRequest node, which is numbered 10. The entry URI is the previous row's exit URI, and the exit URI uses the same value with

'_ER_eRes' concatenated on the end. The function is named 'checkeTag', and the function value uses the event response: '_ER_eRes'.

Rows 3 through 2+i, where i is the number of guard value pairs: For guard value pair GM=gs, a row uses the node named CheckGM, and node number computed using that mode state machine's sequence number. The entry URI is the previous row's exit URI, and the exit URI concatenates on _C_gs. The function is named 'checkGM' and its value is '_C_gs'.

Rows 3+i through 2+i+j, where j is the number of action value pairs: For action value pair AM=ts, a row uses the node named SetAM, and node number computed using that mode state machine's sequence number. The entry URI is the previous row's exit URI, and the exit URI concatenates on _S_ts. The function is named 'setAM' and its value is '_S_ts'.

Row 3+i+j: This row uses the PrepareResponse node, which is numbered 9990. The entry URI is the previous row's exit URI, and the exit URI is the transit-to state pageB plus "Screen".

Row 4+i+j: The transit-to state pageB is mapped to a node named pageBScreen, with sequence number "10000". This node has an entry port $ExitP_{3+i+j}$ with URI pageB-Screen.

There is problem with assigning an arbitrary ordering for the mode state machines. This arises when a state in the Page Navigation state machine has two or more outgoing transitions. If each of the outgoing transition guards checks a different mode state machine, then the ordering of these checks does not matter. If each guard checks exactly the same set of mode state machines, again the ordering does not matter. But for the third case, where the guards check different but overlapping sets of mode state machines, the ordering is critical: the first mode checked must be the shared one. Because different sets of transitions could impose conflicting restrictions on the ordering of mode state machines, we instead "normalize" transitions for the third case, by making each transition from the same state check exactly the same set of mode state machines. Thus a transition that originally did not check a mode must be split into several transitions, one for each value of the added mode.

The Function Name column of Table 2 contains modes that this function checks or sets, and we define a set of check-tags and a set of set-tags to specify these mode state machine and states. The tags were not originally part of the request routing model since they are needed only for maintaining traceability with the navigation model. The tags are kept in a table format. A function may contain multiple check-tags and set-tags. Each tag contains an argument and a set of possible values.

Table 1. A General Page Navigation Transition

Transit-from	Event	Guard	Action	Transit-to
pageA	eTag.eRes	$GM_0 = gs_0$, $GM_1 = gs_1$, ..., $GM_i = gs_i$	$a_0(AM_0=ts_0)$, $a_1(AM_1=ts_1)$, ..., $a_i(AM_i=ts_i)$	pageB

Table 2. The Corresponding Request Route (Result of CoreFARtoRR)

Entry Port	Exit Port	Node Name (seq. num)	Entry URI	Exit URI	Function Name	Func Ret. Val.
-	Exit P_0	pageA (1)	-	pageA_eTag	-	-
Entry P_1	Exit P_1	ProcessRequest (10)	pageA_eTag	pageA_eTag _ER_eRes	check eTag	_ER_eRes
$0<k\leq i$ Entry P_{1+k}	Exit P_{1+k}	CheckGM$_k$ (20+i*10, $M_i = GM_k$)	pageA_eTag _ER_eRes _C_gs$_0$... _C_gs$_{i-1}$	pageA_eTag _ER_eRes _C_gs$_0$... _C_gs$_i$	check GM$_k$	_C_gs$_k$
$0<k\leq j$ Entry P_{1+i+k}	Exit P_{1+i+k}	SetAM$_k$ (30+m*10+i*10, $S_i = AM_k$)	pageA_eTag _ER_eRes_ C_gs$_0$..._C_gs$_i$_S_ts$_0$... _S_ts$_{k-1}$	pageA_eTag _ER_eRes_ C_gs$_0$..._C_gs$_i$_S_ts$_0$... _S_ts$_k$	setAM$_k$	_S_ts$_k$
Entry P_{2+i+j}	Exit P_{2+i+j}	Prepare Response (9990)	pageA_eTag _ER_eRes_C _gs$_0$..._C_gs$_i$_ S_ts$_0$..._S_ts$_k$	pageBScreen	pageA_eTag _ER_eRes_C _gs$_0$..._C_gs$_i$_ S_ts$_0$..._S_ts$_k$ _proRes	-
Entry P_{3+i+j}	-	pageBScreen (10000)	pageBScreen	-	-	-

CoreRRtoFAR

The reverse mapping (from Table 3 to Table 4) is simpler. The first row of Table 4 maps to the transit-from state and the event tag. Then the check values of the middle rows are combined and mapped to the event response and guard of the transition. Similarly the set values of these rows are combined and mapped to the action. The last row maps to the transit-to state.

Table 3. A General Request Route

Entry Port	Exit Port	Node Name	Entry URI	Exit URI	Function Name	Func. Ret.Val.
/	ExitP$_0$	Node$_0$	/	Node$_0$_Event	/	/
$0<i<n$ Entry P_i	ExitP$_i$	Node$_i$	URI$_{i-1}$	URI$_i$	Func$_i$ [checkTag arg.: event"]$^+$ [checkTag arg.: GM$_{i0}$, ..., GM$_{ik}$]$^+$ [setTag arg.: AM$_{i0}$, ..., AM$_{il}$]$^+$	[_ER_erV]$^+$ [_C_cV$_{i0}$..._ C_cV$_{ik}$]$^+$ [_S_sV$_{i0}$..._ S_sV$_{il}$]$^+$
EntryP$_n$	/	Node$_n$	URI$_{n-1}$	/	/	/

Table 4. The Corresponding Page Navigation Transition (Result of CoreRRtoFAR)

Transit- From	Event	Transit-to
$Node_0$	Event.erV	$Node_n$

Guard	Action
$\bigcup\limits_{i=1}^{n-1}(\bigcup\limits_{j=0}^{k}GM_{ij}=cV_{ij})\,\mathbf{cV_{ij}} \neq \text{"ANY"}$	$\bigcup\limits_{i=1}^{n-1}(\bigcup\limits_{j=0}^{k}tosV_{ij}(AM_{ij}=sV_{ij}))\,\mathbf{sV_{ij}} \neq \text{"NOT"}$

The core mapping from a FARNav model to a Request Routing model satisfies Property 1 because according to the mapping steps provided earlier, each page navigation transition is mapped to a request route. It starts from an exit port of a node representing the transit-from state and the event tag; goes through functions checking a mode with a state as return value as all guard value pairs; goes through functions setting a mode with a state as return value as all action value pairs; ends at an entry port of a node representing the transit-to state. Thus it is not possible for a page navigation transition to have no corresponding request route.

Also it is not possible to have a request route with no corresponding page navigation transition. The reason is the request routing model is the combination of all "correct" request routes that are mapped from the page navigation transitions. Because no two request routes can join before the exit port of PrepareResponse, it is not possible to have extra request routes that partially follow the route of a "correct" request route. Thus each request route maps to a page navigation transition in the FARNav model.

A similar argument applies when showing that the core mapping from a Request Routing model to a FARNav model satisfies Property 1.

Since both core mappings satisfy Property 1, it is obvious that Property 2 holds: for both core mappings, the number of request routes and the number of page navigation transitions is the same and there is 1 to 1 mapping. So applying the inverse core mapping gives the original source model.

4.2 FARtoRR and RRtoFAR

The Request Routing model generated using the core mappings can be directly used to guide implementation. However, a developer may want to refine the generated Request Routing model before implementation. We provide two model transformation algorithms: node reordering and node combining. Although we do not have space to show it here, after applying any sequence of these operations, the resulting Request Routing model still maps to the same FARNav model as the original.

Node reordering changes a node's sequence number and updates the request routing model. Node reordering may be used to reduce the complexity of the request routing model or may be used before combining two nodes since only adjacent nodes can be combined. A developer may want to combine adjacent nodes, for example when following the Model2 architecture, which has just one servlet acting as a front controller.

The mappings CoreFARtoRR and CoreRRtoFAR are the core mappings and they are bijections. However, if any transformation operations have been applied to the request routing model, there are now two request routing models that map to the same navigation model, and we lose the bijective property. Thus the transformation operations must be carried along by both models. With this approach, any number of transformation operations can be applied and the core mappings remain bijections. The complete mappings FARtoRR and RRtoFAR are:

```
FARtoRR(N, ops) = executeOps(ops, CoreFARtoRR(N))
RRtoFAR (R, ops) = attachOps (ops, CoreRRtoFAR(R))
```

Although the operations are included in the FARNav model, they are treated as a comment. Thus FARtoRR and RRtoFAR satisfies Property 1 because the core mappings satisfy Property 1 and neither attaching nor applying the operations affects the "correctness" of the result model.

The intuition behind our proof that Property 2 holds for both mappings is as follows. If R is the result of executing a sequence of operations on the core request routing model coreR, then applying RRtoFAR followed by FARtoRR regenerates the original request routing model and reapplies the sequence of operations, so the result is the same R. Similarly, if we start with the FARNav model N, convert it to coreR then to R, we have applied only the operations in N, and these are carried along when converting back to the FARNav model.

5 Tool Support

We developed a Model Helper tool that implements the mappings and transformation operations described in Section 4. For the forward engineering task of the round-trip engineering, Model Helper generates a request routing model including the function tags from a FARNav model. If there are existing transformation operations (OPS), these are applied on the generated request routing model, in the order specified.

Then a developer can use the tool to transform the request routing model by reordering the nodes or combining the nodes. A record of the operations applied is added to OPS. Then the request routing model can be used in Jaza to trace and analyze request routes.

For the reverse engineering task, Model Helper uses the request routing model with embedded function tags and saved OPS in order to generate a FARNav model. The function tags are used to recreate the mode state machines, but the OPS are simply stored as a comment in the FARNav model. Then the developer can check the navigation design using the SMV tool.

We applied the Model Helper in a number of case studies using existing web applications. The goal was to test the Model Helper tool and the round-trip engineering process. Next we give an overview of one of these case studies. A portion of the FARNav model for the PetStore [14] application appears in Fig. 2. This was created by observing the application behavior. Fig. 3 shows the corresponding Request Routing model, which was created by examining the source code.

Starting from the FARNav model in Fig. 2, we applied Model Helper to generate the Request Routing model shown in Fig. 4. The port URIs generated are quite long, because they record the event tag, event response, and all check and set values used in the adaptive navigation. We abbreviated these URIs in the diagram.

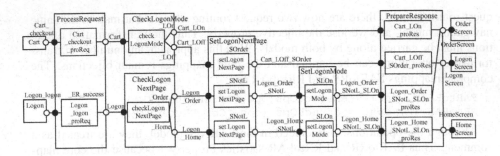

Fig. 4. Generated Request Routing Model for the Example Application

Next we applied some transformation operations, in order to reach a request routing model like the one in Fig. 3, which reflects the current implementation. We applied the node combining operation four times to combine nodes ProcessRequest, CheckLogonNextPage, CheckLogonMode, SetLogonNextPage and SetLogonMode. This combines the processing of requests and the handling of adaptive navigation into one node. Model Helper also records the operations in OPS. The result is shown in Fig. 5.

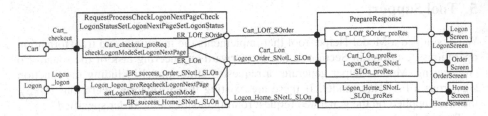

Fig. 5. The Request Routing Model after Node Combining

The structure of the model is now close to our target, and we used Model Helper again to confirm that the new model conforms to the FARNav model, and that it can be regenerated from the FARNav model. However, node names, function names, and URIs are meaningful but quite lengthy, and not all functions are needed, so the next step is to edit the model manually:

- rename the RequestProcess… node to SignOnFilter;
- rename the functions in the SignOnFilter node to "testSignOn" and "validate-SignOn" (also in the function tag table);
- rename the Cart_LOn_… function to "doCheckout";
- remove the other two functions in the PrepareResponse node, and their related entry and exit ports (since no processing is needed for these response pages);
- rename the PrepareResponse node to MainServlet;
- rename the events and URIs with simpler names.

The last change is to add a new Encoding node and encoding function right after the request web page nodes.

The resulting request routing model looks exactly like **Fig. 3**, except for the function return values, which still have their generated long form. Changing these is currently not allowed since that would affect the regeneration of the FARNav model. These manual changes are not currently supported as transformation operations. We can still use the Model Helper to generate the corresponding FARNav model, but the manual changes are not saved as OPS, so they are lost if we then regenerate the request routing model.

After applying the manual changes to the request routing model, we used it to regenerate a FARNav model, then compared that to the original and saw that it does match. So the developer is assured that the manual changes do not affect the navigation design.

6 Related Work

As Henriksson described in [8], for round-trip engineering it is necessary for model transformations to have the two properties we state in Section 4: correctness and bijection. However, examining whether these properties hold is not commonly done for model transformations (for example, [11]).

While nearly every existing web engineering approach includes a navigation model, many are simply descriptive models. There are some formal models for navigation, but none explicitly describe adaptive navigation. For example, Draheim et. al. use bipartite graphs to describe navigation, with one set of nodes representing the web pages and the other set the actions taken on the server side [4]. This is equivalent to our Page Navigation state machine, with transition labels representing the server actions. Their model has no equivalent to our Mode state machines, nor for describing the interaction among them.

Several existing navigation models can be used to generate an implementation. Fraternali et. al use HDM-lite notation to specify structure, navigation and presentation of web application semantics, then generate database schema and web pages from the model [5]. Merialdo et. al generate code from design artifacts [9]. Navarro et. al. use the Pipe approach for characterizing navigational maps then map to UML-Conallen class diagrams with the web pages [10]. Book et. al use the Dialog Flow Notation to represent the dialog flow within a web application, then translate this into an object-oriented dialog flow model for run-time lookups [1]. Vilain et. al created the User Interaction Diagram (UID) as a basis for generating a preliminary class diagram using heuristic guidelines. [14]

However, most of these approaches simply map to web pages (except for [1] and [14]). None support adaptive navigation nor explicitly show how the navigation impacts the request routing. None support traceability between navigation and implementation, nor do they provide an inverse transformation back to the navigation model.

7 Discussion and Conclusion

A FARNav model and a Request Routing model describe different features of a web application and should be used at different stages of development and maintenance.

The FARNav model represents the navigation design so it is related with user requirements and high-level design. The request routing model represents the request routes in a web application, so it is closer to the implementation: the elements in a request routing model directly map to the source code elements.

Earlier we developed these models along with operations that automatically analyze properties of the models. While each is useful independently, their utility is greatly improved by relating the models. This is the contribution described in this paper. We provide mappings (model transformations) between a FARNav model and a Request Routing model and vice versa. We show that the mappings are correct, meaning that they preserve navigation behavior, and that they are bijective.

Because the mappings have these properties, they support traceability between the models, and they allow the models to be used in round-trip engineering. We also gained practical benefit by examining whether the mappings had these properties. We started by writing algorithms for the model transformations in pseudo-code, then discovered errors after explicitly addressing these properties for the mappings.

These mappings are implemented in a new Model Helper tool. In a forward engineering scenario, as when developing a new application, the developer can:

1. Create a FARNav model to represent and verify adaptive navigation.
2. Run the Model Helper to generate an initial Request Routing model.
3. If desired, trace and analyze request routes; reorder and combine nodes.
4. Follow coding conventions to create code skeletons from the Request Routing model.

For reverse engineering, the developer can:

1. Extract the Request Routing model from the source code.
2. If desired, trace and analyze request routes to assist maintenance tasks.
3. Run the Model Helper to generate the FARNav model.
4. Use the FARNav model for maintenance tasks.

When putting these together for round-trip engineering, changes to the FARNav model are propagated to the Request Routing model and vice versa. Thus the Model Helper saves effort for the developer by generating the initial Request Routing model, and by ensuring that the two models are consistent. In this paper we described one of a number of case studies we have done, using the Model Helper in a round-trip engineering scenario.

Although we had a general idea of how we would relate the two models as we developed each, the models have changed as a result of defining the mappings between them. One change was to refine the format of the events on transitions in the Page Navigation state machine. Another change affected the way adaptive navigation is exhibited in the Request Routing model. An earlier version had a special notation for adaptive navigation, but now we generate an assembly of nodes containing check and set functions to handle the adaptive navigation. We had to add function tag tables to the Request Routing model in order to be able to recreate the details of the mode state machines. Finally, we had to carry the Request Routing model's transformation operations along in the FARNav model, in order to reapply those operations to a generated Request Routing model.

The next step for the mappings and Model Helper tool is to define additional transformation operations for a Request Routing model. Once we have a richer set of operations, we can investigate the use of heuristics to generate more customized Request Routing models, for example when the developer specifies that the front controller pattern should be used.

References

[1] de Alfaro, L.: Model Checking the World Wide Web. In: Berry, G., Comon, H., Finkel, A. (eds.) CAV 2001. LNCS, vol. 2102, pp. 337–349. Springer, Heidelberg (2001)

[2] Book, M., Gruhn, V.: Modeling Web-Based Dialog Flows for Automatic Dialog Control. In: ASE 2004. Proceedings of 19th IEEE International Conference on Automated Software Engineering, Linz, Austria, pp. 100–109. IEEE Computer Society Press, Los Alamitos (2004)

[3] McMillan, K.L.: A Methodology for Hardware Certification using Compositional Model Checking. Cadence Berkeley Labs Technical Report 1999-03, http://www.kenmcmil.com/papers/1999-03.ps.gz

[4] Draheim, D., Lutteroth, C., Weber, G.: A Source Code Independent Reverse Engineering Tool for Dynamic Web Sites. In: CSMR 05. Proceedings of the 9th European Conference on Software Maintenance and Reengineering, pp. 168–177 (2005)

[5] Fraternali, P., Paolini, P.: Model-driven development of Web Applications: The AutoWeb system. ACT Trans. Inf. Syst. 18(4), 323–382 (2000)

[6] Han, M., Hofmeister, C.: Modeling Request Routing in Web Applications. In: Proceedings of the 8th IEEE International Symposium on Web Site Evolution (WSE 2006), IEEE Computer Society Press, Los Alamitos (2006) (to appear)

[7] Han, M., Hofmeister, C.: Modeling and Verification of Adaptive Navigation in Web Applications. In: Proceedings of the 6th International Conference on Web Engineering,California, USA, July 11-14, pp. 329–336 (2006)

[8] Henriksson, A., Larsson, H.: A Definition of Round-trip Engineering. Technical report, Linkopings Universitet, Sweden (2003)

[9] Merialdo, P., Atzeni, P.: Design and Development of Data-intensive Web-Sites: The Araneus Approach. ACM Transactions on Internet Technology 3(1), 49–92 (2003)

[10] Navarro, J.L.S., Fernandez-Valmayor, A., Fernández-Manjón, B.: Conceptualization of Navigational Maps for Web Applications. In: Proceedings of Workshop on Model-driven Web Engineering (MDWE2005) (July 26, 2005)

[11] Nickel, U.A., Niere, J., Wadasck, J.P., Zündorf, A.: Roundtrip Engineering with FUJABA. In: Proceedings of 2nd Workshop on Software-Reengineering (WSR), Fachberichte Informatik, Universität Koblenz-Landau (August 2000)

[12] Ricca, F., Tonella, P.: Analysis and testing of web applications. In: Proc. of the International Conference on Software Engineering (ICSE 2001), Toronto, Ontario, Canada, pp. 25–34 (May 2001)

[13] Sendall, S., Küster, J.: Taming Model Round-Trip Engineering. In: Proceedings of Workshop on Best Practices for Model-Driven Software Development (2004)

[14] Sun Java Center. Java Petstore 1.3.01, http://java.sun.com/developer/releases/petstore/petstore1_3_01.html

[15] Vilain, P., Schwabe, D.: Improving the Web Application Design Process with UIDs. In: Proceedings of the 2nd International Workshop on Web-Oriented Software Technology (2002)

A UML2 Profile for Service Modeling

Vina Ermagan and Ingolf H. Krüger

[1] University of California San Diego
9500 Gilman Drive, Mail Code 0404, La Jolla, CA 92093-0404, USA
{vermagan, ikrueger}@ucsd.edu
http://sosa.ucsd.edu

Abstract. In this article we provide an embedding of an interaction-based service notion into UML2. Such an embedding is needed, because to this date, UML2 has only limited support for services – they are certainly not first-class modeling elements of the notation. This is despite the ever increasing importance of services as an integration paradigm for ultra large scale systems. The embedding we provide rests on two observations: (i) services are fundamentally defined by component collaborations; (ii) to support a seamless development process, the service notion must span both logical and deployment architecture. To satisfy (i) and (ii) we introduce modifications to the UML that focus on interaction modeling, and the mapping from logical to deployment service architectures. The result is a novel and comprehensive UML2 profile for service-oriented systems.

Keywords: Rich Services, Service-oriented Architectures, Web Services, Model Driven Architectures.

1 Introduction

A major challenge in the development of ultra large scale software intensive systems is the controlled integration of multiple subsystems, such that the resulting system fulfills a wide spectrum of integration requirements ranging from authentication to security to policy management and governance. *Web services* have proven useful as a lightweight deployment and implementation mechanism for system integration; support for many of these integration challenges is, however, still under development in the Web services community. Furthermore, little guidance exists to date on how to model and design service-oriented architectures such that they leverage the emerging standards, such as WS-Security (authentication and security) and WS-BPEL (business process modeling and execution), as part of an integration solution. However, service-orientation is quickly gaining ground also in other domains with increasing software complexity; the automotive domain is one example, where service-orientation is a declared goal [21] but the deployment architectures are quite removed from a Web services flavor.

Contributions: This paper addresses this challenge by introducing a UML2 profile for the specification of service-oriented architectures that can be deployed on a variety of different object-, component- and service-oriented platforms. In particular,

G. Engels et al. (Eds.): MoDELS 2007, LNCS 4735, pp. 360–374, 2007.

to mention two extremes, service-oriented models according to our profile can be directly mapped not only into a Web service-enabled environment, but also into purely component-oriented deployment environments such as in automotive or avionics.

To that end, we develop a modest set of stereotypes with associated structural and behavioral rules. To address the integration challenges of the system class we target, we place the interplay of the constituent services in the center of concern. Therefore, a major means for specifying services in our approach is by means of interaction diagrams. However, we allow the full set of structural and behavior specification techniques to describe service interfaces and detailed service behaviors.

Figure 3 shows a generic example of the decomposition of a service-oriented architecture according to the *Rich Services Profile* we define in this paper. Intuitively, the profile introduces services as having an interface to their environment and, if they are composite, a predefined internal structure. This internal structure is modeled after two major successful architectural patterns: (1) the emerging Enterprise Service Bus (ESB) and Message-Oriented Middleware (MOM) technologies, such as the increasingly popular Mule/ActiveMQ [22] combination; (2) bus-oriented industrial communication architectures, such as they are found in production plants, cars and airplanes. The basic idea is that every service consists internally of a messaging component, a router, and a set of internal services. Any call upon the service (which we model as a message sent to the service via its Service/Data adapter) is intercepted by the Router, which – using the Messenger as its communication infrastructure – exposes the message to a prescribed set of internal services. Each such internal service can alter or transform the message on its path to its final destination. Analogously, calls made by the Rich Service are also intercepted by the Router before they leave via the Service/Data Adapter. This architectural blueprint provides a rich control framework for service composition.

Benefits: The immediate benefits of this profile are as follows: (i) The concept of service is introduced as a first-class modeling citizen into the UML – in particular, service interfaces and service behavior can be modeled using the well-known description techniques provided by the UML. (ii) The profile defines a structural and behavioral blueprint for controlled service composition and refinement – each composite service can define a set of interaction protocols that govern the interplay of its constituent services, so that the interplay addresses the functional and non-functional integration requirements. Each service, in turn, can be hierarchically decomposed according to the same blueprint to support scalability of the modeling approach. (iii) The distance between a logical service-oriented architecture following the blueprint and a suitable deployment architecture is minimal, resulting in improved traceability from requirements to implementation.

Outline: The remainder of this paper is structured as follows. In Section 2, we introduce the Rich Service Profile in detail by introducing a structural and behavioral domain model for Rich Services modeled after Figure 3. In particular, we describe the stereotypes we introduce, their interplay in terms of behavioral constraints, and our rationale for selecting the design decisions we made. Along the way we also mention the description techniques available to the engineer in specifying systems according to the profile. Section 3 presents a case study illustrating both the modeling approach enabled by the profile, and the use of the stereotypes; the context of the case study is a large-scale system-of-systems integration architecture in the domain of ocean observatories. In Section 4, we discuss our approach in the context of related work. Section 5 contains our conclusions and outlook.

2   Rich Service Profile

As service-oriented modeling and implementation technologies become more popular, so does the need for systematically designing large-scale systems of systems integration solutions based on services. The UML is a common and widely used set of notations providing visual modeling languages, valuable for modeling, design and comprehension of requirements and architectural designs. Currently, the UML supports specific notations for development of object- and component-oriented software, but to date, no explicit notion of service, as a first class modeling entity, is defined in the UML.

The profile mechanism has been specifically defined for providing a lightweight extension mechanism to the UML standard for tailoring UML for various domains or different target platforms. Stereotypes, tagged values, and constraints are the main extension mechanisms available in a profile. To complete the previous versions of UML, the UML2 infrastructure and superstructure specifications have defined the profile mechanism as a specific meta-modeling technique, where stereotypes are specific metaclasses, tagged values are metaattributes, and profiles are specific packages [23]. In this section, we take advantage of the UML2 profile package to create a profile as a metamodel for complex service-oriented architectures.

The goal of the Rich Service Profile we propose here is to provide a common language for describing the central aspects of service-oriented systems. This includes specification of the syntactic *and* semantic interface of individual services, behavior specifications for services, service composition, and the mapping of services to deployment architectures. As mentioned in Section 1, our particular focus is the controlled aggregation of individual services into composite service architectures, such that the resulting architecture, by construction, observes a wide spectrum of crosscutting requirements. The profile we present supports a variety of deployment platforms for implementation of the modeled service(s) – including traditional web service-based approaches, emerging Enterprise Service Bus technologies, and general message-oriented middlewares.

In this section, we utilize the standard mechanism for tailoring the UML, *profiles*, to provide the core of a common language supporting the mentioned goals.

The Rich Service Profile references the UML metamodel as its reference model. It extends Components to specify *Rich Services* and further constructs, including Router, Messenger, and Service Interfaces, needed for supporting them. In essence, a Rich Service serves as a Wrapper around traditional services, including web services, within an architectural framework that supports hierarchical service decomposition, as well as addressing composition and integration concerns within and across hierarchical levels.

The profile also includes collaborations that define the general behavior of the main entities of the profile. These collaborations serve as guidelines for designers who can further refine the general behaviors present in the profile to create a more detailed deployment model.

In the following subsections, we introduce the stereotypes of the Rich Service Profile together with the relevant collaborations in detail.

2.1 Rich Service Profile Stereotypes

The stereotypes of the Rich Service Profile and their base classes are described in Table 1. Figure 1 illustrates the metamodel that the Rich Service Profile provides using the stereotypes of Table 1.

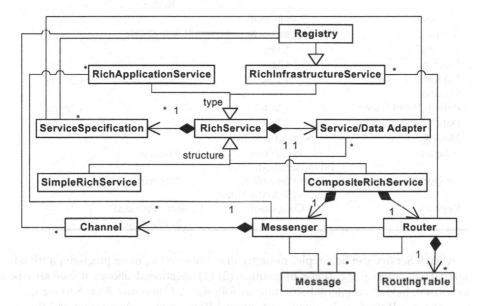

Fig. 1. The Rich Services metamodel

The central entity of the profile is the *Rich Service*. It serves to model individual services, as well as their integration into composite services. Intuitively, a Rich Service consists of the following entities: (1) a *Service/Data Adapter*, which serves as the interface of the Rich Service to its environment, (2) a *Messenger*, which is responsible for message transmission among the sub-services of a Rich Service, (3) a *Router*, which is responsible for intercepting inbound and outbound messages to and from the Service/Data Adapter and routing these messages through the correct set of sub-services, and (4) the *sub-services* themselves, which are also Rich Services that communicate using the Messenger and Router.

A Rich Service is modeled as a stereotype extending Component from the UML BasicComponents package. A Rich Service is active, meaning that it has an associated behavior; it has precisely one externally visible port stereotyped as the Service/Data Adapter. A Rich Service defines provided and required Service Specifications. Service Specification stereotypes the Interface from the

ProtocolStateMachines package and has a tag named Protocol, which is a protocol state machine that defines the external view of the sequence of operation calls that can occur on the interface.

Table 1. Stereotypes of the Rich Service Profile

Stereotype	Base Class	Tags	Parent
RichService	Component	Adapter	
SimpleRichService	Component		RichService
CompositeRichService	Component	Messenger, Router	RichService
Messenger	Component		
Router	Component	RoutingTable	
RoutingTable	Class		
Channel	Class		
PublishSubscribe Channel	Class		Channel
PointToPointChannel	Class		Channel
DataTypeChannel	Class		Channel
Message	NamedElement		
Adapter	Port (From ProtocolStateMachines)	Protocol	
ServiceSpecification	Interface (from ProtocolStateMachines)	Protocol	
Registry	Component	Publish: (Provided Interface)	

A Rich Service can be simple, meaning that it has no (or, more precisely, a trivial) internal structure in the sense of entities (2)-(4) mentioned above; it can also be composite, having an internal structure as follows. A Composite Rich Service has a Messenger, a Router, and a number of internal Rich Services. Messenger and Router are stereotypes extending Component. Multiple internal Rich Services can be attached to the Messenger via ports. Messenger is responsible for Message transmission between the connected Rich Services and between the Rich Services and the Service/Data Adapter. Messenger has a number of Channels to implement the messaging. A Channel is a stereotype extending Class. This allows the profile to support various types of channels, including Publish_Subscribe Channel, DataType Channel, and Point_to_point Channel. A Messenger is always associated with a Router. The Router is responsible for routing the messages through the correct set of Channels based on its Routing Tables. Routing Table is a stereotype extending Class. Intuitively, the router is the mechanism that allows us to inject monitoring and transformation services into a composite service. The idea is that the router intercepts inbound messages at the Service/Data adapter, before they are accessible to the internal Rich Services. The router then follows the configuration stored within its Routing Table to steer the processing of these messages from one internal Rich Service to another. This mechanism can be used, for instance, to encrypt or decrypt messages, to log them, to persist them, etc, without the sender being aware of the intermediate services. Similarly, outbound messages are exposed to the routing scheme before they leave the Rich Service via the Service/Data Adapter.

Rich Services can be of two types: Rich Infrastructure Service (RIS), or Rich Application Service (RAS). Rich Application Services are only aware of the Messenger, while Rich Infrastructure Services can manipulate the Routing Tables and therefore have access to the Router. Rich Infrastructure Services are Rich Services that can directly access the routing tables in order to provide services to the messaging infrastructure, while Rich Application Services provide application-specific services to the system. A specific example of a Rich Infrastructure Service is a Registry where other Rich Services can publish their Service Specification, i.e. their interfaces including the protocol state machines. The Registry associates Channels to published Service Specifications; other Rich Services can subscribe to Channels based on their Service Specifications. The information on subscription of Rich Services to Channels is kept as part of the Routing Table and the Router is responsible for routing the messages sent by the provider Rich Service to the subscribing Rich Services. When a Service/Data Adapter puts a Message on the Messenger, the Router intercepts the Message and routes it based on the Routing Table information through a set of Rich Infrastructure and Application Services.

2.2 Behavior

Collaborations are particularly useful as a means for capturing standard design patterns. Since a Collaboration in UML2 is a kind of classifier, any kind of behavioral description can be attached to it. By extending Collaborations from the UML2 Collaborations package we can form prototypical collaborations to define behavioral pattern of some of the Rich Service Profile entities as part of the profile. These Collaborations can have associated interactions to achieve a more detailed behavior specification. The Stereotyped Collaborations can be used as guidelines for designers on how to use and integrate the profile entities to form meaningful system models.

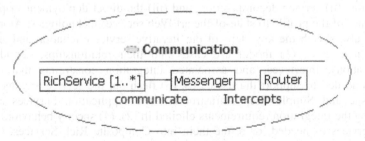

Fig. 2. Communication collaboration

A Communication Collaboration for a Composite Rich Service (see Figure 2) is a stereotype that has a Messenger, a Router, and multiple Rich Services as its parts (tags). Every Composite Rich Service instantiates such a collaboration. The bindings of the collaboration roles to the Rich Service's parts are trivial due to the shared names of the roles and Rich Service's parts. An interaction can be attached to this collaboration, specifying the behavior of the Router as an interceptor (Smart Proxy

[29]). Every Rich Service can send a Message to Messenger. The Router works as an interceptor and picks up the Message, routs it through any specified intermediate Rich Services before sending it to the destination Rich Service. These intermediate Rich services can be Rich Infrastructure Services, or they can be other Rich Application Services. This describes the generic behavior for service composition. A composite Rich Service implements this behavior via the respective role bindings.

Designers can capture the overall behavior of a Composite Rich Service as an interaction. Such an interaction will have the internal Rich Services, the Messenger, and Router as its lifelines. This high level behavior will specify the order in which Rich Services communicate, and can be used to populate the Routing Table. To further refine the model behavior, one can use PartDecomposition from the UML2 Interactions package to decompose the internal Rich Services (modeled as lifelines) to capture the internal behavior of these internal Composite Rich Services. Of course, the internal behavior is visible from outside the Composite Rich Services only to the degree it is specified in the corresponding Service/Data Adapter. The Formal Gates on the decomposed interaction form the interfaces for the Composite Rich Service.

The high level behavior can also be represented as UML2 Protocol State Machines, which can be further redefined to form the internal behavior of encapsulated Composite and Simple Rich Services. This allows us to model service behavior with all the behavior description techniques provided by UML2.

3 Case Study

We demonstrate the utility of the proposed profile and metamodel by using a case study from the domain of global ocean observatories, namely the federated Ocean Research Interactive Observatory Networks (ORION) program [24]. This case study is an elaboration of the ORION-CI conceptual architecture available at [24]. Clearly, here we can only scratch the surface of the complexity of building an architecture of the scale of ORION. However, it allows us to show (i) modeling of services and their integration, (ii) service decomposition, and (iii) the direct deployment mapping from an instance of the profile to state-of-the art Web services technologies. Along the way we will also sketch the key steps of our iterative service elicitation and architecture definition process: (1) model use cases and their relationships, (2) identify the collaborations, interfaces, and associated integration constraints that define the services needed to support the use cases, (3) flesh out the service architecture using Composite and Simple Rich Infrastructure and Application Services as needed, following the integration requirements elicited in (2), (4) specify behaviors of Simple Rich Services as needed, or refine them into Composite Rich Services, (5) specify mapping from the entities in the Rich Service Profile to deployment entities to create an instance of the architecture. Iterate over (1)-(5) until the desired degree of detail is reached.

A system satisfying the goals of ORION would support scientific discovery by providing eligible oceanographers with ubiquitous access to instrument networks for sensing and actuation, computational resources, and modeling and simulation facilities, as well as means for distributed data storage and access. A traditional SOA approach would quickly reach its limits in the face of the challenges induced by the diverse requirements of supporting governance of the different authority domains, access policies, and concerns of the multiple stakeholders involved in such a complex system-of-systems. To capture the requirements for and manage the complexity of the resulting cyber-infrastructure we exploit the Rich Service Profile as defined above; we directly benefit from its disentanglement of logical and deployment architectures for services because the various subsystems indeed rest on a wide spectrum of deployment technologies.

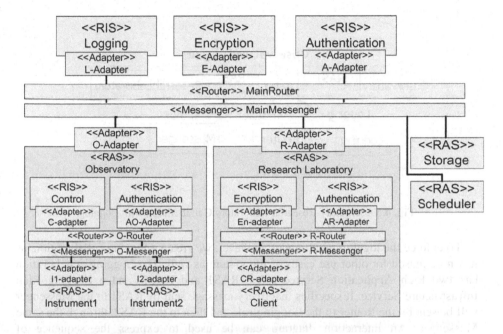

Fig. 3. Orion case study model based on our profile

The hierarchical nature of Rich Services supports creating traceable views for various stakeholders of the system, and a decomposition methodology that supports operation and distributed management of thousands of independently owned taskable resources (modeled as services) of various types (e.g., sensors, sensor platforms, processes, numerical models and simulations) across a core infrastructure operated by independent stakeholders. This also enables hierarchical structuring of the stakeholders' logical roles into the cyber-infrastructure, and encapsulation of crosscutting concerns according to their individual policies.

Figure 3 represents a possible subset of stakeholders as high-level Rich Services such as an *Observatory* and a *Research Laboratory*. Such a decomposition allows us to reason about their role in the cyber-infrastructure without dealing directly with their internal deployment models. Steps (1)-(3): In order to illustrate the steps involved in modeling such a system based on the proposed profile, we will consider *one* use case of the system, namely an oceanographer accessing a remote ocean instrument and retrieving the experimental data from the instrument. As a requirement for this use case, all of the conversations between an oceanographer and an instrument must be logged. The oceanographer and instrument are parts of different authority domains, each with its own set of requirements and policies. At a very high level view, we can abstract from the Adapters and concentrate on the communication between Rich Services. The use case can be modeled as a Communicate collaboration (see Figure 2) use where Research Laboratory and the Observatory play the roles of Rich Services, and the Messenger and the Router play their respective roles.

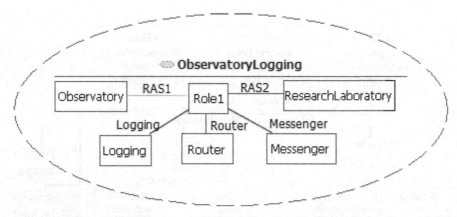

Fig. 4. Observatory-Research Laboratory Collaboration with Logging

To enforce the logging requirement, however, we create a logging collaboration, as it can be reused for other use cases of the system as well. The Logging collaboration has two Rich Application Service roles: RAS1 and RAS2, and a Logging Rich Infrastructure Service. It specifies that every message sent by RAS1 to the Messenger will be sent by the Router to the Logging role and then to the RAS2, through the same Messenger. An interaction diagram can be used to express the sequence of interactions for this collaboration in more detail. Now, we can create a new collaboration for our use case that uses the Logging collaboration to capture the communication of the Observatory and the Research Laboratory. This collaboration is shown in Figure 4. In order to capture the detailed behavior of this collaboration we use a UML interaction diagram shown in Figure 5.

In this interaction, the Research Laboratory, Observatory, Router, and Messenger are captured as lifelines, because they are connectable elements (parts) of their container, i.e. the Rich Service modeling the overall system. The Research Laboratory sends the request to the Messenger, destined for the Observatory. The Router intercepts this communication, sending the request to the Logging Rich Service, also via the Messenger. After being processed by the Logging service, the Router routes the request to the final destination, the Observatory. Note that by using UML interactions, we can further impose time and duration constraints on the occurrences of partial interactions. This interaction model captures the essential behavior of the system to fulfill the use case and its integration requirements and constraints.

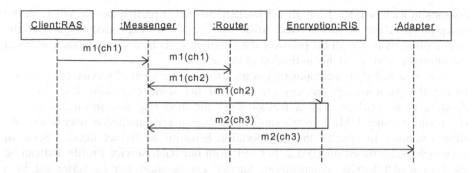

Fig. 5. Interaction diagram for the Observatory-Research Laboratory collaboration

Step (4) - behavior specification and refinement: The Research Laboratory itself is a composite Rich Service and may have many internal services such as oceanographer client service, identification, authentication, and encryption of the outbound messages. The identification and the management of the remote *Instrument* also concerns another stakeholder, as the *Instrument* is located deeper in the hierarchy of the *Observatory*, and within the local control domain of a *Regional Cabled Observatory* near the seashore. We can use PartDecomposition from the UML Interactions package, which is a new concept added in UML2, to further decompose the model capturing the interactions occurring between the internal parts (Rich Services) of each of these composite Rich Services. As an example, Figure 6 shows how the Oceanographer Client's outgoing requests should be intercepted by the internal Router and processed by the Encryption Rich Service before reaching the Adapter, which acts as a gateway for outbound messages.

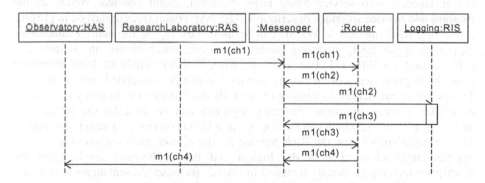

Fig. 6. Research Laboratory internal interaction for Encryption

Note that the only formal gates for the internal interactions of a composite Rich Service exist on the Adapter lifeline of the interaction. Following the semantics of the PartDecomposition, these formal gates must match the actual gates on the decomposed lifeline in the higher level interaction. Together these gates define the partial interfaces of the composite Rich Service. The union of the gates of the composite Rich Service from all the interactions that it participates in will form the complete interface of this service. Since these gates are expressed as parts of the interactions, partial or global protocol state machines can be assigned to their union in the interface, giving a richer definition of the service.

Also note that PartDecomposition is used to model a form of service composition where the participating services are parts of the same composite Rich Service. Sanders et al. [14] propose a methodology for modeling and specifying service composition using UML2 collaborations. They also use interaction overviews, and state machines to specify the collaboration behavior in further details. Such an approach, being based on UML2, fits well with our Rich Service Profile and can be used to model service composition. Service composition can be addressed in a centralized way by adding a coordinating Rich Service that will orchestrate the participating services and with the help of the Router/Interceptor. BPEL4WS is the standard choice for such an approach if the Web Services is the target domain. Service composition can also be addressed in a distributed way, where choreography based languages such as WSCL or WS-CDL can be used when targeting the Web Services domain. Rich Services support these composition approaches while their encapsulation and hierarchy guide developers to focus on one hierarchical level at a time[28].

Step (5) – deployment mapping: A possible deployment plan for such a system might include classic Web services or a more general Enterprise Service Bus (ESB)-based technology. ESBs combine the strengths of message-oriented middleware; a flexible plugin architecture for processing messages to handle crosscutting concerns for a set of connecting Rich Services; and a rich set of data adapters/connectors to facilitate rapid connections between emerging and legacy data sources, applications and services. Examples of ESB implementations include Architect's Toolbox, Cape Clear's ESB, Fiorano ESB, Sonic ESB, SpiritSoft's Spiritwave, and CodeHaus' Mule. For instance, a web service based target platform, might consider WSDL as the Adaptor and Service Interface description. Also conversion rules described in [15] for converting UML models to WDSL can be used. Leveraging the many technologies supported by an ESB, including Mule as transport mechanisms, an Adaptor can publish itself via JMS, HTTP, SOAP, etc. Also, in this example we have abstracted from the Registry services, assuming that the binding is hard-coded into the Routing Tables, which are actually a *feature* of the ESB itself. Since the Registry is modeled as a Rich Service, the same modeling approach can be used for the interactions including a Registry, which can be mapped to a UDDI service as a target technology. This example shows how the Rich Service Profile allows specification of a service-oriented architecture at *both* the logical and the deployment level, using the description techniques already included in UML2. Because it disentangles logical and deployment aspects of services, the profile lends itself to the modeling of complex service architectures with heterogeneous deployment infrastructures.

4   Discussion and Related Work

Model driven design and development of systems is a well-established practice [19]. However, model driven approaches to service-oriented design of systems are still in their early stages. Although UML [23] is a commonly used and widely accepted modeling language, it still has no explicit support for services and their auxiliary constructs. The work presented here leverages the experience we have gained in earlier research, where we have built a comprehensive Service Architecture Definition Language (Service-ADL) with services as first-class modeling citizens [25], and makes these concepts accessible within the UML. In particular, this allows us to also carry over the interaction modeling, behavior synthesis, and architecture exploration techniques we have built for Service-ADL into the UML context [1] [2].

Several other attempts exist to use existing UML constructs to model service-oriented applications. [3] uses UML class diagrams to design a general model for service-oriented architectures (SOAs) and uses collaboration diagrams and graph transformation rules for dynamic architecture reconfigurations. [4] proposes use of UML 2.0 collaboration diagrams for modeling web service collaboration protocols along with activity and interaction diagrams as more detailed modeling levels. Kim [26] investigates how UML diagrams can be used to graphically specify collaboration protocols with an automated mapping to BPSS. UMM [27] provides rich set of UML-based modeling concepts for business to business collaboration protocols and methodological guidance to move from requirements gathering to implementation design. All of these approaches suggest the use of UML modeling techniques for SOA modeling. Our work complements these efforts by embedding an explicit metamodel for SOAs into the UML.

Profiles, as the only lightweight means for extending UML, were leveraged in many approaches to address the lack of support for services. All of these profiles are based on UML 1.x, while the changes from previous UML versions to UML 2.0 have important benefit and impact on service modeling, which is leveraged in our Rich Service Profile. Electronic Services are proposed in [5] as services that are enriched with content and provision. A UML profile for Electronic Service Management Systems [5] is created as a framework for representing operational logic of e-services, providing a conceptual infrastructure for e-services development and management. However, ESMS does not address the business definition and engineering of services. The Enterprise Collaboration Architecture (ECA) defined as part of the UML profile for distributed object computing (EDOC) [6] provides a comprehensive framework for modeling of enterprise systems, while still no explicit notion for services exists in this profile. UML activity models are recommended for service composition modeling in this profile. A metamodel for WSDL is proposed in [8] along with a mapping to UML. Our Rich Service Profile, while based on UML 2.0, has explicit notions for services and adds an architectural pattern to service modeling, while it maps well to currently used Web services technologies, such as WSDL, UDDI and BPEL. For instance, we can use BPEL specifications not only for individual services, but also for the interaction model of composite Rich Services, and then derive the corresponding routing and interaction constraints from these specifications. Gardner [7] describes a UML profile for automated business processes and a mapping of the profile to BPEL4WS. He uses UML activity graphs for specifying the business processes. This profile can be used along with the Rich Service Profile if a mapping of the model to BPEL as a target technology is desired. The profile proposed in [20] is based on UML2. It models services as first class elements; however, as mentioned before, the Rich Services Profile goes beyond by adding a scalable architectural pattern enabling the managed integration of multiple existing service composition and coordination approaches.

In the web services domain, there are several other approaches such as [9] suggesting the use of activities from UML to model web service composition, while [10] proposes a service composition model based on UML class diagrams. Thöne et al. [11] create a UML profile for web service composition and propose the UML_WSC language as a replacement for BPEL4WS. A different approach to service composition leverages rich ontologies that describe service characteristics. For example, the Semantic Web [12] community uses semantic annotations to reason about Web services by using languages such OWL-S [13]. In our Rich Service Profile, services are enriched by having hierarchies and following the specific architectural pattern as part of the metamodel. The use of a Router as an interceptor allows for modeling dynamic reconfiguration at runtime, if needed. Specifically, we can model a wide range of service composition operators, including sequencing, alternatives, repetitions, and parallelism, by means of the Routing Table of the composing Rich Service. More complex operators, such as the ones modeling interrupts, or service synchronization can be modeled using the Routing Table in conjunction with an additional Rich Infrastructure Service that monitors and manages the composition result. Service interfaces are already augmented by suggesting the addition of protocol state machines on them as part of the metamodel, and the profile has the capability of exposing richer interfaces and communication reconfiguration at run time, thereby enabling the use of ontology-based composition techniques.

Work on Web Services composition has highlighted its tight coupling with interaction modeling. [16] explores the use of Message Sequence Charts (MSC) to define interactions. [17], for example, presents a tool that transforms MSC to BPEL specifications to allow Web services composition. We leverage similar techniques to compose Rich Services. Toward this goal we have already experimented with the definition and composition of services based on MSC in [18].

5 Conclusions and Outlook

Service-oriented modeling and implementation are the centerpieces of modern system-of-systems integration approaches. Web services and related technology standards address many important issues of service *deployment*. The *modeling* of service-oriented integration architectures, independently from Web Services deployments, however, is still an area of active research and experimentation. To date there is no widely accepted modeling language for that purpose – specifically, the UML2 proper has not assigned "first-class modeling status" to the notion of service.

In this paper, we have identified the need for having an interaction-based, hierarchical service model that disentangles logical architecture from deployment concerns. We have introduced an UML2 profile for Rich Services. Rich Services introduce an explicit integration architecture, consisting of a messaging and routing component, which allows controlled composition of the internal sub-services that implement a service's behavior. This provides support for a wide spectrum of service composition operators and allows the designer to manage crosscutting aspects of an integration task – examples are: encryption, governance, and policy management.

Furthermore, the hierarchic decomposition of Rich Services allows us to scale service models to any desired level of detail. Using a systems-of-systems integration challenge from the domain of oceanography, we have demonstrated utility of the profile, as well as the direct mapping of Rich Service models to current Web Service-based technologies.

By construction, we leverage all of the UML's description techniques for system specifications based on our profile; tailoring these description techniques further to address dynamic architecture changes, as they are supported by our profile, is one interesting area of future work.

Acknowledgments. Our work was partially supported by the NSF within the project "ITR: Collaborative Research: Looking Ahead: Designing the Next Generation Cyber-infrastructure to Operate Interactive Ocean Observatories" (award OCE/GEO #0427924), as well as by funds from the California Institute for Telecommunications and Information Technology (Calit2). We are grateful to the anonymous reviewers for insightful comments.

References

1. Deubler, M., Krüger, I., Meisinger, M., Rittmann, S.: Modeling Crosscutting Services with UML Sequence Diagrams. In: Briand, L.C., Williams, C. (eds.) MoDELS 2005. LNCS, vol. 3713, pp. 522–536. Springer, Heidelberg (2005)
2. Ermagan, V., Krueger, I., Menarini, M.: Towards Model-Based Failure-Management for Automotive Software. In: Proceedings of the ICSE 2007 Workshop on Software Engineering for Automotive Systems (SEAS) (2007)
3. Baresi, L., Heckel, R., Thöne, S., Varró, D.: Modeling and validation of service-oriented architectures: application vs. style. In: Proceedings of the 11th ACM SIGSOFT Symposium on Foundations of Software Engineering 2003, ESEC/FSE, pp. 68–77 (2003)
4. Kramler, G., Kapsammer, E., Kappel, G., Retschitzegger, W.: Towards Using UML 2 for Modelling Web Service Collaboration Protocols. In: Proceedings of the First International Conference on Interoperability of Enterprise Software and Applications (INTEROP-ESA'05) (2005)
5. Piccinelli, G., Emmerich, W., Williams, S., Stearns, M.: A Model-Driven Architecture for Electronic Service Management Systems. In: Proceeding of International Conference on Service Oriented Computing, pp. 241–255 (2003)
6. Enterprise Collaboration Architecture: (ECA) Specification. Version 1.0. formal/04-02-01 (February 2004), http://www.omg.org/docs/formal/04-02-01.pdf
7. Gardner, T.: UML Modelling of Automated Business Processes with a Mapping to BPEL4WS. In: Cardelli, L. (ed.) ECOOP 2003. LNCS, vol. 2743, Springer, Heidelberg (2003)
8. Bézivin, J., Hammoudi, S., Lopes, D., Jouault, F.: Applying MDA Approach for Web Service Platform. In: Proceedings 8th International Enterprise Distributed Object Computing, pp. 58–70 (2004)
9. Skogan, D., Gronmo, R., Solheim, I.: Web Service Composition in UML. In: Proceedings of the 8th IEEE Intl Enterprise Distributed Object Computing Conference (EDOC), IEEE Computer Society Press, Los Alamitos (2004)

10. Orriëns, B., Yang, J., Papazoglou, M.: Model Driven Service Composition. In: Orlowska, M.E., Weerawarana, S., Papazoglou, M.M.P., Yang, J. (eds.) ICSOC 2003. LNCS, vol. 2910, pp. 75–90. Springer, Heidelberg (2003)
11. Thöne, S., Depke, R., Engels, G.: Process-Oriented, Flexible Composition of Web Services with UML. In: Proceedings of the International Conference on Conceptual Modeling (Workshops), pp. 390–401 (2002)
12. Berners-Lee, T., Hendler, J., Lassila, O.: The semantic web. Scientific American 284(5), 34–43 (2001)
13. OWL-S: Semantic Markup for Web Services (2004), http://www.w3.org/Submission/OWL-S/
14. Sanders, R., Castejón, H., Kraemer, F., Bræk, R.: Using UML 2.0 Collaborations for Compositional Service Specification. In: Proceedings of the 8th International Conference of Model Driven Engineering Languages and Systems, pp. 460–475 (2005)
15. Gronmo, R., Skogan, D., Solheim, I., Oldevik, J.: Model-driven Web services development. In: EEE'04, pp. 42–45. IEEE, Los Alamitos (2004)
16. Krüger, I.H.: Distributed System Design with Message Sequence Charts, Ph.D. dissertation, Technische Univer-sität München (2000)
17. Foster, H., Uchitel, S., Magee, J., Kramer, J.: Tool support for model-based engineering of Web service compositions. In: ICWS 2005, pp. 95–102. IEEE, Los Alamitos (2005)
18. Broy, M., Krüger, I.H., Meisinger, M.: A Formal Model of Services. ACM Transactions on Software Engineering and Methodology (TOSEM) 16(1), 5 (2007)
19. Mellor, S., Clark, A., Futagami, T.: Special Issue on Model-Driven Development. IEEE Software 20(5) (2003)
20. IBM: UML 2.0 Profile for Software Services, http://www-128.ibm.com/developerworks/rational/library/05/419_soa/
21. Krüger, I.H., Nelson, E.C., Prasad, K.V.: Service-Based Software Development for Automotive Applications. In: Proceedings of the CONVERGENCE 2004. Convergence Transportation Electronics Association (2004)
22. http://mule.mulesource.org/wiki/display/MULE/Home
23. Object Management Group: UML 2.1.1 Superinfrastructure version 07-02-03, http://www.omg.org/cgi-bin/doc?formal/07-02-05
24. ORION Program Cyber Infrastructure, http://www.orionprogram.org/organization/committees/ciarch/
25. Ermagan, V., Huang, T.-J., Krüger, I., Meisinger, M., Menarini, M., Moorthy, P.: Towards Tool Support for Service-Oriented Development of Embedded Automotive Systems. In: Proceedings of the Dagstuhl Workshop on Model-Based Development of Embedded Systems (MBEES'07), Informatik-Bericht 2007-01 (2007)
26. Kim, H.: Conceptual Modeling and Specification Generation for B2B Business Process based on ebXML. In: SIGMOD Record vol. 31
27. Hofreiter, B., Huemer, C., Naujok, D.: UN/CEFACT's Buisness Collaboration Framework- Motivation and Basic Concepts. In: Proceedings of the MKWI (2004)
28. Arrott, M., Demchak, B., Ermagan, V., Farcas, C., Farcas, E., Krüger, I.H., Menarini, M.: Rich Services: The Integration Piece of the SOA Puzzle. In: Proceedings of the IEEE International Conference on Web Services (ICWS), Salt Lake City, Utah, IEEE Computer Society Press, Los Alamitos (2007)
29. Hohpe, G., Woolf, B.: Enterprise Integration Patterns: Designing, Building, and Deploying Messaging Solutions. Addison-Wesley, Reading (2003)

Automatic Generation of
Workflow-Extended Domain Models

Marco Brambilla[1], Jordi Cabot[2], and Sara Comai[1]

[1] Dipartimento di Elettronica e Informazione, Politecnico di Milano
Piazza L. Da Vinci, 32. I20133 Milano, Italy
{mbrambil, comai}@elet.polimi.it
[2] Estudis d'Informàtica, Multimèdia i Telecomunicació, Universitat Oberta de Catalunya
Rbla. del Poblenou, 156 E08018 Barcelona, Spain
jcabot@uoc.edu

Abstract. The specification of business processes is becoming a more and more critical aspect for organizations. Such processes are specified as workflow models expressing the logical precedence among the different business activities (i.e. the units of work). Up to now, workflow models have been commonly managed through specific subsystems, called workflow management systems. In this paper we advocate for the integration of the workflow specification in the system domain model. This workflow-extended domain model is automatically derived from the initial workflow specification. Then, model-driven development methods may depart from the extended domain model to automatically generate an implementation of the system enforcing the business processes in any final technology platform, thus avoiding the need of basing the implementation on a dedicated workflow engine.

1 Introduction

Software development processes for complex business applications usually require the definition of a workflow model to express logical precedence and process constraints among the different business activities (i.e. the units of work).

Currently, workflow models are usually implemented with the help of dedicated workflow management systems (e.g., [15], [21]) which are heavy-weight applications focused on the control aspects of the workflow enactment. Alternatively, some approaches focus on the implementation of the workflow model in a specific technology platform, as relational databases (generally in the form of triggers [2]), web applications (by means of hypertextual links and buttons properly placed in Web pages, thus restricting the user navigation [5]) or web services (through transformation into BPEL4WS [18]). These *adhoc* approaches are hardly generalizable to other technologies.

In this paper we adopt a formalized model-driven development process for workflow-based applications and advocate for the automatic integration of the workflow model within the (platform-independent) domain model. Given a domain model d and a workflow model w, it is possible to automatically derive a full fledged domain model d' enriched with the types needed to record the required workflow information

G. Engels et al. (Eds.): MoDELS 2007, LNCS 4735, pp. 375–389, 2007.

in *w* (mainly its activities and the enactment of these activities in the different workflow executions) and with a set of process constraints over such types to control the correct workflow execution. We refer to this resulting model as the *workflow-extended domain model*. We will represent it using UML class diagrams annotated with OCL constraints to represent the process constraints. The whole process is sketched in Fig. 1.1. Note that, if necessary, several workflow models can be integrated within the same domain model. This approach has been implemented in a prototype tool.

The main characteristic of a workflow-extended domain model is that it automatically ensures a consistent behavior of all enterprise applications with respect to the business process specification. As long as the applications properly update the workflow information in the extended model, the generated process constraints enforce that the different tasks are done according to the initial workflow model.

Another advantage of a workflow-extended domain model is that it is platform-independent. Indeed, our workflow-extended model can benefit from any method or tool designed for managing a generic domain model, no matter the target technology platform or the purpose of the tool, spawning from direct application execution, to verification/validation analysis, to metrics measurement and to automatic code-generation in any final technology platform. Those methods do not need to be extended to cope with our workflow-extended models, since our workflow-extended domain model is a completely standard UML model.

Moreover, our workflow-extended models enable the definition of more expressive business constraints, including timing conditions [8] or involving both workflow and domain information. These constraints are generally not allowed by workflow definition languages.

The rest of the paper is structured as follows: in Section 2 the basic workflow concepts and our case study are illustrated. In Sections 3 and 4 we provide the definition of the workflow-extended domain model and of the OCL process constraints, respectively. Section 5 sketches possible implementation strategies for this extended model. Section 6 compares our approach with related work and in Section 7 we draw our conclusions, provide some details about our tool support and discuss future work.

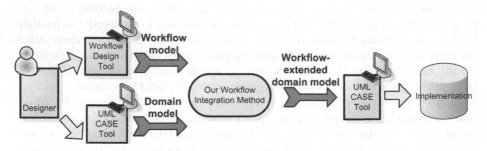

Fig. 1.1. MDD process for workflow-based applications

2 Basic Workflow Concepts

Several visual notations and languages have been proposed to specify workflow models, with different expressive power, syntax, and semantics. Without loss of generality, in our work we have adopted the *Workflow Management Coalition* terminology and the BPMN [20] OMG standard notation[1].

The workflow model is hence based on the concepts of *Process* (the description of the business process), *Case* (a process instance, that is, a particular workflow execution), *Activity* (the elementary unit of work composing a process), *Activity instance* (an instantiation of an activity within a case), *Actor* (a user role intervening in the process), *Event* (some punctual situation that happens in a case), and *Constraint* (logical precedence among activities and rules enabling activities execution). Processes can be internally structured using a variety of constructs: sequences of activities; gateways implementing AND, OR, XOR splits, respectively realizing splits into independent, alternative and exclusive threads; gateways implementing joins, i.e., convergence point of two or more activity flows; conditional flows between two activities; loops among activities or repetitions of single activities. Each construct may involve several constraints over the activities.

Our approach covers a large subset of the full expressive power of BPMN; we do not cope with the concepts of nested subprocesses (which can be easily tackled by flattening the process representation), transactions (which can exploit implementation features), and a few combinations of primitive constructs, such as the direct concatenation of several gateways (which can be handled by introducing fake activities between them).

In the sequel, we will exemplify the proposed approach on a case study consisting of a workflow implementing a simplified purchase process, as illustrated in Fig. 2.1.

According to the BPMN semantics, the depicted diagram specifies a process involving two actors (represented by the two swimlanes): a customer and a seller. The customer starts the workflow by asking for a quotation about a set of products (*Ask Quotation* activity). The seller provides the quotation (*Provide Quotation* activity) and the customer may decide (exclusive choice) to modify the request (and hence the quotation request and response are repeated) or to accept it (then the order is submitted and the seller takes care of it). For simplicity, it is not modeled what happens if they do never reach an agreement. The order management requires two parallel activities to be performed: the choice of the shipment options and the internal management of each order line. The latter is represented by the multi-instance activity called *Process OrderLine*: a different instance is started for each order line included in the order. Once all order lines have been processed and the shipment has been decided (i.e., after the AND merge synchronization), the order is shipped and the customer pays the corresponding amount.

---

[1] The results of our approach when using Activity Diagrams would have been quite similar. See [23] for a correspondence between BPMN and Activity Diagrams.

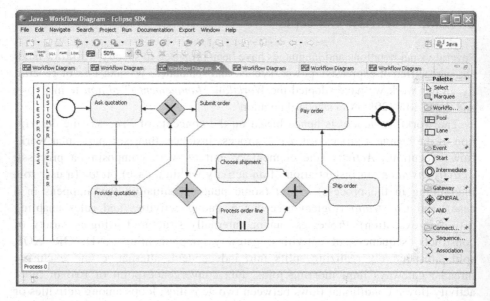

Fig. 2.1. Example of a workflow schema

3 Extending Domain Models with Workflow Information

Given an initial domain model, the workflow-extended domain model of the work-flow-based application is obtained by extending the domain model with some additional elements derived from the workflow specification. This extension can be regarded as a kind of model weaving between the workflow and domain models [13]. We will focus on the case of a single workflow; however, our extensions to the domain model suffice when considering different workflows on the same domain.

Clearly, the workflow-extended domain model is more complex than the original domain model. However, we believe that this increased complexity is compensated by the fact that it may be automatically generated (with our method) and processed (with, for instance, code-generation tools) and thus, the designer does not need to manipulate it. Moreover, the size of the extension is constant regardless the size of the domain model and linear with respect to the number of activities in the workflow.

The workflow-extended model contains the minimum set of concepts required to manage the workflow and to easily specify the needed process constraints. However, richer schemas with further relationship types and/or attributes could be defined, according to the requirements of the specific workflow application (for example, we could have used a more complex pattern for the specification of the role-user relationship [6]). Similarly, simpler extensions could be used instead but then, as a trade-off, the process constraints would become much more complex.

To illustrate the process we will use the workflow model of Fig. 2.1 and we will assume that the initial domain model is the one shown in the bottom part of Fig. 3.1, consisting in the types *Product, Quotation, QuotationLine,* and *Order* (note that when accepted by the customer, a *Quotation* generates an *Order* and then, its quotation lines are referred to as order lines).

The workflow-extended domain model must include at least: *(i)* the original domain model, *(ii)* *user*-related information, *(iii)* *workflow*-related information, *(iv)* a set of possible relationships between the domain schema, the workflow information and the user information, and *(v)* a set of process constraints guaranteeing a consistent state of the whole model with respect to the workflow definition (see the next section). Due to lack of space, the extensions required for *Event* management are only provided in [3].

More formally, we define a workflow-extended domain model as follows. Given an initial domain model with entity types (i.e. classes) $E=\{e_1,...,e_n\}$, representing the knowledge about the domain, and a workflow model w with activities $A=\{a_1,...,a_m\}$, the workflow-extended domain model is obtained in the following way:

i) *Domain subschema*: All entity types in E and their relationships (i.e. associations) remain unchanged in the workflow-extended model (bottom part of Fig. 3.1).

ii) *User subschema:* User-related information is added to the extended model by means of two entity types (see the top-left part of Fig. 3.1): entity type *User* represents individual workflow actors; entity type *Role* represents groups of users, having access to the same set of tasks. A user may belong to different roles.

iii) *Workflow subschema:* Workflow-related information (top-right part of Fig. 3.1) includes several fixed types (i.e. independent of the particular workflow model):
   - Entity type *Process* represents the supported workflows. As an example, an instance of the *Process* type would be our *Purchase* workflow. Other instances would represent additional workflows over the same domain subschema.
   - Entity type *Case* denotes an instance of a process, which has a name, a start time, an end time and a status, which can be: ready, active, cancelled, aborted, or completed [20]. Every execution of a process results in a new instance of this type. This new instance is related with the appropriate *process* instance.
   - Entity type *ActivityType* represents the different activities that compose a process. Activity types are assigned to roles, which are responsible of executing them. In our case study, *AskQuotation*, *ProvideQuotation*, etc. would be instances of *ActivityType*.
   - Entity type *Activity* denotes the occurrence of a particular activity within a *Case*, described by the start time, the end time, and the current status, which can be: ready, active, cancelled, aborted, or completed. Only one user can execute a particular activity instance, and this is recorded by the relationship type *Performs*. The *Precedes* relationship keeps track of the execution order between activities.

and a set of workflow-dependent subtypes:
   - For each activity $a \in A$, a new subtype s_a is added to the entity type *Activity* (*ActivityType* is a *powertype* for this set of generalization relationships). The name of the subtype is the name of a (e.g., in Fig. 3.1 we introduced *ProcessOrderLine, AskQuotation, ShipOrder,* and so on). These subtypes record the information about the specific activities executed during a workflow case. For instance, the action of asking a quotation for the purchase X in a case C of a workflow W would be recorded in the system as an instance of the *AskQuotation* subtype related with the corresponding instance *"C"* in the *Case* type (in its turn related with the *"W"* instance in the *Process* type)

iv) *Relationships between workflow subschema and domain subschema*: each subtype s_a is related with a (possibly empty) set of entity types $E_a \subseteq E$. These new relationship types are useful to record the objects modified/managed during the execution of a certain activity. Also, they are required to evaluate conditions appearing in some process constraints. In the case study (see Fig. 3.1), a set of relationship types are established: *Quotations* are associated to the activities *Ask Quotation* and *Provide Quotation*; *QuotationLines* are associated to the *ProcessOrderLine* activity; and *Orders* are associated to the activities *Submit Order*, *Choose Shipment*, *Process OrderLine*, *Ship Order*, and *Pay Order*. When necessary, these associations between the domain and the workflow subschemata may be automatically generated if the workflow specification includes auxiliary primitives for describing the data flow between activities and/or when the designer defines some pattern-matching among the names of the activities and of the entity types. Otherwise, they must be manually specified.

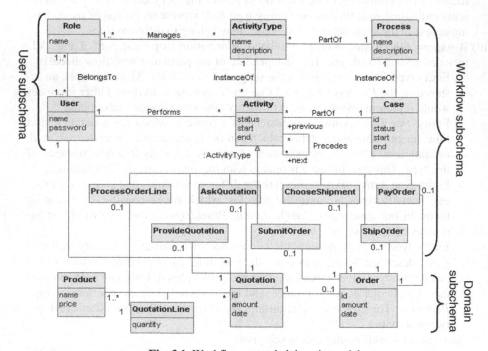

Fig. 3.1. Workflow-extended domain model

4   Translation of Process Constraints

The structure of a workflow model implies a set of constraints regarding the execution order of the different activities, the number of possible instances of each activity in a given case, the conditions that must be satisfied in order to start a new activity, and so forth. These constraints are usually referred to as *process constraints*. The behavior of all enterprise applications must always satisfy these constraints. Thus, the generation of the workflow-extended model must consider all process constraints.

Process constraints are translated as constraints over the population of the $s_{a1},...,s_{am}$ subtypes of *Activity* (see previous section). The generated constraints guarantee that any update event over the population of one of these subtypes (for instance, the creation of a new activity instance or the modification of its status) will be consistent with the process constraints defined in the workflow model.

We specify process constraints by means of invariants written in the OCL language. Invariants in OCL are defined in the context of a specific type, the *context type*. The actual OCL expression stating the constraint condition is called the *body* of the constraint. The *body* is always a boolean expression and must be satisfied by all instances of the context type, that is, the evaluation of the body expression over every instance of the context type must return a *true* value. Constraints are defined to restrict only the execution of the workflow they are created for. Therefore, no interferences among different workflows occur, even if they are defined over an overlapping subset of the domain model.

The complexity of the constraints is of relative importance since all of them are automatically generated from the workflow model, and thus, they do not need to be manipulated (nor even necessarily understood) by the designer but for other tools. However, to simplify its presentation in the extended model, we could easily define an stereotype for each constraint type, as done in [9].

Next subsections define a set of patterns for the generation of the process constraints corresponding to the different constructs appearing in workflow models (sequences, split gateways, merge gateways, conditions, loops, and so on). The patterns can be combined to produce the full translation of the workflow model. As an example, we provide in Section 4.6 the translation of the workflow model of Fig. 2.1. Patterns for the generation of process constraints for *Events* are only shown in [3].

Note that some constructs admit several graphical representations equivalent to the ones used in this paper (see [20] for details). Moreover, the workflow language defines some complex constructs that can be derived from the basic ones, such as complex gateways and event-based gateways, not addressed here due to lack of space.

4.1 Sequences of Activities

A sequence flow between two activities (Fig. 4.1) indicates that the first activity (*A*) must be completed before starting the second one (*B*). Moreover, if *A* is completed within a given case, *B* must be eventually started before ending the case (we do not require *B* to be completed since, for instance, it could be interrupted by the trigger of an intermediate exception event). This behavior can be enforced by means of the definition of three OCL constraints.

Fig. 4.1. Sequence flow

The first constraint (*seq₁* constraint) is defined over the entity type corresponding to the destination activity (*B* in the example) stating that for all activity instances of type *B* the preceding activity instance must be of type *A* and that it must have been already completed. Its specification in OCL is the following:

> *context B inv seq₁: previous->size()=1 and previous->exists(a| a.oclIsTypeOf(A) and a.status='completed')*

This OCL definition enforces that *B* instances (since *B* is the context type of the constraint) have a previous activity (because of the *size* operator over the value of the navigation through the role *previous*) and that such activity is of type *A* (enforced by the *exists* operator). *B* and *A* are *Activity* subtypes as defined in Section 3.

The other two required constraints are:

- A constraint *seq₂* over the second activity to prevent the creation of two different *B* instances related with the same *A* activity instance

> *context B inv seq₂: B.allInstances()-> isUnique(previous)*

- A constraint *seq₃* over the *Case* entity type verifying that when the case is completed there exists a *B* activity instance for each completed *A* activity instance. This *B* instance must be the only instance immediately following the *A* activity instance.

> *context Case inv seq₃: status='completed' implies self.activity-> select(a| a.oclIsTypeOf(A) and a.status='completed')->forAll(a|a.next->exists(b| b.oclIsTypeOf(B)) and a.next->size()=1)*

4.2 Split Gateways

A split gateway is a location within a workflow where the sequence flow can take two or more alternative paths. The different split gateways differ on the number of possible paths that can be taken during the execution of the workflow. For *XOR-split* gateways only a single path can be selected. In *OR-splits* several of the outgoing flows may be chosen. For *AND-splits* all outgoing flows must be followed.

For each kind of BPMN split gateway, Table 4.1 shows the process constraints required to enforce the corresponding behavior.

Besides the process constraints appearing in the table, we must also add to all the activities $B_1...B_n$ the previous constraints *seq₁* and *seq₂* to verify that the preceding activity *A* has been completed and that no two activity instances of the same activity B_i are related with the same preceding activity *A*. We also require that the activity instance/s following *A* is of type B_1 or … or B_n.

Table 4.1. Constraints for split gateways

Split gateway	Process constraints
B1 A XOR Split Bn	- Only one of the $B_1..B_n$ activities may be started *context A inv: next->select(a| a.oclIsTypeOf(B₁) or … or a.oclIsTypeOf(Bₙ))->size()<=1* - If *A* is completed, at least one of the $B_1..B_n$ activities must be created before ending the case *context Case inv: status='completed' implies activities-> select(a|a.oclIsTypeOf(A) and a.status='completed')-> forAll (a|a.next->exists(b|b.oclIsTypeOf(B₁) or..or b.oclIsTypeOf(Bₙ)))*

Table 4.1. (*continued*)

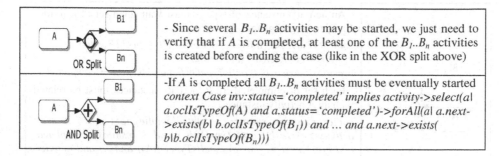

	- Since several $B_1..B_n$ activities may be started, we just need to verify that if A is completed, at least one of the $B_1..B_n$ activities is created before ending the case (like in the XOR split above)
	-If A is completed all $B_1..B_n$ activities must be eventually started *context Case inv:status='completed' implies activity->select(a| a.oclIsTypeOf(A) and a.status='completed')->forAll(a| a.next->exists(b| b.oclIsTypeOf(B_1)) and ... and a.next->exists(b|b.oclIsTypeOf(B_n)))*

4.3 Merge Gateways

Merge gateways are useful to join or synchronize alternative sequence flows. Depending on the kind of merge gateway, the outgoing activity may start every time a single incoming flow is completed (*XOR-Merge*) or must wait until all incoming flows have finished in order to synchronize them (*AND-Merge* gateways). The semantics of the *OR-Merge* gateways is not so clear. If there is a matching *OR-split*, the *OR-Merge* should wait for the completion of all flows activated by the split. If no matching split exists several interpretations are possible, being the simplest one to wait just till the first incoming flow. This is the interpretation adopted in this paper. For a complete treatment of this construct see [24].

Table 4.2 presents the different translation patterns required for each kind of merge gateway. Besides the constraints included in the table, a constraint over A should be added to all gateways to verify that two A instances are not created for the same incoming set of activities (i.e. the intersection between the *previous* instance/s of all A instances must be empty).

Table 4.2. Constraints for merge gateways

Merge gateway	Process constraints
XOR Merge	- All A activity instances have as a previous activity instance a completed activity instance of type B_1 or ... or B_n *context A inv: previous->size()=1 and previous->exists(b| (b.oclIsTypeOf(B_1) or ... or b.oclIsTypeOf(B_n)) and b.status='completed')* - Each $B_1..B_n$ activity instance is followed by an A activity *context Case inv: status='completed' implies activity->select(b| b.oclIsTypeOf(B_1) or ... or b.oclIsTypeOf(B_n))-> forAll(b|b.next->exists(a| a.oclIsTypeOf(A)))*
OR Merge	- An A activity instance must wait for at least an incoming flow *context A inv: previous->select(b| (b.oclIsTypeOf(B_1) or ... or b.oclIsTypeOf(B_n)) and b.status='completed')->size()>=1*

Table 4.2. (*continued*)

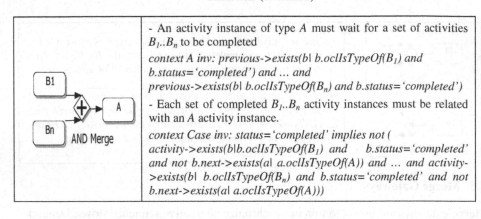

	- An activity instance of type *A* must wait for a set of activities $B_1..B_n$ to be completed *context A inv: previous->exists(b| b.oclIsTypeOf(B$_1$) and b.status= 'completed') and ... and previous->exists(b| b.oclIsTypeOf(B$_n$) and b.status= 'completed')* - Each set of completed $B_1..B_n$ activity instances must be related with an *A* activity instance. *context Case inv: status= 'completed' implies not (activity->exists(b|b.oclIsTypeOf(B$_1$) and b.status= 'completed' and not b.next->exists(a| a.oclIsTypeOf(A))) and ... and activity->exists(b| b.oclIsTypeOf(B$_n$) and b.status= 'completed' and not b.next->exists(a| a.oclIsTypeOf(A)))*

4.4 Condition Constraints

The sequence flow and the *OR-split* and *XOR-split* gateways may contain condition expressions to control the flow execution at run-time. As an example, Fig. 4.2 shows a conditional sequence flow. In the example, the activity *B* cannot start until *A* is completed and the condition *cond* is satisfied. The condition expression may require accessing the entity types of the domain subschema related to *B* in the workflow-extended model. Through the *Precedes* relationship type, we can also define conditions involving the previous *A* activity instance and/or its related domain information.

To handle these condition expressions we must add, for each condition defined in a sequence flow or in an outgoing link of *OR* and *XOR* gateways, a new constraint over the destination activity. The constraint ensures that the preceding activity satisfies the specified condition, according to the following pattern:

context B inv: previous->forAll(a\| a.cond)

Note that these additional constraints only need to hold when the destination activity is created, and thus, they must be defined as *creation-time constraints* [19].

Fig. 4.2. A conditional sequence flow

4.5 Loops

A workflow may contain loops among a group of different activities or within a single activity. In this latter case we distinguish between *standard* loops (where the activity is executed as long as the loop condition holds) and *multi-instance* loops (where the activity is executed a predefined number of times). Every time a loop is iterated a new instance of the activity is created. Fig. 4.3 shows an example of each loop type.

Multi-Instance   Standard   External

Fig. 4.3. Loop examples

Management of *external loops* does not require new constraints but the addition of a temporal condition in all constraints stating a condition like "an instance of type *B* must be eventually created if an instance of type *A* is completed". The new temporal condition on those constraints ensures that the *B* instance is created *after* the *A* instance is completed (earlier *B* instances may exists due to previous loop iterations).

Standard loops may be regarded as an alternative representation for conditional sequence flows having the same activity as a source and destination. Therefore, the constraints needed to handle standard loop activities are similar to those required for conditional sequence flows. We need a constraint checking that the previous loop instance has finished and another one stating that the loop condition is still true when starting the new iteration (again, this is a creation-time constraint). The loop condition is taken from the properties of the activity as defined in the workflow model. Moreover, we need also to check that the activity/ies at the end of the outcoming flows of the loop activity are not started until the loop condition becomes false. To prevent this wrong behavior we should treat all outgoing flows from the loop activity as conditional flows with the condition *'not loopCondition'*. Then, constraints generated to control the conditional flow will prevent next activity/ies to start until the condition *'not loopCondition'* becomes true.

Multi-instance loop activities are repeated a fixed number of times, as defined by the loop condition, which now is evaluated only once during the execution of the case and returns a natural value instead of a boolean value. At the end of the case, the number of instances of the multi-instance activity must be an exact multiple of this value. Assuming that the multi-instance activity is called *A*, the OCL formalization of this constraint would be:

context Case inv: (activity->select(a|a.oclIsTypeOf(A))->size() mod loopCondition)=0

For multi-instance loops the different instances may be created sequentially or in parallel. Besides, we can define when the workflow shall continue. It can be either after each single activity instance is executed (as in a normal sequence flow), after all iterations have been completed (similar to the *AND-merge* gateways), or as soon as a single iteration is completed (similar to the basic *OR-merge* gateway).

4.6 Applying the Translation Patterns

As an example, Table 4.3 summarizes the process constraints resulting from applying the translation over the workflow schema of Fig. 2.1.

For sake of brevity, we do not include here the complete set of constraints, but we exemplify in Table 4.4 the full definition of the constraints involved in the *Provide Quotation* activity (the rest of the specifications can be found in the extended version of the paper at [3]). The *Provide Quotation* activity involves a set of constraints due to the sequence constraint with *Ask Quotation* activity and a set due to the subsequent *XOR* split.

Table 4.3. Process constraints for the workflow running example

Activity	Constraints
Ask Quotation	- When the activity instance comes after a *Provide Quotation*, the latter must have been completed (a single new ask quotation activity can be generated). Otherwise, it must have been created in response to the occurrence of a start event (due to the implicit XOR merge gateway corresponding to the two incoming arrows).
Provide Quotation	- A quotation cannot be provided until the *Ask Quotation* activity has finished. Moreover, if an instance of *Ask Quotation* is completed, a single *Provide Quotation* instance must eventually be created - After providing a quotation we can either ask for a new quotation or submit an order, but not both. At least one of them must be executed.
Submit Order	- The previous *Provide Quotation* activity must be completed. Besides, only a single *Submit Order* instance must be created for the same *Provided Quotation* instance - After submitting an order, both the *Choose Shipment* and the *Process OrderLine* activities must be executed
Choose Shipment	- The preceding *Submit Order* activity instance must be completed. Besides, a single *Choose Shipment* activity must be executed for each *Submit Order* activity instance
Process OrderLine	- The preceding *Submit Order* activity must be completed - The system must exactly execute as many *Process OrderLine* activity instances as the number of order (quotation) lines for the related order
Ship Order	- The order cannot be shipped until the shipment has been chosen and all order lines have been processed. Then, a *Ship Order* activity instance must be executed before ending the case
Pay Order	- An order cannot be paid until it has been shipped. A single *pay order* activity shall be created in response to each order shipment

5 Code-Generation of the Workflow-Extended Domain Model

A workflow-extended domain model is a completely standard domain model. No new modeling primitives have been created to express the extension of the original model with the required workflow information. Therefore, any method or tool able to provide an automatic implementation of the initial domain model can also cope with the automatic generation of our workflow-extended model in any final technology platform using general-purpose MDD techniques and frameworks.

For instance, activity classes (as *AskQuotation* or *ProvideQuotation*) could be implemented as database tables or Java classes while process constraints could be implemented as triggers and method preconditions respectively. Note that a translation from OCL into SQL or Java is already provided by several tools (e.g., [10], [16]), covering also efficient implementation of OCL constraints [7].

Table 4.4. Constraint definitions for the Provide Quotation activity

Constraints due to the sequence with *Ask Quotation*	The preceding activity must be of type *Ask Quotation* and must be completed
	context ProvideQuotation inv: previous->size()=1 and previous->exists(a\ a.ocIIsTypeOf(AskQuotation) and a.status='completed')
	No two instances may be related with the same *Ask Quotation* instance
	context ProvideQuotation inv: ProvideQuotation.allInstances()-> is-Unique(previous)
	A Provide Quotation instance must exist for each completed *Ask Quotation*
	context Case inv: status='completed' implies activity-> select(a\ a.ocIIsTypeOf(AskQuotation) and a.status='completed')->forAll(a\a.next->exists(b\b.ocIIsTypeOf(ProvideQuotation) and a.end<=b.start) and a.next->size()=1)
Constraints due to the *XOR split*	The next activity must be either another *Ask Quotation* instance or a *Submit Order* instance, but not both
	context ProvideQuotation inv: next->select (a\ a.ocIIsTypeOf(AskQuotation) or a.ocIIsTypeOf(ProvideQuotation))->size()<=1
	If the *Provide Quotation* instance is completed, an *Ask Quotation* or a *Submit Order* must be created before ending the case.
	context Case inv: status='completed' implies activity->select(a\a.ocIIsTypeOf(ProvideQuotation) and a.status='completed')-> forAll (a\ a.next-> exists(b\ b.ocIIsTypeOf(AskQuotation) or b.ocIIsTypeOf(SubmitOrder)))
	Only *Ask Quotation* activity instances or *Submit Order* instances may follow a *Provide Quotation* instance
	context ProvideQuotation inv: next->forAll(b\ b.ocIIsTypeOf(AskQuotation) or b.ocIIsTypeOf(SubmitOrder)

6 Related Work

Research on business process in software engineering has mainly addressed the correctness of the design of the workflow model (see [12] as an example) or its direct implementation in specific final technology platforms (see [2] for an implementation over a relational database and [5] for an implementation using web technologies). Integration of workflows and MDD approaches has only been explored from a general framework perspective [14].

As far as we know, ours is the first proposal where both workflow information and process constraints are automatically derived from a workflow model and integrated within a platform-independent domain model. As we have seen in the previous section, this integration permits to generate workflow applications in any final technology without requiring to develop an specific treatment for the workflow model.

Moreover, ours is also the first translation of a workflow model into a set of OCL declarative constraints. Such a translation is necessary regardless how these constraints are to be enforced in the final workflow implementation.

Very few examples of translations to other declarative languages exist (e.g., see [4] for a translation to LTL temporal logics). In literature, workflow metadata and OCL constraints have only been used in [11] to manually specify workflow access control constraints and derive authorization rules, in [1] to express constraints with respect to the distribution of work to teams, in ArgoUWE [17] to check for well-formedness in the design of process models, in [22] to manually specify business models with UML and in [18] to specify the contracts for the transformation of activity diagrams into BPEL4WS.

7   Conclusions

In this paper we presented an automatic approach to integrate the semantics of business process specifications within domain models.

Once the designer has specified both the workflow and the domain models separately, we build an integrated workflow-extended domain model by means of adding to the domain model *(i)* the definition of a set of new entity and relationship types for workflow status tracking and *(ii)* the rules for generating the integrity constraints on such types, needed for enforcing the business process specification.

The integration of both the domain and the workflow aspects in a single extended domain model permits a homogeneous treatment of the workflow-based application. For instance, we can apply the usual model-driven development methods over our extended model to generate its automatic implementation in any technology platform.

To make the proposed approach viable, we have developed a visual editor prototype that allows to design BPMN diagrams (see the tool of Fig. 2.1) and to automatically generate the corresponding workflow subschema (Fig. 3.1) and its process constraints, according to the guidelines presented in this paper. In particular, given the XML representation of the workflow model and the XMI representation of the initial domain model (in particular the XMI version used by *MagicDraw*), our tool generates a new XMI file containing the workflow-extended model and the process constraints.

Future work will include the extension of our translation patterns to directly cover the full expressivity of the BPMN notation and the study and comparison of different implementation options for the workflow-extended models depending on application-specific requirements. Also, we would like to explore the possibility of using our extended model as a bridge to facilitate reverse-engineering of existing applications into their original workflow models and to ease keeping them aligned. Finally, we plan to develop a method that, from the generated process constraints, is able to compute the list of activities that can be enacted by a user in a given case (i.e. those activities that can be created without violating any of the workflow constraints according to the case state at that specific time.

Acknowledgments

This work has been partially supported by the Italian grant FAR N. 4412/ICT, the Spanish-Italian integrated action HI2006-0208, the grant BE 00062 (Catalan Government) and the Spanish Research Project TIN2005-06053.

References

1. van der Aalst, W.M.P., Kumar, A.: A reference model for team-enabled workflow management systems. Data & Knowledge Engineering 38, 335–363 (2001)
2. Bae, J., Bae, H., Kang, S.-H., Kim, Y.: Automatic Control of Workflow Processes Using ECA Rules. IEEE Transactions on Knowledge and Data Engineering 16, 1010–1023 (2004)
3. Brambilla, M., Cabot, J., Comai, S.: Automatic Generation of Worfklow-extended Domain Models (extended version), Available: http://www.elet.polimi.it/upload/mbrambil/WFdomainmodels/

4. Brambilla, M., Deutsch, A., Sui, L., Vianu, V.: The Role of Visual Tools in a Web Application Design and Verification Framework: a Visual Notation for LTL Formulae. In: Lowe, D.G., Gaedke, M. (eds.) ICWE 2005. LNCS, vol. 3579, pp. 557–568. Springer, Heidelberg (2005)
5. Brambilla, M., Ceri, S., Fraternali, P., Manolescu, I.: Process Modeling in Web Applications. ACM Transactions on Software Engineering and Methodology 15, 360–409 (2006)
6. Cabot, J., Raventós, R.: Conceptual Modelling Patterns for Roles. Journal on Data Semantics V, 158–184 (2006)
7. Cabot, J., Teniente, E.: Incremental Evaluation of OCL Constraints. In: Dubois, E., Pohl, K. (eds.) CAiSE 2006. LNCS, vol. 4001, pp. 81–95. Springer, Heidelberg (2006)
8. Combi, C., Pozzi, G.: Temporal Conceptual Modelling of Workflows. In: Song, I.-Y., Liddle, S.W., Ling, T.-W., Scheuermann, P. (eds.) ER 2003. LNCS, vol. 2813, pp. 59–76. Springer, Heidelberg (2003)
9. Costal, D., Gómez, C., Queralt, A., Raventós, R., Teniente, E.: Facilitating the definition of general constraints in UML. In: Nierstrasz, O., Whittle, J., Harel, D., Reggio, G. (eds.) MoDELS 2006. LNCS, vol. 4199, pp. 260–274. Springer, Heidelberg (2006)
10. Demuth, B., Hussmann, H., Loecher, S.: OCL as a Specification Language for Business Rules in Database Applications. In: Gogolla, M., Kobryn, C. (eds.) UML 2001. LNCS, vol. 2185, pp. 104–117. Springer, Heidelberg (2001)
11. Domingos, D., Rito-Silva, A., Veiga, P.: Workflow Access Control from a Business Perspective. In: Proc. ICEIS, vol. 3, pp. 18–25 (2004)
12. Eshuis, R., Wieringa, R.: Verification support for workflow design with UML activity graphs. In: Proc. ICSE'02, pp. 166–176 (2002)
13. Ho, W.-M., Jézéquel, J.-M., Pennaneach, F., Plouzeau, N.: A toolkit for weaving aspect oriented UML designs. In: Proc. AOSD'02, pp. 99–105 (2002)
14. Hur, W., Jung, J.-y., Kim, H., Kang, S.-H.: Model-Driven Approach to workflow execution. In: Desel, J., Pernici, B., Weske, M. (eds.) BPM 2004. LNCS, vol. 3080, pp. 261–273. Springer, Heidelberg (2004)
15. IBM: WebSphere MQ Workflow, http://www.ibm.com/software/ts/mqseries/workflow/ v332/
16. KlasseObjecten: Octopus OCL Tool for Precise Uml Specifications, http://www.klasse.nl/ octopus/ index.html
17. Knapp, A., Koch, N., Zhang, G., Hassler, H.: Modeling Business Processes in Web Applications with ArgoUWE. In: Baar, T., Strohmeier, A., Moreira, A., Mellor, S.J. (eds.) Proc. UML 2004. LNCS, vol. 3273, pp. 69–83. Springer, Heidelberg (2004)
18. Koehler, J., Hauser, R., Sendall, S., Wahler, M.: Declarative techniques for model-driven business process integration. IBM Systems Journal 44, 47–65 (2005)
19. Olivé, A.: A method for the definition of integrity constraints in object-oriented conceptual modeling languages. Data & Knowledge Engineering 58, 243–262 (2006)
20. OMG/BPMI: Business Process Management Notation v.1. OMG Adopted Specification
21. Oracle. Workflow 11i, http://www.oracle.com/appsnet/technology/products/docs/workflow.html
22. Takemura, T., Tamai, T.: Rigorous Business Process Modeling with OCL. In: Nierstrasz, O., Whittle, J., Harel, D., Reggio, G. (eds.) MoDELS 2006. LNCS, vol. 4199, Springer, Heidelberg (2006)
23. White, S.A.: Process Modeling Notations and Workflow Patterns. BPTrends (2004)
24. Wynn, M.T., Edmond, D., van der Aalst, W.M.P., ter Hofstede, A.H.M.: Achieving a general, formal and decidable approach to the OR-join in Workflow using Reset nets. In: Ciardo, G., Darondeau, P. (eds.) ICATPN 2005. LNCS, vol. 3536, pp. 423–443. Springer, Heidelberg (2005)

A Practical Perspective on the Design and Implementation of Service-Oriented Solutions

Alan W. Brown, Marc Delbaere, and Simon K. Johnston

IBM Software Group
awbrown@us.ibm.com, delbaere@be.ibm.com,
skjohn@us.ibm.com

Abstract. Business-driven development is an approach that focuses on automating the path from business understanding to IT solution. IBM's experiences with customers taking a business-driven approach to develop services-oriented solutions are highlighting a number of best practices that are important to share and discuss. This paper focuses on how companies adopting a service-oriented approach are assembling the appropriate environment to be successful. The paper identifies three design techniques for SOA and describes when each of them can be used in practice, depending on the business and IT drivers and the organization's maturity. We then highlight how to use structured enterprise models together with the tools and methods to automate the design of service-oriented solutions. These scenarios and examples are playing an important role in the development of future method content and tooling requirements for IBM Rational tools.

1 Introduction

Intense pressures on developing software-intensive solutions in many business domains are the result of several convergent factors, including [1,2]:

- End users want more timely, synchronized information to be available everywhere, on demand, with no downtime.
- To be competitive, enterprise IT organizations are taking advantage of lower labor rates around the world, integrating components from a variety of suppliers, and reusing solutions across product lines and solution families.
- The fast pace of business change is leading to a greater number of enterprises focused on agile delivery of on time solutions that are "good enough" rather than miss key business deadlines.
- Today's distributed solutions platforms offer a rich collection of distributed infrastructure technologies allowing greater flexibility in how core application capabilities are deployed, optimized, and redeployed as the business needs change.

As a result, existing software design practices are under pressure to support different design styles, architectures, and team structures. The new solutions, often described under the broad umbrella of Service-Oriented Architectures (SOA), exhibit

G. Engels et al. (Eds.): MoDELS 2007, LNCS 4735, pp. 390–404, 2007.

common characteristics: a focus on business/IT alignment, support for assembly of components, and recognition that flexibility in design is paramount [3]. A great deal of attention is being given to technology support for SOA.

However, while many technology aspects of SOA solutions have been discussed in detail, there are remarkably few analyses of the kinds of design techniques organizations are adopting to create an SOA. Indeed, where these topics are addressed, they often devolve into broad generalization of interface-based design approaches (e.g., [4, 5]), or technology-specific tutorials linked to a specific implementation technology (e.g., [6, 7]).

This paper describes real-world design techniques applied by IBM customers when developing business applications in a service-oriented style. Following a discussion on design principles for SOA, this paper describes three categories of design techniques that span the services-oriented solutions we have seen. We then further elaborate services-oriented design approaches by describing a model-driven approach to SOA that uses these techniques based on a substantial set of industry models used in a range of enterprise application in the financial services domain.

2 Design for SOA

The goal of service-oriented solutions is to provide the flexibility to treat elements of business processes and the underlying IT infrastructure as secure, standardized components – typically Web services – that can be reused and combined to address changing business priorities. They enable businesses to address key challenges and pursue significant opportunities quickly and efficiently.

To achieve this goal solutions adopt a specific architectural style [8] that focuses on collaborations among independent providers of capabilities. By considering this style of solution, and its common realization as web services, we recognize a set of characteristics for services that inform us of their nature and applicability. Furthermore, efficient use of services suggests a few high-level best practices, particularly in relation to more established object and component-based design approaches. In particular, services are typically:

- **Coarse-grained** — operations on services are frequently implemented to encompass more functionality and operate on larger data sets, compared with component-interface design.
- **Defined via their interfaces**— services implement separately defined interfaces. The benefit of this is that multiple services can implement a common interface and a service can implement multiple interfaces.
- **Discoverable** — services need to be found at both design time and run time, not only by unique identity but also by interface identity and by service kind.
- **Instantiated as a single instance** — unlike component-based development, which instantiates components as needed, each service is a single, always running instance that a number of clients communicate with.
- **Loosely coupled** — a SOA is loosely coupled architecture because it strictly separates the interface from the implementation. In addition, runtime discovery further reduces the dependency between service consumers and providers and makes SOA even more loosely coupled. Services are connected to other services

and clients using standard, dependency-reducing, decoupled message-based methods such as XML document exchanges.

- **Asynchronous** — in general, services use an asynchronous message passing approach; however, this is not required.
- **Reusable** — services are assets that can be reused in several contexts regardless of the component architecture.

Some of these criteria, such as a focus on interfaces and ease of discovery, are also used in existing component-based development approaches. However, the major difference between SOA and component-based development is the fact that SOA focuses only on interfaces, their discoverability, and emphasizes loose coupling, particularly over network infrastructures. While in contrast, component-based development focuses on the component execution environment, and the acquisition and deployment of software components in that environment. Collectively, these characteristics differentiate a services-based solution from one developed using component architectures.

2.1 SOA Design Elements

In light of these SOA characteristics, when developing service-oriented solutions, our experiences indicate that there are 5 elements to a solution that are essential to such an architectural style [17]. Each must be modeled using appropriate notations and techniques. Briefly, as illustrated in Figure 1, these 5 elements are:

- *Service* – A service is a software resource (discoverable) with an externalized service specification. This service specification is available for searching, binding, and invocation by a service consumer.
- *Message* – Many service-oriented solutions are inherently message based in that they communicate by passing messages. A message is a subset of an information model which is passed into or out of a service invocation.
- *Interaction* – The behavioral specification of a service is described as an interaction, defining the protocol between the service and the consumer, a service may be stateful, or it may have certain conversational requirements fulfilled by the client.
- *Composition* – A Service Collaboration represents a configurable, externalized flow description sequencing a set of message exchanges between services.
- *Policy* – The policy information specifies constraints and governance regarding the operation of the service. Examples of policies include security, availability, quality of service and so forth; these also represent non-functional requirements on the solution as a whole.

In designing for a service-oriented solution, these 5 elements are the basis for the service model. The service model is used to conceive as well as document the design of the software services. It is a comprehensive, composite artifact encompassing all services, providers, specifications, partitions, messages, collaborations, and the relationships between them. The service model primarily sets the architecture, but is also a vehicle for analysis during the elaboration phase. It is then refined by detailed design decisions during the construction phase.

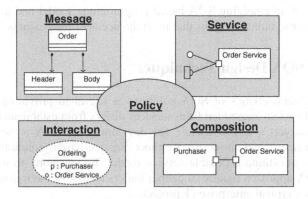

Fig. 1. Elements of an SOA Design Style

2.2 Developing SOA Solutions with the IBM Rational Software Delivery Platform

Our experiences at IBM over the past 3 years in developing and delivering service-oriented solutions have led to the creation of a number of methods, practices, and automated capabilities designed to support their practical realization. The IBM Rational Software Delivery Platform is an integrated set of capabilities for business-driven development of solutions [9]. It supports the design, implementation, testing, deployment, and management of services-based solutions, particularly when those solutions are realized as web services.

From these experiences we can distill some of the key technologies pertaining to SOA design that are often in use, these include the following.

- Business-level tools for modeling and simulation that support description of business-level services in the context of business optimization or business improvement efforts (e.g., WebSphere Business Modeler).
- Business service choreography and assembly using a standards-based business service definition and execution approach, optimized for integration with business integration runtimes (e.g., WebSphere Integration Developer).
- Support for architectural design of services and service assemblies with governance and control of those architectures through the application of SOA patterns. Prepackaged transformations and an open extensibility platforms support a rich set of transformations from analysis and design models to implementation (e.g., Rational Software Architect).
- A set of capabilities for web services creation and validation consisting of a comprehensive set of web services tools to discover, create, build, test, deploy and publish Web services. This includes automated generation of fully functional component tests for Java (classes, interfaces), EJBs, and Web Services. (e.g., Rational Application Developer).

Collectively, these capabilities provide an integrated platform that allows services to be designed, visualized, assembled, constructed, tested, deployed and monitored.

They make use of an emerging SOA-based programming model and design notation that make it easier to build solutions that meet the needs of the business

3   Practical SOA Design Techniques

Analyzing the characteristics of SOA solutions is useful in providing insights into how a delivered system assembled from services differs from established object-based techniques. However, a key challenge for developing an SOA is how to identify appropriate services, and how to architect service-based solutions as service interactions that constitute a flexible service model. Furthermore, to manage risk in adopting an SOA approach requires a deeper understanding of the impact on the roles and activities of a typical enterprise IT project.

Our experiences with users of the IBM Rational Software Delivery Platform have allowed us to observe three primary design patterns being used in the creation of service-oriented solutions: message-centric, services-centric, and collaboration-centric. Here we outline these three design patterns, and illustrate them with specific examples abstracted from real customer experiences. We focus on the impact of these design techniques on the roles and tasks of an enterprise IT project.

3.1   Message-Centric Design

3.1.1   Main Characteristics
Many useful applications of an SOA style involve integrating several existing solutions with overlapping capabilities. A key element of any solution is synchronization and mediation between data sources. Taking a message-oriented view of these situations can frequently be very useful. As a result, the most useful services may be those that act as data management services and the identification of services will tend to focus more on the data model, domain model, or business analysis model.

3.1.2   Practical Example
A large insurance provider was developing new applications to support one of their lines of business. The alignment between the business and supporting IT in this company was reasonably good, with a good focus on delivery of value to the business through IT.

The project in this scenario is the extension of two claims processing systems to provide service-oriented interfaces that will enable their tactical integration. The strategic objective was to be able to retire one system and move all claims processing to a single system.

The line of business folks expressed the desire that the services be in support of some existing business processes, but since these processes change over time, coupling the services to the process would have been impractical. In response, the IT architect for this business area suggested that a more appropriate approach might be to focus on the business artifacts in the process, primarily of course the claim itself. These artifacts were already defined as part of their enterprise architecture efforts based on IBM Insurance Application Architecture (IAA) [10].

The main effort was then in understanding the mappings between these enterprise artifacts and the system-specific representations.

3.1.3 Key Activities

The Business Analyst documented requirements, both functional and non-functional, to support current and planned integration of systems. Process models were developed for both current systems and their intersection.

The IT Architect took these requirements and worked with the Business Analyst to develop more detailed process models that focused on the flow of artifacts in common between the two systems. The IT Architect also worked with the Business Analyst in understanding the data requirements of the current systems and their relationship to the IAA data model.

The Data Architect worked with the IT Architect in drawing up a schema for the "canonical" representations of the artifacts in the new processes.

The IT Architect developed an initial service model that partitioned the process activities into a set of service operations. This model included the message requirements for each operation and the relationship of the messages to the artifact data model.

The IT Architect worked with Integration Specialists to define the mappings required from this new model to the current implementations. These mappings were to be implemented in the middleware but must not be so onerous as to introduce major performance issues.

The IT Architect worked with Senior Developers in elaborating requirements for adapters that may have been needed in the existing applications. The Developers were responsible for the delivery of these adapters to support the new process.

3.2 Service-Centric Design

3.2.1 Main Characteristics

Adapting a common design technique from both Object-Oriented and Component-based Development, a useful way to begin service identification involves identifying the services in some business or technical domain through a domain analysis technique such as noun-verb analysis, or CRC cards. Then, as collaborations between services are analyzed the operations (responsibilities of the service) are identified and added to the service specification.

In addition, some refactoring takes place by aggregating services and operations into more meaningful units. Driven by business-level use cases or typical user scenarios, this helps to ensure that the services defined are not too fine grained and represent more reusable business-focused capabilities.

3.2.2 Practical Example

An on-line provider of components for the chemical process industry wanted to provide access to the purchasing portal via more programmatic means than through the web itself. Previously, the business provided a web portal through which customers could buy parts as well as phone and fax; at some point they had attempted to implement an EDI purchasing solution but the cost of initial deployment as well as the required infrastructure proved too costly. It was felt that a number of current customers were moving to other providers that offer such access as they are able to integrate them better into their more automated processes.

In this case the project was far more driven by the IT organization, as the business goals were simply to provide a mechanism to perform the same ordering operations without having either the customer spend time entering into the web portal or the business entering orders from faxes. So, this left the IT organization to make many decisions independently, and obviously to produce a solution that did not require the cost of the previous EDI attempt.

So, two important design decisions/observations were made by the senior team in the IT organization.

1. The set of capabilities provided must *not* be any less than the customer is able to do via the web or manual ordering.
2. The set of operations should not be driven by *the organization's* business processes, but be used by a customer as a part of their processes with the result that any implementation will have the least possible set of assumptions about their use.

So in this case the IT architect produced a set of use case documents describing the kinds of capabilities expected by customers and agreed them with the business team. The use cases were then broken down into a set of discrete operations and the data requirements (messages) specified. These operations were then aggregated into a set of three services for search, ordering and account management; this step was driven more by experience and judgment than any precise metrics and so was necessarily iterative in nature.

Once the services and the set of operations were known the team defined the service specifications more formally, including the protocols for each interface and any policy information that was required by service consumers. This step was extremely important as it provided the consumers with information that could be understood from simply reading the WSDL deployed by the business once the services were ready for use.

3.2.3 Key Activities

The Business Executive communicated a need to provide the business' clients access to *business* services programmatically as they felt the business was losing revenue to larger providers that currently provided such capabilities.

The IT Architect introduced the proposal to develop a set of services that provided the core capabilities of the web ordering system. The proposal included a set of use cases describing the activities customers were already performing via the web (developed with a Business Analyst from the on-line business). Also included was an early model demonstrating the services and their relationship to the existing back-end system (demonstrating services, providers and partitions representing the service *zone* and back-end application zone).

The IT Architect, on approval of the project, iterated on the initial model to include gateways on the back-end application zone that reflected the services provided by the applications and the permissible access to those services.

In parallel the IT Architect had a Senior Developer look at the existing web application and how it interacted with the back-end system, mining out any required business logic and requirements to be implemented by the services.

The IT Architect and Business Analyst agreed the set of required actions to be provided as operations provided by the services and any constraints/policies required. This iteration of the model then showed three concrete services as well as initial service specifications, although these only specified the interfaces at that time.

The IT Architect worked with Developers on the web application team to understand the specific data requirements for the identified set of operations. This was handed to a Data Architect with XML skills to develop a message model that generated the schemas for the messages to be consumed and produced by the operations.

The Senior Developer began to connect the service model with an implementation model (J2EE) that would be used to describe the components that provide business logic and connect with the existing back-end applications.

The IT Architect and Senior Developer finally refined the service model with details on the service specifications; primarily providing protocol details that enabled a consumer to understand how to interact with the services.

That approach resulted in a complete set of models:

1. A use-case model describing the customer expectations.
2. A service model describing both the consumer view (service specifications) and provider view (services, partitions, etc.).
3. A set of implementation models describing the software components realizing the service specifications.

It was then possible for the Developers to implement the pilot system ready for a select set of customers to use.

3.2.4 Follow-Up Actions

Following the successful deployment of the pilot project it was noted that the provision of the services to clients involved a parallel implementation of certain functionality to the web system. A new project was proposed, and included in some regularly planned updates to the web application, to reconfigure the web application to reuse these services. Such a reconfiguration moved certain business logic from the web application into the services and so a single set of components were then used for both the web- and service-centric customers.

To address performance concerns by the web application team the services were also reconfigured to provide more efficient Java bindings when accessed from the web application in addition to the SOAP/HTTP bindings provided to partners.

3.3 Collaboration-Centric Design

3.3.1 Main Characteristics

Many organizations have a mature view of the business processes that allow them to focus on the major roles and activities that form their business, and how the major stakeholders interact to perform a particular business-driven task. This perspective provides an "outside in" view into the service design, highlighting the collaborations among services, and thereby allowing identification of specific behavior required of each service.

In particular, this approach provides a very tight connection between business stakeholders and the IT organization implementing services by allowing for service operations to directly support the tasks identified in process models. In general, business process models focus on tasks performed by roles and/or resources in an organization to accomplish some task, usually to provide value in the form of product or service to an external party such as a customer or partner.

3.3.2   Practical Example

In this scenario we describe a mid-tier banking organization that decided to pursue a strategy of growth through acquisition and an organization for whom Information Technology is not just a must-have but a competitive advantage in absorbing acquired operations.

The bank developed a leading portfolio of applications, and a portfolio that is carefully managed to ensure both cost and capability efficiency. However, it was clear that certain technical and cultural inhibiters existed in the vision to provide a common set of applications across the whole organization. These inhibiters directed the company to look at the provision of IT capabilities through services and to develop all new capabilities in such a style. The organization has, over the years, developed a very process-centric mind set, and it is these well understood and common processes that allow them to effectively integrate acquired organizations but also to direct their IT in support of these processes. The IT organization has been organized to be very close to the business teams that define the processes that run the banking group – i.e. for this organization the well documented *Business/IT Gap* does not exist simply because the group has organized to eliminate it.

The bank brought one of its legacy applications into the service portfolio, which gave them the opportunity to look at one of the process models. The process model for account consolidation, a business service whereby the organization assists customers in making more efficient use of their accounts, contained a number of "black-box" activities. In the models, it was common for an activity to be performed either by systems that are not enabled as part of the service portfolio, or in ways that were not yet detailed. In this case the models denote these as "black-boxes" and no more details were provided. In this example it was noted that certain legacy applications participate not only in three activities in the account consolidation process, but an analysis of other process models uncovered two other processes where the application was being used.

In the method that the organization has come to rely on, the process models are the primary artifact for analysis and requirements gathering. The process itself represents the complete embodiment of the functional requirements of the business on the IT (and personnel) and attached and formally defined documents describe non-functional requirements. Such process models are carefully maintained under configuration management and published for each operational process so they not only guide the development of the process but also provide material for training the bank's staff in the processes.

The service portfolio mentioned earlier was a RAS [13] repository that allows for searches based on criteria defined by the RAS manifest in each asset. The assets were the deployed services and the manifest consisted of a core profile for describing services and extensions describing business information such as ownership, policy and so forth.

In developing the services to provide the capabilities of the legacy application, the team (combined business and IT representation) focused very much on the collaboration between services as the realization of a business process. In this way the business experts were able to see the result as a traditional business process model whereas the IT side saw the *same model* as a collaboration between services.

3.3.3 Key Activities

The Business Analyst either created a new or updated an existing business process model. This model described the new process as envisioned by the business; this process update also included the definition of business documents managed, consumed or produced by the process activities.

The IT Architect created a gap analysis document that compared the requirements of the new business process against the existing service portfolio.

The Business Analyst defined Key Performance Indicators as a part of the business process itself. This implies that the services participating in the process supported metrics that could be queried to ensure conformance with the KPIs.

The Data Architect refined the business document definitions from the business process model, ensuring commonality with existing schema and then developed a message model that could be used to generate the XML schemas required for the services.

The IT Architect updated the services models with either new services or added new specifications to existing services. The message models developed by the data architect were used (or reused) for these service specifications.

The IT Architect and Business Analyst updated the process model defined earlier to map activities to the new or updated services. The model was therefore complete with respect to the service update, and it could be versioned and published.

The IT Architect and Senior Developer developed detailed service specifications that included the protocol and policy information. Such a detailed specification acted as a contract between the provider and consumer, and so could not be broken by either party.

The Developers created adapters and/or transforms against the legacy application for implementation of the detailed specifications.

The Developers developed the metric required for monitoring, these were generally specific operations required for the monitoring infrastructure to query the service for general state and state of the metrics.

The Integration Specialist updated process choreographies with new services, in this case the business process was transformed into a choreography language (BPEL). In some cases the generated choreography needed to be updated with some additional information before deployed.

The Business Analyst updated the monitoring middleware with the business KPI definitions, and defined their relationship to the detailed metrics. As a result, the new services could then be monitored.

4 Use of Enterprise Models to Automate Realization of Service-Oriented Solutions

Enterprise models can provide an efficient context for designing a service-oriented solution. The three identified SOA design techniques can each leverage a different facet of the enterprise models.

Enterprise models can be designed internally or can customize existing industry models such as IBM's Insurance Application Architecture (IAA) or IBM's Information Framework (IFW) for the banking industry. [10, 11, 12]. The decision of building or customizing these models is out of scope of this paper but this section will assume that the enterprise models are built in a style compatible with the out-of the box IBM Industry Models (and will use their terminology).

4.1 Enterprise Models Relevant to SOA

A variety of enterprise models are relevant in an SOA context, as represented in Figure 2. Here we see:

- The **process models** provide an enterprise-wide definition of business processes. They provide a natural fit for the top layer of an SOA.
- The **use-case models** describe the possibilities of automating parts of the process layer.
- The **business model** is a normalized domain model that describes all of the main concepts of the business domain.
- The **message model** (although not a separate model per se) defines how the data structures from the business model are aggregated and denormalized to represent all data transfers across collaborating services.
- Finally, the **service model** provides the design of the services to support the use cases as well as the collaborations and dependencies between these services.

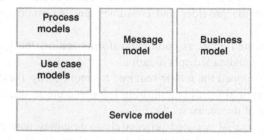

Fig. 2. Enterprise Models relevant to SOA

4.2 Application of Enterprise Models to SOA

This section describes at quite a high level the generic method for deploying enterprise models in SOA engagements. We will then examine in the next sections how to apply enterprise models in the context of the three previously described design techniques.

As shown in Figure 3, business-driven SOA engagements should ideally start with process analysis. In order to achieve efficiency, the process analysis should be conducted across the enterprise. This point is particularly important as it drives the analysis from the very beginning in the direction or re-usable enterprise services.

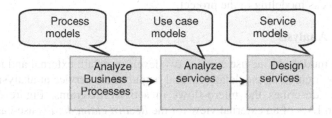

Fig. 3. A Simplified Method for SOA

The service analysis step consists of customizing the use case models to support the automation of the customized processes. It focuses on service identification, data aggregation and definition of the micro-flows at analysis level in activity diagrams.

Finally, the service design consists of fully defining the service interfaces and collaborations.

4.3 Process Analysis

Analyzing the business processes is typically an activity performed by business analysts in conjunction with subject matter experts. The goal is to obtain a representation of how the business is (or should be) run.

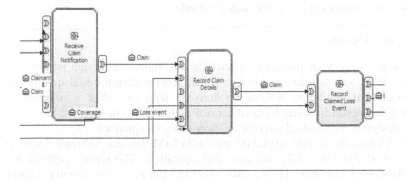

Fig. 4. A Fragment of a Business Process in IAA

Figure 4 shows a small subset from the Insurance Application Architecture of a business process in IBM's WebSphere Business Modeler describing a small number of steps in the claims process.

Although from a theoretical method point of view, one should always start business-driven SOA engagements with process analysis, we have seen in section 3 that in practice it not always the case. Indeed, this requires a certain level of maturity both in the way an enterprise can start to think horizontally at the business level (enterprise-wide view of the processes) and in the business-IT alignment.

As we have seen in the first two design techniques (message-centric and service-centric), it is possible to already derive a lot of benefits from SOA without fully applying process modeling in the project.

4.4 Service Analysis

The industry models define use cases at two levels of detail: external and internal. The external view focuses on the data input and output of the service at analysis level. The internal view describes the micro-flows in activity diagrams. Figure 5 shows an example from IAA of an external view for the *Record claim details* use-case.

Fig. 5. A Sample Use Case in IAA

In most cases, it is a necessary step to go through system use-case analysis but as we have seen, when the main focus is on service rationalization (message-centric design), it is possible to skip that step and to start from the existing service designs as the basis for rationalizing their message payloads.

4.5 Service Design

A key step in any SOA project is service design. Indeed, although purely business-oriented initiatives that do not involve actual service design could qualify as SOA engagements, it would be difficult to derive a path to realize a solution without activities that result in some form of service architecture. In fact, we have seen that service design is an essential step for all three design techniques.

The IAA models provide service definitions in IBM Rational Software Architect. The services are defined by UML interfaces with operations. The service payload design is the application at design level of the data boundaries defined in the system use cases.

5 Summary and Conclusions

Flexibility is essential to organizations today as they seek to react more quickly to the changing demands of their customers, announcements by competitors, and the evolving business environment. The role of software in many businesses is now seen as central to their ability to compete effectively and efficiently. Taking a service-oriented approach to the systems being developed helps to focus businesses on what is essential to them – the services they offer to customers. It also helps IT professionals to look at the systems that support the business in a different way – as composable solution fragments that must be assembled to meet evolving business needs.

In this paper we have focused on the importance of designing for and with services to create quality service-oriented solutions that meet the needs of organizations for flexibility and agility of their enterprise IT systems. We described three design techniques which we believe to be representative of the needs and concerns of our customers.

We have seen that choosing among these techniques is based on the particular domain context in which they are required, and will be influenced by the specific alignment between the business and IT drivers. We have also identified that all of these techniques are important in a model-driven approach to SOA through use of structured industry content as provided by IBM industry models.

These practical lessons are playing an important role in how IBM is delivering tools, technologies, and methods in support of service-oriented design. The recent release of a UML profile for Software Services [14] supported by a Rational Unified Process plug-in for service-based design [15] builds upon these ideas and encourages their direct application in IBM's commercial tooling. These technologies will be updated as further experience with these techniques emerges.

References

1. Friedman, R.: The World is Flat: A Brief History of the 21st Century, Farrar, Straus and Giroux (2005)
2. Bhagwati, J.: In Defence of Globalization. Oxford University Press, Oxford (2004)
3. Krafzig, D., Banke, K., Slama, D.: Enterprise SOA. Prentice Hall, Englewood Cliffs (2005)
4. Bieberstein, N., et al.: Service-Oriented Architecture (SOA) Compass: Business Value, Planning, and Enterprise Roadmap, IBM Press (2005)
5. Herzum, P., Sims, O.: Business Component Factory: A Comprehensive Overview of Component-Based Development for the Enterprise. Prentice Hall, Englewood Cliffs (2002)
6. Erl, T.: Service-Oriented Architecture: A Field Guide to Integrating XML and Web Services. Prentice Hall, Englewood Cliffs (2005)
7. Barry, D.: Web Services and Service-Oriented Architectures. Morgan Kauffman, Seattle (2005)
8. Shaw, M., Garlan, D.: Software Architecture: Perspectives on an Emerging Discipline. Addison Wesley, Reading (1998)
9. Brown, A.W., Delbaere, M., Eeles, P., Johnston, S., Weaver, R.: Realizing Service oriented Solutions with the IBM Software Development Platform. IBM Systems Journal 44(4), 727–752 (2005)
10. IBM Insurance Application Architecture, http://www.ibm.com/industries/financialservices/doc/content/solution/278918103.html

11. Delbaere, M., Dermody, P.: IAA General Information Manual, IBM Financial Services Centre publications, http://www.ibm.com/industries/financialservices/doc/content/bin/fss_iaa_gim_june_2005.pdf?g_type=rhc
12. Huschens, J., Rumpold-Preining, M.: IBM Insurance Application Architecture (IAA). In: Bernus, P., Mertins, K., Schmidt, G. (eds.) Handbook on Architectures of Information Systems, Springer, Heidelberg (1998)
13. OMG: Reusable Asset Specification (RAS), Version 2.2 (2005), http://www.omg.org/technology/documents/formal/ras.htm
14. Johnston, S.K.: UML2.0 Plug-in for Software Services, IBM developerworks (April 2005), http://www.ibm.com/developerworks/rational/library/05/419_soa/
15. Johnston, S.K.: Modeling Service Oriented Solutions, IBM Developerworks (July 2005), http://www.ibm.com/developerworks/rational/library/jul05/johnston/index.html
16. Carlson, D.: Eclipse Distilled. Addison Wesley, Reading (2005)
17. Johnston, S.K., Brown, A.W.: A Model-driven Development Approach to Creating Service-oriented Solutions. In: Dan, A., Lamersdorf, W. (eds.) ICSOC 2006. LNCS, vol. 4294, pp. 624–636. Springer, Heidelberg (2006)

Constructive Techniques for Meta- and Model-Level Reasoning

Ethan K. Jackson and Janos Sztipanovits

Institute for Software Integrated Systems,
Vanderbilt University, Nashville, TN 37235, USA
ejackson@isis.vanderbilt.edu
janos.sztipanovits@vanderbilt.edu

Abstract. The structural semantics of UML-based metamodeling were recently explored[1], providing a characterization of the models adhering to a metamodel. In particular, metamodels can be converted to a set of constraints expressed in a decidable subset of first-order logic, an extended Horn logic. We augment the constructive techniques found in logic programming, which are also based on an extended Horn logic, to produce constructive techniques for reasoning about models and metamodels. These methods have a number of practical applications: At the meta-level, it can be decided if a (composite) metamodel characterizes a non-empty set of models, and a member can be automatically constructed. At the model-level, it can be decided if a submodel has an embedding in a well-formed model, and the larger model can be constructed. This amounts to automatic model construction from an incomplete model. We describe the concrete algorithms for constructively solving these problems, and provide concrete examples.

1   Preliminaries - Metamodels, Domains, and Logic

This paper describes constructive techniques, similar to those found in logic programming, for reasoning about *domain-specific modeling languages* (DSMLs) defined with metamodels. Before we proceed, we must describe how a metamodel can be viewed as a formal object that characterizes the well-formed models adhering to that metamodel. We will refer to the models that adhere to metamodel X as *the models of metamodel X*. In order to build some intuition for this view, consider the simple *DIGRAPH* metamodel of Figure 1. The models of *DIGRAPH* consist of instances of the *Vertex* and *Edge* classes such that *Edge* instances "connect" *Vertex* instances. In other words, *DIGRAPH* characterizes

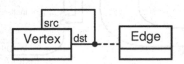

Fig. 1. DIGRAPH: A simple metamodel for labeled directed graphs

G. Engels et al. (Eds.): MoDELS 2007, LNCS 4735, pp. 405–419, 2007.

a class of labeled directed graphs. Thus, a model might be formalized as a pair $G = \langle V \subseteq \Sigma, E \subseteq V \times V \rangle$, where Σ is an alphabet of vertex labels. If Σ is fixed, then the set $\mathscr{G}$ of all models of $DIGRAPH$ is: $\mathscr{G} = \{(V, E) | V \subseteq \Sigma, E \subseteq V^2\}$. This is the classic description of labeled digraphs, and at first glance it might appear possible to extend this description to characterize the models of arbitrary metamodels. Unfortunately, UML-like metamodels[2][3] contain a number of constructs that deny a simple extension of graph-based descriptions. The *UNSAT* metamodel of Figure 2 illustrates some of these constructs. First, classes

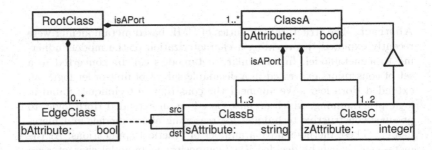

Fig. 2. UNSAT: A complex metamodel with no finite non-trivial models

may have non-trivial internal structure. For example, classes of *UNSAT* have typed member fields (called *attributes*). An instance of *ClassA* has a **boolean** field named *bAttribute*. Classes also inherit this structure, e.g. an instance of *ClassC* has two attributes, *bAttribute* and *zAttribute*, via inheritance. Instances may contain other instances with constraints on the type and number of contained instances. An instance of *ClassA* must contain between 1 and 3 instances of *ClassB*. Second, internal instance structure can be "projected" onto the outside of an instance as *ports*. The containment relation from *ClassA* to *RootClass*

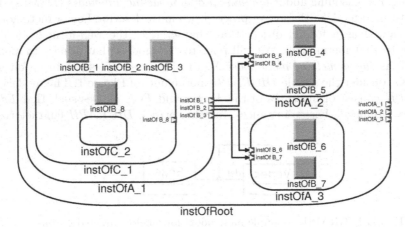

Fig. 3. Model that (partially) adheres to the UNSAT metamodel

has the *isAPort* rolename, requiring that all contained instances of *ClassA* appear as interfaces on the outside of the containing instance of *RootClass*. Figure 3 shows a model with containment and ports. The hollow oblong shapes denote instances that can contain other instances, and the small squares with white arrows on the oblongs' borders denote ports. For example, the outermost container *instOfRoot* is an instance of the *RootClass* and contains three instances of *ClassA*. Each *ClassA* instance appears as a port on the far right-hand side of *instOfRoot*. Containment and ports are a useful form of information hiding, but they also complicate matters because ports permit edges to cross hierarchy. For example, the edges in Figure 3 connect instances of *ClassB* together even though these instances are not contained in the same parent. Furthermore, the edges are actually contained in the *RootClass* instance, even though the endpoints are not. The third major complication arises because edges are not simple binary relations. In UNSAT, edges are instances of *EdgeClass*, and so each edge has a member field named *bAttribute*. In general, edges must be distinguishable (i.e. labeled), otherwise it would not be possible to reliably determine the values of member fields. In fact, the UML-notation (correctly) implies that edges are ternary associations between an edge label, source label, and destination label.

Graph-based formalisms have been used extensively by the model transformation community, and provide reasonable approximations of model structure for the purpose of transformation. However, in this paper we do not focus on model transformation, but rather we explore techniques for reasoning about all the details of metamodel and model structure. One approach to characterizing realistic model structure might be to combine all existing graph extensions and consider models to be *hierarchical*[4], *typed, attributed*[5] *hypergraphs* with labeled edges. However, even this would not handle all aspects of modern metamodeling languages, and it would produce a brittle and unwieldy formalism. In [1] we present an alternative approach to model structure based on formal logic, which we briefly outline now. In order to present our view, we begin with the concept of a *domain* (in the sense of *domain-specific modeling languages*). A domain $D = \langle \Sigma, \Upsilon, \Upsilon_C, C \rangle$ is a quadruple where Σ is an (infinite) alphabet for distinguishing model elements, Υ is a finite signature for encoding model concepts, Υ_C is a finite signature for encoding model properties, and C is a set of logical statements (constraints) for deriving model properties. A *model realization* is set of terms from the *term algebra*[6] $T_\Upsilon(\Sigma)$ over signature Υ generated by Σ. The set of all possible model realizations is $\mathcal{P}(T_\Upsilon(\Sigma))$, i.e. all subsets of terms. We will use the notation $(f, n) \in \Upsilon$ to indicate that *function symbol f* of arity n is a member of the signature Υ.

Example 1. The domain of labeled digraphs $D_{\mathcal{G}}$ has the model realizations given by the signature $\Upsilon = \{(v, 1), (e, 2)\}$ and a countably infinite alphabet ($|\Sigma| = |\aleph_0|$). These two symbols encode the concepts of vertex and edge. Vertices are encoded using the unary function symbol v and edges are encoded using the binary function symbol e. Some model realizations include:

1. $M_1 = \{\ v(c_1), v(c_2), e(c_1, c_2)\ \}$, a 2-path from a vertex c_1 to a vertex c_2.
2. $M_2 = \{\ v(c_3), e(c_3, c_4)\ \}$, a dangling edge starting at vertex c_3.

3. $M_3 = \{\, v(e(c_5, c_6)), v(v(c_7))\}$, a structure that is not a graph at all.

where the symbols written in `typewriter` font indicate members of the alphabet.

The term algebra easily captures arbitrary n-ary concepts and permits concepts to be combined in complex ways. Item 3 of Example 1 shows that function symbols can be arbitrarily nested. Notice also that not all model realizations combine the modeling concepts in ways that match our intentioned meaning of the symbols. Item 1 describes a simple 2-path, but Item 2 describes a dangling edge because vertex c_4 is not in the model. Finally, Item 3 does not correspond to a graph in any obvious way, but is still a legal member of $\mathcal{P}(T_T(\Sigma))$.

The set of model realizations of a domain contains all possible ways that the concepts can be used together. In fact, with a single operator f of arity greater than or equal to one, and an alphabet with at least one element, a countably infinite number of terms can be generated. (Consider a successor operation $succ$ and $\Sigma = \{0\}$.) Thus, for all non-trivial cases the number of possible model realizations is uncountably infinite. Therefore $\mathcal{P}(T_T(\Sigma))$ will typically contain many model realizations that use the function symbols contrarily to our intentions. In order to counteract this, we must define a set of model properties, characterized by another signature T_C, and a set C of logical statements for deriving model properties. For simplicity, we assume that T_C simply extends the signature of T (i.e. $T_C \supset T$). For example, the property of directed paths could be captured by: $T_C = \{(v, 1), (e, 2), (path, 2)\}$ and $C = \{\forall x, y, z\ (e(x, y) \vee path(x, y)) \wedge e(y, z) \Rightarrow path(x, z)\}$. The symbol $path(\cdot, \cdot)$ encodes the concept of a directed path between two vertices. The single logical statement in C defines how to derive the paths in a digraph. The keyword *derive* is important, and there are some subtle points to be made about derivation.

Classically, the notion of a derivation is represented by a *consequence operator*, written $\vdash$, which maps sets of terms to sets of terms $\vdash\colon \mathcal{P}(T_{T_C}(\Sigma)) \to \mathcal{P}(T_{T_C}(\Sigma))$. A consequence operator encapsulates the inference rules of a particular style of logic, and may make use of additional *axioms* to derive terms. In our framework, the set C is the set of axioms that the consequence operator may use. Given a model M (i.e., a set of terms), $M \vdash_C M'$ denotes the set of terms M' that can be discovered from the terms M and the axioms C. A term t can be derived from a model M if $t \in M'$. We will simply write $M \vdash_C t$ to denote that $t \in M'$. Notice that using consequence operators we can extend the notation of a derivation beyond predicate logic. For example, given the simple graph M_1 (Item 1 of Example 1), we can derive the term $path(c_1, c_2)$ without make any claims about the value of this term. Classical consequence operators, in the sense of Tarski, correspond to *closure operators* and are *extensive, isotone*, and *idempotent*[6]. Later, we will discuss the consequence operators of nonmonotonic logics where the isotone property does not hold. The history of mathematical logic is rich and diverse; we will not summarize it here. Instead, we will focus on particular applications and limit our discussion to those applications. For the reader unfamiliar with this area, it suffices to remember these two points: First, consequence operators capture the derivation of terms. Second, terms are not predicates.

Among the properties that can be encoded using Υ_C and C, we require at least one property to be defined that characterizes if a model is well-formed. We permit well-formedness to be defined either positively or negatively. A *positive domain* includes the function symbol $wellform(\cdot)$ in Υ_C, and a model M is well-formed if $\exists x \in T_{\Upsilon_C}(\Sigma)$, $M \vdash_C wellform(x)$. In other words, a model is well-formed if a term of the form $wellform(x)$ can be derived for some x. A *negative domain* is characterized by the function symbol $malform(\cdot)$ such that a model is well-formed if $\forall x \in T_{\Upsilon_C}(\Sigma)$, $M \nvdash_C malform(x)$. In other words, a model is well-formed if it is not possible to prove $malform(x)$ for any x. At first glance it may appear that the positive domains have weaker definitions than negative domains. In fact, this depends on the expressiveness of the underlying logic of $\vdash$. For example, if the logic has a "negation" (which is not the usual propositional negation) then we can define $wellform(x) \Leftrightarrow \forall y \ \neg malform(y)$ for some arbitrary x. On the other hand, if the logic is restricted, then the positive domains may be strictly weaker than the negative domains.

A domain captures the set of possible model realizations and provides a mechanism to discern the good models from the bad ones. From this perspective, the set of all metamodels also defines a domain D_{meta} that characterizes all well-formed metamodels. Let the set $\mathcal{V}$ be a fixed vocabulary of function symbols and the sets Σ and Σ_v be two fixed disjoint countably infinite alphabets. Let $SIG(\mathcal{V}) = \{\Upsilon | \Upsilon : \mathcal{V} \to \mathbf{Z}_+\}$, be the set of all partial functions from $\mathcal{V}$ to the positive integers, i.e., the set of all possible signatures. Finally, let $\mathcal{F}(\Upsilon, \Upsilon_C)$ be the set of all formulas that can be defined over terms composed from function symbols of Υ, Υ_C with constants from Σ and variables from Σ_v. These parameters allow us to characterize the set of all domains $\Delta_{\mathcal{F}}$ that can be defined with a particular style of logic[1]:

$$\Delta_{\mathcal{F}} = \bigcup_{\Upsilon \in SIG(\mathcal{V})} \bigcup_{\Upsilon \subset \Upsilon_C \in SIG(\mathcal{V})} \bigcup_{C \subseteq \mathcal{F}(\Upsilon, \Upsilon_C)} (\Sigma, \Upsilon, \Upsilon_C, C)$$

A *metamodeling language* is a pair (D_{meta}, τ_{meta}) where $\tau_{meta} : D_{meta} \to \Delta_{\mathcal{F}}$ maps metamodels to domains. In [1] we show how the mapping can be constructed for realistic metamodel languages. With this approach, we can extract a precise set of domain concepts and constraints from a metamodel by applying the mapping τ_{meta}. Here we overload the notation D to also represent the set of all well-formed models characterized by the domain D.

Given these preliminaries, we now turn our attention to the analysis of domains. For example, we might like to know: *Does a domain contain any non-trivial finite models?*. It turns out that this fundamental question is difficult to answer for UML-like metamodels. Consider the *UNSAT* metamodel of Figure 2. If a model of *UNSAT* contains anything at all, then it contains an instance of *RootClass*. However, an instance of *RootClass* must contain at least one instance of *ClassA*, which in turn must contain at least one instance of *ClassC*. So far the

---

[1] Technically, we should include the property that all Υ_C signatures contain $wellform(\cdot)$ or $malform(\cdot)$. We have left this out as it unnecessarily complicates the definition of $\Delta_{\mathcal{F}}$.

constraints pose no problem. However, the inheritance operator declares that *ClassC* is a subclass of *ClassA*, so *ClassC* inherits the property that each instance must also contain at least one instance of *ClassC*. This leads to an infinite regress, so there exists no non-trivial finite model of *UNSAT*. This can be seen in Figure 3, which is a finite model that almost adheres to *UNSAT*, except that the instance *instOfC_2* does not contain another instance of *ClassC*. The degree to which we can reason about metamodels depends on the expressiveness of the constraint logic. We now turn our attention to a well-known decidable subset of first-order logic, Horn Logic.

2   Analysis of Nonrecursive Horn Domains

2.1   The Membership Problem

The simplest class of logic we examine is nonrecursive Horn logic[7]. Admittedly, this class is too small for characterizing most realistic domains, but the algorithms for manipulating this logic serve as a foundation for the more expressive logic that we describe in the next section. We begin by recalling some definitions. *Formulas* are built from terms with *variables* and logical *connectives*. There are different approaches for distinguishing variables from constants. Following the notation of the previous section, let Σ_v be an alphabet of variable names such that $\Sigma \cap \Sigma_v = \emptyset$. The terms $T_{\Upsilon_C}(\Sigma)$ are called ground terms, and contain no variables. This set is also called the *Herbrand Universe* denoted $\mathcal{U}_H$. The set of all terms, with or without variables, is $T_{\Upsilon_C}(\Sigma \cup \Sigma_v)$, denoted $\mathcal{U}_T$. Finally, the set of all *non-ground* terms is just $\mathcal{U}_T - \mathcal{U}_H$. A *substitution* ϕ is term endomorphism $\phi : \mathcal{U}_T \to \mathcal{U}_T$ that fixes constants. In other words, if a substitution ϕ is applied to a term, then the substitution can be moved to the inside $\phi f(t_1, t_2, \ldots, t_n) = f(\phi t_1, \phi t_2, \ldots, \phi t_n)$. A substitution does not change constants, only variables, so $\forall g \in \mathcal{U}_H$, $\phi(g) = g$. We say two terms $s, t \in \mathcal{U}_T$ *unify* if there exists substitutions ϕ_s, ϕ_t that make the terms identical $\phi_s s = \phi_t t$, and of finite length. (This implies the *occurs check* is performed.) We call the pair (ϕ_s, ϕ_t) the unifier of s and t. The variables that appear in a term t are $vars(t)$, and the constants are $const(t)$.

A *Horn clause* is a formula of the form $h \Leftarrow t_1, t_2, \ldots, t_n$ where h is called the *head* and $t_1, \ldots, t_n$ are called the *tail* (or body). We write T to denote the set of all terms in the tail. The head only contains variables that appear in the tail, $vars(h) \subseteq \bigcup_i vars(t_i)$. A clause with any empty tail $(h \Leftarrow)$ is called a *fact*, and contains no variables. Recall that these clauses will be used *only* to calculate model properties. This is enforced by requiring the heads to use those function symbols that do not encode model structure, i.e. every head $h = f(t_1, \ldots, t_n)$ has $f \in (\Upsilon_C - \Upsilon)$. (Proper subterms of h may use any symbol.) This is similar to restrictions placed on declarative databases[8]. We slightly extend clauses to permit *disequality* constraints. A Horn clause with disequality constraints has the form $h \Leftarrow t_1, \ldots, t_n, (s_1 \neq s_1'), (s_2 \neq s_2'), \ldots, (s_m \neq s_m')$, where s_i, s_i' are terms with no new variables $vars(s_i), vars(s_i') \subseteq \bigcup_i vars(t_i)$. We can now define the *meaning* of a Horn clause. The definition we present incorporates the *Closed*

World Assumption which assumes all conclusions are derived from a finite initial set of facts (ground terms) I. Given a set of Horn clauses Θ, the operator $\widehat{\vdash}_\Theta$ is called the *immediate consequence operator*, and is defined as follows:

$$M \widehat{\vdash}_\Theta = M \cup \{\, \phi(h_\theta) \mid \exists \phi, \theta,\ \phi(T_\theta) \subseteq M \text{ and } \forall(s_i \neq s'_i)_\theta \in \theta,\ \phi s_i \neq \phi s'_i \,\}$$

where ϕ is a substitution and θ is a clause in Θ. It can be proved that $I \vdash_\Theta I_\infty$ where $I \widehat{\vdash}_\Theta I_1 \widehat{\vdash}_\Theta \ldots \widehat{\vdash}_\Theta I_\infty$. The new terms derivable from I can be calculated by applying the immediate consequence operator until no new terms are produced (i.e. the least fixed point). Notice that the disequality constraints force the substitutions to keep certain terms distinct. *Nonrecursive Horn logic* adds the restriction that the clauses of Θ can be ordered $\theta_1, \theta_2, \ldots, \theta_k$ such that the head h_{θ_i} of clause θ_i does not unify with any tail $t \in T_{\theta_j}$ for all $j \leq i$. This is a key restriction; without it, the logic can become undecidable. Consider the recursive axiom $\Theta = \{f(f(x)) \Leftarrow f(x)\}$. Then $\{f(c_1)\} \vdash_\Theta \{f(c_1), f(f(c_1)), \ldots, f(f(f(\ldots f(c_1) \ldots))))\}$ includes an infinite number of distinct terms. Let $\mathcal{F}_{NH}(\Upsilon, \Upsilon_C)$ be the set of all sets of Horn clauses defined over signatures Υ, Υ_C with alphabets Σ, Σ_v. We call domains specified with formulas from $\mathcal{F}_{NH}$ *nonrecurive Horn domains* (abbreviated NHD). The first problem we wish to solve is the membership problem for positive NHDs.

Definition 1. *The membership problem for positive NHDs: Given a positive NHD D, does there exists a finite model $M \subset \mathcal{U}_H(D)$ such that $M \vdash_C wellform(x)$ for some x. The notation $\mathcal{U}_H(D)$ indicates the set of ground terms defined by the signature Υ of D.*

The membership problem for positive NHDs is the easiest problem to solve. We will solve it by actually constructing a model M for which a *wellform*$(\cdot)$ term can be derived. This is possible because nonrecursive Horn logic has an important property called *monotonicity*: If a model M derives terms M', and another model N contains M, then N must derive at least M'. In symbols, $M \subseteq N$ and $M \vdash_\Theta M'$, $N \vdash_\Theta N'$, then $M' \subseteq N'$. This property implies that an algorithm only needs to examine the "smallest" models that could derive a *wellform*$(\cdot)$ term. Our algorthims are similar to those found in logic programming, but with some necessary augmentations. Typically, logic programs are provided with a set of initial facts that form the closed world. Our task is to determine the set of facts such that if the logic program were initialized with these facts, then the desired outcome (e.g. deriving a *wellform*$(\cdot)$ term) would occur. This distinction means that our algorithms cannot rely on the fact that the closed world contains a finite number of ground terms, because these terms are not yet known. It turns out that although there are an infinite number of "small" models, these models can be partitioned into a finite number of equivalence classes; these classes can be exhaustively examined.

2.2   Finding Well-Formed Members

We have developed a theorem prover called *FORMULA* (FORmal Modeling Using Logic Analysis) which implements these techniques. Figure 4 shows a

412   E.K. Jackson and J. Sztipanovits

positive NHD of directed graphs, called *CYCLE*, using FORMULA syntax. Line 1 declares the two function symbols v (for vertex) and e (for edge). The keyword **in** marks these as input symbols, i.e. elements of the signature Υ. The remaining symbols are used to calculate properties of an input model, and are marked **priv** for private symbols, i.e. elements of Υ_C. The theorem prover will never return a model that contains a private symbol. Well-formed models of the CYCLE

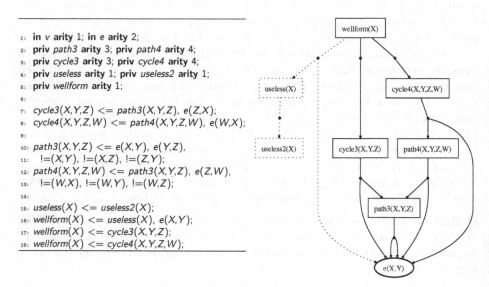

```
1:  in v arity 1; in e arity 2;
2:  priv path3 arity 3; priv path4 arity 4;
3:  priv cycle3 arity 3; priv cycle4 arity 4;
4:  priv useless arity 1; priv useless2 arity 1;
5:  priv wellform arity 1;
6:
7:  cycle3(X,Y,Z) <= path3(X,Y,Z), e(Z,X);
8:  cycle4(X,Y,Z,W) <= path4(X,Y,Z,W), e(W,X);
9:
10: path3(X,Y,Z) <= e(X,Y), e(Y,Z),
11:   !=(X,Y), !=(X,Z), !=(Z,Y);
12: path4(X,Y,Z,W) <= path3(X,Y,Z), e(Z,W),
13:   !=(W,X), !=(W,Y), !=(W,Z);
14:
15: useless(X) <= useless2(X);
16: wellform(X) <= useless(X), e(X,Y);
17: wellform(X) <= cycle3(X,Y,Z);
18: wellform(X) <= cycle4(X,Y,Z,W);
```

Fig. 4. (Left) CYCLE: a positive NHD in FORMULA syntax. (Right) Backwards chaining graph generated from goal $wellform(X)$.

domain must contain either a directed 3-cycle or 4-cycle. Lines 7,8 define the properties of 3-cycles and 4-cycles based on the properties of 3-paths and 4-paths. For example, a 3-cycle exists if there is a 3-path on vertices X, Y, Z and there is an edge from Z to X. (Note that the variable names are local to each clause.) Notice the use of disequality contraints in the definition of 3-paths and 4-paths in Lines 10-13. These constraints ensure that the paths contain unique vertices. Finally, Lines 16-18 define the derivation of $wellform(\cdot)$ terms.

The first step towards generating a well-formed model is to determine the derivation steps that lead to $wellform(\cdot)$ terms. This is done via an augmented form of *backwards chaining*. First, some definitions are necessary. We call two terms s, t *isomorphic* if there exists a substitution ϕ such that ϕ is a term monomorphism (one-to-one map), $\phi s = t$, and ϕ^{-1} is also a substitution. Clearly it holds that $s = \phi^{-1} t$. Given a set of terms T, let I_T be an equivalence relation on terms such that $(s, t) \in I_T$ if s and t are isomorphic. It is easy to see that I_T is an equivalence relation, because composition of monomorphisms yields another monomorphism. A *goal term* g is a term (with variables), and a solution M is a set of ground terms such $M \vdash_\Theta M'$ and $\exists \phi, \exists t \in M'$ $(\phi g = t)$. In other words, a solution is a model that derives a ground term unifying with the goal. The terms derived from the solution M are all ground terms, so, without lost of generality, it can be assumed that the unifier is $(\phi, id_{\mathcal{U}_T})$. Let $terms(D)$ be the union of all terms in the domain definition, (i.e. union of all heads and tails). Given a set of goals G and a domain D, let $[t]$ be the equivalence class of t in $I_{terms(D) \cup G}$. A *backwards chaining graph* $B(G)$ over a set of goal terms G is defined inductively as follows:

1. For each $g \in G$, $[g] \in V_{B(G)}$.
2. For all clauses $h_{\theta_i} \Leftarrow t_1, \ldots, t_m$ in Θ such that $[h_{\theta_i}] \in V_{B(G)}$, then $[t_i]_{1 \leq i \leq m} \in V_{B(G)}$ and there exists a directed "AND" edge $([h_{\theta_i}], \{[t_i]\}_{1 \leq i \leq m}) \in E_{B(G)}$.
3. For all clauses $h_{\theta_i} \Leftarrow t_1, \ldots, t_m$ in Θ such that h_{θ_i} unifies with some tail t_{θ_j} and $[t_{\theta_j}] \in V_{B(G)}$ then $[h_{\theta_i}] \in V_{B(G)}$ and there exists a directed edge $([t_{\theta_j}], [h_{\theta_i}]) \in E_{B(G)}$.

The right-hand side of Figure 4 shows the backwards chaining graph generated by the single goal term $wellform(X)$. There are significantly fewer vertices in the graph than terms in the domain definition, because many terms are isomorphic. $B(G)$ has several properties, though we will not prove them here. $B(G)$ is finite because the domain D has a finite number of clauses, and $B(G)$ is acyclic because D is nonrecursive. Unlike typical backwards chaining, the sinks in the graph are not ground terms, but are terms with function symbols completely in Υ. (The ground terms must be discovered.) Any sinks without this property are pruned from the graph. For example, the $useless(\cdot)$ and $useless2(\cdot)$ terms are pruned, because there are no ways to derive these terms from Υ terms. The vertices and edges in dotted lines are the pruned part of the graph. If a solution exists then there must be a directed path from every $[g]_{g \in G}$ vertex to a non-pruned sink using non-pruned edges. This holds because a solution contains only ground terms, which impose stronger restrictions on the unifier morphisms, than those imposed by the construction of $B(G)$.

The backwards chaining graph captures the various paths from the goal to possible solutions, and each path must be walked until a solution is found or it is confirmed that no solution exists. A path can be "unrolled" one at a time (as in SLD resolution[9]), or a tree can be constructed capturing every possible walk. We choose the latter in order to support other uses of FORMULA. The left-hand side of Figure 5 shows the unrolling of the Figure 4 into a *solution tree*. The tree has a root with a single AND edge having an endpoint on each goal term g. Every goal term g attached to the root receives an edge for each $v \in B(G)$ such that g unifies with v. For example, Figure 5 shows the vertex $wellform(V0)$ connected to the $wellform(V1)$. This edge indicates that $wellform(V0)$ unifies with $wellform(V1)$. The tree construction algorithm always *standardizes apart* unifying terms by instantiating them with unique variables. $wellform(V1)$ has two distinct paths in the backwards chaining graph, and each of these are unrolled into two subtrees of the $wellform(V1)$ vertex. If a clause has disequality constraints, then these appear as constraints on the edges in the solution tree.

The solution tree is viewed as a constraint system over terms. As the tree is walked, equations concerning terms are collected. A unification of terms s, t can be converted to a system of equations over variables. For example $g(X, Y)$

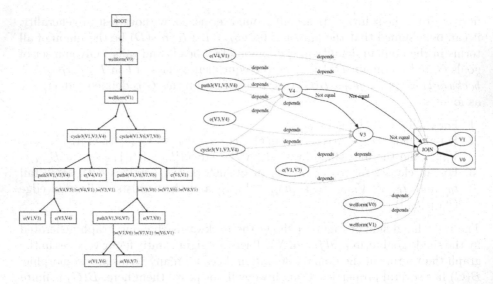

Fig. 5. (Left) Solution tree generated from backwards chaining graph of Figure 4 (Right) Constraint system shown as a forest of union-find trees

unifies with $g(Z, Z)$ if $X = Y = Z$. Clearly any unifier (ϕ_s, ϕ_t) must have $\phi_s(X) = \phi_s(Y) = \phi_t(Z)$. The correct equations are calculated by an inductive procedure as motivated in [9]. The constraint system is represented as a forest of union-find trees; a unification s, t yields a set of equations $\{s_i = t_i\}$, which is converted to operations on the forest: for each equation $s_i = t_i$ perform $join(find(s_i), find(t_i))$ where the $find(x)$ operation creates the vertex labeled x if x does not already exist. For example, there is one non-trivial union-find tree in Figure 5 resulting from the unification of $wellform(V0)$ with $wellform(V1)$, which joins $V0$ and $V1$. As terms are added to the forest, so are their subterms. Dependency links are maintained between vertices, where a term t is dependent on a term s if s is a subterm of t. An operation fails if the dependency edges form a cycle, essentially indicating that a multi-step unification fails. The dependency edges in Figure 5 are gray and labeled "depends". Disequality constraints are implemented as "Not equal" edges between vertices. Notice that all terms in the same union-find tree share the same constraints and dependencies. As trees are joined, all the constraints are moved up to the root. For example, in Figure 5 all constraint edges terminate on the *JOIN* vertex. Thus, a disequality constraint fails if a vertex is deemed unequal to itself, or a join operation moves the source and destination of a disequality edge onto the same join vertex. As the algorithm walks the solution tree, it performs operations on the constraint system. As soon as the constraint system becomes inconsistent, the algorithm restarts on an unexplored combination of subtrees. FORMULA maintains all possible restart configurations, and only fails after all restarts have been tried. Let W be the sequence of vertices visited in a walk of the solution tree. Then $CS(W)$ is the constraint system produced by that walk.

After a consistent walk W has been found, the constraint system $CS(W)$ can be converted into a set of ground terms. Notice that the sinks (ignoring disequality edges) in the constraint system are those terms for which all other terms are dependent. In fact, our construction guarantees that the sinks are just variables or ground terms. Let $sinks(CS)$ be the sinks of a consistent constraint system CS defined as follows: A union-find tree $T \in CS$ is a *sink tree* if the root has no outgoing edges, or only has outgoing disequality edges. If no leaves of the sink tree are ground, then pick a leaf and place it in $sinks(CS)$. Choose any substitution ϕ_{min} such that $\phi_{min}(X) \mapsto c_X \in (\Sigma - const(D))$, where c_X is a unique constant not appearing anywhere in the domain definition. If a variable X is in the same union-find tree as a ground term t_g, then $\phi_{min}(X) \mapsto t_g$. The values of all other variables are calculated transitively to form the full substitution ϕ_{sol}. Finally, the candidate solution M_W for walk W is

$$M_W = \left(\bigcup_{v \in W} \phi_{sol}(t_v) \right) \cap T_\Upsilon(\Sigma)$$

where t_v is the term of a vertex v in the walk W of the solution tree. M_W is a *proper solution* if no model terms of the form $f(t_1, \ldots, t_n)$, where $f \in \Upsilon$, are removed by the intersection with $T_\Upsilon(\Sigma)$. Such a term would be thrown out if it contains a subterm t_i built from symbols of $\Upsilon_C - \Upsilon$. In this case, the candidate solution is discarded and another walk through the solution tree is attempted. Applying this algorithm to the constraint system of Figure 5 gives $sinks(CS) = \{V0, V3, V4\}$. Let $\phi_{min}(V0) \mapsto c_0, \phi_{min}(V3) \mapsto c_1, \phi_{min}(V4) \mapsto c_2$. By transitivity, $\phi_{sol}(V1) \mapsto c_0$, and all variables are accounted for. Applying ϕ_{sol} to each vertex on the left-hand walk of Figure 5 gives a candidate model $M_W = \{e(c_0, c_1), e(c_1, c_2), e(c_2, c_0)\}$, which is a correctly constructed 3-cycle. It is not difficult to prove:

Theorem 1. *A positive NHD has a non-trivial finite model iff there exists a walk W such that $CS(W)$ is consistent and the candiate model M_W is proper.*

2.3   Generating Well-Formed Embeddings

These algorithms can also be used to construct well-formed models with particular *embeddings*. Let $\gamma : \mathcal{U}_H \mapsto \mathcal{U}_H$ be a term endomorphism (i.e. a homomorphism over model terms). A model M' can be *embedded* into a model M, written $M' \leq M$, if there exists a one-to-one term endomorphism (i.e. a monomorphism) such that $\gamma(M') \subseteq M$. Constructive techniques that can produce embeddings allow us to sketch a model that might be malformed, but produce a well-formed version that still contains the original model. This can be quite useful for users who do not understand all of the particular constraints of a modeling language, and would like the computer to correct mistakes. Consider the top-left graph of Figure 6. This *star* graph (S_4) is malformed with respect to the $CYCLE$ domain, because it contains neither a 3-cycle nor 4-cycle. However, with a slight modification to the algorithms above, a new model can be built that is well-formed

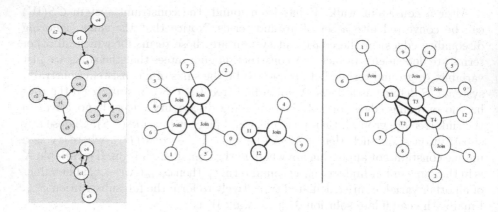

Fig. 6. (Left) From top to bottom: A malformed input model, a well-formed embedding, and a minimal embedding. (Middle) Initial constraint system showing only sink trees and disequality constraints. (Right) Minimized constraint system.

and contains an embedding of the *star* graph. Let D be a domain and let an *input model* M_I be a finite subset of model terms $T_\Upsilon(\Sigma)$. Choose any one-to-one map $\alpha : \Sigma \to \Sigma_v$ that uniquely relates constants to variables in Σ_v. Clearly α induces a monomorphism $\phi_\alpha : T_\Upsilon(\Sigma) \to T_\Upsilon(\Sigma_v)$ from terms without variables to terms that only have variables. We will use this monomorphism to encode the input model as a Horn clause. Pick any function symbol $f \notin \Upsilon_C$ and add it Υ_C with arity $|consts(M_I)|$, i.e. the arity of f is equal to the number of constants in the input model M_I. Add the following clause θ_{M_I} to D:

$$\theta_{M_I} \doteq f(\alpha(c_1), \alpha(c_2), \dots, \alpha(c_n)) \Leftarrow \bigwedge_{t_m \in M_I} \phi_\alpha(t_m) \bigwedge_{i \neq j} (\alpha(c_i) \neq \alpha(c_j))$$

where $1 \leq i, j \leq n = |consts(M_I)|$. Recall from the previous algorithms, that a solution is constructed by defining a substitution ϕ_{sol} that is determined by the sink variables $sinks(CS(W))$. Consider any solution to any goal set G where $f(\alpha(c_1), \dots, \alpha(c_n)) \in G$. By construction, the restriction of ϕ_{sol} to $sinks(CS(W))$ yields a one-to-one map. In the construction above, all pairs of variables induced by M_I have disequality constraints, so $\alpha(consts(M_I)) \subseteq sinks(CS(W))$ for any consistent walk W^2. Therefore, the restriction of any ϕ_{sol} to the terms $T_\Upsilon(\alpha(consts(M_I)))$ must be a monomorphism. Thus, $\gamma = (\phi_{sol} \circ \phi_\alpha)$ gives the embedding of M_I in any proper solution M_W for a consistent walk W.

Theorem 2. *Given an input model M_I and a positive NHD D, augmented with f and θ_{M_I}. Any proper solution to a goal set G, where $f(\alpha(c_1), \dots, \alpha(c_n)) \in G$, contains an embedding of M_I.*

---

[2] This is a slight simplification. There will be some representative sink variable for each variable in the image of α.

In particular, let the goal set $G = \{f(\alpha(c_1), \ldots, \alpha(c_n)), wellform(X)\}$, where the variable X is not in the image of α, then any solution to G contains M_I and is well-formed. The middle-left graph of Figure 6 shows FORMULA's construction of a well-formed version of the star graph in the $CYCLE$ domain.

The default embedding produced by FORMULA is not particularly elegant. It contains a star juxtaposed with a 3-cycle. This solution was constructed because ϕ_{sol} assigns a unique constant to each sink variable, yielding a *maximal solution* with respect to the number of constants. A smaller solution can be found by manipulating the final constraint system $CS(W)$ so that the number of sink variables are reduced. This can be accomplished by merging sink trees, which is legal if the trees do not have disequality constraints between them. The middle graph of Figure 6 shows the sink trees of the constraint system after producing the middle-left embedding. The root of each tree is in bold, and disequality constraints between trees are shown as bold edges. These are the only types of edges between trees, because sink trees do not have dependency edges between them. A *minimal solution* can be formed by partitioning the root vertices into a minimal number of independent sets. This is a computationally hard optimization problem related to the *independent set problem*. The right side of Figure 6 shows the optimized constraint system, which contains only four trees (and four sink variables). The roots of the optimized constraint system form a clique, therefore no further optimization is possible. The bottom-left graph shows the optimized solution generated by FORMULA, wherein the star and 3-cycle have been merged in an ideal fashion. Note that this process yields a minimal, but not neccessarily *minimum* model. Finding a minimum model requires minimizing all possible consistent walks of the solution tree.

3   Extensions, Tools, and Future Directions

We have shown that the constructive reasoning of UML-like metamodels is a rich area of study, both theoretically and algorithmically. In the interest of space we have used directed graphs as our toy example. However, these techniques can be applied to much more complicated metamodels, and with practical applications: *Metamodel composition* is the process of constructing new domain-specific languages by combining existing metamodels. Two metamodels, mm_1 and mm_2, can be syntactically combined with an operator $\circ$, such as class equivalence[10], and the syntactic composition can be converted into a domain $D_{comp} = \tau_{meta}(mm_1 \circ mm_2)$. The membership problem for the domain can then be solved, thereby deciding if the metamodel composition is *semantically meaningful*. Other problems, like the construction of embeddings, correspond to the automatic construction of useful models that satisfy the domain constraints. Model transformations can also be incorporated into our framework, and then constructive techniques can be used to prove that the transformation always produces well-formed output models from well-formed input models. This is the weakest form of correctness one could imagine, but checking these properties has remained mostly open. There is already precedent for the use of Prolog engines to transform a particular input model M_I to an output model M_O, as is done by Viatra2[11]. A particular input/output pair (M_I, M_O) can then be compared to check for mutual consistency (e.g. via bisimulation). However, checking properties of the overall transformation is more difficult, though our approach can handle it as long as the transformation is restricted to an appropriate class of logic. The verification goal resembles Hoare's notion of a *verifying compiler*[12].

This brings us to questions of expressiveness. How expressive is Horn logic and how far can it be taken? This question has driven our development of FORMULA, which we now summarize. Positive NHDs are not particularly expressive, but they are an essential starting point for developing constructive techniques for more expressive domains. The next step in the progression is to solve the membership problem for *negative* NHDs. Recall that negative domains characterize the malformed models with the symbol $malform(\cdot)$, and a model M is wellformed if $\forall x, M \nvdash_C malform(x)$. Negative NHDs can express domains not representable by positive NHDs, because of the universal quantification over $malform(x)$. Notice that the solution tree for a goal $G = \{malform(x)\}$ contains all equivalence classes of malformed models, and the *malformed-ness* property is monotonic in models. With these observations, the membership problem can be solved by repeating this procedure: Prune all leaves in $B(\{malform(x)\})$, except for one symbol $f \in \Upsilon$. If $malform(x)$ can be proved on the corresponding pruned solution tree, then by monotonicity, no wellformed model can contain a term unifiying with f. If $malform(x)$ cannot be proved, then a wellformed model $M = \{f(\cdot, \ldots, \cdot)\}$ has been found. This test is repeated (at least once) for each $f \in \Upsilon$; due to unification issues, it may be repeated multiple times for non-unifying f-terms. This procedure is also implemented in FORMULA.

A further increase in expressiveness can be obtained by extending the Horn logic so that a tail can contain a "negated" term $\neg t_i$. (For example, the *UNSAT* domain (Figure 2) can be defined with this extension.) Loosely, a negated term is a constraint requiring that a solution $M \nvdash_C t_i$. Theoretically, this extension approximates the power of full first order logic, but remains decidable (under additional restrictions on its use). It turns out that this simple extension corresponds to a *nonmonotonic* logic, and has deep theoretical and algorithmic repercussions. Our major challenge has been the development of constructive techniques for domains written in Horn logic extended with negation. These techniques are also implemented in FORMULA, and extend existing work on nonmonotonic inference[7][13] to deal with the particulars of UML-like metamodels. Theoretically, these extensions must be handled carefully in order to maintain the soundness and completeness of the theorem prover. Algorthmically, our approach combines the aforementioned algorithms with state-of-the-art SAT solvers to construct models. In conclusion, a reasonable level of expressiveness can be obtained.

A common criticism of theorem proving is the requirement of the user to understand the underlying mathematics. We have addressed this issue by developing an automated conversion from metamodels to domain definitions. This approach is described in [1], and supports metamodeling in the well-known

Generic Modeling Environment (GME) toolsuite[14]. Furthermore, because the theorem prover is constructive, the results of the prover are concrete models that can be automatically imported back into the GME modeling environment. This closes the loop, providing constructive reasoning about models and metamodels without leaving the comfort of the modeling toolsuite (for most of the common queries). Our future work is to apply these techniques to analyze model transformations, including those specified with the Graph Rewriting and Transformation (GReAT) language[15] that is also part of the GME toolsuite.

References

1. Jackson, E.K., Sztipanovits, J.: Towards a formal foundation for domain specific modeling languages. In: EMSOFT'06. Proceedings of the Sixth ACM International Conference on Embedded Software, pp. 53–62. ACM Press, New York (2006)
2. Object Management Group: Meta object facility specification v1.4. Technical report (2002)
3. Object Management Group: Unified modeling language: Superstructure version 2.0, 3rd revised submission to omg rfp. Technical report (2003)
4. Drewes, F., Hoffmann, B., Plump, D.: Hierarchical graph transformation. J. Comput. Syst. Sci. 64(2), 249–283 (2002)
5. Ehrig, H., Prange, U., Taentzer, G.: Fundamental theory for typed attributed graph transformation. In: Ehrig, H., Engels, G., Parisi-Presicce, F., Rozenberg, G. (eds.) ICGT 2004. LNCS, vol. 3256, pp. 161–177. Springer, Heidelberg (2004)
6. Burris, S.N., Sankappanavar, H.: A Course in Universal Algebra. Springer, Heidelberg (1981)
7. Dantsin, E., Eiter, T., Gottlob, G., Voronkov, A.: Complexity and expressive power of logic programming. ACM Comput. Surv. 33(3), 374–425 (2001)
8. Minker, J.: Logic and databases: A 20 year retrospective. In: Logic in Databases, pp. 3–57 (1996)
9. Chan, D.: Constructive negation based on the complete database. In: Proc. Int. Conference on LP'88, pp. 111–125. MIT Press, Cambridge (1988)
10. Emerson, M., Sztipanovits, J.: Techniques for metamodel composition. In: OOP-SLA 2006 Domain-Specific Languages Workshop (2006)
11. Csertan, G., Huszerl, G., Majzik, I., Pap, Z., Pataricza, A., Varro, D.: Viatra: Visual automated transformations for formal verification and validation of uml models (2002)
12. Hoare, T.: The verifying compiler: A grand challenge for computing research. J. ACM 50(1), 63–69 (2003)
13. Gelfond, M., Lifschitz, V.: The stable model semantics for logic programming. In: ICLP/SLP, pp. 1070–1080 (1988)
14. Karsai, G., Sztipanovits, J., Ledeczi, A., Bapty, T.: Model-integrated development of embedded software. Proceedings of the IEEE 91, 145–164 (2003)
15. Karsai, G., Agrawal, A., Shi, F.: On the use of graph transformations for the formal specification of model interpreters. Journal of Universal Computer Science 9(11), 1296–1321 (2003)

A Metamodel-Based Approach for Analyzing Security-Design Models

David Basin[1], Manuel Clavel[2], Jürgen Doser[1], and Marina Egea[2]

[1] Information Security Group, ETH Zurich
{basin,doserj}@inf.ethz.ch
[2] Computer Science Department, U. Complutense, Madrid
clavel@sip.ucm.es, marina_egea@fdi.ucm.es

Abstract. We have previously proposed an expressive UML-based language for constructing and transforming security-design models, which are models that combine design specifications for distributed systems with specifications of their security policies. Here we show how the same framework can be used to analyze these models: queries about properties of the security policy modeled are expressed as formulas in UML's Object Constraint Language and evaluated over the metamodel of the security-design language. We show how this can be done in a semantically precise and meaningful way and demonstrate, through examples, that this approach can be used to formalize and check non-trivial security properties of security-design models. The approach and examples presented have been implemented and checked in the SecureMOVA tool.

1 Introduction

Model driven development [9] holds the promise of reducing system development time and improving the quality of the resulting products. Recent investigations [2,6,7,8] have shown that security can be integrated into system-design models and that the resulting *security-design models* can be used to generate systems along with their security infrastructures. Moreover, when the models have a formal semantics, they can be reasoned about: one can query their properties and understand potentially subtle consequences of the policies they define.

In previous work [2], we presented a UML-based security modeling language, called SecureUML, closely related to Role Based Access Control (RBAC) [5]. We showed how to systematically combine different design modeling languages with SecureUML in a way that allows users to formalize authorization restrictions on systems implementing the design. The combination scheme was defined both syntactically and semantically and we also described translators that automatically generate distributed, middleware-based systems with complete, configured, access control infrastructure from security-design models.

Our focus in this paper is on formalizing and automatically analyzing security properties of security-design models. In our setting, security-design models constitute formal objects with both a concrete syntax (or notation) and an abstract

G. Engels et al. (Eds.): MoDELS 2007, LNCS 4735, pp. 420–435, 2007.
© Springer-Verlag Berlin Heidelberg 2007

syntax. Security-design models themselves are described by a metamodel that formalizes the structure of well-formed models. We show that, in this setting, security properties of security-design models can be expressed as formulas in the Object Constraint Language (OCL) [11] over this metamodel. We can formalize queries in this language that ask questions about the relationships between users, roles, permissions, and actions. An example of a typical query (taken from Section 5) is: are there two roles such that one includes the set of actions of the others, but the roles are not related in the role hierarchy? Such queries can be answered by evaluating the OCL expressions over the metamodel of the security modeling language.

The idea of formulating OCL queries about role-based access control policies is not new. Our work is inspired by [1,12], who first explored the use of OCL for querying RBAC policies, and we make comparisons in Section 7. Given this previous work, we see our contributions as follows. First, we clarify the metatheory required to make query evaluation formally well-defined. This requires, in particular, precise definitions of both the metamodel of the modeling language and the mapping from models to the corresponding instances of this metamodel. Second, we show the feasibility of this approach and illustrate some of its key aspects on a nontrivial example: a security-design modeling language from [2] that combines SecureUML and a component modeling language named ComponentUML. Finally, we provide evidence that OCL expressions, evaluated in the context of such a metamodel, can be used to formalize and check non-trivial security properties of security-design models. The approach presented here has been implemented and tested in SecureMOVA, a security-design modeling tool whose implementation is directly based on our metamodel-based approach for analyzing security-design models.

2   General Approach

Background: models and meaning. A modeling language provides a vocabulary (concepts and relations) for building models, as well as a notation to graphically depict them as diagrams. Diagrams have to conform with the metamodel of the modeling language. The precise definition of well-formed diagrams is based on the underlying mapping from diagrams (or graphical models) to instances of the metamodel (or abstract models): well-formed diagrams are those that are mapped to instances of the metamodel that satisfy the metamodel's invariants.

Some modeling languages explain the meaning of the diagrams using natural language. In this situation, analyzing the models represented by the diagrams can only be done informally and no rigorous tool support can be expected. Other modeling languages explain the meaning of the diagrams using a formal semantics: that is, they define an interpretation function [_] that associates mathematical structures to well-formed diagrams, or, more precisely, to the instances of the metamodel that correspond to well-formed diagrams. In this case, properties of the models represented by the diagrams can be formally proven, possibly with the assistance of automated tools. In the following, let M be a graphical model (for a modeling language $\mathcal{M}$), $\overline{M}$ be the corresponding abstract model, and $[\overline{M}]$ be the mathematical structure associated to the abstract model by the interpretation function.

Problem statement: rigorously analyzing security models. Given a language with a formal semantics, one can reason about models by reasoning about their semantics. That is, a security model M has a property P (where P is expressed in some logical language) if and only if $[\overline{M}] \models P$. While this approach is standard, it either requires deductive machinery for reasoning about the semantics of models (i.e., a semantic embedding [3] and deduction within the relevant semantic domains) or an appropriate programming logic for reasoning at the level of the models. These are strong requirements and a hurdle for many practical applications. Hence, the question we address is whether there are other ways of formally analyzing security policies modeled by M, but in a more familiar setting.

Approach taken. Our approach for analyzing properties of security-design models M reduces deduction to evaluation: we formalize the desired properties as OCL queries and evaluate these queries over instances $\overline{M}$ of the metamodel. Observe that these queries are formulated over the abstract models, not the (graphical) models that the modeler sees and works with. Hence, for the results to be meaningful, we require that the mapping relating graphical models to abstract models, along with the interpretation function $[\_]$, correctly interacts with the evaluation of OCL expressions. If the mapping is not explicitly given or the requirements are not satisfied, then the validity of the results may be open, or even wrong (for examples, see the related work section).

To be more precise, we state the following requirements. Let f be a function on the semantic domain and let exp_f be an expression intended to formalize f in OCL. We require the following diagram to commute:

$$
\begin{array}{ccccc}
\text{graphical} & & \text{abstract} & & \text{semantic} \\
\text{Model} & & \text{Model} & & \text{Domain} \\
M & \mapsto & \overline{M} & \mapsto & [\overline{M}] \\
& & \downarrow & & \downarrow \\
& & ev(exp_f, \overline{M}) & \mapsto & f([\overline{M}])
\end{array}
$$

In this diagram, the downward arrow on the left side denotes the evaluation of the OCL expression exp_f (the result of which, denoted by the function $ev(\_,\_)$, constitutes another abstract model). The downward arrow on the right side corresponds to the evaluation of the function f in the semantic domain. The requirement says that the OCL expression exp_f can be used to analyze the behavior of f if and only if $[ev(exp_f, \overline{M})] = f([\overline{M}])$. Roughly speaking, this means that an OCL expression can be correctly used for checking a property P if and only if, for arbitrary models M, the result of evaluating this expression over $\overline{M}$ corresponds to the value of the property P in $[\overline{M}]$.

Rigorously proving this correspondence requires detailed metareasoning that involves the semantics of the underlying formal system, the formal semantics of

OCL, and the translation scheme from terms in the semantic domain to OCL expressions. This is a large undertaking and outside the scope of this paper. In many practical cases however, one may settle for the next best thing: it may be sufficient to have a careful understanding of the metamodel of the modeling languages, its invariants, and of the underlying mapping from models to the corresponding instances of the metamodel. Note that this is already a necessary condition for stating meaningful OCL expressions on models in the first place.

Overall, our approach has a number of advantages over more traditional deductive approaches. First, OCL is a formal language defined as a standard add-on to UML. Hence, as noted in [14], "it should be easily read and written by all practitioners of object technology and by their customers, i.e., people who are not mathematicians or computer scientist." Second, there are tools that can automatically evaluate OCL expressions. The limitations are also clear: there may be interesting properties that cannot be naturally expressed using OCL or that cannot be proved by simply evaluating OCL expression over the metamodel.

3   SecureUML+ComponentUML

In this section, we describe SecureUML and Component UML, the security and design modeling languages that we use to illustrate our approach and some of its key aspects, like the mapping from models to instances of the metamodel.

3.1   The SecureUML+ComponentUML Metamodel

SecureUML [2,10] is a modeling language for formalizing access control requirements that is based on RBAC [5]. In RBAC, permissions specify which *roles* are authorized to perform given operations. RBAC additionally allows one to organize these roles in a hierarchy, where roles can inherit permissions along the hierarchy. In this way, the security policy can be described closely following the hierarchical structure of an organization. Users are then granted permissions by being assigned to the appropriate roles, based on their competencies and responsibilities in the organization.

SecureUML provides a language for specifying access control policies for actions on protected resources. However, it leaves open what the protected resources are and which actions they offer to clients. These are specified in a so called *dialect* and depend on the primitives for constructing models in the system-design modeling language. Figure 1 shows the SecureUML metamodel. The system-design modeling language that we consider here, ComponentUML, is a simple language for modeling component-based systems. Essentially, it provides a subset of UML class models: *Entities* can be related by *Associations* and have *Attributes* and/or *Methods*. The metamodel of ComponentUML is shown in the right part of Figure 2. The dialect definition, shown in the left part of Figure 2, additionally specifies:

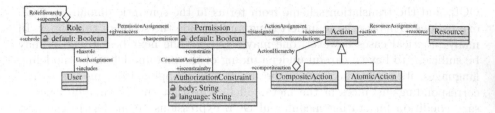

Fig. 1. SecureUML Metamodel

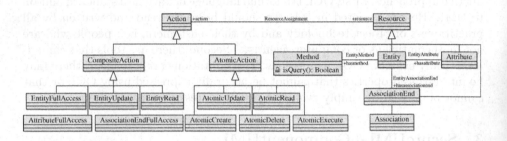

Fig. 2. ComponentUML Dialect Metamodel

- The model element types of the system-design modeling language that represent protected resources. Here, *Entities*, as well as their *Attributes*, *Methods*, and *AssociationEnds* (but not *Associations* as such) are protected resources.
- The actions these resources types offer and hierarchies classifying these actions. The actions offered here are shown in the following table:

Resource	Actions
Entity	create, read, update, delete, full access
Attribute	read, update, full access
Method	execute
Association end	read, update, full access

The atomic actions are intended to directly map onto actual operations of the modeled system. The composite actions are used to group more primitive actions into a hierarchy of more higher-level ones. Here, for example, the composite action *AttributeFullAccess* contains both the read and the update action of the attribute. The precise definition of the actions offered by the different resources, and their hierarchical relationship, is made by adding OCL invariants to the metamodel. The interested reader can find the complete list of these constraints in the references given at `http://maude.sip.ucm.es/securemova`.

- the default access control policy for actions where no explicit permissions is defined (i.e., whether access is allowed or denied by default). Here, by default, access is allowed.

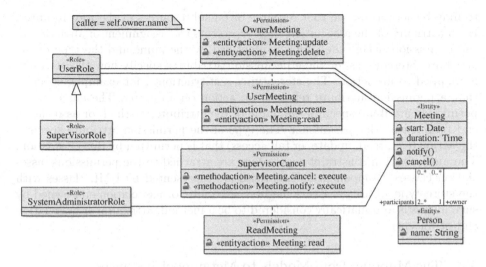

Fig. 3. Example Security Policy

3.2   The SecureUML+ComponentUML Models

We use Figure 3 as a running example to illustrate the concrete syntax of SecureUML and ComponentUML. In this example, the system should maintain a list of users and records of meetings. A meeting has an owner, a list of participants, a time, and a place. Users may carry out standard operations on meetings, such as creating, reading, editing, and deleting them. A user may also cancel a meeting, which deletes the meeting and notifies all participants by email. The system should obey the following (here informally given) security policy:

- All users of the system are allowed to create new meetings and read all meeting entries.
- Only the owner of a meeting is allowed to change meeting data and cancel or delete the meeting.
- A supervisor is allowed to cancel any meeting.
- A system administrator is allowed to read meeting data.

Figure 3 formalizes this security policy using the UML profile for SecureUML and ComponentUML defined in [2]. In this profile, a role is represented by a UML class with the stereotype «Role» and an inheritance relationship between two roles is defined using a UML generalization relationship. The role referenced by the arrowhead of the generalization relationship is considered to be the superrole of the role referenced by the tail. A permission, along with its relations to roles and actions, is defined in a single UML model element, namely an association class with the stereotype «Permission». The association class connects a role with a UML class representing a protected resource, which is designated as the *root resource* of the permission. The actions that such a permission refers to may be actions on the root resource or on subresources of the root resource. Each attribute of the association class represents the assignment of an action to the permission, where the action is identified by the name and the type of the attribute. Stereotypes for these permission attributes specify how the attribute is mapped to an action. The stereotype «entityaction», for example, specifies that a permission attribute refers to an action on an entity. The name of the permission attribute specifies the name of the attribute, method, or association end targeted by this permission. The type of the permission attribute specifies the action (e.g., read, update, or full access) that is permitted by this permission. The authorization constraint expressions are attached to the permissions' association classes. ComponentUML entities are represented by UML classes with the stereotype «Entity». Every method, attribute, or association end owned by such a class is automatically considered to be a method, attribute, or association end of the entity.

3.3   The Mapping from Models to Metamodel Instances

Recall that, in our approach, the specification of security properties using OCL directly depends on the mapping from models to instances of the metamodel, since the expressions formalizing the properties will not be evaluated over the graphical models, but over the corresponding instances of the metamodel. To a large extent, this mapping is straightforward: UML model elements with appropriate stereotypes are mapped to instances of the corresponding metamodel elements, and associations between UML model elements are mapped to appropriate links between the instances of the corresponding metamodel elements.

In some cases, however, this mapping is less straightforward, in particular, where the notation provides the modeler with convenient "syntactic sugar". We list below some examples of such subtleties. Let M be a model, then $\overline{M}$ contains (among others) the following elements:

- "Default" objects of type Role, AuthorizationConstraint, and Permission, which, however, do not correspond to roles, authorization constraints, or permissions depicted in M.
- Objects of subtypes of Action, which correspond to the actions offered by the resources, although they may not be mentioned in the attributes of the permissions depicted in M.
- Links between the "default" objects of type Role, AuthorizationConstraint, and Permission, and between the "default" object of type Permission and the objects of subtypes of Action, which correspond to the default access control policy defined in the metamodel.
- Links between the objects of subtypes of Action, which correspond to the hierarchy of actions defined in the metamodel.

The reader can find the complete definition of this mapping in the references given at http://maude.sip.ucm.es/securemova.

4 Analyzing SecureUML+ComponentUML Models

In this section, we define OCL query operations over the metamodel of SecureUML+ComponentUML that capture different aspects of the access control information contained in the models. These operations will be part of an OCL-based language for analyzing access control decisions that depend on static information, namely the assignment of users and permissions to roles.[1] The approach we take not only allows us to formalize desired properties of models, but also to automatically analyze models by evaluating the corresponding OCL expressions over the instances of the metamodel that corresponds to the models.

4.1 Semantics

We recall here the semantics of SecureUML+ComponentUML models [2], with respect to which we claim that our OCL-operations correctly capture access control information. Let $\Sigma_{RBAC} = (\mathcal{S}_{RBAC}, \geq_{RBAC}, \mathcal{F}_{RBAC}, \mathcal{P}_{RBAC})$ be an order-sorted signature that defines the type of structures specifying role-based access control configurations. Here $\mathcal{S}_{RBAC}$ is a set of sorts, $\geq_{RBAC}$ is a partial order on $\mathcal{S}_{RBAC}$, $\mathcal{F}_{RBAC}$ is a sorted set of function symbols, and $\mathcal{P}_{RBAC}$ is a sorted set of predicate symbols. In detail, let

$$\mathcal{S}_{RBAC} = \{\,Users, Roles, Permissions, AtomicActions, Actions\,\},$$

where $Actions \geq_{RBAC} AtomicActions$. Also, let $\mathcal{F}_{RBAC} = \emptyset$ and

$$\mathcal{P}_{RBAC} = \left\{ \begin{array}{ll} \geq_{Roles} : Roles \times Roles\,, & \geq_{Actions} : Actions \times Actions, \\ UA \quad : Users \times Roles\,, & PA \quad\ : Roles \times Permissions \\ AA \quad : Permissions \times Actions \end{array} \right\}.$$

Given a SecureUML+ComponentUML model M, one defines a Σ_{RBAC}-structure $\mathfrak{S}_{RBAC}$ in the obvious way: the sets $Users$, $Roles$, $Permissions$, $AtomicActions$, and $Actions$ each contain entries for every model element in $\overline{M}$ of the corresponding metamodel types User, Role, Permission, AtomicAction, and Action. The relation $\geq_{Roles}$ is given by the reflexive and transitive closure of the association RoleHierarchy on Role, and the relation $\geq_{Actions}$ by the reflexive and transitive closure of the association ActionHierarchy. Finally, the relations UA, PA, and AA contain tuples for each instance of the associations UserAssignment, PermissionAssignment, and ActionAssignment.

Note that the SecureUML metamodel and its semantics mention "users" and user assignments to roles. These are not usually modeled in the security-design model (e.g., they are not depicted in Figure 3) because this is configuration data that is typically not known at modeling time. For analysis or illustrative purposes, such configuration data can be given as additional input.

---

[1] Programmatic access control decisions that depend on dynamic information, namely the satisfaction of OCL authorization constraints in concrete system states, can be then analyzed using OCL evaluators.

Remark 1. Let $\Im_{RBAC}$ be the Σ_{RBAC} structure defined by a model M. Then, for any u in *Users*, p in *Permissions*, and a in *Actions*, the following table shows the basic correspondence between satisfaction in $\Im_{RBAC}$ and evaluation of OCL expressions in $\overline{M}$:

is satisfied in $\Im_{RBAC}$	evaluates to true over $\overline{M}$
$UA(u,r)$	u.hasrole$->$includes(r)
$PA(r,p)$	r.haspermission$->$includes(p)
$AA(p,a)$	p.accesses$->$includes(a)

4.2   Analysis Operations

In this section, we define a number of OCL *query operations* that are useful for analyzing security properties of security-design models formalized using SecureUML+ComponentUML. We also mention other OCL analysis operations, whose definitions we omit here.

To analyze the relation $\geq_{Roles}$, we define Role::superrolePlus():Set(Role), which is an operation that returns the collection of roles (directly or indirectly) *above* a given role in the role hierarchy.

```
context Role::superrolePlus():Set(Role) body:
  self.superrolePlusOnSet(self.superrole)
```

```
context Role::superrolePlusOnSet(rs:Set(Role)):Set(Role) body:
  if rs->collect(r1|r1.superrole)->exists(r|rs->excludes(r))
  then self.superrolePlusOnSet(rs->union(rs.superrole))
  else rs->including(self)
  endif
```

Similarly, we define the operation Role::subrolePlus():Set(Role) returning the roles (directly or indirectly) *below* a given role in the role hierarchy. Also, we use these operations to define the operation Role::allPermissions():Set(Permission) that returns the collection of permissions (directly or indirectly) assigned to a role.

```
context Role::allPermissions():Set(Permission) body:
  self.superrolePlus().haspermission->asSet()
```

Conversely, we define the operation Permission::allRoles():Set(Role), returning the collection of roles (directly or indirectly) assigned to the given permission.

To analyze the relation $\geq_{Actions}$, we define Action::subactionPlus():Set(Action) that returns the collection of actions (directly or indirectly) subordinated to an action.

```
context Action::subactionPlus():Set(Action) body:
  if self.oclIsKindOf(AtomicAction)
  then Set{self}
  else self.oclAsType(CompositeAction).subordinatedactions.subactionPlus()
  endif
```

Similarly, we define the operation Action::compactionPlus():Set(Action) returning the collection of actions to which an action is (directly or indirectly) subordinated. In addition, we define the operation Permission::allActions():Set(Action) that returns the collection of actions whose access is (directly or indirectly) granted by a permission.

---

context Permission::allActions():Set(Action) **body**:
    self.accesses.subactionPlus()−>asSet()

Conversely, we define the operation Action::allAssignedPermissions():Set(Permission), returning the collection of permissions that (directly or indirectly) grant access to an action. Finally, we define the operation User::allAllowedActions():Set(Action) that returns the collection of actions that are permitted for the given user, subject to the satisfaction of the associated constraints in each concrete scenario.

---

context User::allAllowedActions():Set(Action) **body**:
    self.hasrole.allPermissions().allActions()−>asSet()

Remark 2. Let $\Im_{RBAC}$ be the Σ_{RBAC} structure defined by a model M. Then, for any u in *Users*, r, r_1, r_2 in *Roles*, p in *Permissions*, and a, a_1, a_2 in *Actions*, the following table shows the additional correspondence between satisfaction in $\Im_{RBAC}$ and evaluation of OCL expressions in $\overline{M}$.

is satisfied in $\Im_{RBAC}$	evaluates to true in $\overline{M}$
$r_1 \geq_{Roles} r_2$	r_2.superrolePlus()−>includes(r_1)
	r_1.subrolePlus()−>includes(r_2)
$\exists r_2 \in Roles.\ r_2 \geq_{Roles} r_1 \wedge PA(r_2, p)$	r_1.allPermissions()−>includes(p)
	p.allRoles()−>includes(r_1)
$a_1 \geq_{Actions} a_2$	a_1.subactionPlus()−>includes(a_2)
	a_2.compactionPlus()−>includes(a_1)
$\exists a_2 \in Actions.\ a_2 \geq_{Actions} a_1 \wedge AA(p, a_2)$	p.allActions−>includes(a_1)
	a_1.allAssignedPermisssions()−>includes(p)
$\phi_{RBAC}(u, a)$	u.allAllowedActions()−>includes(a)

Here, $\phi_{RBAC}(u, a)$ is the formula that states whether a user u has a permission to perform action a:

$$\phi_{RBAC}(u, a) = \exists r_1, r_2 \in Roles.$$
$$\exists p \in Permissions. \exists a' \in Actions.$$
$$UA(u, r_1) \wedge r_1 \geq_{Roles} r_2 \wedge PA(r_2, p)$$
$$\wedge AA(p, a') \wedge a' \geq_{Actions} a .$$

5   Analysis Examples

In this section, we give a collection of examples that illustrates how one can analyze SecureUML+ComponentUML models M using the OCL operations defined in Section 4. The questions are formalized as queries over objects in $\overline{M}$, possibly with additional arguments. Note that, with the exception of Example 3, the queries refer to static information about the access control configuration, which is independent of the system state. In contrast, in Example 3 we explicitly query about the circumstances under which a user can perform an action.

The first three examples address the basic question of who can do what, under which circumstances. These functions can provide an elementary sanity check of the access control policy.

Example 1. Given a role, which atomic actions can a user in this role perform?

```
context Role::allAtomics():Set(Action) body:
    self.allPermissions().allAction()−>asSet()
    −>select(a|a.oclIsKindOf(AtomicAction))
```

Example 2. Given an atomic action, which roles can perform this action?

```
context AtomicAction::allAssignedRoles():Set(Roles) body:
    self.compactionPlus().isassigned.allRoles()−>asSet()
```

Example 3. Given a role and an atomic action, under which circumstances can a user in this role perform this action?

```
context Role::allAuthConst(a:Action):Set(String) body:
    self.permissionPlus(a).isconstraintby.body−>asSet()

context Role::permissionPlus(a:Action):Set(Permission) body:
    self.allPermissions()−>select(p|p.allActions()−>includes(a))
```

The next two examples address the question of whether there are possibilities for refactoring or simplifying the role hierarchy. If we have two roles with the same set of allowed actions, one of them may be redundant and could therefore be removed. Similarly, consider two roles where one role is allowed everything the other role is allowed. In this case, the policy could be simplified by letting the second role inherit from the first.

Example 4. Are there two roles with the same set of atomic actions?

```
context Role::duplicateRoles():Boolean body:
    Role.allInstances()−>exists(r1, r2| r1.allAtomics = r2.allAtomics)
```

Example 5. Are there two roles such that one includes the set of actions of the other, but the roles are not related in the role hierarchy?

---

context Role::virtualSubroles():Boolean **body**:
    Role.allInstances()−>exists(r1,r2| r1.allActions()−>includesAll(r2.allActions())
      and not(r1.superrolePlus()−>includes(r2)))

The next example addresses the question of which role a user should be assigned, given that he is supposed to perform a particular action. According to the least-privilege-principle, the user should have no more privileges than absolutely required.

Example 6. Given an atomic action, which roles allow the least set of actions including the atomic action? This requires a suitable definition of "least" and we use here the smallest number of atomic actions.

---

context AtomicAction::minimumRole():Set(Role) **body**:
    self.allAssignedRoles()−>select(r1|self.allAssignedRoles()
      −>forAll(r2| r1.allAtomics()−>size() <= r2.allAtomics()−>size()))

The next two examples address the question of whether there are possibilities for refactoring permissions. Given two permissions that share allowed actions, it may be useful to refactor the common actions into a new, separate permission.

Example 7. Do two permissions overlap?

---

context Permission::overlapsWith(p:Permission):Boolean
body: self.allActions()−>intersection(p.allActions())−>notEmpty()

Example 8. Are there overlapping permissions for different roles?

---

context Permission::existOverlapping():Boolean **body**:
    Permission.allInstances()−>exists(p1,p2| p1 <> p2 and p1.overlapsWith(p2)
      and not(p1.allRoles−>includesAll(p2.allRoles)))

The next example provides another way of detecting opportunities for refactoring permissions. Suppose the policy default is to allow access and, moreover, there is an action that is allowed by every role. The policy can then be simplified by removing this action from all permissions, effectively assigning it the default permission.

Example 9. Are there atomic actions that every role, except the default role, may perform?

---

context AtomicAction::accessAll():Boolean **body**:
    AtomicAction.allInstances()−>exists(a| Role.allInstances−>forAll(r|
      not(r.default) implies r.allAtomics()−>includes(a)))

The above examples provide evidence that OCL expressions can be used to formalize and check non-trivial security properties. This expressiveness is due to the fact that, in our applications, the OCL language is *enriched* with the types provided by the metamodel of SecureUML+ComponentUML, (e.g, Role, Permission, Set(Action)) and vocabulary (e.g., hasrole, givesaccess, isassigned).

6   The SecureMOVA Tool

As [12] observed, although there have been various proposals for specifying role-based authorization constraints, there is a lack of appropriate tool support for analyzing role-based access control policies. In response to this need, [12] shows how to employ the USE system to validate and test access control policies formulated in UML and OCL. We comment on this work in Section 7.

As part of our work, we have implemented a prototype tool called Secure-MOVA for analyzing SecureUML+ComponentUML models. SecureMOVA is an extension of the ITP/OCL tool, a text-input mode validation and analysis tool for UML diagrams with OCL constraints. SecureMOVA extends the ITP/OCL tool with commands for building SecureUML+ComponentUML diagrams and for evaluating OCL queries using, among others, the analysis operations introduced in Sections 4.2 and 5 (the users may, of course, add their own analysis operations to the system). Importantly, SecureMOVA implements the mapping from models to instances of the metamodel introduced in Section 3.3. Thus, the users can work directly with the models (as they are used to), but their queries are evaluated over the corresponding instances of the metamodel, which are automatically generated by SecureMOVA. For reason of space, we omit here the complete definition of the SecureMOVA commands. The interested reader can find it at `http://maude.sip.ucm.es/securemova` along with a collection of examples, including the example in Figure 3.

7   Conclusion

Related Work. As mentioned in the introduction, our work is inspired by [1], who first explored the use of OCL for querying RBAC policies (see also [13,12]). A distinct characteristic of our work is that we spell out and follow a precise methodology, which guarantees that query evaluation is formally meaningful. This methodology requires, in particular, precise definitions of both the metamodel of the modeling language and the mapping from models to the corresponding instances of this metamodel. These definitions make it possible to rigorously reason about the meaning of the OCL expressions used in specifying and analyzing security policies.

To underscore the importance of such a methodology, consider a simple example: specifying two mutually exclusive roles such as "accounts payable manager" and "purchasing manager". *Mutual exclusion* means that one individual cannot have both roles. In [1,13,12] this constraint is specified using OCL as follows:

---

context User **inv**:
 let M : Set = {{accounts payable manager, purchasing manager}, ...} in
 M−>select(m | self.role−>intersection(m)−>size > 1)−>isEmpty()

This constraint correctly specifies mutual exclusion *only if* the association-end role returns all the roles assigned to a user. This should include role assignments explicitly depicted as well as those implicitly assigned to users under the role hierarchy. The actual meaning of the association-end role depends, of course, on the mapping between models and the corresponding instances of the metamodel. Since the precise definition of this mapping is not given in [1,13,12], readers (and tool users) must speculate on the meaning of such expressions and thereby the correctness of their OCL specifications. (Notice that, if the mapping used in [1,13,12] is the " straightforward" one, the association-end role will only return the roles explicitly assigned to a user.)

In our setting, mutual exclusion can be specified using OCL as follows:

---

context User **inv**:
 let M : Set = {{accounts payable manager, purchasing manager}, ...}
 in M−>select(m | self.hasrole.superrolePlus()−>intersection(m)−>size > 1)
   −>isEmpty()

From our definition of superrolePlus in Section 4.2, it is clear that this expression denotes all the roles assigned to a user, including those implicitly assigned to the user under the specified role hierarchy.

OCL has also been used to analyze models of other modeling languages, not only security modeling languages. For example, consider the use of OCL to define metrics, originally proposed by [4]. These approaches share the problems we elaborated in Section 2: without a precise relation between the graphical models and the corresponding metamodel, and a precise relation to the semantic domain, the meaning and validity of OCL formulas is unclear.

Future Work. One direction for future work is tool support for handling queries involving system state. SecureUML includes the possibility of constraining permissions with authorization constraints (OCL formulas), which restrict the permissions to those system states satisfying the constraints. An example of a stateful query for a design metamodel that includes access to the system date is "which operations are possible on week days that are impossible on weekends?" Alternatively, in a banking model, we might ask "which actions are possible on overdrawn bank accounts?" Such queries cannot currently be evaluated as they require reasoning about consequences of OCL formulas and this involves theorem proving as opposed to model checking, i.e., determining the satisfiability of formulas in a concrete model.

Another interesting direction would be to use our approach to analyze the consistency of different system views. In [2] we showed how one can combine

SecureUML with different modeling languages (i.e., ComponentUML and ControllerUML) to formalize different views of multi-tier architectures. In this setting, access control might be implemented at both the middle tier (implementing a controller for, say, a web-based application) and a back-end persistence tier. If the policies for both of these tiers are formally modeled, we can potentially answer question like "will the controller ever enter a state in which the persistence tier throws a security exception?" Again, carrying out such analysis would require support for theorem proving.

References

1. Ahn, G.J., Shin, M.E.: Role-based authorization constraints specification using Object Constraint Language. In: WETICE '01. Proceedings of the 10th IEEE International Workshops on Enabling Technologies, pp. 157–162. IEEE Computer Society, Washington, DC, USA (2001)
2. Basin, D.A., Doser, J., Lodderstedt, T.: Model driven security: From UML models to access control infrastructures. ACM Trans. Softw. Eng. Methodol. 15(1), 39–91 (2006)
3. Boulton, R.J., Gordon, A., Gordon, M.J.C., Harrison, J., Herbert, J., Van Tassel, J.: Experience with Embedding Hardware Description Languages in HOL. In: Proceedings of the IFIP TC10/WG 10.2 International Conference on Theorem Provers in Circuit Design, pp. 129–156. North-Holland, Amsterdam (1992)
4. Brito e Abreu, F.: Using OCL to formalize object oriented metrics definitions. Technical Report ES007/2001, FCT/UNL and INESC, Portugal (June 2001), available at http://ctp.di.fct.unl.pt/QUASAR/Resources/Papers/others/MOOD_OCL.pdf
5. Ferraiolo, D.F., Sandhu, R.S., Gavrila, S., Kuhn, D.R., Chandramouli, R.: Proposed NIST standard for Role-Based Access Control. ACM Trans. Inf. Syst. Secur. 4(3), 224–274 (2001)
6. Georg, G., Ray, I., France, R.: Using aspects to design a secure system. In: ICECCS '02. Proceedings of the Eighth International Conference on Engineering of Complex Computer Systems, pp. 117–126. IEEE Computer Society Press, Washington, DC, USA (2002)
7. Jürjens, J.: Towards development of secure systems using UMLsec. In: Hussmann, H. (ed.) ETAPS 2001 and FASE 2001. LNCS, vol. 2029, pp. 187–200. Springer, Heidelberg (2001)
8. Jürjens, J.: UMLsec: Extending UML for secure systems development. In: Jézéquel, J.-M., Hussmann, H., Cook, S. (eds.) UML 2002. LNCS, vol. 2460, pp. 412–425. Springer, Heidelberg (2002)
9. Kleppe, A., Bast, W., Warmer, J.B., Watson, A.: MDA Explained: The Model Driven Architecture–Practice and Promise. Addison-Wesley, Reading (2003)
10. Lodderstedt, T., Basin, D.A., Doser, J.: SecureUML: A UML-based modeling language for model-driven security. In: Jézéquel, J.-M., Hussmann, H., Cook, S. (eds.) UML 2002. LNCS, vol. 2460, pp. 426–441. Springer, Heidelberg (2002)
11. Object Management Group. Object Constraint Language specification, version 2.0 (May 2006)

12. Sohr, K., Ahn, G.J., Gogolla, M., Migge, L.: Specification and validation of authorisation constraints using UML and OCL. In: di Vimercati, S.d.C., Syverson, P.F., Gollmann, D. (eds.) ESORICS 2005. LNCS, vol. 3679, pp. 64–79. Springer, Heidelberg (2005)
13. Wang, H., Zhang, Y., Cao, J., Yang, J.: Specifying Role-Based Access Constraints with Object Constraint Language. In: Yu, J.X., Lin, X., Lu, H., Zhang, Y. (eds.) APWeb 2004. LNCS, vol. 3007, pp. 687–696. Springer, Heidelberg (2004)
14. Warmer, J., Kleppe, A.: The Object Constraint Language: Getting Your Models Ready for MDA, 2nd edn. Addison-Wesley, Reading (2003)

UML2Alloy: A Challenging Model Transformation

Kyriakos Anastasakis[1], Behzad Bordbar[1], Geri Georg[2], and Indrakshi Ray[2]

[1] School of Computer Science, University of Birmingham, Edgbaston, Birmingham,
UK
{K.Anastasakis,B.Bordbar}@cs.bham.ac.uk
[2] Computer Science Department, Colorado State University, Fort Collins, Colorado,
USA
{georg,iray}@cs.colostate.edu

Abstract. Alloy is a formal language, which has been applied to modelling of systems in a wide range of application domains. It is supported by Alloy Analyzer, a tool, which allows fully automated analysis. As a result, creating Alloy code from a UML model provides the opportunity to exploit analysis capabilities of the Alloy Analyzer to discover possible design flaws at early stages of the software development. Our research makes use of model based techniques for the automated transformation of UML class diagrams with OCL constraints to Alloy code. The paper demonstrates challenging aspects of the model transformation, which originate in fundamental differences between UML and Alloy. We shall discuss some of the differences and illustrate their implications on the model transformation process. The presented approach is explained via an example of a secure e-business system.

1 Introduction

The Unified Modelling Language (UML) [1] is the de-facto language used in the industry for specifying the requirements and the design of software systems. Detecting faults during the early stages of the software development lifecycle, instead of the later stages, provides a significant saving in cost and effort. This necessitates analysing the requirements and design specifications for potential errors and inconsistencies before the system has been developed. Manual analysis is error-prone and tedious. A number of approaches have been proposed in the literature [2,3,4] for analysing UML specifications. These analyses rely mainly on using theorem provers. Theorem provers are hard to use, require expertise, and the analysis requires manual intervention. Consequently, such approaches are not very suitable for use in real-world applications.

In this paper, we advocate the use of Alloy [5] for analysing UML specifications. Alloy is a modelling language for expressing complex structural constraints and behaviour. It has a well-designed syntax most suitable for Object Oriented modelling. Moreover, Alloy is supported by a software infrastructure [6], which

G. Engels et al. (Eds.): MoDELS 2007, LNCS 4735, pp. 436–450, 2007.
© Springer-Verlag Berlin Heidelberg 2007

provides fully automatic analysis of models in the form of simulation and checking the consistency of specifications. Alloy has received considerable attention in the research community. For example, it has been successfully applied to modelling and analysis of protocols in distributed systems [7], networks [8] and mission critical systems [9].

There are clear similarities between Alloy and UML languages such as class diagrams and OCL. From a semantic point of view both Alloy and UML can be interpreted by sets of tuples [5,10]. Alloy is based on first-order logic and is well suited for expressing constraints on Object Oriented models. Similarly, OCL has extensive constructs for expressing constraints as first-order logic formulas. Considering such similarities, model transformation from UML class diagrams and OCL to Alloy seems straightforward. However, UML and Alloy have fundamental differences, which are deeply rooted in their underlying design decisions.

For example, Alloy makes no distinction between sets, scalars and relations, while the UML makes a clear distinction between the three. This has grave consequences in the transformation between the two languages. The current state of model transformation techniques is not dealing with such issues. In this paper we reflect on such differences and their effect on the transformation.

We have incorporated the ideas presented in this paper in a tool called UML2Alloy. UML2Alloy which has been applied to the analysis of discrete event systems [11] and the architecture of enterprise web applications [12].

The next section provides an overview of basic concepts used in this paper.

2   Preliminaries

This section provides a brief introduction to the basic concepts of the MDA and Alloy, which will be used in the rest of the paper.

Model Driven Architecture: The method adopted in this paper makes use of Model Driven Architecture (MDA) [13] techniques for defining and implementing the transformations from models captured in the UML class diagram and OCL into Alloy. Central to the MDA is the notion of *metamodels* [14]. A metamodel defines the elements of a language, which can be used to represent a model of the language. In the MDA a model transformation is defined by mapping the constructs of the metamodel of a *source* language into constructs of the metamodel of a *destination* language. Then every model, which is an instance of the source metamodel, can be automatically transformed to an instance of the destination metamodel with the help of a model transformation framework [15].

Alloy: Alloy [5] is a textual modelling language based on first-order relational logic. An Alloy model consists of a number of *signature* declarations, *fields*, *facts* and *predicates*. Each signature denotes to a set of *atoms*, which are the basic entities in Alloy. Atoms are *indivisible* (they can not be divided into smaller parts), *immutable* (their properties remain the same over time) and *uninterpreted* (they do not have any inherent properties) [5]. Each field belongs to a signature and represents a relation between two or more signatures. Such a relation denotes to a set of tuples of atoms. In Alloy *facts* are statements, which define constraints on the elements of the model. Parameterised constraints, which are referred to as *predicates*, can be invoked from within facts or other predicates.

Alloy is supported by a fully automated constraint solver, called Alloy Analyzer [6], which allows analysis of system properties by searching for instances of the model. It is possible to check that certain properties of the system (*assertions*) are satisfied. This is achieved by automated translation of the model into a Boolean expression, which is analysed by SAT solvers embedded within the Alloy Analyzer. A user-specified *scope* on the model elements bounds the domain. If an instance that violates the assertion is found within the scope, the assertion is not valid. However, if no instance is found, the assertion might be invalid in a larger scope. For more details on the notion of scope, please refer to [5, Sect. 5].

One important characteristic of Alloy is that it treats scalars and sets as relations. For example, a relation between two atoms $A1$ and $A2$ is represented by the pair: $\{(A1, A2)\}$. A set like: $\{A1, A2\}$ is represented by a set of unary relations: $\{(A1), (A2)\}$. Finally a scalar, is represented as a singleton unary relation. For example, the scalar $A1$, will be represented in Alloy as: $\{(A1)\}$.

Treating both scalars and sets as relations, is an interesting property of Alloy, which makes it distinguishable from other popular modelling notations and particularly UML. Hence it introduces additional complexity into the definition of the transformation rules. The following section discusses our MDA based approach to transform UML class diagrams annotated with OCL constraints to Alloy.

3   Model Transformation from the UML to Alloy

This section presents a brief description of our work. We use an MDA compliant methodology to transform a subset of UML class diagram models enriched with OCL constraints to Alloy.

Figure 1 depicts an outline of our approach. Using the EBNF representation of the Alloy grammar [5], we shall first generate a MOF compliant [14] metamodel for Alloy. We then select a subset of the class diagrams [16] and OCL [17] metamodels. To conduct the model transformation, a set of transformation rules has been defined. The rules map elements of the metamodels of class diagrams and OCL into the elements of the metamodel of Alloy. The rules have been implemented into a prototype tool called UML2Alloy. If a UML class diagram, which conforms to the subset of UML we support is provided as input to UML2Alloy, an Alloy model is automatically generated by the tool.

The next section illustrates our work on transforming the EBNF representation of Alloy's grammar into a MOF compliant metamodel.

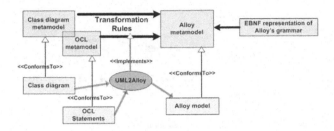

Fig. 1. Outline of the transformation method

3.1 EBNF to MOF

Alloy is a textual language and its syntax is defined in terms of its EBNF [18] grammar [5, Ap. B]. The grammar represents the concrete syntax of the Alloy language. In order to use the MDA, we need to convert the concrete syntax of the Alloy language to a MOF compliant abstract syntax representation. Wimmer and Kramler [19] have already proposed a method for generating a metamodel of a language, based on the EBNF representation of its syntax. We utilised their approach with some simplifications, since some of their proposals were not required in the case of Alloy. For example, we did not use annotations to give additional semantics to the Alloy metamodel that was generated.

Figure 2 depicts a portion of the Alloy metamodel we constructed for signature declarations. A signature declaration (*SigDecl*) is an abstract metaclass. It can either be an *ExtendSigDecl* or an *InSigDecl*, used for subtyping and subseting signatures respectively. A *SigDecl* has a signature body (*SigBody*). It can contain a sequence of constraints (*ConstraintSequence*). A signature declaration can also specify a number of declarations (*Decl*). Declarations are used to define signature fields. They declare one or more variables (*VarId*) and are related to a declaration expression (*DeclExp*). A declaration expression can either declare a binary relation between signatures (*DeclSetExp*) or a relation that associates more than two signatures (*DeclRelExp*). Similarly, we have defined the parts of the Alloy metamodel which represent *expressions*, *constraints* and *operations*.

Since the construction of the metamodel was an intermediate step to utilise the MDA technology, we did not use OCL to specify well-formedness rules on the elements of the metamodel, an approach which is adopted by the UML specification. Instead well-formedness rules were embedded in the transformation, ensuring that the generated Alloy models are well-formed.

3.2 Mapping Class Diagram and OCL to Alloy

This section presents a brief introduction on the transformation rules from UML to Alloy. It provides an informal correspondence between elements of the UML and Alloy metamodels, as a basis on which to present the challenges of the transformation. A more detailed description of the transformation rules can be found in [11]. Due to space limitations the UML and OCL metamodels are

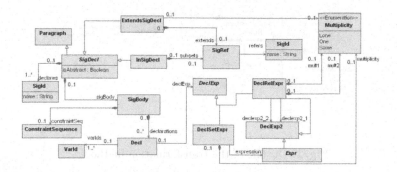

Fig. 2. Part of the Alloy metamodel used to represent signature declarations

not presented here. An extensive explanation can be found in the respective specification documents [16, p. 29] [17].

Table 1 provides an informal correspondence between the most crucial elements of the UML and OCL metamodels and Alloy. More specifically a UML *Class* is translated to an Alloy signature declaration (*ExtendsSigDecl*), which defines a *SigId* with the same name as the class name. If the class is not a *specialization* the Alloy signature is not related to any *SigRef*. Otherwise it will be related to a *SigRef*, which references the signature it might extend.

For example, the *Client* class in the UML model of Fig. 3 is transformed to an *ExtendsSigDecl*, which *declares* a *SigId*, whose *name* is *Client*. Because it doesn't represent a subclass, it is not related to any *SigRef*. Similarly the *SoftwareClient* and *WebClient* are transformed to an *ExtendsSigDecl*. Unlike the *Client* class though, they are related to a *SigRef*, which refers to the *SigId* generated to represent the *Client* class.

The next section presents an example UML class diagram, which will be used to illustrate the challenges of the transformation from UML to Alloy.

Table 1. Informal mapping between UML and Alloy metamodel elements

UML+OCL metamodel element	Alloy metamodel element
Class	ExtendsSigDecl
Property	DeclExp
Operation	Predicate
Parameter	Decl
Enumeration	ExntedsSigDecl
EnumerationLiteral	ExtendsSigDecl
Constraint	Expression

4   Example UML Class Diagram

Figure 3 depicts a UML class diagram that represents the login service of an e-commerce application. The e-commerce system allows clients (i.e. Client) to

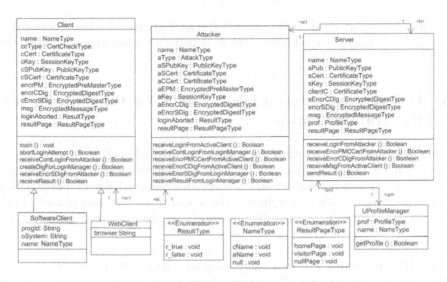

Fig. 3. Partial model of the SSL protocol included in the e-commerce system login service

```
context Client::abortLoginAttempt (): Boolean
post abortLoginAttempt:
    self.loginAborted = ResultType::r_true and
    self.resultPage = ResultPageType::nullPage
```

Fig. 4. OCL specification of the abortLoginAttempt operation of the *Client* class

purchase goods over the internet. It is therefore susceptible to various attacks, including a man-in-the-middle attack that allows an attacker to intercept information that may be confidential. The login service has therefore been augmented with the SSL (Secure Sockets Layer [20]) authentication and confidentiality protocol. We model the man-in-the-middle attack in our example by adding an Attacker class that intercepts all communications between the Client and the e-commerce server, Server, possibly changing message content prior to passing them onto the intended recipient. The SSL protocol works the same whether the Client is a SoftwareClient or a WebClient, but the SoftwareClient provides some extra functionality to the user. If the SSL handshake completes successfully, a secret session key that can be used for message encryption and decryption will have been exchanged between Client and Server. All further communication between them will be encrypted, and thus confidential. If the handshake fails, all communication is aborted between Client and Server.

Figure 3 depicts a high level representation of the system, where attributes of the classes hold the values of the messages exchanged between the entities that participate in the interactions. Due to space limitations some user defined types, such as the *EncryptedDigestType* are not represented in this diagram. For an extended study of this model, please refer to [21].

Figure 4 depicts an excerpt of the OCL specification of the *abortLoginAttempt* method, which will be used later on to demonstrate the differences between OCL and Alloy.

5   Differences Between UML and Alloy Which Influence the Transformation

Although both UML and Alloy are designed to be used in Object-Oriented (OO) paradigm, the two languages have different approaches to some of the fundamental issues of OO such as inheritance, overloading and predefined types [22]. Some of these differences directly influence the model transformation process. In this section, we shall discuss such differences and explain how our approach deals with them.

Inheritance: Both UML and Alloy support inheritance. In the UML, a child class inherits and can specialise the properties of one or more classes [16, p. 126]. The UML standard uses the term '*redefines*' to denote attribute or operation overriding.

In Alloy a *signature* can extend another signature and the elements of the subsignature are a subset of the elements of the supersignature. However, a subsignature can't declare a field whose name is the same as the name of a field of its supersignature. Thus a subsignature can't override the fields of a supersignature. In our transformation we have resolved this shortcoming, by renaming all Alloy fields which have naming conflicts. This is explained with the help of an example.

Consider the SoftwareClient class of Fig. 3. It has an attribute *name*, which overloads the attribute *name* of the Client class. In order to transform this model, we need to create a field with a unique name in Alloy and change all references, which refer to the *name* attribute of the SoftwareClient class, to reference the uniquely named field. However, this brings additional complications. According to the UML specification constraints of a superclasses are propagated to the subclasses. In particular it is mentioned that: '*A redefining element... can add specific constraints or other details that are particular to instances of the specializing redefinition context that do not contradict invariant constraints in the general context.*' [16, p. 126]. Let us assume the following constraint: *self.name* <> *NameType::null* exists in the Client class and the constraint *self.name* <> *NameType::aName* exists in the SoftwareClient. During the transformation the *name* attribute of the SoftwareClient is renamed to *name1*. The original constraint is then translated to a signature constraint: *name1* ! = *aName* in Alloy. However, this allows for the field *name1* to have a *null* value, which is not acceptable in the original UML model. Therefore another constraint (i.e. *name1* ! = *null*) needs to be injected in the translated Alloy model to reflect the constraints applied to the *name* attribute of the *Client* class. A similar approach is followed when dealing with operation overriding.

Namespace: All UML model elements are defined in a *namespace* [17, p. 72]. For example, classes in a class diagram are usually defined in the namespace of the package, while attributes are defined in the namespace of the class they belong to.

Model elements of an Alloy model also belong to a namespace [5, p. 254]. However, the notion of a namespace in Alloy and UML are slightly different. For example, the UML specification defines that: '*The set of attribute names and class names need not be disjoint*' [17, p. 178]. In Alloy on the other hand signature names, have to be distinct from their field names.

Therefore we need to ensure that during the transformation a unique name is created for Alloy elements that belong to the same namespace. In our approach we first identify elements in the UML metamodel, which belong to different UML namespaces, but are translated to the same Alloy namespace (e.g. class operations). If those elements do not have a unique name, we make sure to assign to them a unique name during the transformation to Alloy. All references to those elements in the original UML model, are changed during the transformation to reference the unique names in the generated Alloy model.

Another issue is that in OCL the instance of the class on which the operation is applied, can be accessed using the *self* keyword. In Alloy there is no such concept for predicates, making it difficult to reference the instance of the signature on which the Alloy predicate is applied. As a solution we pass the instance as a parameter to the predicate. For an example of a solution to this problem, consider the *abortLoginAttempt()* operation of the Client. Its OCL specification makes use of the *self* keyword and is depicted in Fig. 4. Following our transformation rules, we translate it in Alloy to the following predicate, where *act*, which is an instance of the signature Client, is passed as a parameter to the predicate:

```
pred abortLoginAttempt(act:Client){
act.loginAborted = r_True && act.resultPage = nullPage }
```

Sets, Scalars, Relations and Undefinedness: Alloy treats sets and scalars as relations. In particular in Alloy a relation denotes to a set of tuples. The number of elements in each tuple depends on the arity of the relation. For example, a binary relation is represented by a 2-tuple. A set is represented as a unary relation and a scalar is a singleton unary relation [5, p. 45].

In UML on the other hand, sets and scalars have the standard meaning they have in set theory. The equivalent of relations in UML is an association between classes, which is represented as a set of tuples [17, p. 184].

These differences in the two languages stem from the fact that UML and Alloy have different design philosophies. More specifically one of the purposes of UML is to represent Object Oriented programming concepts, where the distinctions between scalars and sets is clear. On the other hand Alloy was designed for analysing abstract specifications and the uniform way it deals with sets, scalars and relations contributes to its succinct syntax and leverages its expressiveness [23].

To explain this consider the navigation dot (.). In Alloy it is treated as the relational join [5, p. 59]. As a result navigating over an empty relation denotes to an empty set. Consequently Alloy doesn't need to address the problem of partial functions by introducing a special *undefined* value, like in UML [17, Ap. A.2.1.1]. More specifically let us assume in the model of Fig. 3 we have the following OCL statement.

$$\textit{context Client inv: self.at.name} = \textit{self.at.lm.name} \qquad (1)$$

In UML if the instance of the Client in which this OCL invariant is evaluated is related to no Attacker, the part *self.at.name* of this statement will denote to an *undefined* value. The result of the invariant will then be *undefined*. In an equivalent Alloy model, however, if the Client was related to no Attacker, such a constraint would always denote to true! This is because the left hand side part of the expression, that is *self.at.name*, would denote to an empty set relation. Similarly the right hand side, *self.at.lm.name* would evaluate to an empty set. Therefore the invariant will always evaluate to true.

This has serious implications because the statement will produce a different outcome in Alloy than in OCL. To overcome the problem, we check if an OCL statement can evaluate to an undefined value. To check for undefinedness, the *oclIsUndefined()* OCL operation is used. For example, the OCL statement of (1) will become:

```
context Client
    inv: if not self.at.name.oclIsUndefined() and
        not self.at.lm.name.oclIsUndefined() then
            self.at.name = self.at.lm.name
        else false   endif
```

In this case the modeller has specified that if any part of the expression is *undefined*, the invariant should evaluate to false. This ensures that the OCL statement will either evaluate to true or false, but not undefined. Such an expression is transformed to Alloy using our standard UML to Alloy transformation rules.

Predefined Types: The UML specification defines a number of primitive types (e.g. String, Real, etc.). Those types can be used when developing UML models. For example, the attribute *browser* of the *WebClient* class in Fig. 3 is of type String.

On the other hand, Alloy has a simple type system and the only predefined type it supports is Integers. However, some of the rest of UML's predefined types, can be modelled in Alloy. For example, a String, can be modelled as a sequence of characters and each character can be represented by an *atom*.

One consequence of this approach, is that while in UML primitive types and their operations are part of the metamodel, in Alloy they need to be defined on the model level (i.e. a String has to declared as an Alloy signature). Our transformation rules do this automatically for certain attribute types (e.g. String).

UML's extension mechanism: UML provides two extensions mechanisms [1, p. 11]. One is to create a *profile* and another one is to extend the UML meta-model. If UML has been extended, we need to incorporate the rules involving the new elements into the transformation.

Our current transformation deals with a subset of the standard UML and OCL metamodels. If the metamodels have been extended, the new semantics need to be incorporated into the transformation. For example, let's assume that UML has been extended with the ability to define a *Singleton* stereotype. This stereotype, when used on a class, restricts the class to have only one instance. This is expressed with the following invariant in OCL: *self.allInstances() → size() =1*. In such a case the transformation rules need to be adjusted accordingly. In particular, whenever a *Singleton* stereotype is found on a class, a constraint needs to be injected in the produced Alloy model, to impose that the transformed signature will have only one instance. The implementation of our transformation rules, is modular and uses the SiTra [15] transformation engine, which can be easily augmented to accommodate for any extensions.

Aggregation and Composition: The UML treats aggregation and composition as special kinds of associations [1, p. 112]. Alloy doesn't directly support notions like aggregation and composition. Fortunately [24] present a methodical way of refactoring aggregation and composition as an association with additional OCL constraints that represent the semantics of aggregation and composition. We utilise this approach, as it allows us to use transformation rules for binary associations and OCL constraints we have already defined.

Static vs Dynamic Models: Models in Alloy are static, i.e. they capture the entities of a system, their relationships and constrains about the system. An Alloy model defines an instance of a system where the constraints are satisfied. However, Alloy models do not have an inherent notion of states. In particular, Alloy does not have any built in notion of statemachine [5, Ap. B.5.1]

In UML the term 'static' is used to describe a view of the system, that represents the structural relations between the elements as well as the constraints and the specification of operations with the help of pre and post conditions. In UML, unlike Alloy, static models have an inherent notion of states. A *system state* is made of the values of objects, links and attributes in a particular point in time [17, p. 185].

Hence UML has an implicit notion of states, while Alloy does not support it directly. This introduces additional complexity in the transformation. Let us assume the following OCL statement is the definition of the *receiveResult()* operation of the Client:

```
context Client::receiveResult():void
pre: self.resultPage = ResultPageType::nullPage
post: self.resultPage = ResultPageType::homePage
```

To evaluate this expression two consecutive states are required, one to represent the state before the execution of the operation (precondition) and another

to represent the state after the execution of the operation (postcondition). The
OCL standard formally specifies the *environment* on which pre and postconditions are evaluated [17, p. 210].

If the specification of the *receiveResult()* operation, was directly translated to
Alloy it would translate to:

```
pred receiveResult(act:Client){ act.resultPage = nullPage
act.resultPage = homePage }
```

However, such an Alloy specification leads to an inconsistent model. This is
because the value *nullPage* and *homePage* are assigned to the *resultPage* field,
at the same time. This leads to a logical inconsistency, as both statements can
not be true (i.e. resultPage will either be the *nullPage* or *homepage*, but not
both at the same time).

One solution, which has been proposed is to introduce the notion of a *state*
at the model level [5,25]. This is a standard way of modelling dynamic systems
in Alloy. Our approach uses this pattern of modelling dynamics in Alloy, to
translate UML models to Alloy. This allows us to have two consecutive states and
evaluate the preconditions of each operation on the first state, while evaluating
the postcondition of an operation on the next state.

6   Analysis, Discussion and Future Work

This section presents a brief overview on the results of the analysis we conducted
on the example UML model. It also provides a discussion on further issues that
were encountered and suggests directions for future work.

6.1   Analysis Via Alloy

We applied our model transformation rules from UML to Alloy on the example
model presented in Sect. 4. We checked the produced Alloy model, using the
Alloy Analyzer. The assertion that must be validated is that if the Attacker
obtains the secret session key, the handshake should always fail. This assertion
can be specified using OCL:

```
context Client
inv sameKeySuccess: Client.allInstances() -> forAll(ac:Client |
    ac.loginAborted = ResultType::r_false implies (
    ac.cKey = SessionKeyType::symmKey and
    ac.at.sKey = SessionKeyType::symmKey
    and ac.at.aKey <> SessionKeyType::symmKey))
```

This OCL statement was automatically transformed to the following Alloy
assertion:

```
assert sameKeySuccess{ all ac:Client | ac.loginAborted = r_false
implies (ac.cKey = symmKey && ac.at.lm.sKey = symmKey &&
ac.at.aKey != symmKey) }
```

This assertion was checked for a scope [5, p. 140] of six. A scope of six means that the Alloy Analyzer will attempt to find an instance that violates the assertion, using up to six instances for each of the entities defined in the class diagram of Fig. 3 (for example, Client, Attacker, Server). The assertion produced no counterexample, meaning that it is valid for the given scope.

6.2   Discussion and Future Work

This section briefly presents a discussion on further practical issues we had to deal with, when defining the transformation rules. Moreover it suggests directions for future work.

A difficulty that was encountered when defining the transformation rules, was that parts of the UML specification are inconsistent with the UML *informative semantics* section of the specification [17, An. A]. For example, even though the UML standard allows for overloading of attributes and operations [16, Sect. 7.3.46], the UML formal semantics part of the specification seems to adopt a different stance [17, p. 182]. In particular it doesn't allow for attributes or operations of a subclass to have the same name as the attributes and operations of its superclass. As explained in Sect. 5 we made use of the informal semantics in our transformation rules, since attributes and operations overriding is an important facility provided in object oriented modelling.

Another issue we had to overcome, originates in the nature of OCL. In particular the transformation rules had to be invoked recursively. For example, the definition of the abstract syntax of *If* expressions [17, p. 45] allows for any type of an OCL expression to be part of the *condition* clause. As a result, when defining the transformations someone needs to check the type of the *condition* expression and invoke the corresponding transformation rule, which will be used to transform that specific kind of expression to Alloy. This problem is dealt by the SiTra [15] transformation framework we used for our implementation. SiTra allows for recursive calls of transformation rules with dynamic type checking. Therefore depending on the type of the expression, the corresponding rule is automatically invoked.

The UML specification defines a number of concepts (e.g. *ordered* and *subsets* annotations of association ends, *package import*), which are not formally defined. For some of those concepts the semantics are not clear (e.g. package merge) [26]. Transformation rules for such concepts have not been defined yet. Modellers can define their interpretation of the semantics of such concepts on the model level using OCL.

While Alloy is a purely declarative language, OCL has an imperative flavour. For instance, the OCL standard allows for recursion: '*We therefore allow recursive invocations as long as the recursion is finite*' [17, p. 205]. Alloy on the other hand does not directly support recursion. It might be possible to represent some cases of recursion, using the expressiveness of the Alloy language as suggested in [22]. How this might be incorporated into our transformation rules in order to provide support for recursion, remains an issue for future research.

As we use Alloy to formalise UML, our approach admits some of the inherent limitations of the Alloy language. Some UML primitive types (such as Real numbers) can not be directly transformed to Alloy. Therefore it is not feasible to check whether certain properties involving real values are satisfied. Additionally the UML offers a number of collection types (e.g. *Sequence, Bag*), which can not be directly represented in Alloy. Moreover, since Alloy is a first-order language, it does not support nested collections. The possibility of representing the capabilities of UML collection types in Alloy remains to be investigated further.

7   Related Work

Formalising UML for the purpose of analysis is a popular approach. Evans et al [4] propose the use of Z [27] as the underlying semantics for UML. Marcano and Levy [2] advocate the use of B [28], while Kim [3] makes use of an MDA method to translate a subset of UML to Object-Z. These methods rely on theorem provers to carry out the analysis, which complicates the process.

A number of UML tools also provides support for analysis. For example, the USE tool (UML Specification Environment) [10] is a powerful instance evaluator with the ability of simulation.

Using Alloy to formalise UML has also received considerable attention. More specifically Denis et al [9] use Alloy to expose hidden flaws in the UML design of a radiation therapy machine. Georg et al [8] have used Alloy to analyse the runtime configuration of a distributed system. Unlike our work, those approaches conduct the translation from UML to Alloy manually, a procedure which is tedious and error prone.

Additionally there have been studies on the comparison of languages of UML and Alloy [22,29]. However, they do not use model driven approaches to demonstrate the differences.

Finally transformations from a three-valued logic language to a two-valued logic language, such as ours from UML to Alloy have been applied to the field of database semantics. For example, [30] propose the use of an interpretation operator to treat statements with undefined values in databases either as true or false.

8   Conclusions

Model transformations in the context of MDA are predominantly used for code generation. Model transformations can be used for the creation of analysable models, allowing the discovery of possible flows in the design of a system. Languages used for creation of analysable models have strong formal foundations. Hence, a model transformation from the UML to such languages is highly non trivial.

In this paper we have reflected on the lessons learned from the model transformation from UML class diagrams to Alloy. We discussed some of the differences between UML and Alloy. For example, the different perspectives on inheritance, functions, static and dynamic models. We also studied the implications of such differences in the model transformation. Our proposed solutions to such challenges were presented. The method is implemented in a prototype tool called UML2Alloy. The approach is illustrated with the help of an example from the security domain.

References

1. OMG: UML Infrastructure Document: formal/05-07-05, http://www.omg.org
2. Marcano, R., Levy, N.: Using B formal specifications for analysis and verification of UML/OCL models. In: Workshop on consistency problems in UML-based software development. 5th International Conference on the Unified Modeling Language, Dresden, Germany (2002)
3. Kim, S.K.: A Metamodel-based Approach to Integrate Object-Oriented Graphical and Formal Specification Techniques. PhD thesis, University of Queensland, Brisbane, Australia (2002)
4. Evans, A., France, R., Grant, E.: Towards Formal Reasoning with UML Models. In: Proceedings of the OOPSLA'99 Workshop on Behavioral Semantics (1999)
5. Jackson, D.: Software Abstractions: Logic, Language, and Analysis. The MIT Press, London, England (2006)
6. Jackson, D.: Alloy Analyzer website, http://alloy.mit.edu/
7. Taghdiri, M., Jackson, D.: A lightweight formal analysis of a multicast key management scheme. In: König, H., Heiner, M., Wolisz, A. (eds.) FORTE 2003. LNCS, vol. 2767, pp. 240–256. Springer, Heidelberg (2003)
8. Georg, G., Bieman, J., France, R.B.: Using Alloy and UML/OCL to Specify Run-Time Configuration Management: A Case Study. In: Evans, A., France, R., Moreira, A., Rumpe, B. (eds.) Practical UML-Based Rigorous Development Methods - Countering or Integrating the eXtremists. LNI, German Informatics Society, vol. P-7, pp. 128–141 (2001)
9. Dennis, G., Seater, R., Rayside, D., Jackson, D.: Automating commutativity analysis at the design level. In: ISSTA '04: ACM SIGSOFT international symposium on Software testing and analysis, pp. 165–174. ACM Press, New York (2004)
10. Richters, M.: A Precise Approach to Validating UML Models and OCL Constraints. PhD thesis, Universitaet Bremen, Logos Verlag, Berlin, BISS Monographs, No. 14 (2002)
11. Bordbar, B., Anastasakis, K.: UML2Alloy: A tool for lightweight modelling of Discrete Event Systems. In: Guimarães, N., Isaías, P. (eds.) IADIS International Conference in Applied Computing 2005, vol. 1, pp. 209–216. IADIS Press, Algarve, Portugal (2005)
12. Bordbar, B., Anastasakis, K.: MDA and Analysis of Web Applications. In: Draheim, D., Weber, G. (eds.) TEAA 2005. LNCS, vol. 3888, pp. 44–55. Springer, Heidelberg (2006)
13. Kleppe, A., Warmer, J., Bast, W.: MDA Explained: The Model Driven Architecture-Practice and Promise. The Addison-Wesley Object Technology Series. Addison-Wesley, Reading (2003)
14. OMG: MOF Core v. 2.0 Document Id: formal/06-01-01, http://www.omg.org
15. Akehurst, D.H., Bordbar, B., Evans, M.J., Howells, W.G.J., McDonald-Maier, K.D.: SiTra: Simple transformations in java. In: Nierstrasz, O., Whittle, J., Harel, D., Reggio, G. (eds.) MoDELS 2006. LNCS, vol. 4199, pp. 351–364. Springer, Heidelberg (2006)

16. OMG: UML: Superstructure. Document id: formal/05-07-04, http://www.omg.org
17. OMG: OCL Version 2.0 Document id: formal/06-05-01, http://www.omg.org
18. Aho, A.V., Sethi, R., Ullman, J.D.: Compilers: principles, techniques, and tools. Addison-Wesley, Reading (1986)
19. Wimmer, M., Kramler, G.: Bridging grammarware and modelware. In: Bruel, J.M. (ed.) MoDELS 2005. LNCS, vol. 3844, pp. 159–168. Springer, Heidelberg (2006)
20. TLSWG: SSL 3.0 specification (1996), http://wp.netscape.com/eng/ssl3
21. Georg, G., Anastasakis, K., Bordbar, B., Houmb, S.H., Ray, I., Toahchoodee, M.: Verification and trade-off analysis of security properties in UML system models. Transaction. In: Software Engineering. Special Issue on Security (submitted)
22. Jackson, D.: A Comparison of Object Modelling Notations: Alloy, UML and Z (August 1999), Available at http://sdg.lcs.mit.edu/publications.html
23. Vaziri, M., Jackson, D.: Some Shortcomings of OCL, the Object Constraint Language of UML. In: Technology of Object-Oriented Languages and Systems (TOOLS 34'00), Santa Barbara, California, pp. 555–562 (2000)
24. Gogolla, M., Richters, M.: Transformation rules for UML class diagrams. In: Bézivin, J., Muller, P.-A. (eds.) The Unified Modeling Language. UML 1998: Beyond the Notation. LNCS, vol. 1618, pp. 92–106. Springer, Heidelberg (1999)
25. Wallace, C.: Using Alloy in process modelling. Information and Software Technology 45(15), 1031–1043 (2003)
26. Zito, A., Diskin, Z., Dingel, J.: Package Merge in UML 2: Practice vs. Theory? In: Nierstrasz, O., Whittle, J., Harel, D., Reggio, G. (eds.) MoDELS 2006. LNCS, vol. 4199, pp. 185–199. Springer, Heidelberg (2006)
27. Woodcock, J., Davies, J.: Using Z: Specification, Refinement, and Proof. Prentice Hall, Upper Saddle River, NJ (1996)
28. Abrial, J.R.: The B-book: assigning programs to meanings. Cambridge University Press, New York (1996)
29. He, Y.: Comparison of the modeling languages Alloy and UML. In: Arabnia, H.R., Reza, H. (eds.) Software Engineering Research and Practice, SERP 2006, Las Vegas, Nevada, vol. 2, pp. 671–677 (2006)
30. Negri, M., Pelagatti, G., Sbattella, L.: Formal semantics of SQL queries. ACM Transactions on Database Systems (TODS) 16(3), 513–534 (1991)

i²MAP: An Incremental and Iterative Modeling and Analysis Process*

Sascha Konrad,[1],** Heather J. Goldsby,[2] and Betty H.C. Cheng[2],***

[1] Siemens Corporate Research, Inc.
755 College Road East
Princeton, NJ 08540 USA
[2] Software Engineering and Network Systems Laboratory
Department of Computer Science and Engineering
Michigan State University, 3115 Engineering Building
East Lansing, Michigan 48824 USA
sascha.konrad@siemens.com,{hjg,chengb}@cse.msu.edu

Abstract. Detecting errors early within the development process for an embedded system assists a developer in avoiding excessive error correction costs and minimizing catastrophic losses resulting from failures in deployed systems. Towards that end, this paper presents i²MAP, an iterative and incremental goal-driven process for constructing an analysis-level UML model of an embedded system. The UML model is formally analyzed for adherence to the behavioral properties captured in a companion goal model. The process uses goal modeling to capture the requirements of the system, and uses UML to capture analysis-level structural and behavioral information. Both types of i²MAP models can be used to drive a rigorous approach to model-driven development of embedded systems. In this paper, we illustrate the i²MAP process and the accompanying tool suite in the development of an embedded system model for an adaptive light control system.

1 Introduction

The cost of correcting errors introduced early in the development process is significantly more expensive than the cost of correcting errors introduced in later stages [1]. To further exacerbate this problem, the increasingly popular model-driven development, such as that promoted for use in the model driven-architecture (MDA) by the OMG [2], successively refines models from analysis to design and eventually to code. Thus, undetected errors may be propagated from

---

* This work has been supported in part by NSF grants EIA-0000433, CDA-9700732, CCR-9901017, CNS-0551622, CCF-0541131, Department of the Navy, Office of Naval Research under Grant No. N00014-01-1-0744, Eaton Corporation, Siemens Corporate Research, and a grant from Michigan State University's Quality Fund.
** A significant portion of this work was completed while this author was a doctoral student at Michigan State University.
*** Please contact this author for all correspondences.

G. Engels et al. (Eds.): MoDELS 2007, LNCS 4735, pp. 451–466, 2007.
© Springer-Verlag Berlin Heidelberg 2007

analysis models created early in the development process to code, potentially resulting in expensive correction costs. Within the embedded systems domain, the detection of errors is particularly important, not only because of the high costs of error detection and correction, but also because failures of embedded systems may lead to catastrophic losses [3]. In addition to correctness, embedded systems also need to balance several, often contradictory non-functional system qualities, such as performance and development cost [4]. This paper presents i^{2}MAP (iterative and incremental Modeling and Analysis Process), an iterative and incremental process for constructing an analysis-level UML model of an embedded system, which is guided by an accompanying goal model. While the UML model contains structural and behavioral information, the goal model captures high-level requirements and behavioral properties to which the model should adhere. As part of i^{2}MAP, the UML model is formally analyzed for adherence to the behavioral properties.

Iterative and incremental development practices are advocated by numerous software development processes, such as the Spiral Model [5] and the Rational Unified Process (RUP) [6]. Generally, iterative development constructs increasingly more complex versions of software development artifacts [7], such as UML models or code. Using this strategy, these processes enable developers to detect errors and conceptual misunderstandings of requirements sooner, and incorporate feedback from previous iterations. Several researchers have attempted to combine iterative development practices with the use of formal methods [8, 9, 10]. In contrast to our approach, they have either focused on specification languages other than UML [8] or do not support formal analysis [9, 10]. In addition, none of the aforementioned approaches includes support for goal modeling.

i^{2}MAP uses two different types of models to capture complementary information. A goal model is used to capture high-level functional and non-functional requirements and their rationale. Behavioral properties to which the system should adhere are captured as constraint goals in the goal model, which are specified in terms of formally-analyzable natural language properties. In addition, a UML model is used to capture an operationalization of the requirements in terms of structural and behavioral UML diagrams. Both of these models are realized in increments, where each increment represents the addition of specific functionality to the models. Specifically, the goal model is augmented with goals to capture the functionality that is required for the current increment, while the UML model is extended with structural and behavioral diagram elements realizing the increment's functionality. These modifications to the models are performed in iterations, each of which comprises three phases: (1) the goal model is augmented, (2) the UML model is extended to adhere to the goal model, and (3) both models are formally analyzed for behavioral consistency. In addition, traceability techniques are used to check the models for syntactic consistency [11]. If errors in either the goal or UML model are uncovered during the analysis, then these errors need to be corrected in a subsequent iteration before progressing to the next increment. The process is complete when all requirements for the system have been captured in the models. Requirements that have not been realized yet are evident as goals missing from the goal model.

i²MAP leverages and extends several previously developed methods and tools to create a comprehensive modeling and analysis process supported by a research prototype tool suite. To assist the developer in augmenting the models for realizing an increment, we have developed COBRA (Constraints and Objects for Requirements Analysis) patterns [12] that provide goal model templates and UML diagram templates to capture analysis-level structural and behavioral information of the system. The details of the COBRA patterns are described in [12]; here we only overview the patterns. The *goal model template* declaratively specifies the functional and non-functional requirements of the embedded system using softgoals, and refines the requirements into system constraints using a structured natural language. The *UML model template* operationally specifies, at the requirements-analysis level, UML elements that satisfy the requirements. Structural information is captured using UML class diagrams, while behavioral information is captured using UML state diagrams. *Syntactic consistency* can be established between the goal model and the UML structural diagrams using traceability techniques [11]. *Behavioral consistency* can be established between the goal model and the UML behavioral diagrams by formally analyzing the UML models for adherence to the constraints from the goal model. In order to translate UML models to formal specifications, we use Hydra [13], a previously developed metamodel-based UML formalization framework. To translate the natural language constraints into formal specifications, we leverage SPIDER [14], a customizable natural language specification environment, to generate the corresponding temporal logic formulae. In this paper, we illustrate the i²MAP approach and tool suite through the development of an adaptive light control system.

Overall, i²MAP combines UML and goal modeling, the accessibility of natural language descriptions, and the rigor of formal analysis to support an incremental process for creating and refining requirements-level models. The remainder of the paper is organized as follows. Section 2 briefly overviews COBRA patterns. Section 3 describes the i²MAP process and presents our supporting tool suite. Section 4 applies the process to the development of an adaptive light control system. Section 5 examines related work. Finally, in Section 6 we present conclusions and discuss directions for future work.

2 COBRA Patterns

COBRA patterns have been designed to assist developers in the creation of complementary UML and goal models for the requirements analysis of embedded systems. Specifically, the goal model captures non-functional requirements and declaratively specifies functional requirements and constraints and the UML model operationally specifies behavior that satisfies the requirements. To date, the COBRA patterns have focused on capturing sensors and actuators, specifying the behavior of important system components, and capturing interaction with the user. While this section briefly overviews the COBRA patterns, details of the patterns are described in [12].

Goal models, in general, specify both functional and non-functional system objectives and the relationships between objectives. Developers can use goal models to evaluate alternative solutions and to document the rationale behind requirements [15]. Our COBRA patterns contain goal model templates specified using the Non-Functional Requirements (NFR) modeling notation [16]. Four types of goals are specified with NFR: *softgoals* (non-functional goals whose satisfaction cannot be fully achieved usually), *functional goals* (describing functional objectives of the system that can be achieved), *softgoal operationalizations* (identifying an applicable COBRA pattern and its instantiation), and *constraint goals* (describing system objectives to which the system should adhere). Contribution relationships (depicted as a line with the word "helps" or "hurts") connect an element to a softgoal and indicate whether the element helps or hurts, respectively, the realization of the softgoal. Due to space constraints, this paper focuses on the constraint goals and their relationships to the UML model. Other elements of the goal model are described in more detail in [12]. In order to enable the formal analysis of constraint goals, these goals are specified as structured natural language properties that are amenable to formalization using SPIDER [14].

Thus far, the COBRA pattern repository consists of eleven patterns. Table 1 overviews the main functional goals of the patterns and shows how the COBRA patterns can be evaluated for their contribution to some non-functional requirements. Each COBRA pattern has a *primary functional goal* that describes the basic functionality captured by the pattern, which is further elaborated by a number of a softgoals that describe the impact of applying the pattern. In general, the developer can select an appropriate pattern to apply based on the primary functional goal. However, if two patterns, such as the *Passive Sensor* and *Active Sensor*, have similar functional goals, then the softgoals can be used to assist the developer in evaluating the alternatives. For example, the primary functional goal of both the *Passive Sensor* and *Active Sensor* COBRA patterns is to Monitor environment (as seen in Table 1). However, in the *Active Sensor* pattern, a sensor always sends the information to the computing component when the value changes (information push). As such, the computing component always has the most current value when performing a computation. Nevertheless, the system has to be able to handle the potentially increased number of messages sent by the sensors when a value changes often, which may affect the system cost. In the *Passive Sensor* pattern, the computing component only requests a sensor value when it is actually needed (information pull). However, the decision process for making the request and executing the request takes time and may impact the performance of the system.

Figure 1 depicts an elided portion of the goal model template for the *Passive Sensor* pattern. As denoted in Table 1 and Figure 1, the softgoal (*A*) Performance is hurt, while the softgoals (*B*) Affordability and (*C*) Reusability are helped by the (*D*) Passive sensor pattern. This pattern is AND-refined by (*E*) Passive sensor name, which provides context by naming the pattern instance. This goal is AND-refined by its functional goals, which include (F) Sense environment and (G) Restrict to legal values. Each of the functional goals are, in turn, AND-refined by constraint goals that describe specific, analyzable properties that the UML model should realize. AND-refinement requires all child goals to hold for the parent goal to be satisfied, while OR-refinement only requires one child goal to hold.

Table 1. COBRA pattern evaluation table

Pattern Name	Classification	Primary Functional Goal	Performance	Modifiability	Affordability	Reusability
Active Sensor	Structural	Monitor environment; broadcast of environment information (push)	+		-	+
Passive Sensor	Structural	Monitor environment; requires explicit request for environment information (pull)	-		+	+
Actuator	Structural	Influence environment by setting an actuator value				+
Control	Structural	Receive information from user	+		-	+
Controller Decompose	Structural	Decompose system into components				+
Indicator	Structural	Provide information to user	+		-	+
Communication Link	Behavioral	Interact with external device, e.g., for fault diagnostics			-	+
Computing Component	Behavioral	Distribute computational tasks			-	+
Corrector	Behavioral	Correct faults	-	+	-	
Detector	Behavioral	Detect faults	-	+	-	
Fault-Handler	Behavioral	Centralized handling of faults	-	+	-	

(NFR columns: Performance, Modifiability, Affordability, Reusability)

Instantiating a goal model template has two steps. First, the developer should customize the template by replacing the generic text of the goal model elements with information about the specific embedded system under development.[1] We use underlining in the goals to highlight generic text that needs to be customized. Second, the developer should relate the customized goal model template to the goal model for the overall embedded system by establishing a *hurt* or *help* relationship between the softgoal describing the pattern and a softgoal of the goal model (*e.g.*, the *hurt* relationship between (D) Passive sensor pattern and softgoal (A) Performance in Figure 1).

The COBRA patterns also provide UML model templates. Specifically, the COBRA patterns use UML class diagrams to capture structural information, and sequence and state diagrams to capture behavioral information. Due to space constraints, we do not present UML model templates for the COBRA patterns (please refer to [12] for more details), but we do include example instantiations of UML model templates in the case study description.

---

[1] *Customizing* a natural language template refers to replacing the underlined text with free-form text to apply the goal to a specific system. The property can then be *instantiated* by replacing the underlined text with Boolean propositions in terms of UML model elements.

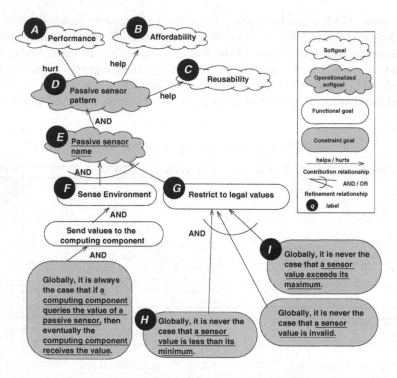

Fig. 1. *Passive Sensor* pattern goal model template

3   Process

In this section, we present i²**MAP** and describe the supporting i²**MAP** tool suite. Specifically, i²**MAP** guides a developer in the cooperative construction of a goal model and an analysis-level UML model of the embedded system in increments, where an *increment* is a unit of functionality. An increment is realized by declaratively specifying the functionality, including constraints, in the goal model and operationally specifying the functionality with structural and behavioral elements in the UML model. Realizing an increment may require several *iterations*, each of which is a single attempt to declaratively and operationally specify the increment. The realization is complete when the UML model adheres to the constraints, specified as part of the goal model. If discrepancies are detected, then these discrepancies need to be corrected during a subsequent iteration within the current increment. An increment may correspond to the application of a COBRA pattern, which provides goal and UML model templates to assist the developer in the specification of the increment functionality. In order to identify suitable patterns, a developer reviews the COBRA pattern repository to look for goals that capture the functional and non-functional requirements needed for the current increment.

To begin the process, the developer seeds the goal model with a high-level objective of the system. Specifically, the primary goal of the system is identified and then decomposed into subgoals until each child goal describes the unit of functionality realizable in one increment, such as what can be matched by the objective of a COBRA pattern. The relationship between the softgoals and operationalized softgoals added during the refinement process is denoted by contribution relationships that are labeled with the keywords "helps" or "hurts". The developer has two options for realizing an increment: (1) using one or more COBRA patterns or (2) without COBRA patterns. In general, the process is simplified if an increment is realized using a COBRA pattern. But it is also possible to use i²MAP without the COBRA patterns. In those cases, there is more burden placed on the developer to develop UML model refinements and goal model refinements without guidance from patterns and templates. For simplicity, we assume in the remainder of the paper that an increment corresponds to the application of one COBRA pattern.

In addition to seeding the goal model, the developer uses the child goals to identify a preliminary list of increments and an initial realization order. Empirical studies [17] have shown that the order in which increments are completed (*i.e.*, specific pattern applications) may affect the complexity of formal analysis, since there may be dependencies between increments. Resolving these dependencies in a suboptimal order could result in large and complex models earlier than necessary.

Once a preliminary increment realization order has been determined, the developer begins to realize increments, where each increment has three phases: (1) planning (augmenting the goal model), (2) modeling (extending the UML model to adhere to the goal model), and (3) analysis (formally analyzing the goal and UML model to determine whether the increment has been completed successfully). Figure 2 gives a data flow diagram overviewing the i²MAP tool suite supporting the process, where ovals represent processes, arrows denote data flow, two parallel lines depict data stores, and external entities are represented by rectangles.[2] Numbers are used to indicate which components are used in which phase. We next explain each phase in detail and show how it is supported in our tool suite.

3.1 Planning

In the planning phase, the developer augments the goal model by including operationalized softgoals describing the non-functional requirements affected by the current increment; and functional and constraint goals are used to capture functional requirements. Since we assume a developer realizes an increment by applying a COBRA pattern, the goal model template provided by the pattern is used to guide the planning phase. Instantiating a goal model template has two steps. First, each goal in the template is customized by replacing the generic underlined text with information about the specific embedded system under development. The customized constraint goals specify properties to be satisfied by the UML model. In the i²MAP tool suite, this specification is created in structured natural language (NL) [14] using the *Increment Specifier* (Process 1 in Figure 2).

---

[2] We use a data flow diagram rather than a UML activity diagram to highlight the flow of data between the different phases of the process.

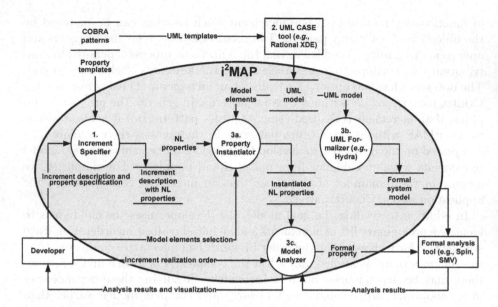

Fig. 2. Abstract view of i²MAP key steps

Second, the developer links the instantiated goal model template to the goal model for the overall embedded system by establishing a *helps* contribution relationship between the softgoal naming the pattern instance (e.g., (**D**) **Passive sensor name**) and a softgoal in the system goal model. Currently, this association needs to be done manually in the companion goal model and is not supported by our tool suite. Techniques exist for establishing the correctness of goal models [18, 19], however, it is outside the scope of our current work.

3.2   Modeling

Next, the UML model is augmented to include the functionality required by the goals. To create the UML model, the developer uses a *UML CASE tool* (Entity 2 in Figure 2). For instance, our i²MAP tool suite currently uses Rational XDE [20]. The structure of the system is captured using UML class diagrams and behavior is captured in terms of UML state and sequence diagrams. To facilitate this process, the COBRA patterns provide templates to guide the creation and refinement of the UML diagrams. These diagram templates are designed to realize the properties specified by the constraint goals for a selected COBRA pattern. To denote a specific instance of the system that should be analyzed, an object diagram is created instantiating the class diagram.

3.3   Analysis

Once the modeling phase is complete, the analysis phase begins. In order to determine whether an increment has been completed successfully, we formalize the constraint goals and analyze the UML model for adherence to these goals. An increment is considered successful if sufficient constraint goals are satisfied so that (1) for each goal with AND child goals, all child goals are satisfied and (2) for each goal with OR child goals, at least one child goal is satisfied. Two tasks need to be performed before the properties specified by the constraint goals can be analyzed. First, constraint goals of the goal model need to be instantiated with model-specific elements. For this purpose, the developer replaces the underlined text in the placeholders of the structured natural language specifications with Boolean expressions describing specific system conditions (extracted from the class and state diagrams). In the i²MAP tool suite, the natural language properties specified in the constraint goals are instantiated by the *Property Instantiator* (Process 3a in Figure 2), which leverages SPIDER [14] to perform the instantiation and formalization (*i.e.*, translate the natural language properties to LTL). To facilitate this process, SPIDER provides user assistance to generate Boolean expression fragments based on information automatically extracted from a UML model. For details on these capabilities, please refer to previously published work [14, 21]. Second, the UML model is translated to a formal system model for the targeted formal analysis tool. In our tool suite, this translation is performed using the *UML Formalizer* (Process 3b in Figure 2). Here, Hydra [13] translates the UML diagrams into a formal model to be analyzed with the Spin model checker [22].

After the properties are instantiated and formalized and the formal system model is created, the UML system model is analyzed for adherence to the constraint goals. The analysis is initiated by the *Model Analyzer* (Process 3c in Figure 2), which iteratively performs the analysis of the formal model for adherence to all constraint goals. The *Model Analyzer* also processes the output received from the formal analysis tool and creates sequence diagram visualizations of the generated error traces. The visualizations can assist a developer in determining the source of each error. If a more interactive visualization is desired, then the developer may alternatively use Theseus [23], a tool that visualizes the model checker error trace in terms of animations on the original UML diagrams. If errors are uncovered during the analysis, then the source of the error needs to be determined and the error must be corrected in another iteration before the increment can be considered complete. As such, this analysis enables developers to find and correct errors quickly, whether the errors were newly introduced in the most recent increment or were introduced in earlier increments but revealed by the latest increment. However, since the size of the system model (in UML) and the number of properties that need to be analyzed is increasing with each iteration, the analysis phase also represents the greatest overhead of our process.

Finally, if no errors are found in the analysis phase, the developer has specified all of the properties for the goal model, as well as constructed all the UML elements for an increment, then the process proceeds to the next increment, or ends when the goal model captures all requirements and, therefore, all increments have been completed.

4   Example Application

Next, we describe the application of i²MAP to the modeling and analysis of an adaptive light control system (ALCS). The ALCS attempts to achieve a user-specified brightness in a single room. The user can select a desired brightness value between 0 and 1000 lux, and the system then attempts to maintain this brightness level. In order to achieve this goal, the system is equipped with a brightness sensor that measures the brightness in the room; and should the brightness be insufficient, the system can activate a dimmer to achieve the desired brightness level. The system is also equipped with a motion sensor. Using the motion sensor, the system can determine whether the controlled room is occupied. If the room is vacant, then the controller will turn off the light after a timeout period. In addition to the automatic mode, the controller also offers a manual mode. When a user activates the manual mode, the controller will set the dimmer to the maximum value for a period of time, and then switch back to automatic mode.

To begin the process, the developer seeds the goal model with the primary goal **Provide adaptive light control** and child goals. Each child goal corresponds to an increment that can be realized using a COBRA pattern. Note that a given COBRA pattern may be used for realizing more than one child goal. Table 2 enumerates the ALCS increments. Specifically, it shows one possible realization order, the increment description, and the name of the COBRA pattern used to realize the increment. In the first few increments, the ability to sense and influence the environment via sensors and actuators is modeled. Next, the *Indicator* pattern provides the functionality needed to indicate the current state of the system to the user. Finally, automatic and manual control are realized using the *Computing Component* pattern.

Table 2. Increment realization order for ALCS

No	Increment Name	Applied COBRA Pattern
1	Detect motion	*Passive Sensor* pattern
2	Measure brightness	*Passive Sensor* pattern
3	Control dimmer	*Actuator* pattern
4	Display status information	*Indicator* pattern
5	Automatic control	*Computing Component* pattern
6	Manual control	*Computing Component* pattern

For illustration purposes, we describe next the realization of an intermediate increment, **(2) Measure brightness**, instead of the initial or final increment.

Planning. At this point, the developer has successfully realized one increment: **(1) Detect motion** (to model the motion sensing functionality). For the second increment **(2) Measure brightness**, the developer needs to model the brightness sensor to measure the brightness level of a room. For this increment, the developer has decided to apply the *Passive Sensor* COBRA pattern (refer to Table 2).

Modeling. The UML model of the ALCS is refined to include the functionality required by the current increment **Measure brightness**. As a result, the class diagram (depicted in Figure 3(a)) is augmented to include a **BrightnessSensor**, which will monitor the brightness level of a room. The **BrightnessSensor**'s behavior is elaborated by a state diagram (depicted in Figure 3(b)).

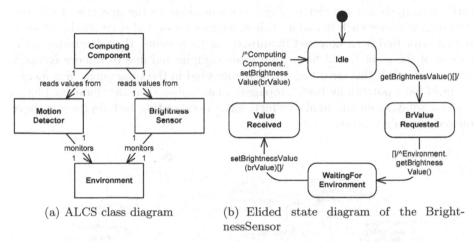

(a) ALCS class diagram   (b) Elided state diagram of the BrightnessSensor

Fig. 3. UML models for the ALCS

Analysis. Once the modeling for the increment **Measure brightness** is complete, analysis begins. The developer uses Hydra to formalize the UML model and SPIDER to instantiate and formalize the constraint goals specified by the goal model. However, formal analysis of the second increment **Measure brightness** reveals that two constraint goals for this increment (shown in Expression 1(a) and 2(a), which are instantiations of the constraint goals (I) Globally, it is never the case that a sensor value exceeds its maximum. and (H) Globally, it is never the case that a sensor value is less than its minimum. in Figure 1) are violated:

> 1(a) Globally, it is never the case that `BrightnessSensor.brightnessValue >` `1000`.

and

> 2(a) Globally, it is never the case that `BrightnessSensor.brightnessValue < 0`.

After inspection of the UML model and the violation trace provided by the analysis tool, the cause for the violation was determined to be the following UML model error: "When the model of the brightness sensor was created, the developer forgot to have the brightness sensor v alidate the value returned by the **Environment**, enforcing upper and lower bounds." Therefore, it was possible for the **Environment** to return a brightness value greater than 1000 lux or less than

0 lux. This error can be seen in the elided state diagram of the **BrightnessSensor** depicted in Figure 3(b), where the **BrightnessSensor** sends the **ComputingComponent** the value that was received in state **ValueReceived**, without performing any validity checks on the value. The corrected state diagram for the **BrightnessSensor** is depicted in Figure 4, where the shaded transitions, state, and variable name indicate corrections. Once these changes were made to the model, further analysis did not detect any violations. Due to the analysis of all constraint goals after each iteration, this error was uncovered early in the process. If this error had been detected in subsequent increments, then determining the source of the error would have been more complicated due to a more complex system model. If this error had not been detected in the subsequent increments, it could have potentially been propagated into subsequent development phases and, possibly, even the final product, since even testing techniques may have missed the subtle error.

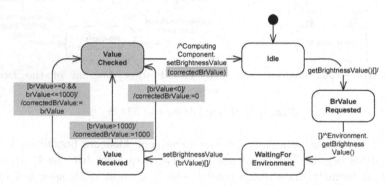

Fig. 4. Elided revised state diagram for the **BrightnessSensor**

5   Related Work

A wide variety of incremental and iterative development processes exists, such as the Spiral Model [5], the Rational Unified Process [6], the Agile Unified Process [24], Agile Model Driven Development [25], Agile Model Driven Architecture [26], and Extreme Programming (XP) [27]. Our iterative and incremental modeling and analysis process shares several characteristics with these approaches, such as incremental development and rapid feedback. However, i²MAP is tailored to the construction of analysis-level goal and UML models for embedded systems and to ensuring structural and behavioral consistency between the UML model and accompanying goal model.

   Several complementary techniques have been developed to combine the use of formal methods and iterative development approaches. Suhaib *et al.* [8] created a modeling and analysis process termed *extreme formal modeling* (XFM). Although their approach is similar to ours, they focus on prescriptive formal hardware models specified in Promela [22] instead of UML models, and do not include patterns or goal modeling. Suhaib *et al.* [8] also investigated the effect of different increment orderings on models created using XFM and report results similar to what we have obtained in our empirical studies [17]: Some orders may result in a larger state space earlier than necessary, hindering the performance of an iterative modeling and analysis approach.

Other approaches exist that integrate iterative and incremental software development and formal methods. SLAM [28] is a technique that uses an object-oriented formal specification language, termed SLAM-SL, to model sequential software programs. Formulae can be used to assert conditions during the execution of a program and, thereby, replace unit tests. In addition, pre- and post-conditions can be used to prove that a program still adheres to all specified requirements after an incremental development step. Henzinger *et al.* [29] have developed an approach termed *extreme model checking* and implemented the approach into an Eclipse plug-in [30]. In this approach, they use the BLAST model checker [29] to implement an efficient regression verification environment for C programs. In contrast to our approach, neither SLAM nor Extreme Model Checking applies to UML models.

Other incremental modeling approaches using UML have been proposed. Boger *et al.* [9] propose *extreme modeling*, in which they apply principles of extreme programming to the modeling phase of software development. To make their models executable and testable, they use a mapping of UML structural and dynamic diagrams to Petri nets. Related to extreme modeling, Hayashi *et al.* [31] introduce the SMART modeling system, which uses action semantics and offers test-driven behavioral guidance. Differing from our approach, extreme modeling and SMART focus on executability and testing of UML diagrams, while our goal is the formal analysis of adherence to properties.

In summary, none of the aforementioned approaches combines the use of goal models to capture functional and non-functional properties, the rigor of formal analysis, guidance of modeling and analysis, and an analysis process that facilitates systematic incremental change and rapid feedback.

6   Conclusions

This paper described i²MAP, a process for the iterative construction of conceptual UML models and goal models for embedded systems. The process has been implemented in the i²MAP tool suite and we illustrated its use in the development of models for an embedded system. Overall, i²MAP enables the incremental development of UML models and accompanying goal models. In order to support the process, COBRA patterns [12] may be used. Behavioral consistency is achieved by formally analyzing the UML model for adherence to constraint goals.

Numerous directions for future work are possible. We are currently working on facilitating the identification of suitable COBRA patterns for realizing increments by developing a browser for the patterns. The process and tool suite could also be extended to include real-time information, where we will leverage real-time extended versions of our property templates [32] and UML formalization framework [33]. could lead to new COBRA patterns, as well as promote the generalization of already existing patterns. Additionally, further empirical evaluation of the process could be performed, in which its impact could be assessed by comparing models created by different user groups using different processes. Such a study could also be used to investigate how the usability of the tool support could be enhanced. Finally, the formalization framework could be extended to support model checkers other than Spin [22], such as Kronos [34], UPPAAL [35], and Hytech [36].

References

[1] Boehm, B.W., Papaccio, P.N.: Understanding and controlling software costs. IEEE Transactions on Software Engineering 14(10), 1462–1477 (1988)
[2] Object Management Group (2007), http://www.omg.org
[3] Lutz, R.R.: Targeting safety-related errors during software requirements analysis. In: SIGSOFT'93 Symposium on the Foundations of Software Engineering (1993)
[4] Cheng, B.H., Atlee, J.M.: Research directions in requirements engineering. In: Briand, L., Wolf, A. (eds.) Future of Software Engineering 2007, IEEE Computer Society, Los Alamitos (2007)
[5] Boehm, B.: A spiral model of software development and enhancement. SIGSOFT Software Engineering Notes 11(4), 14–24 (1986)
[6] IBM Rational: Rational Unified Process (RUP) (2006), http://www-306.ibm.com/software/awdtools/rmc/features/index.html
[7] Pressman, R.S.: Software engineering: a practitioner's approach, 5th edn. McGraw-Hill, Inc., New York (2001)
[8] Suhaib, S., Mathaikutty, D., Berner, D., Shukla, S.: XFM: An incremental methodology for developing formal models. ACM Transactions on Design Automation of Electronic Systems (TODAES) 10(4), 589–609 (2005)
[9] Boger, M., Baier, T., Wienberg, F., Lamersdorf, W.: Extreme modeling. In: Extreme Programming and Flexible Processes in Software Engineering - XP 2000, pp. 175–189. Addison-Wesley Longman Publishing Co., Inc, Boston, MA (2000)
[10] Gargantini, A., Heitmeyer, C.: Using model checking to generate tests from requirements specifications. In: Proceedings of the 7th European Software Engineering Conference, London, pp. 146–162. Springer, Heidelberg (1999)
[11] Fletcher, J., Cleland-Huang, J.: Softgoal traceability patterns. In: 17th International Symposium on Software Reliability Engineering, pp. 363–374. IEEE Computer Society, Los Alamitos, CA (2006)
[12] Goldsby, H.J., Konrad, S., Cheng, B.H.C.: Goal-oriented patterns for UML-based modeling of embedded systems requirements. Technical Report MSU-CSE-07-19, Computer Science and Engineering, Michigan State University, East Lansing, Michigan (March 2007), (Submitted for publication) Available at:http://www.cse.msu.edu/~hjg/goldsby07Goal.pdf
[13] McUmber, W.E., Cheng, B.H.C.: A general framework for formalizing UML with formal languages. In: Proceedings of the IEEE International Conference on Software Engineering (ICSE01), Toronto, Canada (May 2001)

[14] Konrad, S., Cheng, B.H.C.: Facilitating the construction of specification pattern-based properties. In: Proceedings of the IEEE International Requirements Engineering Conference (RE05), Paris, France (August 2005)

[15] van Lamsweerde, A.: Goal-oriented requirements engineering: A guided tour. In: RE '01: Proceedings of the 5th IEEE International Symposium on Requirements Engineering, p. 249. IEEE Computer Society, Washington, DC (2001)

[16] Chung, L., Nixon, B., Yu, E., Mylopoulos, J.: Non-Functional Requirements in Software Engineering. Kluwer Academic, Dordrecht (2000)

[17] Konrad, S.: Model-driven Development and Analysis of High Assurance Systems. PhD thesis, Michigan State University (2006)

[18] Ponsard, C., Massonet, P., Rifaut, A., Molderez, J.F., van Lamsweerde, A., Van Tran, H.: Early verification and validation of mission critical systems. Electr. Notes Theor. Comput. Sci. 133, 237–254 (2005)

[19] Darimont, R.: Process Support for Requirements Elaboration. PhD thesis, Louvain-la-Neuve, Belgium (1995)

[20] IBM: Rational Rose XDE Developer (2005), http://www-306.ibm.com/software/awdtools/developer/rosexde/

[21] Konrad, S., Cheng, B.H.C.: Automated analysis of natural language properties for UML models. In: Bruel, J.-M. (ed.) MoDELS 2005. LNCS, vol. 3844, pp. 48–57. Springer, Heidelberg (2006)

[22] Holzmann, G.: The Spin Model Checker, Primer and Reference Manual. Addison-Wesley, Reading, Massachusetts (2004)

[23] Goldsby, H., Cheng, B.H.C., Konrad, S., Kamdoum, S.: Enabling a roundtrip engineering process for the modeling and analysis of embedded systems. In: Proceedings of the ACM/IEEE 8th International Conference on Model Driven Engineering Languages and Systems, Genova, Italy (October 2006)

[24] Ambler, S.W.: Unified – and agile (January 2006), http://www.sdmagazine.com/documents/s=9947/sdm0601g/0601g.html

[25] Ambler, S.W.: Agile model driven devlopment (AMDD) (2006), http://www.agilemodeling.com/essays/amdd.html

[26] Mellor, S.J.: Agile MDA (2005), http://www.omg.org/mda/mda_files/Agile_MDA.pdf

[27] Beck, K.: Extreme programming explained: Embrace change. Addison-Wesley Longman Publishing Co., Inc., Boston, MA (1999)

[28] Herranz-Nieva, Á., Moreno-Navarro, J.J.: Formal extreme (and extremely formal) programming. In: Marchesi, M., Succi, G. (eds.) XP 2003. LNCS, vol. 2675, pp. 88–96. Springer, Heidelberg (2003)

[29] Henzinger, T.A., Jhala, R., Majumdar, R., Sanvido, M.A.A.: Extreme model checking. In: Dershowitz, N. (ed.) Verification: Theory and Practice. LNCS, vol. 2772, pp. 332–358. Springer, Heidelberg (2004)

[30] Beyer, D., Henzinger, T.A., Jhala, R., Majumdar, R.: An Eclipse plug-in for model checking. In: Proceedings of the 12th IEEE International Workshop on Program Comprehension (IWPC 2004), pp. 251–255. IEEE Computer Society Press, Los Alamitos (CA) (2004)

[31] Hayashi, S., YiBing, P., Sato, M., Mori, K., Sejeon, S., Haruna, S.: Test driven development of UML models with smart modeling system. In: Baar, T., Strohmeier, A., Moreira, A., Mellor, S.J. (eds.) UML 2004. LNCS, vol. 3273, pp. 395–409. Springer, Heidelberg (2004)

[32] Konrad, S., Cheng, B.H.C.: Real-time specification patterns. In: Inverardi, P., Jazayeri, M. (eds.) ICSE 2005. LNCS, vol. 4309, Springer, Heidelberg (2006)

[33] Konrad, S., Campbell, L.A., Cheng, B.H.C.: Automated analysis of timing information in UML diagrams. In: Proceedings of the Nineteenth IEEE International Conference on Automated Software Engineering (ASE04), Linz, Austria, pp. 350–353 (September 2004) (poster summary)

[34] Bozga, M., Daws, C., Maler, O., Olivero, A., Tripakis, S., Yovine, S.: Kronos: A model-checking tool for real-time systems. In: Vardi, M.Y. (ed.) CAV 1998. LNCS, vol. 1427, pp. 546–550. Springer, Heidelberg (1998)

[35] Pettersson, P., Larsen, K.G.: Uppaal2k. Bulletin of the European Association for Theoretical Computer Science 70, 40–44 (2000)

[36] Henzinger, T.A., Ho, P.H., Wong-Toi, H.: HYTECH: A model checker for hybrid systems. International Journal on Software Tools for Technology Transfer 1(1-2), 110–122 (1997)

A Model-Driven Measurement Procedure for Sizing Web Applications: Design, Automation and Validation*

Silvia Abrahão[1], Emilia Mendes[2], Jaime Gomez[3], and Emilio Insfran[1]

[1] Department of Computer Science and Computation
Valencia University of Technology
Camino de Vera, s/n, 46022, Valencia, Spain
{sabrahao, einsfran}@dsic.upv.es
[2] Department of Computer Science, University of Auckland
Private Bag 92019, Auckland, New Zealand
e.mendes@auckland.ac.nz
[3] Department of Information Systems and Languages
University of Alicante
Campus de San Vicente del Raspeig. Apartado 99. 03080 Alicante, Spain
jaime.gomez@ua.es

Abstract. This paper introduces the Object-Oriented Hypermedia Function Points (OO-HFP), which is a functional size measurement procedure for Web projects developed using the Object-Oriented Hypermedia (OO-H) method. This method provides model-driven and transformation-based support for the development of Web applications. Using OO-HFP, a size measure is obtained once a Web application's conceptual model is completed. We follow the steps of a process model for software measurement in order to detail the design and automation of OO-HFP. Finally, we present the validation of OO-HFP for Web effort estimation by comparing the prediction accuracy that it provides to the accuracy provided by another set of validated size measures (the Tukutuku measures) that was found to be a good effort predictor. The results of a study using industrial data show that the effort estimates obtained for projects that are sized using OO-HFP were similar to those using the Tukutuku measures, thus suggesting that the OO-HFP is a suitable effort predictor.

Keywords: Model-driven development, Web Engineering, Functional Size Measurement, Web Effort Estimation, OO-H.

1 Introduction

Developing Web applications is significantly different from traditional software development. The nature of Web development forces project managers to focus primarily on the time variable in order to achieve the required short cycle times.

In this context, Model-Driven Architecture (MDA) approaches [16] seem to be very promising since Web development can be viewed as a process of transforming a

---

* This research is supported by the Generalitat Valenciana, with ref. ACOMP07-216.

G. Engels et al. (Eds.): MoDELS 2007, LNCS 4735, pp. 467–481, 2007.
© Springer-Verlag Berlin Heidelberg 2007

model into another model until it can be executed in a development environment. Over the last few years, several Web development methods that provide some support to MDA have been proposed (i.e., OOHDM [18], WebML [2], and OO-H [5]). Adopting such methods, however, poses new challenges to the Web project manager, in particular with respect to resource estimation and project planning.

A fundamental problem in this context is the size estimation of the future Web application based on its conceptual model. The functional size measurement (FSM) methods used in industry (mainly Function Point Analysis (FPA) [7]) date from a pre-Web era. None of the ISO-standardized FSM methods (e.g., FPA, COSMIC [4]) were designed taking the particular features of Web applications into account. Hence, existing FSM methods need to be adapted or extended to cope with Web projects.

Some approaches for sizing Web sites and applications [6] [3] [17] have been proposed. The main limitation of these approaches is that they cannot be used early in the Web development lifecycle as they rely on implementation decisions. In addition, for project estimation purposes, measurements of this type come too late. Also, some approaches are not automated [1] [3] [6] [17], which limit their use and adoption.

To address these limitations, this paper introduces a model-driven *measurement procedure*[2] for Web applications called Object-Oriented Hypermedia Function Points (OO-HFP). The aim of OO-HFP is to size Web projects that are developed or maintained using the Object-Oriented Hypermedia (OO-H) method [5].

We present the definition and automation of OO-HFP using a process model for software measurement (see Fig. 1) [10]. According to this process, a measurement method is designed, that is, the concept to be measured is defined and the measurement rules are devised. Then, the measurement method is applied. The results of the method are then presented and verified. Such verification includes determining whether the value that is produced is the result of a correct application and interpretation of the measurement rules. Finally, the results are used to build different types of models (e.g, productivity analysis models, effort estimation models).

Fig. 1. Measurement process steps (Source: [10])

We detail how the four steps of the process model were conducted to design and automate OO-HFP. The verification of the measurement results (e.g, evaluation of the reproducibility and accuracy of the results) is out of the scope of this work. This kind of validation is less relevant for automated measurement procedures. More important is the conformity evaluation and the validity of the results for building productivity or effort estimation models. Thus, we analyze the validation of OO-HFP for Web effort estimation by comparing the prediction accuracy that it provides to the accuracy provided by another set of Web size measures (the Tukutuku measures) [12] [13] that has already been validated in the past and that was found to be a good effort predictor.

---

[2] It is defined as a *"set of operations, described specifically, used in the performance of particular measurements according to a given method of measurement"* [9].

The reason to validate OO-HFP lies in the need to provide an accurate size measure that is automatically obtained once the Web application conceptual model is completed. Different from the Tukutuku measures, OO-HFP is compliant to a widely used ISO-standard FSM method. Hence, the size obtained is directly comparable to a large amount of projects existing in industry (e.g., ISBSG repository, www.isbsg.org).

This paper is organized as follows. Section 2 presents an overview of the OO-H development process. Section 3 presents the design of OO-HFP followed by its automation in Section 4. Section 5 shows the validation of OO-HFP against the Tukutuku measures. Finally, section 6 presents the conclusions and further work.

2 The OO-H Development Process

The OO-H (Object-Oriented Hypermedia) method [5] provides designers with the semantics and notation for developing Web applications. The method includes: a design process, a pattern catalog, a Navigation Access Diagram (NAD), and an Abstract Presentation Diagram (APD). The VisualWADE tool automates the entire OO-H development process.

The OO-H design process defines the phases that must be carried out to build a functional interface that fulfills user requirements. The main feature of the OO-H method is the model-driven and transformation-based support for the development of Web applications. Following the approach presented in [11], Fig. 2 depicts the OO-H development process as a stereotyped UML activity diagram. Models are represented with object flow states, and transformations are represented as stereotyped activities (circular icon). The control flow is defined by a set of transformations. The process starts by defining a *Requirements Model* to represent the system's functionality. Platform-independent models (PIMs) are then obtained from these requirements.

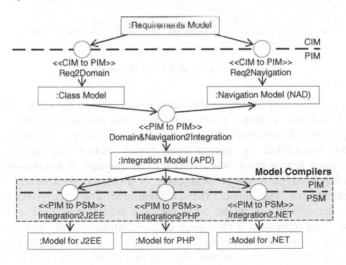

Fig. 2. The OO-H development process

The set of conceptual models represents the different concerns of a Web application: the specification of content requirements (Class Model) and the specification of functional requirements in terms of navigation needs (Navigation Model, NAD). A merge between the class and navigation models results in an integrated PIM model (APD), which covers the relevant aspects of a Web application. Finally, platform-specific models (PSMs) are automatically obtained from the Integration Model (APD), from which source code can be generated. The aim of such a process is to automate model transformations based on a set of transformation rules. The approach follows a *translationist* MDA approach (i.e., the code is generated from PIM models using model compilers).

2.1 The OO-H Metamodel

Fig. 3 shows an excerpt of the OO-H Metamodel [5] including the main concepts used to represent navigational properties. The navigation model of a Web application is defined by a set of navigational maps.

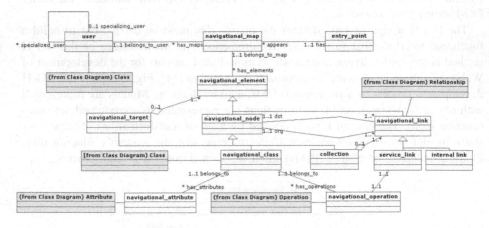

Fig. 3. A portion of the OO-H metamodel

Each *navigational map* structures the navigational view of the system for a specific kind of *user* and has a unique *entry point* that indicates the starting point of the navigation process. A navigational map is made up of a set of *navigational elements* that can be specialized as navigational nodes and/or navigational links. A *navigational node* represents a view over the class diagram. A navigational node can be a navigational_target, a navigational_class, or a collection. A *navigational target* groups elements of the model (i.e. navigational classes, navigational links, collections) that collaborate in the coverage of a user navigational requirement. A *navigational class* represents a view over a set of attributes (*navigational_attribute*) and operations (*navigational_operation*) of a class from the class diagram. A *collection* is a hierarchical structure that groups a set of navigational links. A *navigational link* defines valid node attainability inside the navigational map. Navigational links define the navigation paths that the user can follow through the user interface. Each navigational link has an origin (org) node and its corresponding destination (dst) node. There are two types of links: *internal links* to define the navigation path inside a navigational map, and *service links* to show the services available to the user associated to that navigational map.

In addition, OO-H allows the use of the Object Constraint Language (OCL) to define constraints over the class diagram and the NAD. In the NAD, such constraints are expressed by means of *filters* defined upon links.

3   The Design of OO-HFP

In the first step of the process model described in Fig. 1, four activities are suggested for a complete design of a measurement method [10]: definition of the objectives, characterization of the concept to be measured, definition or selection of the metamodel, and definition of the numerical assignment rules.

The goal of this work is: to **design** a FSM procedure **for the purpose of** automatically sizing OO-H conceptual models **with respect to their** functional size **from the point of view of** the researcher. The **context** is that this procedure should conform to the IFPUG method version 4.1 (ISO/IEC 20926 [7]).

The **entity** to be measured consists of an OO-H conceptual model. It is composed of a *UML Class diagram* and a *Navigation Access Diagram (NAD)* that capture the structural and the navigational views of a Web application, respectively. These views contain all the elements that contribute to the functional size of a Web application. The **attribute** to be measured is *functional size*, which is defined in ISO/IEC 14143-1 [8] as the size of the software derived by quantifying the functional user requirements.

The selection of the metamodel and the definition of the numerical assignment rules are described in the following sections.

3.1   Selection of the Metamodel

Metamodeling is a key concept of the MDA paradigm and is used in Software Engineering to describe the basic abstractions that define the models and their relationships. The metamodel of a FSM method provides a precise basis to design the measurement rules that identify and measure these concepts.

As OO-HFP is intended to conform to the IFPUG method [7], it assumes the same metamodel as FPA. Fig. 4 shows the FPA metamodel for the IFPUG method. It illustrates the information that must be captured in order to size a software project.

Measuring a project basically includes the following activities: i) the determination of the type of count (new project, enhanced project or running application), ii) the identification of the counting scope and application boundary, iii) the identification of data and transactional functions (FunctionType), and iv) the determination of the complexity of each identified function type. The IFPUG method [7] provides several tables to determine the complexity levels of each function type. The *counting scope* defines the functionality that is included in a particular function point count. It defines a (sub) set of the project being sized.

The *boundary* separates the software system being measured from its environment, which contains the users of the system and may include external systems.

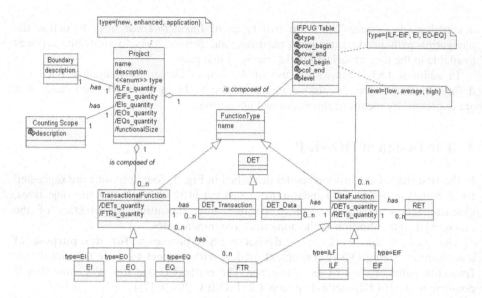

Fig. 4. The FPA metamodel for the IFPUG method

A *data function* consists of both internal logical files (ILF) that are maintained by the system and external interface files (EIF) that are referenced by the system. A *transactional function* consists of external input functions (EI), external output functions (EO), and external inquiry functions (EQ) that use or modify data functions.

The complexity of a data function is a function of the number of Data Element Types (*DET_Data*) and the number of Record Element Types (*RET*). A DET is a unique, user recognizable, non-repeated field. Note that only the functions that have at least one DET must be identified. The complexity of a transactional function is a function of the number of Data Element Types (*DET_Transaction*) and the number of File Types Referenced (*FTR*). A FTR is a data function that is referenced during the execution of a transactional function. The amount and complexity of data and transactional functions determine the amount of functionality that this piece of software delivers, hence its *functional size*.

3.1.1 Definition of the Mapping Rules

The mapping rules help identify the elements in an OO-H conceptual model that contribute to the functional size of the Web application. These rules (see Table 1) are defined as a mapping between the concepts of the FPA metamodel onto the concepts in the UML/OO-H metamodel [5] [15].

First, Rules 1 to 4 are applied to establish the counting scope and the boundary of the Web application. Then, the data (ILF and EIF) and transactional (EI, EO, and EQ) functions are identified by applying Rules 5 to 11.

An ILF *is a user identifiable group of logically related data or control information maintained within the boundary of the system*. In OO-H, this corresponds to a *class*. that encapsulates a set of data items (attributes). Aggregation and composition hierarchies can also be considered an ILF since they represents a unique group of logically related data.

Table 1. Mapping rules

FPA Metamodel	UML/OO-H Metamodels
Counting scope	**Rule 1.** The scope corresponds to an OO-H conceptual model including a UML class diagram and a Navigation Access Diagram (NAD).
Boundary	**Rule 2.** Accept each *class* with the «actor» stereotype as a user of the application.
	Rule 3. Accept each *class* with the «legacy» stereotype as an external application. This class contains the expected object behavior or the connection with existing applications.
	Rule 4. The *boundary* corresponds to an imaginary line traced in the UML class diagram. The «actor» and «legacy» classes are considered to be outside the boundary, whereas the other classes are considered to be inside the boundary.
ILF	**Rule 5.** Accept each *aggregation hierarchy* as an ILF, since aggregation represents "is part of" relationships.
	Rule 6 Accept each *composition hierarchy* as an ILF. Composition is stronger than aggregation since the whole and the parts have coincident lifetimes.
	Rule 7. Accept each *class* of the class diagram that does not participate in an aggregation or composition hierarchy as an ILF.
EIF	**Rule 8:** Accept each *class with the «legacy» stereotype* of the class diagram as an EIF.
EI	**Rule 9:** Accept each method of a class in the class diagram as an EI.
EQ	**Rule 10:** Accept each *navigational target* of a NAD (and each abstract page derived from it) as an EQ.
EO	**Rule 11:** Accept each *navigational target* of a NAD (and each abstract page derived from it) as an EO. It must contain a class with a derived attribute or method.

An EIF is *a user identifiable group of logically related data or control information referenced by the system, but maintained within the boundary of another system.* In OO-H, this is a special type of *class with the «legacy» stereotype* that allows the connection with existing application logic modules.

An EI *is an elementary process that manipulates data or control information that comes from outside the system boundary.* Its goal is to maintain one or more ILFs and/or to alter the behavior of the system. In OO-H, this corresponds to a *method* class since it always changes the state of the objects of the class. A method must be identified only once (in the declared class), even if it is inherited.

An EQ *is an elementary process that sends data or control information outside the system boundary.* Its goal is to present information to a user through the retrieval of data. In OO-H, this corresponds to a *navigational target* defined in the NAD. A navigational target may derive one or several *abstract pages* depending on the value of the property 'effect' of a navigational link. The value *source* (a light arrow) indicates that the information specified in the navigational target will be presented in the current abstract page, while the value *target* (a dark arrow) indicates that the information will be presented in a different abstract page.

Finally, an EO has the same definition as an EQ, but its processing logic must contain at least one mathematical formula or calculation, or it has to create derived data. In OO-H, this also corresponds to the concept of *navigational target*. However, a class in the navigational target must include at least one derived attribute or method.

3.2 Definition of the Numerical Assignment Rules

The purpose of this phase is to produce a quantitative value that represents the Web application functional size. This is accomplished by applying two sets of rules that are introduced in the following subsections.

3.2.1 Definition of the Measurement Rules

According to the FPA metamodel, the complexity of a class (ILF) or «legacy» class (EIF) is determined by counting the number of data element types (DET) and record element types (RET). Table 2 describes the measurement rules proposed to identify the DETs and RETs of a class.

A DET is identified for each single attribute of a class, and a RET is identified for the class itself. In addition, we consider both aggregations/compositions and generalization/specialization relationships as contributing to the complexity of a class. This is because the IFPUG method suggests counting a DET for each piece of data that indicates a relationship with another ILF or EIF.

Associations, aggregations, and compositions are measured according to the multiplicity property of the relationship. This property specifies the lower/upper number of objects that can be associated to a single object of the class. If the number of instances in the target relationship is one, a DET is counted. Otherwise, if the number of instances in the target relationship is greater than one, a RET is identified.

Table 2. Measurement rules for the complexity of a class and class hierarchies

Rules for DET	Rules for RET
Rule 12. Count 1 DET for each *single attribute* type of the class (i.e., String, Integer).	**Rule 17.** Count 1 RET for the *class*.
Rule 13. Count 1 DET for each *association* with a class or «legacy» class having a maximum multiplicity of 1.	**Rule 18.** Count 1 RET for each *association* relationship with a class or «legacy» class having a maximum multiplicity of *.
Rule 14. Count 1 DET for each *aggregation* with a class having a maximum multiplicity of 1.	**Rule 19.** Count 1 RET for each *aggregation* relationship with a class having a maximum multiplicity of *.
Rule 15. Count 1 DET for each *composition* with a class having a maximum multiplicity of 1.	**Rule 20.** Count 1 RET for each *composition* relationship with a class having a maximum multiplicity of *.
Rule 16. Count 1 DET for each super class in an *inheritance* relationship.	**Rule 21.** Count 1 RET for each class in an aggregation, composition or inheritance *hierarchy*.

Specialization relationships are measured in the subclasses by counting a DET for each super class. Finally, the complexity of aggregation, composition, or inheritance hierarchies is measured by counting a RET for each class in the hierarchy.

The complexity of a method (EI) or navigational target (EO and EQ) is determined by counting the number of data element types (DET) and file types referenced (FTR).

Table 3 shows some representative measurement rules for the complexity of a class method. Each single argument of a method is counted as a DET, and each object-valued argument is counted as a FTR. Two DETs are counted in order to be compliant to the IFPUG method: one DET for the capability of the application to send a response message outside the boundary (error, confirmation, control), and another DET for the ability to specify an action to be taken. An FTR is also counted for the class where the method is declared. Other FTRs are also counted for the classes that appear in the OCL formula (e.g., precondition, postcondition) specified in the method class.

Table 3. Some representative measurement rules for the complexity of a class method

Rules for DET	Rules for FTR
Rule 30. Count 1 DET for each *single argument* of the method.	**Rule 33.** Count 1 FTR for the *class* in which the method is declared.
Rule 31. Count 1 DET for the capability of the application to send a *message* outside the boundary.	**Rule 34.** Count 1 FTR for each *complex argument* (object of a class*).
Rule 32. Count 1 DET for the ability of the application to specify an *action* to be taken.	**Rule 35.** Count 1 FTR for each class* in the OCL formula of a *precondition*.
	Rule 36. Count 1 FTR for each new class* in the OCL formula of a *postcondition*.
(*) Only if the class has not been counted yet.	

Table 4 shows some representative measurement rules for the complexity of a navigational target. A navigational target groups elements (navigational classes, navigational links, and collections) to fulfill a user navigational requirement.

A DET is counted for each element in a collection that implies navigating to another collection. Additional DETs are counted for each visible attribute or method of a navigational class. A DET is counted for each link between navigational classes. This is because the IFPUG method suggests counting a DET for each piece of data that indicates a relationship with another group of data. Two other DETs are counted to be compliant to the IFPUG method: one DET for the capability of the application to send a response message and another for the ability to specify an action to be taken.

A FTR is counted for each navigational class of the navigational target. Additional FTRs are counted for the classes that appear in the OCL formula of a navigational class or navigational link (precondition, filter). There are also other measurement rules for navigational patterns (e.g., index, guided tour) that can be specified in a navigational link. The complete set of rules is described in the OO-H measurement guideline available at: www.dsic.upv.es/~sabrahao/OO-HFP.

Table 4. Some representative measurement rules for the complexity of a navigational target

Rules for DETs	Rules for RETs
Rule 46. Count 1 DET for each element (link) of a *collection* that implies navigating to a collection.	**Rule 58.** Count 1 FTR for each *navigational class*.
Rule 47. Count 1 DET for each *visible attribute* of a navigational class.	**Rule 59.** Count 1 FTR for each class* in the OCL formula of a *filter* defined upon navigational links.
Rule 48. Count 1 DET for each *method* of a navigational class.	**Rule 60.** Count 1 FTR for each class* in the OCL formula of a *precondition* defined upon links.
Rule 50. Count 1 DET for each *navigational link* between two navigational classes.	**Rule 61.** Count 1 FTR for each class* in the OCL formula of a navigational class.
Rule 56. Count 1 DET for the capability of the application to send a message.	
Rule 57. Count 1 DET for the ability of the application to specify an action to be taken.	
(*) Only if the class has not been counted yet.	

3.2.2 FPA Counting Rules

Once the DETs, RETs and FTRs have been counted, the FPA counting rules are applied to do the following: to classify the function complexity (low, average, high); to assign weights to the functions; and to aggregate the assigned values into an overall functional size value for the Web application.

The IFPUG method [7] provides several tables to determine the complexity levels of a function type. Table 5 shows the table to determine the complexity of an EO.

Table 5. Complexity weights for EO (navigational target)

FTRs	1-5 DETs	6-19 DETs	20 or more DETs
0-1	Low	Low	Average
2-3	Low	Average	High
4 or more	Average	High	High

Then, a value in Function Points (FP) is assigned to each function depending on its type and complexity level. The IFPUG method also provides several tables to assign FP values for the identified functions. For instance, the values assigned to an EO are the following: **low = 4 FP**, **average = 5 FP**, and **high = 7 FP**.

The function points assigned to each ILF (class or class hierarchy) and EIF (legacy class) are summed to obtain the functional size for the data functions (OO-HFP$_D$). The function points assigned to each EI (class method), EO and EQ (navigational target) are summed to obtain the functional size for the transactional functions (OO-HFP$_T$).

Finally, the values are summed to produce the functional size of the Web application in unadjusted function points: OO-HFP = OO-HFP$_D$ + OO-HFP$_T$

3.3 A Measurement Example

Fig. 5 shows an example of a navigational target (*Concerts*) that is part of an e-commerce application for concert ticket sales. It allows the user to see a list of concerts including information about their sessions, artists, and the room where the concert will be held. It also allows navigation to three other navigational targets.

By applying the OO-H mapping rules (see section 3.1.1), the *Concerts* navigational target is classified as an EO since the *Room* class includes the *capacity* derived attribute. As all the navigational links end in light arrows, only one EO is identified.

Then, the OO-H measurement rules (see section 3.2.1) are applied to determine the complexity of the navigational target. All the attributes of the navigational classes and the relationships among them (dotted arrows) are counted as DETs. Since the relationship between classes and navigational targets (dark arrows) indicates navigation among navigational targets (allows the user to access another EQ), one DET for each link is also counted. Finally, two additional DETs are counted in order to be compliant to the FPA counting rules. A FTR for each class in the navigational target is counted. In total, the *Concerts* navigational target has **16 DETs** and **4 FTRs**.

Once the DETs and FTRs have been counted, the FPA counting rules (see section 3.2.2) are applied to classify the function complexity (low, average or high) and to assign a function point value. According to Table 5, the *Concert* navigational target has a **high complexity**, which corresponds to **7 FP**.

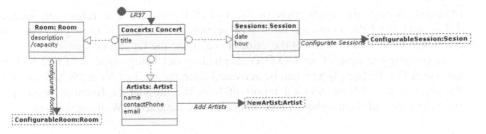

Fig. 5. *Concerts* Navigational Target

4 The Automation of OO-HFP in VisualWADE

In the second step of the process model described in Fig. 1, three activities are suggested for the manual application of a measurement method or procedure: software documentation gathering, construction of the software model, and application of the numerical assignment rules. The automation of OO-HFP was conducted in VisualWADE by automating these three activities. VisualWADE is extensible by means of plug-ins, which allow the analyst to package the customization and automation of several features through the OO-H metamodel.

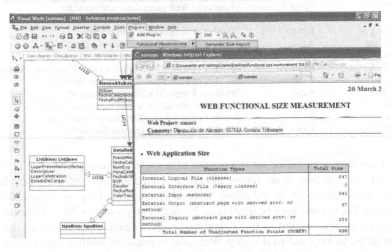

Fig. 6. The OO-HFP plug-in in the VisualWADE tool suite

In the *software documentation gathering* activity, a conceptual model is specified using the OO-H method. The tool stores "on the fly" all the relevant information that is needed to get the functional size. In the *construction of the software model* activity, the OO-HFP mapping rules (see section 3.1.1) are used to obtain the FPA view on the functional size for the conceptual model. This step consists in identifying the elements in the OO-H conceptual model that contribute to functional size. Since the metamodel is expressed in python language, the tool can access the stored metamodel. Finally, in the *application of the numerical assignment rules* activity, a functional size value is obtained through the application of the OO-HFP measurement rules (see section 3.2.1) and the FPA counting rules (see section 3.2.2). The plug-in performs the queries to the metamodel to get the values to calculate the functional size.

A running example of an OO-H conceptual model being sized with OO-HFP is shown in Fig. 6. The plug-in can be activated from the GUI of VisualWADE, giving the analyst a real-time detailed report of both the number of function points per function type and the number of function points for the overall Web application.

5   The Validation of OO-HFP for Effort Estimation

The validation of OO-HFP for Web effort estimation is done by comparing the prediction accuracy that it provides to the accuracy provided by another set of validated size measures (the Tukutuku measures) [13]. This corresponds to the "Exploitation of the measurement result" step of the process model shown in Fig. 1.

The Tukutuku size measures (see Table 6) can be used early on in a Web development life cycle whenever needed, or during the requirements gathering phase.

Table 6. Tukutuku Size measures

Variable name	Scale	Description
TotWP	Ratio	Number of Web pages (new and reused).
NewWP	Ratio	Number of new Web pages.
TotImg	Ratio	Number of images (new and reused).
NewImg	Ratio	Number of new images created.
Fots	Ratio	Number of features reused without any adaptation.
HFotsA	Ratio	Number of reused high-effort features/functions adapted.
Hnew	Ratio	Number of new high-effort features/functions.
totHigh	Ratio	Number of high-effort features/functions
FotsA	Ratio	Number of reused low-effort features adapted.
New	Ratio	Number of new low-effort features/functions.
totNHigh	Ratio	Number of low-effort features/functions

The measures were based on the results of a survey investigation [12] [13], using data from 133 on-line Web forms aimed at giving quotes on Web development projects. In addition, these measures were also confirmed by an established Web company and a second survey involving 33 Web companies in New Zealand. These Web size measures are part of the Tukutuku project, which aims to collect data about Web projects, to be used to develop Web cost estimation models and to benchmark productivity across and within Web Companies. These Web size measures are a subset of those described in Mendes et al. [13]. These measures have been empirically validated as good Web effort predictors (see e.g. [12]).

5.1 Selecting the Estimation Technique

Case-Based Reasoning (CBR) was the technique chosen to obtain effort estimates because it has provided good effort estimates in the past. The results presented here were obtained using the commercial tool CBR-Works. CBR is a branch of Artificial Intelligence where knowledge of similar past cases is used to solve new cases [19].

CBR predicts the effort of a new project by considering similar projects that were previously developed. Specifically, completed projects are characterized in terms of a set of *p* size features (e.g. *totWP*) and make up the case base. The new project is also characterized in terms of the same *p* size attributes and is referred to as the target case. Then, the similarity between the target case and the other cases in the p-dimensional feature space is measured. The most similar cases are used (possibly with adaptations) to obtain a prediction for the target case. To apply the method, the following were selected: the relevant project features, the appropriate similarity function, the number of analogies to select the similar projects to be estimated, and the analogy adaptation strategy for generating the estimation. Our effort estimates were based on the effort of the most similar project in the case base, which was identified using the Euclidean distance, with no different weights for attributes or adaptations of the estimated effort. This seemed to provide the best results when applying different CBR combinations [14].

5.2 Results

The analysis presented in this paper was based on data coming from 12 Web projects volunteered by a Spanish Web development company. All the projects were developed using the OO-H method. Projects had their size measured using OO-HFP and the Tukutuku size measures. Their descriptive statistics are presented in Table 7. They also include the OO-HFP and the Tukutuku size measures. This dataset size has also been previously used in other Web cost estimation studies.

Table 7. Descriptive statistics for the 12 Web Projects

Variable name	Mean	Median	Std. Dev.	Min.	Max.
OO-HFP	675.25	200.50	788.853	24	2400
TotEff	1304.33	132	1557.00	16	3644
TotWP	71.08	23.5	76.25	3	200
NewWP	6.83	3.5	9.50	0	30
TotImg	63.17	12.5	118.94	1	405
NewImg	6.67	0	12.31	0	40
Fots	7.50	7.5	2.61	5	10
HFotsA	148.50	28	196.09	2	611
Hnew	0.00	0	0.00	0	0
totHigh	148.50	28	196.09	2	611
FotsA	3.00	0	5.72	0	20
New	6.08	4.5	5.48	0	15
totNHigh	9.08	5	10.28	0	35

The statistics for OO-HFP show that the functional size of the projects ranges from 24 FP to 2400 FP. The medians for *TotEff, TotWP, TotImg* and *totNHigh* suggest that most projects were small with regard to the number of pages and images and their duration. However, despite being small with short durations, they reused (with adaptation) a comparatively large number of high effort features/functions. This trend is of no surprise given that many of the Web applications developed today reuse components and are focused on functionality. Note that these applications were automatically generated using VisualWADE, hence their short duration.

Next, the effort estimation models were built using CBR. The prediction **accuracy** of models was checked by omitting a group of projects (8 projects) and predicting the effort for the group of omitted projects (4 projects). The rationale was to use different sets of projects to build and to validate a model. Finally the prediction accuracy of each model was measured using the Mean MRE (MMRE), Median MRE (MdMRE), and Pred(25). We also compared each model with predictions obtained using the mean and median efforts of the training set in order to have a benchmark. Statistical significance was checked using the Wilcoxon Signed Ranks test, with $\alpha = 0.05$.

Our case base contained the eight projects in our training set, and used the *11 Tukutuku size measures* and *TotEff* as attributes. Each of the four projects in the validation set was used in turn to obtain its equivalent most similar project, which was measured using the Euclidean distance. The prediction accuracy measures for CBR (see Table 8) indicated that the CBR-based predictions were poor. CBR-based accuracy was not significantly better than predictions based on the median (p = 0.461) or mean (p = 0.461) of the data set.

Table 8. Accuracy statistics for Tukutuku size measures and the OO-HFP

Accuracy measures (%)	Tukutuku size measures			OO-HFP		
	CBR	Mean Effort	Median Effort	CBR	Mean Effort	Median Effort
MMRE	3,614.4	1,532.5	156.51	3,514.4	1,532.5	156.51
MdMRE	1,864.2	786.4	94.8	1,763.2	786.4	94.8
Pred(25)	25	0	0	25	0	0

When CBR was applied to OO-HFP, the same procedure as described above was used. However, the attributes used were *OO-HFP* and *TotEff*. The prediction accuracy measures for CBR (see Table 8) indicated a very similar pattern to those obtained using the Tukutuku size measures. CBR-based accuracy was not significantly better than predictions based on the median or mean of the data set. What these results suggest is that using effort estimates based on early Web size measures (OO-HFP or Tukutuku size measures) provides similar accuracy to using estimates based on the mean or median of past projects.

6   Conclusions and Further Work

This paper has introduced a model-driven measurement procedure for sizing Web applications (OO-HFP). The procedure is compliant to the IFPUG method, which is a widely used FSM method in industry. Since the OO-HFP was designed as a mapping between the IFPUG and the OO-H metamodels, a conformity evaluation was made during the design stage assuring that all the concepts were properly dealt with.

OO-HFP was also automated in VisualWADE. Thus, a size measure of a Web project can be easily calculated when the OO-H conceptual model is specified. This avoids the ambiguity of interpreting the FPA counting rules and the need for special training to count function points in an accurate and repeatable way. In addition, MDA-based methods provide good support for size estimation given the explicit traceability between model elements. Therefore, the size estimated at the PIM level can be representative of the size of the final system obtained. Finally, we validated the use of OO-HFP for Web effort estimation by comparing the prediction accuracy that it provides to the accuracy provided by the Tukutuku size measures. The results showed that both measures presented similar accuracy, which was also not different from the predictions obtained using the mean and median effort of past projects. The main limitation of the study was the small data set used (only 20 projects). We plan to replicate this study using a broad data set and other effort estimation techniques (e.g., Stepwise regression) and size measures.

References

1. Cândido, E.J.D., Sanchez, R.: Estimating the size of web applications by using a simplified function point method. In: Proc. of the WebMedia/LA-WEB 2004, pp. 98–105 (2004)
2. Ceri, S., Fraternali, P., Bongio, A.: Web Modeling Language (WebML): a modeling language for designing Web sites. In: WWW 2000. Proc. of the 9th Word Wide Web Conference, Amsterdam, The Netherlands, pp. 137–157 (2000)
3. Cleary, D.: Web-Based Development and Functional Size Measurement. In: Proceedings of IFPUG Annual Conference, San Diego, USA (2000)
4. COSMIC-FFP Measurement Manual version 2.2 (2003)
5. Gomez, J., Cachero, C., Pastor, O.: Conceptual Modeling of Device-Independent Web Applications. IEEE MultiMedia 8(2), 26–39 (2001)
6. IFPUG: Hints to Counting Web Sites: IFPUG White Paper (1998)
7. ISO: ISO/IEC 20926: Software engineering - IFPUG 4.1. Unadjusted functional size measurement method - Counting practices manual (2003)
8. ISO, ISO/IEC 14143-1- Information Technology - Software measurement - Functional Size Measurement. Part 1: Definition of Concepts (1998)
9. ISO: International Vocabulary of Basic and General Terms in Metrology, 2nd edn. (1993)
10. Jacquet, J.P., Abran, A.: From Software Metrics to Software Measurement Methods: A Process Model. In: ISESS'97. 3rd International Standard Symposium and Forum on Software Engineering Standards, Walnut Creek, USA (1997)
11. Koch, N., Zhang, G., Escalona, M.J: Model transformations from requirements to web system design. In: ICWE 2006. Proc. of the 6th International Conference on Web Engineering, Palo Alto-CA, USA, pp. 281–288 (2006)
12. Mendes, E., Mosley, N., Counsell, S.: Investigating Web Size Metrics for Early Web Cost Estimation. Journal of Systems and Software 77(2), 157–172 (2005)
13. Mendes, E., Mosley, N., Counsell, S.: Investigating Early Web Size Measures for Web Cost Estimation. In: EASE 2003. Proc. of the 7th Conference on Evaluation & Assessment in Software Engineering, Keele, UK (2003)
14. Mendes, E., Watson, I., Triggs, C., Mosley, N., Counsell, S.: A Comparative Study of Cost Estimation Models for Web Hypermedia Applications. Empirical Software Engineering 8(2), 163–196 (2003)
15. OMG: UML 2.1 Unified Modeling Language™ (2006)
16. OMG: MDA Guide, Version 1.0.1 (2003), http://www.omg.org/docs/omg/03-06-01.pdf
17. Reifer, D.: Web Development: Estimating Quick-to-Market Software. IEEE Software 17(6), 57–64 (2000)
18. Schwabe, D., Rossi, G.: The Object-Oriented Hypermedia Design Model. Communications of the ACM 38(8), 45–46 (1995)
19. Shepperd, M.J., Kadoda, G.: Using Simulation to Evaluate Prediction Techniques. In: Proceedings IEEE Metrics Symposium, pp. 349–358. IEEE Computer Society Press, Los Alamitos (2001)

Model-Driven Engineering for Software Migration in a Large Industrial Context

Franck Fleurey[1,2], Erwan Breton[2], Benoit Baudry[1], Alain Nicolas[2],
and Jean-Marc Jézéquel[1]

[1] IRISA/INRIA, Rennes, France
{ffleurey, bbaudry, jezequel}@irisa.fr
[2] Sodifrance, Nantes, France
{ebreton, anicolas}@sodifrance.fr

Abstract. As development techniques, paradigms and platforms evolve far more quickly than domain applications, software modernization and migration, is a constant challenge to software engineers. For more than ten years now, the Sodifrance company has been intensively using Model-Driven Engineering (MDE) for both development and migration projects. In this paper we report on the use of MDE as an efficient, flexible and reliable approach for a migration process (reverse-engineering, transformation and code generation). Moreover, we discuss how MDE is economically profitable and is cost-effective over the migration through out-sourced manual re-development. The paper is illustrated with the migration of a large-scale banking system from Mainframe to J2EE.

1 Introduction

Positioned from the mid 80's on IT servicies dedicated to Banks and Insurance Companies, Sodifrance has developed a strong legacy modernization expertise based on software solutions to industrialize transformation projects. Since 1994, Sodifrance has adopted and promoted model-driven engineering (MDE) approaches for modernization projects. It has industrialized model-driven techniques for reverse-engineering, code analysis and transformation and for representing and manipulating information systems. These solutions allow the company to propose efficient and profitable solutions for migration and modernization of software legacy systems.

In this paper we first present an original model-driven process, developed at Sodifrance, for software migration. This process includes automatic analysis of the existing code, reverse engineering of abstract high-level models, model transformation to target platform models and code generation. We detail the different meta-models and transformations that are produced for the automation of these steps. We also discuss what artefacts can be directly reused and which ones need to be adapted from one project to another. Sodifrance has developed a tool suite for model manipulation called Model-In-Action (MIA) that is used as a basis for automating the migration.

G. Engels et al. (Eds.): MoDELS 2007, LNCS 4735, pp. 482–497, 2007.

A second contribution of this paper is an industrial feedback on the benefits and issues of MDE for migration. First, we present data for a migration project of a large-scale banking system from Mainframe to J2EE. These data are used to discuss the improvements with respect to efficiency, flexibility and reliability that are introduced with a model-driven solution migration. Moreover, we compare MDE and complete manual re-development, and discuss how MDE is economically profitable and is cost-effective.

2   Model-Driven Migration Process

The constant evolution of software technology leads to continuous migrations of software components. These projects may be motivated by different reasons such as the obsolescence of a technology, the pressure of users, or the need to build a single coherant information system when merging companies. Most of the time software migration is achieved through the full re-developement of the legacy application. Model-driven software developement offers an oportunity for increasing the automation in software migration.

The full automation of migration is difficult to achieve not only because of the distance between the legacy platform and the new platform but also in order to ensure the quality of the new application. Most of the time, the objective of migration is not to simply "compile" the legacy application to a new platform but to create a new version of the application using state of the art development techniques. This is necessary to ensure the maintainability of the new application and to leverage the latest technologies in terms of graphical user interfaces, distribution and mobility.

In the following, section 2.1 first presents the general process developed by Sodifrance for model-driven migration, section 2.2 discusses the automation of the process and section 2.3 details how this process is adapted in practice along the phases of a migration project.

2.1   Migration General Process

Figure 1 presents the general process developed by Sodifrance for model-driven migration. This process is mainly divided in four steps.

The first step is the parsing of the code of the legacy application, to build a complete model of the code of the application. This step can be divided into two stages: first a parser builds an abstract syntax tree from the code and, then this syntax tree is processed by a transformation to build an actual model that conforms to the meta-model of the legacy language. During the second stage, all the symbols such as types, variables or function calls are resolved and properly bound to the appropriate model elements. This is a necessary step to allow for a efficient analysis of the legacy system. The meta-model denoted L on figure 1 corresponds to the meta-model of the legacy application implementation language.

The second step is a reverse-engineering from the code model to a platform independent model. The role of this step is to abstract high-level views from the

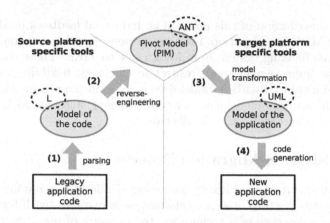

Fig. 1. Model-driven migration priciple

model of the code. This step is implemented by model transformations from the legacy language meta-model (L) to a pivot meta-model. The pivot meta-model used by Sodifrance is a platform independent meta-model called ANT which contains packages to represent:

- Static data structures (close to the UML class diagram).
- Actions and algorithms (it includes an imperative action language).
- Graphical user interfaces and widgets.
- Application navigation.

The navigation is the most high level view of the ANT meta-model. Figure 2 shows an excerpt of this meta-model. It connects dialog elements which correspond to GUI forms, transitions between forms and their GUI events with operations in the class model.

All ANT views have to be created through model transformations from the model of the code of the legacy application. In order to be able to create high-level views, such as a model of the graphical user interface of the legacy application, the model transformations have to rely on a knowledge of the libraries of the legacy platform and on coding conventions (or code patterns introduced by tools) that were used during the development of the legacy application. This is the reason why, even if the legacy platforms for several migration projects are similar, the legacy code must be carefully studied in order to properly adapt the migration tools to every single project.

The third step is the transformation of the ANT model into a platform specific model of the application. This step is implemented using model transformations from the ANT meta-model to the UML meta-model. These transformations are design transformations which refine the platform independant views of the pivot model to fit the target platform. Again at this stage, it is important to adapt the transformation to meet the requirements of every customer. This issue is discussed with more details and illustrated on a specific project in section 4.

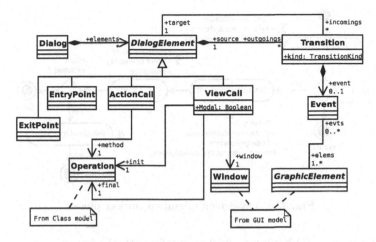

Fig. 2. Excerpt of the ANT navigation meta-model

The last step is the generation of the code of the new application from the platform specific model. To implement this step, Sodifrance uses template-based text generation tools in order to be able to easily customize code generation acording to the customers requirements. The specific tools used by Sodifrance for the implementation of model-transformations and code generation are presented section 3.

2.2 Automation in the Migration Process

To reduce the cost of migration the goal is to achieve an optimum automation in the migration process. However, this should not impact the quality in terms of design, performances or maintainability of the resulting application. Since the legacy application is fully-executable and the target platform is usually powerful enough, one could argue that the migration should be completely automated. It is theoretically possible: it would be the equivalent of writing a compiler for the legacy language that targets the new platform.

However, as stated in the previous section, migration, and especially in the context of modernization, is more than just creating an executable version of the application on top of the new platform. The goal is to design the application for the new platform in order to make it more efficient, more reliable, easier to maintain or easier to extend than the legacy application. In practice this means that the new code should respect the coding standards and best practices of the target platform languages, it should take into account the specific requirements related to the software development process used by the customer company, there should be models for the new application, etc.

In the migration process implemented by Sodifrance the first two steps (as presented on figure 1) are usually completely automated, i.e. all the information from the legacy system is represented in the pivot model. This is to concentrate the manual effort on the transformation from the pivot model to the new

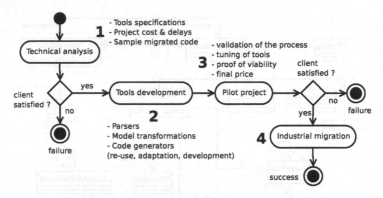

Fig. 3. Model-driven migration project phases

application and avoid having to deal manually with the legacy code as a whole. If some elements of the legacy code cannot fit properly in the pivot model, these elements are captured as notes or tags and presented to the developer when the corresponding parts of the application are transformed or generated.

To maximize the efficiency of the migration process, the tasks that are left to the developer have to be clearly identified and the developer should be provided with all the information he or she needs. This is taken into account in the design of the transformations and code generators. For example in the case of a Java code generator, *TODO* directives can be generated for every piece of code that requires manual inspection, re-factoring or completion. This *TODO* directive can contain the kind of work that has to be done and references to the model elements that are relevant to it. The *TODO* directives are summarized into a task list which gives the developer a clear view of what has to be done.

2.3 Migration Project Phases

Prior to the actual migration and implementation of the new application, the design, the implementation and the validation of a project specific migration process must be completed. This includes the parsing of legacy languages, reverse engineering transformations, high-level design of the new application and mappings between the structures of the legacy application and the concepts of the target platform. All these tasks require some effort due to their complexity and their overall influence on the migration project. In the project structure used by Sodifrance, as represented on figure 3, there are three project phases before the actual migration can start.

The first phase represented on figure 3 is a technical analysis. Its objective is to study the legacy platform, define the target platform and specify the tools that are needed by the migration process. This phase is crucial for the migration project. It is used to estimate the effort that would be required for the development of the tools and the total effort that would be required for the migration. At the end of the technical study a total contractual price is proposed to the customer. During the technical study a small component of the legacy application is usually migrated using generic tools and manually completed to match the code that would be produced using the final tools. This serves as a test for the tool specifications and as a demonstration of the resulting code the customer can expect. If both the price proposed by Sodifrance and the quality of the migrated code are satisfactory to the customer, the project can carry on.

The second phase represented on figure 3 is a tool development phase. The objective is to develop all the tools that have been specified for the migration process. Most of the time the tools do not have to be developed from scratch but are rather re-used or adapted from previous projects. However, most of the time even if the language is the same, the language version and the coding style might be different and require some adaptation.

The third phase represented on figure 3 is a pilot project. The objective of the pilot project is to validate and fine tune the migration process and the tools it uses. It also serves as a demonstration of the viability of the process and allows measuring its efficiency precisely. During this phase, a component of the legacy application is used as a benchmark for the migration process. This component has to be chosen to be as representative as possible of the components of legacy application. In practice the development of the pilot project is truly a testing and debugging phase for the migration tools. For this reason it is usually a lot longer than the migration of a comparable component once the migration process is fully-functional. At the end of the pilot project, the customer is provided with a final price for the project and has a sample of how the new application would look like.

Projects seldom have to stop after the pilot project: the actual migration usually starts shortly afterwards. The preparation of a model-driven migration process can be quite long (the three phases described previously usually require around 6 months to complete but can last up to a year on specific projects such as the one described in section 4), but once the process is up and running, the migration rate can be far more rapid than with any competing techniques. This is discussed in section 5, but before that, the next section presents the model-driven engineering tools used by Sodifrance to practically implement model-driven migration.

3   Model-In-Action (MIA) Tool Suite

Implementing the migration process presented in the previous section requires advanced, scalable and reliable tools for model transformation and code generation. For both the needs of migration project and development projects, Sodifrance has developed Model-In-Action (MIA) [1], a suite of model-driven engineering tools. This section gives a quick overview of these tools.

Figure 4 presents a simplified architecture diagram for the MIA tools. One of the essential requirement for a company like Sodifrance is to be able to adapt to any specific modeling technology used by their clients. In the design of MIA this has been taken into account by creating a generic modeling platform that can connect through various drivers to existing repositories and modelers. On

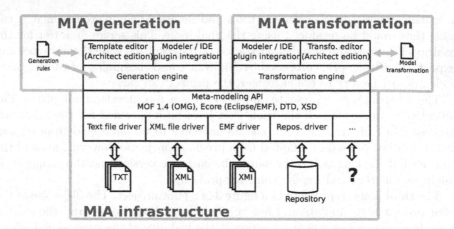

Fig. 4. Model-In-Action tool suite architecture

top of this generic modeling layer the suite is composed of two main products: *MIA-Transformation* for model-to-model transformation and *MIA-Generation* for code generation. Each of these tools is divided in three types of components:

- Core engines for model transformations and code generation. These components are on top of the meta-modeling API and do not have any user interface. They are responsible for the execution of model transformations and code generators.
- Development environments for model transformations and code generators (MIA Architect environments). These environments are used by software architects to design and implement the model transformations and code generators required by MDE projects.
- User environments for model transformation and code generators (MIA developer environments). There are not only standalone versions of these tools but also plug-in versions that integrate directly in the IDEs and modelers of the software developers.

MIA-Transformation is a rule-based model-to-model transformation engine. A model transformation is defined by a set of rules defined between some input meta-models and some output meta-models. Each rule is composed of three elements:

- A context: it corresponds to the set of declared variables and parameters.
- A query: it is an expression that calculates the set of model elements to be processed by the rule.
- An action: it can be a creation, a modification or a deletion of model elements and is performed for each model element returned by the query.

When using MIA-Transformation, alternative languages may be used for expressing transformation rules. MIA-Transformation includes both a fully declarative language (close to the declarative form of QVT) and an imperative language.

The two languages can even be mixed in a single transformation rule: the query can be written using the declarative language and the action implemented imperatively. In addition, as rule based transformation has some limitations, it is possible to define transformation services in Java and use them in transformation rules.

MIA-Generation is a template based model-to-text transformation engine. The idea of MIA-Generation is to attach text generation scripts directly in meta-models in order to define how each model element should be generated. There are two kinds of scripts:

- Templates that textually describe the piece of code to be generated.
- Macros that allow more complex operations such as string handling or model navigation.

The macros are defined directly in Java and can be called from the template. The fact that the generation scripts are directly attached to the meta-model makes MIA text generators easy to understand, adapt and maintain. In addition, the generation engine can keep track of the execution of each generation script and the text it has produced. This provides the developer with all the information required to tune or fix a code generator.

4   Migration of a Large-Scale Banking Application

This section reports on how the migration process described in section 2 is applied in the context of a large-scale banking application. The migration of this application is part of the modernization of the information systems of a French bank[1]. The objective of the project was to migrate a mainframe system made of around a million lines of code to J2EE in order to ease the maintenance and future evolutions of the system. The overall system is composed of:

- 42 applications (for a total of 800 forms and 7500 events)
- 99 prints and exports using Cristal Report
- 990 server services
- 20 batch processes

Sodifrance (and their model-driven migration approach) was chosen by the bank for the migration of this system not only because of the quality assurance provided by the use of automation but also for pricing reasons. After an initial study of the project by Sodifrance and several competing companies, the price proposed by Sodifrance was significantly lower than the price of any brute-force redevelopment strategy (out-sourced or not). In the following, section 4.1 presents the customer's requirements and the migration process that has been developed, section 4.2 details the project schedule and section 4.3 discusses the problem of the validation of the migrated application.

---

[1] For confidentiality reasons, and for the protection of Sodifrance customers, this paper does not provide specific details on the migrated application.

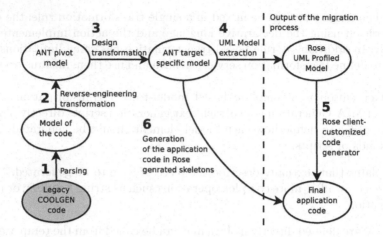

Fig. 5. Banking application migration process

4.1   Specific Requirements and Migration Process

For the modernization of its information system both the servers and the client applications of the bank had to be migrated. The whole legacy application had been developed using the COOLGEN IDE. COOLGEN provides an intermediate programming language and produces executable application by compiling this language to a combination of C code and COBOL code. For the modernization of the system, the servers had to be migrated to plain COBOL because the code generated by COOLGEN was difficult to maintain and had some performances issues. The 42 client applications had to be migrated from COOLGEN to J2EE web applications. The applications and the servers would communicate through a COBOL/Java middleware. The following focuses on the migration of the 42 client applications from COOLGEN to J2EE.

An important requirement of the customer for this project was the strict respect of its internal development standards. All the new applications developed by this bank are generated from Rational Rose UML models. All the models conform to a UML profile developed by the bank itself and specific code generators are used. As a result of the migration process the bank expected to be able to round-trip between models and code using its usual profiles, tools and code generators. The model-driven migration process had to be adapted to take this specific requirement into account.

Figure 5 presents the migration process that is applied to each of the 42 applications of the legacy system. Steps 1 and 2, which correspond to the parsing and reverse-engineering of the application, are similar to the two first phases of the general process presented in section 2. These two phases produce an ANT model of the legacy application which includes all the information contained in the code of the applications (windows, widgets, statements, expressions). Step 3 (also quite similar to the third step of the general process) does the mapping

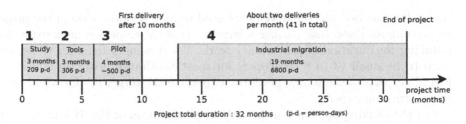

Fig. 6. Banking application migration time schedule and cost breakdown

between the source architectural concepts and the target ones to produce a complete platform specific model of the target application.

Steps 4, 5, and 6 of the process presented figure 5 are specific to the banking system migration and designed to produce customer-specific synchronized UML models and source code for the target application. Firstly, step 4 is a model transformation that extracts a UML-profiled model from the ANT application model. The elements of the target application, such as statements, that do not fit in the UML-profiled model are ignored. Then, step 5 consists in using the regular code generator used in all the development projects of the bank to produce code skeletons from the UML model. In regular projects these skeletons have to be filled manually but here the role of step 6 is to automatically generate the final application code in the code skeletons. The manual phase of the migration can then be carried out: the model transformations and the code generators have left notes in the UML model and comments in the code wherever a manual migration task has to be accomplished.

4.2 Project Time Schedule and Cost Breakdown

This section details the organization and cost breakdown for the banking system migration. The overall project required a total of 9315 days of work including 7815 for the migration of the 42 client applications. As discussed in section 2 any model-driven migration project has several mandatory initialization phases to design a specific migration process and adapt or develop the required tools. Figure 6 presents the scheduling and the cost (in terms of days of work) for each phase of the banking system migration project.

The first preliminary phase of the project is the technical study. In the case of the banking system it took 3 months and required a total of 209 days of work (which represents about 2.5% of the project effort). Then, the tool development phase and the pilot project took 7 months to complete and required an approximate effort of 800 days of work (around 10% of the project effort). For the pilot project, a representative client application has been chosen among the 42 application that had to be migrated. The delivery of the pilot project occurred 10 months after the beginning of the project and after about 12% of the project effort has been spent.

The important investment and delay before the first delivery is specific to model-driven migration. Moreover, because the preliminary tasks are difficult to

492   F. Fleurey et al.

parallelize and because the developers need to have a global view of the project to accomplish these tasks, using a large team of developers cannot really help reducing the duration of preliminary work. The developer team for these tasks have to be small (3 to 8 developers for most Sodifrance projects) and should include experts of the source platform, experts of the target platform and model transformation experts.

On the banking application the industrial migration of the 41 remaining application started 3 months after the end of the pilot project. This phase required a total of 19 months to complete. During this industrial phase of the project the migration is performed in parallel by three independent teams of around 15 developers each. Sodifrance migrates three applications at a time, and, during the 19 month period of the industrial migration an average of 2 deliveries are made per month. Contrary to the project preliminary phases that require a small developer team, the industrial migration duration can easily be shortened by increasing the number of developers.

4.3 Validation and Quality Assurance

Even with the use of automation, since there is still a significant part of the work done by hand, the migrated application has to be carefully validated in order to check its correctness, performance and integration in its new environment. In practice this is achieved thanks to a strict non-regression testing process. This test process is costly for the customers because they have to provide test cases together with the legacy applications and they have to perform acceptance testing[2]. It is also costly for Sodifrance who perform unit testing for the new application and uses the test cases provided with the legacy application to do regression testing. In the case of the banking application the total testing cost is around 3500 days of work (around 1000 days for unit testing and 2500 for regression testing). This represents 45% of the total project cost.

5   Discussion

This section compares *model-driven migration* with brute-force *re-development* migration strategies. The most significant difference between the two approches is the significative *preliminary tasks* required by model-driven techniques. This section especially discusses the influence of these preliminary tasks on the schedule and cost breakdown of migration projects and shows that for projects over a critical size, the model-driven approach is more profitable than re-development.

Complete re-development has some advantages over automated migration. Firstly the development process is similar to the development of any application except that it has a fixed and non-ambiguous specification. This allows using efficient software engineering techniques which is reassuring and unsurprising

---

[2] The numbers provided in this section do not include this cost. Only the cost for Sodifrance is taken into account.

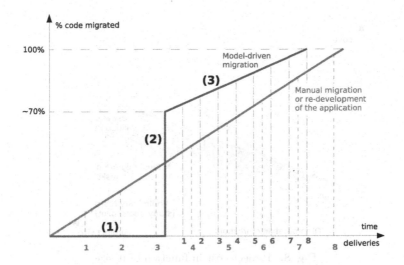

Fig. 7. Migrated code percentage in function of time

to the customer. Secondly, the target application can easily be re-designed, re-factored and adapted to the new platform. Thirdly, evolutions to the legacy application can be taken into account in the design of the new application. All these advantages, most of the time combined with out-sourcing to cut workforce cost, allow full re-development to be a common option for modernization.

In this context, thanks to model-driven migration, Sodifrance has managed to provide a comparable quality of service at lower prices to its customers on a number of modernization projects.

5.1 Migration Time Schedule

Figure 7 compares re-developpment and model-driven migration with respect to the percentage of code migrated over time. For re-development the model we use is linear: the components of the legacy application are re-developed one by one. For model-driven migration the process is a little different: during the first stage of the project (1) the objective is to develop the tools that will be used to partially automate the migration. During this first stage no code of the new application is produced at all but once the tools are fully functional they typically allow generating about 70 percent of the final application code (2). The actual migration can then begin (3), each component of the legacy application is manually completed and delivered to the customer.

The most important difference between the two approaches is the first phase of the model-based process which is an investment in specific tools that will make the migration faster. One of the drawback of model-driven migration is that for an initial period of time, no final code is produced and thus nothing can be delivered to the customer. In the example of figure 7 the legacy application has been divided in 8 components. Using a re-development strategy, the first

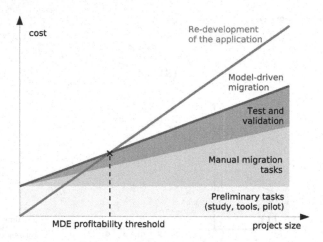

Fig. 8. Project cost in function of it size

component can be delivered to the customer just after the beginning of the project. On one the hand the first component is delivered after quite a long period of time: using model-driven migration, when the first component is delivered, 3 components have already been finished with re-development. But on the other hand, using model-driven migration, once the production of the new application has started, the delivery rate can be faster than for re-development. Eventually, the delivery of components developed using the model-driven approach can catch up with the delivery of re-developed components (this is the case for components 7 and 8 on the figure).

In the case of the banking application described previously, the preliminary tasks of the migration project required 10 months which represents about a third of the total project duration. From an economical point of view, more than 10% of the total migration cost was spent on these preliminary tasks. The next subsection discusses the profitability of this investment.

5.2   Migration Cost Repartition

To be profitable, the model-driven migration process must be applied on legacy applications that have a sufficient size. Indeed, the effort that has to be invested for developing migration tools mostly depends on the complexity of the input and output platforms but not on the volume of code that has to be migrated. Figure 8 presents an estimation of the cost of a migration project using both model-driven migration and re-development.

In the case of re-development the cost is directly proportional to the size of the legacy application. In the case of model-base migration, there is a fixed initial cost related to the development of tools which is complemented by a linear cost corresponding to manual migration efforts. The gradient of the function corresponding to model-driven migration is lower than the gradient for re-development because using migration tools reduces the manual effort that has to be provided.

A general profitability threshold for model-driven migration cannot be estimated accuratly because it really depends on the ratio of tools that have to be developed for each project. In practice, the experience of Sodifrance on model-driven migration shows that the profitability threshold for MDE in the context of migration is quite low. Even for projects that require about 1000 days of work, the initial overhead of developing tools pays off. On the migration of the large banking application described previously, Sodifrance estimated that the cost of re-development whould have been around twice the price of model-driven migration.

5.3 Benefits and Limitation of Model-Driven Migration

The primary advantage of model-driven migration is to partly automate the migration process. As discussed in the previous sections this allows Sodifrance to significantly lower the prices and duration of migration projects. This is the reason why Sodifrance is often chosen over competing companies that propose full re-development.

The second advantage is to allow for reuse between migration project. This is another element that allows cutting the cost of migration. All the transformations and tools that have been developed for a migration projects can be adapted to future project that have similar input or output platforms.

The first limitation of model-driven migration is a commercial limitation related to the cost and time consumed by preliminary tasks. In the project presented in section 4 the first delivery of migrated code occured after 10 months. This is a commercial issue because after the begining of the project the customer has to wait for a long time without sceing the progression of the project. To mitigate this issue, possible solution is to works in close collaboration with the IT department of the customer and, if possible, to include members of the customer company in the development team.

The second limitation is related to the cost of testing. This is not specific to model-driven migration but is a general problem in software migration. For the banking application discussed previously, testing represents 45% of the total migration cost. This cost does not include the cost of the production of regression tests and acceptance testing which are the responsability of the customer [3]. One of the reasons of the important cost of testing tasks is that, for most migration project, they are mostly handled manually. In the same way model-based techniques has been applied to smartly automate repetitives migration tasks, Sodifrance is now studying model-based regression testing using meta-modeling languages such as Kermeta [2] to reduce the cost of testing.

---

[3] The production of the tests is a cost that has to be taken into account by the client. However, this cost is usually far lower than the cost of providing the complete specifications required for full re-development.

6   Related Works

Software modernizaton has been identified by the OMG as an important application field for model-driven architecture. The Architecture-Driven Modernization (ADM) is an OMG task force dedicated to this topic [3] that aims at building standard metamodels and tools for software modernization. Reus et al. in [4] propose a MDA process for software migration that is quite similar to ours. They parse the text of the original system and build a model of the abstract syntax tree. This model is then tranformed into a pivot language that can be translated into UML. A prototype automates parts of this process using Arc-Styler [5]. Bordbar et al. [6] propose a model-based approach for maintenance of data-centric systems. Their MDA approach improves the evolution and maintenance of databases in applications developed with java and modelled with UML. In [7], Zou transforms legacy code towards object-oriented languages. Parts of this process are implemented with automatic program transformations.

Another type of works related to the study presented in this paper concerns feedbacks from industrial projects that have applied model-driven approaches. In the last two editions of the MoDELS conference, two studies gave such feedback. In [8], Baker et al report on the significant improvements in productivity and reliability gained with MDE techniques and also present the remaining issues to profit more from those approaches. In [9], Staron presents a study on the requirements for the adoption of MDE in software industries. The paper reports on the observations of two companies that tried to MDE in their development process. In [10] M1_Global_solution compares MDE and off-shore development. They conclude that MDE tool increased developer productivity by over 50 percent and advocate a combination of MDE for automatic production of a large part of the system and off-shore development for the parts that need to be manually developed.

7   Conclusion

This paper has precisely presented a model-driven migration and modernization process developed by the Sodifrance company. We have detailed the process and the tools that automate this process. The paper has also dicussed the benefits introduced by MDE in terms of reuse and automation, and also the issues that are introduced to fully benefit from reusable transformations and generators. Finally we have showed that, even if the process is not fully automated and requires manual adaptation from one project to the other as well as manual implementation of some parts of the final application, it is still viable compared to manual re-development.

Even if model-driven engineering is already economically profitable for migration, there are still some important challenges that need to be tackled. A major issue in terms of human effort is testing. Today, regression test is used to validate the migration. However, the production of efficient regression test cases is manual, ad-hoc and difficult to evaluate. Future work consists in adding, in the reverse-engineering phase, a step to reverse a model for high-level control flow in the application in order to eveluate test coverage at use-case level. Moreover, unit and integration test for the migrated code is also very expensive. A possible solution here could consist in generating test objectives when generating the code.

References

1. Sodifrance: Model-in-action tool-suite (2007), http://www.mia-software.com/
2. Muller, P.A., Fleurey, F., Jézéquel, J.M.: Weaving executability into object-oriented meta-languages. In: MODELS/UML 2005, Montego Bay, Jamaica. LNCS, Springer, Heidelberg (2005), http://www.kermeta.org/
3. OMG: Architecture-driven modernization (2006)
4. Reus, T., Geers, H., Deursen, A.v.: Harvesting software systems for mda-based reengineering. In: Rensink, A., Warmer, J. (eds.) ECMDA-FA 2006. LNCS, vol. 4066, pp. 213–225. Springer, Heidelberg (2006)
5. Objects, I.: Arcstyler (2007)
6. Bordbar, B., Draheim, D., Horn, M., Schulz, I., Weber, G.: Integrated model-based software development, data access, and data migration. In: Briand, L.C., Williams, C. (eds.) MoDELS 2005. LNCS, vol. 3713, pp. 382–396. Springer, Heidelberg (2005)
7. Zou, Y., Kontogiannis, K.: Migration to object oriented platforms: a state transformation approach. In: ICSM'02 (International Conference on Software Maintenance), pp. 530–539 (2002)
8. Baker, P., Loh, S., Weil, F.: Model-driven engineering in a large industrial context - motorola case study. In: Briand, L.C., Williams, C. (eds.) MoDELS 2005. LNCS, vol. 3713, pp. 476–491. Springer, Heidelberg (2005)
9. Staron, M.: Adopting model driven software development in industry - a case study at two companies. In: Nierstrasz, O., Whittle, J., Harel, D., Reggio, G. (eds.) MoDELS 2006. LNCS, vol. 4199, pp. 57–72. Springer, Heidelberg (2006)
10. M1 Global Solutions: Model driven software development and offshore outsourcing (2004)

Introducing Variability into Aspect-Oriented Modeling Approaches

Philippe Lahire[2], Brice Morin[1], Gilles Vanwormhoudt[3], Alban Gaignard[2], Olivier Barais[1], and Jean-Marc Jézéquel[1]

[1] IRISA Rennes, Projet Triskell, IRISA - Campus de Beaulieu F-35042 Rennes Cedex
[2] I3S Nice-Sophia Antipolis, Equipe Rainbow, I3S-UNSA
Les algorithmes, 2000 route des lucioles BP 121 F-06903 Sophia-Antipolis Cedex
[3] GET Telecom-Lille 1/ LIFL, Université de Lille 1, F-59655 Villeneuve d'Ascq Cedex

Abstract. Aspect-Oriented Modeling (AOM) approaches propose to model reusable aspects, or cross-cutting concerns, that can be composed in different systems at a model or code level. Building complex systems with reusable aspects helps managing software complexity. But in general, reusability of an aspect is limited to a particular context. On the one hand, if the target model does not match the template point-to-point, the aspect cannot be applied. On the other hand, even when it is actually applied, it is woven into the target model always in the same way. In this paper[1], we point out the needs of variability in the AOM approaches and introduce seamless variability mechanisms in an existing AOM approach to improve reusability. Our aspects can fit various contexts and can be composed into the base model in different ways. Introducing variability into AOM approaches will turn standard aspects into highly reusable aspects.

1 Introduction

The Aspect Oriented Software Development (AOSD) paradigm first appeared at the code level a decade ago [7] with the most famous AOP language AspectJ [6]. The aspect paradigm offers a new way to construct complex systems by composing crosscutting concerns with the base system. In the earlier stages of the software life-cycle, several Aspect-Oriented Modeling approaches (AOM) already exist [1,2,4,16], with various levels of abstraction (requirement, design, architecture). In general, these approaches decrease the complexity of systems by composing models that represents the different concerns of the system (business, security, persistence ...). To help developers saving time designing systems and therefore reduce the time-to-market of these systems, models should be reusable.

Currently, AOM approaches provide some means to design reusable and flexible aspects. But, reusability and flexibility are often limited. In general, they

---

[1] This work was partially supported by the French National Research Agency (RNTL FAROS Project).

G. Engels et al. (Eds.): MoDELS 2007, LNCS 4735, pp. 498–513, 2007.

describe one possible variant of an aspect and propose one possible way to integrate it. For example, a designer cannot model a design pattern in its full genericity with these approaches: he can only model one specific implementation choice for this design pattern. Consequently, aspects are only reusable in similar or very related contexts. In this paper, we argue that aspects must be reusable in various contexts. Designing context independent aspects requires seamless variability mechanisms for specifying the weaving, the pointcut expression, *etc...* . Such mechanisms will turn standard aspects into highly reusable and flexible pieces of models. The contribution of this paper is to point out the needs of variability in the AOM approaches, to provide some mechanisms to support variability in one particular AOM approach and to illustrate these new mechanisms on a concrete example. To address variability in software development, Software product lines (SPL) offer some mechanisms to support functional variability[2] and to derive products that match the user's needs. However, this variability only concerns the software module specifications. In the case of AOM approaches, variability should also be applied onto the composition mechanisms.

The remainder of this paper is organized as follows. Section 2 points out the needs of variability in the AOM approaches with a motivating example. Section 3 presents an overview of an AOM approach. This approach is extended in the section 4 to support variability mechanisms. Section 5 describes a metamodel for this approach and the implementation of a modeling tool. Section 6 presents related works and section 7 concludes and discusses future work.

2   Motivating Example

To illustrate the needs of variability in the AOM approaches, we use the example of a mobile phone device. Figure 1 shows a simplified class diagram presenting the main functionalities of an accountancy package for a mobile phone.

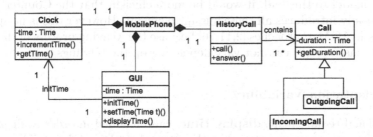

Fig. 1. A Simplified class diagram of the mobile phone

When the user is calling (resp. is called by) someone, the *HistoryCall* class creates a new outgoing (resp. incoming) call and saves the duration. The class *GUI* can display its local variable *time* which is initialized when switching on the phone. The class *Clock* only contains a variable *time* which is incremented every minute.

---

[2] see Software Product Line Conferences : http://www.splc.net

2.1 Matching Variability

Two optional requirements, **total calls** and **total outgoing calls**, can be added to our mobile phone in order to compute the total duration of the (outgoing) calls. We will use the Counter pattern [11] to realize these two requirements.

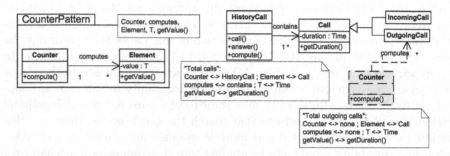

Fig. 2. The Counter pattern realizing the **total calls** requirement

In most of the AOM approaches [2,11,16], a template specifies the model elements of the concern that have to be bound with target model elements. Reusability is then limited to iso-structural target models because if the structure does not match the template point-to-point, the aspect cannot be applied.

Figure 2 shows the Counter pattern composition into the mobile phone model. In order to realize the **total calls** requirement, we use the existing *HistoryCall* and *Call* classes to respectively act as *Counter* and *Element*. We now want to realize the **total outgoing calls** requirement in a separate Counter class. This requires the creation of a new class acting as a Counter and the insertion of a new association between this class and the *OutgoingCall* class. Instead of modifying the base model to this end, it would be more efficient that the Counter pattern automatically introduces all these missing elements. However this is not possible with classic AOM approaches [2,11,16] because the weaving process of the aspect upon the base system can not vary depending on the bindings.

2.2 Adaptation Variability

One optional requirement, **display time**, can be added in order to display and update the time every minute, when the internal clock is updated. The Observer pattern will realize this requirement, notifying the *GUI* (Observer) that the *Clock* (Subject) has been updated.

In most of the AOM approaches [2,11,16], aspects are composed into the target model using one composition rule at a time, offering poor flexibility. Depending on the context, it would be very useful to easily switch between different composition rules. In the context of embedded systems it may be preferable to reduce the number of classes because of memory limitations, and completely merge the aspect while in some other cases, it may be preferable to compose the aspect

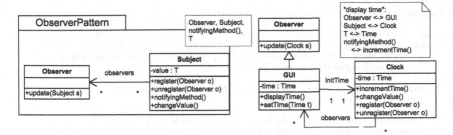

Fig. 3. The Observer pattern merged into the base model

by inheritance in order to improve readability. Figure 3 illustrates another composition rule where *Subject* is merged into *Clock* whereas *GUI* inherits from *Observer*.

This motivating example has shown the needs of variability in two contexts i.e. matching and adaptation. There is also a need for functional variability e.g. how to design many versions of the Counter pattern (total or average for example). Since Software Product Line approaches [17] can help modeling this kind of variability, we do not cover them into this paper.

3   An AOM Approach Overview

The approach which is presented in this paper is only one among many possible approaches for addressing AOM [3,11]. It focuses on providing capabilities for concerns (functional or extra-functional) to be reused. In this context, the expressiveness of the concern modeling is not a primary objective. For example, contrary to other non aspect-oriented approaches like [17], we do not offer more capabilities for expressing the variability of concerns than the one provided by the underlying metamodel used for the concern specification. The approach called SMARTADAPTERS had been applied first to Java programs [8] and more recently to EMF models. It leverages the notions of subject [14] and aspect programming [6,7]. Its key concepts are **concerns, adapters, adaptations** and **adaptation target**. The main idea is the following: each concern identified as reusable should go with an adapter which specifies a **composition protocol**, that is a set of adaptations and adaptation targets describing how the concern should be composed with other concerns when it is reused. This protocol will guide the designer to identify the specific parts for reuse when composing a reusable concern into a target concern.

We propose to explain this approach through the reuse of the *Observer* design pattern. First we define its composition protocol (see Figure 4). For better readability, we use a concrete textual syntax in order to specify this composition protocol. Details in the concrete syntax are not important and the syntax might be slightly modified in the future.

```
01 concern designpattern.observer
02 abstract adapter ObserverAdapter {
03
04    abstract Class target "class(es) representing an observer" : observerClass
05    abstract Class target "class(es) representing a subject " : subjectClass
06    abstract Method target " method(s) notifying  changes " : notifyingMethod
07      require notifyingMethod in subjectClass.*
08
09    adaptation becomeObserver "Modify class to make it an observer" :
10        inherit Observer in observerClass
11
12    adaptation becomeSubject "Modify class in order to make it  a subject " :
13        merge class subjectClass  with Subject
14
15    adaptation  introduceLink "introduce an association (subject to observer)  " :
16          introduce Association observers (subjectClass -¿ observerClass)
17
18    adaptation notifyingObserver
19       " Alter notifyingMethods to tell observers about modification" :
20        extend method notifyingMethod( ... ) with after { changeValue(); }
21
22    abstract adaptation updateObserver "add an update facility  to observers " :
23        introduce method public void update(subjectClass s) in Observer
24
25    ... Protocol includes also :object initialization,observers registration,...

26 }
```

Fig. 4. Snippet of the composition protocol for the Observer design pattern

Let us now detail this example illustrated in Figure 4. Line *01* specifies the concern to be reused. The adapter called *ObserverAdapter* describes its composition protocol (Line *02*). When the composition protocol is defined the concern(s) that may reuse it are not known so that we do not know the classes corresponding to the objects acting as *subjects* and those acting as *observers*. The only thing that we may assume is that there are classes that act as observers and subjects. They are represented by the two abstract targets of type **class**: *observerClass* and *subjectClass* (Lines *04* and *05*). Each of these targets may be associated to one or several classes at composition time.

Considering the design pattern *Observer* of Figure 3, any *subject* must inform an *observer* that its content has been modified by calling the method *changeValue*. For the same reasons that the classes mentioned above are not known the method(s) playing this role are also not known but they should exist and be declared in the *subjectClass* (Lines *06* and *07*). To ensure that the call to *changeValue* is performed by the method(s) *notifyingMethod*, the composition protocol specifies an adaptation of type *interception* which adds this call at the end of the corresponding method(s) (Lines *18* to *20*).

More generally this kind of adaptation deals with some actions to be taken when a classifier member (attributes, methods...) is accessed or called. These adaptations allow the designer to add behavior at the beginning, end or around some existing methods but also to add some treatment when an exception is triggered. For attributes, interception may occur when the attribute is read or modified.

Let us continue with our example. To be able to call *changeValue* or any other feature of class *Subject*, it is necessary to have access to it from within the classes corresponding to *subjectClass*. This means that we have to specify another adaptation. Two possibilities could be chosen: to merge all the features of class *Subject* into *subjectClass*, or to make *subjectClass* inherit from *Subject*. Here we choose an adaptation of type **Merging** (Lines *12* and *13*).

Such adaptations deal mainly with packages, classifiers features and associations. Method merging is particularly interesting if there is a support for describing the behavior (programming constructs in KERMETA [13], Sequence diagrams in UML, etc.). At present time merging policies are mainly execution of one method before the other; the handling of interlaced method bodies could be inspired by approaches like [9]. Merging classifiers is either straightforward (no conflict, name of features to be merged are identical, feature appear only in one of the classes,...) or may need more information in order to relate the features of the classifiers that need to be merged [2,16].

All these adaptations were dealing with the *subjects*. It is then necessary to address *observers* and to also insert class *Observer* at the right place(s) in the target concern. We chose here to inherit from it (Lines *09* and *10*). Such adaptation is of type **Introduction**. It deals not only with superclass introduction as it is the case here but also with adding classifier members (new attributes or methods), as well as association. It is also possible to add a classifier invariant or a method assertion (Precondition or postcondition).

We use the same type of adaptation to insert the association between *subject* and *observer* classes as specified in the design pattern. Depending on the association to be introduced we may provide additional information. For example, in the current case the association is unidirectional from *subjects* to *observers* (Lines *15* and *16*).

It only remains one thing to do: to add to the *observerClass* class(es) a method *update* (also an adaptation of type **Introduction**), that reacts to the changes made in the subject object. At this time we do not know the content of this feature because we do not know what the purpose of the target concern is. This is why the adaptation is **abstract**. The advantage to plan this adaptation in the composition protocol is to guide and control the reuse of the design pattern.

This composition protocol continues with the description of the initialization and the registration of observers but for space reasons we do not include it.

Let us suppose now that this concern is reused by the concern described in Figure 1 (Section 2) dealing with mobile phones. So we need to compose these two concerns. The information which is imcomplete into the composition protocol (abstract targets and adaptations) is described into a concrete adapter *ApplicationPhone* which specializes the adapter *ObserverAdapter* as it is shown in Figure 5. Please note that, in this example, the insertion is *in situ*. It means that adaptations are performed within the concern *application.phone*. In some cases, it is better to make the composition *ex situ* that is to say to compose the two concerns into a new one.

In the above composition protocol (Figure 4) we made several assumptions about the target concern. For example, we suppose that the association does not yet exist between the classes *GUI* (the observer) and *Clock* (the subject). This is a drawback because if the composition does not deal with a concern which satisfies these assumptions, it will be impossible to reuse the composition

```
01 concern application.phone
02 compose designpattern.observer with application.phone
03 adapter ApplicationPhone extends ObserverAdapter {
04
05    target typeOfValue = Time
06    target subjectClass = application.phone.Clock
07    target observerClass = application.phone.GUI
08    target notifyingMethod = application.phone.Clock.incrementTime()
11
12    adaptation observerUpdate :
13      introduce method public void update (subjectClass s) in observerClass {
14        setTime(time++)
14        displayTime()
15      }
16 }
```

Fig. 5. Reuse of Design Pattern Observer for a mobile phone

protocol in another context. Thus, we reach the conclusion that we need to introduce some variability within the composition protocol. This is the purpose of section 4.

4   Extension to Support Variability

In Section 3 we proposed an overview of the SMARTADAPTERS approach. We now consider the needs of variability pointed out in Section 2. Our objective in this section is to introduce matching and adaptation variability into the composition protocol in order to make it more reusable and as consequence to make the concern itself more reusable. SMARTADAPTERS is a support for explaining our approach but we plan to address other AOM approaches. Variability mechanisms introduced are inspired by Software Product Lines approaches, especially [17].

Figure 6 shows what we should introduce in an adapter to better customize the composition protocol. In Section 5 we will describe the metamodel containing the capabilities that are suggested here.

We may note first that adapter *ObserverAdapter* is now preceded by the keyword **derivable** (Line *02*). This means that it may present several alternatives to implement the composition and may consider some adaptation targets or adaptations as optional. This adapter acts as a template where some information should be given in order to choose between possible variants or options.

A first possible customization is dealing with the insertion of the features provided by classes *Subject* and *Observer*. Depending on the target concern or more generally on the context of reuse, it may be interesting to have the choice between inheriting from those classes or merging their features into *observerClass* and *subjectClass*. In Figure 4 a choice is made *a priori*. In Figure 6, the choice is described by the Lines *09* to *24* through a clause **Alternative** *InsertionChoices* which specifies here two variants (more variants could be defined if needed). A variant may contain several adaptation target declarations and adaptations. Implicitly this means that these targets and adaptations are dependent from each others.

Now, we can introduce the *update* method. If we merge the *Subject* and the *Observer*, we need to introduce the *update* method in the class where the *Observer* is merged i.e., *observerClass* (Lines *22* and *23*). *Subject* is also merged in a target

```
01 concern designpattern.observer
02 derivable adapter ObserverAdapter  {
03
04     abstract Class target ''class(es) representing an observer'' : observerClass
05     abstract Class target ''class(es) representing a subject '' : subjectClass
06     abstract Method target '' method(s) notifying  changes '' : notifyingMethod
07        require notifyingMethod in subjectClass.*
08
09     Alternative InsertionChoices '' Choice between inheritance and merging'' {
10        [Vinheritance] '' Inheritance variant '' :
11        adaptation becomeSubject ''Modify class in order to make it a subject '' :
12           inherit class Subject  in subjectClass
13        adaptation becomeObserver ''Modify class to make it an observer'' :
14           inherit Observer in observerClass
15        abstract adaptation updateObserver ''add an update facility  to observers '' :
16           introduce method public void update(Subject s) in Observer
17     or else [Vmerge] '' Merging variant '' :
18        adaptation becomeSubject '' Modify class in order to make it a subject '' :
19           merge class subjectClass  with Subject
20        adaptation becomeObserver ''Modify class to make it an observer'' :
21           merge  class observerClass with Observer
22        abstract adaptation updateObserver ''add an update facility  to observers '' :
23           introduce method public void update(subjectClass s) in observerClass
24     }
25
26     Alternative NotificationTime '' Choice of notification time'' {
27        [Vbegin] '' Method beginning variant'' :
28        adaptation notifyingObserver
29           '' Alter notifyingMethods to tell observers about modification'' :
30           extend method notifyingMethod( ... ) with before { changeValue(); }
31     or else [Vend] '' Method ending variant'' :
32        adaptation notifyingObserver
33           '' Alter notifyingMethods to tell observers about modification'' :
34           extend method notifyingMethod( ... ) with after { changeValue(); }
35     }
36        ...Protocol  includes also :object initialization,observers registration,...
37 }
```

Fig. 6. Composition protocol for the Observer with variability

class, therefore the parameter of the *update* method has the type of this target class i.e. *subjectClass*. If the pattern is composed by inheritance, the *update* method is introduced in the *Observer* class itself, and the parameter has the type *Subject* (Lines *15* and *16*). The *update* method is very related to the composition variant, so we integrate its introduction in the InsertionChoices alternative. Depending on the chosen composition variant, the right *update* method will be introduced. In both cases the contents of this method is not already known, that is why this method is **abstract**.

A second possible customization is related to the location of the call to method *changeValue* within *notifyingMethod*. It may be useful depending on the target concern to notify the subject changes to observers either at the beginning or at the end of the execution of *notifyingMethod*. The corresponding variants are described by the Lines *22* to *31* through a second clause **Alternative**. Each variant corresponds to a unique adaptation of type Interception.

In figure 7 we extend this protocol to experiment the combination of **optional** and **constraint** clauses. We now address the association between observers and subjects (called *observers* in the design pattern of Figure 2). It is very likely that

```
01 concern designpattern.observer
02 derivable adapter ObserverAdapter  {
03    ...
04
05    is optional AssociationExist  '' association (observers to subject)  may exist '' {
06         abstract  Association target '' handling association mapping'' :
07              subjectObserverAssociation
08         adaptation  mergeLink ''merge association with the Observer pattern one  '' :
09              merge association subjectObserverAssociation with observers
10              require subjectObserverAssociation ⊆ (subjectClass -> observerClass)
11    }
12
13    is optional LinkModification1  '' Existing association may be renamed ''   {
14         abstract adaptation renameLink ''rename association-end of association  '' :
15              rename association  subjectObserverAssociation
16    is optional LinkModification2  '' Existing association may be redefined '' :
17         adaptation  alterLink ''add an association-end to association   '' :
18              add association observers (subjectClass  -> observerClass)
19              ...
20    constraint AssociationHandling ''working on association implies it exists ''  {
21    LinkModification1 depends on {AssociationExist}
22    LinkModification2 depends on {AssociationExist}
23    {LinkModification1, LinkModification2} are exclusive
24    }
25 }
```

Fig. 7. Options and matching variability

depending on the target concern this association may already exist in it. In order to authorize both situations we propose some optional adaptations (Lines *05* to *18*). A first optional clause assumes that the association exists in the target concern and is identified by the target *subjectObserverAssociation*; it must be merged with *observers*. Then it may be possible to specify a renaming adaptation because nothing can ensure that it has the same association-end name in the target concern. It is also possible to add an association-end when the association exists but in the opposite way in this concern.

The example developed in Figures 6 and 7 especially illustrates the needs for optional parts and variant definitions. In order to insure the consistency of the composition protocol, the user can define mutual exclusion and dependency constraints. These constraints restrict the number of possible combinations to sensible ones. In our example, we want to ensure that *i)* renaming and redefinition may not be performed if the association between observers and subjects does not exists in the target concern and, *ii)* renaming its association-end is incompatible with adding *observers*. These contraints are expressed (Figure 7 - Lines *20* to *23*) by introducing dependencies between *L*inkModification1, *L*inkModification2 and *A*ssociationExists options and a mutual exclusion between the first two options.

Now, we can compose the variable "Design Pattern Observer" into the mobile phone base model. In addition to the tasks described in figure 5 it is necessary to select options and variants (adaptation targets and adaptations) which are suitable for the concern "mobile phone". Of course the abstract adaptation targets and adaptations to concretize in the adapter *ApplicationPhone* depends on the variants and options which are selected (Figure 8).

The selection is made through a clause **derive** (Lines *05* to *08*). No association can match the *observers* association in the target model, so the optional clauses

```
01 concern application.phone
02 compose designpattern.observer with application.phone
03 adapter ApplicationPhone derives ObserverAdapter {
04
05    derive designpattern.observer with {
06       options: none
07       alternatives: InsertionChoices#[Vinheritance], NotificationTime#[Vend]
08    }
09
10    target typeOfValue = Time
11    target subjectClass = application.phone.Clock
12    target observerClass = application.phone.GUI
13    target notifyingMethod = application.phone.Clock.incrementTime()
14
15    adaptation observerUpdate :
16       introduce method public void update (Subject s) in observerClass {
17          setTime(time++)
18          displayTime()
19       }
20 }
```

Fig. 8. Reuse of Design Pattern Observer for a mobile phone

are not selected (note that an association exists in application.phone but in the opposite way so that it would be possible to keep only one association selecting *AssociationExist* and *LinkModification2*). We also select the two variants associated to the alternative clauses *InsertionChoices* and *NotificationTime*. Finally, we have to concretize the update method, specifying that the GUI has to increment its variable *time* and refresh the screen. Concretizing abstract methods in a concrete adapter is close to the mechanism defined in the AOP approach of Hannemann *et al.* [5]. Mandatory targets and adaptations of Figure 6 are processed normally in the same way as it is done in Figure 5.

Figure 9 shows two types of composition i.e, merging and inheritance, in order to realize the **display time** requirement. **Inheritance** corresponds to the adapter we have derived above, while **Merging** corresponds to another possible derivation provided by the protocol.

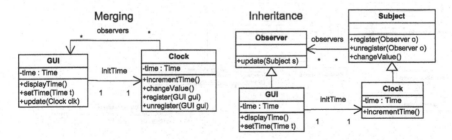

Fig. 9. Two possible compositions of the Observer pattern

In the motivating example, we were not able to realize the **total outgoing calls** with the standard Counter pattern because the template approach was not flexible enough. We can now realize the **total calls** and the **total outgoing calls** requirements using the same Counter pattern. Indeed, the Counter pattern now can be applied either if the class acting as *Counter* is present or not in the base model. For space limitation, the derivable adapter and the concrete adapter are not shown but the principle is similar to the Observer protocol (Figures 6, 7 and 8).

Finally, it is interesting to note that introducing variability did not affect the guidance and the controls when reusing a derivable concern. On the contrary, the choices induced by the addition of variability is also controlled and guided thanks to the expressiveness of the composition protocol.

5   Metamodeling and Implementing AOM with Variability

This section proposes a metamodel of concerns that includes concepts for adapters and variability illustrated in sections 3 and 4. This metamodel aims at giving a precise formulation of concerns and make it possible their integration into modeling tools. Figure 10 shows an excerpt of the metamodel where concepts introduced to handle variability are identified with a circle at the upper left. The key concepts of the metamodel are concern, adapter, target and adaptation.

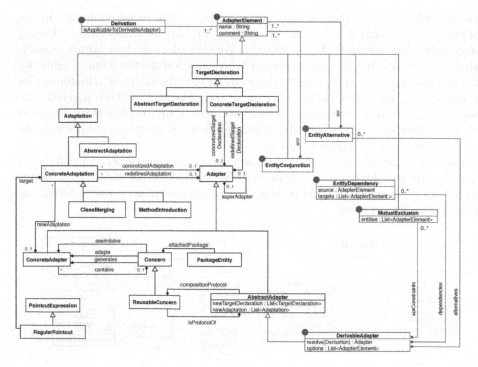

Fig. 10. Metamodel of SmartAdapter with variability

A reusable concern (*class ReusableConcern*) is associated to a package (*class PackageEntity*) which contains the concern description[3] and its protocol of composition (*class AbstractAdapter*). Concerns are not always reusable (*class Concern*).

---

[3] We assume that a concern is described by a package of classes similarly to a UML class diagram.

For example the concern which describes the GUI of an application is rather specific and may not be reusable; such concerns do not have a composition protocol but could be composed with other concerns. A concern refers to as many concrete adapters (*ConcreteAdapter*) as there are concern to be integrated with it.

An adapter (class *Adapter*) is identified by a name and may inherit (i.e. specialize) from another adapter. An adapter may be abstract (class *AbstractAdapter*), concrete (class *ConcreteAdapter*) or derivable (class *DerivableAdapter*). Each adapter contains adaptations (class *Adaptation*) and adaptation targets (class *TargetDeclaration*). A derivable adapter is an abstract adapter which supports variability: it allows the designer to customize the set of adaptations or/and the set of adaptation targets by expressing options, alternatives, dependencies and exclusions. Such an adapter is not intended to be used directly for composing concerns but serves to derive an adapter. A derived adapter is obtained using the method *resolve* of class *DerivableAdapter* which takes a derivation (class *Derivation*) parameter to select the adaptations and the adaptation targets among the options and variants. This adapter may be concrete, abstract or derivable depending on what is resolved by the derivation parameter.

A target declaration (class *TargetDeclaration*) specifies an adaptation target that matches the entities on which the adaptations relies on. An adaptation target may identify just one required element (class *AbstractTargetDeclaration*) (like the *observers* or the *subjects* in the design pattern *Observer*) or be fully specified (class *ConcreteTargetDeclaration*) by referencing the real element (class, method, ...) to adapt.

An adaptation (class *Adaptation*) specifies the action to be taken for an element of the reusable concern when it is composed. The metamodel includes a hierarchy of adaptation classes that are typed according to the types of target entities (package, classifier, method, attributes and association) and reflect the four kinds of adaptation currently proposed: interceptions, introduction, merging and redefinitions. Figure 10 shows two of the adaptation classes used in the previous examples (class *ClassMerging* and class *MethodIntroduction*).

To be able to take into account several variants for the integration of the concern, the metamodel includes the concept of alternative entity (*EntityAlternative*). An alternative entity may refer to several adaptations or adaptation targets (see *xor* link) but only one will be selected at composition time.

Adaptations, adaptation targets and even alternatives can be optional in a derivable adapter, that is to say that they are planned in the composition protocol but they could be retained or not when the concern is composed with another one. Optional elements of a derivable adapter are referenced by its *options* link.

Practically several adaptations or adaptations targets may be described in a given variant or be declared as an optional block. For this purpose we propose a way to group thoses entities (class *EntityConjunction*).

In a derivable adapter, classes *EntityDependency* and *MutualExclusion* allows designers to specify that an *AdapterElement* (variant or option) may not be selected with other ones or on the contrary must be selected if some others are selected. These classes define constraints that are checked before deriving a derivable adapter, in order to insure the consistency of the derived adapter. If a derivation does not respect these constraints then an exception is raised that asks the user to modify the derivation.

The metamodel described above has been used to build a modeling tool integrated in the Eclipse environment. This tool currently provides two main functionalities: designing models of concerns and adapters; composing concerns from their models. This tool has been implemented using the Eclipse Modeling Framework (EMF) and the Kermeta language [13]. We have exploited EMF to define a Ecore version of our metamodel, reusing the Ecore metamodel for the description of concerns. The Kermeta language has been exploited to extend the Ecore version of our metamodel with operational behavior. This behavior performs several tasks related to the design and composition of concerns: it checks the consistency of adapters, computes derived adapters and compose elements of concerns from a set of adaptations. At this time, we are investigating the design of a concrete textual syntax for our metamodel like the one used in the previous section and we plan to build the concrete syntax tool using a meta-model centric approach as [12].

6   Related Work

There exists numerous AOM approaches but few of them support variability mechanisms at the composition level [4,16,1]. In [2], Clarke *et al.* model an aspect in a template package specifying the structure and the behavior of the aspect with a class diagram and sequence diagrams. The template is composed of model elements present in the concern's class diagram and specifies the elements that have to be matched in the target model. There is no functional or matching variability mechanism. The composition relationship authorizes multiple bindings i.e. it is possible to match several target model elements to the same concern model element. Adaptation lacks variability: concerns are indeed always merged into the target model. Note that it is possible to generate AspectJ code to postpone the weaving at code level. Our adaptation protocol allows the designer to define different variants of how the concern will be integrated in the target model. All the variability mechanisms we have identified may be adapted to Theme.

Muller *et al.* [11] also propose an approach to compose a reusable model expressed as a template package with an existing model. To express this composition, they introduce an apply operator that specifies the mapping between template parameters and elements of the target model. Their approach addresses variability at the composition level by giving the capacity to annotate the apply operator with different strategies such as "merge" or "view". Strategies are only provided to get different resulting models. Compared to our proposal, this solution does not offer any mechanism to express options and variants for the reusable model. It is also less flexible as it does not offer finer grain mechanisms to control how elements of reusable and target models must be composed.

France *et al.* [16] have developed a systematic approach for composing class diagrams in which a default composition procedure based on name matching can be customized by user-defined composition directives. These directives constrain how class diagrams are composed. The framework automatically identifies conflicts between models that have to be composed and it solves them thanks to the composition directives. Contrary to Theme, composition directives address the weaving only from the structural point of view. They consider the composition as a model transformation. The variability can be addressed by designing several composition directives depending on the integration context. However, the definition of the composition directive would then become messy and error-prone. Besides, it is a symmetric AOM approach in which they do not differentiate between aspect model and base model. Consequently, they do not currently provide a pointcut language to manage the composition.

In [5], Hannemann *et al.* propose an AOP approach to implement design patterns with AspectJ. They propose up to seven different implementations for each design pattern. The only variability mechanism is the generalization relationship between an abstract aspect and an aspect. For example, the *update* method of the *Observer* is declared abstract in an abstract aspect and its contents will be specified in a concrete aspect. We also use this mechanism but the variability mechanisms we introduced allow a concern to be applied in multiple contexts whereas we would have to create a new aspect depending on the context with the Hannemann *et al.* approach. Option and variant notions do not exist, reducing the reusability of the aspects. Our concerns are adaptable and do not need modifications to be applied, but only customization. Introducing the same variability mechanisms at the code level code could enhance the expressiveness of AOP language such as AspectJ.

7   Conclusion

In this work, we propose an approach for introducing variability in aspect-oriented modeling (AOM). To achieve this goal, two important parts of such an AOM approach were needed: A concern model and a weaver that support variability. In this paper we mainly focus on the second one. Indeed, the variability in the concern specification depends on the expressiveness of the meta-model dedicated to concern modeling. Consequently, a reasonable solution to integrate variability in the concern model can be inspired by product lines researches and more precisely by [17].

To introduce variability in the weaving process, the composition meta-model of our AOM approach has been extended. These extensions concern the adaptations primitives and the pointcut specification. They are composed of a set of entities specifying optional parts, alternatives, dependencies and mutual exclusion constraints. These extensions allow the user to design a family of aspects at the design level that can be derived to be applied in a particular context.

One of the main benefits of building a composition protocol is the capability to control and guide the software architect when he designs new applications. The variability introduction does not affect the guidance and the control when reusing a derivable concern. On the contrary, the choices induced by the addition of variability are also controlled and guided thanks to the expressiveness of the composition protocol.

In the SMARTADAPTERS platform, we plan to improve the pointcut language and the target identification. One possible solution is to describe the pointcut with a template model and to use pattern matching [15] to identify targets. We also want to generalize the SMARTADAPTERS to various metamodels, not only class diagrams or Java programs. In [10], we have proposed and implemented a metamodel-driven approach to generate domain-specific AOM frameworks that uses the aforementioned pointcut language. Finally, AOM approaches can be used to manage variability in software product line. Our work can be merged to these approaches to show why variability is also needed in the aspects in order to use an AO approach to build software product line.

References

1. Aldawud, O., Elrad, T., Bader, A.: UML Profile for Aspect-Oriented Software Development. In: 3rd International Workshop on Aspect Oriented Modeling (In conjunction of AOSD'03), Boston, Massachusetts (March 2003)
2. Baniassad, E., Clarke, S.: Theme: An Approach for Aspect-Oriented Analysis and Design. In: ICSE '04. Proceedings of the 26th International Conference on Software Engineering, pp. 158–167. IEEE Computer Society, Washington, DC, USA (2004)
3. Barais, O., Le Meur, A.F., Duchien, L., Lawall, J.: Safe integration of new concerns in a software architecture. In: ECBS '06. Proceedings of the 13th Annual IEEE International Symposium and Workshop on Engineering of Computer Based Systems, pp. 52–64. IEEE Computer Society, Washington, DC, USA (2006)
4. Elrad, T., Aldawud, O., Bader, A.: Aspect-Oriented Modeling: Bridging the Gap between Implementation and Design. In: Batory, D., Consel, C., Taha, W. (eds.) GPCE 2002. LNCS, vol. 2487, pp. 189–201. Springer, Heidelberg (2002)
5. Hannemann, J., Kiczales, G.: Design Pattern Implementation in Java and Aspectj. In: OOPSLA '02. Proceedings of the 17th ACM SIGPLAN conference on Object-oriented programming, systems, languages, and applications, pp. 161–173. ACM Press, New York, NY, USA (2002)
6. Kiczales, G., Hilsdale, E., Hugunin, J., Kersten, M., Palm, J., Griswold, W.G.: An Overview of Aspectj. In: Knudsen, J.L. (ed.) ECOOP 2001. LNCS, vol. 2072, pp. 327–353. Springer, Heidelberg (2001)
7. Kiczales, G., Lamping, J., Menhdhekar, A., Maeda, C., Lopes, C., Loingtier, J-M., Irwin, J.: Aspect-Oriented Programming. In: Aksit, M., Matsuoka, S. (eds.) ECOOP 1997. LNCS, vol. 1241, pp. 220–242. Springer, Heidelberg (1997)
8. Lahire, Ph., Quintian, L.: New Perspective To Improve Reusability in Object-Oriented Languages. Journal Of Object Technology (JOT) 5(1), 117–138 (2006)
9. Mens, T., Kniesel, G., Runge, O.: Transformation Dependancy Analysis, a Comparison of two Approaches. In: Rousseau, R., Urtado, C., Vauttier, S. (eds.) Proceedings of LMO 2006, Langages et Modèles à Objets, Nîmes, France, pp. 167–182. Hermes-Lavoisier (Mars 2006)
10. Morin, B., Barais, O., Jézéquel, J.M., Ramos, R.: Towards a Generic Aspect-Oriented Modeling Framework. In: 3rd International Workshop on Models and Aspects (In conjunction of ECOOP'07), Berlin, Germany (2007)
11. Muller, A., Caron, O., Carré, B., Vanwormhoudt, G.: On Some Properties of Parameterized Model Applications. In: Proceedings of ECMDA'05: First European Conference on Model Driven Architecture - Foundations and Applications, Nuremberg, Germany (November 2005)

12. Muller, P.A, Fleurey, F., Fondement, F., Hassenforder, M., Schneckenburger, R., Gérard, S., Jézéquel, J.M: Model-driven analysis and synthesis of concrete syntax. In: Nierstrasz, O., Whittle, J., Harel, D., Reggio, G. (eds.) MoDELS 2006. LNCS, vol. 4199, pp. 98–110. Springer, Heidelberg (2006)
13. Muller, P.A., Fleurey, F., Jezequel, J.M.: Weaving Executability into Object-Oriented Meta-languages. In: Briand, L.C., Williams, C. (eds.) MoDELS 2005. LNCS, vol. 3713, Springer, Heidelberg (2005)
14. Ossher, H., Tarr, P.: Hyper/J: Multi-Dimentionnal Separation of Concern for Java. In: Ghezzy, C. (ed.) Proceedings of ICSE'00, Limerick, Ireland, ACM Press, New York (2000)
15. Ramos, R., Barais, O., Jézéquel, J.M.: Matching model-snippets. In: MoDELS '07. Model Driven Engineering Languages and Systems, 10th International Conference, Nashville, Tennessee (2007)
16. Reddy, Y.R., Ghosh, S., France, R.B., Straw, G., Bieman, J.M., McEachen, N., Song, E., Georg, G.: Directives for Composing Aspect-Oriented Design Class Models. In: Rashid, A., Aksit, M. (eds.) Transactions on Aspect-Oriented Software Development I. LNCS, vol. 3880, pp. 75–105. Springer, Heidelberg (2006)
17. Ziadi, T., Jézéquel, J.M.: Families Research Book. In: Product Line Engineering with the UML: Products Derivation. LNCS, pp. 557–588. Springer, Heidelberg (2006)

An Expressive Aspect Composition Language for UML State Diagrams

Jon Whittle[1], Ana Moreira[2,*], João Araújo[2,*], Praveen Jayaraman[1], Ahmed Elkhodary[1], and Rasheed Rabbi[1]

[1] Dept. of Information & Software Engineering,
George Mason University, Fairfax VA 22030
[2] Dept. of Informatics, FCT, Universidade Nova de Lisboa,
2829-516 Caparica, Portugal
jwhittle@gmu.edu, {amm, ja}@di.fct.unl.pt,
{pjayara1,aelkhoda,rrabbi}@gmu.edu

Abstract. The goal of aspect-oriented software development is to maintain a clear separation of concerns throughout the software lifecycle. Concerns that are separated, however, must be composed at some point. The hypothesis in this paper is that existing aspect-oriented modeling composition methods are not expressive enough for composing state-dependent behavioral models. The paper presents a new aspect composition language, SDMATA, for UML state diagrams. SDMATA supports a richer form of model composition than previous approaches to aspect-oriented modeling. Firstly, pointcuts are given as patterns which allows for sequence pointcuts, loop pointcuts, etc. Secondly, SDMATA supports rich forms of composition including parallel composition and alternative composition. The language is applied to the use case slice technique of Jacobson and Ng. The findings are that it is possible to maintain the separation of state-dependent models during software design and that expressive model composition methods are necessary to do this in practice.

Keywords: aspect-oriented development, state machines, use cases.

1 Introduction

In software engineering, a concern is anything that is of interest to one or more stakeholders, such as a feature, a component or a non-functional requirement. Separation of concerns is the process of isolating key parts of a software system so that these elements can be developed and reasoned about independently. Separating concerns has been recognized as a way of tackling complexity in requirements engineering (e.g., viewpoints [1], aspect-oriented requirements engineering [2]), software architecture [3], software design (e.g., multi-dimensional separation of concerns MDSOC [4]) and coding (e.g., AspectJ [5], Hyper/J [6]).

---

* This author was partially supported by the AMPLE project (STREP IST-33710).

G. Engels et al. (Eds.): MoDELS 2007, LNCS 4735, pp. 514–528, 2007.

A key problem to solve is: How to specify the composition of a set of separated concerns? Concern composition (called *weaving* in aspect-oriented programming) is necessary so that the entire set of concerns can be executed as a whole.

This paper focuses on concern composition during software design. We introduce a new aspect composition language, SDMATA[1], for specifying the composition of aspect-oriented behavioral models given as UML state diagrams. Aspect-oriented behavioral modeling (AOBM) works as follows. An aspect model defines behavior that crosscuts a base model. A specification of an aspect includes three parts—a definition of the model elements in the base model affected by the aspect (often called the joinpoint model), a definition of the aspect model elements, and a definition of how behavior from the aspect model affects the base model elements (often called advices).

SDMATA brings two advances over previous work on aspect-oriented modeling with state machines (e.g., [8-10]). Firstly, the joinpoint model is defined by a state diagram pattern which allows for very expressive joinpoints. For example, a joinpoint may define a sequence of transitions, a loop, or a set of orthogonal regions. This is in contrast to most previous approaches to AOBM that only allow pointcuts to be single model elements, such as a transition or an event. Secondly, SDMATA supports rich composition types. For example, an aspect state diagram can be composed in parallel with the base, as an alternative to the base, or, in fact, in any way allowed by the underlying state diagram syntax. Previous work has often been limited to the before, after, around advices of AspectJ.

An additional contribution of this paper is that we apply SDMATA to composition specification in the use case slice technique by Jacobson and Ng [11]. Use case slices are a way of maintaining a use case-based decomposition throughout the development lifecycle. For state diagrams, this means that each use case maintains its own state diagram and these diagrams are composed to obtain the overall design.

To evaluate both SDMATA and the use case slice approach, we used SDMATA to specify state diagram compositions for seven student design solutions. These solutions were UML designs that were created as part of a graduate course on software design. They were originally produced following a traditional use case-based methodology. The solutions were re-engineered to keep separate the state diagrams derived from different use cases and our composition technique was applied to compose the state diagrams. This exercise provided evidence that an expressive composition language is necessary for practical UML designs and that our composition language is sufficiently expressive.

The remainder of this paper is structured as follows. Section 2 introduces motivation for our work. Section 3 presents SDMATA. Section 4 applies SDMATA to the use case slice method. Section 5 compares related work and is followed by suggestions for future work in Section 6.

---

[1] MATA (Modeling Aspects Using a Transformation Approach) is a more general aspect-oriented modeling approach that includes class diagrams, sequence diagrams and state diagrams. This paper describes the language features for state diagrams, hence the name SDMATA. [7] describes other language features applied to modeling software product lines.

2 Motivation

Figure 1 is an example of using UML state diagrams to maintain separation of concerns in a distributed application. The left-hand-side (LHS) is a use case for calling a remote service and consists of a state dependent class ServiceController and a state diagram that defines its behavior. Similarly, the right-hand-side (RHS) is a use case for handling a network failure, which contains the same class ServiceController, but with a different set of attributes and a different state diagram. This second use case describes a limited number of attempts to retry a service.

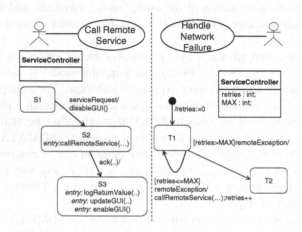

Fig. 1. Aspect-Oriented Modeling with State Diagrams

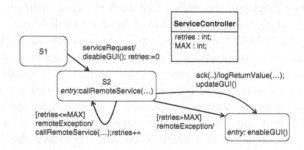

Fig. 2. Desired Composition of State Diagrams from Figure 1

The RHS crosscuts the LHS in the sense that whenever callRemoteService appears on the LHS, the RHS behavior should be used to handle a failure. This is a non-trivial example of crosscutting behavior. Prior to calling the remote service on the LHS, a GUI is disabled (via the action disableGUI). The GUI is only re-enabled (via the action enableGUI) once the remote service has been called *successfully*—the service call succeeds, a log is taken, and the GUI is updated before the GUI is re-enabled.

Now consider the desired result from composing the RHS with the LHS—this is shown in Figure 2. Note that when failure-handling is incorporated, what is now needed is that the GUI should be re-enabled whether the calling of the remote service succeeds or not. That is, even if the maximum number of retries is exceeded, enableGUI must still occur. Furthermore, logging and updating must only occur if the service call succeeds.

Capturing this composition is very difficult if a composition model based on that of AspectJ is used. Existing work on aspect-oriented modeling might, for example, define a pointcut as the occurrence of the action callRemoteService. But there is then no way to insert behavior after or around this pointcut in such a way that enableGUI is called whether or not the service call succeeds, and that logging/updating is not called in the failure scenario.

The example illustrates that it is not enough to consider a single action as a pointcut. In fact, the pointcut must be a *sequence* of actions and events. If one could specify that the pointcut is the sequence of actions/events between callRemoteService and enableGUI, then we get closer to the desired composition of Figure 2. Sequence pointcuts are not currently possible with most aspect-oriented modeling (AOM) approaches[2], although some AOP languages do support them [12].

More generally, when composing crosscutting state diagrams, it may be desirable to use advices that are more expressive than before, after or around. For example, an aspect state diagram may need to be composed *in parallel* with a base state diagram, or an aspect state diagram may need to be inserted *inside* a state in the base diagram (i.e., the base state becomes a composite state). It may also be useful to insert the aspect state diagram after a base state but then return to the base state once the aspect has completed—in other words, the aspect and base are composed in a loop. In fact, composition should allow two diagrams to be composed using any of the syntactic constructs of the modeling language. In the case of state diagrams, for example, composition could be achieved using orthogonal regions, composite states, or even history states.

In other words, aspect-oriented model composition may require models to be composed in complex ways rather than just before or after each other. It is for this reason that we propose a new model composition language in this paper.

3   SDMATA: Aspects for UML State Diagrams

In this section, we present SDMATA, a language for composing aspect-oriented UML state diagrams. Although we apply this language to UML state diagrams in this paper, the approach can be adapted to other UML diagrams (see, for example, [7]). SDMATA is given semantics in terms of graph transformations and so we briefly present background on graph transformations.

3.1   Graph Transformations

A graph consists of a set of nodes and a set of edges. A graph transformation is a graph rule $r: L \rightarrow R$ from a left-hand-side (LHS) graph L to a right-hand-side (RHS) graph R. The process of applying r to a graph G involves finding a graph

---

[2] The only known approach that does allow this is join point designation diagrams (JPDDs) [9] but JPDDs do not support expressive advices.

monomorphism, *h*, from *L* to *G* and replacing *h(L)* in *G* with *h(R)*. Graph transformations may also be defined over attributed typed graphs. A typed graph is a graph in which each node and edge belongs to a type. Types are defined in a type graph. An attributed graph is a graph in which each node and edge may be labeled with attributes where each label is a (value, type) pair giving the value of the attribute and its type. In a graph rule, variables may be used to capture a set of possible values and/or a set of possible types.

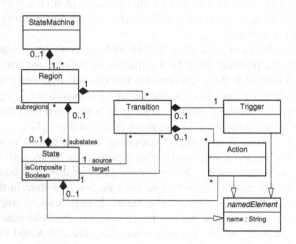

Fig. 3. UML State Machine Metamodel

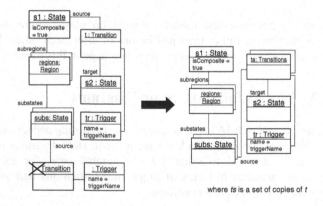

Fig. 4. Graph Rule to Move Down Transitions

Graph rules have previously been used for transforming UML models (e.g., UML refactorings [13]). Such work requires that UML models be represented as graphs. The usual approach is to define node types as the metaclasses in the UML metamodel. Graph rules can then be shown graphically using object diagrams.

As an example, Figure 3 shows a (simplified) fragment of the UML state machine metamodel. A state machine contains 1 or more (orthogonal) regions, each of which contains states. Each transition is from a source to a target state and has a trigger and actions. States may also have actions. A state may contain 0 or more regions. A state is composite if it contains 1 or more regions. If it contains 2 or more regions, then the regions in this state are orthogonal. The State metaclass has an attribute isComposite indicating whether or not the state is composite. Finally, states, triggers and actions have names (as represented by a generalization relationship to namedElement).

Figure 4 is an example graph transformation which moves all outgoing transitions from a composite state to its substates. The notation used to define this graph transformation is that of [13]. (We defer to [13] for the subtleties of this notation.) Nodes in the graph are given as rectangles. Nodes are attributed and typed so UML object diagram notation can be used to represent them. There are two additional notations. First, a set of nodes of a certain type is shown by a stacked rectangle. For example, *regions* is a set of Regions associated with a composite state. Secondly, the cross in the figure is a negative application condition and says that any match against the LHS graph cannot have a substate with a transition trigger called *triggerName*. The LHS in Figure 4 matches any graph with at least one composite state with an outgoing transition. Furthermore, there should not be a transition on any of the substates with the same trigger. The RHS redirects the matched transition to all substates (by creating copies) thus moving the transition down in the state hierarchy.

3.2 SDMATA

Given a base state diagram, S_B, crosscut by an aspect state diagram, S_A, an SDMATA composition rule merges S_A and S_B to produce a composed model S_{AB}. The merge could be specified simply using a graph rule, r: S_B S_A $\rightarrow$ S_{AB}. However, SDMATA does not take this approach because graph rules are specified at the meta-level. A requirement of SDMATA is that it can be used by working software modelers. Modelers typically do not have the in-depth knowledge of the UML metamodel required to specify a graph rule—Figure 4, for example, is hard to get right. Therefore, although SDMATA is based on graph transformations, it uses the concrete syntax of UML wherever possible. This means that SDMATA is not as expressive as graph transformations over UML (since metaclasses cannot be accessed) but it is tailored to aspect composition in a way that is accessible to software engineers. The semantics of SDMATA is given by mapping the language to the equivalent meta-level graph rule. Standard graph matching and rule execution semantics can then be used.

A SDMATA composition rule consists of two parts. Firstly, the rule specifies which part of S_B is crosscut by S_A. This is done by a *state diagram pattern* that identifies a sub-diagram of S_B. Secondly, *composition operators* specify how crosscutting behavior from S_B is combined with the behavior from S_A. State diagram patterns can be thought of as the joinpoint model of SDMATA and the composition operators can be thought of as the advices.

3.3 State Diagram Patterns

State diagram patterns resemble the concrete syntax of UML state diagrams very closely. This concrete syntax is familiar to developers and is therefore more accessible. A state diagram pattern is an abstract representation of a family of state diagrams and contains pattern variables. Pattern variables are typed over state diagram metaclasses and are marked with multiplicities. Pattern variables are regular expressions prefixed with a vertical bar 'I' to denote that they are variables. A pattern variable IX has a multiplicity of one. A pattern variable IX$^+$ has a multiplicity of one or more. A state diagram pattern matches a state diagram if all the pattern variables can be instantiated to elements of the state diagram in a way that preserves the variable's metaclass and multiplicity.

State Diagram Pattern Syntax. We denote the type of a pattern variable by (IX : T). Only the metaclasses in the list below are allowed to be types for pattern variables. We assume the metamodel of Figure 3 in the remainder of this paper.

1. (IX : State) matches against a single state. (IX$^+$: State) matches against one or more states and also matches the transitions between these states.
2. (IX : StateMachine) matches a single state machine. (IX$^+$: StateMachine) is not allowed.
3. (IX : Action) matches a single action. (IX$^+$: Action) matches a sequence of one or more actions.
4. (IX : Trigger) matches a single event. (IX$^+$: Trigger) matches a sequence of one or more events.
5. (IX : Region) matches a single orthogonal region. (IX$^+$: Region) matches one or more regions within the same composite state.

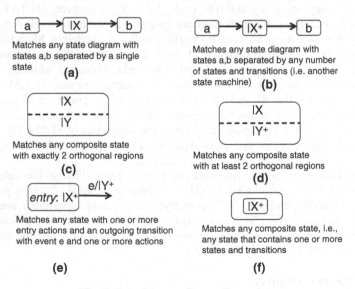

(a) Matches any state diagram with states a,b separated by a single state

(b) Matches any state diagram with states a,b separated by any number of states and transitions (i.e. another state machine)

(c) Matches any composite state with exactly 2 orthogonal regions

(d) Matches any composite state with at least 2 orthogonal regions

(e) Matches any state with one or more entry actions and an outgoing transition with event e and one or more actions

(f) Matches any composite state, i.e., any state that contains one or more states and transitions

Fig. 5. State Diagram Pattern Examples

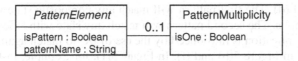

Fig. 6. Pattern Elements

Whenever possible, the concrete syntax of a pattern variable is the same as the UML concrete syntax of its type. See Figure 5 for examples.

Figure 5(a), for example, matches any sequence of states starting with a state named a, ending with a state named b, and with another state in between (different from a and b). In contrast, the variable $|X^+$ in Figure 5(b) matches one or more states in between a and b as well as any transitions between those states. This means that $|X^+$ represents any number of states and transitions with at least one of those states connected to the incoming transition shown, and at least one state connected to the outgoing state shown. In a similar way, Figures 5(c) and 5(d) show how to match against a specific number of regions and one or more regions, respectively. Figure 5(e) is self-explanatory. Figure 5(f) matches a state which contains a state machine $|X$, i.e., there must be at least one substate, but the composite state may contain any number of substates and transitions.

New notation is introduced only to match composite states (see Figure 5(f)). The latter is necessary because composite states are given by a meta-attribute in the UML state machine metamodel (*isComposite* in Figure 3) so it is not possible to distinguish a simple or composite state based purely on the concrete syntax of state. Note also that state diagram patterns need not be valid state diagrams—Figure 5(e), for example, has no target state.

The abstract syntax of state diagram patterns is defined by an extension of the metamodel in Figure 3 and their semantics is given by mapping them to the notation used in Figure 4—this notation is more or less standard for graph rules and so existing graph matching algorithms can then be used. A metamodel for state diagram patterns can be defined in terms of Figure 3. (Due to space limitations, we limit the discussion to syntactic elements defined in Figure 3). We introduce a new abstract metaclass, PatternElement, which is a generalization of StateMachine, State, Region, Trigger and Action. Note that Transition is not a PatternElement. Also, we introduce a PatternMultiplicity metaclass. PatternElement has two attributes. *isPattern* denotes whether a model element is a pattern or not. If it is, it has a pattern name which is a regular expression prefixed by '|'. PatternMultiplicity has a single attribute, *isOne*, that denotes whether the multiplicity is exactly one, or one or more. The new metaclasses are defined in Figure 6.

State Diagram Pattern Semantics. The pattern matching semantics for state patterns is given by mapping each pattern to a typed graph consisting of instances of the appropriate metaclasses. Standard graph matching algorithms (e.g., [14]) can then be used. If a pattern element has a multiplicity of one, it maps to a single instance of its metaclass. If it has multiplicity one or more, it maps to a set of instances. To illustrate, Figure 7 shows the mapping to metaclass instances for the patterns given in

Figure 5(c) and (d). The first pattern will match any composite state with exactly two orthogonal regions. The second pattern will match any state with at least 2 regions.

A slight complication is introduced by the use of IX^+ to match against a set of states and transitions in Figure 5(b) and (f). In Figure 5(f), for example, instead of mapping IX^+ to a set of instances of State, it must be mapped to an instance of Region containing any number of instances of State and Transition. This issue arises because of the peculiarities of the UML metamodel. The full details of the mapping to metaclass instances cannot be given here due to space limitations.

3.4 State Diagram Pattern Example

Figure 8 shows the state diagram pattern required in the example from Section 2. Recall that a sequence pointcut was required. The figure illustrates how to specify a sequence from callRemoteService to enableGUI. The pattern variable IX^+ matches against any number of actions in the target state of the transition but will not match against enableGUI(). The effect is that the state diagram pattern matches any sequence starting with the callRemoteService(...) action, followed by a transition, followed by one or more entry actions, and ending with the action enableGUI().

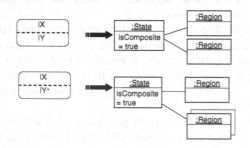

Fig. 7. Metaclass Instance Representation of Patterns

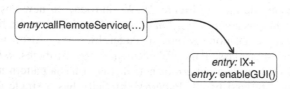

Fig. 8. State Diagram Pattern for Figure 1

3.4 Composition Operators

State diagram patterns identify model elements in the base model that are crosscut by the aspect model. SDMATA also gives a way to define how model elements from the aspect should be composed with model elements from the base. This is done using a language based on graph transformations. SDMATA represents a graph rule $r: L \rightarrow R$, where L and R are UML models, as a single UML model in which model elements may be annotated with one of three stereotypes—<<create>>, <<delete>> or <<context>>.

Given a pointcut definition given as a state diagram pattern, model elements from the aspect that should be added to the state diagram pattern are marked with the <<create>> stereotype. Similarly, elements may be removed using the <<delete>> stereotype. Simple examples are shown in Figure 9. In (a), the pointcut is any state and the aspect elements added are a state, *a*, and a transition to *a*. In (b), the pointcut is any pair of states and the aspect element is a superstate that is added so that it contains these states. In general, <<create>> and <<delete>> can be used to add (or remove) any kind of aspectual model element. For example, an aspect could be added as an orthogonal region to an existing base model that matches a state pattern—see Figure 9(c). The use of <<create>> is "optimized" in the sense that if a state is stereotyped as <<create>> then any of its substates or transitions are also created. Hence, in Figure 9(a), the transition is created but does not need to explicitly be given a stereotype. This optimization reduces the number of stereotypes a user must specify. However, in Figures 9(b) and 9(c), the user wants to wrap a composite state around existing states. To stop these substates from being created, they are stereotyped as <<context>>.

<<create>> and <<delete>> are a way of representing both the LHS and RHS of a graph rule on the same diagram—the familiar LHS → RHS notation can be obtained by taking the LHS as all elements without a stereotype and the RHS as the LHS but with the <<create>> elements added and the <<delete>> elements removed. This idea comes from VIATRA [15]. The <<context>> stereotype is unique to MATA.

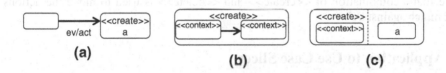

(a)   **(b)**   **(c)**

Fig. 9. Examples of Composition Operators

3.5 Applying SDMATA Rules

An SDMATA rule consists of (apart from the models themselves) a state diagram pattern and composition operators specified as above. An SDMATA rule can be translated into a graph rule. Hence, existing graph rule execution engines can be used to actually compose an aspect state diagram with its base. Currently, we use AGG [16] to execute aspect composition. We have implemented a tool to allow users to specify aspects and aspect composition rules in IBM Rational Software Modeler. AGG is then used to seamlessly execute the composition and the composed models are presented to the user for inspection or refinement.

3.6 SDMATA Example

Figure 10 shows the SDMATA rule definition that will merge the two models from Figure 1 and produce the composed model as given in Figure 2. To make it easier to read, elements that are created or deleted are in bold italics. Note that a SDMATA rule contains the state diagram pattern to match against, the aspect model elements, and the composition operators that detail how those aspect elements are merged with the base. The effect of applying this rule is that: (1) a match is found in the base model with the state diagram pattern, and (2) the matched submodel of the base is modified by creating and deleting elements according to the <<create>> and <<delete>> composition operators.

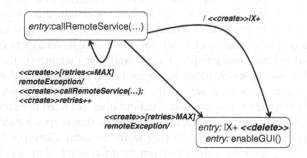

Fig. 10. SDMATA Example

The SDMATA rule in Figure 10 contains both the aspect (the RHS of Figure 1) as well as a specification of how the aspect should be composed with the base (the LHS of Figure 1). The rule essentially defines a sequence pointcut to match against and then creates and deletes elements, which, in effect, composes the aspect with the base. Note that a combination of <<create>> and <<delete>> is used to move the actions that match against IX^+.

4 Application to Use Case Slices

SDMATA is more expressive than existing aspect-oriented behavioral modeling (AOBM) techniques. Broadly, there have been two types of approaches to AOBM. The first is to reuse the joinpoint model and advices from AspectJ. The second is to use a generic merge algorithm (that can be tailored) to compose an aspect and a base model. The second approach will be referred to as the MDSOC-like approach because it is inspired by MDSOC programming models [4]. The most notable example of this approach is Theme/UML [17].

For SDMATA, the question remains whether the additional expressiveness is actually required in practice. To answer this question, we undertook an investigation of existing design solutions to see which kinds of compositions are needed in practice. Our experiment attempted to answer the following question. In practical examples, are model composition mechanisms based on AspectJ-like or MDSOC-like approaches expressive enough? Moreover, is our approach expressive enough? The investigation was undertaken for the use case slice technique of Jacobson and Ng [11]. Use case slices are a way of maintaining a use case-based decomposition throughout the development lifecycle. For state diagrams, this means that each use case maintains its own state diagram and these state diagrams are composed during late design or implementation to obtain the overall design.

In [11], Jacobson and Ng do not adequately address how to compose use case slices during design. Their approach is to apply AspectJ-like composition operators. The hypothesis of this paper is that such operators are not expressive enough. To test this hypothesis, we examined existing UML designs, refactored those designs to reflect the use case slice technique of Jacobson and Ng, and then investigated the level of expressiveness required to compose state diagrams from different use case slices. Because of availability of the models, we chose to study seven student team design solutions, each expressed in UML consisting of use cases, class diagrams, interaction diagrams and state diagrams. Only the use cases and state diagrams were relevant to the study. Projects were conducted by teams of three to four students. Each of the seven projects tackled the same problem statement using the same set of use cases. The scale of the student solutions is clearly not industrial in size and the results offered here are meant to be just the first step.

Based on an analysis of the compositions required in the state-dependent use case slices, we identified four categories of composition.

C1: One-to-One State Matching. The first category includes model compositions that can be expressed using simple matching of states. In other words, for two state diagrams, $S1$ and $S2$, with state sets $\Sigma1$ and $\Sigma2$, the composed state diagram $S1 \cdot S2$, can be obtained by defining a one-to-one mapping $\theta: \Sigma1 \rightarrow \Sigma2$. Figure 11(a) gives an example. In the student solutions, this case occurred typically when two state diagrams defined sequences that were joined together into a loop.

C2: Many-to-many state matching. This category is an extension of the previous one whereby states in the two state diagrams have a many-to-many relationship, i.e., $\theta(\sigma)$ is a set for any state σ. This allows a much richer form of composition. In particular, it allows for the creation of composite states (see Figure 11(b)).

C3: State diagram refactoring. In this category, one or more of the state diagrams must be refactored to enable composition to take place. In other words, one state diagram cannot be inserted in its entirety into the other. Rather, it must be broken up before being inserted in multiple places. This type cannot be handled by state matching because matching cannot refactor a state diagram. Figure 11(c) illustrates.

C4: State diagram refinement. In this type of composition, additional behavior (i.e., states and transitions) must be added when composition takes place. Clearly, state matching does not apply because state matching cannot refine behavior. This type of composition is necessary in cases where two use case slices have been developed independently but where there are dependencies between the slices that must be resolved when the slices are composed. A typical example concerns access to data. If a single use case slice reads from a data object, then no data access synchronization is required. However, if another use case slice writes to this data object, when the two use case slices are composed, an access synchronization mechanism such as mutual exclusion must be added. Figure 11(d) gives an example.

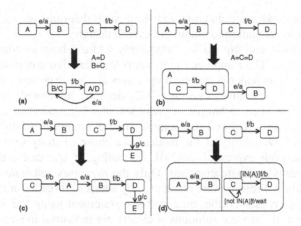

Fig. 11. Composition Categories

Based on the student design solutions, we found that all four categories of composition occur for use case slice development. The relative frequency for the four categories was as follows: 13%, 39%, 46%, 2%. Of the three approaches considered—AspectJ-like, MDSOC and SDMATA—only SDMATA is expressive enough to support all four categories. The MDSOC approach supports only category C1 although it can be easily extended to support C2 (as was done in [18]). The AspectJ-approach does not support either category C1 or C2 because both categories allow complex interleavings that cannot be expressed using just before/after advices. Some compositions in C3 could be supported by the AspectJ approach if the aspect state diagram is first refactored into multiple state diagram fragments. Each fragment can then be inserted at a different place. However, we view this as a non-optimal approach to composition because it involves representing fragments of a state diagram separately which leads to problems in reusability and readability.

SDMATA supports all categories because the entire state machine diagram syntax is available. For example, two use case slices can be merged in parallel using UML orthogonal regions. The results of the investigation reveal that, at least for use case slice composition, a greater degree of expressiveness is required in practice. Further investigation is required, of course, to see if these results are true for other aspect-oriented software development methods.

5 Related Work

There is a large body of work on aspect-oriented modeling, although much of this has been restricted to structural models. Work of note that considers behavioral models is the Motorola WEAVR tool for state machines [10], Song et al.'s work on weaving security aspects into sequence diagrams [19], and Jézéquel et al.'s work on semantic composition for interaction diagrams [20]. Although expressive pointcut mechanisms, such as sequence pointcuts, have been considered for aspect-oriented programming (e.g., [12]), to the authors' knowledge, this paper is the first work to bring expressive pointcuts to behavioral models. Furthermore, the vast majority of existing work on aspect-oriented behavior modeling (e.g., [9]) considers a limited set of composition operators based on those of AspectJ.

This paper considers joinpoints to be static in the sense that the runtime semantics of state diagrams is not taken into consideration. Dynamic join points can also be defined for state diagrams. However, since models are most commonly used for communication and documentation, and are not necessarily executed, static joinpoints are perhaps more useful in current modeling practices. It would be interesting to extend SDMATA to dynamic join points, however. A related issue is that of semantic-based aspect composition in which the fragile pointcut problem is addressed by capturing pointcuts according to their semantics rather than their syntax. The first work applying this idea to behavioral models is [20], although the concept has also been applied in early aspect requirements engineering [21]. The pointcuts in this paper are purely syntactic and the authors recognize the limitations of this.

6 Conclusion

This paper presents a novel language, SDMATA, for defining aspect compositions for UML state diagrams in a way that offers very expressive composition. As shown by the study in this paper, SDMATA is more expressive than existing model composition techniques in the AOSD field. Furthermore, this richer expressiveness *is* required in practice—even relatively small design solutions would require it.

We have implemented an execution engine for SDMATA, although the support for patterns is still under development. This tool allows users to specify UML models and aspects in IBM Rational Software Modeler. Composed models may be viewed in the same tool. The tool is being used in a number of research projects for applying aspects in product line development and for adaptive systems. The use of graph transformations as an underlying formalism also provides some support for the aspect interaction problem—whereby application of multiple compositions may have unexpected effects. We are investigating the use of critical pair analysis to assist users by detecting some kinds of aspect interactions automatically.

References

1. Nuseibeh, B., Kramer, J., Finkelstein, A.: ViewPoints: meaningful relationships are difficult! In: ICSE. International Conference on Software Engineering, Portland, Oregon, pp. 676–683 (2003)
2. Moreira, A., Rashid, A., Araújo, J.: A Multi-Dimensional Separation of Concerns in Requirements Engineering. In: RE. International Conference on Requirements Engineering, pp. 285–296. IEEE Computer Society, Paris, France (2005)
3. Clements, P., Bachmann, F., Bass, L., Garlan, D., Ivers, J., Little, R., Nord, R., Stafford, J.: Documenting Software Architectures: Views and Beyond. Addison-Wesley, Reading (2002)
4. Tarr, P.L., Ossher, H., Harrison, W.H., Sutton, S.M.: N Degrees of Separation: Multi-Dimensional Separation of Concerns. In: ICSE. International Conference on Software Engineering, Los Angeles, CA, pp. 107–119 (1999)

5. Kiczales, G., Hilsdale, E., Hugunin, J., Kersten, M., Palm, J., Griswold, W.G.: An Overview of AspectJ. In: Knudsen, J.L. (ed.) ECOOP 2001. LNCS, vol. 2072, pp. 327–353. Springer, Heidelberg (2001)
6. Ossher, H., Tarr, P.L.: Hyper/J: Multi-Dimensional Separation of Concerns for Java. In: ICSE. International Conference on Software Engineering, Limerick, Ireland, pp. 737–737 (2000)
7. Jayaraman, P., Whittle, J., Elkhodary, A., Gomaa, H.: Model Composition and Feature Interaction Detection in Product Lines using Critical Pair Analysis. In: Engels, G., Opdyke, B., Weil, F. (eds.) MODELS. International Conference on Model Driven Engineering, Languages and Systems, Springer, Nashville, TN (2007)
8. Katara, M., Katz, S.: Architectural Views of Aspects. Aspect-Oriented Software Development (AOSD), Boston, Massachusetts 1–10 (2003)
9. Stein, D., Hanenberg, S., Unland, R.: Expressing Different Conceptual Models of Join Point Selections in Aspect-Oriented Design. Aspect-Oriented Software Development (AOSD), Bonn, Germany, 15–26 (2006)
10. Cottenier, T., van den Berg, A., Elrad, T.: Motorola WEAVR: Model Weaving in a Large Industrial Context. Aspect-Oriented Software Development (AOSD), Vancouver, Canada (2007)
11. Jacobson, I., Ng, P.-W.: Aspect Oriented Software Development with Use Cases. Addison-Wesley Professional, Reading (2004)
12. Douence, R., Fritz, T., Loriant, N., Menaud, J.-M., Segura-Devillechaise, M., Sudholt, M.: An Expressive Aspect Language for System Applications with Arachne. Aspect-Oriented Software Development (AOSD), Chicago, Illinois, 27–38 (2005)
13. Markovic, S., Baar, T.: Refactoring OCL Annotated UML Class Diagrams. In: Briand, L.C., Williams, C. (eds.) MoDELS 2005. LNCS, vol. 3713, pp. 280–294. Springer, Heidelberg (2005)
14. Zuendorf, A.: Graph Pattern Matching in PROGRES. In: Cuny, J., Engels, G., Ehrig, H., Rozenberg, G. (eds.) Graph Grammars and Their Application to Computer Science. LNCS, vol. 1073, pp. 454–468. Springer, Heidelberg (1996)
15. Balogh, A., Varro, D.: Advanced Model Transformation Language Constructs in the VIATRA2 Framework. In: ACM Symposium on Applied Computing (Model Transformation Track), pp. 1280–1287. ACM Press, Dijon, France (2006)
16. Taentzer, G.: AGG: A Graph Transformation Environment for Modeling and Validation of Software. In: Pfaltz, J.L., Nagl, M., Böhlen, B. (eds.) AGTIVE 2003. LNCS, vol. 3062, pp. 446–453. Springer, Heidelberg (2004)
17. Clarke, S., Baniassad, E.: Aspect-Oriented Analysis and Design: The Theme Approach. Addison-Wesley, Reading (2005)
18. Araújo, J., Whittle, J., Kim, D.-K.: Modeling and Composing Scenario-Based Requirements with Aspects. In: RE. International Conference on Requirements Engineering, Kyoto, Japan, pp. 58–67 (2004)
19. Song, E., Reddy, R., France, R.B., Ray, I., Georg, G., Alexander, R.: Verifiable Composition of Access Control and Application Features. In: SACMAT. ACM Symposium on Access Control Models and Technologies, Stockholm, Sweden, pp. 120–129. ACM Press, New York (2005)
20. Klein, J., Helouet, L., Jézéquel, J.-M.: Semantic-Based Weaving of Scenarios. Aspect-Oriented Software Development (AOSD), Vancouver, Canada, 27–38 (2006)
21. Chitchyan, R., Rashid, A., Rayson, P., Waters, R.: Semantics-Based Composition for Aspect-Oriented Requirements Engineering. Aspect-Oriented Software Development (AOSD), Vancouver, Canada, 36–48 (2007)

Enhancing UML State Machines with Aspects

Gefei Zhang, Matthias Hölzl, and Alexander Knapp[*]

Ludwig-Maximilians-Universität München
{gefei.zhang,matthias.hoelzl,alexander.knapp}@pst.ifi.lmu.de

Abstract. Separation of Concerns (SoC) is an important issue to reduce the complexity of software. Recent advances in programming language research show that Aspect-Oriented Programming (AOP) may be helpful for enhancing the SoC in software systems: AOP provides a means for describing concerns which are normally spread throughout the whole program at one location. The arguments for introducing aspects into programming languages also hold for modeling languages. In particular, modeling state-crosscutting behavior is insufficiently supported by UML state machines. This often leads to model elements addressing the same concern scattered all over the state machine. We present an approach to aspect-oriented state machines, which show considerably better modularity in modeling state-crosscutting behavior than standard UML state machines.

1 Introduction

Separation of Concerns (SoC) is an important issue in software engineering. A clear SoC could improve the modularity of software artefacts, reduce the complexity of software systems, and thus make them less error-prone and better maintainable.

Recent advances in programming language research propose the use of Aspect-Oriented Programming (AOP [11]) to achieve a better SoC in programs. AOP helps in particular to solve the problem of "scattered code": code, that would be otherwise scattered all over the program, may be collected together in a new language construct called *aspect*. This way, AOP is particularly helpful for separating cross-cutting concerns, i.e. concerns that are involved in other concerns, like logging or transaction management.

However, not only code may be scattered all over a program, but also model elements all over a software design model. In particular, the Unified Modeling Language (UML [16]), the *lingua franca* in object-oriented software analysis and design, lacks aspect-like language constructs to address the problem of "scattered model elements" by centralizing model elements involved in one concern at a dedicated location.

We propose to extend the UML with aspect-oriented language concepts. In particular, we present a design of aspect-oriented state machines. Aspect-oriented state machines show considerably better modularity in, among others, the design of state-crosscutting behavior, such as state synchronization and trace-based behavior.

The remainder of this work is organized as follows: in Sect. 2 we summarize the syntax and semantics of UML state machines and in Sect. 3 we demonstrate some of

---

[*] This work has been partially sponsored by the EU project SENSORIA (IST-2005-016004) and the DFG project MAEWA (WI 841/7-1).

G. Engels et al. (Eds.): MoDELS 2007, LNCS 4735, pp. 529–543, 2007.

their weaknesses w.r.t. SoC. In Sect. 4 we present our proposal to aspect-oriented state machines and show how they may be used to achieve a better SoC by modularizing state interaction and trace-based behavior. An algorithm for for translating aspects into UML state machines is presented in Sect. 5. Related work is discussed in Sect. 6. Finally, we conclude and outline some future work.

2   UML State Machines

A UML state machine provides a behavioral view of its context. Figure 1 shows a state machine of a process which contains two parallel threads. After creation, the process is first initialized in the state Init, then the two threads run parallel in the state Running, until they receive the events astop and bstop while they are in the states A3 and B3, respectively. In this case each thread waits for the other to receive his stop signal in waiting states A4 and B4, respectively, before the threads terminate conjointly.

We briefly review the syntax and semantics of UML state machines according to the UML specification [16] by means of Fig. 1. A UML state machine consists of *regions* which contain *vertices* and *transitions* between vertices. A vertex is either a *state*, where the state machine may dwell in and which may show hierarchically contained regions; or a *pseudo state* regulating how transitions are compound in execution. Transitions are triggered by *events* and describe, by leaving and entering states, the possible state changes of the state machine. The events are drawn from an *event pool* associated with the state machine, which receives events from its own or from different state machines. The *context* of a state machine, a UML classifier, describes the features, in particular, the attributes, which may be used and manipulated in execution.

A state of a state machine is *simple*, if it contains no regions (such as Init and all states contained in Running in Fig. 1); a state is *composite*, if it contains at least one region; a composite state is said to be *orthogonal* if it contains more than one region, visually separated by dashed lines (such as Running). Each state may show an *entry* behavior (like actB2 in B2), an *exit* behavior (like actA2 in A2), which are executed on activating and deactivating the state, respectively; a state may also show a *do activity* (like in Init) which is executed while the state machine sojourns in this state. Transitions are triggered by events (a12, a23), show guards (condB), and specify actions to be executed when a transition is fired (actA23). Completion transitions (transition leaving

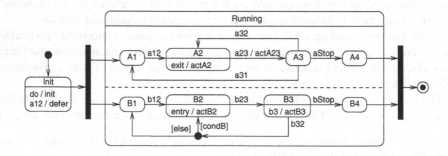

Fig. 1. State machine of a process containing two parallel threads

Init) are triggered by an implicit *completion event* emitted when a state completes all its internal activities. Events may be *deferred* (as a12 in Init), that is, put back into the event pool, if they are not to be handled currently but only later on. By executing a transition its source state is left and its target state entered; transitions, however, may also be declared to be *internal* to a state (b3 / actB3), thus skipping the activation-deactivation scheme. An *initial* pseudo state, depicted as a filled circle, represents the starting point for the execution of a region. A *final* state, depicted as a circle with a filled circle inside, represents the completion of its containing region; if all regions of a state machine are completed the state machine terminates. *Junction* pseudo states, also depicted as filled circles (see lower region of Running), allow for case distinctions. Transitions to and from different regions of an orthogonal composite state can be synchronized by *fork* and *join* pseudo states, presented as bars. For simplicity, we omit the other pseudo state kinds (entry and exit points, shallow and deep history, and choice).

During runtime, a state gets active when entered and inactive when exited as a result of a transition. The set of currently active states is called the active *state configuration*. When a state is active, so is its containing state. The active state configuration is thus a set of trees starting from the states in the top-level regions down to the innermost active substates. The execution of a state machine consists in changing its active state configuration in dependence of the current active states and a *current event* dispatched from the event pool. We call the change from one state configuration to another an *execution step*. First, a maximally consistent set of prioritized, enabled compound transitions is chosen. Transitions are combined into *compound transitions* by eliminating their linking pseudo states; for junctions this means to combine the guards on a transition path conjunctively, for forks and joins to form a fan-out and fan-in of transitions. A compound transition is *enabled* if all of its source states are contained in the active state configuration, its trigger is matched by the current event, and its guard is true. Two enabled compound transitions are consistent if they do not share a source state; an enabled compound transition takes priority over another enabled compound transition if its source states are below the source states of the other transition in the active state configuration. For each compound transition in the set, its least common ancestor (LCA) is determined, i.e. the lowest composite state containing all the compound transition's source and target states. The compound transition's main source state, i.e., the direct substate of the LCA containing the source states, is deactivated, the transition's actions are executed, and its target states are activated.

Given a state machine M, we denote by the sets $states(M)$, $transitions(M)$, $compounds(M)$ and $events(M)$ the states, transitions, compound transitions and events of M. Given a state $s \in states(M)$, we write $compounds(M, s)$ for the compound transitions leaving s. Given a transition $t \in states(M)$ we write $trigger(t)$ for the event triggering t and $guard(t)$ for the guard of t.

3   Pervasive Modification of State Machines

UML state machines work fine as long as the only form of communication among states is the activation of the subsequent state via a transition. More often than not, however, an active state has to know how often some other state has already been active and/or if

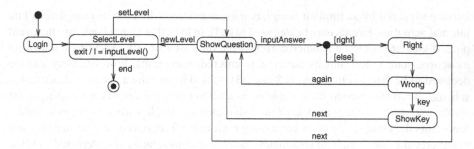

Fig. 2. State machine of an e-learning system

other states (in other regions) are also active. Unfortunately, behavior that depends on such information cannot be modeled modularly in UML state machines.

For an example, consider an e-learning system as modeled in Fig. 2: the user first logs in, selects a level of difficulty, and then proceeds to answering questions of this level. The system tells him whether his answer is right or not, if it is right, the user may proceed to the next question; if not, he may choose to try again, to see the key to the question, or to proceed to the next question. Instead of answering the current question, the user may also choose to go to another level.

Suppose the selection of a level $l > 0$ should be allowed only if the user has already answered minRight questions of level $l - 1$ in a row, otherwise the system should give an error message. Figure 3 shows how this restriction might be modeled in a standard UML state machine: a new attribute variable crir is introduced for counting the length of the current stroke of correct answers (current right in a row), it is incremented in state Right and reset to 0 once Wrong is active. An array r is introduced to store the maximal length of crir at each level. Once the user gives a wrong answer, the system has to check if this record should be updated. In order to know whether the user has selected in SelectLevel a different level than the current one or just continues with the same level, another new variable cl stores the current level each time SelectLevel is entered. The transition from SelectLevel to ShowQuestion is split into two to handle the cases whether the level selected by the user is selectable or not. Finally, the variable crir has to be reset to 0 when the user has successfully changed to another level. Obviously, it is unsatisfactory

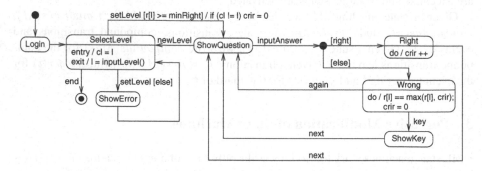

Fig. 3. Modeling the level selection restriction using standard UML

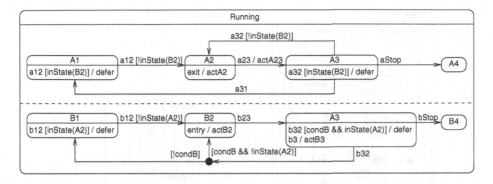

Fig. 4. Mutual exclusion using standard UML

that the model elements involved in this one feature are scattered all over the state machine, switching the feature on and off thus is difficult and requires modifications all over the state machine. The state machine gets rather hard to understand and maintain.

Besides history-based behavior, state synchronization is also difficult to model in UML state machines. Suppose for our state machine shown in Fig. 1 an additional mutual exclusion rule that the states A2 and B2 must not be active at the same time: before any thread entering any of these two states, the process must check if the other critical state is active; the thread may only proceed if this is not the case, otherwise it must wait for the other thread to leave the "blocking" state.

Figure 4 extends the state Running of the state machine shown in Fig. 1 with this rule: The OCL [15] predicate inState is used for checking whether a particular state is active; every transition leading to A2 and B2 must be guarded by such a predicate in order to ensure that the A2 and B2 do not become active at the same time. Simultaneously, the source states of these offending transitions defer the triggering event such that it will be reconsidered if the activation condition of either A2 or B2 changes. (Note that we have taken the liberty of adding a guard to the deferring of an event in order to make a peek to the current event, which is not covered by the UML metamodel. However, this construct can be replaced by adding an internal transition involving the negated deferring guard.) Again, it is a tedious and error-prone task to model a simple thread synchronization rule including no more than two states like this, since invasive modifications to the original state machine are necessary almost everywhere in it.

In both examples, having to model interaction between several states by the pervasive usage of sometimes intricate state machine features makes the resulting state machine rather complex and no longer intuitively understandable, in contrast to what models in a graphical modeling language are supposed to be.

4   Aspect-Oriented State Machines

Our answer to the problem of scattered model elements in UML state machines is, motivated by the success of aspect-oriented programming in solving the problem of scattered code, to enhance UML state machines with aspects. We introduce a new language construct, *aspect*, to the UML, which contains a *pointcut* and a piece of *advice*. The

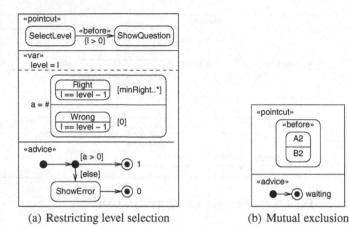

(a) Restricting level selection   (b) Mutual exclusion

Fig. 5. Aspects for a better separation of concerns

pointcut specifies some special point of time in the execution of a UML state machine, which is called the *base state machine*. The advice defines some additional or alternative behavior. An aspect is thus a (graphical) statement expressing that at the special point of time specified by the pointcut, the behavior defined by the advice should be performed. The complete behavior of the context of the base state machine is therefore given by the composition of the base state machine and the aspect. The process of composition is called *weaving*.

4.1   Concrete Syntax and Informal Semantics

We first define a new language construct, *superstate*, to be used in aspects. A super-state contains a subset of the states of the base state machine and may be guarded by constraints.

A pointcut is either a *configuration pointcut* or a *transition pointcut*. A configuration pointcut consists of a superstate, stereotyped by «before» or «after». It specifies, depending on the stereotype, the point of time just before or just after any state configuration that contains all states in this superstate is active. A transition pointcut consists of two superstates, connected by a transition with a stereotype «before» or «after». It specifies, depending on the stereotype, the point of time just before or just after this transition is fired. In the aspect shown in Fig. 5(a), the transition from SelectLevel to ShowQuestion is stereotyped «before» and guarded by a constraint $l > 0$. The pointcut thus specifies the point of time just before the base state machine is about to change from an active state configuration, which contains the state SelectLevel, to a new active state configuration which would contain the state ShowQuestion, where the value of the variable l is greater than zero.

The behavior to execute at the point of time specified by the pointcut is defined in the advice. A piece of advice is a UML state machine enhanced with some pre-defined final states. The variables of the base state machine may be used, the aspect may also define local variables, in particular *trace variable*s for counting the occurrences of a certain sequence of active state configurations in the execution history. The local variables may also be used in the advice.

In Fig. 5(a), two variables are introduced: a normal variable level which stores the current value of the variable l each time the advice is executed, and a trace variable a. The value of a is the number of occurrences (indicated by #) of the sequence specified by the pattern on the right hand side: it is a sequence that contains at least minRight ([minRight..*]) active state configurations in which Right is active while the variable l has the value of level − 1, and that does not contain any active state configuration ([0]) where Wrong is active while l has the value of level − 1. The terms l == level−1 are added as constraints to the respective states. The advice in Fig. 5(a) checks whether such a sequence can be found in the trace ([a > 0]) or not ([else]), and resumes the execution of the base state machine (final state named 1) or prevents the base state machine from firing the specified transition by making it stutter at the current state configuration (final state named 0). As a whole, this aspect implements additively to the base state machine that changing to level $l > 0$ is only allowed when the user has already answered $l − 1$ questions correctly in a row. Note that no more invasive modification to the base state machine is necessary and that the model elements used are now gathered at a dedicated location instead of scattered all over the state machine.

Figure 5(b) shows another aspect, which is applied to the state machine given in Fig. 1 and implements the desired mutual exclusion. Its configuration pointcut contains only one superstate and specifies the point of time just before («before») any state configuration containing A2 and B2 gets active (this configuration should be avoided according to the mutual exclusion requirement). The advice does not need any variable, but simply makes the base state machine stutter (final named waiting). The difference between 0 and waiting is that returning to the base state machine after 0 the event that kicked off the aspect is consumed and the base state machine therefore needs explicitly another instance of this event if it should try again, while after waiting the base state machine does not need such an event but will be waiting to complete the interrupted transitions as soon as possible, e.g., as soon as the next active state configuration would not be caught by the pointcut again. Figure 5(b) thus models the logic of mutual exclusion highly modularly and non-invasively, switching on and off this feature is now a very easy task.

4.2   Resuming from Advice

Note that waiting and 0 are not the only "stuttering" final states that may be used in the advice. When the base state machine is told to stutter, it needs not only information on whether it should try to resume the interrupted transition "automatically" or only upon an explicit event, but also whether it should react to other events or not. For example, suppose the base state machine shown in Fig. 1 tries to enter the state A2 from A3, but has to stay in A3 since B2 is active, then what should it do when it now receives an a31 event? Should it proceed to A1 or not?

Therefore, we could distinguish four stuttering strategies:

1. The base state machine tries automatically to resume the interrupted transition without any explicit event and reacts to other events; this is our case waiting.

2. The base state machine needs an explicit event to make another try of the interrupted transition and reacts to other events; this is our case 0.
3. The base state machine tries automatically to resume the interrupted transition without any explicit event and does not react to other events. We say the base state machine is in this case "pinned" to the interrupted transition and model this case with
   a final state with the name pinned.
4. The base state machine needs an explicit event to make another try of the interrupted transition but does not react to other events.

We currently have no examples that require case 4 and therefore do not include it
in our aspect language. However, both assigning this case a name and extending our
implementation (see Sect. 5) are straightforward so that a simple extension would make
our language to cover this case as well.

We call final states with label 0 or 1 *progressing* and final states with labels waiting
or pinned *inhibiting*.

5   Translation of Aspects

The weaving process transforms a state machine and an aspect into a new state machine.
As combining several aspects presents a number of additional challenges we concentrate on weaving a single aspect into a state machine. Weaving proceeds in the following
stages which we describe in more detail in the rest of the section.

1. Normalize the state machine to eliminate syntactic variation.
2. Identify the states and transitions which have to be modified.
3. Construct the finite automata that track the relevant history.
4. Insert variables and actions.
5. Insert advice.

Normalization. UML state machines allow a number of notational variations. To simplify the translation process we transform the state machine into an equivalent canonical state machine which is well-suited for the next stages of the translation process. We
require that the normalization process transform the state machine into a form where
all transitions that can lead to a state configuration in which the pointcut applies are
explicit. Hence the normalization process ensures that all states and transitions which
may potentially be modified by the introduction of the aspect can be determined from
the advice or a variable declaration.

Identification of Relevant States and Transitions. There are two different kinds of
relevant state machine elements: Some elements are necessary for inserting advice. We
call these elements *advice-relevant* and write $\mathsf{arel}(M, \mathcal{A})$ for the set of advice-relevant
elements of state machine M and aspect $\mathcal{A}$. Other elements are relevant for keeping
track of the history of active state configurations; these are called *history-relevant*. The
set of all history-relevant elements is written as $\mathsf{hrel}(M, \mathcal{A})$. The set of elements which
are history-relevant for a single trace variable a is written $\mathsf{hrel}(M, a)$. It is, of course,
possible for an element to be both advice-relevant and history-relevant.

The advice-relevant elements can be found from the pointcut specification:

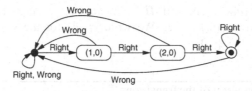

Fig. 6. Finite automaton for trace variable a

- If the stereotyped element of the pointcut is a state configuration and the stereotype is «before», then all simple states in this state configuration and all transitions with such a state as target are advice-relevant. If the stereotype is «after», then all simple states in the state configuration and all transitions with one of these states as source are advice-relevant.
- If the stereotyped element of the pointcut is a transition, then all the transitions in the state machine which go directly from the source state of the stereotyped transition to its target state are advice-relevant; furthermore all simple states contained in the source state of the stereotyped transition are advice-relevant.

A state is history relevant if and only if it appears on the right hand side of a variable declaration. Transitions are never history-relevant.

Construction of History Automata. Aspect-oriented state machines can take decisions based on their history of active state configurations. It would obviously be prohibitively expensive to store and analyze the complete history of the state machine in order to decide which transition to take. Therefore our aspect language is designed such that the whole relevant history for unbounded executions can be stored in a finite amount of memory unless the context variables referenced by trace variables in some aspect take an infinite number of values.

This is possible because a trace pattern for a trace variable a defines a regular language over the alphabet $\mathrm{hrel}(M, a)$. For example, the definition of the trace variable a with minRight $= 3$ in Fig. 5(a) is translated into the non-deterministic finite automaton shown in Fig. 6. The initial state of this automaton accepts the whole alphabet $\mathrm{hrel}(M, a)$; this represents the fact that we are looking for occurrences of the pattern anywhere in the history. The automaton for recognizing a sequence of trace patterns can be obtained by connecting the automata for the individual trace patterns with ϵ-transitions.

The rest of this section is rather technical as it describes the construction of the non-deterministic finite automaton (NFA) for integer trace variables of the form $a = \#(s)$ for some superstate s.

We assume that the states in $\mathrm{hrel}(M, a)$ are numbered from 0 to n and write s_i for the state with number i. If the constraint on a state s is of the form $n_1..n_2$ we call $\min(s) := n_1$ the *minimal* and $\max(s) := n_2$ the *maximal number of occurrences* of that state. If the constraint on s is of the form $n..*$ we define $\min(s) := \max(s) := n$. in this case we define infinite?$(s) = $ true, otherwise infinite?$(s) = $ false.

To distinguish states of the finite automaton from states of the state machine we refer to the former as *hstates*. We define a hstate for each tuple $(a_0, \ldots, a_n)$ with $0 \leq a_i \leq \max(s_i)$ and write $hstate(a_o, \ldots, a_n)$ for this hstate. Furthermore we write $tuple(H) = (a_0, \ldots, a_n)$ iff $H = hstate(a_0, \ldots, a_n)$. For all other tuples we define $hstate(a_0, \ldots, a_n) = \bot$. We write $\mathcal{H}(\mathcal{A})$ for the set of all hstates, i.e., $\mathcal{H}(\mathcal{A}) := hstate[\mathbb{N}^{n+1}] \setminus \{\bot\}$.

Algorithm 1. Computation of the transitions $\mathcal{T}$

1: $H \leftarrow \mathcal{H}(\mathcal{A})$
2: $\mathcal{T} \leftarrow \emptyset$
3: $h_0 \leftarrow hstate(0, \ldots, 0)$
4: **for all** i, $\quad 0 \le i \le n$ **do**
5: $\quad \mathcal{T} \leftarrow \mathcal{T} \cup \{h_0 \xrightarrow{s_i} h_0\}$
6: **end for**
7: **for all** $h \in H$ **do**
8: $\quad (a_0, \ldots, a_n) \leftarrow tuple(h)$
9: $\quad$ **for all** $k \leftarrow (a_0, \ldots, a_i + 1, \ldots, a_n)$, $\quad 0 \le i \le n$ **do**
10: $\quad\quad$ **if** $hstate(k) \ne \bot$ **then**
11: $\quad\quad\quad \mathcal{T} \leftarrow \mathcal{T} \cup \{h \xrightarrow{s_i} hstate(k)\}$
12: $\quad\quad$ **else if** infinite?(s_i) **then**
13: $\quad\quad\quad \mathcal{T} \leftarrow \mathcal{T} \cup \{h \xrightarrow{s_i} h\}$
14: $\quad\quad$ **else**
15: $\quad\quad\quad \mathcal{T} \leftarrow \mathcal{T} \cup \{h \xrightarrow{s_i} h_0\}$
16: $\quad\quad$ **end if**
17: $\quad$ **end for**
18: **end for**

The set of transitions $\mathcal{T}$ is generated as described in Algorithm 1. This algorithm inserts a transition with label s_i between hstates of the form $hstate(a_0, \ldots, a_i, \ldots, a_n)$ and $hstate(a_0, \ldots, a_i + 1, \ldots, a_n)$; it inserts a self-transition with label s_i for hstates of the form $hstate(a_0, \ldots, a_n)$ if $a_i = \max(s_i)$ and s_i allows infinitely many repetitions, otherwise it inserts a transition into the initial hstate h_0. By inserting (non-deterministic) transitions $h_0 \xrightarrow{s} h_0$ for all states $s \in \mathrm{hrel}(M, a)$ we ensure that the automaton keeps track of all sequences of states in its history.

A hstate $hstate(a_0, \ldots, a_n)$ is accepting if $\min(s_i) \le a_i \le \max(s_i)$ for all i with $0 \le i \le n$. To achieve the desired semantics for the aspect-oriented state machine we execute a transition of the automaton for each run to completion step of the state machine. By construction the finite automaton enters an accepting state from a non-accepting state for the n-th time precisely when the state machine is in the corresponding superstate for the n-th time. The construction of the automaton described above takes into account overlapping patterns, a slight modification of the construction of the NFA and the functions described in the next section would disallow them.

Depending on the space/time trade-off for the state machine we can either convert the NFA into a deterministic finite automaton (DFA) or simulate the execution of the DFA with the NFA. For the sake of concreteness we assume that convert the NFA into a DFA which we call the *history automaton for a*, $\mathfrak{H}(a)$, with hstates $states(\mathfrak{H}(a))$, transitions $transitions(\mathfrak{H}(a))$, initial hstate h_0^a (corresponding to the translation of $hstate(0, \ldots, 0)$), and accepting hstates $accepting(\mathfrak{H}(a))$. If h is a hstate of $\mathfrak{H}(a)$ we write $\mathfrak{H}(a)(h,s)$ to denote the result of executing the transition labeled s starting from hstate h of the automaton $\mathfrak{H}(a)$.

Insertion of Variables and Actions. Having defined the history automata for all trace variables we are now in a position to introduce the necessary modifications for deciding the applicability of aspects into the context of the base state machine. Let a be a trace variable. We write vars(a) for the set of context variables in the trace patterns of a. For example, in Fig. 5(a), vars(a) $= \{l\}$. We need to keep track of a separate history for each value of each variable in vars(a). This is also the case when manually modifying the state machine, as can be seen in Fig. 3 where an array $r[l]$ is used to store the number of correct answers for each level.

For each trace variable we therefore introduce the following elements into the context of the base state machine:

- *history*: This is a finite function with domain vars(a) $\times E(a)$, where $E(a)$ is the set of environments for vars(a), i.e., the set of all assignments of a value to each variable in vars(a). The value for each key is a pair $\langle \mathfrak{H}(a), h \rangle$, where h is the current state of the state machine $\mathfrak{H}(a)$ (for the values in e). The initial value for *history* is the function mapping each value $\langle a, e \rangle$ in its domain to $\langle \mathfrak{H}(a), h_0^a \rangle$.
- *val*: This is a finite function with domain vars(a) $\times E(a)$ as described above and range $\mathbb{N}$, evaluating the current value of a in environment e. The default value is a function mapping each value in its to domain to 0
- *updateHistory*: This is a function with arguments $\langle a, s, e \rangle$, where a and e are as described above and s is a state of the base state machine. This function updates the result $\langle \mathfrak{H}(a), h \rangle$ of *history*(a, e) by executing transition s of $\mathfrak{H}(a)$ in hstate h, i.e., *history*(a, e) is updated to $\langle \mathfrak{H}(a), \mathfrak{H}(a)(h, s) \rangle$. If h is not accepting and the result of this transition is an accepting hstate of $\mathfrak{H}(a)$, i.e., if $h \notin accepting(\mathfrak{H}(a))$ and $\mathfrak{H}(a)(h, s) \in accepting(\mathfrak{H}(a))$, then *val*($a, e$) is incremented by one.

To keep track of the state machine's history, it is now sufficient to add an entry action *updateHistory*(a, s, e) (where e is the local environment for vars(a)) to every state s in hrel(M, a) for each trace variable a. To evaluate an expression X in which a occurs we replace a by *val*(a, e) (with the current environment e for vars(a)) in X, i.e., we evaluate $X[a/\ val(a, e)]$.

Inserting Advice. The final step in the weaving process is the insertion of the advice itself. We write init($\mathcal{A}$) for the initial transition of the advice, final($\mathcal{A}, l$) for the set of final transitions of the advice with label l, source(t) for the source state of a transition t, target(t) for the target state of a transition, and $guard(t)$ for the guard of a transition. We write source(t) $\leftarrow s$ to denote the operation of replacing the source state of transition t with state s, similar for the target node. The operation copy($\mathcal{A}$) copies the advice of an aspect; all states and transitions of the copy are disjoint from the original. We write inhibited(s) for the set of all compound transitions leading from s to an inhibited final state.

To simplify the translation algorithm we place the following restrictions on advice: the advice itself may only contain a single region; for each compound transitions all final states must either be progressing or inhibiting. The first restriction ensures that

Algorithm 2. Inserting Advice

1: $\mathcal{T}_{old} \leftarrow transitions(\mathrm{arel}(M, \mathcal{A}))$
2: $transitions(M) \leftarrow transitions(M) \setminus \mathcal{T}_{old}$
3: **for all** $t \in \mathcal{T}_{old}$ **do**
4:     $A \leftarrow copyAdvice(\mathcal{A})$
5:     $\mathsf{source}(\mathrm{init}(A)) \leftarrow \mathsf{source}(t); guard(\mathrm{init}(A)) \leftarrow guard(t)$
6:     ADD BEHAVIOR FOR INHIBITED FINAL STATES(A)
7:     UPDATE TRANSITION TARGETS(A, t)
8: **end for**
9: $transitions(M) \leftarrow transitions(M) \cup transitions(A)$
10: $states(M) \leftarrow states(M) \cup states(A)$

Algorithm 3. Adding behavior for inhibited final states

1: **function** ADD BEHAVIOR FOR INHIBITED FINAL STATES(A)
2:     **for all** $s \in states(A)$ **do**
3:         $ew \leftarrow \emptyset; ep \leftarrow \emptyset$
4:         $\forall e \in events(A). (g_e^w \leftarrow \mathsf{false}; g_e^p \leftarrow \mathsf{false})$
5:         **for all** $t' \in inhibited(s)$ **do**
6:             $g \leftarrow \bigwedge \{\mathsf{inState}(s') \wedge \mathsf{constraints}(s') \mid s' \in states(\mathrm{arel}(M, A)) \setminus target(t')\}$
7:             **if** $t' \in final(A, \mathsf{waiting})$ **then**
8:                 $g_e^w \leftarrow g_e^w \vee (guard(t') \wedge g); \quad ew \leftarrow ew \cup \{\mathsf{event}(t')\}$
9:             **else**
10:                 $g_e^p \leftarrow g_e^p \vee (guard(t') \wedge g); \quad ep \leftarrow ep \cup \{\mathsf{event}(t')\}$
11:             **end if**
12:             $guard(t') \leftarrow guard(t') \wedge \neg g$
13:         **end for**
14:         **for all** $e \in ew$ **do**
15:             Add behavior $e[g_e^w]$/defer to s
16:         **end for**
17:         **for all** $e \in ep$ **do**
18:             Add entry behavior $v_e^s = \mathsf{false}$ to s
19:             Add the following behavior to s:
20:             $e[g_e^p]/v_e^s = \mathsf{true}; \quad e[g_e^p \wedge v_e^s]$/defer
21:             $\forall e' \in events(A) \setminus \{e\}. e'[g_e^p \wedge v_e^s]/$
22:         **end for**
23:     **end for**
24: **end function**

Algorithm 4. Updating transition targets

1: **function** UPDATE TRANSITION TARGETS(A, t)
2:     **for all** $t' \in final(A, 1) \cup final(A, \mathsf{waiting}) \cup final(A, \mathsf{pinned})$ **do**
3:         $target(t') \leftarrow target(t)$
4:     **end for**
5:     **for all** $t' \in final(A, 0)$ **do**
6:         $target(t') \leftarrow \mathsf{source}(t)$
7:     **end for**
8: **end function**

compound transitions cannot have multiple final states, the second restriction simplifies the introduction of guards in the weaving process.

Algorithms 2, 3 and 4 describe the process of inserting advice into the state machine for configuration pointcuts: For each advice-relevant transition t we remove t from the base state machine, attach the advice to the source state of t and connect the final states of the advice with appropriate states of the state machine. We also modify the advice relevant states of the resulting state machine to defer events for advice ending in a final state with label waiting or pinned and disable the outgoing transitions once the advice arrives in a pinned state. The algorithm for transition pointcuts is similar.

As an example of the weaving process consider Fig. 4 which is the result of weaving the mutual exclusion advice in Fig. 5(b) into the base state machine in Fig. 1. The result of manually extending the e-learning example in Fig. 3, however, differs from the result of the automatic weaving process, as this process uses a history automaton for the trace variables instead of explicitly manipulating counters.

6 Related Work

Our idea of dynamic aspects of state machines has been inspired by dynamic aspect-oriented programming languages such as JAsCo [20] and Object Teams [9]. Such languages are recognized as useful for separation of concerns (cf. [7]); a recent example of using control flow based aspects to build powerful debuggers is given in [4]. In particular, using trace variables to quantify over the trace is reminiscent to the trace aspects of Arachne [6]. Aspect-oriented modeling, aiming at a better separation of concerns on the level of software design, is still in its infancy. Most existing work (e.g. [1,17,19]) focuses on modeling aspect-oriented programs rather than making modeling languages aspect-oriented.

In the realm of modeling languages and state machines in particular, Altisen et al. [2] propose aspects for Mealy automata. Pointcuts are also defined as automata. In comparison, the weaving algorithm of our approach is due to the richer language constructs of the UML much more elaborate. Thanks to the wider acceptance of the UML, our approach is more tightly connected to common practice. Mahoney et al. [13] propose to combine several state machines into one orthogonal composite state and to relate by textual notations triggering events in different regions so that related transitions can be fired jointly. This approach can be used to modularize the synchronization of state machines, although having to declare all events of the wrapping state machine to be executed before triggering transitions in the base state machine may lead to quite complicated annotations. JPDD [8] is a pointcut language that facilitates the definition of trace-based pointcuts. In comparison, our approach also allows the modeler to define state machine synchronization modularly. Moreover, we have also defined a translation semantics for our aspects including both pointcuts and advice.

Theme/UML [5] models different features in different models (called themes) and uses UML templates to define common behavior of several themes. It does not contain a pointcut language, model elements have to be bound to formal parameters explictly by textual notations (A first step towards using JPDD as the pointcut language is presented in [10], although there are still compatibility problems between these two approaches).

The definition of history-based and modular modeling of state machine synchronization does not seem possible.

7 Conclusions and Future Work

We have defined a syntax and a translation semantics of aspect-oriented state machines. Using aspects may improve the modularity of UML state machines considerably by separating state interaction and trace-based behavior from other concerns. Our weaving algorithm works with a bounded amount of memory in realistic cases, i.e., as long as the variables used in the constraints of superstates do not take infinite many values.

Both the pointcut language and the advice language may be extended. In particular, it is expected to be straightforward yet useful to allow the pointcut to contain more than two superstates and/or to quantify over variable traces (and thus allow data-oriented aspects as proposed in [18]). It would also be interesting to allow the advice to make the base state machine not only stutter, but jump into the past. This might be useful for modeling compensations in long-running transactions in service-oriented systems. Moreover, in order to allow the application of several aspects to a state machine, a notation should be designed for defining the order of weaving in case of conflicts; the implementation described in Sect. 5 should be extended as well.

Another important issue of future work is model validation. We plan first to extend an existing UML model checker, such as Hugo/RT [12], to validate the weaving product. In a second step, it would be interesting to investigate techniques of compositional validation, in order to allow validation of larger models. Finally, extending aspect-orientation to other UML diagrams and generating aspect-oriented programs from aspect-oriented models are also part of our future research.

References

1. Aldawud, O., Elrad, T., Bader, A.: UML Profile for Aspect-Oriented Software Development. In: AOM. Proc. 3rd Int. Wsh. Aspect-Oriented Modeling, Boston (2003)
2. Altisen, K., Maraninchi, F., Stauch, D.: Aspect-Oriented Programming for Reactive Systems: Larissa, a Proposal in the Synchronous Framework. Sci. Comp. Prog. 63(3), 297–320 (2006)
3. Barry, B., de Moor, O. (eds.): AOSD'07. Proc. 6th Int. Conf. Aspect-Oriented Software Development, ACM Press, New York (2007)
4. Chern, R., De Volder, K.: Debugging with Control-Flow Breakpoints. In: Barry, de Moor [3], pp. 96–106 (2007)
5. Clarke, S., Baniassad, E.: Aspect-Oriented Analysis and Design. Addison-Wesley, Reading (2005)
6. Douence, R., Fritz, T., Loriant, N., Menaud, J.-M., Ségura-Devillechaise, M., Südholt, M.: An Expressive Aspect Language for System Applications with Arachne. In: Mezini, Tarr [14], pp. 27–38
7. Filman, R.E., Haupt, M., Hirschfeld, R. (eds.): Proc. 2nd Dynamic Aspects Wsh. (DAW'05). Technical Report 05.01. Research Institute for Advanced Computer Science (2005)
8. Hanenberg, S., Stein, D., Unland, R.: From Aspect-Oriented Design to Aspect-Oriented Programs: Tool-Supported Translation of JPDDs into Code. In: Barry, de Moor [3], pp. 49–62

9. Herrmann, S.: Object Teams: Improving Modularity for Crosscutting Collaborations. In: Aksit, M., Mezini, M., Unland, R. (eds.) NODe 2002. LNCS, vol. 2591, pp. 248–264. Springer, Heidelberg (2003)

10. Jackson, A., Clarke, S.: Towards the Integration of Theme/UML and JPDDs. In: Proc. 8th Wsh. Aspect-Oriented Modeling, Bonn (2006)

11. Kiczales, G., Lamping, J., Menhdhekar, A., Maeda, C., Lopes, C., Loingtier, J.-M., Irwin, J.: Aspect-Oriented Programming. In: Aksit, M., Matsuoka, S. (eds.) ECOOP 1997. LNCS, vol. 1241, pp. 220–242. Springer, Heidelberg (1997)

12. Knapp, A., Merz, S., Rauh, C.: Model Checking Timed UML State Machines and Collaborations. In: Damm, W., Olderog, E.R. (eds.) FTRTFT 2002. LNCS, vol. 2469, pp. 395–416. Springer, Heidelberg (2002)

13. Mahoney, M., Bader, A., Elrad, T., Aldawud, O.: Using Aspects to Abstract and Modularize Statecharts. In: Proc. 5th Wsh. Aspect-Oriented Modeling, Lisboa (2004)

14. Mezini, M., Tarr, P.L. (eds.): AOSD'05. Proc. 4th Int. Conf. Aspect-Oriented Software Development, ACM Press, New York (2005)

15. Object Management Group. Object Constraint Language, version 2.0. Specification, OMG (2006), http://www.omg.org/cgi-bin/doc?formal/06-05-01.pdf

16. Object Management Group. Unified Modeling Language: Superstructure, version 2.1.1. Specification, OMG (2007), http://www.omg.org/cgi-bin/doc?formal/07-02-05.pdf

17. Pawlak, R., Duchien, L., Florin, G., Legond-Aubry, F., Seinturier, L., Martelli, L.: A UML Notation for Aspect-Oriented Software Design. In: AOM'02. 1st Int. Wsh Aspect-Oriented Modeling, Enschede (2002)

18. Rashid, A., Chitchyan, R.: Data-Oriented Aspects. In: Hannemann, J., Baniassad, E., Chen, K., Chiba, S., Masuhara, H., Ren, S., Zhao, J. (eds.) AOASIA'06. Proc. 2nd Asian Wsh. Aspect-Oriented Software Development, pp. 24–29. National Institute of Informatics, Tokyo (2006)

19. Stein, D., Hanenberg, S., Unland, R.: A UML-based Aspect-Oriented Design Notation for AspectJ. In: AOSD'02. Proc. 1st Int. Conf. Aspect-Oriented Software Development, pp. 106–112. ACM Press, New York (2002)

20. Vanderperren, W., Suvée, D., Verheecke, B., Cibrán, M.A., Jonckers, V.: Adaptive Programming in JAsCo. In: Mezini, Tarr [14], pp. 75–86

Complementary Use Case Scenario Representations Based on Domain Vocabularies

Michał Śmiałek, Jacek Bojarski, Wiktor Nowakowski,
Albert Ambroziewicz, and Tomasz Straszak

Warsaw University of Technology,
Warsaw, Poland
{smialek,bojarsj1,nowakoww,ambrozia,straszat}@iem.pw.edu.pl

Abstract. Use cases are commonly used as notation for capturing functional requirements through scenarios. The problem is that there is no universal notation for use case contents which is capable of accommodating all the needs of software project participants. Business analysts and stakeholders need understandability and informality, while for architects and designers, precision and unambiguity are the most crucial features. In this paper we propose a metamodel and concrete syntax for three complementary representations of use case scenarios. These representations present the same information, but put emphasis on different aspects of it thus accommodating for different readers. This metamodel utilises the idea of separation of requirements as such from their representations as well as the idea of clear distinction between description of the system's behaviour and of the problem domain.

Keywords: use cases, requirements, scenarios, activity diagrams, interaction diagrams.

1 Introduction

In a typical software development project, several roles with sparse background have to maintain and read the requirements specification. For business people, requirements need to be understandable. Designers necessitate precision and unambiguity. Unfortunately, most notations for requirements do not offer both of these characteristics. Notations that are capable of being handled by machines for transforming into design are usually hard to read by "ordinary people". Natural language notations are human readable, but leave too much space for interpretation and lack rigour needed by technical people. The ideal notation should allow for getting as diverse group of people as possible better involved in the process of eliciting, modelling, communicating and agreeing requirements, thus improving the quality of the resulting system. This issue has been widely discussed in the literature (see [1,2,10] for example).

It seems that an ideal notation should be a model – one that is understandable for the users (diagrammatic or in simple structured language) and precise enough to be handled by developers or even machines.

G. Engels et al. (Eds.): MoDELS 2007, LNCS 4735, pp. 544–558, 2007.

A commonly used notation for requirements in the modelling world are use cases. Unfortunately, there are numerous problems with use case notation – mainly with their representations. Use cases and relationships between them, as defined in UML [11] have quite vague semantics (see [4] for a discussion). This results in multitude of notations for their contents (see [6] for a survey) and is source for confusion and misunderstandings (as explained in [15]). Lack of clear separation of use case scenarios from problem domain description causes inconsistencies in requirements specifications (see [13,16]).

Thus, in this paper we propose a notation that would unify sparse approaches to use case representation and allow for comprehension by different participants in a software project. This notation consists of three separate but complementary scenario-based representations of use cases. At the basis of the notation lies the idea of separation of requirements as such from their representations as well as the idea of clear distinction between description of the behaviour and the domain vocabulary. In the following sections, concrete and abstract syntax as well as semantics of the notation are explained.

2   Use Case Scenarios Based on Domain Vocabulary

Before we present the various use case scenario representations, we shall start with providing a definition of a use case that seems most suitable for our purpose. From among tens of definitions which can be found in the literature, we shall use the one that is most representative and does not relate to any concrete notation. Such a definition of a use case was introduced by Cockburn in [3]. According to this definition a use case is:

> "A collection of possible scenarios between the system under discussion and external actors, characterized by the goal the primary actor has toward the system's declared responsibilities, showing how the primary actor's goal might be delivered or might fail."

Now, we need a definition for a scenario as a use case component. Again, referring to [3], a scenario is:

> "A sequence of interactions happening under certain conditions, to achieve the primary actor's goal, and having a particular result with respect to that goal. The interactions start from the triggering action and continue until the goal is delivered or abandoned, and the system completes whatever responsibilities it has with respect to the interaction."

Above definitions are the basis on which we have designed three complementary use case scenario representations suitable for people having various roles in a software project and thus looking at a use case from different points of view. While designing these representations we took into account two important issues. First, we had to resolve the problem of precise control flow semantics for use cases. It can be argued that the semantics of "include" and "extend" relationships in UML is disadvantageous ([14], [9]). Thus, we substituted UML relationships with "invoke" relationship. Its semantics and modified use case metamodel have been described in details in [16]. In the following section we summarise briefly this solution as it influences scenario representations introduced in this paper. The second important issue that applies to our approach is separation of use case scenario representations describing the system's behaviour from the description of problem domain. This allows to eliminate many inconsistencies from requirements specifications ([16,17]). In-depth research in above mentioned areas is being carried on as a part of the ReDSeeDS project (www.redseeds.eu). Below we present some results of this work concerning use case representations.

2.1 Redefined Use Case and Types of Scenario Sentences

As defined above, use case is a set of scenarios with the same goal. There must be at least one scenario that reaches the goal (basic scenario). There can be also any number of alternate scenarios which end either with success or failure in reaching the goal. Every scenario is a sequence of actions forming a dialogue between the primary actor and the system. Every such action can be expressed by a single sentence in a simple SVO grammar [1] (see [5] for an original idea). In addition to action sentences, we need to introduce two additional sentence types: condition and control sentences. They are used in a scenario to express the flow of control between alternative scenarios of the same use case as well as between scenarios of different use cases (see [8,18]).

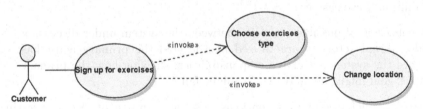

Fig. 1. Simple use case model with "invoke" relationships

Introducing control sentences needs prior redefinition of vague control flow semantics for use cases presented in UML. Figure 1 shows a simple use case model where one use case invokes two another use cases. In this model the "invoke" relationship denotes that another use case (actually, one of its scenarios) can be invoked from within currently performed use case. After performing one of the final actions in the invoked use case, the flow of control returns to the invoking use case right after the point of invocation to perform the remaining actions of the base use case scenario. There are two types of invocation: conditional and unconditional. A use case is invoked conditionally when explicitly requested

---

[1] A sentence in SVO grammar consists of a subject, a verb and an object. Optionally it can have a second indirect object.

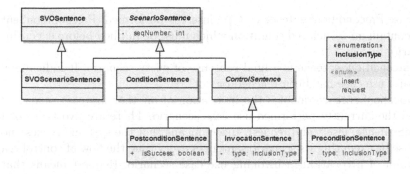

Fig. 2. Metamodel for scenario sentences

by an actor or under a certain condition on the system state. A use case is invoked unconditionally every time the scenario of the base use case containing appropriate invocation sentence is performed. The type of the invocation, the name of a use case to be invoked and the exact point of invocation is defined by a special kind of control sentence in the invoking use case scenario.

Now we will describe the semantics of specific scenario sentences that is common for all use case representations. Figure 2 shows a fragment of the metamodel that deals with scenario sentences. There are three main types of sentences that can be used in a scenario: SVOScenarioSentence, ConditionSentence and ControlSentence. All these sentences are subtypes of an abstract Scenario-Sentence which has a sequence number that defines the sentences position in the scenario. SVOScenarioSentence describes a single scenario step (an action) in the form of a sentence in the SVO grammar – it derives the whole syntax from SVOSentence described in details in the following section. A scenario step represented by this sentence can be performed by an actor or by the system.

As every action can be performed under certain condition, we need to introduce ConditionSentence which is a special kind of ScenarioSentence that controls the flow of scenario execution. ConditionSentence is a point of conditional control flow: the following scenario step can be executed only when the condition expressed by the sentence is met, otherwise a sentence from alternative scenario is executed.

Another type of a scenario sentence we need is a ControlSentence. It is a type of ScenarioSentence which controls the flow of scenario execution in situations when the control enters or leaves a use case. This abstract class is a superclass for concrete classes: PreconditionSentence, PostconditionSentence and InvocationSentence. PreconditionSentence is an initial sentence of every use case scenario indicating where the flow of control of every use case scenario starts. There are two types of PreconditionSentences: insert and request defined in InclusionType enumeration. PreconditionSentence of type request is always performed when the actor triggers a use case directly or requests invoking a use case (see InvocationSentence) from another use case scenario through initial actor action (first SVOScenarioSentence in the scenario). When a use case is invoked by inserting its scenario into the flow of invoking use case, the initial action is omitted. In this case PreconditionSentence of type insert is performed. PreconditionSentence may contain an associated condition which must be fulfilled before executing the scenario.

PostconditionSentence is a final sentence of every scenario. It indicates if the goal of a use case has been reached or not.

InvocationSentence denotes the invocation of another use case scenario from within the currently performed use case scenario. There are two types of InvocationSentence: insert and request. Insert means that the system invokes another use case by executing its scenario sentences every time the flow of control reaches the point of invocation in invoking use case scenario. Request means that the actor or the system explicitly requests invoking another use case – it depends either on the actor's decision or system state (a certain condition is met) whether scenario sentences of invoked use case will be executed or not. After performing all scenario steps of the invoked use case, the flow of execution returns to the invoking use case scenario to execute the remaining sentences. InvocationSentences are related to appropriate PreconditionSentences in invoked use case's scenario.

2.2 Sentences in SVO Grammar with Separated Domain Vocabulary

Sentences in the SVO grammar, as mentioned above, are used in scenarios to express actions performed by the actor or by the system. Experience shows that this simple grammar is sufficient to express interactions in a scenario, eg. "Customer submits sign-up for exercises" or "System signs up customer for exercises". This grammar combines informality with necessary precision (see [15]). The biggest strength of this grammar, is that SVO sentences only allow for describing the behaviour – no interleaving of domain element definitions are allowed. Such a separation makes requirements specifications unambiguous and consistent.

Considering the above, we need means for creating a separate specification of the domain and the way to link notions used in sentences with their definitions in the domain vocabulary. This issue has been resolved in the Requirements Specification Language (RSL) which has been developed recently as a part of the ReDSeeDS project (see [7]). Below we will explain how the separation is done by presenting a simplified metamodel.

A domain element from the domain vocabulary is usually a noun along with its definition in the context of the problem domain. A noun can be preceded by a modifier which can change the meaning of the noun, e.g. "registered user". In addition to nouns, the domain vocabulary can also contain verbs. Verbs do not have their own definitions - they are related to nouns as their meaning depends on the context of a noun. Verbs are treated as behavioural features of related nouns. For example, "choose exercise" has a different meaning than "choose time from time schedule".

In order to use such constructs in SVO sentences, we introduced the concept of phrases. Figure 3 shows an SVOSentence composed of one Subject and one Predicate. These two classes are kind of hyperlinks that can be embedded in SVOSentences linking it with phrases which are part of domain specification.

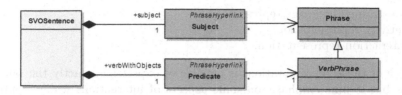

Fig. 3. The structure of SVOSentences

Phrases are building blocks of SVOSentence and depending on the type, they are composed of nouns (in the role of objects), modifiers, verbs and prepositions (see Figure 4 for phrases metamodel).

Every Phrase consists of at least one noun, optionally preceded by a modifier. Phrases occurs in SVOSentences in the role of 'subject's. In the role of a 'verbWithObjects' there can be used a ComplexVerbPhrase or a SimpleVerbPhrase. SimpleVerbPhrase extends Phrase by adding a verb which precedes the noun (with an optional modifier). This makes it possible to express constructs like "shows registered customers". ComplexVerbPhrase contains a preposition which links the SimpleVerbPhrase pointed by the phrase with a contained noun (again, with an optional modifier). This type of phrase can express constructs like "adds registered customers to customer list", where "registered customer" is a direct objects and "customer list" is an indirect object.

All phrases that refer to the same noun are grouped within a domain element, where the noun is the element's name. Every phrase grouped in a domain element has its own definition. Elements together with their relationships form the domain specification, which should be partially created during the problem domain analysis. While writing scenarios, the writer should be able to browse through the domain specification to search appropriate domain elements and their phrases and insert them directly into scenario sentences. The writer should also be able to add new elements to the domain specification as needed. Such an approach significantly improves the quality of the final requirements specification.

2.3   Introducing Complementary Scenarios Representations

Now, as we have precisely defined use case scenarios and all necessary types of scenario sentences, we can introduce complementary representations of use case scenarios:

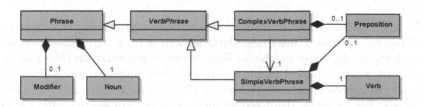

Fig. 4. Phrases metamodel

- constrained language representation,
- activity representation,
- interaction representation.

Each of these representations is capable of expressing exactly the same scenarios but it puts emphasis on some aspects of interactions while suppressing some others, thus making it usable for diverse groups of people having different roles in the software project. Due to the precise metamodel, every representation is directly transformable one into another.

The constrained language representation is a purely textual representation, where scenarios are written as sequences of numbered sentences in the SVO grammar, interlaced with condition and control sentences. A single scenario represents a single story without alternative paths. This representation is most readable for "ordinary people" like users or stakeholders who are usually reluctant to any technical notation. Some people, usually analysts, prefer precise structure for use case scenarios in a graphical form. Activity representation shows all scenarios of a single use case (main path and all alternative paths) as one activity diagram. This precisely reflects the flow of control in a use case as a single unit of functional requirement. Interaction representation presents a single scenario as a sequence of messages send between the system, the actors and other use cases in the form of a sequence diagram. This representation clearly reflects temporal interaction of the actors with the system as well as actions performed by the system in response to the actors' interaction. It seems to be most suitable for designers as it prepares them for transformation into design level interaction diagrams.

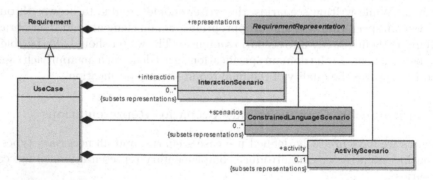

Fig. 5. Three complementary representations of use case scenarios

The metamodel in Figure 5 shows metaclasses representing the three introduced representations of use cases. UseCase derives from the Requirement metaclass, which can have any number of RequirementRepresentations (abstract metaclass). Such a separation of requirements and their representations gives us the possibility of representing requirements in different forms depending on the current needs. For example, draft requirements can be represented with a natural language description while detailed requirements can be represented in more

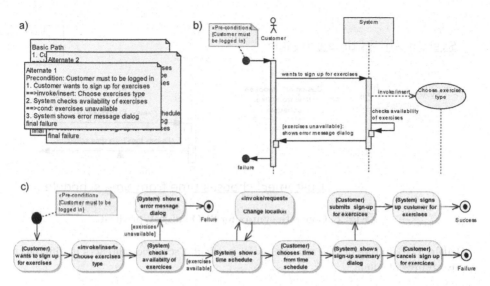

Fig. 6. Example of concrete syntax for all complementary representations of the same scenario: a) constrained language b) interaction c) activity

formal way, e.g. a constrained language description with relation to the domain vocabulary. Representations of a use case are concrete subclasses of Requirement-Representation and they subset representations pointed to from a Requirement.

Figure 6 shows examples of concrete syntax for these three representations. Figure 6a presents one of scenarios of a use case in constrained language. In Figure 6b, the interaction representation of the same scenario is shown. We can see that all sentences from the constrained language scenario have their equivalents in the interaction scenario in the form of messages between lifelines. The activity scenario representation in Figure 6c shows all scenarios of the use case in one diagram. Scenario sentences are presented as activity nodes or edges (in case of condition sentences). For comparison of concrete syntax for particular scenario sentence types for all three representations please refer to Table 1.

Figure 7 shows details of SVO sentence's subjects and predicates. In constrained language representation, both subject and predicate, while using an appropriate tool, could be shown as hyperlinks leading to their descriptions in the domain vocabulary. In activity representation predicate is delineated as an activity action's name, while subject is represented as an activity partition – subject name is placed in parenthesis above the action name. In interaction representation subject of a sentence is depicted as a lifeline (see Customer's and System's lifeline in the example). Predicate of every SVO sentence, in turn, is depicted as a message with its name above. It starts from the subject's lifeline and goes either to another subject's lifeline or comes back to the same lifeline it starts from.

Condition and control scenario sentences have also different concrete syntax in different representations (see Table 1). A condition sentence in textual representation is a text expressing the condition preceded with "⇒cond:" prefix. In activity and interaction representation it is shown as a text in square brackets

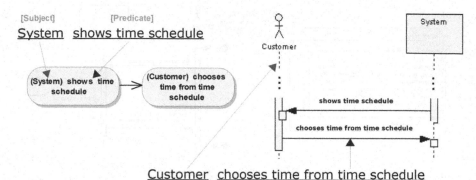

Fig. 7. Concrete syntax for SVO subject/predicate

Table 1. Examples of concrete syntax for particular scenario sentences in different use case representations

Sentence	Textual	Activity	Interaction
SVO	System shows time schedule	(System) shows time schedule	shows time schedule
condition	⇒cond: exercises available	[exercises available]	[exercises available]: shows time schedule
invoke/insert	⇒invoke/insert: Change location	«invoke/insert» Choose exercises type	invoke/insert Choose exercises type
invoke/request	⇒invoke/request: Change location	«invoke/request» Change location	invoke/request Change location
precondition	precondition: Customer must to be logged in	«Pre-condition» {Customer must to be logged in}	«Pre-condition» {Customer must to be logged in}
postcondition	final: success postcondition: Customer is registered for choosen exercises	«Post-condition» {Customer is registered for choosen exercises} Success	«Post-condition» {Customer is registered for choosen exercises} final: success

attached to a control flow edge in activity diagram or to a message in sequence diagram. Sentences of invoke type are presented in textual representation as invoking use case name preceded with a prefix denoting the type of invocation: "⇒invoke/request:" or "⇒invoke/insert:". In activity representation, this type of sentences is depicted as an action node with the name of invoking use case inside and an appropriate stereotype: "invoke/request" or "invoke/insert". In

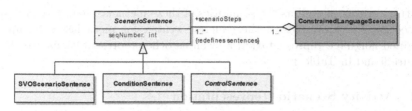

Fig. 8. Constrained language scenario representation

case of insertion, the incoming control edge comes from preceding action and the outgoing control edge goes to the action representing consecutive scenario sentence. In case of request, the outgoing control edge goes back to the action the incoming edge came from. In interaction representation, both invoke sentences are presented as "create" message from the actor's or system's lifeline to the invoking use case's lifeline. The type of invocation is shown as the message's text. Precondition and postcondition sentences are presented in textual representation as a text with "precondition:" or "postcondition:" prefix. In two another representations, sentences of these types are shown as constraints attached to start/final node in case of activity or lost/found message in case of interaction representation. In all representations, a postcondition sentence contains also "success" or "failure" text, indicating whether the goal of a use case has been reached or not.

3 Metamodel of Complementary Scenarios Representations

In the following sections we introduce the metamodel defining the abstract syntax for three complementary representations of a use case scenarios. This metamodel is expressed by the means of MOF ([12]). It refers to elements defined in UML 2.0 Superstructure ([11]), mainly from BasicActivities and BasicInteractions packages. Due to the scope of this paper, we present only the most essential constructs from our metamodel. Some high-level elements of this metamodel have already been introduced in previous sections. For more details, please refer to the language specification ([7]).

3.1 Constrained Language Scenario Representation

Constrained language scenario representation can be treated as a basis representation. Its syntax, both concrete and abstract, is the least complex in comparison with two remaining representations which extend the basic syntax mainly by specialising from UML elements.

As shown in Figure 8, ConstrainedLanguageScenario is composed of ScenarioSentences as its scenarioSteps. ScenarioSentence is an abstract metaclass that defines seqNumber which is an order number of a sentence in a scenario. This general metaclass is a base for subclasses representing scenario sentences of specific types like SVOScenarioSentence, ConditionSentence and ControlSentence. The abstract syntax as well as the semantics of these metaclasses has been described in section 2.1. Examples of the concrete syntax for metaclasses forming constrained language representation as well as other representations are shown in Figure 6 and in Table 1.

3.2 Activity Scenario Representation

Activity scenario representation utilises UML activity diagrams to present use case scenarios. In this way, it allows for showing all possible scenario paths of a single use case in one picture. In order to utilise activity diagrams for this purpose, elements of the metamodel for this representation specialise from appropriate UML elements defined in BasicActivities and IntermediateActivities packages (see [11]).

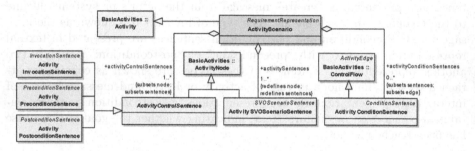

Fig. 9. Activity scenario representation

ActivityScenario extends Activity and it can contain three types of scenario sentences. Metaclasses representing these sentences combine syntax and semantics derived from sentences metaclasses from constrained language representation and from appropriate UML metaclasses (Figure 9).

An ActivitySVOScenarioSentence represents a single scenario action in the form of ActivityNode which it derives from. It means that these actions can have incoming and outgoing ControlFlows showing possible execution sequences. ActivitySVOScenarioSentence also derives from SVOScenarioSentence what means that the action it represents is described by SVO sentence (see Figure 10). ActivitySVOScenarioSentence redefines subject and verbWithObjects derived from SVOSentence with ActivitySubject and ActivityPredicate respectively. Redefined ActivitySubject derives from UML's ActivityPartition which is a kind of activity group for identifying actions that have some characteristic in common. In case of scenarios, it identifies whether an action is performed by an actor or by the system. In the diagram, subject is represented as a text in parentheses, placed inside action above its name (see Figure 7).

An ActivityControlSentence represents ControlSentence in ActivityScenario in the form of ActivityNode which is its superclass. It has three concrete subclasses: ActivityPreconditionSentence, ActivityPostconditionSentence and ActivityInvocationSentence. Each of these three subclasses corresponds to the appropriate ControlSentences subclass.

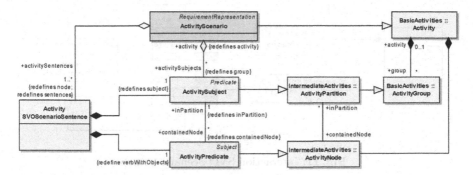

Fig. 10. Realisation of Subject and Predicate in SVO sentence in activity representation

An ActivityPreconditionSentence shows the starting point of a scenario on an activity diagram. ActivityPreconditionSentence can have the precondition of a scenario attached as a constraint. It has semantics similar to PreconditionSentence. Additionally it has the semantics of UMLs InitialNode.

The end point of a scenario on activity diagram is represented by an Activity-PostconditionSentence. It shows, if a scenario ends with success or failure. It can also have the postcondition of a scenario attached as a constraint. It has semantics similar to PostconditionSentence and additionally it derives the semantics from UML's FinalNode.

In activity representation, points where another use cases scenarios are invoked are shown as ActivityInvocationSentence. ActivityInvocationSentence is a subtype of ActivityControlSentence and InvocationSentence. In addition to that, it indirectly specialises from ActivityNode which makes it possible to be presented on activity diagrams.

3.3 Interaction Scenario Representation

Interaction representation of use case scenarios aims at emphasising the temporal sequence of interaction messages exchanged between an actor and the system as described in section 2.3. To achieve this, interaction representation utilises UML's sequence diagrams to represent scenarios. This means that elements of the metamodel for this representation specialise from appropriate UML elements defined in BasicInteractions packages (see [11]).

Figure 11 presents a fragment of the metamodel, handling InteractionSVO-ScenarioSentence. It derives from SVOScenarioSentence and is composed of SubjectLifeline and PredicateMessage. Both derive from Subject and Predicate respectively and redefine subject and verbWithObjects. Realisation of scenario sentences is based on UML's Interaction model. Interaction has a set of Messages connected to Lifelines by MessagesEnds (see Figure 12). Every message is depicted as an arrow pointing from one lifeline to another or to the same lifeline.

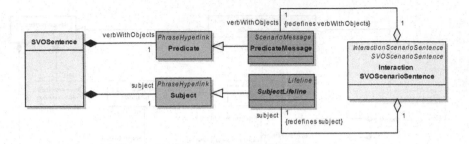

Fig. 11. Interaction SVO Scenario Sentence

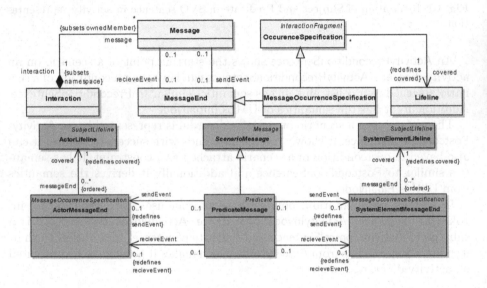

Fig. 12. Realisation of Predicate in interaction representation

Subject of InteractionSVOScenarioSentence acts as a lifeline in a sequence diagram. SubjectLifeline derives from UML's interaction lifeline. This metaclass is abstract and it is the base class for two concrete subclasses: ActorLifeline and SystemElementLifeline. The first metaclass represents subjects of sentences performed by use case's Actor. The latter, corresponds to subjects of sentences performed by SystemElement. Both are kind of DomainElement (see Figure 13 for more details).

Predicate of InteractionSVOScenarioSentence acts as a message. It indirectly derives from Interaction's Message. Predicate message can be connected to SystemElementLifeline or actor's lifeline (through respective MessagesEnds). As it is shown in Figure 12, Predicate has sendEvent and receiveEvent redefined with SystemElementMessageEnd and ActorMessageEnd. Those two message ends are covered by ActorLifeline and SystemLifeline.

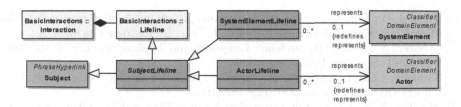

Fig. 13. Realisation of SVO Subject in interaction representation

4 Conclusions and Future Work

Presented notations for use case scenarios as part of the requirements specification language, has already been validated by students during software engineering courses at the Carlos III University of Madrid and Warsaw University of Technology. The results of case studies prepared by students showed that proposed approach makes capturing and specifying of system requirements easier. Students had no problems with understanding and discussing the resulting models. Created requirements specifications appear to be much more consistent and precise than specifications written in natural language. They also appear to be a good basis for further development process. The next step to prove usability of the language is validation in real-life projects by the industrial partners involved in the ReDSeeDS project. This validation is currently in progress.

Full utilisation of capabilities of proposed language calls for a tool support. A tool supporting a simplified concept behind the RSL was already developed (see [16]). Development of a tool covering the whole RSL language is in the scope of the ReDSeeDS project.

Acknowledgments. This work is partially funded by the EU: Requirements-driven Software Development System (ReDSeeDS) (contract no. IST-2006-33596 under 6FP). The project is coordinated by Infovide-Matrix, Poland with technical lead of Warsaw University of Technology and with University of Koblenz-Landau, Vienna University of Technology, Fraunhofer IESE, University of Latvia, HITeC e.V. c/o University of Hamburg, Heriot-Watt University, PRO DV, Cybersoft and Algoritmu Sistemos. Initial interaction representation metamodel was proposed by University of Koblenz-Landau team (J. Ebert, V. Riediger, D. Bildhauer, H. Schwarz).

References

1. Alexander, I.: A taxonomy of stakeholders, human roles in system development. International Journal of Technology and Human Interaction 1(1), 23–59 (2005)
2. Alexander, I., Maiden, N.: Scenarios, Stories, Use Cases Through the Systems Development Life-Cycle. John Wiley, New York, NY (2004)
3. Cockburn, A.: Structuring use cases with goals. Journal of Object-Oriented Programming 5(10), 56–62 (1997)

4. Genova, G., Llorens, J., Metz, P., Prieto-Diaz, R., Astudillo, H.: Open issues in industrial use case modeling. In: Nunes, N.J., Selic, B., Rodrigues da Silva, A., Toval Alvarez, A. (eds.) UML Modeling Languages and Applications. LNCS, vol. 3297, pp. 52–61. Springer, Heidelberg (2005)

5. Graham, I.M.: Task scripts, use cases and scenarios in object-oriented analysis. Object-Oriented Systems 3(3), 123–142 (1996)

6. Hurlbut, R.R: A survey of approaches for describing and formalizing use cases. Technical Report XPT-TR-97-03, Expertech Ltd. (1997)

7. Kaindl, H., Śmiałek, M., Svetinovic, D., Ambroziewicz, A., Bojarski, J., Nowakowski, W., Straszak, T., Schwarz, H., Bildhauer, D., Brogan, J.P., Mukasa, K.S., Wolter, K., Krebs, T.: Requirements specification language definition. Project Deliverable D2.4.1, ReDSeeDS Project (2007), www.redseeds.eu

8. Metz, P., O'Brien, J., Weber, W.: Specifying use case interaction: Types of alternative courses. Journal of Object Technology 2(2), 111–131 (2003)

9. Metz, P., O'Brien, J., Weber, W.: Against use case interleaving. In: Gogolla, M., Kobryn, C. (eds.) UML 2001. LNCS, vol. 2185, pp. 472–486. Springer, Heidelberg (2001)

10. Nuseibeh, B., Easterbrook, S.: Requirements engineering: a roadmap. In: ICSE - Future of SE Track, pp. 35–46 (2000)

11. Object Management Group. Unified Modeling Language: Superstructure, version 2.0, formal/05-07-04 (2005)

12. Object Management Group. Meta Object Facility Core Specification, version 2.0, formal/2006-01-01 (2006)

13. Som, S.S.: Beyond scenarios: Generating state models from use cases. In: Scenarios and state machines: models, algorithms and tools - ICSE 2002 Workshop, Orlando. Florida (2002)

14. Simons, A.J.H.: Use cases considered harmful. In: TOOLS Europe'99. Proceedings of the 29th Conference on Technology of Object-Oriented Languages and Systems, Nancy, France, pp. 194–203. IEEE Computer Society Press, Los Alamitos (1999)

15. Śmiałek, M.: Accommodating informality with necessary precision in use case scenarios. Journal of Object Technology 4(6), 59–67 (2005)

16. Śmiałek, M., Bojarski, J., Nowakowski, W., Straszak, T.: Scenario construction tool based on extended UML metamodel. In: Briand, L.C., Williams, C. (eds.) MoDELS 2005. LNCS, vol. 3713, pp. 414–429. Springer, Heidelberg (2005)

17. Śmiałek, M., Bojarski, J., Nowakowski, W., Straszak, T.: Writing coherent user stories with tool support. In: Baumeister, H., Marchesi, M., Holcombe, M. (eds.) XP 2005. LNCS, vol. 3556, pp. 247–250. Springer, Heidelberg (2005)

18. van den Berg, K.G, Simons, A.J.H.: Control flow semantics of use cases in UML. Information and Software Technology 41(10), 651–659 (1999)

Modeling Time(s)

Charles André, Frédéric Mallet, and Robert de Simone

I3S, Université de Nice-Sophia Antipolis, CNRS, F-06903 Sophia Antipolis
INRIA, F-06902 Sophia Antipolis
{candre,fmallet,rs}@sophia.inria.fr

Abstract. Time and timing features are an important aspect of modern electronic systems, often of embedded nature. We argue here that in early design phases, time is often of logical (rather than physical) nature, even possibly multiform. The compilation/synthesis of heterogeneous applications onto architecture platforms then largely amounts to adjusting the former logical time(s) demands onto the latter physical time abilities. Many distributed scheduling techniques pertain to this approach of "time refinement".

We provide extensive Time and Allocation metamodels that open the possibility to cast this approach in a Model-Driven Engineering light. We give a UML representation of these concepts through two subprofiles, parts of the foundations of the forthcoming OMG UML Profile for Modeling and Analysis of Real-Time and Embedded systems (MARTE). Time modeling also allows for a precise description of time-related entities and their associated timed properties.

Keywords: UML profile, real-time embedded.

1   Introduction

Modeling of Time should be a central concern in Model-Driven Engineering for Real-Time Embedded systems. Nevertheless, (too?) many modeling frameworks consider Time annotations as to be considered in timing/schedulability/performance, and accordingly build uninterpreted stereotypes and label locations with insightful names only for the future analysis tool (and no meaning at all for the time augmented profile). Given that the default operational semantics of the UML is inherently *untimed*, and rightfully so since there is no Time information in the ground metamodel, one can reach the situation where the same model can be understood differently depending on whether it is considered from the UML causality model or the intended timed analysis viewpoint. Our primary goal here is to lay the foundation for a Time model which could be *deeply* embedded in UML as a profile allowing the subsequent clean and precise definition of a *timed causality* model enforcing timed operational semantics of events and actions.

Following some works on dedicated *Models of Computation and Communication* (MoCCs) for real-time embedded systems [1,2,3], we view *Time* in a very

G. Engels et al. (Eds.): MoDELS 2007, LNCS 4735, pp. 559–573, 2007.
© Springer-Verlag Berlin Heidelberg 2007

broad sense. It can be *physical*, and considered as *continuous* or *discretized*, but it can also be *logical*, and related to user-defined clocks. For instance, durations could be counted in terms of numbers of execution steps, or clock cycles on a processor or even more abstract time bases, without a strong relation to the actual physical duration (which may not be known at design time, or fluctuate, or be a parameter that allows the same model to be instantiated under different contexts and speeds). With modern embedded designs where, for low-power reasons, the actual processor clock can be shut down and altered at times, such usage of logical time in the application design will certainly become customary. In our approach, time can even be *multiform*, allowing different time threads to progress in a non-uniform fashion.

This approach looks certainly non-standard, but is getting increasing interest from a number of directions. A mostly untimed concurrent application can be considered as comprising several unrelated (or loosely coupled) time threads (thereafter called "clocks", not to be confused with the physical measurement device which we will never consider). The process of allocating the various operations/functions/actions of such a concurrent model onto an embedded execution platform comprises aspects of spatial distribution and temporal scheduling. This is accomplished by resolving the mutual sets of timing constraints imposed by the designer of the time scales of the application, the target architecture, and the real-time requirements to be met. We call the approach one of *Time refinement*.

A number of existing transformation techniques can be cast in this framework. Nested loop scheduling and parallelization [4,5] in high-performance computing, software pipe-lining, SoC synthesis phases from so-called *transactional level (TLM)* down to cycle-accurate *RTL* level, to mention a few. In all cases, the purpose is to progressively refine the temporal structure, which starts with a number of degrees of freedom, to attain a fully scheduled and precisely cycle-allocated version, with predictable timing. In that sense our model allows, and it is in fact its primary aim, to describe formal clock relations in a simple mathematical way.

We provide a UML model for Time in its different guises, physical/logical, dense/discrete, single/multiple, and some useful basic operators and relations to combine timed events or clocks. From this set of primitives, we hope to build explicit representation of MoCCs, and to provide a Timed causality model to endow the timed models with a timed semantics, according to the one that would be considered by analysis tools. When the relation are simple enough (periodic or regular), the system of contraints imposed by these relations can be solved, and the schedule itself becomes an explicit modeling element, traceable to the designer. In other, more complex cases, the constraints embody a given scheduling policy, which can be analyzed with corresponding schedulability analysis techniques when applicable.

After describing some existing time and allocation models (Section 2), Section 3 introduces our contribution, the MARTE[1] subprofiles for time and allocation. Section 4 briefly illustrates their use.

---

[1] a preliminary version is available at www.promarte.org

2 Existing Time and Allocation Models

2.1 Time Modeling

This subsection focuses on time models and time-related concepts in use in the UML and some of its profiles.

UML. In UML [6] Time is seldom part of the behavioral modeling, which is essentially untimed (by default, events are handled in the same order as they arrive in event handlers). UML describes two kinds of behaviors [7]: the intra-object behavior—the behavior occurring within structural entities—and the inter-object behavior, which deals with how structural entities communicate with each other. The CommonBehaviors package defines the relationship between structure and behavior and the general properties of the behavior concept. A subpackage called SimpleTime adds metaclasses to represent *time* and *duration*, as well as actions to observe the passing of time. This is a very simple time model, not taking account of problems induced by distribution or by clock imperfections. In particular the UML *causality model*, which prescribes the dynamic evaluation mechanisms, does never refer to time (stamps). Instead, the UML specification document explicitly states that *"It is assumed that applications for which such characteristics are relevant will use a more sophisticated model of time provided by an appropriate profile"*. Our contribution can be seen as providing the means for building such sophisticated time models.

SPT. The UML Profile for Schedulability, Performance, and Time (SPT) [8] aimed at filling the lacks of UML 1.4 in some key areas that are of particular concern to real-time system designers and developers. SPT introduces a *quantifiable* notion of time and resources. It annotates model elements with quantitative information related to time, information used for timeliness, performance, and schedulability analyses.

SPT only considers *(chrono)metric* time, which makes implicit reference to physical time. It provides time-related concepts: concepts of instant and duration, concepts for modeling events in time and time-related stimuli. SPT also addresses modeling of timing mechanisms (clocks, timers), and timing services. SPT, which relies on UML 1.4, had to be aligned with UML 2.1. This is one of the objectives of the MARTE profile, presented in Section 3.

Non OMG Profiles. Several "unofficial" UML profiles are also considering time modeling. We mention a few, developed for different purposes, as work related to ours.

EAST-EEA is an ITEA project on Embedded Electronic Architecture [9]. It provides a development process and automotive-specific constructs for the design of embedded electronic applications. Temporal aspects in EAST are handled by requirement entities. The concepts of Triggers, Period, Events, End to End Delay, physical Unit, Timing restriction, can be applied to any behavioral EAST elements. It is compliant with UML2.0, the intent is to deliver a UML2 profile.

The UML profile Omega-RT [10] focuses on analysis and verification of time and scheduling related properties. It is a refinement of the SPT profile. The profile is based on a specific concept of event making it easy to express duration constraints between occurrences of events. The concept of *observer*, which is a stereotype of state machine, is a convenient way for expressing complex time constraints. It would have to be aligned with UML2.0.

Summary. The abovementioned profiles introduce relationships between Time and Events or Actions. They annotate the UML model with quantitative information about time. None consider logical and multiform time.

2.2   Allocation Models

These are concerned with the mapping of application elements onto architectural platform resources and services. The following frameworks are currently untimed. It is in fact our main goal that a Time Model can be used to select and optimize such mapping according to the timing demands of both sides (and possibly additional real-time requirements).

UML Deployments. UML deployments consist in assigning concrete software elements of the physical world (artifacts) to nodes. Nodes can represent either hardware devices or software execution environments. Artifacts are physical piece of information—a file or a database entry—and model elements are stored in resources specified by artifacts. The MARTE allocation mechanism is complementary to the UML deployment mechanism, the differences are described in section 3.2.

SysML Allocation. SysML[11] provides a mechanism to represent, at an abstract level, cross-associations among model elements with the broadest meaning. A SysML allocation is expected to be the precursor of more concrete relationships. It differentiates three of the many possible and not exclusive categories: behavior, flow and structure allocations. Behavior allocations separate the functions from the structure; they provide a way to allocate a behavior to a behavioral feature. Flow allocations have many usages; they include allocations of activity transitions (SysML flows) to connectors of structured activities (SysML blocks). Structure allocations acknowledge the needs for a mapping relation of logical parts to more physical ones. The MARTE allocation is inspired from the SysML allocation and the differences are described in section 3.2. One reason for this choice is that we want to be able to define, in the most convenient way, how various durations and clock streams are connected in the course of the allocation. This can easily fit some of SysML *constraints/parametrics* and *requirements* modeling features, which were originally used to model physical constraints or uninterpreted requirement engineering information respectively.

2.3   Timed Allocation Models

We believe that suitable models for real-time and embedded systems design and analysis should support both time and allocation. We give here a brief insight of the Society of Automotive Engineer(SAE)'s Architecture Analysis & Design Language(AADL) standard [12].

The temporal semantics of AADL concepts is defined using "hybrid automata". These automata are hierarchical finite state machines with real-valued variables that denote the time. Temporal constraints, expressed as state invariants and guards over transitions, define when the discrete transitions occur. Concurrent executions are modeled using threads managed by a scheduler. The dispatch protocol (periodic, aperiodic, sporadic and background) determines when an active thread executes its computation. AADL supports multiform time models. However, it lacks model elements to describe the application itself, independently of the resources. UML activities allow for a description of the application, actions executed sequentially or concurrently, without knowing, at first, whether actions are executed by a periodic thread or a subprogram. This important information is brought by an orthogonal process, the allocation. After several iterations, analysis, the threads are eventually deployed (or bound) to the execution platform.

AADL offers a binding mechanism to assign software components (data, thread, process, etc.) to execution platform components (memory, processor, buses, etc.). Each software component can define several possible bindings and properties may have different values depending on the actual binding. This binding mechanism encompasses both the UML deployment and the MARTE allocation, while sometimes it is useful to separate the two concepts.

3   MARTE

MARTE is a response to the OMG RFP to provide a UML profile for real-time and embedded systems [13]. MARTE is a successor of SPT, aligned with UML 2, and with a wider scope. MARTE introduces a number of new concepts, including time and allocation concepts, which are central to this paper.

3.1   MARTE Time Model

Time in SPT is a *metric* time with implicit reference to physical time. As a successor of SPT, MARTE supports this model of time. However, MARTE goes beyond this quantitative model of time and adopts more general time models suitable for system design. In MARTE, Time can be *physical*, and considered as *dense* or *discretized*, but it can also be *logical*, and related to user-defined clocks. Time may even be *multiform*, allowing different times to progress in a non-uniform fashion, and possibly independently to any (direct) reference to physical time.

Concept of time structure. Figure 1 shows the main concepts introduced in MARTE to model time. This is a conceptual view, or in the UML profile terminology, a *domain view*. The corresponding UML representations will be presented later. The building element in a time structure is the *TimeBase*. A time base is a totally ordered set of instants. A set of instants can be *discrete* or

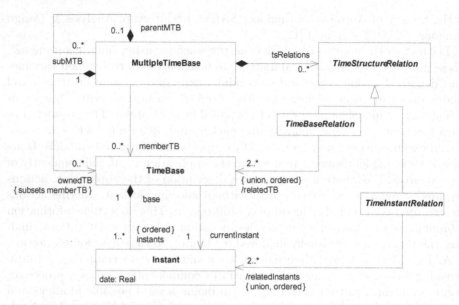

Fig. 1. Time structure (Domain view)

dense. The linear vision of time represented by a single time base is not sufficient for most of the applications, especially in the case of multithreaded or distributed applications. Multiple time bases are then used. A MultipleTimeBase consists of one or many time bases. A time structure contains a tree of multiple time bases.

Time bases are *a priori* independent. They become dependent when instants from different time bases are linked by relationships (coincidence or precedence). The abstract class TimeInstantRelation in Figure 1 has CoincidenceRelation and PrecedenceRelation as concrete subclasses. Instead of imposing local dependencies between instants, dependencies can be directly imposed between time bases. A TimeBaseRelation (or more precisely one of its concrete subclasses) specifies many (possibly an infinity of) individual time instant relations. This will be illustrated later on some time base relations. TimeBaseRelation and TimeInstantRelation have a common generalization: the abstract class TimeStructureRelation. As a result of adding time structure relations to multiple time bases, time bases are no longer independent and the instants are partially ordered. This partial ordering of instants characterizes the *time structure* of the application.

This model of time is sufficient to check the logical correctness of the application. Quantitative information, attached to the instants, can be added to this structure when quantitative analyses become necessary.

Clock. In real world technical systems, special devices, called clocks, are used to measure the progress of physical time. In MARTE, we adopt a more general point of view: a clock is a *model* giving access to the time structure. Time may be logical or physical or both. MARTE qualifies a clock refering to physical time as a *chronometric* clock, emphasizing on the quantitative information attached to this model. A Clock makes reference to a TimeBase. Clocks and time structures have mathematical definitions introduced below. This formal modeling is transparent to the user of the profile.

The mathematical model for a clock is a 5-tuple $(\mathcal{I}, \preccurlyeq, \mathcal{D}, \lambda, u)$ where $\mathcal{I}$ is a set of instants, $\preccurlyeq$ is an order relation on $\mathcal{I}$, $\mathcal{D}$ is a set of labels, $\lambda : \mathcal{I} \to \mathcal{D}$ is a labeling function, u is a symbol, standing for a *unit*. For a chronometric clock, the unit can be the SI time unit s (second) or one of its derived units (ms, us...). The usual unit for logical clocks is tick, but user-defined units like clockCycle, executionStep ...may be chosen as well. To address multiform time, it is even possible to consider other physical units like angle degrees (this is illustrated in an application of our time model to an automotive application [14]). Since a clock refers to a TimeBase, $(\mathcal{I}, \prec)$ is an ordered set.

A *Time Structure* is a 4-tuple $(\mathcal{C}, \mathcal{R}, \mathcal{D}, \lambda)$ where $\mathcal{C}$ is a set of clocks, $\mathcal{R}$ is a relation on $\bigcup_{a,b \in \mathcal{C}, a \neq b} (\mathcal{I}_a \times \mathcal{I}_b)$, $\mathcal{D}$ is a set of labels, $\lambda : \mathcal{I}_\mathcal{C} \to \mathcal{D}$ is a labeling function. $\mathcal{I}_\mathcal{C}$ is the set of the instants of a time structure. $\mathcal{I}_\mathcal{C}$ is not simply the union of the sets of instants of all the clocks; it is the quotient of this set by the coincidence relation induced by the time structure relations represented by $\mathcal{R}$.

Time-related concepts. Events and behaviors can be directly bound to time. The occurrences of a (timed) event refer to points of time (instants). The executions of a (timed) behavior refer to points of time (start and finish instants) or to segments of time (duration of the execution). In MARTE, Instant and Duration are two distinct concepts, specializations of the abstract concept of Time. TimedEvent (TimedBehavior, resp.) is a concept representing an event (a behavior, resp.) *explicitly* bound to time through a clock. In this way, time is not a mere annotation: it changes the semantics of the timed model elements.

MARTE Time profile. The time structure presented above constitutes the semantic domain of our time model. The UML view is defined in the "MARTE Time profile". This profile introduces a limited number of powerful stereotypes. We have striven to avoid the multiplication of too specialized stereotypes. Thanks to the sound semantic grounds of our styereotypes, modeling environments may propose patterns for more specific uses.

The main sterotypes are presented in figures 2 to 4. ClockType is a stereotype of the UML Class. Its properties specifies the kind (chronometric or logical) of clock, the nature (dense or discrete) of the represented time, a set of clock properties (*e.g.*, resolution, maximal value...), and a set of accepted time units. Clock is a sterotype of InstanceSpecification. An OCL rule imposes to apply the Clock stereotype only to instance specifications of a class stereotyped by ClockType. The unit of the clock is given when the stereotype is applied. Unit is defined in the Non Fonctional Property modeling (NFPs) subprofile of MARTE, it extends EnumerationLiteral. It is very convenient since a unit can be used like any user-defined enumeration literal, and conversion factors between units can be specified (*e.g.*, $1ms = 10^{-3}s$). TimedElement is an abstract stereotype with no defined metaclass. It stands for model elements which reference clocks. All other *timed* stereotypes specialize TimedElement.

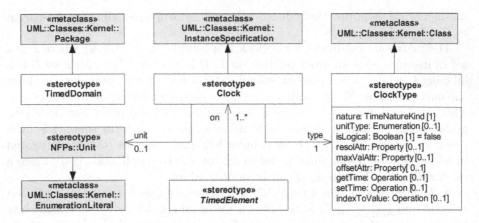

Fig. 2. MARTE TimeModeling profile: Clock

Clock constraints. ClockConstraint is a stereotype of the UML Constraint. The clock constraints are used to specify the time structure relations of a time domain. In turn, these relations characterize the $\mathcal{R}$ relation of the underlying mathematical model of the time structure.

The *context* of the constraint must be a TimedDomain. The *constrained elements* are clocks of this timed domain and possibly other objects. The *specification* of a clock constraint is a set of declarative statements. This raises the question of choosing a language for expressing the clock constraints. A natural language is not sufficiently precise to be a good candidate. UML encourages the use of OCL. However, our clocks usually deal with infinite sets of instants, the relations may use many *mathematical* quantifiers, which are not supported by OCL. Additionnally, OCL [15] is made to be evaluatable, while our constraints often have to be processed altogether to get a set of possible solutions. So, we have chosen to define a simple constraint expression language endowed with a mathematical semantics. The specification of a clock constraint is a UML::OpaqueExpression that makes use of *pre-defined* (clock) relations, the meaning of which is given in mathematical terms, outside the UML. Our *Constraint Specification Language* is not normative. Other languages can be used, so long as the semantics of clocks and clock constraints is respected.

TimedEvent and TimedProcessing. In UML, an Event describes a set of possible occurrences; an occurrence may trigger effects in the system. A UML2 TimeEvent is an Event that defines a point in time (instant) when the event occurs. The MARTE stereotype TimedEvent extends TimeEvent (Figure 3). Its instant specification *explicitly* refers to a clock. If the event is recurrent, a repetition period—duration between two successive occurrences of the event—and the number of repetitions may be specified.

In UML, a Behavior describes a set of possible executions; an execution is the performance of an algorithm according to a set of rules. MARTE associates a duration, an instant of start, an instant of termination with an execution,

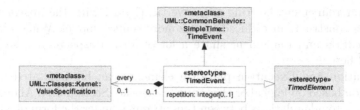

Fig. 3. MARTE TimeModeling profile: TimedEvent

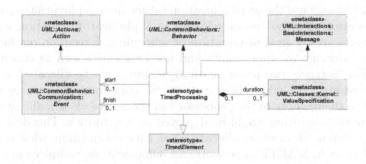

Fig. 4. MARTE TimeModeling profile: TimedProcessing

these times being read on a clock. The stereotype TimedProcessing (Figure 4) extends the metaclasses Behavior, Action, and also Message. The latter extension assimilates a message tranfer to a *communication* action.

Note that, StateMachine, Activity, Interaction being Behavior, they can be stereotyped by TimedProcessing, and thus, can be bound to clocks.

3.2   MARTE Allocation Model

Allocation of functional application elements onto the available resources (the execution platform) is main concern of real-time embedded system design. This comprises both spatial distribution and temporal scheduling aspects, in order to map various algorithmic operations onto available computing and communication resources and services.

The MARTE profile defines *application* and *execution platform* models. A MARTE allocation is an association between a MARTE application and a MARTE execution platform. Application elements may be any UML element suitable for modeling an application, with structural and behavioral aspects. An execution platform is represented as a set of connected resources, where each resource provides services to support the execution of the application. So, resources are basically structural elements, while services are rather behavioral elements. Application and Execution platform models are built separately, before they are paired through the Allocation process. Often this requires prior adjustment (inside each model) to abstract/refine its components so as to allow a direct match. Allocation can be viewed as a "horizontal" association, and abstraction/refinement layering as a "vertical" one, with the abstract version relying on constructs introduced in the more refined model. While different in role, allocation and refinement share a lot of formal aspects, and so both are described here.

Application and Execution platform elements can be annotated with time information based on logical or chronometric clocks (Section 3.1). Allocation and refinement provide relations between these timings under the form of constraints between the clocks and their instants. Other similar non-functional properties such as space requirement, cost, or power consumption are also considered.

In MARTE, we use the word allocation rather than deployment (as in UML) since allocation does not necessarily imply a physical distribution and could simply represent a logical distribution or scheduling. Execution platform models can be abstract at some points and not necessarily seen as concretization models. For instance, two pieces of an algorithm could be allocated to two different processor cores, while the executable file containing both pieces would be deployed on the memory of the processor and the source file containing the specification of the algorithm would be deployed on a hard disk. This dual function was recognized in SPT, where allocation was called realization, while refinement was used as such. MARTE allocation and refinement are complementary to the UML deployment; we prefer to keep the three concepts separated. This is not the case of AADL that provides a single mechanism—the binding—for all three concepts. The allocation mechanism proposed by MARTE is actually very close to the structure allocations of SysML because it allocates logical parts to more physical ones. However, MARTE makes it explicit that both the logical and physical parts could be either of a behavioral or structural nature. Contrary to SysML, MARTE makes a difference between allocation—from application model elements to execution platform model elements—and refinement of an abstract model elements (logical or physical) into more specific elements.

The stereotype Allocate. A MARTE allocation is materialized by the stereotype Allocate (Figure 5), which extends the UML metaclass Abstraction, and can be associated with NFP constraints. Allocation can be structural, behavioral, or hybrid. Structural allocation associates a group of structural elements and a group of resources. Behavioral allocation associates a set of behavioral elements and a service provided by the execution platform. When clear from context, hybrid allocations are allowed (*e.g.*, when an implicit service is uniquely defined for a resource). At the finer level of detail, behavioral allocation deals with the mapping of UML actions to resources and services.

The stereotype Allocated. MARTE advocates the need to differentiate the potential sources of an allocation from the targets. Each model element involved in an allocation is annotated with the stereotype Allocated (as in SysML), which extends the metaclass NamedElement or rather one of its specializations (Figure 6). The stereotype ApplicationAllocationEnd, noted by the keyword «ap_allocated», denotes a source of an allocation. The stereotype ExecutionPlatformAllocationEnd, noted by the keyword «ep_allocated», represents the

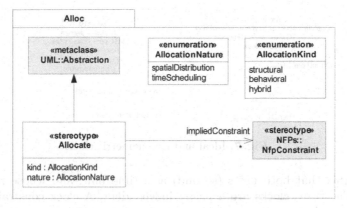

Fig. 5. The stereotype «allocate»

target of an allocation. The stereotype Allocated is not abstract to ensure compatibility with SysML, but one of its specializations should be preferred. The property allocatedTo, respectively allocatedFrom, is a derived property resulting from the process of creating the abstraction (allocation); they facilitate the identification of the targets, respectively the sources, of the allocation when all model elements cannot be drawn on the same diagram.

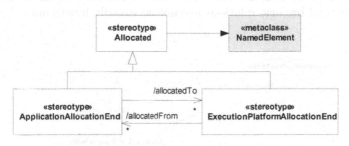

Fig. 6. The stereotype «allocated»

4   Illustrative Examples

4.1   Chronometric Clocks

The MARTE TimeLibrary provides a model for the *ideal time* used in physical laws: idealClk, which is an instance of the class IdealClock, stereotyped by ClockType (Fig. 7). idealClk is a dense time clock, its unit is the SI time unit s.

Starting with idealClk, the user can define new discrete chronometric clocks (Fig. 8). First, the user specifies Chronometric—a class stereotyped by ClockType—which is discrete, not logical (therefore chronometric), and with a read only attribute (resolution). Clocks belong to timed domains. In Fig. 8, a single time domain is considered. It owns 3 clocks: idealClk, cc1 and cc2, two instances of

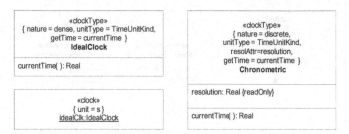

Fig. 7. Ideal and Chronometric clocks

Chronometric that both use s (second) as a time unit; and whose resolution is 0.01 s. The three clocks are *a priori* independent. A clock constraint specifies relationships among them.

The first statement of the constraint defines a clock c local to the constraint. c is a discrete time clock derived from idealClk by a *discretization* relation. The resolution of this clock is 1 ms. The next two statements specify that cc1 and cc2 are subclocks of c with a rate 10 times slower than c. The fourth and fifth statements indicate that cc1 and cc2 are not perfect clocks. Flaws are characterized by *non functional properties* like stability and offset. Their rate may have small variations (a stability of 10^{-5} implicitly measured on idealClk). The last statement claims that the two clocks are out of phase, with an offset value between 0 and 5 ms measured on idealClk. Note that even if cc1 and cc2 look alike, they are not identical because relations are not necessarily functional.

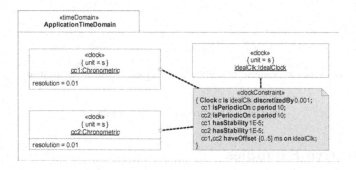

Fig. 8. Clock constraints

4.2   AADL Communication

To explain its *port-based* communication semantics, the AADL specification takes the example of a thread Read that captures a sample and sends it to a second thread Control. The two threads are assumed to be dispatched at the same time. Several cases are studied, the case where the two threads are dispatched with the same period, the case where Read is dispatched twice faster than Control (undersampling), and the case where Read is dispatched twice slower than Control (oversampling). These cases are studied first with an immediate communication (the output value is available as soon as the thread Read completes) and then with a delayed communication (the output value is available only at the next dispatch of the thread Read).

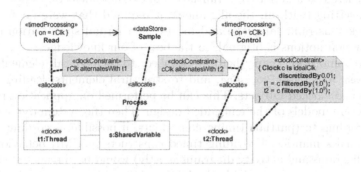

Fig. 9. Clock constraints

To compare our approach to AADL, we take the case of an immediate communication, which is the more challenging, with undersampling. As said before, the main difference of our approach is that we separate the application models from the execution platform models. The application is described with a UML activity diagram (Fig. 9) using purely logical clocks and stereotyped by «TimedProcessing». The behavior of Read and Control are executed repetitively, they communicate through a datastore object node that allows for multiple readings of the same sample.

In a second step, the application is allocated to the model of an execution platform, a process containing two threads that communicate through a shared variable. In our model, t1 and t2, or rather their dispatch time, are considered as clocks. These two harmonic clocks are defined using a local 100Hz-clock c. Then t1 and t2 are derived from c with respective frequency 20Hz and 10Hz.

Additional clock constraints are associated with the allocate dependency to map the application clocks to the platform clocks. All these constraints define a partial order. In the case of a delayed communication, these would have been enough to have an equivalence of all possible schedulings. With an immediate communication, an additional constraint is required to guarantee the same behavior than AADL. This constraint would follow a greedy scheduling in order to execute Control as soon as possible. Our constraint model allows for delaying subjective choices as much as possible in order to avoid overspecification.

5   Conclusion

We presented a UML profile for comprehensive Time Modeling. Time here can be of discrete or dense, physical or logical. Logical time allows to model various time threads sustaining asynchronous or loosely time-related concurrent processes. This philosophy (of assigning logical clocks in order to explicitly handle time rates) borrows to foundational notions in embedded MoCC design. To this we add a kernel language of clock constraint relations, as well as timed events constraint relations. This constraint language, while currently simple, allows to define most useful clock relations (such as being periodic). While the profile can be considered as a "creative" translation of existing ideas on tagged systems to a UML setting (with all the alignments it required that were far from trivial), the clock constraint language and its use as a formal specification of classical time relation notions is original, to the best of our knowledge.

Time annotation can then be applied to behavioral elements, leading TimedEvent and TimedProcessing, and to structural elements, leading to clocked Classes and clocked Objects. This can be performed on application models and architecture models of the embedded design. Then the system dynamics should run according to (partial) timing constraints, if possible, according to a timed operational semantics. Providing timed constructs in UML behavioral models (state diagrams and activity diagrams mostly) would be the next step here. Numerous examples exist (outside the UML) of timed languages and calculi under the form of MoCC constructors inside the proper time domain. Model transformation tools could extract the time properties, feed them into timing analysis tools and bring the result back into UML within a UML simulator. In that sense, our profile would give the time semantics of UML models.

Clock constraints provide partial scheduling information, and an actual schedule can be obtained by solving such a set of constraints, some of which originate from the application model, some from the execution platform model, and some from the system's real-time requirements. The same formalisms of clock relations can also be used in some case to represent the *result* of the scheduling decisions, and display them to the designer.

We provided modeling instances and case studies to illustrate and motivate the modeling framework. We showed how it allows to introduce a number of useful time predicates on events. We also showed the intent behind logical time by considering examples with various clocks running at unrelated speeds.

References

1. Lee, E.A., Sangiovanni-Vincentelli, A.L.: A framework for comparing models of computation. IEEE Transactions on Computer-Aided Design of Integrated Circuits and Systems 17(12), 1217–1229 (1998)
2. Buck, J., Ha, S., Lee, E., Messerschmitt, D.: Ptolemy: A framework for simulating and prototyping heterogeneous systems. International Journal of Computer Simulation, special issue on Simulation Software Development 4, 155–182 (1994)
3. Jantsch, A.: Modeling Embedded Systems and SoCs - Concurrency and Time in Models of Computation. Morgan Kaufman, San Francisco (2003)
4. Darte, A., Robert, Y., Vivien, F.: Scheduling and Automatic Parallelization. Birkhaüser (2000)
5. Feautrier, P.: Compiling for massively parallel architectures: a perspective. Microprogramming and Microprocessors (41), 425–439 (1995)

6. OMG: UML 2.1 Superstructure Specification, Object Management Group, Inc., 492 Old Connecticut Path, Framing-ham, MA 01701. OMG document number: ptc/2006-04-02 (2006)

7. Selic, B.: On the semantic foundations of standard uml 2.0. In: Bernardo, M., Corradini, F. (eds.) SFM 2004. LNCS, vol. 3185, pp. 181–199. Springer, Heidelberg (2004)

8. OMG: UML Profile for Schedulability, Performance, and Time Specification, Object Management Group, Inc., 492 Old Connecticut Path, Framing-ham, MA 01701. OMG document number: formal/05-01-02 (v1.1) (2005)

9. ITEA: EAST-ADL — The EAST-EEA Architecture Description Language. ITEA Project Version 1.02 (2004)

10. Graf, S., Ober, I., Ober, I.: A real-time profile for UML. STTT, Software Tools for Technology Transfer 8(2), 113–127 (2006)

11. OMG: Systems Modeling Language (SysML) Specification. OMG document number: ad/2006-03-01 (2006)

12. SAE: Architecture Analysis and Design Language (AADL). document number: AS5506/1 (2006)

13. OMG: UML profile for Modeling and Analysis of Real-Time and Embedded systems (MARTE), Request for proposals, Object Management Group, Inc., 492 Old Connecticut Path, Framing-ham, MA 01701. OMG document number: realtime/2005-02-06 (2005)

14. André, C., Mallet, F., Peraldi-Frati, M.A.: A multiform time approach to real-time system modeling: Application to an automotive system. Technical Report ISRN I3S/RR–2007-14–FR, I3S laboratory, Sophia-Antipolis, France (2007)

15. OMG: Object Constraint Language, version 2.0, Object Management Group, Inc., 492 Old Connecticut Path, Framing-ham, MA 01701. OMG document number: formal/06-05-01 (2006)

A UML Profile for Developing Airworthiness-Compliant (RTCA DO-178B), Safety-Critical Software

Gregory Zoughbi[1,2], Lionel Briand[2], and Yvan Labiche[2]

[1] General Dynamics Canada, 3785 Richmond Road, Ottawa, ON K2H 5B7, Canada
[2] Carleton University, 1125 Colonel By Drive, Ottawa, ON K1S5B6, Canada
Gregory.Zoughbi@gdcanada.com, briand@sce.carleton.ca,
labiche@sce.carleton.ca

Abstract. Many safety-related, certification standards exist for developing safety-critical systems. System safety assessments are common practice and system certification according to a standard requires submitting relevant software safety information to appropriate authorities. The airworthiness standard, RTCA DO-178B, is the de-facto standard for certifying aerospace systems containing software. This research introduces an approach to improve communication and collaboration among safety engineers and software engineers by proposing a Unified Modeling Language (UML) profile that allows software engineers to model safety-related concepts and properties in UML, the de-facto software modeling language. Key safety-related concepts are extracted from RTCA DO-178B, and then a UML profile is defined to enable their precise modeling. We show that the profile improves the line of communication between safety engineers and software engineers, for instance by allowing the automated generation of certification-related information from UML models. This is illustrated through a case study on developing an aircraft's navigation controller subsystem.

Keywords: UML, UML Profile, Airworthiness, RTCA DO-178B, Safety, Safety-Critical, Safety Assessment, Certification, Certification Authority.

1 Introduction

Safety-critical software must exhibit safe behaviour that does not contribute to hazards within the context in which it is used. For example, an aircraft must only allow the pilot to lift up the landing gears if it is airborne. If the landing gears were lifted while the aircraft is on ground, then there would be a hazard, which is likely to result in damaging the aircraft and hurting its occupants. A hazard is a state of the system that could ultimately lead to an accident that may result in a loss in human life.

Many standards require that a safety assessment be performed when designing or modifying a safety-critical system. Safety assessments, which have some similarities with risk assessments [8] and are performed using similar methods, produce a list of safety requirements and constraints that the system must fulfil. RTCA DO-178B [14], also known as the "airworthiness" standard, is the de-facto safety-related standard for commercial and military aerospace software.

G. Engels et al. (Eds.): MoDELS 2007, LNCS 4735, pp. 574–588, 2007.

Due to safety-related requirements, and the need for safety assessment and thorough software analysis, developing safety-critical software is more expensive than developing non-safety-critical software (in the order of 20 to 30 times more expensive [10]). One of the challenges, besides actually designing and implementing safety and certification requirements of the system, is to be able to accurately communicate safety aspects among the different stakeholders involved in software development. This has already been documented in [4], where the authors reported on an extensive survey, performed by the NASA Langley Research Center (in collaboration with the Federal Aviation Administration (FAA)), to identify the challenges in developing safety-critical software for aerospace systems. The authors identified "the challenge of accurately communicating requirements between groups of people" as "the root of many of the current challenges" in software aspects during certification. They presented this challenge as a combination of the following two major communication channels: (a) Among regulatory people (e.g. certification authorities) and systems people (e.g. systems engineers and safety engineers); (b) Among systems people and software people (e.g. software engineers). The authors also found out that "requirements definition is difficult", which undoubtedly contributes to the communication challenges among the various groups. For example, safety engineers, who are rarely software engineers, may define requirements that software engineers find infeasible or expensive. If software engineers better understand the needs behind the requirements, then they may be able to propose solutions that are more practical and cost effective. On the other hand, software engineers may misinterpret the requirements due to their lack of experience in safety.

In this paper, we report on an approach to address these communication challenges. Acknowledging that the Unified Modeling Language (UML) [13] is now the de-facto standard for developing (object-oriented) software systems, and is actually used in the aerospace industry, we aim at extending the UML notation with safety concepts by defining a new profile. In this research, because of our focus on aerospace software, the airworthiness standard [14] is analyzed to extract a list of key safety-related concepts that are of interest to both safety engineers and software engineers. We show that if those concepts are properly represented in UML models of software, then software engineers can document their decisions (that relate to safety) and a tool can automatically generate reports containing safety and certification-related information about the software. This provides safety engineers with better insights into the software compliance with safety requirements, which they can easily track over time. Those reports could also be used as evidence of software compliance with the airworthiness requirements, which can then be presented to the external certification authority. Furthermore, such safety UML profile will increase the software engineers' awareness of safety-related issues, which will enable them to implement safer software and better communicate with safety engineers. Using stereotypes and tagged values, our focus in this paper is to first model safety and certification information in class diagrams. However, future work will tailor the stereotypes and tagged values to other diagrams such as sequence diagrams and state machines.

The remainder of the paper is structured as follows. We shortly introduce safety assessment for aerospace systems in Section 2. We then present the requirements that a UML profile should have to adequately facilitate the definition and use of safety information in an aerospace context (Section 3). Section 4 discusses existing

UML-based techniques to specify safety concepts, in the light of the information requirements specified in Section 3. We then discuss our profile in Section 5 on a case study, using a number of representative examples, and show how it can indeed address the communication challenges reported in [4]. Conclusions are drawn in Section 6.

2 Safety Assessment of Aerospace Systems

Many industrial standards exist for system and software safety. Some are common to all industry sectors (e.g., IEC 61508-3 [6]) whereas others are industry specific (e.g., CENELEC 50128 for Railway applications [1]). Hermann provides a high-level summary of those standards in [5], and we discuss in [15] those that relate directly or indirectly to safety. One of those standards, RTCA DO-178B [14], also known as the airworthiness standard, is the de-facto safety-related standard for developing aerospace software systems.

RTCA DO-178B takes into account the fact that not all software components in an airborne system have the same impact on the safety of the aircraft and its occupants. For example, the failure of software that controls the altitude of an aircraft is less acceptable than the failure of the software that stores and displays geographical information maps for navigation, since the failure of the former may significantly reduce the aircraft's chances of a safe flight and landing. The failure of the latter, however, introduces inconveniences that the pilots may be able to cope with. As a result, DO-178B classifies software failure conditions into the following five categories [14]: *Catastrophic*, *Hazardous/Severe-Major*, *Major*, *Minor*, and *No Effect*; depending on the impact of the failure conditions on the safety of the flight or landing of the aircraft. According to these categories, a software component is determined to belong to a specific software or airworthiness level: A, B, C, D, or E.

It should be noted that there exists a difference between the concepts of "airworthiness" and "safety". The airworthiness standard defines failure condition categories and software levels based on the "severity of failure conditions on the aircraft and its occupants" [14]. This is more specific than Leveson's definition of safety, which was stated as "the freedom from accidents or losses" [8]. Airworthiness is concerned with accidents or losses regarding the aircraft and its occupants and is a subset of safety: safe software is airworthy but airworthy software is not necessarily safe.

3 Requirements for an Airworthiness Profile

Based on the communication challenges reported in [4], we first identify in Section 3.1 how safety information, if included in analysis and design (UML) models, could be used by the different stakeholders involved in the development of safety-critical software (e.g., safety engineers, software engineers, certification authority). In Section 3.2, we perform a careful analysis of the safety concepts involved in the airworthiness standard, thereby identifying and precisely modeling the information requirement that an airworthiness UML profile should satisfy.

3.1 Usage Scenarios of Safety Information

Stakeholders use safety information as depicted by the use case diagram in Fig. 1.

Usage 1—*Provide Safety Requirements*: Safety engineers perform a safety assessment of the system being developed. This results in safety requirements, a subset of which is allocated to software and communicated to software engineers.

Usage 2—*Design Safety Requirements in Systems*: Software engineers design the software system according to the safety requirements allocated to software.

Usage 3—*Record and Justify Design Decisions*: Software engineers record and justify their design decisions. Traditionally, architectural and major design decisions have been recorded in documents separate from the software model and, furthermore, detailed design decisions have often appeared as plain text comments in the source code. In practice, this makes it hard to automatically retrieve justifications for design decisions.

Usage 4—*Monitor Safety*: Safety engineers continuously monitor the safety of the system, including the software, over the project's lifecycle. In order to do so, they need to identify how software engineers designed the software (Usage 2) according to the safety requirements they were provided with (Usage 1). The software engineers' justifications for the design decisions (Usage 3) are also considered. Safety engineers can then assess this information and discuss any issue with software engineers, thus ensuring that the software's safety is continuously improving during the software's lifecycle so that it meets the final safety objectives of the system.

Usage 5—*Get Safety Information*: Safety and certification information is submitted to the certification authorities for certification, which usually occurs towards the end of the development lifecycle. This information includes the safety requirements (Usage 1), the software design (Usage 2), the justification of the software design (Usage 3) given the safety requirements of the software, and the process used to continuously monitor the system and software safety over the development lifecycle (Usage 4).

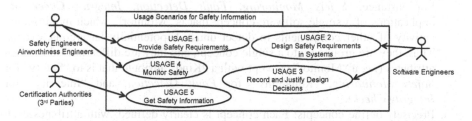

Fig. 1. Usage scenarios of safety information

Those usage scenarios would be greatly facilitated if safety information is captured in the lifecycle UML models already in use to support development. In other words, a UML model could play a central role in the communication of safety information across engineering groups. Software engineers record safety information in UML models. Safety engineers can monitor safety information by automatically generating reports, using appropriate tool support to query UML models. Therefore, they need not understand the specifics of UML because any tool that extracts safety information from the model can present it in a model-independent format. To certify the system, safety and certification information can be automatically extracted from the UML models, using a tool that could produce the required information in a format suitable for submission to the certification authorities. Our approach is therefore to extend the UML notation with safety information, through the definition of a profile [13].

3.2 Safety Information Requirements

We identify airworthiness-related concepts from the airworthiness standard [14] that should be part of the UML profile to be defined. Since airworthiness is a subset of safety (Section 2), all of the airworthiness-related concepts we identified are also safety-related concepts and are referred to as such in the remainder of this document. Though not restricted to airworthiness, those concepts are however clearly not the only ones needed for other safety-critical applications including transportation, medical, and nuclear.

A careful analysis of the airworthiness standard led us to identify 65 safety-related concepts that were found relevant for developing airworthiness-compliant software. Those safety-related concepts are not solely safety concepts, and hence the rationale behind using the term "safety-related concepts" rather than simply "safety concepts" – in fact, many of those concepts are primarily non-safety concepts, such as reliability concepts (e.g. fault-tolerance), performance concepts, and certification concepts.

In order to prepare for the definition of the profile, this set of concepts was then refined and restructured to:

1. Remove duplicate concepts: Seemingly different concepts sometimes appear in the airworthiness standard where, in reality, they represent a single concept or term. For instance, *Multiple-Version Dissimilar Software* and *Software Redundancy* describe the same concept or idea of using multiple software components that have the same functionality but different implementations.

2. Group concepts: Some concepts are in fact examples of a more general concept. For instance, *Safety Monitoring*, *Fault Detection*, *Integrity Check* are applications of a single software-concept that is "Monitor", which monitors the activity of other components to detect unusual, potentially hazardous, events. Thus, they are grouped into a single concept, and they are differentiated through attributes (i.e. a "Monitor" concept with a "Kind" attribute that is to "Safety" for *Safety Monitoring*, "Reliability" for *Fault Detection*, and "Integrity" for *Integrity Check*).

3. Precisely define concepts: Each concept is clearly defined, with attributes each describing a single aspect of the concept. For instance, when defining the "Safety Requirement" concept, we give it an attribute called "Specification" that can be used to specify the details of the requirement. We use a precise template for concept definition, adapting templates used in existing UML profiles (e.g., [11, 12]): concept definition, attribute definition, relations to other concepts, reference to the original airworthiness concept.

This resulted in the definition of 27 safety-related concepts, 48 attributes, and inter-concept relationships. Their various combinations can represent all of the original 65 safety-related concepts plus additional details. These refined concepts and their relationships were defined using the abovementioned template and formalized under the form of a conceptual model, i.e., a UML class diagram (Fig. 2). Broadly speaking, the concepts in Fig. 2 cover the notions of safety criticality, event/reaction/handler/monitor, replication, configurable component, partitioning and requirement. It is worth noting from the list of concepts in Fig. 2 that they are not specific to airworthiness, which suggests that the profile we describe later in this paper could be used (and perhaps extended) in other safety-critical domains.

The 27 safety concepts are not detailed here due to space constraints but can be found in [15]. Nevertheless, Fig. 3 shows the definition of the "Safety Critical" concept as an example. Notice that the concept has a "Criticality Level" attribute whose definition refers to the airworthiness standard as well as the IEC 61508 standard [6]. This illustrates our conscious effort to be as general as possible while focusing on the airworthiness standard. Additionally, the attributes' values are not fixed and can be organization (or project) specific, e.g., the "High", "Medium" and "Low" values for attribute "Confidence Level" can be assigned specific meanings. This will translate into a flexible UML profile, with customizable tagged values.

As discussed and illustrated in Section 5, these concepts correspond to stereotypes in our airworthiness profile. Some of those concepts will therefore be presented in Section 5, as we discuss the corresponding stereotypes such as Event, Reaction, Rationale, Monitor, Handler, and Safety Critical.

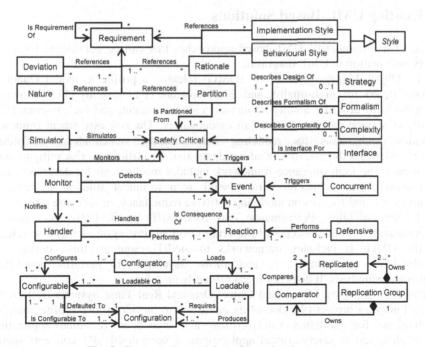

Fig. 2. Conceptual model for safety-related concepts

After defining the safety concepts and their inter-concept relationships in detail, we developed a list of information requirements [15]. These information requirements identified the information that a suitable mechanism, or profile, must be able to model

Definition:
  The "Safety Critical" concept represents a safety-critical design (element) that impacts safety. It also identifies the safety or airworthiness level of design elements.
Attributes:

Name	Description	Examples
Criticality Level	Indicates the level of criticality (e.g. airworthiness level, Safety Integrity Level (SIL)), on some pre-defined scale, such as the software level or the failure condition category	For RTCA DO-178B: "A", "B", "C", "D", "E" For IEC 61508: "SIL 1", "SIL 2", "SIL 3", "SIL 4"
Confidence Level	Indicates the level of confidence, on some pre-defined scale, that the criticality level is satisfied	"High", "Medium", "Low", "80%", "50%", ... etc

Relationships:

Name	Description
Triggers	Identifies zero or more "Event" instance that the "SafetyCritical" instance may trigger

Original Safety-Related Concepts
  Safety-Critical, Software Level, Level of Confidence, Failure Condition Category

Fig. 3. Template description of the "Safety Critical" concept

in UML. Therefore, they became the basis upon which a mechanism, or profile, to model safety concepts in UML was assessed.

4 Existing UML-Based Solutions

There exist a number of UML-based approaches that can be considered for adding safety information to UML diagrams.

The Object Management Group (OMG) released a profile to model Quality of Service (QoS) for high-quality and Fault-Tolerant (FT) systems [11]. It includes frameworks to describe quality of service, risk assessment, and fault tolerance. The quality of service part can be used to model safety. The risk assessment framework provides mechanisms for modeling risk contexts, stakeholders, weaknesses, opportunities and threats, unwanted incidents, risk quantification, risk mitigation and treatments. The fault tolerance framework includes mechanisms for describing fault-tolerant software architectures in general as a technical solution to reliability requirements, and focuses on modeling software redundancy, or software replication.

The Schedulability, Performance, and Time (SPT) profile [12], which provides mechanisms to model concepts of importance to real-time systems was also released by the OMG. It includes frameworks to model resources, time, concurrency, schedulability, performance ... and allows developers to perform model-based performance analysis. It does not focus primarily on safety.

The High Integrity Distributed Object-Oriented Real-Time Systems (HIDOORS) was a European project [9]. One of its goals was to introduce SPT profile-compliant mechanisms for modeling safety-critical and embedded real-time applications. Although aimed at safety-critical applications, it specialised SPT concepts such as triggers, actions, resources, and scheduling jobs without strongly focusing on safety.

Jan Jürjens presented a UML profile [7] that aimed at modeling reliability aspects regarding transmitting messages (e.g., maximum failure rates for message communication). Jürjens argued that since failures related to lost, delayed, or corrupted messages have an impact on safety in safety-critical applications, the profile can be used for developing safety-critical applications. It included mechanisms to model risks, crashes, guarantees, redundancy, safe links, safe dependencies, safety critical elements, safe behaviours, containment, and error handling.

Based on the IEC 61508 standard [6], Hansen and Gullesen presented a series of UML patterns that can be used to model some aspects of safety-critical systems [3]. The patterns allowed modeling safety quality of service, software diversity and voting, partial diversity with built-in diagnostic or monitoring, "safe" communication protocols, and some other topics such as testing, hazard analysis and quality development. They include mechanisms to model, in use cases, redundancy, monitoring, and voting based on multiple output comparisons.

We evaluated how the UML-based solutions discussed above score with respect to addressing the information requirements identified in Section 3.2. Each profile's score was calculated based on how many information requirements it met (Section 3.2). We observed (see details in [15]) that none of the existing techniques that were evaluated achieved more than 31% of the maximum score. In fact, all of the profiles combined only met 44% of the information requirements. For instance: reactions and events can be modeled in HIDOORS and SPT [9, 12], but not in the other three approaches; monitors can only be modeled in [7, 12]; a model element's contribution to failure conditions (i.e., criticality levels in the airworthiness standard) is only handled in [3, 7, 12]; the implementation style (e.g., recursion, dynamic memory allocation), hardware/software interfaces, and configurable components cannot be modeled by any of the existing approaches. We therefore decided to develop our own UML profile, as described in the next section.

5 The Airworthiness Profile by Examples

From the 27 safety-related concepts (and 48 attributes) we identified, we defined a UML profile containing 32 stereotypes[1] (with 80 tagged values). These stereotypes and tagged values clearly cannot be all described in details in this article. Instead, we refer the interested reader to [15] for all the details, and present a number of representative examples of the use of the profile on a realistic case study, thereby illustrating a subset of the profile. The goal is to demonstrate, through a realistic case study, the usefulness of the profile in the context of the usage scenarios we defined (Section 3.1). Stereotypes and tagged values are shown in the figures for illustrative purposes, but modeling tools can capture this without cluttering the diagrams.

We considered, as a safety-critical system, the Navigation Controller (NC) subsystem of an aircraft's navigation system. Note that this case study has been conducted for the purpose of evaluating the profile only and does not correspond to a certified (e.g., by the FAA) system, though the first author is involved in the development of similar systems. We, however, followed every standard practices [14] when designing this controller.

The NC subsystem is used to control the aircraft's flight paths through both automatic pilot and manual input from the pilots. In autopilot mode, the subsystem

---

[1] Each of the 27 safety-related concepts translates into a stereotype and we added stereotypes such as <<ReliabilityContext>> and <<ConcurrencyContext>> that were used to stereotype diagrams containing reliability and concurrency information, respectively.

can choose an appropriate flight path based on the source and destination of the aircraft, and guide the aircraft by generating appropriate commands to the aircraft's ailerons and spoilers (on the wings), rudder (on the vertical tail), and engines to change the speed and heading (i.e. direction) as required. In custom flight mode, the subsystem can accept commands from the pilots such as a specific position's latitude and longitude. Then, it controls the aircraft's speed and heading to get to the desired position that was indicated by the pilot.

In order to perform such functionality, the NC subsystem needs to have continuous input from the aircraft's navigation system, which reports the current position and altitude of the aircraft at all times. In addition, it needs to be able to command the aircraft's ailerons, spoilers, rudders, and engines to change the speed and heading.

When designing the subsystem, we followed the procedures required in the DO-178B [14]. We first identified five high-level functional requirements (main functionalities) for the subsystem, referred to as FREQ1 to FREQ5, e.g., the subsystem shall be able to list pre-determined flight paths for a requested source/destination pair (FREQ1). We then performed a safety assessment, using four standard, complementary methods (namely Action Error Analysis, Failure Modes and effects Analysis, Hazards and Operability Analysis, and Interface Analysis [8]). This generated eleven safety requirements, referred to as SREQ 1 to SREQ 7 (SREQ 6 is further decomposed into SREQ 6.1 to SREQ 6.5), e.g., the subsystem shall disable the autopilot when the pilot is flying the aircraft (using the mechanical and steering stick subsystem), and re-enable it when the pilot stops doing so (SREQ 1). We then designed the subsystem while recording design decisions (in particular safety-related ones) using our profile. Last, we evaluated the resulting design documents with respect to the usage scenarios discussed in Section 3.1.

Below we detail the design of the controller subsystem in two steps. We first identify the events the controller subsystem has to handle, and the reactions it has to take, according to the functional and safety requirements (Section 5.1). Second, we discuss the design of the subsystem itself (Section 5.2). In both cases, we record design decisions using our profile. Last, we show how the safety engineers and third party certification authorities can use this information (Section 5.3).

5.1 Identification of Events and Reactions

As part of the design, we first identified all the events received and the reactions performed by the NC subsystem that could have safety implications. To identify the events, one needs to determine which inputs to the system, or changes in its state, may impact its safety. To identify the reactions, one needs to determine how the system should behave when any of the identified events occurs.

The answers to those questions are found in the safety requirements. For example, one can identify at least two events of interest from safety requirement SREQ 1 above: (1) The event of when the pilot starts using the mechanical and steering stick subsystem; (2) The event of when the pilot stops using this subsystem. Also from this requirement, one can identify at least the following reactions: (1) Disabling the autopilot when the pilot starts using the mechanical and steering stick subsystem; (2) Enabling the autopilot when the pilot stops using this subsystem.

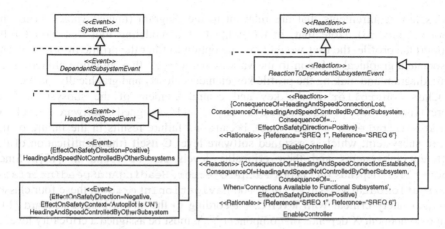

Fig. 4. Excerpt of the <<Event>> and <<Reaction>> hierarchies

Such an analysis of the requirements led to the identification of an inheritance hierarchy of 27 events classes (stereotyped <<Event>>) and an inheritance hierarchy of 12 reactions (stereotyped <<Reaction>>). Excerpts of the two hierarchies appear in Fig. 4, showing the events and reactions identified from SREQ 1. The <<Event>> hierarchy shows two concrete classes, namely HeadingAndSpeedControlledBy-OtherSubsystem and HeadingAndSpeedNotControlledByOtherSubsystem, sent to the NC subsystem when the pilot is using the mechanical and steering stick subsystem, or when she is not using it, to control the aircraft heading and speed, respectively. The <<Event>> stereotype has two tagged values: one provides information on the impact of receiving the event on the safety of the subsystem (EffectOnSafetyDirection), and the other provides some context under which the event can be sent (EffectOnSafetyContext). When the subsystem receives these events, it performs either of two reactions: DisableController or EnableController (Fig. 4). The <<Reaction>> stereotype has a number of tagged values to: refer to the event(s) that the reaction is a consequence of (ConsequenceOf), indicate the impact of the reaction on safety (EffectOnSafetyDirection), and indicate the condition(s) under which the reaction occurs (When). (Reactions DisableController and EnableController are also consequences of events other than the previously mentioned two, which are not shown in Fig. 4 to focus our discussion.) The two <<Reaction>> classes are also stereotyped <<Rationale>> to record the design decisions. The <<Rationale>> stereotype has a tagged value, namely Reference, to refer to the requirements that justify the existence of the reactions (here, safety requirements SREQ 1 and SREQ 6).

5.2 Controller Subsystem Design

The design (class diagram) of the controller subsystem appears in Fig. 5, and selected stereotypes are applied to the model for illustrative purposes. The diagram itself is stereotyped <<SafetyCritical>> to indicate that it contains information that is relevant to safety. It is also stereotyped <<Requirement>> to indicate the functional and safety requirements that are relevant to the diagram (tagged values Kind and Specification). Note that, in the context of a modeling tool that would fully support the profile, the user would have the option to filter the profile information and associate a graphical notation to the various stereotypes, thus avoiding cluttering the class diagram and making the profile key elements clearly and graphically visible.

Class Controller is the key and central element of the subsystem. It is stereotyped with <<SafetyCritical>> and assigned software level C (CriticalityLevel tagged value), because its failure results in the failure of the entire subsystem, which is assigned software level C itself (the identification of the different subsystems and their criticality levels can be found in [15]). Since Controller depends on classes PathProjector, HeadingAndSpeedInterface, NavigationDatabaseInterface, and NavigationInterface, these four classes are also safety critical with level C. (According to the airworthiness standard [14], when component X depends on component Y, Y must be assigned a criticality level at least as high as the level of X.)

Class Controller is also stereotyped with <<Handler>> with tagged value HandleableEvent equal to PilotInputEvent to indicate that it handles all concrete events that are subclasses of class PilotInputEvent (stereotyped <<Event>> but not shown in Fig. 4). In addition to normal code execution (e.g., changing the flight path in response to a ChangeFlightPath input event), Controller also executes the InvestigateFuelShortage reaction to decide whether the changes requested by the pilots will result in a fuel shortage or not.

Class Controller is also stereotyped <<state dependent control>>, indicating it is a control class with a state-dependent behaviour. This stereotype, as well as the <<algorithm>>, <<coordinator>>, and <<system interface>> are not part of our profile. They were defined in [2].

The controller subsystem interacts with the HeadingAndSpeedSubsystem, the NavigationDatabaseSubsystem, and the NavigationSubsystem (subsystem decomposition not shown here). Therefore, the Controller class is associated with three <<system interface>> classes, namely HeadingAndSpeedInterface, NavigationDatabaseInterface, and NavigationInterface. These interface classes are stereotyped <<Rationale>>, thereby relating the design decision of having those subsystems to functional and safety requirements. They are also stereotyped <<Interface>> to indicate the kind of interface they correspond to (i.e., software-software, or software-hardware). (Note that identifying the software-hardware interfaces is a requirement of the DO-178B [14].) Class Controller can receive events from these subsystems (through the corresponding interface classes). These events are monitored by three <<Monitor>> classes (i.e., HeadingAndSpeedMonitor, NavigationDatabaseMonitor, Navigation-Monitor), which all interact with one handler to perform the required reactions, namely ExternalSubsystemsEventHandler (bottom of Fig. 5). The <<Monitor>> stereotype indicates that those monitors are safety monitors (Kind tagged value equals to Safety). It also specifies the monitored entity (tagged value MonitoredEntity), the detectable event from this entity (tagged value DetectableEvent), and the handler for those events (tagged value EventHandler). In particular, referring to the already discussed safety requirement SREQ 1, class HeadingAndSpeedMonitor monitors class

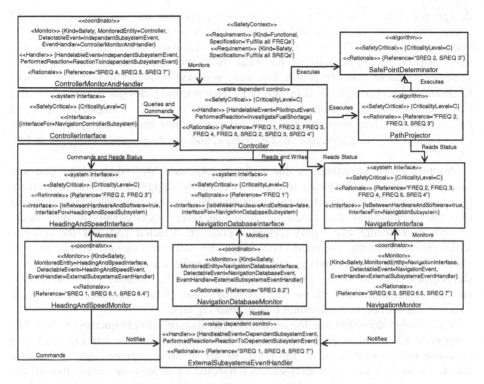

Fig. 5. Controller Subsystem Design

HeadingAndSpeed-Interface to detect HeadingAndSpeedControlledBy
OtherSubsystem and HeadingAndSpeedNotControlledByOtherSubsystem
events and then notifies ExternalSubsystemsEventHandler that triggers the
DisableController or EnableController reactions accordingly. (The
PerformedReaction tagged value of the <<Handler>>'s stereotype indicates that
the handler performs actions that are subclasses of
ReactionToDependentSubsystemEvent, which is the case of
DisableController and EnableController—Fig. 4). Classes HeadingAnd-
SpeedMonitor and ExternalSubsystemsEventHandler are therefore
stereotyped <<Rationale>>, and the Reference tagged value shows SREQ 1.

5.3 Safety Monitoring and Certification

In this section, we revisit the usage scenarios discussed in Section 3.1 (Fig. 1) with
the help of the NC subsystem case study. When designing a safety critical system
with our airworthiness profile, software engineers have to relate their design decisions
to functional as well as non-functional (safety) requirements. This is done in our
profile by using the <<Rationale>> stereotype. Each model element that
implements at least one safety (or functional) requirement is stereotyped with
<<Rationale>> in Fig. 5. Thus, the design is explicitly and precisely related to the
safety requirements, which then supports usage scenario 3 in Fig. 1 (record and justify

design decisions). The <<Rationale>> stereotype has an Explanation tagged value. The events, reactions, and monitors classes, as well as links between classes and requirements (Reference tagged value of stereotype <<Rationale>>) clearly show which class is responsible for what. Note that other stereotypes and tagged values were defined in the profile but were not used in Fig. 5 though they further facilitate usage scenario 3 (Fig. 1) [15]. Documenting design decisions is also possible for replications (see the Replicated, Comparator, and ReplicationGroup concepts in Fig. 2) and configurable parts of the system (see the Configurator, Configuration, Loadable, and Configurable concepts in Fig. 2).

Another major advantage is that software engineers will likely better consider safety requirements if they have to explicitly relate the design model elements to them (usage scenario 2: design safety requirements). By decomposing safety requirements into events, reactions and monitors, designing safety requirements in the system reduces to ensuring that safety-related events are detected and that the relevant reactions are properly executed. This helps the software engineer ensure that all identified safety requirements are accounted for in the design. The <<Rationale>> stereotype and its Reference tagged value, as well as the <<Reaction>> stereotype and its EffectOnSafetyDirection tagged value (which can be Positive or Negative), also play an important role in ensuring this.

Since design decisions are documented in the diagram, simple queries on the UML diagram can help safety engineers monitor safety (usage scenario 4 in Fig. 1). Using adequate tool support to query the UML model, a safety engineer can for instance check whether all safety requirements are handled by at least one class, that all the safety-related events correspond to some reaction and are handled in the design. Thus, safety engineers can determine how safety requirements are designed into the system (usage scenario 2 in Fig. 1), and how design decisions are justified (usage scenario 3 in Fig. 1). It is worth mentioning that reports from such query do not need to refer to UML terminology and concepts. This is important as safety engineers are usually not experienced in UML, object-orientation or programming. Additionally, although we have not developed a prototype tool that supports the profile and implements such queries, we ensured that the technology exists. For instance, the Rational Software Architect tool provides an OCL query engine to query a UML model, which could also benefit software engineers, e.g., they could use them to ensure for instance that all safety requirements are handled.

Similar querying mechanisms could also be developed for certification purposes, to prove compliance with airworthiness requirements (usage scenario 5 in Fig. 1). Here are a number of examples, extracted from the standard [14]. Certifying authorities require for instance that each safety requirement be implemented. This was discussed above (events/reactions). Certification authorities also require that software levels and hardware/software interfaces be clearly specified. Software levels can be obtained by querying the model for all the model elements with the <<SafetyCritical>> stereotype and reading its CriticalityLevel tagged value. Hardware/software interfaces can be obtained by querying the model for all model elements with the <<Interface>> stereotype and reading its IsBetweenHardwareAndSoftware tagged value. Certification authorities require that partitions be clearly specified, and our profile has a <<Partition>> stereotype. More complex queries could also be implemented. For instance, a query could ensure that the software level of a component be equal to at least that of the highest software level for all the components that depend on it.

In conclusion, our profile clearly supports all but one (usage scenario 1) of the usage scenarios discussed in Section 3.1, which described how safety information is used in practice.

6 Conclusion

Communication and collaboration among safety stakeholders, namely safety engineers and software engineers, has been found to be a major challenge when building safety-critical software. To address this issue, we have first focused on one specific domain, specifically commercial and military aerospace systems. We first identified and modeled, from the RTCA DO-178B standard (a.k.a. the airworthiness standard), the information that both engineering groups have to exchange and how they would use such information in practice (usage scenarios): safety engineers communicate safety requirements to software engineers, software engineers design software with safety in mind and record design decisions, safety engineers monitor safety during design and prepare reports for certification. We modeled this information in terms of 27 safety-related concepts (and 48 attributes) and their relationships thus leading to a conceptual model (class diagram). Although primarily focussing on aerospace software systems, we made sure that those concepts would also be usable for other safety critical software systems.

Because we focus on embedding safety information in UML models, the de-facto standard for designing object-oriented software, we used the defined safety conceptual model to compare existing UML-based approaches (e.g., existing profiles). Our conclusion was that existing UML-based approaches are far from being adequate for our purpose: they only capture a small portion of the safety concepts in our conceptual model. We therefore defined our own UML profile, composed of 32 stereotypes and 80 tagged values, focusing first on adding safety information to class diagrams.

This profile was applied on a realistic case study, a navigation controller subsystem that can control an aircraft's flight paths through both automatic pilot and manual input from the pilots. We designed the controller subsystem using our profile, i.e., embedding safety information in the model, following standard procedures required by the airworthiness standard (e.g., we performed a safety analysis of the controller subsystem). We then showed that our profile can indeed facilitate communication between software engineers and safety engineers and the certification process (the usage scenarios).

Future work should first focus on (1) providing full automation for the profile, (2) extending it to other diagrams such as sequence diagrams and state machines, and (3) evaluating its usefulness in practice through additional case studies.

Acknowledgements. This work was performed within the framework of Gregory Zoughbi's Master's thesis. This work was partly supported by a Canada research Chair (CRC) grant. Lionel Briand and Yvan Labiche were further supported by NSERC discovery grants.

References

[1] CENELEC EN 50128: Railway Applications: Software for Railway Control and Protection Systems, Version (1997)
[2] Gomaa, H.: Designing Concurrent, Distributed, and Real-Time Applications with UML, Object Technology. Addison Wesley, Reading (2000)
[3] Hansen, K.T., Gullesen, I.: Utilizing UML and Patterns for Safety Critical Systems. In: Proc. Workshop on Critical Systems Development with UML, in conjunction with the International Conference on the UML (2002)
[4] Hayhurst, K.J., Holloway, C.M.: Challenges in Software Aspects of Aerospace Systems. In: Proc. Annual NASA Goddard Software Engineering Workshop (2001)
[5] Herrmann, D.S.: Software Safety and Reliability: Techniques, Approaches, and Standards of Key Industrial Sectors. Wiley, Chichester (2000)
[6] International Electrotechnical Commission (IEC), Functional Safety of Electrical/ Electronic/Programmable Electronic Safety-Related Systems, IEC 61508 (1998)
[7] Jürjens, J.: Developing Safety-Critical Systems with UML. In: Stevens, P., Whittle, J., Booch, G. (eds.) UML 2003. LNCS, vol. 2863, pp. 360–372. Springer, Heidelberg (2003)
[8] Leveson, N.G.: Safeware - System Safety and Computers. Addison-Wesley, Reading (1995)
[9] Meunier, J.-N., Lippert, F., Jadhav, R.: RT Modeling with UML for Safety Critical Applications - the HIDOORS Project Example. In: Proc. Workshop on Specification and Validation of UML Models for Real-Time and Embedded Systems, in conjunction with the International Conference on the UML (2003)
[10] Nilsen, K.: Certification Requirements for Safety-Critical Software. RTC Magazine (2004)
[11] OMG: UML Profile for Modeling Quality of Service and Fault Tolerance Characteristics and Mechanisms (2005), Final Adopted Submission, http://www.omg.org/docs/ptc/05-05-02.pdf
[12] OMG: UML Profile for Schedulability, Performance, and Time Specification, Adopted Specification (2005), http://www.omg.org/docs/formal/05-01-02.pdf
[13] Pender, T.: UML Bible. Wiley, Chichester (2003)
[14] RTCA: Software Considerations in Airbone Systems and Equipment Certification, Radio Technical Commission for Aeronautics (RTCA), European Organization for Civil Aviation Electronics (EUROCAE), Standard Document no. DO-178B/ED-12B, (December 1992)
[15] Zoughbi, G., Briand, L.C., Labiche, Y.: A UML Profile For Developing Airworthiness-Compliant (RTCA DO-178B) Safety-Critical Software. Carleton University, Technical Report SCE-05-19 (December 2006)

Forensic Debugging of Model Transformations

Mark Hibberd, Michael Lawley, and Kerry Raymond

Queensland University of Technology, Brisbane, Australia
mt.hibberd@student.qut.edu.au,{m.lawley,k.raymond}@qut.edu.au

Abstract. Software bugs occur in model-driven development, just as they do with traditional development techniques. We explore the types of bugs that occur in model transformations and identify debugging approaches that can be applied or adapted to a model-driven context. Investigation shows that the detailed source-to-target traceability available with model transformations enables effective post-hoc, or forensic, debugging. Forensic debugging techniques are introduced for automated bug localisation in model transformations. The methods discussed are grounded with examples using the Eclipse Modeling Framework (EMF) and Tefkat, a declarative model transformation engine.

1 Introduction

As model-driven engineering techniques have become widely adopted in commercial environments the need for related high quality, pragmatic engineering processes has become very apparent. A key aspect of this is the need for efficient and effective debugging techniques.

Model transformations form the backbone of model-driven engineering, and correspondingly become a primary point of failure. Transformation development faces similar challenges to traditional programming; specifically, the possibility of human error in any stage of the development life-cycle, thus the need for debugging.

Debugging is readily classified into three parts, identifying the existence of a problem, fault localisation and the actual correction of the problem [1]. Generally, once a problem is located, a developer who is adequately experienced with the technology can correct the problem with minimal effort. Traditionally the majority of effort is spent on bug localisation [1,2,3] and the case in model transformations is no different. Based on the premise that automation of bug localisation will provide the greatest benefit to the developer or modeler, we describe debugging primarily in terms of this localisation.

In this paper we address bug localisation for model transformations in four key parts: the identification of questions that are asked when debugging model transformations; the classification of model transformation bugs into a set of bug classes and patterns; the exploration of debugging approaches that can be applied or adapted to these types of bugs; and, the demonstration of these approaches to automate bug localisation in model transformations.

G. Engels et al. (Eds.): MoDELS 2007, LNCS 4735, pp. 589–604, 2007.

2   Concepts and Context

The goal of a model transformation is to produce one or more target models, from an input of one or more source models. When talking about source and target models in the context of a model transformation, there are two parts, the *meta-model* that describes, or defines, the model and the model *instance*, which is a specific occurrence of the meta-model.

2.1   Model Transformation Tools

There are a number of model transformation tools available which utilise different techniques to solve the transformation problem. The defining characteristics [4,5,6] of these techniques include:

- how the transformation is specified; as a general purpose language or as a problem specific language.
- the transformation approach; either an imperative or declarative approach.
- what needs to be transformed; the types, described by a meta-model or free text; the number of models involved, 1..m source and 1..n target models
- the traceability between transformation artifacts; firstly the detail of the traceability (if any) and secondly the directionality, i.e. can the trace be followed source-to-target, target-to-source or both.
- the level of automation; applied programatically or manually.

To look at debugging problems, the transformation approach and traceability of the transformation are the most important characteristics. It is assumed that most practical transformations will achieve an adequate level of automation and that both source and target models are instances of a well defined meta-model. The number of models involved in the transformation is disregarded as the debugging concepts should extend to any number of models.

Using these characteristics as a guide, we have utilised the Eclipse Modeling Framework (EMF) [7] and the Tefkat [8] model transformation engine. Tefkat uses a declarative approach to model transformations. It has a formal trace model, which links the target, source and transformation. Another important feature of Tefkat is that the abstract syntax of a transformation is represented as a model, with corresponding meta-model. This allows the trace model to accurately reference the transformation as well as the source and target models.

Declarative approaches, like Tefkat, concentrate on *what* relationships exist between the source and target, compared with imperative approaches which concentrate on *how* to explicitly transform from the source to target. By defining only the relationships, declarative transformations allow for complete and correct transformations to occur without concern for execution order, source traversal and target creation. The use of a declarative approach does introduce complex concepts that make traditional imperative debugging techniques difficult. The most obtrusive of these is the lack of a pre-defined execution order. This makes interactive or step-through debugging difficult as the execution order is independent from the concrete syntax.

2.2   The Model Transformation Environment

To identify and localise bugs in model transformations, the model transformation environment must be understood. This environment may vary depending on the specific model transformation technologies, however similar concepts are applicable to most types of transformation. The core artifacts in the model transformation environment are the source model(s), the target model(s), the transformation and any available trace information. The Tefkat model transformation environment (figure 1) is comprised of:

- Source extents: One or more source model instances and their meta-models.
- Target extents: One or more target model instances and their meta-models.
- The transformation: The model transformation and its meta-model.
- The trace extent: A trace model instance and meta-model that links target objects to the source objects that contributed to their creation and the transformation rule involved (see figure 2).

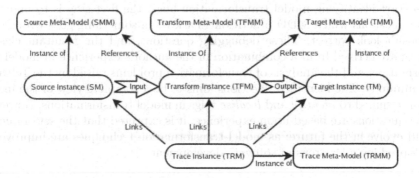

Fig. 1. The model transformation environment

The trace extent is a key enabling factor for the techniques presented in this paper. Source-to-target traceability was identified as a key requirement in early model research [9,10,11]. Gerber et al. [12] even noted the possibility of utilising trace information for tasks such as debugging, change propagation and round-trip engineering. As discussed further in section 4, this trace information is yet to be fully exploited for debugging.

The information contained in the trace extent can be leveraged for more effective post-hoc debugging techniques than is possible with traditional languages. Figure 2 provides a visual representation of the information contained by the Tefkat trace model. Each trace object references a target object, the transformation rule which created the target object and the source object(s) which contributed to its creation.

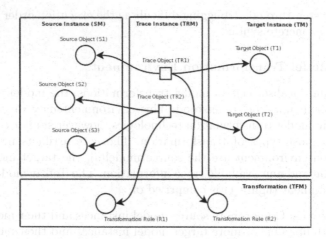

Fig. 2. Model trace environment

3   Model Transformation Bugs

To start identifying model transformation bugs, the first step is to understand the questions which MDD practitioners ask when something goes wrong or just doesn't look correct. These debugging questions, and the resultant classes of bugs are derived from a combination of the author's experience in model transformations and the analysis of transformation problems raised by the Tefkat user community. The identified questions aims to be a complete view of the information required to *identify* and *localise* bugs in model transformations. However, as the questions are based upon experience, it is expected that the set of questions will evolve in the future; as model-transformation techniques are improved and adopted in wider areas of software development.

3.1   The Debugging Questions

The debugging questions commonly asked by a modeler can be divided into two high-level categories. These categories are characterised by model transformations that produce *incorrect* output, logical bugs, compared with those that produce *invalid* output, well-formedness bugs.

Logical bugs, category A, can be identified by the violation of a relationship constraint between the source and target of a given transformation. Commonly these are only informal or implicit constraints, such as, we expect a few x's in the target as we know there are some y's in the source. There are several ways which these constraints could be provided. The first, the constraint could be provided by an oracle, the modeler, as a direct input to the debugging process. Next the constraint could possibly be inferred from the transformation itself. Finally, the constraint could be specified as a part of a formal testing or validation framework. This paper only deals with the first case, but recognises the benefits of, and the requirement for, more formal specification and/or automated discovery of this type of constraint. The set of logical bug related questions:

A.1 Why are there no objects of type t in the target?
A.2 Why are there so many objects of type t in the target?
A.3 Why is there only one object of type t in the target?
A.4 Why didn't source type, t, result in any target objects being created?
A.5 Why doesn't object x contain any references?
A.6 Why does a particular reference point to object x?
A.7 Why isn't reference r set?
A.8 Why are references r_1 to r_n out of order?
A.9 Why does attribute a have value v?
A.10 Why isn't attribute a set?

Well-formedness bugs, category B, can be identified by violation of the constraints specified by the target meta-model(s). Handling a model which is invalid with respect to its defining meta-model is a more difficult problem than the incorrect output case. To address this set of questions, debugging tools shall require special case handling and dynamic discovery of the structure of model instances. Dealing with invalid output models is out of scope for this paper, however it is an important direction for future model-transformation debugging research.

The set of well-formedness bugs:

B.1 Why isn't object x contained?
B.2 Why was the single valued reference, r, assigned more than once?
B.3 What violated meta-model constraint c?
B.4 Why is there no target model at all, i.e. no output compared with empty output as described by question A.1?
B.5 Why is there an instance, x, of an abstract class c?
B.6 Why is there an instance, x, that has been created with two different classes c_1 and c_2?

The questions provided for categories A and B are parametrised, indicating the requirement for a debugging context to be well defined and provided as input to the question. The problem of identifying this debugging context results in the definition of a third set of questions, analysis questions. Analysis questions, category C, encompass two sub-groups, *bug smells* or static-analysis questions, category C.I, and information-discovery questions, category C.II. These questions are more about refinement of the problem than debugging questions but they are relevant to localising bugs.

Bug smells represent a pattern or relationship between the source, target and transformation that are commonly the result of a bug. It is important to note that these smells are not always bugs, sometimes there will be a legitimate reason for a bug smell pattern to be found. An example of a bug smells question is question C.I.1.

C.I.1 Which source objects did not contribute to any target objects?

Bug smell questions often need to be refined to produce more meaningful output. In the example (question C.I.1), there may be a lot of cases where it is acceptable for a source object to not contribute to any target objects. An example of this is where a transformation is not completely exhaustive for the source meta-model. Any objects not referenced by the transformation will not contribute to the target, but is clearly not a bug. This process leads to questions aimed at refining the output. Extending the example in question (question C.I.1).

    C.I.2 For all source types, which source objects of the selected type did not contribute to the creation of any target objects?

As these debugging questions evolve, it is apparent that there is a need for supplementary information to use as input to the parametrised debugging questions. This supplementary information is gathered by asking information-discovery questions, category C.II. The information required to answer these questions is often directly available in the trace model, however in large transformations it can still be quite time consuming and error prone to access the information without tool support. The category C.II questions identified are.

    C.II.1 Given a target object, what source objects contributed to its creation?
    C.II.2 Given a source object, what target objects did it contribute to?
    C.II.3 Given an object type, which transformation rules reference the type?
    C.II.4 Given a target object, what are the relevant slices of the transformation that could effect its creation and/or attributes?
    C.II.5 Which source objects contributed to the creation of target objects?

Figure 3 gives an overview of the debugging question categories that have been defined. The next step in identifying model transformations is to turn these debugging questions into a set of bug categories.

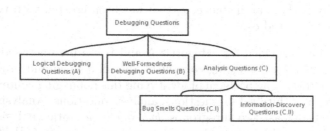

Fig. 3. Question categories

3.2 Classes of Bugs

The debugging questions raised in section 3.1 allow the classification of possible bugs in model transformations. The bug classes identified can be used to facilitate several decision making processes, including allowing appropriate debugging approaches to be linked to specific bug scenarios. Following is a set of bug classes with descriptions to identify the different bug scenarios.

Existence bugs: Existence relationships often exist between the source and target, e.g. for all source objects of type x there will be only one target object of type y. These bugs are characterised by the debugging questions A.1-4. Existence relationships are often specified as informal requirements rather than strict rules, which can make them more difficult to identify.

Containment bugs: Meta-models which define containment references expect a strict set of semantics to be adhered to. There are two bugs which result in a containment reference being violated. An object that should be contained is not, or too many objects are contained by a single container.

Bi-directional reference bugs: A common bug pattern with bi-directional references is where both ends of the reference don't point back at each other. Bi-directional constraints are enforced by EMF; this means that a bi-directional reference bug will only result from a bug in the meta-model.

Range bugs: Range bugs occur where there are invalid values in the target instance with respect to the constraints defined by the target meta-model.

Completeness bugs: Completeness bugs occur when some non-optional part of the target is not generated as a part of the transformation.

Well-formedness bugs: Well-formedness bugs are closely related to the category B debugging questions which result from invalid output. A well-formedness bug occurs when the target model instance does not conform to the target meta-model. Well-formedness bugs are a superset of a number of other bug categories, including completeness bugs and containment reference bugs.

Technology specific bugs: Technology specific bugs occur as a result of the model transformation tools and techniques used. This paper limits the discussion of technology specific bugs as there is limited benefit in approaching technology specific problems from a generic model-driven development perspective.

Using these types of bugs as a reference, debugging techniques will now by analyzed to determine effective approaches to bug localisation that may be applied to model transformations.

4   Debugging Techniques

Localisation is the key facet of any debugging process. Figure 4 visualises the goal of bug localisation. The *before* snapshot represents a situation of a buggy transformation, the developer knows there is a bug. However, there is a large area (represented in white) of unexplored code where the bug may be located. The *after* snapshot shows how a debugging process can be applied to narrow the unknown area, and inturn help pinpoint the bugs location. It is important to note that a realistic goal of bug localisation is not to pinpoint the precise problem, merely to localise the possible causes to a minimum area. It is accepted that there may always be some level of developer interaction required to go from a localised bug to the correct solution.

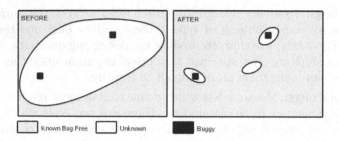

Fig. 4. Before and after localisation of potential problems

To achieve this bug localisation there are two primary categories which can encapsulate most techniques: post-hoc or *forensic* debugging, and interactive or *live* debugging.

4.1 Forensic vs Live Debugging

The key difference between *forensic* and *live* debugging is that live debugging requires access to a complete runtime environment that is not required for forensic debugging. Live debugging may always be required to solve more complex bugs. However in model-driven engineering, the inherent traceability and the well-understood nature of the source and target in model transformations provide a unique opportunity for forensic debugging techniques. Leveraging the additional traceability and artifacts available, a higher level of detail and most importantly automation can be achieved with forensic debugging of model transformations compared with traditional programming languages.

This paper concentrates on a static or *forensic* approach to bug localisation. There are many advantages in pursuing the forensic case over live debugging. These include:

- The time and effort cost for the developer usually associated with interactive debugging.
- In a software development community where automation is continually being pursued, forensic debugging can save valuable resources by using information from failed builds or tests to track down problems rather than relying on developers to re-run the problematic program/transformation to perform live debugging tasks.
- Offline debugging means that bugs can be localised in inaccessible environments, such as production environments where live debugging is often impossible.
- Transient errors or heisenbugs[1] can be difficult or impossible to reproduce in a live debugging environment.

To assist in the development of debugging techniques that can take advantage of these aspects of forensic debugging we look to past debugging research in comparative areas of software development.

---

[1] A bug that disappears or changes its behavior when debugging [13].

4.2 Learning from the Past

There has been no significant research into the post-hoc debugging possibilities specific to model transformation. However, the declarative paradigm used by Tefkat and many other transformation engines is not new to software development. They share similar debugging problems as those seen in fifth generation and logic languages such as Prolog and Mercury. The difficulty in debugging these declarative languages is well-understood, with significant research into debugging techniques such as program slicing [14,15] and algorithm debugging [16,17,18,19,20].

Many traditional automated debugging techniques such as anomaly detection [21], test based fault localisation [3], statistical based fault localisation [22] and nearest neighbour queries [23] struggle with the paradigm shift from an imperative to declarative approach. However, there is also a portion of automated debugging research which can apply equally, or at least be adapted, to suit both imperative and declarative programs. These approaches include using data-flow analysis to help with program slicing [24], predicate switching [25] and knowledge-based localisation [1].

5 Localisation of Model Transformation Bugs

To address the debugging questions, section 3.1, we present two forensic debugging approaches: analysis and re-enactment. These approaches adapt and extends the techniques discussed above in section 4.2, to best suit forensic debugging of model transformations.

5.1 Analysis

Analysis involves gathering evidence from the artifacts available in the normal model transformation environment (see figure 1). At its simplest level this is simply a collation and refinement of the available data. Analysis is best suited to addressing category C debugging questions by identifying bug smells, and gathering evidence to be used as input to the re-enactment processes. All the information is readily available, however the volume and complexity of the output can prevent viable manual processing.

We have experimented in the use of analysis techniques to gather the information required for bug-localisation. To implement and automate the information gathering required, we have used two different methods. Firstly, programatically through the EMF API and secondly, as a Tefkat transformation where the static environment; the original source, target, transform and trace; form the transform inputs and the transform output is a reference or set of references which answer the query. Both techniques have been successful, and the best choice of implementation depends greatly on the specific tools and automation techniques that are being utilised.

The following sections describe, in generic terms, how the model transformation environment can be utilised to answer some of the debugging question using analysis.

Tracing from a target object to its contributers. To address question
C.II.1, the required information is readily provided by the trace model (TRM).
A direct look-up for each target object will find the rule which created it and
the source objects which contributed to its creation.

Tracing from source objects to target objects. Question C.II.2 is effectively addressed by the algorithm specified for question C.II.1, with the source
and target roles reversed.

Source objects that contributed to the creation of target objects. The
source references in the trace model (TRM), ($TRM[source\text{-}references]$), is a
subset of the source model (SM). Using this, question C.II.5 is addressed by
finding the intersection of all the source references in the trace model and the
objects in the source model.

$$x = TRM[source\text{-}references] \cap SM$$

**Source objects that did not contribute to the creation of a target
object.** Similarity, question C.I.1 is addressed by determining the relative com-
plement of all the source references in the trace model and those in the source
model.

$$y = TRM[source\text{-}references] - SM$$

As discussed in section 3.1 and presented in question C.I.2, the output of
this question must be refined to produce a useful result. An additional filter is
applied to reduce the output; the objects found by the first process that have a
type referenced by the transformation (TFM).

$$z = \{o \mid o \in y \wedge o.class \in TFM[MOFInstance]\}$$

Analysis of the model transformation environment has provided enough information to answer straight forward, query based questions. Re-enactment extends
this information to pinpoint specific rules and/or terms in rules that trigger a
bug.

5.2   Re-enactment

Re-enactment involves the selective re-execution of logical parts of the model
transformation in a controlled runtime environment to gather knowledge about
specific problems. Typically there are two parts to the re-enactment, determining
a part of the transformation which could potentially cause a given problem,
and executing that part in isolation. The execution phase of the re-enactment
will utilise program slicing [14,15] and predicate switching [25] to narrow down
possible failures over a number of iterations.

The re-enactment process developed involves executing modified slices of the
transformation in isolation. The transformation slices shall be created using predicate switching to replace irrelevant or at least suspected irrelevant parts of the

transformation. The predicate switching is implemented by replacing conditional terms with an explicit TRUE term. The re-enactment algorithms have been designed with automation in mind. As such, they utilise only information available in the model-transformation environment and they do not rely on any additional knowledge to be provided by the user.

Re-enactment is best suited to answering category A debugging questions. For the following examples it is assumed that the output is always valid, but is not the expected output.

Choosing a slice. The process of choosing a slice is dependent on the constraint being tested. Currently this research assumes that a slice has already been identified. There is space for future work in this area, as it may be possible to choose a slice using a heuristics based approach for selecting rules from the transformation or by interacting with a test framework that uses formal constraints to identify bugs.

An example. The first example, figure 5, shows a Tefkat transformation rule.

```
RULE FindPersistentClasses
  FORALL  UMLClass uml
  WHERE   uml.kind = "persistent" AND uml.parent.kind = "persistent"
  MAKE    UMLClass result
  SET     result.name   = uml.name, result.kind   = uml.kind,
          result.parent = uml.parent;
```

Fig. 5. Simple Tefkat rule containing a bug

This rule is working with a simplified UML meta-model, figure 6. The rule is attempting to locate all persistent classes, that is classes with a kind attribute of "persistent" or a parent with a kind attribute of "persistent".

The rule contains a bug in the conditional logic. The use of AND instead of OR prevents the finding any persistent classes. The modeler knows that there are some persistent classes in his input model, so he asks debugging question A.1,"why are there no *UMLClasses* in my output?". To answer this question re-enactment is used.

A simple slice. To answer the question posed, a *head-first* or *tail-first* predicate switching approach can be applied to this problem. Our experiments have identified benefits to both approaches. An important point to note when evaluating each approach is that the *head* or *tail* is logical only, to re-iterate the point in section 2, there is no explicit execution order so the terms in the rule may be re-ordered by the transformation engine as required. The choice of starting from the *head* or *tail* is arbitrary, but draws on techniques commonly applied by developers attempting to localise a bug.

The head-first switching algorithm is shown in figure 7. This approach replaces all conditional terms with a TRUE term, adding the conditions back one at a time until the transformation output goes from the *correct* output to the *buggy*

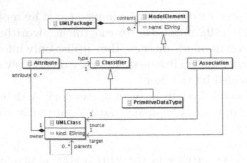

Fig. 6. Simple UML meta-model

output. The last term switched is identified as a potential problem. If no terms made a difference to the output it indicates that the input to the rule actually caused it to produce the unexpected output.

The tail-first switching algorithm is shown in figure 8. Tail-first switching iterates through each conditional term, replacing it with a TRUE term until the transformation goes from the "buggy" output to the "correct" output. Similar to the head-first approach, if no terms made a difference to the output it indicates that the input to the rule actually caused it to produce the unexpected output.

To produce a complete picture it is possible to combine both approaches. Often both approaches will return the same result, but it is possible that two potential problems can be identified. An advantage of using both approaches is the removing terms in different orders helps the elimination of *down-stream* bugs; those which would not occur except for a problem earlier on in the rule.

In the uml example, applying both of these rules highlights the uml.parent .kind = "persistent" term. As highlighted by figure 4 and section 5 this may not be the root cause of the bug, however it localises the problem sufficiently to realise that it doesn't make sense that the term always has to be true and the bug can be corrected by modifying the AND to an OR.

```
1  Select source term for rule (current slice)
2  mofInstances <- new list, conditions  <- new list
3  For each term
   3.1  If source term is a MOFInstance, add to mofInsances
   3.2  Otherwise term is potentially a condition, add to conditions
4  For each condition in conditions
   4.1  Replace condition with TRUE term
5  Execute new version of rule
6  If result contains NO required objects, the input is at fault,
   return mofInstances as potential bug
7  For each condition in conditions (from head to tail)
   7.1  Replace TRUE term with condition
   7.2  Execute new version of rule
   7.3  If result contains NO required objects, return condition
8  return Rule is OK
```

Fig. 7. Head-first predicate switching

```
1  Select source term for rule (current slice)
2  mofInstances <- new list,  conditions <- new list
3  For each term
   3.1  If source term is a MOFInstance, add to mofInsances
   3.2  Otherwise term is potentially a condition, add to conditions
4  For each condition in conditions (from tail to head)
   4.1  Replace condition with TRUE
   4.2  Execute new version of rule
   4.3  If result contains any of required object, return condition
5  No terms effected output, the input is at fault,
   return mofInstances as potential bug
```

Fig. 8. Tail-first predicate switching

There are a number of caveats to this approach. Most importantly, it is not possible to easily differentiate between a source term that will bind a variable and one that acts as a condition or filter. This means that removing the source term could break the injection part of the rule and cause the transformation to flounder[2]. To address this problem the transformation rule can be modified to not depend on any variables possibly bound in the source term. The first step to this is eliminating all target conditions (SET clause in the example). The second step is to eliminate all non-default injections. The example does not have any non-default injections, which take the form of a MAKE/FROM clause. These changes to the transformation rule do not affect the algorithms in figures 7 and 8 as, although some values will differ from the "correct" output, there will be no changes to the objects that are created.

An advanced slice. The algorithms presented in figures 7 and 8 address a simple case where all the effecting logic is encapsulated within a single rule and with no branching. A more realistic example would involve the use of an OR condition, IF/THEN/ELSE statement, PATTERN use or implicit dependencies between rules created by LINKS/LINKING terms. These more complex structures require additional checks that must be made to ensure a complete set of results is determined. For example, in the case of OR, a potential problem could be identified for each branch within the rule.

Figure 9 shows the recursive execution of the predicate tail first switching algorithm on each branch of the OR condition. This algorithm can be inserted at step 7.1 in figure 7 or step 4.1 in figure 8. The first additional check is for an OR condition. If either of these statements are encountered, each of its branches must be traversed separately. It is possible that the conditional statement will result in 0, 1 or 2 additional results.

Other advanced constructs; IF/THEN/ELSE statements, PATTERNs and LINKS/LINKING statements; can be approached with similar predicate switching algorithms. The IF/THEN/ELSE case is identical to the OR case where each branch is replaced then the whole statement is replaced. Patterns can be addressed by identifying and recursively applying the predicate switching to each PATTERN

---

[2] The transformation can not complete execution as a rule is dependent on variable that is never bound.

```
1. If the condition is an 'OR' term
   1.1 Replace the first nested term with FALSE and recursive apply
       predicate switching algorithm to right hand side of 'OR' term
   1.2 Replace the second nested term with FALSE and recursive apply
       predicate switching algorithm to left hand side of 'OR' term
   1.3 Replace entire 'OR' term with TRUE
2. Otherwise continue normal predicate switching algorithm
```

Fig. 9. Handling multiple branches

declaration, and finally to the PATTERN use. This approach can be used to identify any terms inside the PATTERN declaration that effect the output.

Dependencies between rules, normally identified by the LINKS construct in
Tefkat, present additional problems. In the simple slice example it was noted
that floundering could be prevented by modifying the MAKE and SET clauses
to only depend on variables that are not bound by the terms involved in the
predicate switching. A rule containing a LINKS term may not depend on any
other input and as a result the LINKS term can not be switched out. There
are two approaches to handling this situation. Firstly, the LINKS term can be
processed last (similar to the MOFInstances in the simple slice). If none of the
other terms affect the output then it can be said the rule does not produce the
expected output as no dependent objects were created. This information can be
used to identify the rules which create those dependent objects, allowing the
debugging questions to be asked again for the new rule. The second approach is
to ensure that the dependency always exists. This approach is useful when there
is more than one LINKS term that must be processed.

6 Conclusion

The key to addressing the debugging problem, with respect to model transformations, is understanding the types of questions raised when a problem is identified.
In section 3.1, we presented a framework, as a set of questions, to define the goals
of model-tranformation debugging.

Utilising forensic debugging approaches we have addressed a number of the
model transformation debugging questions highlighted. We have demonstrated
the potential that leveraging the trace available in model transformations brings
to forensic debugging. We have also demonstrated the adaptability of previously
live debugging approaches into forensic algorithms.

Analysis techniques do benefit from leveraging the current trace information.
However as the research has progressed it has highlighted the potential for improvements to the information provided by the trace model. Some of these possible enhancements include linking target objects to the specific *injection* that
created them and also the rules that resulted in the objects' attributes being set.

The re-enactment approach is able to greatly extend the value which forensic
debugging can provide. However, it is important to realise that the re-enactments
can rarely (if ever) provide a definitive answer to its queries without help from the

user. That said, it contributes significantly towards solving the original problem of localising the fault and minimising developer debugging effort.

References

1. Sedlmeyer, R., Thompson, W., Johnson, P.: Knowledge-based fault localization in debugging: preliminary draft. In: Proceedings of the ACM SIGSOFT/SIGPLAN software engineering symposium on high-level debugging, vol. 8(4), pp. 25–31. ACM Press, New York (1983)
2. Ducassé, M., Emde, A.M.: A review of automated debugging systems: knowledge, strategies and techniques. In: Proceedings of the 10th international conference on Software engineering, pp. 162–171 (1988)
3. Jones, J., Harrold, M., Stasko, J.: Visualization of test information to assist fault localization. In: Proceedings of the 24th international conference on Software engineering, pp. 467–477 (2002)
4. Mens, T., Czarnecki, K., Van Gorp, P.: A Taxonomy of Model Transformation. In: International Workshop on Graph and Model Transformation (2005)
5. Czarnecki, K., Helsen, S.: Classification of Model Transformation Approaches. Proceedings of the 2nd Object-Oriented Programming, Systems, Languages and Applications (OOPSLA) Workshop on Generative Techniques in the Context of the Model Driven Architecture (2003)
6. Czarnecki, K., Helsen, S.: Feature-based survey of model transformation approaches. IBM Systems Journal 45(3), 622 (2006)
7. Eclipse Foundation: Eclipse Modeling Framework Project (EMF) (2007) (accessed February 20, 2007), Available at: http://www.eclipse.org/modeling/emf/
8. Lawley, M., Steel, J.: Practical Declarative Model Transformation With Tefkat. In: Bruel, J.-M. (ed.) MoDELS 2005. LNCS, vol. 3844, p. 139. Springer, Heidelberg (2006)
9. Object Management Group: MOF 2.0 Query - Views - Transformations RFP. OMG Document ad/2002-04-10 (April 2002)
10. Miller, J., Mukerji, J., et al.: MDA Guide Version 1.0.1. OMG Document omg/2003-06-01 (June 2003)
11. DSTC-IBM-CBOP: MOF 2.0 Query/Views/Transformations, OMG Document ad/2004-01-06 (January 2004) (second revised submission)
12. Gerber, A., Lawley, M., Raymond, K., Steel, J., Wood, A.: Transformation: The Missing Link of MDA. In: Corradini, A., Ehrig, H., Kreowski, H.-J., Rozenberg, G. (eds.) ICGT 2002. LNCS, vol. 2505, Springer, Heidelberg (2002)
13. Bourne, S.: A conversation with Bruce Lindsay. Queue 2(8), 22–33 (2004)
14. Weiser, M.: Programmers use slicing when debugging. Communications of the ACM 25(7), 446–452 (1982)
15. Xu, B., Qian, J., Zhang, X., Wu, Z., Chen, L.: A brief survey of program slicing. ACM SIGSOFT Software Engineering Notes 30(2), 1–36 (2005)
16. Shapiro, E.: Algorithmic program diagnosis. In: Proceedings of the 9th ACM SIGPLAN-SIGACT symposium on Principles of programming languages, pp. 299–308. ACM Press, New York (1982)
17. Shapiro, E.: Algorithmic Program DeBugging. MIT Press, Cambridge, MA, USA (1983)
18. Fritzson, P., Shahmehri, N., Kamkar, M., Gyimothy, T.: Generalized algorithmic debugging and testing. ACM Letters on Programming Languages and Systems (LOPLAS) 1(4), 303–322 (1992)

19. Naish, L.: Declarative Debugging of Lazy Functional Programs. Dept. of Computer Science, University of Melbourne (1992)
20. Naish, L.: A Declarative Debugging Scheme. Department of Computer Science, University of Melbourne (1995)
21. Hangal, S., Lam, M.: Tracking down software bugs using automatic anomaly detection. In: Proceedings of the 24th International Conference on Software Engineering, pp. 291–301 (2002)
22. Liu, C., Yan, X., Fei, L., Han, J., Midkiff, S.: SOBER: Statistical Model-based Bug Localization. In: Proceedings of the 10th European software engineering conference held jointly with 13th ACM SIGSOFT international symposium on Foundations of software engineering, pp. 286–295. ACM Press, New York (2005)
23. Renieres, M., Reiss, S.: Fault localization with nearest neighbor queries. In: Proceedings 18th IEEE International Conference on Automated Software Engineering, 2003, pp. 30–39. IEEE Computer Society Press, Los Alamitos (2003)
24. Agrawal, H., Horgan, J.R., London, S., Wong, W.E.: Fault localization using execution slices and dataflow tests. In: Proceedings Sixth International Symposium on Software Reliability Engineering, pp. 143–151 (1995)
25. Zhang, X., Gupta, N., Gupta, R.: Locating faults through automated predicate switching. In: International Conference on Software Engineering, pp. 272–281 (2006)

Runtime Debugging Using Reverse-Engineered UML

Orest Pilskalns, Scott Wallace, and Filaret Ilas

School of Engineering and Computer Science
Washington State University, Vancouver
{orest, swallace, filas}@vancouver.wsu.edu

Abstract. Finding runtime faults in object-oriented code can be very difficult even with the aid of modern runtime debuggers. Failures may manifest themselves due to decisions in the code that were executed much earlier in the program. Tracing execution paths and values backward from a failure to the faulty code can be a daunting task. We propose a fault finding approach that uses unit tests to exercise source code in order to trace object-method execution paths. This is similar to reverse-engineering techniques used to create Sequence Diagrams from code. It is often too complex to debug a program using a large set of reverse-engineered Sequence Diagrams each obtained from an individual execution. Therefore, our approach partitions and aggregates individual execution paths into into fault and non-fault revealing categories. By examining the differences between fault and non-fault paths, we are left with a simplified graph. The graph can then be transformed into a useful Sequence Diagram that may reveal the location of the faulty code.

1 Introduction

A fault can be defined as missing or incorrect code. A failure is the observable inability of a system or component to perform a function due to a fault [3]. A runtime failure is a failure that is revealed by running the code. Tests may reveal where a runtime failure occurs, but they often do not reveal the location of the fault responsible for the failure. We provide an approach that aids in finding runtime faults by using reverse-engineered UML Sequence Diagrams.

The current approach to finding runtime faults relies on using a debugger usually built into an integrated development environment (IDE). Debuggers allow users to execute and observe the state of their program. Debuggers provide a friendly interface for observing the stack frames of an executing program. Variables can be flagged for observation, which is called setting a watch. A user can choose to execute the program one line at a time known as stepping through code or set a break point which allows the program to execute up to a pre-specified location and then halt. The user can than start the execution again if needed. Usually a combination of breakpoints and stepping are called for. Once a failure manifests itself, the following steps can be taken using a debugging tool:

1. Place a watch on suspicious variables.
2. Set break points.

G. Engels et al. (Eds.): MoDELS 2007, LNCS 4735, pp. 605–619, 2007.
© Springer-Verlag Berlin Heidelberg 2007

3. Execute the suspicious code segment and observe variables.
4. Step through code as needed.

This process is repeated until the error is found or the user gives up.

2   Hard to Find Faults

The traditional approach to finding runtime faults works reasonably well if the failure occurs in close proximity to the underlying fault. This is the situation, for example, when both fault and failure occur within the same method. As the distance between failure and fault increases, however, the traditional approach of setting break points and stepping through executing code becomes much more difficult.

2.1   Notions of Distance

A straightforward notion of distance between the failure and the fault could be measured as lines of executed code; we call this *syntactic* distance. Intuitively, syntactic distance between fault and failure has an obvious relationship on the difficulty of a particular debugging task. The larger this distance, the more code must be examined. Worse, since the developer will be examining source code (as opposed to an execution trace), they must reconstruct (either mentally, or through repeated execution) how decisions made at each branch point in the source code conspired to create the observed failure. As the syntactic distance between fault and failure grow, the number of branches that must be examined will likewise increase. Without accurately knowing which branch has been taken during a particular execution, this effect will lead to an exponential growth in the code base relevant to this search.

A second notion of distance is based on the developer's intuitive beliefs about where a particular fault may be located compared to the actual location of the fault. These intuitions can be viewed as creating a priority-queue of source-code locations that the developer will examine. We call the location of the true fault in this priority-queue *heuristic* distance. That is, as the developer's intuition guides them down incorrect paths, the heuristic distance is large. However, as more and more possibile locations are eliminated, the location of the true fault will necessarily rise toward the top of the priority queue and the heuristic distance will likewise decrease. The heuristic distance is thus critically related to how much time the debugging task is likely to require.

In the traditional approach to debugging, there is an important interplay between syntactic distance and heuristic distance. As syntactic distance increases, the developer will have to incrementally set break points further and further back from the failure. Because the number of branches that may need to be explored will be high, the heuristic distance is also likely to be initially high and will be higher still if the fault occurs in a seemingly unrelated area of the source code. In other words, as the syntactic distance increases, the developer is likely to spend significantly more time looking for the fault.

The correlation between syntatic and heuristic distance is not fundamental. We believe that given an appropriate set of tools it is possible to substantially reduce or even remove the correlation between these two measures. In the following section, we provide a detailed illustration of one failure situation in which the heuristic distance is likely to be relatively high. Then, we present a new tool that can be used to quickly isolate sections of the source code that are likely contributors to the failure. By visually identifying these potentially problematic sections, our tool is able to decouple the relationship between syntactic and heuristic distance, thus making hard-to-find faults readily apparent.

2.2   Cache (In)Consistency: An Example

Consider two objects: a DataSource that provides a volatile dataset based on its own internal state (see Listing 1.1); and a DataConsumer that examines the datasets provided by multiple DataSource objects.

Listing 1.1. DataSource

```
public class DataSource {

  Version version = new Version();

  // two internal state values impact the results
  // returned by getData()
  private int type;           // internal state
  private double initialvalue; // internal state

  public int getVersion() {
    // the version is used to indicate changes to the
    // DataSource's internal state. So long as the
    // version remains unchanged, calls to getData()
    // should return the consistent results.
    return version;
  }
  public DataSet getData(int n) {
    // do something potentially complicated based on the
    // internal state and return a dataset with n elements...

    // source code continues...
```

DataSource's job of generating the dataset may be complex, requiring significant computational overhead. If this is the case, it may make sense for the DataConsumer to store these datasets in a cache. In this way, the DataConsumer can control how storage is allocated, optimizing the caching policy for its own purposes. The object in Listing 1.2 follows this model. Note also that two DataConsumer methods (invert() and setInitialValues()) manipulate the underlying DataSource objects. This, in turn, impacts the datasets that will be produced by the getData() method.

Listing 1.2. DataConsumer

```java
public class DataConsumer {
  DataSource[] srcs;
  // the cache maps DataSets to their DataSource and keeps a
  // Version number to quickly check if the DataSet is stale
  Cache<DataSource,DataSet,Version> cache;

  public void setInitialValues(double d) {
    srcs[0].setInitialValue(d);
    srcs[1].setInitialValue(d);
  }

  public void invert() {
    int t = srcs[0].getType();
    srcs[0].setType(srcs[1].getType());
    srcs[1].setType(t);
  }
  // source code continues...
```

The DataConsumer uses the cache in its getPoint() method. The getPoint() method is illustrated in Listing 1.3 and is intended to retrieve a particular data point for one of the datasets.

Listing 1.3. DataConsumer.getPoint()

```java
public Point2D getPoint(int s, int n) {
  CacheEntry e = cache.get(srcs[s]);
  DataSet ds;
  Version dsVersion = srcs[s].getVersion();

  if ( e == null ) {
    // if the dataset is not in the cache, fetch it
    ds = srcs[s].getData(size);
    cache.put( srcs[s], ds, dsVersion );
  }
  else if ( !e.getVersion().equals(dsVersion) ) {
    // inconsistency detected -- cache is stale, refresh it
    ds = srcs[s].getData(size);
    cache.put( srcs[s], ds, dsVersion );
  }
  else {
    // cache seems consistent
    ds = e.getData();
  }
  return (Point2D)ds.get(n);
}
```

The cache relies on a simple method for determining cache consistency. Each time a dataset is obtained from a `DataSource`, the version is also obtained and stored in the cache. So, to retrieve a data point, `getPoint()` first looks for a cache entry. If an entry is found whose version matches the `DataSource`'s current version, the entry is determined to be consistent and the point is fetched directly from the cache. Otherwise, `DataSource.getData()` is invoked to get a new copy of the dataset thereby refreshing the cache and providing the return results.

This approach relies on `DataSource.getVersion()` to indicate when the `DataSource` has changed in a manner that will affect the dataset it produces. A problem arises if this assumption is violated. Consider the two methods below, both of which change the internal state of a `DataSource` object.

Listing 1.4. DataSource methods

```
public void setInitialValue( double v ) {
  initialvalue = v;
  version = version.next();
}


public void setType( int t ) {
  // BUG: setting the type affects the data that an instance
  // would produce. We should increment the version number
  // to indicate such a change.
  type = t;
}
```

Listing 1.4 illustrates a type of fault that may be quite difficult to identify, especially given the context of how the `DataSource` object will be used. Because the `setType()` method does not appropriately increment the version, it makes any cached dataset susceptible to inconsistency. Moreover, since `setType()` is otherwise correct, a fault will only occur when this method is invoked between calls to the `Data-Consumer`'s `getPoint()` method. The net result is that syntactic distance is likely to be quite high. The fault will not occur in the same method that the stale data is used, nor will backtracing to the `invert()` method reveal the fault. Rather, the developer's search will need to continue back to `setType()`.

In addition to the syntatic distance between fault and failure, there is also likely to be significant heuristic distance in this situation. Consider the two unit tests in Listing 1.5.

In both tests, identical methods are invoked, yet only one test (`testFailure`) produces a failure. This of course, is because `invert()` calls `setType()` and since this method is invoked between calls to `getPoint()` the `DataConsumer`'s cache becomes inconsistent. A developer having noticed the success of the first test, might be inclined to assume that `invert()` and all of the methods it calls all work correctly. This would place a large heuristic distance between failure and fault and may mean that a considerable time is spent before the developer notices that it is `invert()`'s underlying calls to `setType()` which are actually responsible for the erronous behavior.

Listing 1.5. Unit Tests

```java
public void testSuccess() {
  // this test exercises the code
  consumer.invert();
  for( int i = 1; i < nTests; i++ ) {
    consumer.setInitialValues( i );
    // check end points
    assertEquals( new Point2D.Double( 0.0, i ),
            consumer.getPoint( 1, 0 ) );
    assertEquals( new Point2D.Double( 3, i+6 ),
            consumer.getPoint( 1, 3 ) );
    assertEquals( new Point2D.Double( 3, i-9 ),
            consumer.getPoint( 0, 3 ) );
  }
}

public void testFail() {
  // this test exercises the code
  consumer.setInitialValues( 1 );
  assertEquals( new Point2D.Double( 3, 7),
          consumer.getPoint( 0, 3 ) );
  assertEquals( new Point2D.Double( 3, -8),
          consumer.getPoint( 1, 3 ) );
  consumer.invert();
  // fails! cache is out of sync!
  assertEquals( new Point2D.Double( 3, 7 ),
          consumer.getPoint( 1, 3 ) );
  assertEquals( new Point2D.Double( 3, -8),
          consumer.getPoint( 0, 3 ) );

}
```

3   The Approach

Our approach relies on differentiating between successful code execution and fault revealing code execution. By tracing code execution of both successful and fault revealing
unit tests, we can create directed acyclic graphs that show the differences. These graphs
can be transformed into Unified Modeling Language (UML) Sequence Diagrams. UML
Sequence Diagrams are often used by Software Engineers to represent the behavior of
program in the design phase. Here we use Sequence Diagrams to reveal faults while
eliminating the unnecessary clutter of code-level detail. The following steps outline our
approach:

1. Create Unit Tests
2. Instrument the source code so message paths (and associated objects) can be traced.
3. Execute the tests and record objects and message paths.

4. Partition paths into fault and non-fault revealing partitions.
5. Aggregate all paths into a single graph and differentiate based on fault partitions.
6. Generate Sequence Diagram from differentiated graph.
7. Use Sequence Diagram to reason about fault.

3.1   Unit Tests

Our approach relies upon unit tests that provide coverage of the code that produces the fault. Additional unit tests are needed to provide coverage of the code using test cases that do not fail. Therefore, our method is applicable when the application is mature enough that some unit tests succeed, but no so mature as to pass all of the unit tests.

Test coverage is important since our objective is to differentiate between successful code and faulty code. If the coverage is inadequate then a failed unit test may have little in common with successful tests negating the usefulness of differentiating the two. Ideally, adequate coverage would reveal localized differences in the object method traces of successful and failed unit tests. It is reasonable to assume that as test coverage increases so should our success rate. However, there is the possibility that the fault exists at a lower level (statement level) of the code.

3.2   Instrument the Code

Instrumenting the code is the process of inserting tracing code that records the method execution calls between objects. This can be accomplished by inserting code that logs each method call, the calling object's id, and the calling object's class type. Logging could be done at the source code level, but would require tools for both inerting and removing the instrumentation code. We take an alternate approach that simplifies the process for the developer by automatically inserting tracing methods into the Java Byte code. After the debugging process is complete, the Java Byte code can be discarded, and the (unmodified) source simply recompiled.

Since our goal is to create Sequence Diagrams, we chose to track method calls. However, we could choose a lower or high granularity level. For example we could choose to track the execution sequence line by line, or we could only track messages between components. Additional empirical work and the creation of a fault model will be necessary to see the shorting coming or benefits of choosing method calls.

Our analysis tool uses the utilities in *org.apache.bcel* java library in order to accomplish the instrumentation of the byte code. Classes selected for instrumentation are loaded and injected with a reference to a static object named *LumberJack*. *LumberJack* uses a static counter to keep track of method calls and inserts trace code for each method *call* and each method *return*. The tracing code keeps track of the following information in the bytecode:

1. *method* contains information such as class name, method name and method signature (returned type and arguments);
2. *location* tracks the line number where the method occurs in the source code and indicates if the method is a return call or an initial call;
3. *runtime* tracks the order of calls during execution;

In addition, the *LumberJack* class provides tools for printing an XML representation of the trace logs.

3.3   Execute the Tests

During the execution process, the instrumented byte code is traversed using the unit tests. For this example, the unit tests provide branch coverage of the code. Each unit test generates an object-method trace through the code, which is recorded to an XML trace file. Listing 5 shows a sample of the data recorded in the trace file after the unit tests have been executed on the Cache Example. Each unit test trace is tagged as successful or unsuccessful based on the outcome of the test.

Listing 1.6. trace.xml

```
...
<vertex>
   <method>DataConsumer.invert ()V</method>
   <location>called: DataConsumer.java:40
   </location>
   <runtime>myCallCount=1</runtime>
</vertex>

<vertex>
   <method>DataSource.getType () I</method>
   <location>called: DataSource.java:47
   </location>
   <runtime>myCallCount=1</runtime>
</vertex>

<vertex>
   <method>DataSource.getType () I</method>
   <location>return: DataSource.java:47
   </location>
   <runtime>myCallCount=1</runtime>
</vertex>

<vertex>
   <method>DataSource.getType () I</method>
   <location>return: DataSource.java:47
   </location>
   <runtime>myCallCount=1</runtime>
</vertex>
...

<method>DataConsumer.invert ()V</method>
   <location>return: DataConsumer.java:45
   </location>
   <runtime>myCallCount=1</runtime>
</vertex>

...
```

3.4   Partition Paths

The previous step provides enough information to allow us to differentiate between successful and failed code execution. The generated trace files contain information about each test and thus potentially the fault. Given such information it is trivial to partition the paths into what we have named fault and non-fault revealing partitions. Thus, the trace files associated with successful tests are classified as non-fault revealing and likewise unsuccessful tests are classified as fault revealing.

3.5   Aggregate Paths

The merging algorithm aggregates all the trace paths generated during the unit test execution. Merging results in an acyclic graph where the vertices represent the actual method calls and the directed links between vertices specify the order of the method calls. Figure 1 displays the acyclic graph obtained for the Cache example. Every vertex in the graph contains the following information: id, method, location. Before merging every vertex id is named based on the unit test name and the index of the vertex in the trace path. After merging the vertex, the id may be renamed with a unique alpha-numeric symbol beginning with m to indicate that two vertices have been merged.

Trace paths are aggressively merged by looking for object-method calls that coexist between traces. Merging the results of two identical unit tests results in a linear graph with no branching. If two unit tests traverse different object-method calls, however, the process will introduce branches into the graph which may later merge back to the same path.

The merge algorithm iteratively processes object-method call traces. At each step, a new trace t_i is added to the graph G. Note that both t_i and G are directed acyclic graphs but t_i has a branching factor of exactly one. When the algorithm begins, the graph G consists of only a single root node with the label *start*. When the algorithm is complete, G is the aggregation of all execution paths through the unit tests. The process follows five steps:

1. Initially, set m_g to the root of G and m_t to the root of t_i.
2. Place a pointer p_g at m_g and another pointer p_t at m_t.
3. For each child of the node pointed to by p_g, scan foward in t_i for a matching object-method call.
4. If a matching pair is not found, repeat the scan forward in t_i from m_t trying all descendents of p_g in a breadth-first fashion. If no match is found, add the directed graph rooted at m_t as a new child of the node pointed to by m_g. The algorithm is now complete; no new merging has occured.
5. Otherwise, the nodes at p_g and p_t are the same object-method call and represent a "rejoining" of the graph G and the trace t_i. Splice a new branch between m_g and t_g that includes the sequence between m_t and p_t exclusive of these endpoints. Repeat from step 2.

The algorithm above aggressively merges traces to reduce the number of branches in the aggregate representation. This results in a less complex and smaller graph than would be created if braches were not allowed to merge back to one another.

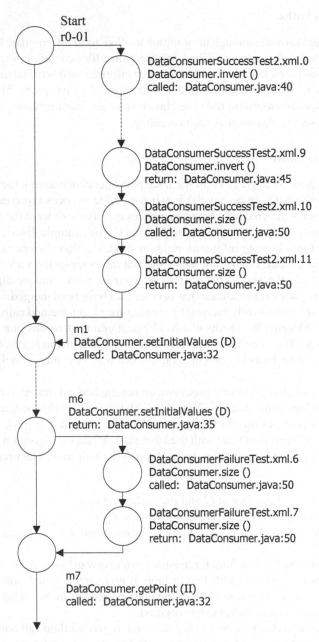

Fig. 1. Acyclic graph

Figure 1 illustrates the result of merging two traces: one successful and one unsuccessful unit test. The resulting graph is rooted at the node labeled *start*. A branch occurs immediately, indicating that the initial execution paths of the successful and failed unit tests differ. A dotted line indicates that a sequence of events has been collapsed and hidden from view for display purposes. The user interface allows us to examine the method calls in details if required. After calls to and returns from `invert()` and `size()` the execution traces merge and execute the method `setInitialValues()` which is the first object-method call in the failed execution trace. Both traces return from that method before once again diverging briefly.

In Figure 2, the trace continues with a new branching after `getPoint()`. During the first four DataConsumer.getPoint() method calls the graphs correspond, and thus the nodes in the two unit tests merge together as expected. A new branching occurs caused by two different method calls in the unit tests as shown in the Figure 2. The right branch corresponds to the successful partition. The left side branch corresponds to the fault revealing partition. The method call DataConsumer.invert() will cause cache inconsistency for the DataConsumerFailureTest unit test. Therefore during the next method call DataConsumer.getPoint() the normal code execution fails after the method DataSource.getVersion() returns a value different from the actual cache version. The successful unit test trace continues with the vertices corresponding to the method calls from the DataConsumerSuccessTest2.

3.6   Generate Sequence Diagrams

A UML Sequence Diagram is a behavioral representation of objects interacting with each other via method calls. In the previous steps we created an acyclic graph representing both successful and fault revealing unit tests. The graph is also a representation of objects interacting with each other. Therefore we can use the graph to generate UML Sequence Diagrams. We generate a Sequence Diagram for each branched segment of the direct acyclic graph that that contains a failed test. Each branch is visualized as a *combined fragment*. A combined fragment is used to visually display the conditional flow in a Sequence Diagram. Thus Sequence Diagrams are systematically generated by traversing each vertex, v, in the graph and using the following steps:

1. When a vertex contains more then one child and at least one child represents a failed test, create a new Sequence Diagram (if not already created) and create a combined fragment for each child vertex. Each child vertex should be represented as an object in the Sequence Diagram.
2. For each newly added child vertex, check its children, if is contains only one child, add the child vertex to the combined fragment and connect to the parent vertex using the method call in the previous vertex (label appropriately). If it contains more than one child return to step one.

Using this algorithm we created the Sequence Diagram in Figure 3 which represents the branched segment in the directed acyclic graph shown in Figure 2.

3.7   Reason About Fault

The Sequence Diagram shows where we can find the section of code responsible for the failure of the unit test. It now seems obvious that the method *invert()* with it's underlying call *setType()* causes the undesired behavior. The DataConsumer's cache becomes inconsistent when this method is invoked between calls to *getPoint()*. Therefore the heuristic

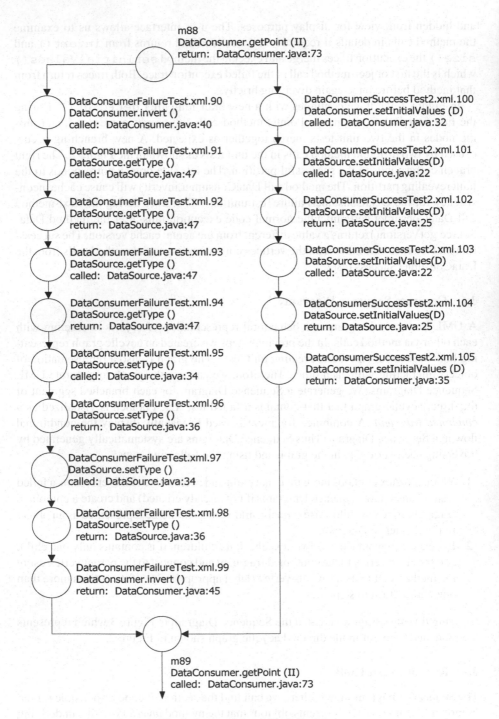

m88
DataConsumer.getPoint (II)
return: DataConsumer.java:73

DataConsumerFailureTest.xml.90
DataConsumer.invert ()
called: DataConsumer.java:40

DataConsumerFailureTest.xml.91
DataSource.getType ()
called: DataSource.java:47

DataConsumerFailureTest.xml.92
DataSource.getType ()
return: DataSource.java:47

DataConsumerFailureTest.xml.93
DataSource.getType ()
called: DataSource.java:47

DataConsumerFailureTest.xml.94
DataSource.getType ()
return: DataSource.java:47

DataConsumerFailureTest.xml.95
DataSource.setType ()
called: DataSource.java:34

DataConsumerFailureTest.xml.96
DataSource.setType ()
return: DataSource.java:36

DataConsumerFailureTest.xml.97
DataSource.setType ()
called: DataSource.java:34

DataConsumerFailureTest.xml.98
DataSource.setType ()
return: DataSource.java:36

DataConsumerFailureTest.xml.99
DataConsumer.invert ()
return: DataConsumer.java:45

DataConsumerSuccessTest2.xml.100
DataConsumer.setInitialValues (D)
called: DataConsumer.java:32

DataConsumerSuccessTest2.xml.101
DataSource.setInitialValues(D)
called: DataSource.java:22

DataConsumerSuccessTest2.xml.102
DataSource.setInitialValues(D)
return: DataSource.java:25

DataConsumerSuccessTest2.xml.103
DataSource.setInitialValues(D)
called: DataSource.java:22

DataConsumerSuccessTest2.xml.104
DataSource.setInitialValues(D)
return: DataSource.java:25

DataConsumerSuccessTest2.xml.105
DataConsumer.setInitialValues (D)
return: DataConsumer.java:35

m89
DataConsumer.getPoint (II)
called: DataConsumer.java:73

Fig. 2. Successful and Fault Partitions

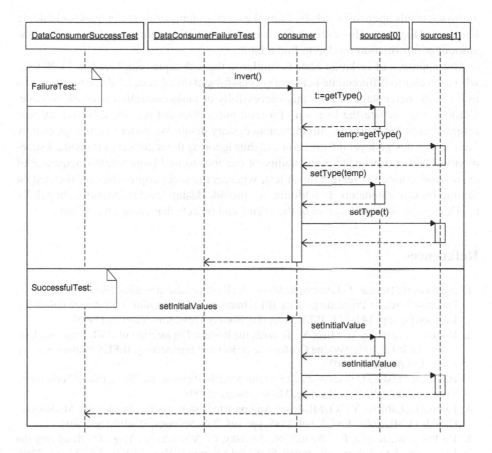

Fig. 3. UML Sequence Diagram

distance between failure and fault is mitigated. We can easily find where the actual fault occurs. The diagrams do not reveal why this method causes the inconsistency. Once the fault location is found, the developer can examine the source code and develop a solution.

4 Related Work

Telles et al. [17] describe debugging as the process of understanding the behavior of a system to facilitate the removal of bugs. Our approach assumes that extracting object level information and reducing it to a set of Sequence Diagram is useful in understanding the behavior of the system. In a case study, Vans et al. [18] have shown that up to 46% of the debugging task was spent trying to understand the software at the design level. Many have created approaches for visualizing code either using graphs [5,8,19] or the UML [7,4,14] for the purpose of better understanding a system. These approaches have focused on round-trip engineering and program comprehension. Our approach builds upon this work, by using the reverse engineering techniques for building Sequence Diagrams. However we focus on differentiating paths in order to reduce the amount of information needed to find a fault.

Our approach of isolating code is similar to the techniques employed by [1,20,6,9], who use color to differentiate between successful and failed code, and use brightness to indicate the percentage of tests that successfully or unsuccessfully traversed the code. Coloring may isolate the location of a fault but it does not provide a behavioral representation of the fault and eliminate unnecessary detail. We differ in our approach by focusing on design level information and thus ignoring the minutiae of individual statements. Another advantage is the ability of our tool to find bugs where a sequence of events collaborate to create a failed test, whereas the other approaches are focused on finding broken statements. In addition, we provide design level information, hopefully resulting better comprehension of the events and objects that caused the failure.

References

1. Agrawal, H., Horgan, J., London, S., Wong, W.: Fault Localization using Execution Slices and Dataflow Tests. In: Proceedings of the IEEE International Symposium on Software Reliability Engineering, pp. 143–151. IEEE Computer Society Press, Los Alamitos (1995)
2. Briand, L., Labiche, Y., Miao, Y.: Towards the Reverse Engineering of UML Sequence Diagrams. In: IEEE 10th Working Conference on Reverse Engineering, IEEE Computer Society Press, Los Alamitos (2003)
3. Binder, R.: Testing Object-Oriented Systems Models, Patterns, and Tools. Object Technology Series. Addison Wesley, Reading, Massachusetts (1999)
4. Briand, L., Labiche, Y.: A UML-based Approach to System Testing. In: Gogolla, M., Kobryn, C. (eds.) UML 2001. LNCS, vol. 2185, pp. 194–208. Springer, Heidelberg (2001)
5. De Pauw, W., Jensen, E., Mitchell, N., Sevitsky, G., Vlissides, J., Yang, J.: Visualizing the Execution of Java Programs. In: Diehl, S. (ed.) Software Visualization. LNCS, vol. 2269, pp. 151–162. Springer, Heidelberg (2002)
6. Eagan, J., Harrold, M.J., Jones, J., Stasko, J.: Technical note: Visually Encoding Program Test Information to Find Faults in Software. In: Proceedings of IEEE Information Visualization, pp. 33–36. IEEE Computer Society Press, Los Alamitos (2001)
7. Jacobs, T., Musial, B.: Interactive Visual Debugging with UML. In: SoftVis '03. Proceedings of the 2003 ACM Symposium on Software Visualization, pp. 115–122. ACM Press, New York (2003)
8. Jerding, D.F., Stasko, J.T., Ball, T.: Visualizing Interactions in Program Executions. In: Proceedings International Conference on Software Engineering, pp. 360–370 (1997)
9. Jones, J., Harrold, M., Stasko, J.: Visualization of Test Information to Assist Fault Localization. In: Proceedings of the 24th International Conference on Software Engineering, pp. 467–477 (2002)
10. Kollman, R., Gogolla, M.: Capturing Dynamic Program Behavior with UML Collaboration Diagrams. In: Proceedings CSMR, pp. 58–67 (2001)
11. Larman, C.: Applying UML and Patterns, 3rd edn. Prentice-Hall, Englewood Cliffs (2005)
12. DeMillo, R., Pan, H., Spafford, E.: Failure and Fault Analysis For Software Debugging. In: Proceedings of the Computer Software and Applications Conference, pp. 515–521 (1997)
13. Object Management Group, UML 2.0 Draft Specification (2005), http://www.omg.org/uml

14. Oechsle, R., Schmitt, T.: JAVAVIS: Automatic Program Visualization with Object and Sequence Diagrams Using the Java Debug Interface (JDI). In: Diehl, S. (ed.) Software Visualization. LNCS, vol. 2269, pp. 176–190. Springer, Heidelberg (2002)
15. Richner, T., Ducasse, S.: Using Dynamic Information for the Iterative Recovery of Collaborations and Roles. In: Proceedings International Conference on Software Maintenance, pp. 34–43 (2002)
16. Systä, T., Koskimies, K., Muller, H.: Shimba – An Environment for Reverse Engineering Java Software Systems. Software – Practice and Experience 31(4), 371–394 (2001)
17. Telles, M., Hsieh, Y.: The Science of Debugging, The Coriolis Group, Scottsdale, AZ (2001)
18. Vans, M., von Mayrhauser, A., Somlo, G.: Program Understanding Behavior during Corrective Maintenance of Large-scale Software. Int. Journal Human-Computer Studies 51, 31–70 (1999)
19. Walker, R.J., Murphy, G.C., Freeman-Benson, B., Wright, D., Swanson, D., Isaak, J.: Visualizing Dynamic Software System Information through High-Level Models. In: Proceedings OOPSLA, pp. 271–283 (1998)
20. xSlice: A Tool for Program Debugging,
    http://xsuds.argreenhouse.com/html-man/coverpage.html

Formally Defining a Graphical Language for Monitoring and Checking Object Interactions

Kai Xu and Donglin Liang

Department of Computer Science and Engineering
University of Minnesota, Minneapolis, Minnesota 55455

Abstract. Monitoring and checking object interactions is an important activity for testing/debugging scenario implementation in an object-oriented system. In our previous work, we proposed behavior view diagrams (BVD) as a graphical language for writing programs that automate such monitoring and checking process. In this paper, we illustrate the formal definition of the syntax and the semantics of an extended version of BVD that can also be used to describe multi-threaded scenarios. This formal definition provides a critical foundation both for understanding the language and for building its tool support.

1 Introduction

Software testing and debugging often require software developers to exercise the system with appropriate inputs, monitor the program execution, and compare the observed behaviors of the software with the expected behaviors to detect and to investigate bugs. In modern software development methodologies (e.g., [9]), the expected behaviors for a system or a subsystem are often identified and documented as *scenarios* during requirements analysis and design. Therefore, testing and debugging at the system or subsystem level should focus on monitoring program actions and verifying properties relevant to the progress of these scenarios. This *scenario-based* monitoring approach allows software developers to effectively utilize their knowledge of scenarios built during analysis and design to detect and pinpoint problems in the program.

Existing testing and debugging techniques provide inadequate support for the scenario-based monitoring. Assertions have been widely used in testing and debugging to check whether the program behaves as intended (e.g., [2,15]). However, because the assertions are often specified independent of the execution history, they are not suitable for specifying properties specific to particular scenarios, especially those that are related to several steps of object interactions in a scenario. Existing debugging techniques provide various supports for execution monitoring. Source level debugging mechanisms, such as breakpoints, allow software developers to interactively inspect the program states when the program control reaches specific code locations. Event-based debugging techniques (e.g., [1,3,13]), on the other hand, allow the software developers to specify the inspections to be automatically performed when specific execution events occur. However, because these techniques do not emphasize on correlating the monitoring of the program actions

G. Engels et al. (Eds.): MoDELS 2007, LNCS 4735, pp. 620–634, 2007.
© Springer-Verlag Berlin Heidelberg 2007

at different points of time during execution, these techniques provide inadequate support for scenario observation and inspection.

The goal of our research is to develop better techniques to support scenario-based execution monitoring. To achieve this goal, we propose an intuitive graphical language, behavior view diagrams (BVD), for specifying expected object interaction scenario, and a tool suite that can automatically check the actual progresses of these scenarios against BVD specifications. For a specific scenario, a BVD precisely specifies the runtime objects that are involved, the sequence of object interactions among these objects, and the properties that must hold when specific object interactions occur. Thus, it is a powerful mechanism for software developers to compare the actual behaviors of the system with their expectation. We have implemented a prototype that can translate simple BVD specifications into execution monitors. These execution monitors can detect relevant execution events and examine the states of the target programs based on the BVD specifications. Our limited experience shows that, with tool support, using BVD may improve the effectiveness of testing and debugging.

In our previous work [11], we have proposed a simplified version of BVD for monitoring and checking sequential scenarios. In this paper, we formally define the syntax and the operational semantics of an extended version of BVD that can also be used to specify multi-threaded scenarios. The formal definition provides a critical foundation for understanding this language and for building tool support.

Formally defining the operational semantics of BVD is challenging. First, it is difficult to build a mathematical model that is required by the operational semantics based on the 2-D BVD drawings. Second, BVD has a hierarchical name space. For a multi-threaded BVD, more than one instance of child name spaces may exist concurrently within the instance of a parent name space. To address these challenges, we first transform a BVD specification into an algebraic expression. We then build a mathematical model on top of the algebraic expression by combining each of its sub-expression that represents a block structure in the BVD drawing with its corresponding name space instance to model the states of an execution monitor generated from the BVD specification. Finally, we define a set of semantic rules over this mathematical model to specify the possible behavior of the execution monitor. We expect that these treatments may provide inspirations for defining other similar graphical languages.

In the rest of the paper, Section 2 gives an overview of testing and debugging with BVDs. Section 3 presents the monitoring profile. Section 4 presents the operational semantics. Section 5 discusses the related work. Section 6 concludes the paper and discusses future work.

2   Testing and Debugging with Behavior View Diagrams

BVD is a high-level graphical language for writing monitoring and checking programs. It extends UML 2.0 sequence diagrams with new elements to facilitate runtime monitoring for investigating the runtime behaviors and the states of the objects in the target program. In a previous work [11], we have introduced a

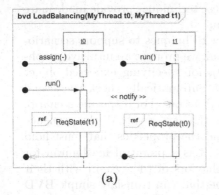
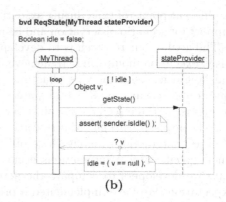

(a)   (b)

Fig. 1. An example BVD specification

preliminary version of this language for monitoring sequential object interaction scenarios. In this section, we will use a simplified example to illustrate the key features of an improved version of this language that can be used to monitor multi-threaded scenarios.

Fig. 1 shows the example BVD specification. The specification contains two bvds that define how to monitor an implementation of a concurrent state-space-exploration algorithm [4]. This implementation uses two threads to explore a state-space tree concurrently. To balance the work-load, one thread will ask the other thread for un-explored states when it finishes its current work. The algorithm terminates when all states have been explored.

The bvds in Fig. 1 shows how the two threads should interact to implement the algorithm.[1] In Fig. 1(a), the two thread objects are represented by the two life-lines, each of which is marked with a formal parameter name. This indicates that the two objects will be identified by the formal parameters. This object-identification mechanism is referred to as the *parameter-based* binding for life-lines. According to Fig. 1(a), `assign()` should be invoked to assign work to thread t_0 before t_0 is started. The two threads can be started by `run()` in any order. Once t_1 is started, it waits for a notification from t_0 indicating that t_0 has been started and is ready to distribute un-explored states. After that, the two threads explore the state space concurrently. When one of the thread objects runs out of states (this mode is called *idle*), it requests an un-explored state from the other thread object by invoking method `getState()`. This interaction is specified by another bvd in Fig. 1(b), and is referred to by the bvd in Fig. 1(a) through interaction use constructs (the constructs marked with `ref`).

Fig. 1(b) illustrates some other features in BVD. The first life-line in Fig. 1(b) has a rounded-rectangle head. This life-line represents the same object as that represented by the life-line covered by the interaction use construct that refers to this bvd. For example, the object represented by the life-line t_0 in

---

[1] Due to the space limitation, the diagrams omit many other important interactions and properties.

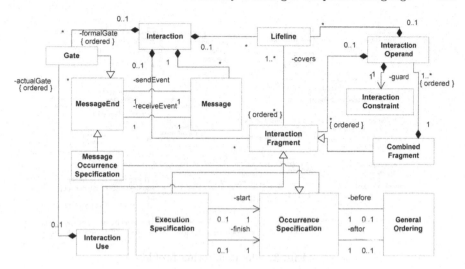

Fig. 2. A meta-model for sequence diagrams

Fig. 1(a) will be passed into the bvd in Fig. 1(b) that is referred to by the through the interaction use construct on the left side of Fig. 1(a). It will be binded with the life-line *MyThread*. This object-identification mechanism is referred to as *position-based* binding. When more than one life-line is covered by an interaction use construct, the position-based bindings are determined by the relative positions of the covered life-lines. The comment box (referred to as the *monitoring block*) associated with getState() in Fig. 1(b) checks, by invoking the pure method isIdle() in the target program, whether the requesting thread object is idle when the getState() invocation is detected. The label "?v" on the return arrow for method getState() instructs the monitor to extract the return value of this method and store it into variable "v". This value can then be used to determine whether the requesting thread should quit the loop.

This example illustrates that BVD provides an effective notation for specifying runtime checkable design intention. With the tools that we are developing, a BVD specification can be translated into execution monitors. The execution monitors can recognize relevant events and perform various inspections based on the BVD specification. Comparing to other testing and debugging techniques, using BVD allows the software developers to focus on checking the important milestones and scenario-specific properties during the scenario progress. It is suitable for debugging when the final output is not useful for determining the root cause. It is also suitable for testing incomplete programs when the final output is not available or meaningful.

3 Defining BVD Syntax with UML Profile

UML 2.0 uses meta-models to define the syntax of UML diagrams. It offers profiles as a mechanism to extend the meta-model for specialize the syntax of

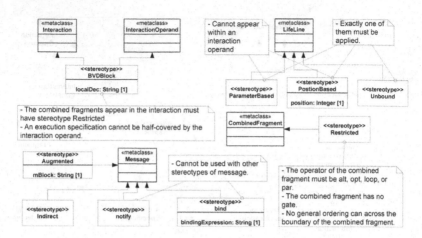

Fig. 3. The monitoring profile

the diagrams for modeling in a specific domain. In this section, we will define the syntax for BVD with a profile to the meta-model of sequence diagrams. We refer to such a profile as the *monitoring* profile.

Figure 2 shows a simplified meta-model defining sequence diagrams (see [14] for a complete description of the meta-model). In the meta-model, a sequence diagram is an *interaction*. It contains a set of life-lines, a set of messages, and an ordered set of interaction fragments. An interaction fragment can be one of the following types. An *occurrence specification* represents the sending or receiving event of a message, or the end of a method invocation. An *execution specification* is the thin box holding together the events belonging to a particular method invocation. A *combined fragment* is a control element. It contains one or more *interaction operands*, each of which contains its own ordered set (or sequence) of interaction fragments. An *interaction use* refers to another interaction. It stands as a place holder for the sequences defined by the referred interaction. An interaction use may contain an ordered set of *actual gates* that are corresponding to the *formal gates* defined in the referred interaction. These gates allow the sending event and the receiving event of a message to be specified in different interactions. A *general ordering* specifies a ordering constraint between two occurrence specifications.

Note that, in UML 2.0, all life-lines are owned by interactions. However, if a life-line is covered only by the interaction fragments within an interaction operand, making this life-line local to this operand may improve the modularity. Thus, in Figure 2, we extend the meta-model specified in UML 2.0 so that interaction operands can contain local life-lines.

The monitoring profile. The monitoring profile defines how the meta-classes for sequence diagrams can be extended to specify BVD. Figure 3 shows the major features of this profile. We extend both `Interaction` and `InteractionOperand` with stereotype `BVDBlock` for holding the declarations of local monitoring variables. The stereotyped construct is thus refers to as a *bvd block*. We extend

`Message` with stereotype `Augmented` for adding an extra property `mBlock` that specifies the monitoring block associated with the message. We also extend `Message` with stereotype `Indirect` for specifying indirect messages, and with stereotype `notify` for specifying the notification from one thread to another. We further define three stereotypes for `LifeLine` to indicate the mechanismsfor identifying the runtime object that will be represented by a life-line. We refer to these approaches as the *binding* approaches.

Stereotypes `PositionBased` or `ParameterBased` are used to annotate life-lines that use position-based binding or parameter-based binding, respectively. Such life-lines have been illustrated in Fig. 1. Stereotype `Unbound` is used to annotate a life-line whose binding will be determined when the first message pointing to this life-line is detected. The first message can be a creation message or a call message. In both cases, the target object of the message will be bound to the life-line. To allow more flexible bindings, we also introduce a a special kind of message, the `binding` message, that is marked with stereotype `bind`. The label of a binding message has a binding expression whose value will be used as a reference to find the object to be bound to the target life-line. To re-use the syntactic constructs of UML sequence diagrams, we define the binding message as a stereotype of the creation message.

The monitoring profile also specifies important constraints that have been imposed on an interaction and its elements to make it a BVD. For example, a combined fragment in a BVD must have stereotype `Restricted`. The operator of such a combined fragment can only be `alt`, `opt`, `loop`, or `par`. Other types of operators are excluded because they are less common or their meanings are still under debate (e.g., `assert` and `neg` [5])[2]. The monitoring profile further specifies constraints for defining well-formed BVDs. E.g., a combined fragment cannot contain gates. The well-formness is critical for defining the semantics.

The discussion above shows that defining BVD syntax with a profile allows us to reuse the UML 2.0 meta-model. Such a reuse simplifies the syntax specification for BVD. Potentially, it would also allow software developers to use a profile-aware UML editor to write BVD specifications. This reuse of tools paves the way for a quick adoption of BVD in the future.

4   The Operational Semantics of BVD

A BVD specification is used for monitoring and checking the object interactions. For this purpose, the monitor created from the specification must be able to recognize the relevant object interactions and to perform various inspection to determine whether the interactions progress as expected. The operational semantics for BVD defines the behavior for such a monitor.

The operational semantics for a specification language is typically defined by a mathematical model that consists of a domain of configurations and a set of rules that define the possible transitions among these configurations. For BVD, the configurations are used to represent the states of the monitor. Intuitively,

---

[2] The `neg` can be simulated in BVD by using `assert(false)` in the monitoring block.

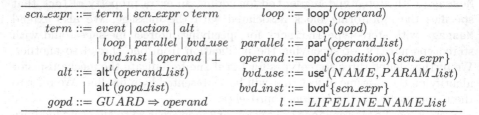

$$scn\_expr ::= term \mid scn\_expr \circ term \qquad loop ::= \mathsf{loop}^l(operand)$$
$$term ::= event \mid action \mid alt \qquad\qquad\quad \mid \mathsf{loop}^l(gopd)$$
$$\mid loop \mid parallel \mid bvd\_use \quad parallel ::= \mathsf{par}^l(operand\_list)$$
$$\mid bvd\_inst \mid operand \mid \bot \qquad operand ::= \mathsf{opd}^l(condition)\{scn\_expr\}$$
$$alt ::= \mathsf{alt}^l(operand\_list) \qquad\qquad bvd\_use ::= \mathsf{use}^l(NAME, PARAM\_list)$$
$$\mid \mathsf{alt}^l(gopd\_list) \qquad\qquad\quad bvd\_inst ::= \mathsf{bvd}^l\{scn\_expr\}$$
$$gopd ::= GUARD \Rightarrow operand \qquad\qquad l ::= LIFELINE\_NAME\_list$$

Fig. 4. The grammar rules for the syntax of scenario expressions, where X_list stands for a list of grammar symbol X. To save space, the opt element is considered as special case of alt, and the rules about *condition* are not listed. Finally, we assume that $\bot$ is automatically eliminated from a scenario expression if this term is connected to other terms in the rest of the paper.

a state of the monitor contains two important pieces of information: the object interactions yet to be observed and the sequencing constraints among them; the current bindings of the life-lines and the values of the monitoring variables. The rules define how the relevant object interactions are recognized and how the monitor state changes when a relevant object interaction is recognized.

In the rest of this section, we first explain how to represent BVDs with expressions in Section 4.1, then explain how to represent data environments and the configurations in Section 4.2. Finally, we present the major operational semantic rules in section 4.3.

4.1   Encoding Sequencing Constraints with Scenario Expressions

In BVD, the sequencing constraints among object interactions are specified by the 2-D spatial relations among different elements. Although spatial relations are easy for human to understand, directly using them in a configuration may complicate the specification of the operational rules. We thus translate these spatial relations into an algebraic expression. We refer to such an expression as a *scenario expression*. With the semantic rules that will be shown in the following, the sequencing constraints defined by the spatial relations will be enforced.

Fig. 4 shows the major grammar rules that define how a scenario expression can be composed. A scenario expression is a sequence of *terms* connected with *weak sequencing connectors*, denoted as $\circ$. Each term in the scenario expression represents a BVD element in a BVD specification. It is associated with a superscript l, which is the set of names of the lifelines that are covered by its corresponding BVD element. Terms have different types. *Event* terms represent message arrows and the terminations of methods. *Action* terms represent variable declarations, assertions, and value-based life-line binding constructs. *Compound* terms represent box elements (e.g., loop, alt, par, and their operands). A compound term may also be used to represent an interaction use construct or a bvd instance that in-lines the content of the used bvd. From the perspective of the operational semantics, an interaction use construct is similar to a function call-site in a functional language, and a bvd instance is similar to the in-lining of the function body in the place of the call-site. Finally, an empty term $\perp$ is introduced to represent an "empty" BVD element.

A BVD specification can be translated into a scenario expression through the following process. Initially, the top-level elements in a BVD are considered. Each of these elements will be represented by a term. The order in which these terms are connected in the expression is determined by a topological sorting among these top-level elements.[3] This ordering approach ensures that, if an element A appears above element B on a common life-line, then A must be sorted before B. As a next step in the process, a term is created for each top-level element. If this element is a box element, a sub-scenario-expression will be created by recursively invoking this process on the internal of this box element.

Thus, BVD LoadBalancing in Fig. 1(a) can be translated as

$$a \circ r_1 \circ r_2 \circ n \circ \mathsf{use}^{t_0}(\mathsf{ReqState}, (t_1)) \circ \mathsf{use}^{t_1}(\mathsf{ReqState}, (t_0)) \circ r_1' \circ r_2' \qquad (1)$$

by following this process, where a represents the call-message arrow assign, r_1 represents the call message arrow run to t_0, r_2 represents the call message arrow run to t_1, n represents the notification, and r_1', r_2' represent the corresponding returns of r_1 and r_2, respectively.

4.2   Encoding the Data Environment

BVD has a hierarchical data environment. The nodes in this hierarchy is the local data environments of BVD blocks. In a multi-threaded BVD, different child BVD blocks of a parent BVD block may be active concurrently. Moreover, the statements in BVD may have side-effects. Hence, it is challenging to maintaining correctly the data environment in the operational semantics.

To deal with this complexity, in the configurations for our operational semantics, we break the bindings between the names and their values into two different kinds of mappings. The first kind of mapping is called an *environment*, denoted as β or α, which is a partial function that maps a set of names to individual locations. The second kind of mapping is called a *global store*, denoted as σ, which is a partial function that maps valid locations to values. This treatment is common for defining semantics for language with side-effects.

We use two different types of environments. A *local* environment for a BVD block maps the names declared inside this block to locations. A *contextual* environment for a BVD block maps, to locations, all the names that are visible from outside of this block. The *actual* environment that contains the names visible for an element in the block can be obtained by composing the local and contextual environments, β and ρ, respectively, in the following way:

$$\rho[\beta] = (\rho \setminus \{n \mapsto v \mid n \in names(\beta)\}) \cup \beta$$

where $names(\beta)$ is the set of names in β. This formula respects the shadowing of names declared in nesting scopes. In the rest of this paper, when we say that

---

[3] This topological sorting can be obtained by simply sorting the top y-coordinates of these elements. Due to the space limitation, we will not show the proof here.

α is a local (or contextual) environment for a scenario expression, we refers to the corresponding environment for the innermost block that contains all the elements represented by this expression.

The two different types of environments for a block are computed differently by the monitor. A local environment for a block is initially empty (denoted as $\emptyset$). It will be populated when the monitor processes the action terms that represent the life-line/variable declarations in this block. The local environment may change when this sub-scenario expression is being processed. In contrast, the contextual environment for the internal of a compound term in a block is computed when the monitor begins to process the compound term. Such an environment will remain unchanged when the internal of a compound term is being processed. For this reason, a transition is represented in the form of $\rho \vdash \langle X : \beta, \sigma \rangle \overset{e}{\to} \langle X' : \beta', \sigma' \rangle$, where X and X' are *annotated* scenario expressions, σ and σ' are global stores, β (β') and ρ are the local and contextual environments, respectively, for X and X'. It reads: under the contextual environment ρ, configuration $\langle X : \beta, \sigma \rangle$ can transit to configuration $\langle X' : \beta', \sigma' \rangle$ by observing runtime event e. In an annotated scenario expression, each sub-scenario expression in a compound term is associated with its local environment (initially $\emptyset$). As we will see in the operational rules, this arrangement allows the monitor to handle the name spaces correctly.

4.3   Operational Semantic Rules

Monitoring with a BVD specification starts when a monitoring controller creates a monitor that is generated from this BVD specification. The state of the monitor, called a configuration, is a tuple of a annotated scenario expression $X : \beta$, a global store σ, and a contextual environment ρ. Initially, the expression part X of the annotated scenario expression is the scenario expression transformed from the BVD specification. The environment part β and the global store σ contain the mapping that binds the names of the formal parameters and the life-lines with their initial values, which are set by the monitoring controller. The contextual environment at this time is the global environment ρ_0 that maps the names of available BVDs to the scenario expressions that represent the bodies of such BVDs. Once the monitor has been created, it will use a set of deduction rules presented in this subsection to process the events. The monitor will stop when the scenario expression in its configuration is empty or when it is terminated by the monitoring controller. During its life-time, when a relevant event is detected, the monitor will attempt to match this event with an event term in the scenario expression. If a match can be found, the monitor will perform the monitoring action associated with the matched term and update its configuration. Otherwise, the monitor will ignore such an event.

In many cases, the first term in a scenario expression is not an event term. In this case, the internal of this term will be explored to find a match, or the first term will be moved across and the next term will be considered if this does not violate the sequencing constraints.[4] In the rest of this subsection, we will discuss

---

[4] An example for the latter case is when one tries to recognize the invocation `t1.run()` using the scenario expression shown in (1).

$$\frac{t \text{ is an event term,}\quad (match(t,e,\rho[\beta],\sigma,\Delta_e)=\sigma')\neq error}{\Delta_e,\rho\vdash\langle t{:}\beta,\sigma\rangle\xrightarrow{e}\langle\bot{:}\beta,\sigma'\rangle}(ME1)$$

$$\frac{\Delta_e,\rho\vdash\langle t{:}\beta,\sigma\rangle\xrightarrow{e}\langle t'{:}\beta,\sigma'\rangle}{\Delta_e,\rho\vdash\langle \mathsf{alt}^L(t,S){:}\beta,\sigma\rangle\xrightarrow{e}\langle t'{:}\beta,\sigma'\rangle}(MAu1)\qquad\frac{\Delta_e,\rho\vdash\langle t:\beta,\sigma\rangle\xrightarrow{e}\!\!\!\!/\;,\quad \Delta_e,\rho\vdash\langle \mathsf{alt}^L(S):\beta,\sigma\rangle\xrightarrow{e}\langle t':\beta,\sigma'\rangle}{\Delta_e,\rho\vdash\langle \mathsf{alt}^L(t,S){:}\beta,\sigma\rangle\xrightarrow{e}\langle t'{:}\beta,\sigma'\rangle}(MAu2)$$

$$\frac{true=eval(g,\rho[\beta],\sigma,\Delta_e),\quad\Delta_e,\rho\vdash\langle t:\beta,\sigma\rangle\xrightarrow{e}\langle t':\beta,\sigma'\rangle}{\Delta_e,\rho\vdash\langle \mathsf{alt}^L(g{\Rightarrow}t,S){:}\beta,\sigma\rangle\xrightarrow{e}\langle t'{:}\beta,\sigma'\rangle}(MAg1)\qquad\frac{false=eval(g,\rho[\beta],\sigma,\Delta_e),\quad\Delta_e,\rho\vdash\langle \mathsf{alt}^L(S):\beta,\sigma\rangle\xrightarrow{e}\langle t':\beta,\sigma'\rangle}{\Delta_e,\rho\vdash\langle \mathsf{alt}^L(g{\Rightarrow}t,S){:}\beta,\sigma\rangle\xrightarrow{e}\langle t':\beta,\sigma'\rangle}(MAg2)$$

$$\frac{\Delta_e,\rho\vdash\langle t{:}\beta,\sigma\rangle\xrightarrow{e}\langle t'{:}\beta,\sigma'\rangle}{\Delta_e,\rho\vdash\langle \mathsf{loop}^L(t){:}\beta,\sigma\rangle\xrightarrow{e}\langle t'\circ\mathsf{loop}^L(t){:}\beta,\sigma'\rangle}(MLu)\qquad\frac{true=eval(g,\rho[\beta],\sigma,\Delta_e),\quad\Delta_e,\rho\vdash\langle t:\beta,\sigma\rangle\xrightarrow{e}\langle t':\beta,\sigma'\rangle}{\Delta_e,\rho\vdash\langle \mathsf{loop}^L(g\Rightarrow t):\beta,\sigma\rangle\xrightarrow{e}\langle t'\circ\mathsf{loop}^L(g\Rightarrow t):\beta,o'\rangle}(MLg)$$

$$\frac{\Delta_e,\rho\vdash\langle t{:}\beta,\sigma\rangle\xrightarrow{e}\langle t'{:}\beta,\sigma'\rangle}{\Delta_e,\rho\vdash\langle \mathsf{par}^L(t,S):\beta,\sigma\rangle\xrightarrow{e}\langle \mathsf{par}^L(t',X):\beta,\sigma'\rangle}(MP1)\qquad\frac{\Delta_e,\rho\vdash\langle t:\beta,\sigma\rangle\xrightarrow{e}\!\!\!\!/\;,\quad\Delta_e,\rho\vdash\langle \mathsf{par}^L(S):\beta,\sigma\rangle\xrightarrow{e}\langle \mathsf{par}^L(S'):\beta,\sigma'\rangle}{\Delta_e,\rho\vdash\langle \mathsf{par}^L(t,S):\beta,\sigma\rangle\xrightarrow{e}\langle \mathsf{par}^L(t,S'):\beta,\sigma'\rangle}(MP2)$$

$$\frac{\Delta_e,\rho[\beta]\vdash\langle X{:}\alpha,\sigma\rangle\xrightarrow{e}\langle X'{:}\alpha',\sigma'\rangle}{\Delta_e,\rho\vdash\langle \mathsf{opd}^L\{X{:}\alpha\}{:}\beta,\sigma\rangle\xrightarrow{e}\langle \mathsf{opd}^L\{X'{:}\alpha'\}{:}\beta,\sigma'\rangle}(MO)$$

$$\frac{\rho_0(n)=bvd(x_1,\cdots,x_n,pl_1,\cdots pl_m)\{X'\},\quad \alpha'=\{x_i\mapsto eval(expr_i,\rho[\beta],\sigma,\Delta_e)\mid 1\le i\le n\}\cap\{pl_i\mapsto\rho[\beta](L[j])\mid 1\le j\le n\}\quad \Delta_e,\rho_0\vdash\langle \mathsf{bvd}^L\{X':\alpha'\}:\beta,\sigma\rangle\xrightarrow{e}\langle \mathsf{bvd}^L\{X:\alpha\}:\beta,\sigma'\rangle}{\Delta_e,\rho\vdash\langle \mathsf{use}^L(n,expr_1,\cdots expr_n):\beta,\sigma\rangle\xrightarrow{e}\langle \mathsf{bvd}^L\{X{:}\alpha\}{:}\beta,\sigma'\rangle}(MU)$$

$$\frac{\Delta_e,\rho_0\vdash\langle X{:}\alpha,\sigma\rangle\xrightarrow{e}\langle X'{:}\alpha',\sigma'\rangle}{\Delta_e,\rho\vdash\langle \mathsf{bvd}^L\{X{:}\alpha\}{:}\beta,\sigma\rangle\xrightarrow{e}\langle \mathsf{bvd}^L\{X'{:}\alpha'\}{:}\beta,\sigma'\rangle}(MB)\qquad\frac{\Delta_e,\rho\vdash\langle X{:}\beta,\sigma\rangle\xrightarrow{e}\langle X'{:}\beta',\sigma'\rangle}{\Delta_e,\rho\vdash\langle X\circ t{:}\beta,\sigma\rangle\xrightarrow{e}\langle X'\circ t{:}\beta',\sigma'\rangle}(MR)$$

Fig. 5. Matching rules for various terms

the rules for these two choices separately. We refer to the first set of rules as the *matching rules* and the second set of rules as the *permission rules*. Such a distinction is inspired by the deduction rules presented in [12].

Matching rules. The matching rules define the matching transition relations among configurations. A matching transition relation between two configurations s and s' is denoted as $\Delta_e,\rho\vdash s\xrightarrow{e}s'$. Δ_e represents the current state of the target program being monitored. Namely, it means the values of the variables visible from the location of the target program where the event e occurred.

Fig. 5 shows the key rules for deriving the matching transition relations. Rule (ME1) defines that a runtime event can be recognized as an event term if their descriptions match. If they do, the match function in rule (ME1) will execute the monitoring action associated with the event term and return the updated global store. The event term, then, will be removed from the scenario expression to reflect the monitoring history. Rule (MR) defines how to propagate the effect of recognizing the runtime event e in the prefix of a scenario expression to its end. The rest of the rules in this figure deal with compound terms. Rules (MAu1) and (MAu2) express how to recognize a runtime event within an alternative combined fragment with no guard conditions. If the runtime event can be recognized in the branches, then the interactions in the chosen branch should be observed in the future. Therefore, the `alt` term in the configuration should be replaced with the `opd` term representing this branch. The rules also enforce that priorities are given to the branches based on the order of appearance. Similarly, rules (MAg1) and (MAg2) express how to recognize a runtime event within an alternative combined fragment with guard conditions. In this case, a particular branch can be chosen to recognize the event only if its guard is evaluated true in addition.

Rules (MLu) and (MLg) are used to explore un-guarded or guarded loops, respectively. In both cases, if the runtime event can be recognized in the internal operand, then a new iteration of the loop will begin. Rules (MP1) and (MP2) express how to recognize a runtime event within a parallel combined fragment. If the event can be recognized in one branch of the parallel, this branch will be updated. If the event can be recognized in more than one branch, a non-deterministic choice will be made as both (MP1) and (MP2) can be applied. Rules (MO) expresses how an interaction operand can be explored to recognize the event. If a match can be found in the sub-scenario expression under the inner environment $\rho[\beta]$, then the `opd` term will be updated accordingly.

Rule (MU) expresses how an interaction use can be replaced by the referred bvd for recognizing an event. This involves several actions. First, the specification of the referred bvd is looked up from the global environment, ρ_0. Second, the local environment of the new instance of the referred bvd is prepared by evaluating the actual parameters of the interaction use and assigning them to the formal parameters, and passing the bindings of the life-lines covered by the interaction use to the life-lines with PositionBased stereotypes in the referred BVD. Third, a bvd term that represents an instance of the referred BVD is explored to recognize the event with the prepared local environment. Note that this exploration is done under ρ_0, instead of the actual environment $\rho[\beta]$ because the referred bvd cannot access the names in the referring bvd. If the event is recognized successfully, then the resulting bvd term replaces the use term in the configuration. Once a bvd term is introduced in the configuration, rule (MB) can be used to continue exploring within the bvd instance for recognizing future events.

Permission rules. The permission rules are used for moving across terms in a scenario expression before finding a matching term. Intuitively, a monitor may move across the terms in the prefix prior to a term t in a scenario expression if the terms in the prefix do not specify an event that must occur before t. That is, this prefix "permits" a matching to be found in t, or term t is "permitted" by its prefix in the scenario expression. Note that, in some cases, moving across a term may require a transformation of this term. For example, when the monitor moves across an action term that represents a variable declaration, it will remove this term from the scenario expression and update the local environment to include the new variable. Therefore, the permission rules are specified as rules

$$\frac{\Delta_e,\rho\vdash\langle X:\beta,\sigma\rangle\not\xrightarrow{e},\quad \Delta_e,\rho\vdash\langle X:\beta,\sigma\rangle\xrightarrow{\cdots\xrightarrow{\ell(c)}}\langle X':\beta'',\sigma''\rangle,\quad \Delta_e,\rho\vdash\langle t:\beta'',\sigma''\rangle\xrightarrow{e}\langle t':\beta',\sigma'\rangle}{\Delta_e,\rho\vdash\langle X\circ t:\beta,\sigma\rangle\xrightarrow{e}\langle X'\circ t':\beta',\sigma'\rangle}(M1)$$

$$\frac{\ell(t)\cap L=\emptyset,\quad \Delta_e,\rho\vdash\langle X:\beta,\sigma\rangle\xrightarrow{L}\langle X':\beta',\sigma'\rangle}{\rho\vdash\langle(X\circ t):\beta,\sigma\rangle\cdots\xrightarrow{L}\langle(X'\circ t):\beta',\sigma'\rangle}(PR1)$$

$$\frac{\begin{array}{c}\ell(t)\cap L\neq\emptyset,\\ \Delta_e,\rho\vdash\langle X:\beta,\sigma\rangle\xrightarrow{L\cup\ell(t)}\langle X':\beta'',\sigma''\rangle,\\ \Delta_e,\rho\vdash\langle t:\beta'',\sigma''\rangle\xrightarrow{L}\langle\bot:\beta',\sigma'\rangle\end{array}}{\Delta_e,\rho\vdash\langle(X\circ t):\beta,\sigma\rangle\cdots\xrightarrow{L}\langle X':\beta',\sigma'\rangle}(PR2)$$

$$\frac{\ell(t)\cap L=\emptyset}{\Delta_e,\rho\vdash\langle t:\beta,\sigma\rangle\cdots\xrightarrow{L}\langle t:\beta,\sigma\rangle}(PT)\qquad \frac{\ell(t)\cap L\neq\emptyset,t\text{ is an action term},(\beta',\sigma')=exec(t,\rho[\beta],\sigma,\Delta_e)}{\Delta_e,\rho\vdash\langle t:\beta,\sigma\rangle\cdots\xrightarrow{L}\langle\bot:\beta',\sigma'\rangle}(PX)$$

$$\frac{\begin{array}{c}true=eval(g,\rho[\beta],\sigma,\Delta_e),\\ \Delta_e,\rho\vdash\langle t:\beta,\sigma\rangle\cdots\xrightarrow{L'}\langle\bot:\beta,\sigma'\rangle\end{array}}{\Delta_e,\rho\vdash\langle\mathsf{alt}^{L'}(g{\Rightarrow}t,S):\beta,\sigma\rangle\cdots\xrightarrow{L}\langle\bot:\beta,\sigma'\rangle}(PAg1)\qquad \frac{\begin{array}{c}false=eval(g,\rho[\beta],\sigma,\Delta_e),\\ \Delta_e,\rho\vdash\langle\mathsf{alt}^{L'}(S):\beta,\sigma\rangle\cdots\xrightarrow{L}\langle\bot:\beta,\sigma'\rangle\end{array}}{\Delta_e,\rho\vdash\langle\mathsf{alt}^{L'}(g{\Rightarrow}t,S):\beta,\sigma\rangle\xrightarrow{L}\langle\bot:\beta,\sigma'\rangle}(PAg2)$$

$$\frac{\begin{array}{c}L'\cap L\neq\emptyset,\\ \Delta_e,\rho\vdash_L\langle t:\beta,\sigma\rangle\cdots\xrightarrow{L}\langle\bot:\beta,\sigma'\rangle\end{array}}{\Delta_e,\rho\vdash\langle\mathsf{alt}^{L'}(t,S):\beta,\sigma\rangle\cdots\xrightarrow{L}\langle\bot:\beta,\sigma'\rangle}(PAu1)\qquad \frac{\begin{array}{c}L'\cap L\neq\emptyset,\\ \Delta_e,\rho\vdash\langle\mathsf{alt}^{L'}(S):\beta,\sigma\rangle\cdots\xrightarrow{L}\langle\bot:\beta,\sigma'\rangle\end{array}}{\Delta_e,\rho\vdash\langle\mathsf{alt}^{L'}(a,S):\beta,\sigma\rangle\xrightarrow{L}\langle\bot:\beta,\sigma'\rangle}(PAu2)$$

$$\frac{\begin{array}{c}L'\cap L\neq\emptyset,\quad true=eval(g,\rho[\beta],\sigma,\Delta_e),\\ \Delta_e,\rho\vdash\langle t:\beta,\sigma\rangle\cdots\xrightarrow{L'}\langle\bot:\beta,\sigma''\rangle,\\ \Delta_e,\rho\vdash\langle\mathsf{loop}^{L'}(g\Rightarrow t):\beta,\sigma''\rangle\\ \cdots\xrightarrow{L}\langle\bot:\beta,\sigma'\rangle\end{array}}{\Delta_e,\rho\vdash\langle\mathsf{loop}^{L'}(g{\Rightarrow}t):\beta,\sigma\rangle\cdots\xrightarrow{L}\langle\bot:\beta,\sigma'\rangle}(PL1)\qquad \frac{\begin{array}{c}L'\cap L\neq\emptyset,\\ false=eval(g,\rho[\beta],\sigma,\Delta_e)\end{array}}{\Delta_e,\rho\vdash\langle\mathsf{loop}^{L'}(g{\Rightarrow}t):\beta,\sigma\rangle\xrightarrow{L}\langle\bot:\beta,\sigma'\rangle}(PL2)$$

$$\frac{L'\cap L\neq\emptyset}{\Delta_e,\rho\vdash\langle\mathsf{loop}^{L'}(t):\beta,\sigma\rangle\cdots\xrightarrow{L}\langle\bot:\beta,\sigma\rangle}(PL3)$$

$$\frac{\begin{array}{c}L'\cap L\neq\emptyset,\Delta_e,\rho\vdash\langle t:\beta,\sigma\rangle\cdots\xrightarrow{L'}\langle\bot:\beta,\sigma''\rangle,\\ \Delta_e,\rho\vdash\langle\mathsf{par}^{L'}(S):\beta,\sigma''\rangle\cdots\xrightarrow{L}\langle\bot:\beta,\sigma'\rangle\end{array}}{\Delta_e,\rho\vdash\langle\mathsf{par}^{L'}(t,S):\beta,\sigma\rangle\cdots\xrightarrow{L}\langle\bot:\beta,\sigma'\rangle}(PP1)\qquad \frac{\begin{array}{c}L'\cap L\neq\emptyset,\\ \Delta_e,\rho\vdash\langle t:\beta,\sigma\rangle\cdots\xrightarrow{L'}\langle\bot:\beta,\sigma'\rangle\end{array}}{\Delta_e,\rho\vdash\langle\mathsf{par}^{L'}(t):\beta,\sigma\rangle\cdots\xrightarrow{L}\langle\bot:\beta,\sigma'\rangle}(PP2)$$

$$\frac{L'\cap L\neq\emptyset,\quad \rho[\beta],\Delta_e\vdash\langle X:\alpha,\sigma\rangle\cdots\xrightarrow{L'}\langle\bot:\alpha',\sigma'\rangle}{\Delta_e,\rho\vdash\langle\mathsf{opd}^{L'}\{X:\alpha\}:\beta,\sigma\rangle\cdots\xrightarrow{L}\langle\bot:\beta,\sigma'\rangle}(PO)\qquad \frac{L'\cap L\neq\emptyset,\quad \Delta_e,\rho_0\vdash\langle X:\alpha,\sigma\rangle\cdots\xrightarrow{L'}\langle\bot:\beta,\sigma'\rangle}{\Delta_e,\rho\vdash\langle\mathsf{bvd}^{L'}\{X:\alpha\}:\beta,\sigma\rangle\cdots\xrightarrow{L}\langle\bot:\beta,\sigma'\rangle}(PB)$$

Fig. 6. Permission relation rules for matching a term in a sequence

for deriving a transformation relationship among configurations. We refer to such a relationship as a *permission* relationship.

A permission relation between two configurations s and s' is denoted as $\Delta_e,\rho\vdash s\cdots\xrightarrow{L}s'$, where L is a set of life-lines. It means that, under the context of (Δ_e,ρ), s can be transformed into s', without violating the sequencing constraints defined in s, such that the life-lines in L are not covered by any term in s'. As stated in Rule (M1) of Fig. 6, if the prefix X can be transformed into X' with respect to the life-lines covered by a term t (i.e., $\ell(t)$), then t can be used to match an event without violating the sequencing constraints defined by $X\circ t$. That is, the monitor can safely move across the terms in the prefix X. (M1) also contains a clause, $\Delta_e,\rho\vdash\langle X:\beta,\sigma\rangle\not\xrightarrow{e}$, to ensure that e cannot be matched with the prefix X. This clause is in place to prevent the monitor from "moving across" the terms in a scenario expression too aggressively.

Rules (PR1) and (PR2) in Fig. 6 specify how the transformation can be applied recursively to a scenario expression w.r.t. a set L of life-lines. Rule (PR1) handles the case in which the right-most term t of the expression does not cover any life-line in L. In this case, if the prefix X can be transformed to X' such that X' does not contain any element that covers life-lines in L, $X' \circ t$ will not contain any element that covers life-lines in L. Thus, there is a permission relation between $X \circ t$ and $X' \circ t$. Rule (PR2) handles the case in which the right-most term t of the expression covers some life-lines in L. Intuitively, if X can be transformed to X', and t contains a path that does not include any event element under the new environment β', then there is a permission relation between $X \circ t$ and X'.

The rest of the rules in Figure 6 define how an individual term may be transformed w.r.t. a set of life-lines L. Rule (PT) deals with the case in which the term t does not cover any life-line in L. In this case, t will not be changed. The other rules deal with the case in which the term does cover some life-lines in L. In this case, there must be a path through the bvd construct represented by this term such that this path does not contain any event specification. The rules can thus be viewed as exploring the paths through the bvd construct while avoiding event specification. If such a path can be found, then this term will be transformed to $\perp$. The rules for various terms are quite self-explained. Due to the space limitation, we will not go through each individual rule in this paper.

General orderings and gated messages. To ease the explanation, the rules discussed in this subsection do not consider the general orderings and messages involving gates (gated messages). The idea of handling general orderings is the following. General ordering can be recorded in the scenario expressions by adding a *toBefore* attribute to event terms. In addition, each event term will have a *id* attribute that contains an id that is unique within the scope of a bvd. A general ordering starting from a and ending with b can be represented by adding $a.id$ to $b.toBefore$. With the additional attributes, the matching rules can be enhanced to ensure that an event term can be used to match a runtime event only if its *toBefore* attribute is empty. When an event term has been matched with an event, additional rules can be introduced to remove the id of this event term from the *toBefore* attributes of other event terms.

In the presence of gates to an interaction use, a message may go from a life-line to an actual gate, a formal gate to an actual gate, or a formal gate to a life-line. The basic idea for handling such messages is to use separate terms to represent the sending and the receiving of such messages and then to use general orderings to connect the sending to the receiving. In this case, both the matching rules and permission rules must be extended to handle these cases. Due to the space limitation, we will not discuss the details in this paper.

5   Related Works

Kiviluoma et. al. [10] proposed a behavior profile of UML diagrams for specifying architecturally significant behaviors that can be automatically checked during runtime. The behavior profile differs from BVD in at least two aspects. First, the behavior profile can only be used to specify the sequencing constraints of object interactions. In contrast, BVD can also be used to specify monitoring actions for inspecting the states of the target program. This broader monitoring capability is important for detecting and locating state-related bugs. Second, the behavior profile applies the monitoring to all the objects that match the role description. BVD allows software developers to precisely specify the runtime objects to be monitored and the period of time in which the monitoring should be deployed. This precision is important for investigating a specific bug in a program.

UML 2 sequence diagrams [14] and MSC [7,8] are widely used to describe object interactions and process communications. Presumingly, by using bisimulation, these notations can be used to check the sequencing constraints of object interactions. However, because the semantics (e.g., [12]) of these notations often emphasize on trace generation, they do not consider the inter-dependencies between states and sequencing constraints. Such inter-dependencies could be quite sophisticated, and thus, must be inspected during testing and debugging. BVD provides several mechanisms for dealing with such inter-dependencies. BVD allows objects of interest to be selected dynamically during monitoring. BVD also allows state information extracted from the target program to be be used to determine scenario paths. BVD further allows to verify properties over the state information extracted from multiple steps during monitoring. Such capability is important for investigating state-related bugs.

Live Sequence Charts (LSC) [6], an extended form of MSCs, offer syntax and semantics for monitoring scenario-specific properties. However, there is a paradigm difference between LSC and BVD. LSC follows a rule-based composition paradigm: different LSCs are specified as relatively independent rules, and are implicitly composed together through event unification. In contrast, BVD follow a call-based composition paradigm: different BVDs are composed explicitly through interaction uses. BVD allows a hierarchical way for specifying complex interaction scenarios. In addition, because testing and debugging OO programs is not the design goal of LSC, it does not provide the same level of support for this task as BVD does (e.g., the value-based binding and monitoring actions).

The permission relations presented in this paper is inspired by the semantics framework proposed for MSC'96 by Mauw and Reniers [12]. However, because BVD must support monitoring actions and states, as well as block-scoped declaration of facility variables, the configurations and the deduction rules presented in this paper are richer than that proposed by Mauw and Reniers.

6 Conclusion

This paper formally defines BVD—a graphical language for automatically monitoring and checking object interaction scenarios. With the tools that we are developing, automatic execution monitors can be generated from BVD specifications. Initial experiences [16] show that this capability has the potential to improve both the effectiveness and the efficiency of testing and debugging.

Providing a formal definition to BVD is one step in our ongoing effort of building a tool suite in supporting scenario-driven monitoring methodologies. We have built a prototype that can read BVD models in the standard XMI format and generate execution monitors for Java programs. The generated runtime monitors can detect method call/return events through Java Debugging Interface, and can check the ordering of the detected events as well as the state of the target Java program when the events are detected. So far, only sequential BVDs that contains no combined fragments can be accepted by the prototype. In our future work, we will extend this prototype to fully support the syntax and semantics presented in this paper, and explore efficient methods of event detection. We will also continue our empirical studies to evaluate the usefulness of this language and the usability of our tools.

References

1. Auguston, M., Jeffery, C., Underwood, S.: A framework for automatic debugging. Technical Report TR-CS-004/2002, New Mexico State University (2002)
2. Binder, R.: Testing Object-Oriented Systems. Addison-Wesley Professional, Reading (1999)
3. Ducasse, M.: Coca: An automated debugger for C. In: ICSE'99, pp. 504–513 (May 1999)
4. Grama, A., Karypis, G., Kumar, V., Gupta, A.: Introduction to Parallel Computing, 2nd edn. Addison Wesley, Reading (2003)
5. Harel, D., Maoz, S.: Assert and negate revisited: modal semantics for uml sequence diagrams. In: ICSE. Proceedings of the 2006 international workshop on Scenarios and state machines: models, algorithms, and tools, pp. 13–20 (2006)
6. Harel, D., Marelly, R.: Come, Let's Play: Scenario-Based Programming Using LSCs and the Play-Engine. Springer, Heidelberg (2003)
7. ITU-T. Recommendation Z.120 (10/96), Message Sequence Chart (MSC). International Telecommunication Union, Geneva (1996)
8. ITU-T. Recommendation Z.120 (11/99), Message Sequence Chart (MSC). International Telecommunication Union, Geneva (1999)
9. Jacobson, I., Booch, G., Rumbaugh, J.: The Unified Software Development Process. Addison-Wesley, Reading (1999)
10. Kiviluoma, K., Koskinen, J., Mikkonen, T.: Run-time monitoring of behavioral profiles with aspects. In: The 3rd Nordic Workshop on UML and Software Modeling (2005)
11. Liang, D., Xu, K.: Monitoring with behavior view diagrams for scenario-driven debugging. In: IEEE Asia-Pacific Software Engineering Conference, IEEE Computer Society Press, Los Alamitos (2005)
12. Mauw, S., Reniers, M.A.: Operational Semantics for MSC'96. Computer Networks 31(17), 1785–1799 (1999)
13. Olsson, R.A., Cawford, R.H., Ho, W.W.: A dataflow approach to event-based debugging. Software - Practice and Experience 21(2), 209–230 (1991)
14. OMG. UML 2.0 superstructure, http://www.omg.org/cgi-bin/doc?ptc/2004-10-02
15. Voas, J., Kassab, L.: Using assertions to make untestable software more testable. Software Quality Professional Journal 1(4) (1999)
16. Xu, K., Liang, D.: Supporting scenario-driven debugging with behavior view diagrams. Technical Report 06-002, Dept. of CSE, Univ. of Minnesota (2006)

Statechart Development Beyond WYSIWYG

Steffen Prochnow and Reinhard von Hanxleden

Real-Time and Embedded Systems Group, Department of Computer Science
Christian-Albrechts-Universität Kiel, Olshausenstr. 40, D-24118 Kiel, Germany
{spr,rvh}@informatik-uni-kiel.de
http://www.informatik.uni-kiel.de/rtsys/

Abstract. Modeling systems based on semi-formal graphical formalisms, such as Statecharts, have become standard practice in the design of reactive embedded devices. Statecharts are often more intuitively understandable than equivalent textual descriptions, and their animated simulation can help to visualize complex behaviors. However, in terms of editing speed, project management, and meta-modeling, textual descriptions have advantages.

As alternative to the standard WYSIWYG editing paradigm, we present an approach that is also graphical but oriented on the underlying structure of the system under development, and another approach based on a textual, dialect-independent Statechart description language. These approaches have been implemented in a prototypical modeling tool, which encompasses automatic Statechart layout. An empirical study on the usability and practicability of our Statechart editing techniques, including a Statechart layout comparison, indicates significant performance improvements in terms of editing speed and model comprehension compared to traditional modeling approaches.

1 Introduction

Statecharts [1] constitute a widely accepted formalism for the specification of concurrent reactive systems. They extend classical finite-state machines and state transition diagrams by incorporating hierarchy, orthogonality, compound events, and a broadcast mechanism for communication between concurrent components. Statecharts provide an effective graphical notation, not only for the specification and design of reactive systems, but also for the simulation of the modeled system behavior. Statecharts have also been incorporated into the Unified Modeling Language (UML) and are supported by several commercial tools, e. g., Rational Rose, Matlab/Simulink/Stateflow, or Esterel Studio. Since the inception of Statecharts some twenty years ago, significant progress has been achieved concerning their semantics, formal analysis and efficient implementation. Concerning the practical handling of Statecharts, however, it appears that comparatively little progress has been made since the very first Statechart modeling tool set [2]. Specifically, the *construction, modification,* and *revision management* of Statecharts tend to become increasingly burdensome for larger models, and we feel that in this respect Statecharts are still at a disadvantage

G. Engels et al. (Eds.): MoDELS 2007, LNCS 4735, pp. 635–649, 2007.
© Springer-Verlag Berlin Heidelberg 2007

relative to other development activities, such as classical programming. This observation, corroborated in numerous discussions with practitioners and modeling experiences ranging from small, academic models to industrial projects, has motivated the work presented in this paper.

A commonly touted advantage of graphical formalisms such as Statecharts is their intuitive usage and the good level of overview they provide—according to the phrase "one picture is worth ten thousand words." However, when moving from toy examples to realistic systems, one is quickly confronted with large and unmanageable graphics originating from a high number of components or from intricate interactions and interdependencies.

As an alternative to the graphical modeling one can also develop reactive systems using textual notations. There exist a couple of languages that either describe Statecharts directly (e. g., SCXML [3], SVM [4]) or indirectly (e. g., Esterel [5,6]). Consequently the developer of reactive systems may choose between the textual and the graphical approach to specify systems. In principle, they offer the same expressiveness and the same level of abstraction. However, there are notable differences in terms of practical use, and both approaches have their benefits. Graphical models benefit from intuitiveness and are good for higher level context. Textual languages can represent precise details very well and they permit powerful macro capabilities (e. g., using generic scripting or preprocessing languages such as perl or m4) and allow a detailed revision management (e. g., applying the UNIX diff utility to compare different versions).

In summary, textual as well as graphical languages have their specific domains and advantages. The traditional model-based design flow starts with entering a graphical model of the *System Under Development* (SUD), from which textual programs are synthesized; however, as we argue here, it would combine the advantages of both techniques to allow the designer to work with textual and graphical representations of the SUD simultaneously.

Statecharts are commonly created using some *what you see is what you get* (WYSIWYG) editor, where the modeler is responsible for the graphical layout, and subsequently a Statechart appears the way a designer has modeled it. We believe that the WYSIWYG construction paradigm, which leaves the task of graphical layout to the human designer, has so far been a limiting factor in the practical usability of Statecharts, or graphical modeling in general. The premise of this paper is that this paradigm may have been justified at some point, but advances in layout algorithms and processing power today make it feasible to free the designer from this burden.

The main contributions of this paper are:

- an analysis of the graphical editing process using WYSIWYG editors and the identification of generic Statechart editing patterns;
- the presentation of two alternative Statechart construction paradigms—a macro-based and a text-based technique—that let the modeler focus on the modification of the Statechart *structure*, rather than their layout;

- a textual Statechart language, called *KIel statechart extension of doT* (KIT), which is concise and Statechart-dialect independent and supports the text-based construction of Statecharts; and
- an empirical study that evaluates the proposed construction techniques and shows their practicability and efficiency.

The rest of the paper is organized as follows. The remainder of this section discusses related work, and introduces the prototypical *Kiel Integrated Environment for Layout* (KIEL) tool, which serves as an evaluation platform for our proposals. Sect. 2 presents the analysis of WYSIWYG editing patterns. The macro-based and text-based Statechart construction approaches and the KIT language are discussed in Sect. 3. Sect. 4 summarizes the findings of our empirical study, Sect. 5 concludes and discusses possible future extensions.

1.1 Related Work

As indicated above, there is to our knowledge little published work that is directly related to the pragmatics of Statechart construction. However, the work presented here cuts across several related areas that have been studied already, namely textual Statechart description languages, the layout of Statecharts, graphical editors, and cognitive studies on the effectiveness of graphical and textual languages. Each of these areas is briefly discussed in the following.

The SCXML [3] Statechart description language has a comprehensible structure, but the required *tags* and their hierarchical dependencies call for specific XML editors. Alternative Statechart descriptions such as SVM [4] and the UMC [7] Statecharts use explicit declarations of Statechart objects, which reduces the readability especially for large Statecharts. This provides the advantages of textual entry, but does not offer the Statechart dialect-independent, concise constructs as available in KIT. These Statechart description languages generally serve as an intermediate format synthesized from manually edited Statecharts; to our knowledge, none of these languages has been used so far for Statechart synthesis, as we propose to do here. An exception is the RSML approach [8], which synthesizes a graphical view of the topology using a very simple, but surprisingly effective layouting scheme, which inspired KIEL's alternating linear layout. However, RSML still keeps much information that is normally part of the graphical model instead in textual AND/OR tables.

Castelló et al. [9] have developed a framework for the automatic generation of layouts of Statecharts based on floor planning. Harel and Yashchin [10] have investigated the optimal layout of *blobs*, which are edge-less hierarchical structures that correspond to Statecharts without transitions. KIEL offers several layout mechanisms, some employ the GraphViz [11] layout framework, others are developed from scratch.

A well-established technique to obtain consistency between model artifacts produced at different stages of the model life-cycle are transformational modeling approaches. DiaGen [12] and AToM³ [13] employ graph grammars to generate graphical editors for visual languages. GenGEd [14] uses graph grammars to modify visual languages using graph productions. The visual language is produced by a priori specified production sequences; we here instead propose interactive manipulations of the model. The graph grammar based tools use graphical constraints for placing graphical elements; we perform an automatic layout from scratch.

Several experimental studies address the comprehensibility of textual and visual programs; e. g., Green and Petre [15] performed an experimental study to evaluate the usability of textual and graphical notations using LabView. They determined that visual programs can be harder to read than textual ones. Purchase et al. [16] have evaluated the aesthetics and comprehension of UML class diagrams. We are not aware of any experimental studies on the effectiveness of *editing* visual languages.

1.2 The KIEL Modeling Environment

The *Kiel Integrated Environment for Layout* (KIEL) is a prototypical modeling environment that has been developed for the exploration of complex reactive system design [17]. As the name suggests, a central capability of KIEL is the automatic layout of graphical models, which makes KIEL a suitable testbed for the construction paradigms presented here. The following paragraph briefly summarizes KIEL's capabilities.

The tool's main goal is to enhance the intuitive comprehension of the behavior of the SUD. While traditional Statechart development tools merely offer a static view of the SUD during simulation, apart from highlighting active states, KIEL provides a simulation based on the *dynamic focus-and-context* visualization paradigm [17]. It employs a generic concept of Statecharts which can be adapted to specific notations and semantics, and it can import Statecharts that were created using other modeling tools. The currently supported dialects are those of Esterel Studio, Stateflow, and the UML via the XMI format, as, e. g., generated by ArgoUML [18]. Alternatively, KIEL can synthesize graphical SSMs from (textual) Esterel v5 programs [6]. KIEL also provides an automated checking framework, which checks compliance to *robustness rules* [19].

2   The WYSIWYG Statechart Editing Process

To analyze and educe improvements in developing Statecharts, we inspected the common WYSIWYG editing process. We identified nine main *editing schemata*, which can be grouped into three categories: Statechart *creation*, *modification* of Statechart elements, and *deletion* of elements. Fig. 1 illustrates some of these editing schemata. For example, to apply the schema "add hierarchical successor state" (Fig. 1d), the modeler has to perform the following steps: (1) select the state to supplement, (2) add a new hierarchical state, (3) insert an inner initial connector, (4) insert an inner state, and (5) insert connecting transitions.

When using conventional Statechart editors, none of the editing schemata can be realized as a single action. Generally, each editing schema using WYSIWYG editors passes the following action sequence:

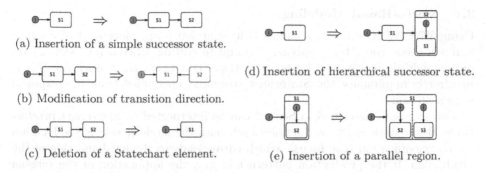

(a) Insertion of a simple successor state.

(b) Modification of transition direction.

(c) Deletion of a Statechart element.

(d) Insertion of hierarchical successor state.

(e) Insertion of a parallel region.

Fig. 1. Exemplary generic editing schemata derived from a typical editing process using WYSIWYG editors

1. If needed, create free space (e. g., expand hierarchical states for new sub-elements, move existing elements for placing new elements).
2. Focus on a Statechart element for modification resp. supplementation (move pointer, select per mouse click).
3. Apply an editing schema.
4. If needed, rearrange the modified chart to improve readability.

It is a common experience that the modeler spends much time with the layout-related activities of steps 1 and 4. For Statecharts developed from scratch, this effort may be small. In contrast, if an existing chart has to be modified, the work for arranging the elements increases roughly with the number of State-chart elements and Statechart complexity. Quoting a practitioner: *"I quite often spend an hour or two just moving boxes and wires around, with no change in functionality, to make it that much more comprehensible when I come back to it"* [20]. Furthermore, each editing schema requires the modeler to perform a sequence of low-level editing steps. The alternative proposals presented in the next section aim to improve both of these points.

3 Proposals for Enhancements in Statechart Editing

The basic idea of our approach is to automate the editing process as far as possible. Specifically, we propose to reduce the effort of re-arranging Statechart elements by applying automatic Statechart layout mechanisms. This produces Statecharts laid out according to a *Statechart Normal Form* (SNF) [17], which is compact and makes systematic use of secondary notations to aid readability. Due to the application of an automatic layout mechanism, the editing action sequence of Sect. 2 is reduced to:

1. Focus on a Statechart element for modification resp. supplementation;
2. Apply an editing schema.

Both editing actions remain under control of the modeler and will be treated by the following editing proposals.

3.1   Macro-Based Modeling

Using WYSIWYG editors, a simple editing action (e. g., placement of a state) scarcely needs time; but applying a complete editing schema (cf. Sect. 2) requires multiple mouse and keyboard actions. Our proposal to optimize this is to directly manipulate the Statechart *structure*, uncoupled from its graphical representation.

The schemata described in Sect. 2 can be interpreted as *Statechart productions*. Before applying a production (a schema), the modeler selects the location for the modification (the focus), which corresponds to the left-hand side of the production. If the production pattern matches, the application of the schema replaces the focus with the right-hand side of the production. The set of productions constitutes a Statechart *grammar*, which has the nice property that every application of a production results in a syntactically correct Statechart. Hence, a design does not go through meaningless intermediate editing stages, which frees the modeler from time-consuming syntax-checking. (An exception to this are productions that delete model elements, which may result in isolated states; KIEL does provide syntax checks that detect these, however.)

Concerning step 1, the setting of the focus, we propose to not only provide the traditional mouse-oriented mechanism, but also to allow a structure-oriented navigation, similar to text editors. E. g., in the KIEL macro editor (see Sect. 3.3), (1) the right/left key navigates through state sequences, (2) the up/down key navigates among sibling elements (e. g., multiple outgoing transitions from a state object), and (3) the page up/down keys navigate up resp. down in state hierarchies. Fig. 2a illustrates some navigation examples.

Concerning step 2, the selection of an editing schema, the designer may select a schema from a pull-down menu or by pressing a keyboard shortcut. E. g. in KIEL Ctrl+I generates a new successor state with a connecting transition and adjusts if necessary the priorities associated with transitions (cf. Fig. 2b). Afterwards a rearrangement of the Statechart elements will be performed automatically, according to the SNF.

3.2   Text-Based Modeling

Macro-based modeling works directly on the Statechart topology, combining several simple editing actions. As another, alternative structure-based Statechart editing technique we propose to employ a textual Statechart structure description. The *KIel statechart extension of doT* (KIT) combines implicit declarations as used in *dot* [11], the hierarchy construction as used in textual Argos [21], and the orthogonal construction as used in Esterel [22] with the ability to describe different dialects of Statecharts.

Fig. 3 presents a KIT example with the equivalent graphical model of a Safe State Machine (SSM), the Statechart-dialect implemented in Esterel Studio. Fig. 3a lists the KIT code, which is shortly described in the following. The Statechart preamble is listed in Line 1, containing the Statechart name and the model type and version, which determine the Statechart dialect and the accompanying graphical Statechart representation of the targeted modeling tool. Lines 2–5 declare the

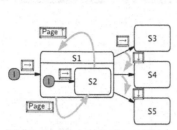

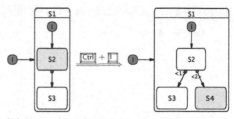

(a) Navigation with key strokes.

(b) Example of applying the "insert simple successor state" schema. Before applying the schema, state S2 is selected; afterwards the inserted state S4 remains selected for further editing operations.

Fig. 2. Editing actions and navigation using the macro-based modeling approach

```
1   statechart abro[model="Esterel Studio";version="5.0"]{
2     input A;
3     input B;
4     input R;
5     output O;
6     {
7       ->ABO;
8       ABO{
9         AB{
10          ->A;
11          A->AF[type=sa;label="A"];
12          AF[type=final];
13          ||
14          ->B;
15          B->BF[type=sa;label="B"];
16          BF[type=final];
17        };
18        ->AB;
19        AB->Program_Terminated[type=nt;label="/ O"];
20        Program_Terminated[type=final];
21      };
22      ABO->ABO[type=sa;label="R"];
23    };
24  };
```

(a) KIT description representation.

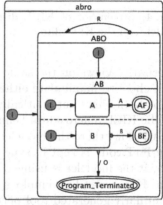

(b) SSM representation.

Fig. 3. Textual and graphical representations of the ABRO example [22]

signal events. Afterwards, Lines 7–23 declare Statechart elements and their relations. State objects are implicitly identified by their state names, (cf. Line 8), curly braces define the scope of hierarchical relations (e. g., state AB, cf. Line 9–17), transitions are written as -> (cf. Line 11), and the || operator denotes parallel regions (cf. Line 13). KIT includes a couple of shorthand notations; e. g., a transition without a source node determines an initial connector (cf. Line 7), a transition of type sa abbreviates the SSMs *strong abortion* (cf. Line 11).

3.3   Implementation in KIEL

We have implemented the above proposed Statechart editing techniques in KIEL, which resulted in the *KIEL macro editor* and the *KIT editor*. Both editors are accessible simultaneously and are arranged side by side so that they allow

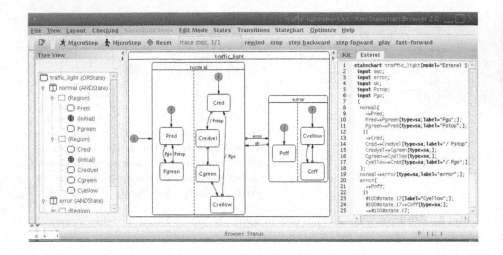

Fig. 4. Screenshot of KIEL displaying the Statechart tree-structure, the graphical model, and the KIT editor

alternative views on the same SUD, as can be seen in Figure 4. The user may thus chose to manipulate either the textual or the graphical view, and the tool keeps both views automatically and continuously consistent.

The KIEL macro editor is implemented as an extension of the graphic display-ing window; there the modeler marks and modifies graphical elements directly. The KIT code is kept in sync with the graphical model. In the opposite direc-tion, if the modeler is using the KIT editor, the graphical model is synthesized from KIT code. This employs a parser/synthesizer generated by SableCC [23]. A similarly generated tool performs the application of the production rules us-ing the KIEL macro editor. The productions are specified using an underlying grammar; the set of production can be easily extended with further production rules. Figure 5 depicts the tool chain of the KIT editor and the KIEL macro editor and their integration within KIEL.

4   Experimental Evaluation

We have used successive generations of KIEL in the classroom since 2005, in-cluding the macro-based and text-based editors presented here. The feedback on these editing approaches has been generally quite positive, also in comparison to the classical editing paradigms employed by the other, commercial modeling tools also used in classes. However, to gain a better, objective insight into the effectiveness of our modeling approaches, we have performed an experiment to investigate differences in editing performance between the conventional WYSI-WYG approach, the KIT editor, and the KIEL macro editor. As mentioned in Sect. 3.3, KIEL provides a mechanism that automatically produces our preferred Statecharts arrangements; in the following we call this the *alternating dot layout*

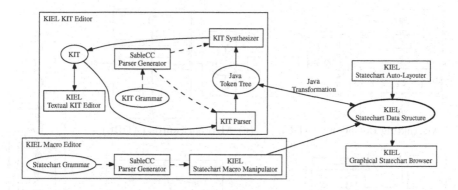

Fig. 5. Integration of the KIT editor and the KIEL macro editor into KIEL. The solid lines characterize the information flow during runtime, the dashed lines represent dependencies during compile-time of KIEL.

(ADL) (see Fig. 6a). Hence, a further goal of the experiment was to compare the readability of the ADL with other layout strategies.

4.1   Experiment design

The participants in the experiment (the *subjects*) were graduate-level students attending the lecture "Model-Based Design and Distributed Real-Time Systems" in the Winter Semester 2006/07[1]. Most of them were not familiar with the Statechart formalism in advance. The experiment consisted of two parts. The first part took place early in the semester, after two lecture units introducing the Statechart formalism. The subjects had by then also solved a first homework on understanding Statechart semantics. The second part proceeded after the final lecture unit at the end of the semester. In the meantime the subjects had gained practical experiences in modeling Statecharts; furthermore, they had learned about the importance of modeling paradigms, such as maintainability and co-notation of Statecharts. 24 students participated in the first experiment, 19 took the second one. In the following we refer to the participants of the first experiment as *novices* and to the participants of the second experiment as *advanced*. Furthermore, we define as *experts* the modelers that have significant practical experience beyond course work. Both experiments have similar design and consist of three parts:

Modeling Technique Evaluation: The subjects had to create Statecharts of varying complexity using different Statechart modeling techniques: a graphical WYSIWYG Statechart editor (we decided to employ Esterel Studio), the KIEL macro editor, and the KIT editor. Afterwards the created Statechart had to be extended and modified. A one-page reference card per modeling

---

[1] URL: http://www.informatik.uni-kiel.de/rtsys/teaching/ws06-07/v-model/

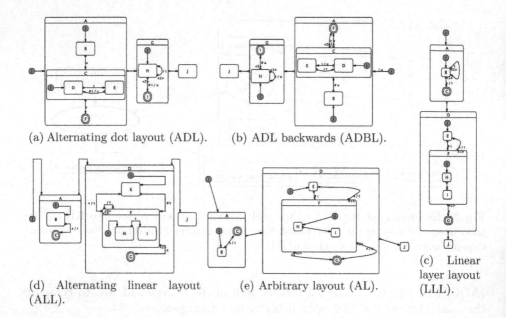

(a) Alternating dot layout (ADL).   (b) ADL backwards (ADBL).

(d) Alternating linear layout (ALL).   (e) Arbitrary layout (AL).   (c) Linear layer layout (LLL).

Fig. 6. Different Statechart layouts for experimental comparison. The layouts in Fig. 6a, 6b, and 6d were automatically generated from the KIEL layout mechanism; the Statechart models in Fig. 6c and 6e were drawn manually.

tool instructed the subjects. As editing performance metric the elapsed time was measured.

Subjective Layout Evaluation: The subjects were asked to score readability and comprehensibility of five different Statechart layouts without understanding (cf. Fig. 6).

Objective Layout Evaluation: In this experiment the subjects had to analyze different Statechart models, constructing a sequence of active states and upcoming signal events according to the semantics of SSMs. The elapsed time of each Statechart reading was measured for performance evaluation.

Each subject had to process a personal randomized experiment assignment (see Sect. 4.3), containing the tasks described above. The study was realized as a controlled experiment, i.e., the experiment leader checked and rejected the solutions of Parts A and C in case of incorrectness. Each of the subject's experiments was performed in a single session of one to two hours; the sessions were videotaped.

4.2 Hypotheses

The main questions asked in this experiment are the following: "Do the macro-based and text-based editing techniques make the Statechart construction process easier and faster than the conventional WYSIWYG method? Are the resulting Statecharts more readable and comprehensible?" To guide the analysis of the results it is useful to formulate some explicit expectations in form of hypotheses about the differences that might occur. Hence, the experiment should investigate the hypotheses as follows:

1. *Statechart Creation:* We expect that novices will need less time to create a Statechart using the WYSIWYG editor compared to the usage of the KIEL macro editor or the KIT editor. However, the Statechart creation times of advanced modelers using the KIT editor should be less than when using the WYSIWYG editor.
2. *Statechart Modification:* We expect that modification of an existing Statechart using the KIT editor or the KIEL macro editor is faster than using the WYSIWYG editor.
3. *Aesthetics:* Statecharts are sensed as aesthetic if their elements are arranged conforming to a certain layout style guide. We expect the best scores for Statecharts laid out according to the ADL (see Fig. 6a).
4. *Comprehension:* We suppose that well arranged Statecharts influence the readability. Hence, we expect a faster comprehension of the ADL compared to other Statechart layouts.

4.3 Validity

Concerning the *internal validity*, all relevant external variables (subjects' Statechart modeling experience, maturation, aptitude, motivation, environmental condition, etc.) were equalized between appropriate groups by randomized group assignment. Regarding the *external validity*, there are several sources of differences between the experimental and real Statechart modeling situations that limit the generalization of the experiments: In real situations, there are modelers with more experience, often working in teams, and there are Statechart models of different size or structure. However, we do not consider this to invalidate the basic findings of the experiment.

4.4 Results

This section presents and interprets the results of the experiments. The analysis is organized according to the hypotheses listed in Sect. 4.2. *Box plots* present the obtained statistical data; the comparison of means will be assisted by the two sample *t-test*. The test compares the difference of sample means from two data series to an hypothesized difference of the series means. It computes the p-value, which indicates statistically significance. We will call a difference significant if $p < 0.05$. The analysis and plots were performed with R v. 2.4.0 [24].

Evaluation of Modeling Techniques. The plots in Fig. 7a corroborate our Hypothesis 1 for novices. Due to the novelty of the KIEL macro editor and KIT editor, the novices sought advice in the reference card (cf. Sect. 4.1); in contrast

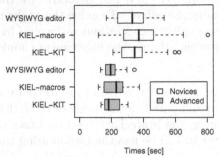

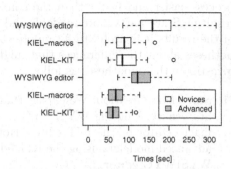

(a) Distribution of times for creating a new Statechart. *Novices:* The Statechart creation times using a WYSIWYG editor are smaller than using the KIEL macro editor (t-test $p = 0.04$) and tend to be smaller using the KIT editor (t-test $p = 0.25$). *Advanced:* The Statechart creation times using a WYSIWYG editor tend to be smaller than using the KIEL macro editor (t-test $p = 0.12$); time differences between WYSIWYG editor and KIT editor are not significant (t-test $p = 0.46$).

(b) Distribution of times for modifying an existing Statechart. *Novices and Advanced:* The needed times for Statechart modification using the KIEL macro editor or using the KIT editor are smaller than the times using the WYSIWYG editor (both: t-test $p = 0.00$).

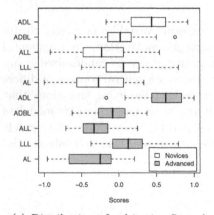

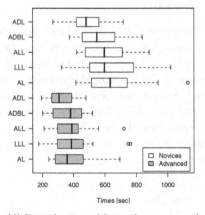

(c) Distribution of subjective Statechart layout scores. *Novices and Advanced:* The ADL scores better than all other Statechart layouts (all layouts: t-test $p = 0.00$). Score meaning: 1.0: strong preference, −1.0: strong rejection.

(d) Distribution of Statechart comprehension times. *Novices:* Less time is needed for comprehending Statechart according to ADL (ADBL: t-test $p = 0.02$, others: t-test $p = 0.00$). *Advanced:* ADL times tend to be smaller than times of the ADBL (t-test $p = 0.1$); less time is needed for other layouts (ALL: t-test $p = 0.04$, LLL: t-test $p = 0.03$, AL: t-test $p = 0.03$).

Fig. 7. Distribution of times for modeling Statecharts and distribution of Statechart layout assessments. The box plots denote quartiles, small circles indicate outliers.

the WYSIWYG editor could be used intuitively and without any reference card. Hence, on average the novices needed less time for creating Statecharts using the WYSIWYG editor than using the KIT editor or the KIEL macro editor. For advanced learners, however, the mean times are slightly less using the KIT editor. We suppose that for experts in Statechart creation this difference would increase further.

Fig. 7b illustrates the efficiency using the KIT editor and the KIEL macro editor in Statechart modification; this corroborates Hypotheses 2. With the KIT editor and the KIEL macro editor the modeler only works on the Statechart structure, while KIEL's Statechart auto-layouter arranges the graphical model. In contrast, using the WYSIWYG editor the subjects spent most of the time with making room for new Statechart elements and re-arranging the existing ones to make the developed chart readable. Despite the fewer operations needed using the KIEL macro editor, the subjects needed more time to modify a State-chart. The time was largely due to frequent consultations of the reference cards. Hence, for experts we suppose that the KIEL macro editor would provide the fastest modeling method.

Evaluation of Statechart Layouts. The scores of subjective Statechart layout assessment (cf. Fig. 7c) clearly show the subjects' preference for Statecharts laid out according to the ADL; hence, hypothesis 3 can be retained. Apparently it is not sufficient that layouts underlie an automatic layout; in fact Statechart layouts have to satisfy certain aesthetics to be assessed as good layouts. Accordingly, subjects stated that "transitions must be short and traceable" and "the element structure has to follow the Statechart meaning". E. g., due to unnecessary long transitions the ALL scores lower than the LLL.

Figure 7d demonstrates that a proper layout enhances the readability of State-charts; Statecharts laid out according to the ADL are faster comprehensible than other Statechart layouts, which corroborates hypothesis 4. This results from the accompanying proper micro layout (e. g., label placement) as well as proper macro layout (e. g., compact and white-space avoiding element arrangement).

5   Conclusion and Future Work

Embedded devices are proliferating, and their complexity is ever increasing. Statecharts are a well established formalism for the description of the reactive behavior of such devices. However, there is evidence that the current use of this formalism is not optimal, in particular as models get more complex.

We have presented a description language called KIT that was developed with the intention to describe topological Statechart structures. The KIEL tool combines the ability of easy textual editing and simultaneous viewing of the resulting graphical Statechart model. As another alternative to the classic, low-level WYSIWYG graphical editing paradigm, the graphical model can be modified using high-level editing schemata. This technique employs Statechart production rules that ensure the syntax-consistency through the whole editing process.

The user feedback on this has been generally very positive, and this has been supported by experimental data.

In the future we intend to experiment further with the simultaneous display of textual and graphical representation of the SUD. E. g., for a better traceability an indexing mechanism between elements of the textual and the graphical models could be useful. Beyond, we intend to apply the graphical model synthesis from a textual description, in combination with layout and simultaneous display, to data-flow languages such as SCADE/LUSTRE.

Acknowledgments. Mirko Wischer has performed the implementation of the KIEL macro editor and KIT editor presented here; the rest of the KIEL development team have also contributed. Prof. Dr. Jürgen Golz, Dept. of Psychology, CAU Kiel, has been an invaluable help in the design and analysis of the validation experiment. We also thank Ken Bell and the anonymous reviewers for helpful comments on this paper. Finally, we thank the students participating in the experimental study for their support and interest.

References

1. Harel, D.: Statecharts: A visual formalism for complex systems. Science of Computer Programming 8(3), 231–274 (1987)
2. Harel, D., Lachover, H., Naamad, A., Pnueli, A., Politi, M., Sherman, R., Shtull-Trauring, A., Trakhtenbrot, M.: STATEMATE: A Working Environment for the Development of Complex Reactive Systems. IEEE Transactions on Software Engineering 16(4), 403–414 (1990)
3. W3C: State Chart XML (SCXML): State Machine Notation for Control Abstraction (February 2007), URL: http://www.w3.org/TR/scxml/
4. Feng, T.H.: An extended semantics for a Statechart Virtual Machine. In: Bruzzone, A., Itmi, M. (eds.) Summer Computer Simulation Conference (SCSC 2003), Student Workshop, The Society for Computer Modelling and Simulation, Montréal, Canada, pp. S147–S166 (July 2003)
5. Berry, G., Gonthier, G.: The Esterel Synchronous Programming Language: Design, Semantics, Implementation. Science of Computer Programming 19(2), 87–152 (1992), URL: http://citeseer.nj.nec.com/berry92esterel.html
6. Prochnow, S., Traulsen, C., von Hanxleden, R.: Synthesizing Safe State Machines from Esterel. In: LCTES'06. Proceedings of ACM SIGPLAN/SIGBED Conference on Languages, Compilers, and Tools for Embedded Systems, Ottawa, Canada, ACM Press, New York (2006)
7. Mazzanti, F.: UMG User Guide (Version 2.5). Istituto di Scienza e Tecnologie dell'Informazione "Alessandro Faedo" (ISTI), Pisa, Italy (2003)
8. Heimdahl, M.P.E., Leveson, N.G.: Completeness and Consistency in Hierarchical State-Based Requirements. Software Engineering 22(6), 363–377 (1996)
9. Castelló, R., Mili, R., Tollis, I.G.: A Framework for the Static and Interactive Visualization for Statecharts. Journal of Graph Algorithms and Applications 6(3), 313–351 (2002)
10. Harel, D., Yashchin, G.: An Algorithm for Blob Hierarchy Layout. The Visual Computer 18, 164–185 (2002)

11. Gansner, E.R., North, S.C.: An open graph visualization system and its applications to software engineering. Software—Practice and Experience 30(11), 1203–1234 (2000), URL: http://www.research.att.com/sw/tools/graphviz/GN99.pdf
12. Minas, M.: Specifying Statecharts with DiaGen. In: HCC '01 – 2001 IEEE Symposia on Human-Centric Computing Languages and Environments, Symposium on Visual Languages and Formal Methods, Statechart Modeling Contest (September 2001), http://www2.informatik.uni-erlangen.de/VLFM01/Statecharts/minas.pdf&e=7%47
13. de Lara, J., Vangheluwe, H., Alfonseca, M.: Meta-modelling and graph grammars for multi-paradigm modelling in AToM^3. Software and Systems Modeling (SoSyM) 3(3), 194–209 (2004)
14. Bardohl, R.: GenGEd – A visual environment for visual languages. Science of Computer Programming (special issue of GraTra '00) (2002)
15. Green, T.R.G., Petre, M.: When Visual Programs are Harder to Read than Textual Programs. In: Human-Computer Interaction: Tasks and Organisation, Proceedings ECCE-6 (6th European Conference Cognitive Ergonomics) (1992), URL: http://citeseer.nj.nec.com/green92when.html
16. Purchase, H.C., McGill, M., Colpoys, L., Carrington, D.: Graph drawing aesthetics and the comprehension of UML class diagrams: an empirical study. In: ACM International Conference Proceeding Series archive, Australian symposium on Information visualisation, pp. 129–137. ACM Press, New York (2001)
17. Prochnow, S., von Hanxleden, R.: Comfortable Modeling of Complex Reactive Systems. In: DATE'06. Proceedings of Design, Automation and Test in Europe, Munich (March 2006)
18. ArgoUML: Tigris.org. Open Source Software Engineering Tools, URL: http://argouml.tigris.org/
19. Prochnow, S., Schaefer, G., Bell, K., von Hanxleden, R.: Analyzing Robustness of UML State Machines. In: Proceedings of the Workshop on Modeling and Analysis of Real-Time and Embedded Systems (MARTES'06), held in conjunction with the 9th International Conference on Model Driven Engineering Languages and Systems, MoDELS/UML 2006, Genua (October 2006)
20. Petre, M.: Why looking isn't always seeing: readership skills and graphical programming. Communications of the ACM 38(6), 33–44 (1995)
21. Maraninchi, F.: The Argos language: Graphical representation of automata and description of reactive systems. In: IEEE Workshop on Visual Languages, IEEE Computer Society Press, Los Alamitos (1991)
22. Berry, G.: The Foundations of Esterel. In: Plotkin, G., Stirling, C., Tofte, M. (eds.) Proof, Language and Interaction: Essays in Honour of Robin Milner (2000)
23. Gagnon, E.M., Hendren, L.J.: SableCC, an object-oriented compiler framework. In: TOOLS (26), pp. 140–154. IEEE Computer Society, Los Alamitos (1998)
24. R Development Core Team: R: A Language and Environment for Statistical Computing. R Foundation for Statistical Computing, Vienna, Austria (2006), URL: http://www.R-project.org

Model-Based Design of Computer-Controlled Game Character Behavior

Jörg Kienzle, Alexandre Denault, and Hans Vangheluwe

McGill University, Montreal, QC H3A 2A7, Canada
{Joerg.Kienzle, Alexandre.Denault, Hans.Vangheluwe}@mcgill.ca

Abstract. Recently, the complexity of modern, real-time computer games has increased drastically. The need for sophisticated game AI, in particular for Non-Player Characters, grows with the demand for realistic games. Writing consistent, re-useable and efficient AI code has become hard. We demonstrate how modeling game AI at an appropriate abstraction level using an appropriate modeling language has many advantages. A variant of Rhapsody Statecharts is proposed as an appropriate formalism. The Tank Wars game by Electronic Arts (EA) is used to demonstrate our concrete approach. We show how the use of the Statecharts formalism leads quite naturally to layered modeling of game AI and allows modelers to abstract away from choices between, for example, time-slicing and discrete-event time management. Finally, our custom tools are used to synthesize efficient C++ code to insert into the Tank Wars main game loop.

1 Introduction

Recently, global sales of the world's computer game industry have grown higher than those of the movie industry. Consequently, there is a growing demand for technology which supports rapid, re-usable game development accessible to non software experts. Computer games can be roughly classified into two categories: turn-based games (such as board games, adventures, and some role playing games) and real-time games (such as action or arcade games, and real-time strategy games). The kind of *artificial intelligence* found in computer games is different for turn-based and real-time games.

Board games are usually computerized versions of existing board games. Real board games typically require 2 or more players, but in a computerized version the computer can play the opponent. A good example of a board game that has seen many successful computerized implementations is *Chess* [6]. In turn-based games and particularly in board games, an artificial intelligence component that plans the moves of a player typically uses advanced search algorithms and heuristics to evaluate many possible future game situations. It then chooses as the current move the one that maximizes the likelihood of winning the game in the future. Timing is not that critical. Since the game is turn-based, the state of the game does not change until a player makes a move. Usually, waiting several seconds for an artificial intelligence component to make a move is acceptable.

G. Engels et al. (Eds.): MoDELS 2007, LNCS 4735, pp. 650–665, 2007.

Real-time games are very different in nature. The state of the game changes continuously (or in tiny increments), and the screen is continuously updated to present the new game state to the player. Modern computer games usually provide at least 30 frames-per-second updates. In real-time games (with the exception of real-time strategy games) the player usually controls one character (or a small number of characters), and plays within a game environment against a set of computer controlled characters (or in multiplayer games against characters controlled by other players).

In such games, the term *artificial intelligence* is used to designate the algorithms that specify the behavior of computer-controlled game characters, often also called *non-player characters* (NPC). The ultimate goal is to make the NPCs' own actions and reactions to game events seem as intelligent and natural as possible. For example, a guard protecting a building might walk back and forth in front of the main door. If he ever hears shots nearby, he should not simply continue this behavior, but for instance seek cover and call for backup. In its simplest form, such AI can be specified with scripts or rules that specify the NPC's behavior case by case. More realism can be achieved if the NPC has the ability to analyze a situation and evaluate different options, taking into account even the game history.

We believe that the specification of such advanced real-time AI should not be done within a programming language, but at a higher level of abstraction using visual modeling formalisms. Since the main focus of the models is to define reactions to game events, an event-based formalism seems to be the most natural choice. We decided to use our own variant of *Rhapsody statecharts* [5], a combination of state diagrams and class diagrams, for our experiments.

Our paper is structured as follows. Section 2 describes our approach to modeling game AI, and explains the details by designing a game AI that controls the behavior of a tank. Section 3 shows how we used our model to generate code that executes within the EA Tank Wars environment. Section 4 presents some related work and section 5 discusses the benefits of our approach and concludes.

2   Modeling Game AI

In games or simulations, a character perceives the environment through his senses or *sensors*, and reacts to it through actions or *actuators*. For instance, a character might observe an obstacle with his eyes, and subsequently decide to turn left. Our AI modeling framework follows this control-inspired philosophy. The transformation from sensor input to actuator output is described by means of simple components. Each component's structure is modeled by a class, and its behavior by a statechart. The main mechanism of communication between the components is the asynchronous sending/receiving of events. This lowers the coupling between components and hence makes reconfiguration and reuse easier. In some situations, a component may also synchronously invoke an operation of another component.

The architecture of our AI models is described in Fig. 1. The first level contains components that represent the *sensors* that allow the character to observe the environment as well as its own state. The sensors filter the abundant information and send events of interest on to the next levels. The second level contains components that *analyze* or correlate the events from individual sensors, which might lead to the generation of further events. The *memorizer* components keep track of the history of events. The *strategic decider* components are conceptually at the highest level of abstraction. They have to decide on a strategy for the character based on the current state and memory. At the next level, the *tactical deciders* plan how to best pursue the current strategy. The *executors* then translate the decisions of the tactical components to low-level commands according to the constraints imposed by the game or simulation. *Coordinator* components understand the inter-relationships of actuators and might refine the low-level commands further. Finally, the *actuators* perform the desired action.

To illustrate the power of our approach, we show in the following sections how we modeled the AI of a computer-controlled tank.

2.1 Modeling the State of a Tank

A tank is a heavy armored fighting vehicle carrying guns and moving on a continuous articulated metal track. When developing a model of a real-world object such as a tank, the modeler abstracts away certain details depending on the context in which the model is going to be used.

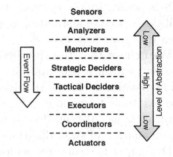

Fig. 1. AI Model Architecture

Some games actually model game objects such as vehicles and their physical interactions with the environment in great detail. These games are typically called *simulators*, such as flight simulators, helicopter simulators and tank simulators. Simulating the physics of a tank requires a detailed model of the physical components of a tank (physical shape, material, mass) and equations describing the physical interactions of these components.

In this paper, our interest is to reason about the behavior of a tank pilot. Therefore we can model a tank at a much higher level of abstraction (see Fig. 2). In the particular game that our AI is going to be playing, a tank has a given physical size, approximated by a bounding rectangle. The gun is mounted on a rotating turret anchored in the middle of the tank. A tank also has a set of sensors that relay information about the state of the tank and the surrounding environment to the pilot. The instruments in the cockpit tell the driver the position of the tank, in which direction the tank is facing, what speed it is going at, and at what angle the turret is currently positioned. A fuel indicator shows the current fuel level of the tank, and a status indicator reports on the current damage. Finally, two radars, one mounted in the front of the tank, and the other one on the turret, scan the environment for enemies and obstacles. The tank has also an advanced weapon detection system, which informs the pilot when the tank is under attack, and from what position the enemy attack is originating.

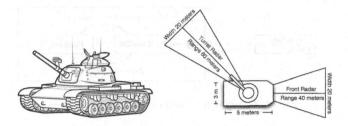

Fig. 2. Tank and it's Abstraction

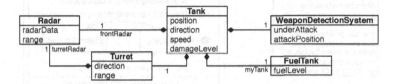

Fig. 3. Modeling the State of a Tank with Class Diagrams

The above mentioned state of a tank can naturally be modeled using class diagrams as shown in Fig. 3. Each sensor of the tank, such as the radar, can be modeled as a stand-alone class. The composition association is then used to connect the different components together to form the complete state of a particular tank. The advantage of using hierarchical composition is easy to see: models of tanks with different components, for example with 2 turrets, can easily be constructed by combining the individual components in different configurations.

2.2 Sensors – Generating Important Game Events

During a game or simulation, the state of the tank and the states of its components evolve (according to the game rules or laws governing the simulation). As mentioned in the introduction, a tank pilot (or a computer player) pursues a specific high level goal and performs actions that work towards the achievement of that goal. High level goals usually remain the same as long as no significant changes in the tank's state or in its environment occur.

We suggest to explicitly model the generation of significant events using state diagrams. The state diagrams are attached to the class that has access to all the state needed to generate the event. The events are generated either by inspecting the values of the attributes of the current class, or by looking at attributes of other classes associated by composition relationships.

A simple example is shown in Fig. 4. The *FuelTank* class encapsulates an attribute that stores the current fuel level of the tank. Fuel is essential for the tank to function, but the exact fuel level is not of great importance. Hence we abstract from the continuous fuel level to two discrete states, *FuelLevelOK* and *FuelLow*. Only when the fuel is low, the tank pilot should take appropriate measures. We can model the generation of a *fuelLow* event in case the fuel level crosses a certain threshold by attaching a state diagram to the *FuelTank* class.

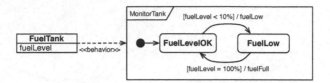

Fig. 4. Generating *FuelLow* and *FuelFull* Events

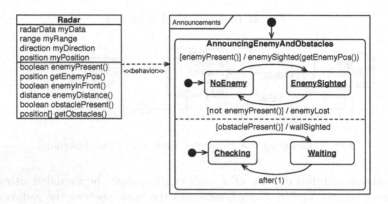

Fig. 5. Generating Events based on Simple Computations

Note that this simple behavior introduces hysteresis: once the fuel level drops below 10%, the *FuelLow* state is entered, to only be exited once the level reaches 100% again.

A more complicated example is shown in Fig. 5. In this case, the *Radar* component wants to signal *EnemySighted* and *EnemyLost* event when an enemy enters/exits the radar surveillance zone. This behavior is described in the first orthogonal component of the statechart *Announcements*. Analyzing the radar data for enemy presence, and calculating the enemy position are both operations that require a small computation. They can be modeled as simple operations such as getEnemyPos() attached to the *Radar* class. The state diagram attached to the *Radar* class can use these operations to trigger the transition that sends the desired *EnemySighted* and *EnemyLost* events. The orthogonal *AnnounceObstacles* component also shown in Fig. 5 performs similar event generation for detected obstacles.

2.3   Analyzers – Correlating Sensor Events

Some significant events can only be detected or calculated based on the state of several tank components. For instance, to determine if the enemy is in range, information from the turret as well as the turret radar is needed. The *InRangeDetector* state diagram shown in Fig. 6 takes care of this. While in the *Seeking* state, if the front radar ray of the turret radar detects an enemy, and the distance is smaller than the turret range, then the *ReadyToShoot* event is sent.

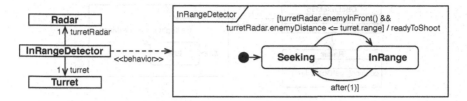

Fig. 6. Generating Events based on the State of Several Components

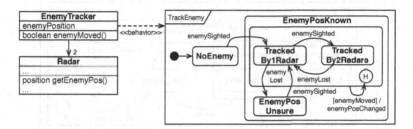

Fig. 7. Remembering the Position of the Enemy

2.4 Memorizers – Modeling Memory

A tank pilot does not only react to current events, but also makes decisions based on events/state from the past. In order to remember interesting state or events for future strategical decisions, we need to add state to the model that acts as the tank pilot's "memory".

Occurrences of events can be remembered using boolean or enumeration fields, or states in a statechart. An example of the latter is shown in Fig. 7, which depicts an *EnemyTracker* class that contains an `enemyPosition` field that remembers at what position the enemy has last been seen, even if the enemy is not within range of one of the radars anymore.

While the enemy is in range of at least one of the radars (# received *enemySighted* events > # received *enemyLost* events), the `enemyMoved` operation compares the enemy position of the *EnemyTracker* with the position obtained from the two radars. If the positions differ by a significant amount, the stored position is updated and an *enemyPosChanged* event is sent.

Remembering complex state, for instance geographical information, is less trivial, and usually requires the construction of an elaborate data structure that stores the state to be remembered in an easy-to-query form. This is done in the *ObstacleMap* class shown in Fig. 8. It reacts to the *WallDetected* events sent by the radars and updating the map data structure accordingly. The actual algorithm is not shown here, but abstracted within the `updateMap()` operation.

2.5 Strategic Deciders – Deciding on a High-Level Goal

Now that we have the event generation (based on environment sensors, current state of the tank and memory) in place, it is possible to model the high level

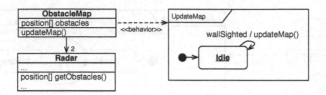

Fig. 8. Creating a Map of the World

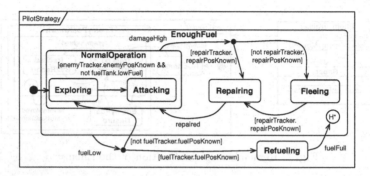

Fig. 9. The Tank Pilot Strategy

strategy of the tank pilot. This is depicted in Fig. 9. At the highest level of abstraction, a tank pilot switches between different operating modes based on events. He starts in *Exploring* mode, and switches to *Attacking* mode once the enemy position is known (and there is still enough fuel). If at any point in time the sustained damage is too high, then, if the location of the repair station is known, he switches to *Repairing* mode. Otherwise, *Fleeing* is the best strategy. In the event that the fuel is low, if the location of the fuel station is known, the tank pilot chooses to switch to *Refueling* mode. Otherwise, it is best to continue *Exploring*, hoping to find a fuel station soon. When the fueltank is full, the pilot switches back to whatever he was doing before he was interrupted.

The mode changes are announced by sending corresponding events: when *Exploring* is entered, the *explore* event is sent, when *Attacking* is entered, the *attack* event is sent, etc.

2.6 Tactical Deciders – Planning how to Achieve the Goal

The high-level goals sent by the pilot strategy component have to be translated into lower-level commands that can be understood by the different actuators of the tank, such as the motor and the turret. This translation is not trivial, since it can require complex tactical planning decisions to be made. In addition, the planning should take into account the history of the game, i.e. consult the memorizers for important game state or events that happened in the past.

Each strategy of the pilot should have a corresponding *planner* component. Fig. 10 illustrates how the *AttackPlanner* decides to carry out an attack:

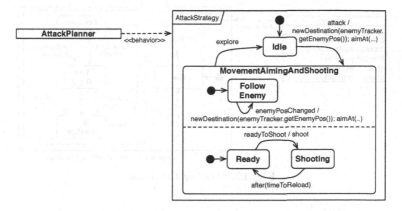

Fig. 10. Attack Movement Strategy

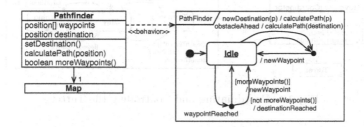

Fig. 11. Pathfinding

whenever the tank is ready to shoot, a *shoot* event is sent. The movement strategy is also simple: the planner chooses to move the tank to the position of the enemy. Whenever the enemy position changes, it sends out a *newDestination* event.

The *Pathfinding* component, shown in Fig. 11, knows how to perform obstacle avoidance. It translates the *newDestination* event into a list of waypoints by analyzing the current world information obtained from the *Map*. The *Pathfinding* component then announces the first waypoint by sending an event. Whenever the tank reaches a waypoint, the next waypoint is announced.

2.7   Executors – Mapping the Decisions to Actuator Commands

The executors map the decisions of the tactical deciders to events that the actuators can understand. The mapping of events is constrained by the rules of the game or simulation. There is typically one executor for each actuator.

In our case the *Steering* component shown in Fig 12translates the waypoints into events that the *MotorControl* understands. Every second, depending on whether the waypoint is ahead of, left of, right of or behind the tank, the corresponding command event is sent to the *MotorControl*. A more sophisticated *Steering* component would take into account the dynamics of the tank such as speed, mass, and acceleration.

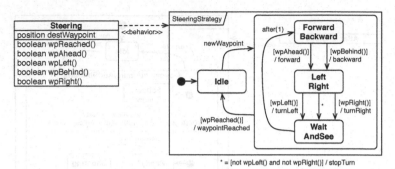

Fig. 12. Steering the Tank

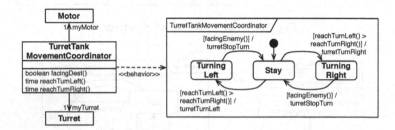

Fig. 13. Coordinating and Controlling the Turret

2.8   Coordinators – Resolving Undesired Actuator Interactions

For modularity and composability reasons, executors individually map tactical decisions to actuator events. This mapping can result in inefficient and maybe even incorrect behavior when the effects of actuator actions are correlated. In such a case it is important to add an additional *coordinator* component that deals with this issue.

For example, while attacking, the turret should turn until it is facing the enemy tank and then shoot. However, the optimal turning strategy depends on whether the tank itself is also turning or not. Fig. 13 illustrates a *Turret-TankMovementCoordinator* class that performs this coordination step. The calculations required to determine if it is faster to turn right or turn left based on the current turning decisions of the motor are done in the operations `reachTurnLeft()` and `reachTurnRight()`.

2.9   Actuators – Signaling the Action to the Game

At our level of abstraction, the tank actuators are very simple. A tank pilot can decide whether to advance or move the tank backwards at different speeds, and whether to turn left or right. Likewise, a turret can be turned left or right, and shots can be fired at different distances. Finally, commands can be given to refuel or repair, if the tank is currently located at a fuel or repair station. We suggest to model each actuator as a separate *Control* class.

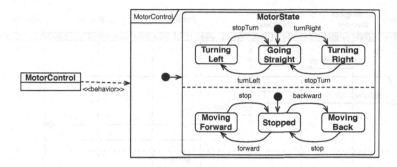

Fig. 14. The *Motor* Actuator

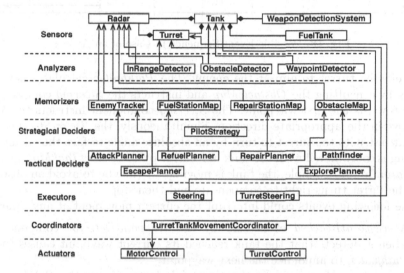

Fig. 15. The Tank AI Components

Fig. 14 shows the *MotorControl* class, an actuator that controls the movement of the tank. The state diagram shows how the motor reacts to *turnLeft, turnRight, stopTurn, forward, backward* and *stop* events. How this action is finally executed within the game or simulation is going to be discussed in section 3.

2.10 Tank AI Model Summary

The detailed architecture that shows all the components of our tank AI is depicted in Fig. 15. Most communication is done using events, and hence the individual components are only loosely coupled. Only when conceptually necessary, i.e. when a component's functionality depends on data from another component, synchronous communications must occur. In this case, the dependency between the involved classes is shown with directed associations.

Fig. 16 shows a possible sequencing of events in case the *PilotStrategy* component decides to attack. The *AttackPlanner* (concurrently) sets a new destination

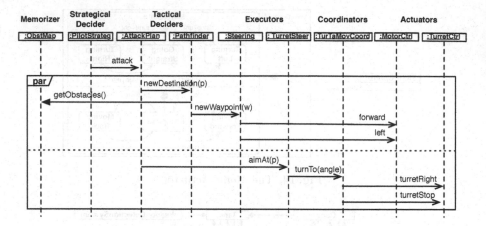

Fig. 16. Possible Event Sequence in Case of an Attack

and tells the turret to aim at the enemy. The *Pathfinder* calculates new waypoints by consulting the *ObstacleMap* and instructs the *Steering* component to move towards the first waypoint. The *Steering* component instructs the Motor to move in the appropriate direction. Simultaneously, the *TurretSteering* component instructs the turret to turn by an absolute angle to face the enemy. The turning is coordinated with the tank movement by the *TurretTankMovementCoordinator*. In our example, the tank is moving left (maybe to avoid an obstacle), and the turret turns right to compensate and then stops.

The following events could interrupt the current movement at any time:

- A *waypointReached* event sent by the *WaypointDetector* component when it detects that the tank reached its current waypoint causes the *Pathfinder* to announce the next waypoint.
- An *obstacleAhead* sent by the *ObstacleDetector* causes the *Pathfinder* to calculate a new path and announce the first waypoint.
- A *fuelLow* event sent by the *FuelTank* causes the *PilotStrategy* to transition into the refueling state and send the *refuel* event, which causes the *RefuelPlanner* to announce the position of the fuel station as the new destination.
- Similarly, a *damageHigh* event sent by the *Tank* can cause the pilot to decide to repair.

3   Mapping to an Execution Platform

3.1   The EA Tank Wars Simulation Environment

In 2005, Electronic Arts announced the EA Tank Wars competition [2], in which Computer Science students compete against each other by writing artificial intelligence (AI) components that control the movements of a tank. EA released a simulation environment written in C++, prepared to compile on Windows,

Linux and MacOS X, in which, two tanks, both controlled by an AI component, fight a one-on-one battle in a 100 by 100 meter world. In Tank Wars, a tank is equipped with sensors and actuators identical to our example model of section 2 (see also Fig. 2). During the simulation, an animation shows the moving tanks, their radar ranges and their state.

The Tank Wars simulation environment is *time-sliced* (as opposed to discrete-event). Every time slice, the AI component of the tank is given the current state of the world as seen by the tank sensors. The AI then has to decide whether to change the speed of the tank, whether to turn, whether to turn the turret, whether to fire and how far, and whether to refuel or repair, if possible. Each turn lasts *50 milliseconds*. If the AI does not make a decision when the time limit elapsed, the tank does not move during that time slice.

3.2   Time-Slicing vs. Continuous Time

The simulation in Tank Wars is built on a time-sliced architecture. Every 50ms, the new state of the environment is sent to the AI component. Statecharts on the other hand are purely event-based. At the modeling level, as well as when the model is simulated, time is continuous, i.e. infinite time precision is available. There is no time-slicing: a transition that is labeled with a time delay such as after(t) means that the transition should fire exactly after the time interval t has elapsed, t being a real number. Continuous time is most general, and is most appropriate at this level of abstraction for several reasons:

- Modeling freedom: The modeler is not unnecessarily constrained or encumbered with implementation issues, but can focus on the logic of the model.
- Symbolic analysis: Using timed logic it is possible to analyze the model to prove properties.
- Simulation: Simulation can be done with infinite accuracy (accuracy of real numbers on a computer) in simulation environments.
- Reuse: Continuous time is the most general formalism, and can therefore be used in any simulation environment.

When a model is used in a specific environment, actual code has to be synthesized, i.e. the continuous time model has to be mapped to the time model used in the target simulation. In games that are event-based such a mapping is straightforward. This is however not the case for Tank Wars, in which an approximation has to take place: the synthesized code can execute at most once every timeslice. Fortunately, if the time slice is small enough compared to the dynamics of the system to be modeled (such as the motion of a tank), the approximation is acceptable and the resulting simulation close to equivalent to a continuous time simulation.

3.3   Bridging the Time-Sliced – Event-Driven Gap

In order to use event-based reasoning in a time sliced environment, a bridge between the two worlds has to be built. In Tank Wars, at every time slice, the

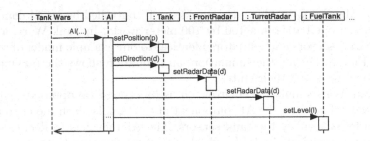

Fig. 17. Converting Time Sliced Execution to Events

Fig. 18. Wall Encounter Execution Trace

framework calls the C++ function `static void AI (const TankAIInput in,`
`TankAIInstructions & out)` of an AI object. We implement the bridge between
the time-sliced game environment and our statechart model in this function.

In section 2.1, we modeled the state of the sensors in separate classes. The
`in` parameter of the AI function contains a data structure that describes the
current state of all sensors. Our custom code within the AI function proceeds by
storing the new sensor states in the appropriate objects (see Fig. 17).

The mapping to events is done at the level of the sensor objects according to
the attached statecharts. After the operation updated the state of the sensor,
the guards in the statechart are evaluated, and the corresponding event is fired,
if any. For instance, according to the statechart shown in Fig. 4, the execution
of the `setLevel` operation of the *FuelTank* might generate a *fuelLow* event in
case the fuel level sinks below 10%.

From then on, propagation of events and triggering of actions is entirely done
within the statechart formalism. After all events have been processed, or at least
just before the 50 ms deadline expires, the state in the actuator objects is copied
into the `out` struct of the AI function and returned to the Tank Wars simulation.

3.4   ATOM3 and Code Generation

To validate our approach, we built our tank model in our AToM³ visual metamodeling and model transformation environment [1]. We compiled the model
into C++ code with our own custom-built Statechart compiler. After inserting
this code into the Tank Wars game (in the `AI` function), realistic behavior is
observed as shown in Fig. 18. The figure shows a trace of a scenario where a
tank encounters a wall, initiates turning until the wall is no longer in the line of
sight, and finally continues on its way.

4   Related Work

The use of visual modeling environments is not new to the gaming industry. Also known under the name of *Visual Scripting Languages*, finite state machines and other formalisms have been used to model various features of games, including cinematics and story narratives [8]. The main objective of developing such systems is to offload work from the programmers to the game designers and the animators, allowing them to participate to the development of the game without requiring any programming or scripting knowledge [4].

More interesting is the use of modeling environment to define the behavior of agents, as proposed by *Simbionic* and its toolset which allow a developer to describe the behavior of intelligent agents using finite state machines [3]. Although similar to our approach, the *Simbionic* toolset represents states as actions, transitioning from one action to another solely through the use of conditions and guards. In addition, the toolset functions exclusively in a time-slice fashion, abstracting time as simple clock ticks.

Viusal modeling environments can also be found in commercial engines. The Unreal Engine 3 [9] includes *UnrealKismet*, a visual scripting system, which provides artists and level designers the freedom to design stories and action sequences for non player characters within a game without the need for programming. One key feature of *UnrealKismet* is the support for hierarchy of components, which makes it possible to structure complicated behavior descriptions nicely. The difference with our approach is that the models in *UnrealKismet* essentially describe the decision making steps of an AI algorithm graphically. Our approach does not model the control flow explicitly. The behavior emerges based on the components that listen for and react to events.

Also worth mentioning is *ScriptEase* [7], a textual tool for scripting sequences of game events and reactions of non player characters. Although it doesn't use a visual formalism, *ScriptEase* introduces a pattern template system – a library of frequently used sequences of events – that allows designers to put together complex sequences with little programming.

5   Discussion and Conclusion

In this section, we situate our efforts into the broader context of Model-Based Design and highlight the benefits of this approach. The core idea of Model-Based Design is to explicitly *model* the structure and behavior of systems under study. Such models can be described at different *levels of abstraction* or detail as well as by means of different *formalisms*. The particular formalism and level of abstraction chosen depends on the background and goals of the modeler as much as on the system modeled.

Working at the Appropriate Level of Abstraction. In general, the process of abstraction can be considered a type of transformation that preserves some invariant properties (usually behavioral) of the system. In the case of our Tank

Wars example, several types of abstraction take place. First of all, there is the explicit *layering* of levels of abstraction. At the lowest levels (most detailed data, closest to the physical entities/game engine) are sensors and actuators. At the highest level is the strategic planning level. Intermediate levels help bridge the information gap between these levels. Different levels of abstraction are crossed quite naturally by means of event aggregation and synthesis.

Abstraction is also applied to data. Sensors filter the large amounts of data and propagate only salient events to higher abstraction layers. As is common in object-oriented design, an abstraction is made by choosing only relevant properties to be modeled as object attributes.

Appropriate Formalism and Visual Notation. Orthogonal to the choice of model abstraction level is the selection of suitable formalisms in which the models are described. The choice of formalism is related to the abstraction level, the intended audience, and the availability of solvers/simulators/code generators for that formalism. In the case of our Tank Wars example, a variant of *Rhapsody Statecharts* was chosen as the main formalism. This formalism allows for modular description of both structure (in the form of Class Diagrams) and behavior (in the form of associated Statecharts) of the different components described above. Statecharts have been used extensively to model behavior of reactive systems. It is hence no surprise that this formalism is a natural choice to model the types of modern computer games we are interested in. The formalism has an intuitive visual notation which makes it suitable for use by non software experts. It also allows us to almost perfectly encapsulate the individual components.

During the modelling (and analysis and possibly simulation) stage of development, Statecharts allow us to ignore implementation details such as whether the game engine uses time-slicing or event-scheduling time management. This issue, albeit very important, is taken care of transparently by the model compiler. Thus, the game AI modeler is no longer burdened with making coding detail decisions, but only with higher-level choices. This shows the power of working at an appropriate level of abstraction, using appropriate formalism(s). Note that complexity does of course never disappear. In Model-Based Design, *accidental complexity* is kept to a minimum and is factored out and encoded in formalism transformation models (the model compiler in this case).

Enhanced Modularization. There is the more detailed de-composition, within each abstraction layer, of structure and behavior into easily identifiable components. These components either correspond to physical entities (such as a *Turret*) or to conceptual entities such as an *AttackPlanner*. This high degree of modularity allows both for independent development and understanding of components. While working on a specific component within a well-defined abstraction level, a developer is maximally focused on the task at hand.

Enhanced Evolution and Reuse. The abstraction layers we presented are commonly found in a variety of modern computer games and provide a conceptual framework within which models can easily be formulated and (re-)used. For instance, the *AttackPlanner* could be easily reused within a game in which a computer-controlled knight has to decide how to attack an enemy soldier.

The elegant breakup into loosely coupled components makes it easy to evolve an AI by simply replacing an existing component with a more sophisticated component that provides similar functionality. For instance, the performance of our tank AI could be enhanced by using a better *Pathfinder* or an enhanced *AttackPlanner* that hides behind an obstacle to ambush the enemy.

Finally, the loose coupling makes it very easy to create AI for tanks with different sensors and actuators by removing or adding individual components. For instance, a tank could have a better radar, or just one radar, or 2 turrets.

References

1. de Lara, J., Vangheluwe, H., Alfonseca, M.: Meta-modelling and graph grammars for multi-paradigm modelling in AToM^3. Software and Systems Modeling (SoSyM) 3(3), 194–209 (2004)
2. Electronic Arts. EA Tank Wars (2005), http://www.info.ea.com/company/company-tw.php
3. Fu, D., Houlette, R.T.: Putting AI in entertainment: An AI authoring tool for simulation and games. IEEE Intelligent Systems 17(4), 81–84 (2002)
4. Gill, S.: Visual Finite State Machine AI Systems. Gamasutra (November 2004), http://www.gamasutra.com/features/20041118/gill-01.shtml
5. Harel, D., Kugler, H.: The rhapsody semantics of statecharts (or, on the executable core of the UML). In: Ehrig, H., Damm, W., Desel, J., Große-Rhode, M., Reif, W., Schnieder, E., Westkämper, E. (eds.) INT 2004. LNCS, vol. 3147, pp. 325–354. Springer, Heidelberg (2004)
6. Newborn, M.: Deep blue's contribution to AI. Ann. Math. Artif. Intell. 28(1-4), 27–30 (2000)
7. Onuczko, C., Cutumisu, M., Szafron, D., Schaeffer, J., McNaughton, M., Roy, T., Waugh, K., Carbonaro, M., Siegel, J.: A Pattern Catalog For Computer Role Playing Games. In: Game-On-NA 2005 - 1st International North American Conference on Intelligent Games and Simulation, pp. 33–38. Eurosis (August 2005)
8. Pickett, C.J.F., Verbrugge, C., Martineau, F.: (P)NFG: A Language and Runtime System for STructured Computer Narratives. In: Game-On-NA 2005 - 1st International North American Conference on Intelligent Games and Simulation, pp. 23–32. Eurosis (August 2005)
9. Unreal Technology. The Unreal Engine 3 (2007), http://www.unrealtechnology.com/html/technology/ue30.shtml

Model-Driven Construction of Certified Binaries

Sagar Chaki[1], James Ivers[1], Peter Lee[2], Kurt Wallnau[1], and Noam Zeilberger[2]

[1] Software Engineering Institute
[2] Computer Science Department
Carnegie Mellon University

Abstract. Proof-Carrying Code (PCC) and Certifying Model Checking (CMC) are established paradigms for certifying the run-time behavior of programs. While PCC allows us to certify low-level binary code against relatively simple (e.g., memory-safety) policies, CMC enables the certification of a richer class of temporal logic policies, but is typically restricted to high-level (e.g., source) descriptions. In this paper, we present an automated approach to generate certified software component binaries from UML Statechart specifications. The proof certificates are constructed using information that is generated via CMC at the specification level and transformed, along with the component, to the binary level. Our technique combines the strengths of PCC and CMC, and demonstrates that formal certification technology is compatible with, and can indeed exploit, model-driven approaches to software development. We describe an implementation of our approach that targets the Pin component technology, and present experimental results on a collection of benchmarks.

1 Introduction

Today, off-the-shelf programs are increasingly available as modules or components that are attached to an existing infrastructure. Often, such plug-ins are developed from high-level component specifications (such as UML Statecharts), but distributed in executable machine code, or "binary" form. In this article we present a framework for generating trustworthy "binaries" from component specifications, and for proving that such binaries satisfy specific policies. A more detailed exposition of this work is available as a technical report [1].

Our approach builds on two existing paradigms for software certification: proof-carrying code and certifying model checking. Proof-Carrying Code (PCC) [2] constructs a proof that machine code respects a desired policy, packages the proof with the code so that the validity of the proof and its relation to the code can be independently verified before the code is deployed. In contrast, Certifying Model Checking (CMC) [3] is an extension of model checking [4] for generating "proof certificates" for finite state models against a rich class of temporal logic policies. In recent years, CMC has been augmented with iterative abstraction-refinement to enable the certification of C source code [5,6].

PCC and CMC have complementary strengths and limitations. Specifically, while PCC operates directly on binaries, its applications to date have been

G. Engels et al. (Eds.): MoDELS 2007, LNCS 4735, pp. 666–681, 2007.

restricted to relatively simple memory safety[1] policies. The progress of PCC has also been hindered by the need for manual intervention, e.g., to specify loop invariants. In contrast, CMC is able to certify programs against a richer class of temporal logic policies (which subsumes both safety and liveness), and is automated. However, CMC is only able to certify source code (for example "C") or other forms of specification languages.

Finally, while PCC and CMC both require a small trusted computing base–usually consisting of a verification condition generator and a proof checker–they both tend to generate prohibitively large proofs. This can pose serious practical obstacles in using PCC in resource constrained environments. Unfortunately, embedded software (e.g., in medical devices) that might benefit from the high confidence obtained with PCC are almost certainly going to be resource constrained. In this context, our approach has the following salient features:

1. **Expanded Applicability:** We generate certified binaries directly from component specifications expressed in a subset of UML Statecharts. The key technique involved is a process of translating "ranking functions", along with the component itself, from one language to the next. Thus, our approach bridges the two domains of model-driven software development and formal software certification.
2. **Rich Policies:** As with CMC, we certify components against a rich class of temporal logic policies that subsumes both safety and liveness. We use the state/event-based temporal logic called SE-LTL [7] developed at the SEI.
3. **Automation:** As with CMC, we employ iterative refinement in combination with predicate abstraction and model checking to generate appropriate invariants and ranking functions required for certificate and proof construction in an automated manner.
4. **Compact Proofs:** We use state-of-the-art Boolean satisfiability (SAT) technology to generate extremely small proofs. Our results indicate that the use of SAT yields proofs of manageable size for realistic examples.

2   Basic Concepts

In this section, we present the basic concepts of components, policies, ranking functions, verification conditions, certificates, etc., that we use later.

Logical Foundation. We assume a denumerable set of variables *Var*, and a set of expressions *Expr* constructed using *Var* and the standard C operators. We view every expression as a formula in quantifier-free first order logic with C interpretations for operators and truth values (0 is false and anything else is true). Thus, we use the terms "expression" and "formula" synonymously and apply concepts of validity, satisfiability, etc. to both expressions and formulas.

Component. We deal with several forms of a component—their Construction and Composition Language (CCL) form, C implementation form, analysis form,

---

[1] Informally, a safety policy stipulates a condition that must never occur, while a liveness policy stipulates a condition that must eventually occur.

and their binary (assembly language) form. The syntax and semantics of CCL have been presented elsewhere [8], and we use the PowerPC assembly language. Hence, we only describe the other two (analysis and C implementation) forms.

In its analysis form, a component is simply a control flow graph (CFG) with a specific entry node. Each node of the component is labeled with either an assignment statement, a branch condition, or a procedure call. The outgoing edges from a branch node are labeled with THEN and ELSE to indicate flow of control. For any component C, we write $Stmt(C)$ to denote the set of nodes of C since each node corresponds to a component statement. Figure 1 shows a component on the left and its representation in C syntax on the right.

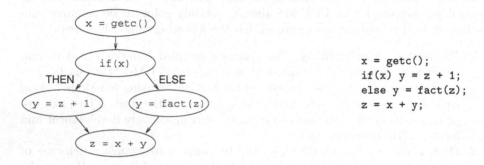

```
x = getc();
if(x) y = z + 1;
else y = fact(z);
z = x + y;
```

Fig. 1. Component in Analysis (Left) and C (Right) Forms

The C implementation is generated from CCL, and contains both the logical behavior specified by Statecharts, and the infrastructure imposed by the Pin [9] component model. However, we impose several strong restrictions on the C code itself. For instance, we disallow recursion so that the entire component is inlined into a single CFG. We also disallow internal concurrency. Variable scopes and return statements are not considered. All variables are assumed to be of integral type, and pointers and other complicated data types are disallowed.

While these are severe restrictions when viewed from the full generality of ANSI-C, they are not so severe when viewed from the more restrictive vantage of CCL specifications. In particular, a CCL specification for a component with a single reaction (the CCL unit of concurrency) obeys the above restrictions by definition. Even when a restriction is violated (e.g., CCL allows statically declared fixed size arrays), simple transformations (e.g., representing each array element by a separate variable) are possible. Since all C programs and binaries we consider are obtained via some form of semantics-preserving translation of CCL specifications, they obey our restrictions as well.

Policy. Policies are expressed in CCL specifications as SE-LTL formulas. Prior to verification, however, the policy is transformed into an equivalent Büchi automaton. Thus, for the purpose of this paper, a policy φ is to be viewed simply as a Büchi automaton. The theoretical details behind the connection between SE-LTL and Büchi automata can be found elsewhere [7], and are not crucial to grasp the main ideas presented here.

Ranking Function. Ranking functions are a technical device used to construct proofs of liveness, which require a notion of progress toward some objective O. The essential idea is to assign ranks—drawn from an ordered set R with no infinite decreasing chains—to system states. Informally, the rank of a state is a measure of its distance from O. Then, proving liveness boils down to proving that with every transition, the rank of the current system state decreases appropriately, i.e., the system makes progress toward O. Since there are no infinite decreasing chains in R, the system must eventually attain O. In our case, it suffices to further restrict R to be a finite set of integers with the usual ordering.

Definition 1 (Ranking Function). *Given a component C, a policy φ, and a finite set of integral ranks R, a ranking function RF is a mapping from Expr to R. The expressions in the domain of RF represent states of the composition of C and φ, using additional variables to encode the "program counter" of C and the states of φ. Given any ranking function RF, C and φ are known implicitly.*

Definition 2 (Verification Condition). *Given a ranking function RF, we can effectively compute a formula called the verification condition of RF, and denoted by $VC(RF)$, using an algorithm called VC-Gen.*

Ranking functions, verification conditions, and software certification are related intimately, as expressed in Fact 1. Note that we write $C \models \varphi$ to mean component C respects policy φ, and that a formula is *valid* if it is true under all possible variable assignments.

Fact 1 (Soundness) *For any component C and policy φ, if there exists a ranking function $RF : Expr \rightarrow R$ such that $VC(RF)$ is valid, **then** $C \models \varphi$.*

We will not go into a detailed proof of Fact 1 since it requires careful formalization of the semantics of C and φ. In addition, proofs of theorems that capture the same idea have been presented elsewhere [2,6].

Definition 3 (Certificate). *For any component C and policy φ, a certificate for $C \models \varphi$ is a pair (RF, Π) where $RF : Expr \rightarrow R$ is a ranking function over some finite set of ranks R, and Π is a resolution proof of the validity of $VC(RF)$.*

Indeed, if such a certificate (RF, Π) exists, then, by the soundness of resolution[2], we know that $VC(RF)$ is valid, and hence, by Fact 1, $C \models \varphi$. This style of certification, used in both PCC and CMC, has several tangible benefits:

- Any purported certificate (RF, Π) is validated by the following effective (i.e., automatable) procedure: (i) compute $VC(RF)$ using VC-Gen, and (ii) verify that Π is a correct proof of $VC(RF)$ using a proof checker.
- Necula and Lee demonstrated that this effective procedure satisfies a fundamental soundness theorem: any program with a valid certificate satisfies the policy for which the certificate is constructed [2]. This fact is not altered

---

[2] More details on resolution can be found in our technical report [1].

even if the binary program, the proof certificate, or both, are tampered with. A binary program may exhibit different behavior in its modified form than in its original form. However, this new behavior will still be guaranteed to satisfy the published policy if its proof certificate is validated.

– The policy, VC-Gen, and proof checking algorithms are public knowledge. Their mechanism does not depend in any way on secret information. The certificate can be validated independently and objectively. The soundness of the entire certification process is predicated solely upon the soundness of the underlying logical machinery (which is time tested), and the correctness of the "trusted computing base" (TCB), as discussed later.

– The computational complexity of the certification process is shouldered by the entity generating the certificate. In the case of software components, this entity is usually the component supplier who has the "burden of proof".

Overall, the existence of a valid certificate implies that $C \models \varphi$ irrespective of the process by which the certified component was created or transmitted. This feature makes our certification approach extremely attractive when incorporating components derived from unknown and untrusted sources.

3   Framework for Generating Certified Binaries

Figure 2 depicts our infrastructure for certified component binary generation. Key elements are numbered for each of reference and are correlated with the steps of the procedure described in this section. The flow of artifacts involved in generating a certified binary is indicated via arrows. Certified component binaries are generated step-wise as follows:

Step 1. A component is specified in CCL [8]. CCL uses a subset of UML 2.0 Statecharts that excludes features that are not particularly useful given the Pin component model as a target. The specification *Spec* contains a description of the component as well as the desired SE-LTL policy φ that the component is to be certified against.

Step 2. *Spec* is transformed ("interpreted" [10]) into a component C, that can be processed by a model checker. C is comprised of a C program along with finite state machine specifications for procedures invoked by the program. This step was implemented by augmenting prior work [11] so that C contains additional information relating its line numbers, variables and other data structures with those of *Spec*. This information is crucial for the subsequent reverse-interpretation of ranking functions in Step 4.

Step 3. C is input to Copper, a state-of-the-art certifying software model checker that interfaces with theorem provers (TP) and boolean satisfiability solvers (SAT). The output of Copper is either a counterexample (CE) to the desired policy φ, or a ranking function $RF1 : Expr \rightarrow R$, over some set of ranks R, such that $VC(RF1)$ is valid.

Step 4. The certificate $RF1$ only certifies C (the result of the interpretation) against the policy φ. It is reverse-interpreted into a certificate $RF2 : Expr \rightarrow R$

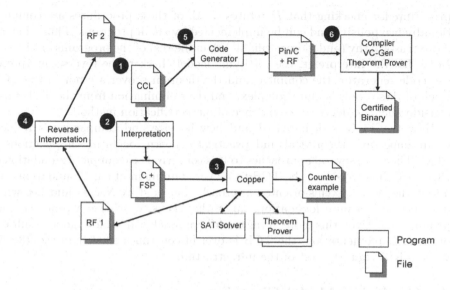

Fig. 2. Framework for Generating Certified Binaries

such that $VC(RF2)$ is valid. This process is enabled by the additional information generated during interpretation to connect *Spec* with C in Step 2.

Step 5. *Spec* and *RF2* are transformed into Pin/C component code that can be compiled and deployed in the Pin runtime environment [9]. We augmented an existing Pin/C code generator to also create a ranking function, using *RF2*, and embed it in the generated code. In essence, we transform the ranking function, and the component, from CCL to the Pin/C formalism.

Step 6. The final step consists of three distinct sub-steps.

Step 6.1. The component with the embedded ranking function is compiled from Pin/C to binary form. In our implementation we use GCC[3] (targeting the PowerPC instruction set) for this step. Let *RF3* be the ranking function embedded in the binary obtained as a result.

Step 6.2. We compute $VC(RF3)$ using VC-Gen.

Step 6.3. We obtain a proof Π of $VC(RF3)$ using a proof-generating theorem prover. In our implementation we use a SAT-based theorem prover for this step. In essence, we convert $\neg VC(RF3)$ (i.e., the logical negation of $VC(RF3)$) to a Boolean formula ϕ. We then check if ϕ is unsatisfiable using ZChaff [12]. If $\neg VC(RF3)$ is unsatisfiable, i.e., if $VC(RF3)$ is valid, then the resolution proof emitted by ZChaff serves as Π. The use of SAT enables us to obtain extremely compact proofs [6] in practice. Finally, the certificate (RF, Π) along with the binary is emitted as the end result—the certified binary for *Spec*.

Trusted Computing Base. It is instructive to discuss the artifacts that must be trusted for our approach to be effective. In essence, the TCB is comprised of: (1) VC-Gen, (2) the procedure for converting $\neg VC(RF3)$ to ϕ, and (3) the

---

[3] http://gcc.gnu.org

procedure for checking that Π refutes ϕ. All of these procedures are computationally inexpensive and can be implemented by small programs. Thus, they are more trustworthy (and more verifiable) than the rest of the programs of Figure 2. Note that the interpreter, the certifying model checker, the reverse-interpreter, the code generator, the compiler, and the theorem prover are not in the TCB. Each of these tools is quite complex, and their elimination from the TCB raises considerably the degree of confidence of our certification method.

How the TCB is demarcated and how its size and complexity is reduced is an important theoretical and practical concern for future applications of PCC. There are several approaches to this concern. For example, "foundational" PCC [13] aims to reduce the TCB to its bare minimum of logic foundations. We adopt the more systems-oriented approach pioneered by Necula and Lee which does not seek a pure foundation, but rather seeks to achieve a practical compromise [14]. Even this more "pragmatic" approach can achieve good results. In our own implementation, the TCB is over fifteen times smaller in size (30 KB vs. 450 KB) than the rest of the infrastructure.

4   Certifying Model Checking

The infrastructure for performing certifying model checking corresponds to steps 1-4 from Figure 2. We begin with component specifications expressed in CCL.

Overview of CCL. CCL is a simple composition language for describing component behavior and how components are wired together into assemblies for deployment. In CCL, a component is viewed as a collection of potentially concurrent units of computation called *reactions*, each of which describes how the component responds to stimuli on its sink pins and under what circumstances it initiates interactions on its source pins. Figure 3 shows a CCL specification for a component comp with a single reaction R. The reaction R reacts to stimuli from its environment on its incr sink pin by incrementing an internal counter (up to a maximum and then reseting to a minimum) and informing its environment of the new value on its value source pin. The semantics of the state machine provided for each reaction is based on the UML 2.0 semantics of Statecharts. Aside from the obvious syntactic differences, CCL differs from Statecharts in two important ways:

1. CCL does not permit some concepts defined in UML Statecharts, most significantly hierarchical states and concurrent sub-states (within a reaction).
2. CCL provides more specific semantics for elements of the UML standard that are identified as semantic variation points (e.g., the queuing policy for events queued for consumption by a state machine). These refined semantics are based on the execution semantics of the Pin component technology, the target of our code generator.

Interpreting CCL to C. CCL specifications are transformed into an equivalent representation in C and FSP [15] for use with Copper, a software model checker. This corresponds to Step 2 from Figure 2. In the interpreted form, each

```
component comp () {
   sink asynch incr ();
   source asynch value (produce int v);
   threaded react R (incr, value) {
      int i = min;
      start -> idle { }
      idle -> incrementing {trigger ^incr;}
      incrementing -> idle {trigger $value; action $incr();}
      state incrementing {
         if (i < max) i++;
         else i = min;
         ^value(i);
      }
   } // end of react R
} // end of component comp
```

Fig. 3. CCL Specification for a Simple Component

state of the specification state machine is implemented in a correspondingly labeled program block; guards are represented by **if** statements; transitions are completed using **goto** statements; and so on. The equivalence is straightforward, particularly given CCL's use of C syntax for actions. Two elements that are less intuitive are the representation of events used for interaction (communication) between components and annotations used to facilitate reverse interpretation (expressing model checking results in terms of the original CCL specification instead of the interpreted C program).

Communication between concurrent units (representations of interacting components) in Copper is primarily handled using event semantics based on FSP. Our interpretation uses events to model message-based interactions between components in the Pin component technology. In Pin, interactions occur in synchronous or asynchronous modes, and the initiation and completion of an interaction are differentiated syntactically by a ^Pin for initiation on a pin Pin, or a $Pin for completion on a pin Pin. These phenomena are mapped to FSP-style events as part of the interpretation.

For example, initiation of an interaction over a source pin (^value) is represented by a **begin_value** event. This event is denoted in the interpreted C program using the __COPPER_HANDSHAKE__() function. Representing a choice among several events, however, is more difficult. Thus, when a component is willing to engage in an interaction over any of several sink pins (i.e., pull the next message from its queue and respond accordingly), this corresponds to a willingness to synchronize over one of several FSP-style events. This concept is not as easily represented in C, and we use Copper's ability to provide specifications of functions. We insert a call to an **fsp_exernalChoice()** function and provide a specification of that function's behavior as an FSP process that allows a choice among a specific set of events and returns an integer indicating the event with which the process synchronized.

The annotations used to simplify reverse interpretation are inserted via CCL_NODE(x) function calls. The parameter passed to each such call denotes the node in the CCL abstract syntax tree (AST) of the CCL specification that corresponds to C statement that follows the annotation. These calls are known to Copper, and are stripped from the program prior to verification. When used for certifying model checking, however, Copper retains the parameter values and includes them in the ranking functions emitted upon successful verification.

Ranking Function Generation by Copper. Copper uses iterative-predicate-abstraction-refinement for verification. This paradigm has been presented in detail elsewhere [16,17,18,1] and we only present its relevant features here. The key idea is that conservative models of the C program are constructed via predicate abstraction, verified, and refined iteratively until either the verification succeeds, or a real counterexample is found. Let M be the model verified successfully. Then each state of M is of the form (l, V) where l is a location in the C program, and V is a valuation of the set of predicates used to construct M. Each valuation V has a concretization $\gamma(V) \in Expr$. Also, for any two distinct valuations V and V', $\gamma(V)$ and $\gamma(V')$ are logically disjoint.

We now describe the ranking function generated by Copper. The ranking function is generated as a set of triples of the form $((l, I), s, r)$ where: (i) I is an invariant, i.e., the concretization of a predicate valuation V such that (l, V) is a reachable state of M^4, (ii) s is a state of the Büchi automaton corresponding to the policy, and (iii) r is a rank. The procedure for constructing an appropriate ranking function is presented elsewhere [6] and we do not describe it further.

Recall, from Definition 1, that a ranking function is a mapping from expressions to ranks. Each triple $((l, I), s, r)$ emitted by Copper corresponds to an entry in this mapping as follows. Let PC and SS be special variables representing the program location (i.e., program counter) and the policy state respectively. Then, the triple $((l, I), s, r)$ denotes a mapping in the ranking function from the expression I && $(PC == l)$ && $(SS = s)$ to the rank r. Note that, for any two triples $((l, I), s, r)$ and $((l', I'), s', r')$ emitted by Copper, either $l \neq l'$ or I and I' are disjoint (since they are the concretizations of two distinct predicate valuations). Hence, the ranking function emitted is always well-formed.

Figure 4 shows an excerpt from the ranking function generated for our example CCL specification and a policy asserting that min <= i <= max is always true (RF 1 from Figure 2). Each line denotes a triple $((l, I), s, r)$. The first field is the CCL AST node number, corresponding to the location l. The second and third fields (which, in the excerpt, are always 8 and 0) correspond to the policy automaton state s and the rank r respectively. The last field is the invariant I.

The final step in certifying model checking is to relate the ranking function back to the original CCL specification. This is achieved via a process of mapping elements from the interpreted C program back to CCL elements. For example, variable names are "demangled" and replaced with references to AST node

---

[4] Strictly speaking, an invariant at location l is the disjunction of the concretizations of all predicate valuations V such that (l, V) is a reachable state of M. We use a slightly looser definition of invariant for simplicity.

```
104 : 8 : 0 [(-1 < P0::R__i ),(P0::R__i < 7 )]
106 : 8 : 0 [(P0::R__i < 7 ),(-2 < P0::R__i ),(P0::R__i != -1 )]
116 : 8 : 0 [(-1 < P0::R__i ),(P0::R__i < 7 )]
```

Fig. 4. Ranking Function in terms of Interpreted C Program

numbers and predicates relating to variables that were introduced during interpretation are stripped or remapped to the appropriate CCL concepts. At the
conclusion of certifying model checking, if a component is known to satisfy all of
its policies, we obtain evidence to that effect in the form of a ranking function
expressed in terms of nodes of the AST for the component's CCL specification.

5 Certified Source Code Generation

The infrastructure for generating certified source code corresponds to Step 5
from Figure 2. We begin with a component specification expressed in CCL and
a ranking function expressed in terms of nodes of its AST. From previous work,
we have a code generator for CCL that generates C code targeted for deployment
in the Pin component technology (Pin/C). To support certified code generation,
we extended this code generator to embed invariants from the ranking function
in the generated Pin/C code. The key decision was choosing how to embed this
information to maintain a correlation between the location of these invariants in
the Pin/C code and the assembly code resulting from compilation.

```
__begin__();                        1:        bl __begin__
                                    2:        li %r0,0
                                    3:        stw %r0,16(%r31)
                                    4:        lwz %r0,8(%r31)
                                    5:        cmpwi %cr7,%r0,0
                                    6:        blt %cr7,.L5
                                    7:        lwz %r0,8(%r31)
                                    8:        cmpwi %cr7,%r0,9
                                    9:        bgt %cr7,.L5
                                    10:       li %r0,1
                                    11:       stw %r0,16(%r31)
                                  .L5:
                                    12:       lwz %r3,16(%r31)
                                    13:       crxor 6,6,6
__inv__((n > =0) && (n < 10));      14:       bl __inv__
```

Fig. 5. Invariants in Pin/C Code (Left) and Assembly Code (Right)

The convention we chose (shown in Figure 5) encodes invariants using a pair
of function calls inserted in the Pin/C code prior to the location associated
with each invariant. The invariant itself is used as the argument to the second
function of the pair. When such code is compiled, pairs of recognizable assembly

call instructions appear in the assembly code and the instructions necessary to represent the invariant appear between these calls.

We extended the Pin/C code generator to insert these pairs of calls at any locations for which the ranking function provides invariants (a short excerpt from the generated code is shown in Figure 6). The code generator also adds an additional predicate to each invariant in the ranking function, an encoding of the current state of the state machine. At the conclusion of certified source code generation, we have C source code that includes the invariants necessary for generating a proof that the binary form of this component satisfies the desired policy. An important point to note is that the generated certified source code contains at least one call to __begin__() and __inv__(...) inside every loop. This is crucial for effective computation of the certified binary, as presented in the next section, without having to supply loop invariants.

```
else if (_THIS_->R_CURRENT_STATE == 1) {
    __begin__();
    __inv__(((__pcc_claim__ == 0 && __pcc_specstate__ == 8 &&
        __pcc_rank__ == 0 && ((-2 < _THIS_->R_i ) &&
        (_THIS_->R_i != -1 ) && (_THIS_->R_i < 7 )) &&
        _THIS_->R_CURRENT_STATE == 1))); /* 52 */

    if (pMessage->sinkPin == 0 /* ^incr */ ) {
        ...
```

Fig. 6. Excerpt from Generated Pin/C Code

6   Certified Binary Generation

In this section we describe the process of obtaining the end-product of our approach, the certified binary code. To this end, we present the procedure for constructing the two components of the certified binary—the binary itself, and a certificate which is essentially the proof of a verification condition.

The certified binary is obtained by simply compiling this C source code with any standard compiler. In our implementation, we used GCC targeted at the PowerPC instruction set for this step of our procedure. The binary generated by the compiler contains assembly instructions, peppered with calls to __begin__() and __inv__(...). Let us refer to an assembly fragment starting with a call to __begin__(), and extending up to the first following call to __inv__(...), as a *binary invariant*. Note that in any binary invariant, the code between the calls to __begin__() and __inv__(...) effectively compute and store the value of the argument being passed to __inv__(...) in register r3.

To construct the certificate, we first construct the verification condition VC. This is done one binary invariant at a time. Specifically, for each binary invariant β, we compute the verification condition for β, denoted by $VC(\beta)$. Let BI be the set of all binary invariants in our binary. Then, the overall verification condition VC is defined as follows: $VC = \bigwedge_{\beta \in BI} VC(\beta)$.

The technique for computing $VC(\beta)$ is based on computing weakest preconditions, the semantics of the assembly instructions, and the policy that the binary is being certified against. It is similar to the VC-Gen procedure used in PCC. The main difference is that our procedure is parameterized by the policy, and is thus general enough to be applied to any policy expressible in SE-LTL. In contrast, the VC-Gen procedure used in PCC has a "hard-wired" safety policy, viz., memory-safety. It is also noteworthy that our procedure does not require loop invariants since every loop in the binary contains at least one binary invariant.

Once we have VC, the certificate is obtained by proving VC with a proof-generating theorem prover. We leverage our previous work on using a theorem-prover based on Boolean satisfiability (SAT) [6] to generate extremely compact certificates as compared to existing non-SAT-based proof-generating theorem provers. In addition, it enables us to be sound with respect to bit-level C semantics, which is crucial when certifying safety-critical software.

Given a binary B and an associate certificate C, we validate B as follows. We first compute the verification condition VC using the technique described above. We then check that C is a correct proof of the validity of VC. Validation succeeds if and only if C turns out to be indeed a proper proof of VC.

Note that once a certified binary has been validated successfully, the embedded binary invariants are stripped off before the binary is actually deployed. This is crucial for both correctness (since what we really certify is the binary *without* the invariants) and performance. Finally, it is noteworthy that our choice of mechanism for carrying invariants from C code to assembly code is sensitive to compiler optimizations. Certain optimizations (e.g., code reordering across the boundaries demarcated by calls to __begin__ and __inv__) *may* break this correspondence. Fortunately, the fundamental soundness theorem still holds. In the worst case, such a reordering might result in a failure in proof checking, but will never validate a proof for a program that violates a policy.

7   Related Work

PCC was proposed by Necula and Lee [19,2,20] for certifying memory safety policies on binaries. PCC works by hard-coding the desired safety policies within the machine instruction semantics, while our approach works at the specification level and encodes the policy as a separate automaton. Foundational PCC [13,21] attempts to reduce the trusted computing base of PCC to include only the foundations of mathematical logic. Bernard and Lee [22] propose a new temporal logic to express PCC policies for machine code. Non-SAT-based techniques for minimizing PCC proof sizes [23,24] have also been proposed. Whalen et al. [25] describe a technique for synthesizing certified code. They augment the AUTO-BAYES synthesizer to add annotations based on "domain knowledge" to the generated code. Their approach is not based on CMC, and generates certified source code rather than binaries.

Certifying model checkers [3,26] emit an independently checkable certificate of correctness when a temporal logic formula is found to be satisfiable by a finite state model. Namjoshi [27] has proposed a two-step technique for obtaining proofs of Mu-Calculus policies on infinite state systems. In the first step, a proof is obtained via certifying model checking. In the second step, the proof is "lifted" through an abstraction. Namjoshi's approach is still restricted to certifying source code while our work aims for low-level binaries. Iterative refinement has been applied successfully by several software model checkers such as SLAM [16], BLAST [17] and MAGIC [18]. While SLAM and MAGIC do not generate any proof certificates, BLAST implements a method [5] for lifting proofs of correctness. However, BLAST's certification is limited to source code and purely safety properties. Assurance about the correctness of binaries can also be achieved by proving the correctness of compilers (which is difficult and yet to be widely adopted) or via translation validation [28] (which still assumes that the source code is correct). In contrast, our approach requires no such correctness assumptions.

In previous work, we developed an expressive linear temporal logic called SE-LTL [7] that can be used to express both safety and liveness claims of component-based software. In the work reported here, we modified SE-LTL to express certifiable policies. Also previously, we developed an infrastructure to generate compact certificates for C *source code* against SE-LTL claims in an automated manner [29]. There, the model checker is used to generate invariants and ranking functions that are required for certificate and proof construction. Compact proofs were obtained via state-of-the-art Boolean satisfiability (SAT) technology [6]. In the current work, we extend this framework to generate certified *binaries* from component specifications. Finally, we build on the PACC infrastructure for analyzing specifications of software component assemblies and generating deployable machine code for such assemblies.

8   Experimental Results

We implemented a prototype of our technology and experimented with two kinds of examples. First, we created a simple CCL specification of a component that manipulates an integer variable and the policy that the variable never becomes negative. Our tool was able to successfully prove, and certify at the assembly code level that the implementation of the component does indeed satisfy the desired claim. The CCL file size was about 2.6 KB, while the generated Pin/C code was about 20 KB. In contrast, the assembly code was about 110 KB while the proof certificate size was just 7.7 KB. The entire process took about 5 minutes with modest memory requirements.

To validate the translation of a certified C component to a certified binary (Step 6 in Figure 2), we conducted additional experiments with Micro-C, a lightweight operating system for embedded real-time applications. The OS source code consists of about 6000 lines of C (97 KB) and uses a semaphore to ensure mutually exclusive access to shared kernel data structures. Using out approach we were able to certify that all kernel routines follow the proper locking order when using the semaphore. The total certification time was about one minute, and the certificate size was about 11 KB, or roughly 11% of the operating system source code size.

We also experimented with the C implementation of the "tar" program in the Plan 95 operating system. Specifically, we certified, using our approach, that a particular buffer will never overflow when the program is executed. The source code was manually annotated in order to generate the appropriate proof certificates. While our experiments show that our approach is viable, we believe that a more robust implementation and more realistic case studies are needed in order to push our technique amongst a wider user base.

9   Conclusion

In this paper, we presented an automated approach for generating certified binaries from software component specifications. Our technique is based on, and combines the strengths of, two existing paradigms for formal software certification—PCC and CMC. It also demonstrates that a model driven approach can be combined effectively with formal certification methodologies. In addition, we developed and experimented with a prototypical implementation of our technique. Our implementation, and our overall approach, does have limitations which we like to classify into the following two broad categories:

Deferred Features. Some of the missing features from our implementation are not difficult conceptually, but are best deferred until a target environment has been selected. For example, we did not define the format of certified binaries—in particular how the proof object is packaged with executable code.

Technical Limitations. CCL currently supports only a primitive assortment of types, and, as a consequence, the implementation supports a limited range of C language features (e.g., pointers, structs, and arithmetic types other than int and float are not supported). We have also not implemented our own proof checker or SAT formula generator, even though these are key elements of a TCB. Instead, we rely on (in principle) untrusted publicly available implementations. However, both of these are relatively simple to implement. Also, Copper is only able to generate ranking functions that involve a finite and strictly ordered set of ranks, and thus is able to certify a restricted set of programs. More general ranking functions are generated by other tools such as Terminator[5].

Nevertheless, we believe that our work marks a positive and important step toward the development of rigorous, objective and automated software certification practices, and the reconciliation of formal and model-driven approaches for software development. Our experiment results are preliminary, but realistic and encouraging, and therefore underline the need for further work in this direction.

References

1. Chaki, S., Ivers, J., Lee, P., Wallnau, K., Zeilberger, N.: Certified binaries for software components. Technical Report CMU/SEI-2007-TR-001 (2007)
2. Necula, G., Lee, P.: Safe Kernel Extensions without Runtime Checking. In: OSDI'96 (1996)

---

[5] http://research.microsoft.com/TERMINATOR/default.htm

3. Namjoshi, K.S.: Certifying Model Checkers. In: Berry, G., Comon, H., Finkel, A. (eds.) CAV 2001. LNCS, vol. 2102, Springer, Heidelberg (2001)
4. Clarke, E., Emerson, A.: Design and Synthesis of Synchronization Skeletons for Branching Time Temporal Logic. In: Proc. of WLP (1982)
5. Henzinger, T.A., Jhala, R., Majumdar, R., Necula, G.C., Sutre, G., Weimer, W.: Temporal-Safety Proofs for Systems Code. In: Brinksma, E., Larsen, K.G. (eds.) CAV 2002. LNCS, vol. 2404, Springer, Heidelberg (2002)
6. Chaki, S.: SAT-Based Software Certification. In: Hermanns, H., Palsberg, J. (eds.) TACAS 2006 and ETAPS 2006. LNCS, vol. 3920, Springer, Heidelberg (2006)
7. Chaki, S., Clarke, E.M., Ouaknine, J., Sharygina, N., Sinha, N.: State/Event-Based Software Model Checking. In: Boiten, E.A., Derrick, J., Smith, G.P. (eds.) IFM 2004. LNCS, vol. 2999, Springer, Heidelberg (2004)
8. Wallnau, K., Ivers, J.: Snapshot of CCL: A language for predictable assembly. Technical note CMU/SEI-2003-TN-025, Software Engineering Institute (2003)
9. Hissam, S., Ivers, J., Plakosh, D., Wallnau, K.C.: Pin Component Technology (V1.0) and Its C Interface. Technical Report CMU/SEI-2005-TN-001 (2005)
10. Ivers, J., Sinha, N., Wallnau, K.: A Basis for Composition Language CL. Technical Report CMU/SEI-2002-TN-026, Software Engineering Institute (2002)
11. Ivers, J., Sharygina, N.: Overview of ComFoRT: A Model Checking Reasoning Framework. Technical Report CMU/SEI-2004-TN-018 (2004)
12. Zhang, L., Malik, S.: Validating SAT Solvers Using an Independent Resolution-Based Checker: Practical Implementations and Other Applications. In: DATE'03 (2003)
13. Appel, A.W.: Foundational proof-carrying code. In: Proc. of LICS (2001)
14. Schneck, R.R., Necula, G.: A gradual approach to a more trustworthy, yet scalable, proof-carrying code. In: Voronkov, A. (ed.) Automated Deduction - CADE-18. LNCS (LNAI), vol. 2392, Springer, Heidelberg (2002)
15. Magee, J., Kramer, J.: Concurrency: State Models and Java Programs (2006)
16. Ball, T., Rajamani, S.K.: Automatically Validating Temporal Safety Properties of Interfaces. In: Bošnački, D., Leue, S. (eds.) Model Checking Software. LNCS, vol. 2318, Springer, Heidelberg (2002)
17. Henzinger, T., Jhala, R., Majumdar, R., Sutre, G.: Lazy Abstraction. In: POPL'02 (2002)
18. Chaki, S., Clarke, E., Groce, A., Jha, S., Veith, H.: Modular verification of software components in C. IEEE Transactions on Software Engineering (TSE) (6) (2004)
19. Necula, G.C.: Proof-Carrying Code. In: Proc. of POPL (1997)
20. Necula, G.C., Lee, P.: Safe, Untrusted Agents Using Proof-Carrying Code. In: Proceedings of Mobile Agents and Security (1998)
21. Hamid, N.A., Shao, Z., Trifonov, V., Monnier, S., Ni, Z.: A Syntactic Approach to Foundational Proof-Carrying Code. In: Proc. of LICS (2002)
22. Bernard, A., Lee, P.: Temporal Logic for Proof-Carrying Code. In: Voronkov, A. (ed.) Automated Deduction - CADE-18. LNCS (LNAI), vol. 2392, Springer, Heidelberg (2002)
23. Necula, G., Lee, P.: Efficient Representation and Validation of Proofs. In: LICS'98 (1998)
24. Necula, G., Rahul, S.: Oracle-Based Checking of Untrusted Software. In: POPL'01 (2001)
25. Whalen, M.W., Schumann, J., Fischer, B.: Synthesizing certified code. In: Eriksson, L.-H., Lindsay, P.A. (eds.) FME 2002. LNCS, vol. 2391, Springer, Heidelberg (2002)

26. Kupferman, O., Vardi, M.: From Complementation to Certification. In: Jensen, K., Podelski, A. (eds.) TACAS 2004. LNCS, vol. 2988, Springer, Heidelberg (2004)
27. Namjoshi, K.S.: Lifting Temporal Proofs through Abstractions. In: Zuck, L.D., Attie, P.C., Cortesi, A., Mukhopadhyay, S. (eds.) VMCAI 2003. LNCS, vol. 2575, Springer, Heidelberg (2002)
28. Pnueli, A., Siegel, M., Singerman, E.: translation validation. In: Steffen, B. (ed.) ETAPS 1998 and TACAS 1998. LNCS, vol. 1384, Springer, Heidelberg (1998)
29. Chaki, S., Wallnau, K.: Results of SEI Independent Research and Development Projects and Report on Emerging Technologies and Technology Trends. Technical report CMU/SEI-2005-TR-020, Software Engineering Institute, Ch. 6 (2005)

Workshops and Symposia
at MODELS 2007

Holger Giese*

Software Engineering Group, University of Paderborn
Warburger Str. 100, D-33098 Paderborn, Germany
hg@upb.de

1 Introduction

Following the tradition of previous instances of the MODELS conference series, also in 2007 a number of workshops and symposia were hosted. 11 workshops and two symposia complemented the main conference by providing room for important subject areas and enabling a high degree of interactivity.

The selection of the workshops has been organized like in former instances of the MODELS conference series by a workshop selection committee. The following well known experts agreed to serve on this committee:

- Gabor Karsai, Vanderbilt University, USA
- Thomas Kühne, Darmstadt University of Technology, DE
- Jochen Küster, IBM Research Zurich, CH
- Henry Muccini, University of L'Aquila, IT
- Sebastian Uchitel, Imperial College London, UK

The workshops provided collaborative forums for particular topics. They enabled a group of participants to exchange recent and/or preliminary results, to conduct intensive discussions, or to coordinate efforts between representatives of a technical community. They served as forums for lively discussion of innovative ideas, recent progress, or practical experience on model-driven engineering for specific aspects, specific problems, or domain-specific needs.

In addition, like in previous editions there have been a Doctoral Symposium and an Educators' Symposium. The Doctoral Symposium provided specific support for Ph.D. students to discuss their work and receive useful guidance for the completion of the dissertation research. The Educators' Symposium addressed how to educate students as well as practitioners to move from the traditional thinking to an engineering approach based on models.

A more detailed description of the workshops and symposia is provided in the following two sections. A corresponding satellite-event proceeding will be published in the LNCS series by Springer after the conference, featuring summaries as well as revised selected papers from the workshops, the Doctoral Symposium, and the Educators' Symposium.

---

* Currently a visiting professor at the Hasso-Plattner-Institut, University of Potsdam, Germany.

G. Engels et al. (Eds.): MoDELS 2007, LNCS 4735, pp. 682–690, 2007.

2   Detailed List of Workshops

W1: Aspect-Oriented Modeling

Organizers: Omar Aldawud (Lucent Technologies, USA), Walter Cazzola (University of Milano, Italy), Tzilla Elrad (Illinois Institute of Technology, USA), Jeff Gray (University of Alabama at Birmingham, USA), Jörg Kienzle (McGill University, Canada), Dominik Stein (University of Duisburg-Essen, Germany)

Abstract: Aspect-orientation is a rapidly advancing technology. New and powerful aspect-oriented programming techniques are presented at many international venues every year. However, it is not clear what features of such techniques are "common aspect-oriented concepts" and what features are rather language-specific specialties. Research in aspect-oriented modeling has the potential to help find such common characteristics from a perspective that is at a more abstract level (i.e., programming language-independent). The Aspect-Oriented Modeling (AOM) Workshop brings together researchers and practitioners from two communities, aspect-oriented software development (AOSD) and model-driven engineering. This workshop provides a forum for presenting new ideas and discussing the state of research and practice in modeling various kinds of crosscutting concerns at different levels of abstraction. The goals of the workshop are to identify and discuss the impacts of aspect-oriented technologies on model engineering to provide aspect-oriented software developers with general modeling means to express aspects and their crosscutting relationships onto other software artifacts.

URL: http://www.aspect-modeling.org/models07/

W2: Language Engineering

Organizers: Jean-Marie Favre (University of Grenoble, France), Dragan Gašević (Athabasca University, Canada), Ralf Lämmel (Microsoft, USA), Andreas Winter (University of Mainz, Germany)

Abstract: The workshop on Language Engineering, which is held as the 4th edition of the ATEM Workshop series, brings together researchers from different communities to study the disciplined engineering and application of various language descriptions in order to further expand frontiers of their synergetic use in model-driven engineering. The importance of metamodels, schemas, grammars, and ontologies (or "language descriptions") is generally acknowledged by the model-driven engineering community, but, as yet, the study of these artifacts lacks a common umbrella. Language Engineering (in the context of software engineering) promotes language descriptions to first class citizens, just like programs, data, and models based on the systematic, programmatic analysis and manipulation of these language descriptions. To have a deeper focus on the Language Engineering perspective of MDE,

ATEM2007 pays attention to the fact that language descriptions, which are used in developing software systems, can be defined in different ways and used to define for different software artifacts, but still they have to be used together in integrated software life cycle. Thus, we need ways that enable us to fully support the Language Engineering life cycle including (but not limiting to), requirement analysis, design, implementation, testing, deployment, application, re-engineering, reverse engineering, and evolution of language descriptions. Since most of language descriptions have rather different technological, research and cultural origins, the synergic use is rather a complex task that requires join efforts of different communities.

URL: http://www.planetmde.org/atem2007/

W3: Model Driven Development of Advanced User Interfaces

Organizers: Andreas Pleuß (University of Munich, Germany), Jan Van den Bergh (Hasselt University, Belgium), Heinrich Hußmann (University of Munich, Germany), Stefan Sauer (University of Paderborn, Germany), Daniel Görlich (University of Kaiserslautern, Germany)

Abstract: The user interface of an application is often one of the core factors determining its success. While model-based user interface development is an important line of research in the human-computer-interaction (respectively human-machine-interaction) community, model-driven application development is an important area in the software engineering community. This workshop aims at integrating the knowledge from both domains, leading to a model-driven development of user interfaces. Thereby, the focus of the workshop lies on advanced user interfaces corresponding to the current state-of-the-art in human-computer interaction, like e.g. multimedia or context-sensitive user interfaces or multimodal interaction techniques.

This is the third workshop in this series, building up on the results of its predecessor held at MODELS 2006. In particular, it addresses challenges identified on the preceding events, such as better integration of model-driven user interface development and creative design, model transformations which allow a better control of the usability of the resulting user interface and advanced usage of models at runtime for adaptation of user interfaces.

URL: http://planetmde.org/mddaui2007/

W4: Model Size Metrics

Organizers: Michel Chaudron [Workshop Chair] (Technische Universiteit Eindhoven), Betty H.C. Cheng (Michigan State University), Christian F.J. Lange (Technische Universiteit Eindhoven), Jacqueline McQuillan (National University of Ireland, Maynooth), Andrij Neczwid (Motorola Software Group), Frank Weil (Motorola Software Group)

Abstract: Within the MODELS community a standardized method of determining sizing concepts for software models that allows the effective base lining and comparison of model concepts is needed. Such metrics are crucial for effective estimation and quality management of model development. Additionally measuring the model size is important to provide context information for empirical studies using models.

One of the most commonly used measures of source code program size is the source lines of code (SLOC) metric. However, the concept of lines of code does not readily apply to modeling languages such as UML and SDL. Furthermore, software models are heterogeneous in nature (consisting of several different types of diagrams), can exist at varying levels of abstraction and can be created using different modeling styles. As a result, researchers face many challenges when trying to define the size of a software model.

The aims of the workshop are to bring together practical experience and research results related to sizing techniques for software models, to build a community of researchers and practitioners that share software design artefacts for the purpose of empirical studies, and to identify a research agenda in the area of model size measurement.

URL: http://www.win.tue.nl/~clange/MSM2007/

W5: Model-Based Design of Trustworthy Health Information Systems

Organizers: Elske Ammenwerth (University for Health Sciences, Medical Informatics and Technology, Austria), Ruth Breu (Universität Innsbruck, Austria), Ruzena Bajcsy (University of California, Berkeley, USA), John C. Mitchell (Stanford University, USA), Alexander Pretschner (ETH Zürich, Switzerland), Janos Sztipanovits (Vanderbilt University, USA)

Abstract: While many information-intensive industries have developed and deployed standards-based information infrastructures, healthcare has been characterized as a "trillion dollar cottage industry" that, in its current state, is heavily dependent on paper records and fragmented, error-prone approaches to service delivery. In response, Health Information Systems (HIS) are emerging as a new and significant application domain for information technologies to capture and promote interactions between patients and healthcare providers. However, the ingratiation of information systems into the complex world of healthcare generates unique technology challenges. A primary concern is that privacy and security requirements for HIS are frequently expressed in vague, as well as contradictory, complex laws and regulations. To address this problem, model-based methods offer a revolutionary way to formally and explicitly integrate privacy and security goals into HIS architectures. End-to-end architecture modeling, integrated with formal privacy and security models, offer new opportunities for HIS system designers and end users. This workshop intends to bring computer scientists, medical experts, and legal policy experts together to discuss research results in the development and application of model-based methods for representing, analyzing and integrating, architectures, privacy and security policies, computer security mechanisms, web authentication, and human factors engineering.

URL: http://mothis.isis.vanderbilt.edu/

WS6: Model-Driven Engineering, Verification and Validation

Organizers: Benoit Baudry (IRISA/INRIA, France), Alain Faivre (CEA-LIST, France), Sudipto Ghosh (Colorado State University, USA), Alexander Pretschner (ETH Zurich, Switzerland)

Abstract: Model-Driven Engineering (MDE) is a development process that extensively uses models and automatic model transformations to handle complex software developments. Many software artifacts, tools, environments and modeling languages need to be developed to make MDE a reality. Consequently, there is a crucial need for effective V&V techniques in this new context. Furthermore, the novelty of this development paradigm gives rise to questions concerning its impacts on traditional V&V techniques, and how they can leverage this new approach. The objective of this workshop is to offer a forum for researchers and practitioners who are developing new approaches to V&V in the context of MDE.

Major questions that cross-cut V&V and MDE include: Is the result of a transformation really what the user intended? Is the model correct with respect to the expected security, time, and structural constraints? What models can be used for validation or verification? Does the implementation generated after several model transformations conform to the initial requirements?

The workshop will discuss V&V of model transformations and code generation; techniques for validating a model or generating test cases from models including simulation, model-checking, and model-based testing; application of MDE to validation, testing, and verification; tools and automation; case studies and experience reports.

URL: http://www.modeva.org/2007

W7: Modeling and Analysis of Real-Time and Embedded Systems

Organizers: Sébastien Gérard (CEA, List, France), Susanne Graf (Verimag, France), Øystein Haugen (SINTEF and University of Oslo, Norway), Iulian Ober (IRIT), Bran Selic (IBM/Rational, Canada)

Abstract: The Model Driven Architecture (MDA) initiative of the OMG puts forward the idea that future process development will be centered on models, thus keeping application development and underlying platform technology as separate as possible. In the area of DRES (distributed, Real-time and Embedded Systems), this model-oriented trend is also very active and promising. But DRES are different from general-purpose systems. The purpose of this workshop is to serve as an opportunity to gather researchers and industrials in order to survey some existing experiments related to modeling and model-based analysis of DRES. Moreover in order to be able to exchange models with the aim to apply formal validation tools and to achieve interoperability, it is important to have also a common understanding of the semantics of the given notations. Other important issues in the domain of real-time are methodology and modeling paradigms allowing breaking down the complexity, and tools which are able to verify well designed systems.

URL: http://www.martes.org/

W8: Ocl4All: Modelling Systems with OCL

Organizers: David H. Akehurst (University of Kent at Canterbury, UK), Martin Gogolla (Technische Universität Bremen, Germany), Steffen Zschaler (Technische Universität Dresden, Germany)

Abstract: The requirements that the modelling community now wants to see supported by OCL go far beyond its initial requirements. When OCL was conceived it was designed as a language for supporting precise modelling. The advent of the MDA (Model Driven Architecture) vision and the rapid acceptance of MDE (Model Driven Engineering) approaches emphasize new application domains (like Semantic Web or Domain Specific Languages). This increase in new modelling languages causes a need for new OCL-like languages for systems modelling, frequently developed as extensions to the original.

This workshop is a continuation of the well-established series of MODELS workshops on OCL. As this year's special focus, we wish to recognise, officially, that OCL will be used as a basis for many text-based navigation languages and to bring together the community that defines these extensions in order to consolidate the experiences, successes and failures involved in doing so. We also hope to discuss the potential for redesigning or at least restructuring the OCL standard definition in order to better facilitate and support third party extensions to the language. This workshop aims to look specifically at how to apply the key software engineering principles of modularity and reuse to the definition of OCL.

URL: http://st.inf.tu-dresden.de/Ocl4All2007/

W9: Models@run.time

Organizers: Nelly Bencomo (Lancaster University, UK), Gordon Blair (Lancaster University, UK), Robert France (Colorado State University, USA)

Abstract: In the design modelling community, research effort has focused on using models at design, implementation, and deployment stages of development. The complexity of adapting software during runtime has spawned interest in how models can be used to validate, monitor and adapt runtime behaviour. The use of models during runtime extends the use of modelling techniques beyond the design and implementation phases of development. Model-driven software development would help providing the infrastructure to reconfigure and adapt a runtime system based on input QoS and context based values. The perspective of models at runtime consists in bringing this model-based capability forward to the runtime. The goal of this workshop is to look at issues related to developing appropriate model-driven approaches to managing and monitoring the execution of systems. The workshop aims to integrate and combine research ideas from relevant areas as model-driven software development, software architectures, reflection, and autonomic and self healing systems, and provide a "state-of-the-research" assessment expressed in terms of research issues, challenges, and achievements.

URL: http://www.comp.lancs.ac.uk/~bencomo/MRT07/

W10: Multi-Paradigm Modeling: Concepts and Tools

Organizers: Pieter Mosterman (The MathWorks, Inc., USA), Tihamér Levendovszky (Budapest University, Hungary), Juan de Lara (Universidad Autónoma de Madrid, Spain)

Abstract: Computational modeling has become the norm in industry to remain competitive and be successful. As such, Model- Based Design of embedded software has enterprise-wise implications and modeling is not limited to isolated uses by a single engineer or team. Instead, it has reached a proliferation much akin to large software design, with requirements for infrastructure support such as version control, configuration management, and automated processing.

The comprehensive use of models in design has created a set of challenges beyond that of supporting one isolated design task. In particular, the need to combine, couple, and integrate models at different levels of abstraction and in different formalisms is posing a set of specific problems that the field of Computer Automated Multiparadigm Modeling (CAMPaM) is aiming to address.

The essential element of multiparadigm modeling is the use of explicit models throughout. This leads to a framework with models to represent the syntax of formalisms used for modeling, models of the transformations that represent the operational semantics, as well as model-to-model transformations for inter-formalism transformation.

These models are then used to facilitate generative tasks in language engineering, such as evolving a domain specific modeling formalism as its requirements change, but also in a tool engineering space, such as automatic generation of integrated development environments. Moreover, given an explicit model of a model transformation allows analyses such as termination characteristics, consistency, and determinism.

URL: http://mpm07.aut.bme.hu/

W11: Quality in Modeling

Organizers: Ludwik Kuzniarz (Blekinge Institute of Technology, Ronneby, Sweden), Jean Louis Sourrouille (INSA Lyon, France), Miroslaw Staron (IT University, Gothenburg, Sweden)

Abstract: Quality assessment and assurance is an important part of software engineering. The issues of software quality management are widely researched and approached from multiple perspectives and viewpoints. The introduction of a new paradigm in software development – Model Driven Development (MDD) – raises new challenges in software quality management, and as such should be given a special attention. The issues of early quality assessment based on models at a high abstraction level and building prediction models for software quality are important from the software engineering perspective.

The workshop is intended to provide a premier forum for discussions related to software quality and MDD. The discussions are to be organized around addressing the following topics: assessment of quality, quality models and best practices, quality checking and ensuring, software processes for ensuring quality, impact of MDD on quality, experience reports and empirical studies of quality in the context of MDD.

The workshop is built upon the experience and discussions during the first workshop on Quality in Modeling at MODELS 2006 and a follow up of a series of consistency workshops held annually at the UML conferences and MDA-FA conference.

URL: http://www.ipd.bth.se/lku/Quality-in-Modeling-2007

3   Detailed List of Symposia

S1: Doctoral Symposium

Organizer: Claudia Pons (Universidad Nacional de La Plata, Argentina)

Abstract: The MODELS 2007 Doctoral Symposium is a forum for Ph.D. students to discuss their goals, methodology and results at an early stage in their research, in a critical but supportive and constructive environment. The symposium will provide an opportunity for student participants to interact with other students at a similar stage in their careers and with established researchers in the broader software modeling community. The symposium aims to provide useful guidance for the completion of the dissertation research and motivation for a research career.

The symposium is intended for students who have already settled on a specific research topic (closely related to model-driven engineering) and have obtained initial results, but still have enough time remaining before their final defense so that they can benefit from the symposium experience. Students should be at least a year from completion of their dissertation (at the time of the symposium), to obtain maximum benefit from participation.

Each student that is invited to attend the symposium will be assigned a specific mentor who will be in charge of leading the discussion after the student's presentation.

URL: http://www.modelsconference.org

S2: Educators' Symposium

Organizer: Miroslaw Staron (IT University of Göteborg, Sweden)

Abstract: Software engineering is progressing towards using higher abstraction levels to increase productivity of development teams and quality of software. Software engineering industry needs to be supported by educating skilled professionals who are able to use advanced modeling techniques directly after graduating the university. Software engineering education is therefore facing a challenge of efficient move from the traditional focus on computer science to the focus on engineering techniques, in particular centered around models rather than source code and textual design specifications.

The purpose of the Educators' Symposium is to serve as a premiere forum for discussions and exchange of experiences between researchers, practitioners, and teachers interested in problems of teaching modeling. This year's focus of the symposium is on the industrial relevance of university courses in the area of modeling. The discussions during the symposium are intended to stimulate exchange of good practices in teaching modeling and requirements for industrially relevant and good courses in modeling. The leading topic for the symposium in 2007 is transitioning from the traditional, programming oriented, curricula and courses to modern, model based, software engineering curricula and courses.

The best paper will be published in Elsevier Information and Software Technology Journal.

URL: http://www.modelsconference.org

Acknowledgements

I am grateful to the members of the selection committee who did a great job in reviewing the workshop proposals and selecting the best workshops. In particular Thomas Kühne (member of the selection committee and my predecessor as workshop chair) was of great help for me and eased my work by generously sharing his experiences from the former year with me.

Tutorials at MODELS 2007

Jeff Gray

Department of Computer and Information Sciences, University of Alabama at Birmingham
1300 University Boulevard, Birmingham, Alabama 35294 USA
gray@cis.uab.edu

Abstract. The MODELS 2007 conference offered four high-quality tutorials from leading experts in the area of model-driven engineering. Each tutorial was presented as a half-day event that was organized during the first two days of the conference. This short overview provides an introduction to the tutorials program and a summary of each tutorial as submitted by the presenters.

1 Introduction

The MODELS 2007 tutorials provided conference attendees with a broad spectrum of opportunities to increase their knowledge about modeling practice and theory. The four tutorials that were selected cover topics on general modeling practice, specific application of modeling to particular domains, and an introduction to popular modeling toolsuites. The collection of tutorials offered something for every attendee – from first time participants and students looking for an overview of current practice, to seasoned practitioners and researchers seeking knowledge on new developments among popular tools.

There were a large number of high quality tutorial proposals that were submitted to the MODELS 2007 conference, but space only permitted four tutorials to be selected. Each of the tutorial proposals received three reviews from the Selection Committee. Members of the MODELS 2007 tutorials Selection Committee were:

> Jean Bézivin, University of Nantes, France
> Jeff Gray, University of Alabama at Birmingham, USA
> Ivan Kurtev, Twente University, Netherlands
> Alfonso Pierantonio, University of L'Aquila, Italy
> Gianna Reggio, Universita' di Genova, Italy
> Antonio Vallecillo, University of Malaga, Spain

The tutorials for the conference occurred during the first two days of the conference. More details about the dates, times, and locations of the tutorials can be found at the main MODELS 2007 conference web page:

www.modelsconference.org

A summary of the four tutorials can be found in the next section.

G. Engels et al. (Eds.): MoDELS 2007, LNCS 4735, pp. 691–694, 2007.
© Springer-Verlag Berlin Heidelberg 2007

2   Tutorial Summaries

This section provides an abstract of each tutorial. The abstracts are taken from the descriptions provided by each presenter during their initial submission.

Tutorial T1: *Effective Model Driven Engineering Patterns, Principles, and Practices*

Presenters: Bruce Trask and Angel Roman (MDE Systems, USA)

Model Driven Engineering (MDE) brings together multiple technologies and critical innovations and formalizes them into the next wave of software development methods. This tutorial will cover the basic patterns, principles and practices of MDE. The three main MDE categories include the development of Domain-Specific Languages, Domain-Specific Editors (including Domain-Specific Visual Languages) and, Domain-Specific Transformation Engines or Generators. Expressed in terms of language development technology, these mirror the development of the Abstract Syntax, Concrete Syntax and Semantics of a new Domain-Specific Language. This tutorial will cover the basic effective patterns, principles and practices for developing these MDE software artifacts. The tutorial will show how to apply these concepts as effective means with which to both raise levels of abstraction and domain specificity and thus increase power and value of tools and languages that allow developers to tackle the complexities of today's software systems. It will also show how to effectively leverage abstraction without sacrificing the ability to robustly and precisely refine these abstractions to solve real world problems. Additionally, this tutorial will cover the exact details of how to leverage the Eclipse Modeling Framework (EMF), the Eclipse Graphical Editor Framework (GEF), and the Eclipse Graphical Modeling Framework (GMF), to support the development of these three areas. These three frameworks provide a unique and integrated platform in which to learn the basics of Model-Driven Engineering (MDE) in full application not just in theory. Conversely, the tutorial provides an effective context in which to learn how to apply the power of these integrated Eclipse Frameworks developed to support MDE.

Tutorial T2: *Model-Driven Engineering for QoS Provisioning in Distributed Real-Time and Embedded Systems*

Presenter: Aniruddha Gokhale (Vanderbilt University, USA)

Distributed real-time and embedded (DRE) systems require multiple, simultaneous quality of service (QoS) properties, such as predictability, reliability and security. Assuring QoS properties for DRE systems is a hard problem due to the conflicting demands imposed by each dimension of QoS. On one hand, DRE domain experts face substantial challenges in defining the desired QoS properties for different parts of their DRE systems since they must first understand the impact of individual QoS dimensions on each other at a conceptual level. On the other hand, DRE system integrators face substantial challenges provisioning QoS properties on the platforms that host the DRE systems due to a lack of proper understanding of how choices of platform configurations they make will impact the overall QoS delivered to the system. Model-driven engineering (MDE) plays a vital role in addressing these challenges. Domain-specific modeling languages (DSMLs) associated with MDE tools provide intuitive abstractions of QoS properties to DRE domain experts, who can use them to express the QoS properties they desire for their systems. Analysis tools associated with the MDE frameworks can provide vital feedback to the domain experts on the severity of conflicts between the QoS dimensions. For the systems integrators, the same MDE framework can provide automated mechanisms to map the QoS properties defined in the problem space to platform configurations in the solution space. This tutorial will illustrate these features of MDE using the CoSMIC MDE framework. A number of case studies and short demonstrations will be used to illustrate the challenges and the solutions provided by MDE tools.

Tutorial T3: *From Rapid Functional Prototypes to Robust Implementations to Product-Lines: A Model-Driven Software Factory Approach*

Presenter: Vinay Kulkarni (Tata Research, India)

We describe an approach wherein one begins by capturing the functional specifications in the form of models and domain-specific languages. An interpretive approach helps quickly build a functional prototype from these specifications, and helps derive correct functional specifications by iterating until all stakeholder concerns are satisfied. An architectural prototyping exercise is carried out in parallel to arrive at a solution architecture that meets the non-functional requirements. The desired solution architecture is then 'codified' into a set of models and model transformation specifications leading to a set of tools. We define a purpose-specific software factory instance wherein these tools transform the functional specifications in successive stages of refinement culminating in a complete implementation that conforms to the desired solution architecture. We share our experience of using this approach, over the last twelve years, to deliver several large enterprise applications on a variety of technology platforms and implementation architectures.

Tutorial T4: *Putting MDA to Work on Eclipse with the AMMA Tool Suite*

Presenters: Mikaël Barbero (University of Nantes, France) and Frédéric Jouault (INRIA, France and University of Alabama at Birmingham, USA)

As part of the OMG process for issuing common recommendations, a Request for Information (RFI) has recently been issued on "MDA Tool Capabilities." The objective is to find what capabilities (e.g., functionalities, methodology definitions, and process guidance) of tools the MDA user community currently uses for their projects, and which new capabilities it would like to have. As part of the Eclipse foundation process, a new project called EMP (for Eclipse Modeling Project) has recently been created to foster the evolution and promotion of model-based development technologies within the Eclipse community. It does so by providing a unified set of modeling frameworks, tools, and standards implementations. This tutorial will investigate the multiple relations between the complementary OMG and Eclipse activities. There are modeling specifications and open source tool solutions that may be deployed to implement the MDA approach, which interact in many ways. Model-based and DSL-based practical solutions to software production and maintenance will be presented. The various aspects of using modeling solutions to implement forward and reverse engineering will be particularly discussed. The tutorial will concretely show how a set of Eclipse open source components (namely: KM3, ATL, AM3, AMW, TCS, MoDisco) can be used to find new solutions to difficult problems. These components are part of a modeling platform named AMMA (ATLAS Model Management Architecture). The tutorial will conclude by revisiting the application scope of model engineering, seven years after the initial proposal of the MDA approach by OMG.

Panels at MODELS 2007

Jean-Michel Bruel

LIUPPA
Université de Pau et des Pays de l'Adour
64000 Pau, France
Jean-Michel.Bruel@univ-pau.fr

Panel 1: The Future of Aspect Modeling: Will MDD Absorb It?

Chair: João Araújo, Universidade Nova de Lisboa, Portugal

Currently there are some young and promising approaches in software development, such as Model-Driven Development (MDD), and Aspect-Oriented Software Development (AOSD). While MDD focuses on the systematic use of models, where the software is built through a chain of model transformations, AOSD aims to offer enhanced mechanisms to identify, model and compose crosscutting concerns. But what AOSD proposes doesn't it involve just a specific kind of modelling? And regarding composition, isn't it a special kind of model transformation? If these questions are true, will MDD absorb AOSD in the end? If not, what are then the true crosscutting relationships between MDD and AOSD?

Panel 2: Evaluating MDE Research: Are We Doing Enough?

Chairs: Sudipto Ghosh and Robert France, CSU, Fort Collins, USA

Research papers typically describe novel results and convey new ideas to a community of researchers and practitioners. Authors should back their claims and convince readers that the results are, in some sense, good. Pilot studies, case studies, and experiments are some of the techniques used for evaluating approaches. Evaluations can range from small-scale, proof-of-concept studies in academia to large-scale formal evaluations in industrial settings.

Since model driven engineering (MDE) is still in its infancy, researchers and practitioners are faced with the dilemma of quickly publishing a novel approach without much evaluation versus performing a formal, lengthy evaluation and consequently delaying publication of results. In this panel we will discuss issues and challenges related to the evaluation of MDE research results. The following two questions will be the starting points for discussions in the panel:

1. What are the expectations on the nature and quality of evaluation for MDE papers to get accepted to a workshop, conference, or journal?
2. What are the challenges in evaluating MDE research and how do we address them?

G. Engels et al. (Eds.): MoDELS 2007, LNCS 4735, p. 695, 2007.
© Springer-Verlag Berlin Heidelberg 2007

Author Index

Lecture Notes in Computer Science

Sublibrary 2: Programming and Software Engineering

For information about Vols. 1– 4066
please contact your bookseller or Springer

Vol. 4405: L. Padgham, F. Zambonelli (Eds.), Agent-Oriented Software Engineering VII. XII, 225 pages. 2007.

Vol. 4401: N. Guelfi, D. Buchs (Eds.), Rapid Integration of Software Engineering Techniques. IX, 177 pages. 2007.

Vol. 4385: K. Coninx, K. Luyten, K.A. Schneider (Eds.), Task Models and Diagrams for Users Interface Design. XI, 355 pages. 2007.

Vol. 4383: E. Bin, A. Ziv, S. Ur (Eds.), Hardware and Software, Verification and Testing. XII, 235 pages. 2007.

Vol. 4379: M. Südholt, C. Consel (Eds.), Object-Oriented Technology. VIII, 157 pages. 2007.

Vol. 4364: T. Kühne (Ed.), Models in Software Engineering. XI, 332 pages. 2007.

Vol. 4355: J. Julliand, O. Kouchnarenko (Eds.), B 2007: Formal Specification and Development in B. XIII, 293 pages. 2006.

Vol. 4354: M. Hanus (Ed.), Practical Aspects of Declarative Languages. X, 335 pages. 2006.

Vol. 4350: M. Clavel, F. Durán, S. Eker, P. Lincoln, N. Martí-Oliet, J. Meseguer, C. Talcott, All About Maude - A High-Performance Logical Framework. XXII, 797 pages. 2007.

Vol. 4348: S. Tucker Taft, R.A. Duff, R.L. Brukardt, E. Plödereder, P. Leroy, Ada 2005 Reference Manual. XXII, 765 pages. 2006.

Vol. 4346: L. Brim, B. Haverkort, M. Leucker, J. van de Pol (Eds.), Formal Methods: Applications and Technology. X, 363 pages. 2007.

Vol. 4344: V. Gruhn, F. Oquendo (Eds.), Software Architecture. X, 245 pages. 2006.

Vol. 4340: R. Prodan, T. Fahringer, Grid Computing. XXIII, 317 pages. 2007.

Vol. 4336: V.R. Basili, D. Rombach, K. Schneider, B. Kitchenham, D. Pfahl, R.W. Selby (Eds.), Empirical Software Engineering Issues. XVII, 193 pages. 2007.

Vol. 4326: S. Göbel, R. Malkewitz, I. Iurgel (Eds.), Technologies for Interactive Digital Storytelling and Entertainment. X, 384 pages. 2006.

Vol. 4323: G. Doherty, A. Blandford (Eds.), Interactive Systems. XI, 269 pages. 2007.

Vol. 4322: F. Kordon, J. Sztipanovits (Eds.), Reliable Systems on Unreliable Networked Platforms. XIV, 317 pages. 2007.

Vol. 4309: P. Inverardi, M. Jazayeri (Eds.), Software Engineering Education in the Modern Age. VIII, 207 pages. 2006.

Vol. 4294: A. Dan, W. Lamersdorf (Eds.), Service-Oriented Computing – ICSOC 2006. XIX, 653 pages. 2006.

Vol. 4290: M. van Steen, M. Henning (Eds.), Middleware 2006. XIII, 425 pages. 2006.

Vol. 4279: N. Kobayashi (Ed.), Programming Languages and Systems. XI, 423 pages. 2006.

Vol. 4262: K. Havelund, M. Núñez, G. Roşu, B. Wolff (Eds.), Formal Approaches to Software Testing and Runtime Verification. VIII, 255 pages. 2006.

Vol. 4260: Z. Liu, J. He (Eds.), Formal Methods and Software Engineering. XII, 778 pages. 2006.

Vol. 4257: I. Richardson, P. Runeson, R. Messnarz (Eds.), Software Process Improvement. XI, 219 pages. 2006.

Vol. 4242: A. Rashid, M. Aksit (Eds.), Transactions on Aspect-Oriented Software Development II. IX, 289 pages. 2006.

Vol. 4229: E. Najm, J.-F. Pradat-Peyre, V.V. Donzeau-Gouge (Eds.), Formal Techniques for Networked and Distributed Systems - FORTE 2006. X, 486 pages. 2006.

Vol. 4227: W. Nejdl, K. Tochtermann (Eds.), Innovative Approaches for Learning and Knowledge Sharing. XVII, 721 pages. 2006.

Vol. 4218: S. Graf, W. Zhang (Eds.), Automated Technology for Verification and Analysis. XIV, 540 pages. 2006.

Vol. 4214: C. Hofmeister, I. Crnković, R. Reussner (Eds.), Quality of Software Architectures. X, 215 pages. 2006.

Vol. 4204: F. Benhamou (Ed.), Principles and Practice of Constraint Programming - CP 2006. XVIII, 774 pages. 2006.

Vol. 4199: O. Nierstrasz, J. Whittle, D. Harel, G. Reggio (Eds.), Model Driven Engineering Languages and Systems. XVI, 798 pages. 2006.

Vol. 4192: B. Mohr, J.L. Träff, J. Worringen, J.J. Dongarra (Eds.), Recent Advances in Parallel Virtual Machine and Message Passing Interface. XVI, 414 pages. 2006.

Vol. 4184: M. Bravetti, M. Núñez, G. Zavattaro (Eds.), Web Services and Formal Methods. X, 289 pages. 2006.

Vol. 4166: J. Górski (Ed.), Computer Safety, Reliability, and Security. XIV, 440 pages. 2006.

Vol. 4158: L.T. Yang, H. Jin, J. Ma, T. Ungerer (Eds.), Autonomic and Trusted Computing. XIV, 613 pages. 2006.

Vol. 4157: M. Butler, C.B. Jones, A. Romanovsky, E. Troubitsyna (Eds.), Rigorous Development of Complex Fault-Tolerant Systems. X, 403 pages. 2006.

Vol. 4143: R. Lämmel, J. Saraiva, J. Visser (Eds.), Generative and Transformational Techniques in Software Engineering. X, 471 pages. 2006.

Vol. 4134: K. Yi (Ed.), Static Analysis. XIII, 443 pages. 2006.

Vol. 4119: C. Dony, J.L. Knudsen, A. Romanovsky, A.R. Tripathi (Eds.), Advanced Topics in Exception Handling Techniques. X, 302 pages. 2006.

Vol. 4111: F.S. de Boer, M.M. Bonsangue, S. Graf, W.-P. de Roever (Eds.), Formal Methods for Components and Objects. VIII, 447 pages. 2006.

Vol. 4089: W. Löwe, M. Südholt (Eds.), Software Composition. X, 339 pages. 2006.

Vol. 4085: J. Misra, T. Nipkow, E. Sekerinski (Eds.), FM 2006: Formal Methods. XV, 620 pages. 2006.

Vol. 4079: S. Etalle, M. Truszczyński (Eds.), Logic Programming. XIV, 474 pages. 2006.

Vol. 4067: D. Thomas (Ed.), ECOOP 2006 – Object-Oriented Programming. XIV, 527 pages. 2006.